OXIDATIVE STRESS AND DISEASE

Series Editors

LESTER PACKER, PhD
ENRIQUE CADENAS, MD, PhD

UNIVERSITY OF SOUTHERN CALIFORNIA SCHOOL OF PHARMACY
LOS ANGELES, CALIFORNIA

W0018483

1. Oxidative Stress in Cancer, AIDS, and Neurodegenerative Diseases, *edited by Luc Montagnier, René Olivier, and Catherine Pasquier*
2. Understanding the Process of Aging: The Roles of Mitochondria, Free Radicals, and Antioxidants, *edited by Enrique Cadenas and Lester Packer*
3. Redox Regulation of Cell Signaling and Its Clinical Application, *edited by Lester Packer and Junji Yodoi*
4. Antioxidants in Diabetes Management, *edited by Lester Packer, Peter Rösen, Hans J. Tritschler, George L. King, and Angelo Azzi*
5. Free Radicals in Brain Pathophysiology, *edited by Giuseppe Poli, Enrique Cadenas, and Lester Packer*
6. Nutraceuticals in Health and Disease Prevention, *edited by Klaus Krämer, Peter-Paul Hoppe, and Lester Packer*
7. Environmental Stressors in Health and Disease, *edited by Jürgen Fuchs and Lester Packer*
8. Handbook of Antioxidants: Second Edition, Revised and Expanded, *edited by Enrique Cadenas and Lester Packer*
9. Flavonoids in Health and Disease: Second Edition, Revised and Expanded, *edited by Catherine A. Rice-Evans and Lester Packer*
10. Redox–Genome Interactions in Health and Disease, *edited by Jürgen Fuchs, Maurizio Podda, and Lester Packer*
11. Thiamine: Catalytic Mechanisms in Normal and Disease States, *edited by Frank Jordan and Mulchand S. Patel*
12. Phytochemicals in Health and Disease, *edited by Yongping Bao and Roger Fenwick*
13. Carotenoids in Health and Disease, *edited by Norman I. Krinsky, Susan T. Mayne, and Helmut Sies*
14. Herbal and Traditional Medicine: Molecular Aspects of Health, *edited by Lester Packer, Choon Nam Ong, and Barry Halliwell*
15. Nutrients and Cell Signaling, *edited by Janos Zempleni and Krishnamurti Dakshinamurti*
16. Mitochondria in Health and Disease, *edited by Carolyn D. Berdanier*
17. Nutrigenomics, *edited by Gerald Rimbach, Jürgen Fuchs, and Lester Packer*
18. Oxidative Stress, Inflammation, and Health, *edited by Young-Joon Surh and Lester Packer*

LIPID OXIDATION IN HEALTH AND DISEASE

EDITED BY

CORINNE M. SPICKETT
HENRY JAY FORMAN

CRC Press
Taylor & Francis Group
Boca Raton London New York

CRC Press is an imprint of the
Taylor & Francis Group, an **informa** business

First published in paperback 2024

First published 2015 by CRC Press
2385 NW Executive Center Drive, Suite 320, Boca Raton FL 33431

and by CRC Press
4 Park Square, Milton Park, Abingdon, Oxon, OX14 4RN

CRC Press is an imprint of Taylor & Francis Group, LLC

© 2015, 2024 Taylor & Francis Group, LLC

ISBN: 978-1-48-220285-4 (hbk)
ISBN: 978-1-03-291986-7 (pbk)
ISBN: 978-0-42-908907-7 (ebk)

DOI: 10.1201/b18138

Visit the Taylor & Francis Web site at
http://www.taylorandfrancis.com

and the CRC Press Web site at
http://www.crcpress.com

Contents

SECTION I   Chemistry and Biochemistry of Lipid Oxidation

SECTION II Sites of Biological Actions of Oxidized Lipids

SECTION III Oxidized Lipids in Pathology and Disease

Series Preface

OXYGEN BIOLOGY AND MEDICINE

Through evolution, oxygen—itself a free radical—was chosen as the terminal electron acceptor for respiration. The two unpaired electrons of oxygen spin in the same direction; thus, oxygen is a biradical. Other oxygen-derived free radicals, such as superoxide anion or hydroxyl radicals, formed during metabolism or by ionizing radiation are stronger *oxidants*, that is, endowed with a higher chemical reactivity. Oxygen-derived free radicals are generated during metabolism and energy production in the body and are involved in regulation of signal transduction and gene expression, activation of receptors and nuclear transcription factors, oxidative damage to cell components, the antimicrobial and cytotoxic action of immune system cells, as well as in aging and age-related degenerative diseases. Conversely, the cell conserves antioxidant mechanisms to counteract the effect of oxidants; these *antioxidants* may remove oxidants either in a highly specific manner as for example, by superoxide dismutases or in a less specific manner (e.g., small molecules such as vitamin E, vitamin C, and glutathione). *Oxidative stress as classically defined is an imbalance between oxidants and antioxidants.* Overwhelming evidence indicates that oxidative stress can lead to cell and tissue injury. However, the same free radicals that are generated during oxidative stress are produced during normal metabolism and, as a corollary, are involved in both human health and disease.

UNDERSTANDING OXIDATIVE STRESS

In recent years, the research disciplines interested in *oxidative stress* have grown and enormously increased our knowledge of the importance of the cell redox status and the recognition of oxidative stress as a process with implications for many pathophysiological states. From this multi- and interdisciplinary interest in oxidative stress emerges a concept that attests to the vast consequences of the complex and dynamic interplay of *oxidants* and *antioxidants* in the cellular and tissue settings. Consequently, our view of *oxidative stress* is growing in scope and new future directions. Likewise, the term "reactive oxygen species"—adopted at some stage in order to highlight nonradical oxidants such as H_2O_2 and $^1O_2$—fails nowadays to reflect the rich variety of other reactive species in free radical biology and medicine, that is, encompassing nitrogen-, sulfur-, oxygen-, and carbon-centered radicals. With the discovery of nitric oxide, nitrogen-centered radicals gathered momentum and has matured into an area of enormous importance in biology and medicine. Nitric oxide or nitrogen monoxide (NO), a free radical generated in a variety of cell types by nitric oxide synthases (NOS) is involved in a wide array of physiological and pathophysiological phenomena, such as vasodilation, neuronal signaling, and inflammation. Of great importance is the radical–radical reaction of nitric oxide with superoxide anion; this is among the most rapid nonenzymatic reactions in biology (well over the diffusion-controlled limits) and yields the potent nonradical oxidant, peroxynitrite. The involvement of this species in tissue injury through oxidation and nitration reactions is well documented.

Virtually all diseases thus far examined involve free radicals. In most cases, free radicals are secondary to the disease process, but in some instances, causality is established by free radicals themselves. Thus, there is a delicate balance between oxidants and antioxidants in health and disease. Their proper balance is essential for ensuring healthy aging.

Both reactive oxygen and nitrogen species are involved in the redox regulation of cell functions. Oxidative stress is increasingly viewed as a major upstream component in the signaling cascade involved in inflammatory responses, stimulation of cell adhesion molecules, and chemoattractant production and as an early component in age-related neurodegenerative disorders, such as Alzheimer's, Parkinson's, Huntington's disease, and amyotrophic lateral sclerosis. Hydrogen peroxide is probably the most important redox signaling molecule that, among others, can activate NF-κB, Nrf2, and other universal transcription factors. Increasing steady-state levels of hydrogen peroxide have been linked to the cell's redox status with clear involvement in adaptation, proliferation, differentiation, apoptosis, and necrosis. The identification of oxidants in regulation of redox cell signaling and gene expression was a significant breakthrough in the field of oxidative stress: the classical definition of oxidative stress as an *imbalance between the production of oxidants and the occurrence of cell antioxidant defenses* proposed by Sies in 1985 now seems to provide a limited concept of oxidative stress, but it emphasizes the significance of the cell's redox status. Because individual signaling and control events occur through discreet redox pathways rather than through global balances, a new definition of oxidative stress was advanced by Dean P. Jones (*Antioxidants & Redox Signaling*, 2006) as a disruption of redox signaling and control that recognizes the occurrence of compartmentalized cellular redox circuits. Recognition of discreet thiol redox circuits led D. P. Jones to provide this new definition of oxidative stress. Measurements of GSH/GSSG or cysteine/cystine, thioredoxin$_{reduced}$/thioredoxin$_{oxidized}$ provide a quantitative definition of oxidative stress. The redox status is thus dependent on the degree to which tissue-specific cell components are in the oxidized state. In general, the reducing environment inside cells helps to prevent oxidative damage. In this reducing environment, disulfide bonds (S–S) do not spontaneously form because sulfhydryl groups are maintained in the reduced state (SH), thus preventing protein misfolding or aggregation. The reducing environment is maintained by metabolism and by the enzymes involved in maintenance of thiol/disulfide balance and substances, such as glutathione, thioredoxin, vitamins E and C, and enzymes such as superoxide dismutases, catalase, and the selenium-dependent glutathione reductase and glutathione and thioredoxin-dependent hydroperoxidases (periredoxins), which serve to remove reactive oxygen species (hydroperoxides). Also of importance is the recognition of the existence of many tissue- and cell compartment-specific isoforms of antioxidant enzymes and proteins.

Compelling support for the involvement of free radicals in disease development originates from epidemiological studies showing that an enhanced antioxidant status is associated with reduced risk of several diseases. Of high significance is the role that micronutrients play in modulation of redox cell signaling; this establishes a strong link between diet and health and disease and is centered on the ability of micronutrients to regulate redox cell signaling and modify gene expression.

These concepts are anticipated notions to serve as a platform for development of tissue-specific therapeutics tailored to discreet, compartmentalized redox circuits. This, in essence, dictates principles of drug development–guided knowledge of mechanisms of oxidative stress. Hence, successful interventions will take advantage of new knowledge of compartmentalized redox control and free radical scavenging.

OXIDATIVE STRESS IN HEALTH AND DISEASE

Oxidative stress is an underlying factor in health and disease. In this series of books, the importance of oxidative stress and disease associated with organ systems of the body is highlighted by exploring the scientific evidence and the clinical applications of this knowledge. This series is intended for researchers in the basic biomedical sciences and clinicians. The potential of such knowledge for healthy aging and disease prevention warrants further knowledge about how oxidants and antioxidants modulate cell and tissue function.

This book examines the chemistry and biochemistry of lipid oxidation from a solid redox biology platform; highlights include the effects of lipid oxidation on diverse cell signaling pathways and inflammatory responses. The role of oxidized lipids in pathology includes chapters on cardiovascular disease, carcinogenesis, and age-related neurodegenerative disorders.

Corinne M. Spickett and Henry J. Forman are to be congratulated for producing this excellent and timely book in an ever-growing field of importance to health sciences research.

Lester Packer
Enrique Cadenas

Editors

Corinne M. Spickett is currently a reader at Aston University, following a move from the University of Strathclyde in January 2011. Her first degree is in biochemistry from Oxford University and she has a DPhil (Oxon) on the application of NMR (nuclear magnetic resonance) to study yeast bioenergetics in vivo. After further postdoctoral work using NMR to investigate stress responses in plants and glutathione metabolism in preeclamptic toxemia, Dr. Spickett became a Glaxo-Jack Research Lecturer in the Department of Immunology at the University of Strathclyde. Since then, she has been working on the analysis of phospholipid oxidation by electrospray mass spectrometry and the biological effects of oxidized lipids, especially as relating to atherosclerosis and inflammation, and has published extensively in this field. More recently, she expanded her research to include analysis of protein oxidation and formation of lipoxidation products during inflammation, and has been developing label-free, semi-targeted approaches to their identification in biological samples.

Dr. Spickett is currently treasurer of the Society for Free Radical Research Europe and a member of the Steering Committee of the International HNE-Club, having been its secretary from 2004 to 2010. She is also very active in European collaborations, and has been a workgroup leader in the COST Actions B35 on lipid peroxidation associated disorders and CM1001 on chemistry of nonenzymatic protein modification. Dr. Spickett is also a member of the Society for Free Radical Biology and Medicine, the Biochemical Society, and a Fellow of the Society of Biology.

Henry Jay Forman is a distinguished professor of biochemistry and chemistry and founding faculty at the University of California, Merced. He is also a research professor of gerontology at the University of Southern California. He earned a BA in chemistry from Queens College of the City University of New York and then a PhD in biochemistry from Columbia University in 1971. After a postdoctoral position at Duke University, he held faculty positions in multiple disciplines at several universities including the University of Pennsylvania, the University of Southern California, and the University of Alabama, where he was the chairman of Environmental Health Sciences.

Dr. Forman's expertise is in the areas of oxidative stress and signal transduction. His major research achievements are pioneering work in mitochondrial superoxide production in which he demonstrated that manganese superoxide dismutase actually pulls the reaction forward, redox signaling in which he first demonstrated signaling by endogenously generated hydrogen peroxide, and mechanisms of induced resistance to oxidative stress in which he demonstrated the induction by quinones of the first enzyme in glutathione biosynthesis. He has over 180 peer-reviewed publications and been an invited lecturer at many national and international symposia. Dr. Forman is currently president of the Society for Free Radical Biology and Medicine and reviews editor of *Free Radical Biology & Medicine*.

Contributors

Naser M. I. Al-Aaswad
Bradford School of Pharmacy
University of Bradford
Bradford, United Kingdom

Fiorella Biasi
Department of Clinical and
    Biological Sciences
University of Turin
Turin, Italy

Valery N. Bochkov
Institute of Pharmaceutical Sciences
University of Graz
Graz, Austria

D. Allan Butterfield
Department of Chemistry
University of Kentucky
Lexington, Kentucky

Sean S. Davies
Department of Pharmacology
Vanderbilt University Medical School
Nashville, Tennessee

Klaus Eder
Institute of Animal Nutrition and
    Nutrition Physiology
Justus-Liebig-University Giessen
Giessen, Germany

Maria Fedorova
Institute of Bioanalytical Chemistry
Center for Biotechnology and
    Biomedicine
Leipzig University
Leipzig, Germany

David A. Ford
Department of Biochemistry and
    Molecular Biology
Saint Louis University
Saint Louis, Missouri

Henry Jay Forman
Life and Environmental Sciences Unit
University of California, Merced
Merced, California

and

Andrus Gerontology Center of the
    Davis School of Gerontology
University of Southern California
Los Angeles, California

Paola Gamba
Department of Clinical and Biological
    Sciences
University of Turin
Turin, Italy

Simona Gargiulo
Department of Clinical and Biological
    Sciences
University of Turin
Turin, Italy

Denise K. Gessner
Institute of Animal Nutrition and
    Nutrition Physiology
Justus-Liebig-University Giessen
Giessen, Germany

Françoise Guéraud
Institut National de la Recherche
    Agronomique
Institut National Polytechnique de
    Toulouse
Université Paul Sabatier
Toulouse, France

Ralf Hoffmann
Institute of Bioanalytical Chemistry
Center for Biotechnology and
    Biomedicine
Leipzig University
Leipzig, Germany

Xueting Jiang
Burnett School of Biomedical
    Sciences
College of Medicine
University of Central Florida
Orlando, Florida

Emilia Kansanen
Department of Biotechnology and
    Molecular Medicine
University of Eastern Finland
Kuopio, Finland

Gregor Leibundgut
Department of Medicine
University of California
La Jolla, California

and

University of Basel
Basel, Switzerland

Norbert Leitinger
Department of Pharmacology
University of Virginia
Charlottesville, Virginia

Gabriella Leonarduzzi
Department of Clinical and Biological
    Sciences
University of Turin
Turin, Italy

Anna-Liisa Levonen
Department of Biotechnology and
    Molecular Medicine
University of Eastern Finland
Kuopio, Finland

Sharon A. Murphy
Manchester Pharmacy School
University of Manchester
Manchester, United Kingdom

Chandrakala Aluganti Narasimhulu
Burnett School of Biomedical
    Sciences
College of Medicine
University of Central Florida
Orlando, Florida

Anna Nicolaou
Manchester Pharmacy School
University of Manchester
Manchester, United Kingdom

Olga Oskolkova
Institute of Pharmaceutical Sciences
University of Graz
Graz, Austria

Sampath Parthasarathy
Burnett School of Biomedical
    Sciences
College of Medicine
University of Central Florida
Orlando, Florida

Nicos A. Petasis
Department of Chemistry
University of Southern California
Los Angeles, California

Maria Philippova
Cardiovascular Research
    Laboratories
Basel University Hospitals
Basel, Switzerland

Giuseppe Poli
Department of Clinical and Biological
    Sciences
University of Turin
Turin, Italy

Tanea T. Reed
Department of Chemistry
Eastern Kentucky University
Richmond, Kentucky

Aladdin Riad
Burnett School of Biomedical
   Sciences
College of Medicine
University of Central Florida
Orlando, Florida

Robert Ringseis
Institute of Animal Nutrition and
   Nutrition Physiology
Justus-Liebig-University Giessen
Giessen, Germany

Homero Rubbo
Departamento de Bioquímica
Universidad de la República
Montevideo, Uruguay

Irene Fernandez Ruiz
Burnett School of Biomedical
   Sciences
College of Medicine
University of Central Florida
Orlando, Florida

Claus Schneider
Department of Pharmacology
Vanderbilt University Medical
   School
Nashville, Tennessee

Zachariah P. Sellers
Department of Chemistry
Eastern Kentucky University
Richmond, Kentucky

Vlad Serbulea
Department of Pharmacology
University of Virginia
Charlottesville, Virginia

Barbara Sottero
Department of Clinical and
   Biological Sciences
University of Turin
Turin, Italy

Corinne M. Spickett
School of Life and Health
   Sciences
Aston University
Birmingham, United Kingdom

Adam Taleb
Department of Medicine
University of California
La Jolla, California

and

Internal Medicine
St. Elizabeth's Medical Center
Brighton, Massachusetts

Michael J. Thomas
Wake Forest School of Medicine
Winston-Salem, North Carolina

Andrés Trostchansky
Departamento de Bioquímica
Universidad de la República
Montevideo, Uruguay

Sotirios Tsimikas
Department of Medicine
University of California
La Jolla, California

Koji Uchida
Laboratory of Food and
   Biodynamics
Graduate School of Bioagricultural
   Sciences
Nagoya University
Nagoya, Japan

Vivek Krishna Pulakazhi Venu
Burnett School of Biomedical Sciences
College of Medicine
University of Central Florida
Orlando, Florida

Kathryn Young
Burnett School of Biomedical Sciences
College of Medicine
University of Central Florida
Orlando, Florida

Introduction

LIPID OXIDATION PRODUCTS IN HEALTH AND DISEASE

OVERVIEW

Lipid oxidation was long regarded as a deleterious process that was responsible for lipid rancidity, loss of function, and generation of toxic products, but in recent years it is becoming understood that the physiological and pathological effects of lipid oxidation products are much more complex than this. Studies of lipid peroxidation began early in the twentieth century and focused on free radical mechanisms of attack, but in biological systems alternative forms of oxidation including nitration and chlorination are also important. In addition to nonenzymatic mechanisms of lipid oxidation, several enzymatic systems contribute to the complexity of the products. Even from a single lipid structure, dozens of different oxidation products can form, depending on the exact site of attack and subsequent reactions. Although the mechanisms of these modifications are quite well understood, new lipid oxidation products continue to be identified and questions remain about the molecular mechanisms responsible.

A number of lipid products, especially small, aldehyde-containing breakdown products of oxidized lipids, show highly toxic effects on cells and tissues, and it is now understood that this is related to their reactivity and ability to undergo further reactions, such as adduct formation, with proteins and nucleic acids. However, other oxidized lipid products, especially oxidized phospholipids, have more physiological effects involving the modulation gene expression and alteration of cell behavior. There are a variety of cell surface receptors (e.g., scavenger receptors and Toll-like receptors [TLRs]) and intracellular receptors (e.g., peroxisome proliferator-activated receptors [PPARs] and liver X receptors [LXRs]) that bind to oxidized lipids and initiate signaling pathways. In many cases, this leads to pro-inflammatory processes and disease pathology, although activation of PPARs can result in anti-inflammatory responses and oxidized phospholipids can interfere with signaling via TLRs in response to pathogen-associated molecular patterns, again leading to anti-inflammatory effects. Some oxidized lipid products are also able to modulate the activity of some transcription factors, most notably Nrf2-KEAP1, which leads to protective cellular responses by upregulation of antioxidant defenses. The balance of these cellular responses depends on the molecular structure and reactivity of the lipid oxidation products, as well as their concentration, and determines the outcome of physiological signaling versus disease pathology.

This book covers both established information and recent findings on the action of oxidized lipid products on cell signaling and gene expression in health and disease. It is composed of three sections: Section I "Chemistry and Biochemistry of Lipid Oxidation"; Section II "Sites of Biological Actions of Oxidized Lipids"; and Section III "Oxidized Lipids in Pathology and Disease." Within each section, chapters by specialists in the field explain the basic principles and address current ideas and recent findings on a selection of critical topics.

1 An Introduction to Redox Balance and Lipid Oxidation

Corinne M. Spickett, Maria Fedorova,
Ralf Hoffmann, and Henry Jay Forman

CONTENTS

REACTIVE SPECIES AND BIOMOLECULE MODIFICATIONS

In all living cells and tissues, there is a delicate balance between oxidizing compounds (oxidants) and reductants, called biological redox balance, which influences many of the biochemical processes in the cell and ultimately determines the cell's behavior. An excess on either side can lead to stress, either oxidative or reductive stress, if sufficiently severe. The relevant oxidizing compounds are most commonly derived from molecular oxygen by partial reduction, and as they are relatively reactive they are grouped together under the imprecise term "Reactive Oxygen Species" (ROS). They can react subsequently with small molecules or ions containing heteroatoms such as nitrogen and halogens, for example, generating other reactive nitrogen or chlorine species. These groups of molecules, often called RNS and RClS, usually contain oxygen and are still oxidizing. As any one abbreviation for all of these reactive species is as vague as another, defining ROS as any reactive species formed from oxygen is what should be used, but only when the actual species involved is uncertain.

1

Formation of any of these compounds at sufficient levels can lead to oxidative stress, which is defined as an imbalance in which the production of these reactive species is favored over the antioxidant defense system, leading to disruption of redox regulation and signaling, or in extreme conditions to molecular damage. In recent years, scientists' perceptions of oxidative stress and the role of small oxidants in cell function have changed considerably, but there is nevertheless good evidence that they play a part in various diseases and pathological conditions (Forman et al., 2014).

ROS, such as the superoxide anion radical ($O_2^{\bullet-}$), hydrogen peroxide (H_2O_2), and the hydroxyl radical ($^{\bullet}OH$), are generated under numerous conditions *in vivo* (Imlay, 2003) and by exogenous factors. Mitochondria are a major endogenous source of cellular $O_2^{\bullet-}$ and H_2O_2, even in nonpathological conditions (Forman and Boveris, 1982; Turrens, 2003; Orrenius et al., 2007), where they are formed during metabolism through leakage of electrons from the electron transport chain. Even under normal conditions up to 1% of the mitochondrial electron flow produces superoxide anions (Ott et al., 2007), which represents the primary source of $O_2^{\bullet-}$ production (Andreyev et al., 2005). Reduction of $O_2^{\bullet-}$ yields further ROS, such as H_2O_2. Subsequent reactions of hydrogen peroxide and superoxide anions (Haber–Weiss reaction) and the Fe^{2+} (or Cu^+)-driven cleavage of H_2O_2 (Fenton reaction) generate $^{\bullet}OH$, which is even more reactive and thus more toxic in cells (Imlay, 2003; Mikkelsen and Wardman, 2003). Moderate levels of H_2O_2, however, play an integral role in several physiological functions of the cell, including gene expression and signal transduction.

Cellular enzymes represent another important endogenous source of $O_2^{\bullet-}$ and H_2O_2, for example, NADPH oxidase enzymes in phagocytic cells and elsewhere (Wu et al., 2003; Qin et al., 2007). Activation of phagocytic cells, such as macrophages and neutrophils, by inflammatory cytokines leads to an oxidative burst with production of high amounts of $O_2^{\bullet-}$ and H_2O_2. Superoxide dismutase is responsible for catalyzing the dismutation of $O_2^{\bullet-}$ superoxide anions to H_2O_2 and O_2. Hydrogen peroxide, along with oxidizing compounds produced from the myeloperoxidase-catalyzed oxidation by H_2O_2 of chloride, bromide, thiocyanate anion or nitrite kill invading pathogens (Hampton et al., 1998). Hypochlorite can react further with hydrogen peroxide to form singlet oxygen, which is a well-established chemical process (Cui et al., 2011). Hypobromous acid reacts even more efficiently than HOCl to produce singlet oxygen (Kanofsky et al., 1988a). However, the production of singlet oxygen is at such low yield that the significance of its contribution to oxidative events in biology is questionable.

The situation is even more complex with regard to reactive species formed from nitric oxide ($NO^{\bullet}$), which is generated *in vivo* by NO synthases (NOS). Although $NO^{\bullet}$ *per se* is not particularly reactive with lipids directly, it reacts with molecular oxygen to give nitrogen dioxide, and very readily with $O_2^{\bullet-}$, yielding peroxynitrite ($ONOO^-$). Peroxynitrous acid ($ONOOH$, $pK_a = 6.8$) is unstable and spontaneously decomposes to a large variety of reactive species. One highly biologically relevant reaction is that with CO_2 (Pavlovic and Santaniello, 2007), which is present at high levels in cellular systems ($\sim$24 mmol/L in plasma). This reaction rapidly produces nitrosoperoxycarbonate ($ONOOCO_2^-$), which subsequently decomposes to nitrogen dioxide ($NO_2^{\bullet}$) and carbonate radical ($CO_3^{\bullet-}$). Interestingly, it has recently been demonstrated that at high $O_2^{\bullet-}$ levels (presumably produced by NADPH oxidases) most of the $NO^{\bullet}$ generated by eNOS is converted into reactive peroxynitrite, which can oxidize the eNOS cofactor tetrahydropterin, causing uncoupling of oxygen reduction from NO synthesis and leading to production of O_2^- by eNOS. The uncoupling of eNOS further amplifies O_2^- production, which may play an important role in oxidative pathogeneses (Thum et al., 2007).

Under any conditions where the oxidants described above are not removed by cellular low molecular weight antioxidants (Kanofsky et al., 1988b) or antioxidant enzymes, they may react with cellular macromolecules, yielding covalent modifications. However, it is important to note that there are no specific scavengers of hydroxyl radicals, owing to their high reactivity and rate of reaction with all biomolecules. If there is constant generation of oxidants by endogenous or exogenous sources, oxidative modifications can accumulate, affecting many cellular functions (Gracy et al., 1999). Historically, such modifications were viewed simply as undesirable damage to the macromolecules that caused dysfunction. However, it is now becoming well-accepted that while this is true in some cases, in others, the oxidatively modified molecules have novel bioactivities, some of which are important in cell signaling and contribute to cellular responses. Oxidation, nitration, or halogenation of biomolecules add new functional groups that influence the chemical, physical, and biological properties of the biomolecules, and this applies equally to lipids as to proteins.

MODIFICATION OF LIPID MOLECULES

Lipids are usually classified into eight major classes: fatty acids (FAs), glycerolipids, glycerophospholipids, sphingolipids, sterols, prenol lipids, saccharolipids, and polyketides; overall, these contribute to several thousand different lipid species. Lipids are comparatively reduced molecules, and are therefore susceptible to oxidation; indeed, this is the reason why they are excellent energy sources for catabolic metabolism. For long hydrocarbon chains, oxidation is usually energetically favoured at bis-allylic sites (Figure 1.1), so polyunsaturated fatty acids (PUFAs) are most easily oxidized, although depending on the reactivity of the oxidant, other sites can also be attacked (Spickett et al., 2010). For example, in phospholipids, some head groups can also undergo adventitious oxidation, such as phosphoethanolamine and phosphoserine (Simoes et al., 2013; Maciel et al., 2014). Lipid oxidation can be mediated by enzymes (e.g., lipoxygenases, cytochrome P450s, and cyclooxygenases) or can occur nonenzymatically resulting from reaction with some of the oxidants and free radicals described above. There are also several types of lipid oxidation: for example, free radical attack commonly causes lipid peroxidation, whereas hypohalites react by electrophilic addition, which is a 2-electron process and not a chain reaction (Spickett, 2007). Lipid nitration is thought to occur both by free radical and electrophilic addition mechanisms, depending on the particular reactive species involved (Rubbo et al., 2009). It is important to be aware of the types of reactions that can occur and their products, in order to understand the significance of the formation of different oxidation products. The lipid modifications triggered by oxidation, nitration, glycation, and glycoxidation increase the lipidome complexity enormously, perhaps by a factor of 100.

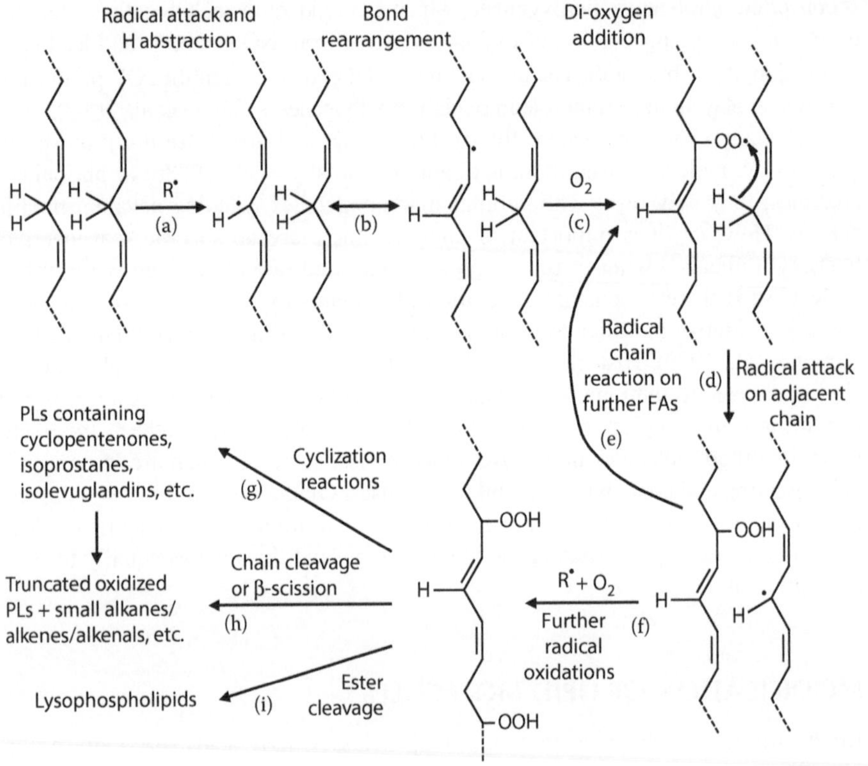

FIGURE 1.1 Radical attack and peroxidation of polyunsaturated fatty acyl chains. (a) A reactive radical (•R), such as hydroxyl radical, abstracts a hydrogen atom to form a carbon-centered radical. (b) The carbon-centered radical can rearrange to form a conjugated diene, which is more stable. (c) Addition of molecular oxygen forms a peroxyl radical (−OO•). (d) This peroxyl radical can attack a bis-allylic site in an adjacent fatty acyl chain to form a hydroperoxide (−OOH) and a new carbon-centered radical. (e) Steps c and d can be repeated many times as a radical chain reaction. (f) The hydroperoxide can also undergo further oxidations through new radical attack (i.e., steps a–d). Both the mono- and bis-hydroperoxides can undergo further reactions as indicated in steps g–i, which generates a wide range of full-chain and truncated products with a variety of different chemical groups.

Enzymatic Lipid Oxidation

Enzymatic lipid oxidation has been mostly studied for PUFAs, and is described by Murphy et al. in Chapter 3. A large variety of enzymatic oxidation products have been identified, including eicosanoids, prostaglandins, thromboxanes, leukotrienes (LT), lipoxins, and hepoxilins as well as hydroxylated, and epoxidized FAs (Massey and Nicolaou, 2011). Mostly, the enzymes act on free PUFAs, which are released from membrane-associated lipid esters by the action of phospholipase A_2 (PLA_2) (Burke and Dennis, 2009). Cyclooxygenases (COXs) can catalyze oxidation and peroxidation of arachidonic acids (AA) to produce prostanoids, such as prostaglandins (PG), prostacyclins (PGI), and thromboxanes (TX), which posses a wide range

of biological functions in mediation and resolution of inflammation and vasoconstriction (Smith et al., 2000a). Additionally, COXs mediate hydroxylation of AA to
(11R)- and (15S)-hydroxyeicosatetraenoic acid (HETE). In contrast to the constitutive COX-1, the COX-2 isoform is inducible and can oxidize linoleic acid (LA) to
9- or 13-hydroxyoxtadecadienoic acid (HODE) and docosahexaenoic acid (DHA) to
13-hydroxydocosahexaenoic acid (HDHA) (Smith et al., 2000b).

In contrast, the enzyme lipoxygenase (LOX) catalyzes hydroperoxidation of FAs
with conjugated double bonds, including AA, DHA, and LA. Hydroperoxides of FA
can be either reduced to the corresponding hydroxides by glutathione peroxidase or
oxidized further to other bioactive lipids. Different isoforms of LOX are classified
by the position of the selective hydroxylation (e.g., 5-, 8-, 12-, and 15-LOX), which
produces different oxygenated FAs (Back, 2009; Duroudier et al., 2009). 5-LOX,
for example, oxidizes AA to 5-hydroperoxyeicosotetreanoic acid (HPETE), which
can be reduced to 5-HETE or metabolized to various LT involved in immune and
inflammatory responses. 12-LOX and 15-LOX can modify not only free PUFA but
also the corresponding PL-esters (Massey and Nicolaou, 2011). Thus, AA-esterified
PL are oxidized to isoprostanes, which can be released by PLA$_2$ and further oxidized
to isolevuglandins (isoLG), which are toxic keto-aldehydes capable of modifying
nucleophilic sites in proteins and nucleic acids, as described further by Schneider
and Davies (Chapter 2) and Murphy et al. (Chapter 3).

Finally, the cytochrome P450 (CYP450) superfamily of monooxygenases can
also epoxidize and hydroxylate PUFAs. CYP450 can form epoxides on virtually any
double bound of AA, resulting in multiple positional isomers and enantiomers of
epoxyeicosatrienoic acid (EET). Further oxidation can produce various HETEs with
mid-chain or omega-hydroxylation, whereas epoxide hydrolase converts EET into
dihydroeicosatetraenoic acid (DHET). Additionally, CYP450 can oxidize cholesterol to oxysterols, with the hydroxyl group being incorporated at different positions
to yield 7α-, 24 or 25-hydroxycholesterols, which can be further oxidized and are
involved in several metabolic (e.g., bile acid synthesis) and pathological processes
(Konkel and Schunck, 2011; Westphal et al., 2011).

LIPID PEROXIDATION BY RADICAL MECHANISMS

Polyunsaturated lipids are very susceptible to reaction with free radicals, owing to
the strong reactivity of the hydrogens at the carbon between two adjacent double
bonds, as the carbon–hydrogen bond of bis-allylic positions is slightly weaker (ca.
75 kcal/mol) than for mono-allylic carbon–hydrogen bonds (88 kcal/mol). When one
of these hydrogens is abstracted by a sufficiently reactive radical, a carbon-centered
radical is formed, a bond rearrangement tends to occur and molecular oxygen can
be added to yield peroxyl radicals (explained by Murphy et al. in Chapter 2 and Poli
et al. in Chapter 15) (Figure 1.1). A fatty acyl peroxyl radical can abstract a hydrogen
atom from the bis-allylic position of an adjacent PUFA, which converts the peroxyl
radical to a hydroperoxide but generates a new carbon-centered radical, thus initiating a chain reaction of lipid peroxidation (Gutteridge and Halliwell, 1990; Moore
and Roberts, 1998). Addition of further oxygen atoms and cleavage of the lipid
hydroperoxide is common, and while if the FA contained more than one bis-allylic

pair of hydrogens, it is susceptible to radical attack at multiple sites; for example, arachidonic acid contains three bis-allylic carbons, and can form tris-hydroperoxides (Spickett et al., 1998). The early peroxyl and hydroperoxide products are relatively unstable (the former more than the latter), and can undergo a variety of rearrangements to form full chain length and cyclized products containing epoxy, keto, and hydroxy groups (Reis and Spickett, 2012). The instability of the hydroperoxy groups can alternatively trigger cleavage of the hydrocarbon chain by several mechanisms to yield chain-shortened oxidized phospholipids and complementary fragments of the original chain, such as small alkanes, alkenes, alkanals and alkenals or hydroxy derivatives thereof. The chemical reactions involved are explained much more fully by Schneider and Davies in Chapter 2 and Thomas in Chapter 7. Thus, the range of potential oxidation products even from a single lipid species is extensive.

Many of the resulting oxidative cleavage products, such as alkenals, hydroxy/oxo-alkenals, epoxy-alkenals, and γ-ketoaldehydes (Benzie, 1996), retain a double bond and can readily react with nucleophilic groups in proteins. The protein adducts are predominantly formed by Michael or Schiff base-type addition of lipid peroxidation products (LPPs) to nucleophilic Lys-, Cys-, and His-residues (Sayre et al., 2008). This process is often referred to as protein lipoxidation, and is described in detail in Chapter 6 by Uchida. The extent of these reactions increases with cellular aging due to their less effective antioxidant defense systems causing structural and functional protein alterations (Ethen et al., 2007). Moreover, secondary oxidants produced during primary lipid peroxidation can also modify phospholipid classes. For instance, 4-hydroxynonenal (HNE), hydroxyhexenal, and hydroxydodecadienal can form stable Michael and Schiff base adducts with the amino group of PE *in vitro*.

In addition to oxidative cleavage to yield chain-shortened phospholipids, oxygen addition by nonenzymatic reaction with free radicals often results in the formation of lysolipids (loss of a FA at the *sn*-1 or *sn*-2 position), diacylglycerides (loss of the head group moiety) or mono-acylglycerides (loss of the head group and one FA), either nonenzymatically or through the action of various phospholipases (Girotti, 1998).

LIPID MODIFICATION BY REACTIVE NITROGEN SPECIES

There is now good evidence that unsaturated lipids can be modified by a variety of reactive nitrogen species, leading to nitration (addition of NO_2) or nitrosation (addition of NO) among other reactions (O'Donnell and Freeman, 2001; Rubbo and Radi, 2008). These reactions can occur by radical mechanisms analogous to those described in the previous section, or by electrophilic addition or substitution, which is explained further in the next section. For example, the radical nitrogen dioxide ($^{\bullet}NO_2$) can abstract a hydrogen from a bis-allylic carbon to form a carbon centered radical, which can then react with a second $^{\bullet}NO_2$ in a radical–radical reaction to yield a nitroalkene (LNO_2) (Figure 1.2). Reactive nitrogen-containing species, including $^{\bullet}NO$, can also react with preformed lipid peroxy radicals; the intermediates are relatively unstable and decompose to give a variety of nitrated lipids products as well as epoxides and hydroxides. The half-life of $^{\bullet}NO$ is several seconds, which provides sufficient time for it to diffuse into hydrophobic membrane bilayers or lipoprotein complexes. As with lipid peroxidation where addition of multiple

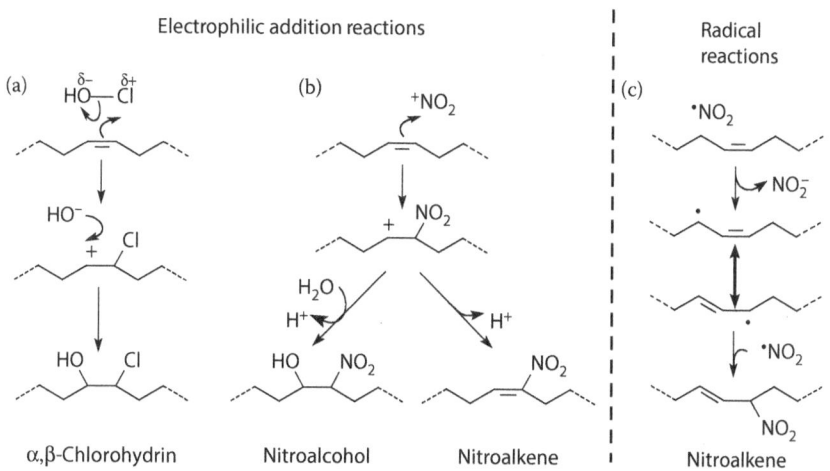

FIGURE 1.2 Halogenation and nitrations of unsaturated fatty acyl chains. (a) Electrophilic attack by hypochlorous acid (HOCl) to form a chlorohydrin. The carbocation is actually stabilized by a chloronium bridge (not shown) and the nucleophilic attack is commonly by water rather than hydroxide. (b) Electrophilic attack by a nitrosonium ion ($^+NO_2$) initially forms a carbocation, which is less stable than the chlorinated form on the left, and tends to lose a proton to form a nitroalkene. However, it can also undergo nucleophilic attack by water to form a nitroalcohol. (c) Radical attack and hydrogen abstraction by nitrogen dioxide ($^\bullet NO_2$) initially forms a carbon-centered radical, which may rearrange, and then reacts with another nitrogen dioxide in a radical–radical quenching reaction to yield a nitroalkene.

di-oxygens can occur, subsequent further reactions with RNS lead to the generation of a complex spectrum of products, including nitro- and dinitroalkenes, alkylnitrites (LONO), nitro alcohols, nitroepoxides, and nitrohydroperoxides (Pryor et al., 1985; O'Donnell and Freeman, 2001). Owing to the reactivity of several reactive nitrogen-containing species with molecular oxygen, and the possibility of peroxidation or nitration following the formation of a carbon-centered radical, the ultimate profile of products is very dependent on the $^\bullet NO$ derived to $O_2^{\bullet-}$ derived species ratio (Baker et al., 2005). Despite the large number of different nitrogen-containing lipid products demonstrated *in vitro*, the main products *in vivo* are nitro-linoleate and nitro-oleate, which have been detected in human plasma and urine (Baker et al., 2005), and found at higher levels in plasma from hypercholesterolemic patients. Nitrated cholesterol esters have also been detected in human plasma and lipoproteins (Lima et al., 2003). The formation and effects of nitrated lipids are explained more thoroughly by Trostchansky and Rubbo in Chapter 5.

ELECTROPHILIC ATTACK ON UNSATURATED LIPIDS

In addition to the radical-dependent mechanism of lipid oxidation described above, unsaturated lipids (as indeed many other types of biomolecule) can also be modified by electrophilic compounds that add across double bonds. The most commonly known oxidants that carry out such reactions are hypohalous acid (HOX), such as hypochlorous acid (HOCl) and hypobromous acid (HOBr). Biologically, these are produced by the enzymes myeloperoxidase (MPO) and eosinophil peroxidase (Spickett, 2007). MPO is present at very high levels in neutrophils and when activated in response to pathogen invasion produces mainly the highly reactive oxidizing and chlorinating agent hypochlorous acid (HOCl) (Prokopowicz et al., 2012). It reacts with unsaturated lipids by an electrophilic addition of the chloronium ion (Cl^+) across a double bond to form chlorohydrins, or α-chloro, β-hydroxy derivatives (Spalteholz et al., 2004). A similar mechanism underlies the formation of nitroalkenes or nitro alcohols by the nitrosonium ion, NO_2^+ (Rubbo and Radi, 2008) (Figure 1.2).

It has been found that plasmenyl phospholipids (containing a vinyl ether linkage) are particularly susceptible to attack by HOCl, which causes formation of α-chlorofatty aldehydes and lysolipids (loss of a FA at the *sn*-1 or *sn*-2 positions) (Ford, 2010). Interestingly, lysolipids have also been observed following chlorohydrin formation from polyunsaturated phospholipids (Arnhold et al., 2002). Finally, phospholipids with primary amino groups in the head group can be modified by HOCl as well, yielding mono- and di-chloroamines, imines, chloroamines and nitriles (Jaskolla et al., 2009). The formation of chlorinated and halogenated lipids is described in detail by Ford in Chapter 4.

LIPID GLYCATION AND GLYCOXIDATION

The term glycation refers to the nonenzymatic reaction of reducing sugars with primary amino groups yielding Schiff bases that can undergo an Amadori rearrangement, which is well studied for proteins and peptides. Glucose-derived glycation sites have been reported and characterized for many proteins over the last few decades, with the glycated N-terminus of hemoglobin A (HbA1c) being a well-established biomarker for diagnosing and controlling diabetes (Zhang et al., 2010). Similarly, primary amino groups of PE and phosphatidylserine (PS) can be glycated. Glycated PE species were detected as a major product of LDL glycation in human plasma and in red blood cells (Ravandi et al., 1996, 2000). Domingues' research group identified recently several glycated PLs when PE and PS were incubated with glucose *in vitro* and also characterized the glycated species by mass spectrometry (MS) (Simoes et al., 2013; Maciel et al., 2013a). Interestingly, glycated lipids appear to enhance lipid peroxidation and can be additionally oxidized at their *sn*-2 PUFA ester, which yields numerous glycated species. Moreover, the glycated head group can be further oxidized in a process known as glycoxidation, which yields the so-called advanced glycation end products (AGE) (Maciel et al., 2013b). The reaction pathways of AGE formation have been intensively discussed but whether they are formed via oxidative degradation of Amadori products or in separate reactions between primary amino groups and reactive products of glucose oxidation, such as dicarbonyls like methylglyoxal, still needs to be determined. AGE-modified proteins are of high biological relevance as they can activate cells of the innate immune system and endothelial cells via specific receptors (RAGE, receptor for advanced glycation end products) and thereby mediate inflammation-related processes (Vistoli et al., 2013). Several AGE-modified PL, such as carboxymethyl- and carboxyethyl-modified PE, were also detected in human plasma and erythrocytes (Shoji et al., 2010). The ability of such products to activate the immune system is discussed in Chapter 8.

WHY IS LIPID OXIDATION IMPORTANT?

Oxidative modification of lipids and phospholipids, including radical damage, halogenation and nitration, result in significant changes to the chemical properties of the molecules, which in turn have a major effect on their biochemical functions. At the most basic level, many phospholipids, as well as cholesterol, form the structural basis of cell membranes, and are responsible for the biophysical properties of membranes, including their barrier function (Kinnunen et al., 2012). Oxidation of the lipids introduces polar groups in the nonpolar region of the membrane, causing disruption of the fluid packing and potentially leading to cell lysis or more subtle structural changes that contribute to apoptosis, depending on the lipids oxidized and the extent of physical disorganization caused (Kinnunen et al., 2012). Oxidized fatty acyl chains are often not thermodynamically stable within the bilayer, and protrude as "lipid whiskers" (Greenberg et al., 2008). This can alter cell signaling and recognition by receptors on neighboring cells. The activity of many membrane proteins is affected by the phospholipids present in the local region, and oxidation of the phospholipids can therefore alter their conformation and activity. There is growing appreciation that oxidized phospholipids have a wide range of biological effects; furthermore, as described in Chapter 3 (Murphy et al.), phospholipases can result in the release of oxidized fatty acyl chains from membrane, which are well known to play a role in many types of inter- and intra-cellular signaling (Calder, 2006 and Chapter 12 by Ringseis et al.). Overall, oxidized phospholipids, oxidized FAs and oxysterols have a wide range of biological effects on cells including altered metabolism, proliferation, cell migration, and apoptosis, and there has been much work to elucidate the signaling mechanisms involved (Leonarduzzi et al., 2005; Niki, 2009; Bochkov et al., 2010), which is described in several subsequent chapters of this book. Some of the effects are beneficial, but many contribute to pathology, especially inflammation (Greig et al., 2012), and have been shown to play roles in several diseases, as discussed in the later chapters of this book (Figure 1.3).

BIOLOGICAL EFFECTS OF OXIDIZED LIPIDS

Much of the early work on effects of oxidized phospholipids was carried out to elucidate their pro-inflammatory effects, especially with relevance to atherosclerosis and the cardiovascular system. Following this, a wide range of specific effects has been reported, which can be classified into several types that are mentioned below. For example, it is well-established that oxLDL and oxPAPC (which mimics its action) can enhance monocyte–endothelial adhesion through upregulation of adhesion molecule expression on both cell types (Witztum and Steinberg, 1991; Vora et al., 1997; Watson et al., 1997). This has been well studied both in cultured cell models and *ex-vivo* tissue models, and adhesion molecules that contribute to the effects have been identified, including CS-1 fibronectin (Leitinger et al., 1999), E-selectin, and VCAM-1 (Subbanagounder et al., 2000). Similar effects have also been reported for

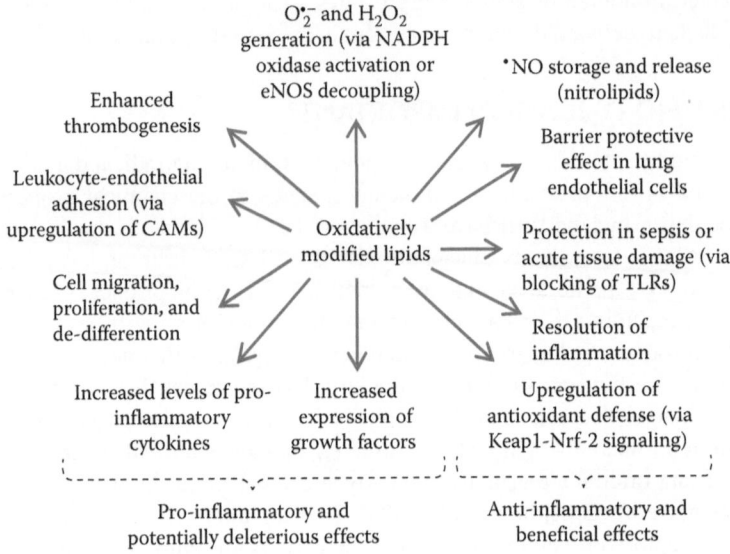

FIGURE 1.3 The diversity of biological effects of oxidatively modified lipids. A generalized scheme showing both the beneficial and deleterious effects that oxidized lipids can have overall, including the effects of oxidized phospholipids, oxidized FAs, oxysterols, halogenated phospholipids, and nitrated lipids. It should be remembered that not all oxidized lipids have these effects and indeed different individual oxidized species may have opposing effects.

the chlorohydrin product of stearoyl-oleoyl phosphatidylcholine (Dever et al., 2008). Activation of monocytes leading to extravasation is probably induced by increased production of several inflammatory cytokines; in particular, upregulation of MCP-1 has been reported and suggested to be part of the monocyte-specific actions of oxidized phospholipids. There has also been a big focus on upregulation of the chemotactic factor IL-8 by oxLDL, oxPAPC and specific short-chain oxidation products, as different signaling pathways have been reported to be involved (Lee et al., 2000; Walton et al., 2003b; Yeh et al., 2004b; Erridge et al., 2007). Other cytokines that are known to be produced at increased levels in response to oxidized lipids include IL-6, macrophage inflammatory proteins MIP-1α and β, TNFα and IL-10 (Fu and Birukov, 2009; Bochkov et al., 2010). Several oxidized lipids and oxysterols also induce expression of some growth factors, such as transforming growth factor-β (TGF-β) (Leonarduzzi et al., 2001), VEGF (Afonyushkin et al., 2010). All of these intercellular signaling molecules can induce pro-atherogenic effects.

Several other biological actions of oxidized lipids have also been described. Oxidized phospholipids can induce migration and proliferation of smooth muscle cells to form an atherosclerotic cap, which is a factor in the development of vascular lesions (Chatterjee et al., 2004). This corresponds to a phenotypic switching from a contractile to a de-differentiated proliferative cell type; this is now understood to represent a response to injury and can be considered in part as a protective response that attempts to repair the vessel (Spin et al., 2012). Also relevant to atherosclerosis are the thrombogenic effects of oxidized phospholipids, which are caused in part by increased endothelial cell expression of tissue factor, a cell-surface receptor for clotting factor VIIa, via activation of early growth response factor 1 (EGR-1) (Bochkov et al., 2002b). Altered redox balance and increased levels of ROS are thought to be important in pathological and atherogenic processes, so it is interesting that several oxidized and chlorinated lipids and phospholipids have been found to increase ROS generation. At a downstream level in endothelial cells, this can be mediated either by activation of the NADPH oxidase (Rouhanizadeh et al., 2005) or uncoupling of the endothelial nitric oxide synthase (Landar et al., 2006). The increased production of cellular $O_2^{\cdot-}$ and H_2O_2 may further contribute to the toxicity of oxidized and chlorinated phospholipids and oxysterols, which can involve either necrosis or apoptosis depending on the concentration. Toxicity of these compounds has been reported for many cell types, including endothelial cells, epithelial cells, smooth muscle cells, leukocytes, and many cultured cell lines (Salvayre et al., 2002; Dever et al., 2003, 2006; Bochkov et al., 2010).

It is critical to recognize that in contrast to these mainly deleterious effects of oxidatively modified lipids, there are also a number of anti-inflammatory and beneficial actions.

Full chain oxidized products of arachidonate-containing phospholipids have protective effects on lung endothelial cells by enhancing their barrier function in situations where this is compromised (Birukov et al., 2004; Birukova et al., 2013). Oxidized PAPC and some of its individual components show anti-inflammatory effects in acute injury or sepsis, which appears to be due to the ability to bind to and interfere with signaling from receptors for pathogen-associated molecular patterns (PAMPs) such as bacterial lipopolysaccharides (LPS) and lipopeptides (Bochkov et al., 2002a). A number of studies have demonstrated the protective effects using animal models of sterile or aseptic lung-injury, endotoxemia, and subcutaneous skin infection (Erridge et al., 2008; Fu and Birukov, 2009; Bochkov et al., 2010). These properties are discussed in detail in the chapters by Fedorova and Hoffmann (Chapter 8), Philippova et al. (Chapter 10) and Serbulae and Leitinger (Chapter 11). Moreover, several lipid oxidation and nitration products (mainly those with electrophilic reactivity) are able to induce upregulation of antioxidant enzymes through phase II responses, involving the Keap1–Nrf-2 system (Wright et al., 2009; Zhang and Forman, 2009), as described by Kansanen and Levonen in Chapter 13. Some nitrated lipids, such as nitro-linoleate and nitro-cholesterol, have beneficial effects through a completely different mechanism: they are able to decompose to release ·NO, which causes vasorelaxation and has anti-inflammatory effects (Lima et al., 2005).

Understanding of the biological activities of oxidatively modified lipids depends, at least in part, on knowledge of the receptors to which they bind and the signaling pathways that are affected. Research over the last three decades has enabled the identification of several important receptors, although it is likely that others will continue to be discovered. Much work has been carried out on the platelet activation factor (PAF) receptor, which recognizes a number of truncated oxidized phospholipids (McIntyre, 2012). A lipoprotein-associated phospholipase A_2, PAF-acetylhydrolase, can hydrolyze oxidized chains at the sn-2 position of phospholipids to yield lysolipids, which in turn are able to bind to and activate some G-protein-coupled receptors (GPCRs), such as the PGE_2 receptor EP2, which is activated by oxPAPC and PEIPC and upregulates β-integrin expression, as well as G2A, GPR4, and LPA1-4 (Fu and Birukov, 2009). Scavenger receptors of macrophages, endothelial cells, and other cells have also received a lot of attention; examples include SRA-I/II, SR-BI, CD36, and lectin-like oxidized low-density lipoprotein receptor-1 (LOX-1). These receptors are responsible for unregulated uptake of oxidized lipoproteins, often leading to foam cell formation as well as changes in cellular gene expression. CD36 has been carefully studied in terms of its specificity for a range of different oxidized phospholipids (Podrez et al., 2002a, b). Toll-like receptors (TLRs), which are responsible for detection of many pathogen- and damage-associated molecular patterns and present on many cells other than immune cells, have also been implicated in responses to oxidized phospholipids, which are structurally similar to the phosphorylated-polysaccharide with FA chains that compose lipopolysaccharide (LPS), also known as lipoglycans and endotoxin). However, there has been some controversy in this area, resulting from differences in behavior between cell types and concentration-dependent effects (Walton et al., 2003a, b; Yeh et al., 2004b; Erridge et al., 2008). These issues are discussed in much more detail in Chapter 10 by Bochkov and Chapter 11 by Leitinger. Finally, peroxisome proliferator-activated receptors (PPARs) are well established as receptors for nonpolar compounds such as oxidized FAs and steroids, but more recently evidence has been presented for their interaction with oxidized phospholipids, which is presented more fully in Chapter 12 by Ringseis et al. In contrast to the membrane-localized receptors above, these receptors are present in the cytoplasm and migrate to the nucleus following ligand binding, thus representing rather different signaling mechanisms.

While much attention has focused on receptor-mediated effects, it must also be recognized that several receptor-independent mechanisms of action are possible. Oxidized lipid products can be divided broadly into those that are reactive (usually electrophilic) and those that are not. The receptor interactions mentioned above are assumed to be mainly non-covalent, depending on structural complementarity and ligand recognition. In contrast and as mentioned in Section 2.2, reactive electrophilic products can undergo covalent reactions with many different targets (proteins, DNA, and other lipids) leading to adduct formation (Catala, 2009; Fritz and Petersen, 2013). Although for many proteins, lipoxidation causes loss of activity, and was for a long time regarded as adventitious damage, it is now emerging that occasionally an increase in activity or gain or function may occur, and consequently lipoxidation may in some cases be considered as a regulatory mechanism (Domingues et al., 2013). Indeed, adduction of either 15-deoxy- Δ-12,14-prostaglandin J2 or 4-hydroxy-2-nonenal to Keap1, activates the Nrf2 transcription factor that results in the induction of a large number of protective and repair enzymes (Levonen et al., 2004). This argument is considered further in Chapter 6, by Uchida. In addition to actions through covalent or noncovalent bonding, biophysical mechanisms may play a role in some oxidized phospholipid effects. This appears to be the situation with the sterol response element binding protein (SREBP), where oxidized phospholipids have been hypothesized to cause loss of membrane cholesterol, resulting in alterations to the structure membrane lipid rafts and triggering activation of

SREBP (Yeh et al., 2004a). Similarly, a recent study demonstrated that a very small amount of iron-catalyzed lipid peroxidation in the macrophage plasma membrane can activate a signaling cascade leading to induction of inflammatory cytokines (Premasekharan et al., 2011).

It is important to distinguish between the types of lipid-containing material that have been tested in studies on effects of oxidized lipids. For example, many studies have been carried out using oxLDL, which contains many different lipids that are oxidizable as well as the protein moiety; therefore it is usually not clear which oxidized component is responsible for any effects observed. Moreover, the different methods of oxidation commonly used (metal-catalyzed, radical initiators, HOCl or MPO, RNS) lead to different profiles of oxidation. Other studies have been carried out using more defined mixtures of oxidized lipids, such as auto-oxidized 1-palmitoyl-2-arachidonyl-3-glycerophosphocholine (oxPAPC), which contains dozens of different full chain and chain-shortened oxidation products, or oxysterol mixtures derived from oxidation of cholesterol, containing a variety of oxidation sites and functional groups. Again, using such systems it is not possible to distinguish which particular oxidized lipid is responsible for any one effect. Ultimately, this can only be achieved by studying individual purified oxidized lipids or phospholipids; many studies have adopted this approach using the (now growing) number of commercially available products, or ones isolated and purified in-house. Even so, it is important to bear in mind the proviso that the test compound may be metabolized by the cells or tissues to other bioactive compounds; this is quite hard to control for, especially if the pathways and products are not characterized for this oxPL or tissue type. This is an important issue, as there have been found to be quite different and even opposing effects of individual oxidized phospholipids: for example, full chain oxygenated phospholipids demonstrate barrier-protective effects whereas truncated oxidized phospholipids do not, and the rather similar compounds POVPC and PGPC have differential effects on adhesion molecule expression and monocyte versus neutrophil activation (Fu and Birukov, 2009). This issue is also considered in Chapter 10 (Philippova et al.). Overall, there are arguments in favor of different approaches, as the complex lipid mixtures could be considered as being more physiologically relevant, at least in cases where adventitious oxidation is the cause. The bottom line is that these approaches to studying the biological effects of oxidatively modified lipids all provide valuable information that is complementary. This section can only give a flavor of the types of biological effects that have been reported to be induced by oxidized lipid products, without detailing all the individual products that contribute to each of the effects. That kind of specific detail is provided within the main chapters of the book that address these topics.

It can be seen already from the outline of biological activities and effects of oxidized, nitrated, and chlorinated lipids above that these compounds can contribute substantially to inflammatory conditions and other pathologies, although they also have some anti-inflammatory and beneficial actions. Unsurprisingly, therefore, they have been shown to play a part in several diseases, and these are considered in greater detail in subsequent chapters on cardiovascular disease (Chapters 14 by Leibundgut et al. and Chapter 15 by Poli et al.), metabolic disorders (Chapter 17 by

Parthasarathy et al.), neurodegenerative diseases (Chapter 16 by Reed et al.), and cancer (Chapter 18 by Guéraud).

MEASURING LIPID OXIDATION

Understanding of the importance of oxidatively modified phospholipids and their products in biological systems is underpinned by their measurement and quantification in cells and tissues. The analysis of lipid oxidation and its products is as old as rubber tires, as much early work was done to study the perishing of rubber, as well as understanding rancidity in food (Gutteridge and Halliwell, 1990). Consequently, there are many relatively simple assays based on colorimetric tests for products of lipid peroxidation (Moore and Roberts, 1998). These generally involve reaction of the oxidation product with a reagent to generate a chromophore or fluorophore. One example is the ferrous oxidation of xylenol (FOX) assay for lipid hydroperoxides (Jiang et al., 1992). Another is the thiobarbituric acid reactive substances (TBARS) assay, which depending on the conditions used may measure malondialdehyde or lipid hydroperoxides (Grintzalis et al., 2013). A wide variety of protocols exist for this assay (Breusing and Grune, 2008; Spickett et al., 2010). A number of alternative aldehyde-reactive reagents exist, such as 2,4-dinitrophenylhydrazine (DNPH) or dicyclohexanedione (CHD); the former can be used either for direct spectrophotometric analysis, or in combination with antibodies against the hydrazone derivative. If possible, it is advisable to separate the analytes by HPLC before detection, in order to limit interference from contaminants and identify the products with more confidence. Indeed, some journal editors are now rejecting manuscripts where claims of lipid peroxidation are made using the TBARS assay without identification of the specific products. HPLC with luminescence detection has also been used for the detection of phospholipid and cholesterol hydroperoxides. All these methods are still in common use, and more information can be found in recent reviews (Niki, 2009; Spickett et al., 2010).

Currently, some of the most popular methods are antibody-based, in the form of ELISAs or Western blotting. Antibodies against several small lipid oxidation products (malondialdehyde, acrolein, and 4-hydroxynonenal) have been developed, but they actually recognize adducts of these reactive aldehydes with proteins (Domingues et al., 2013; Spickett, 2013). The antibodies are usually relatively specific for either histidine or lysine adducts. The application of antibodies for western blot detection of HNE adducts in neurodegenerative diseases is mentioned in Chapter 16 (Reed et al.). In other studies, natural antibodies that recognize oxidized phospholipids have been exploited, such as the antibody E06 (Tsimikas et al., 2004), which is thought to bind to POVPC, although there is some debate over the exact antigens, as mentioned by Philippova et al. in Chapter 10.

Gas chromatography–mass spectrometry (GC–MS) has long been considered the gold standard for analysis of oxidized FAs, oxysterols and isoprostanes (Milne et al., 2013), but electrospray ionization (ESI) MS interfaced with liquid chromatography has gained considerably in popularity recently. This ionization method has the advantage of minimal sample preparation as derivatization to produce volatile compounds is not required, which gives it much greater flexibility; essentially almost any ionizable molecule can be detected. These methods are information-rich and MS-based lipidomics is fast becoming the method of choice for profiling lipids in biological samples (Wenk, 2010). ESI–MS and tandem MS (MSMS) is ideal for the detection of polar and charged phospholipids and their oxidation products (Domingues et al., 2008). LC-MS(MS) has been used in many studies to identify a wide variety of different types of oxidized phospholipid, as mentioned briefly by Thomas in Chapter 7. It is also invaluable for the detection of chlorinated (Wacker et al., 2013) and nitrated lipids (Lima et al., 2002); analysis of the latter is described more fully in Chapter 5 by Trostchansky and Rubbo. The methodology is evolving, and targeted MS routines that are specific for individual oxidized lipid species or types of oxidative modification are now being reported (Morgan et al., 2010; Spickett et al., 2011; Gruber et al., 2012). These help to increase confidence in the identification, while allowing detection of a large number of products in a single chromatographic run. Although MS is not intrinsically a quantitative technique, relative quantification can be achieved using multiple reaction monitoring, and if isotopically labeled standards of the analytes are available, absolute quantification is also possible. For a recent compendium of reviews on methods for measuring lipid oxidation, we refer readers to a special issue of *Free Radical Biology & Medicine* (Forman and Moore, 2013).

SUMMARY AND PERSPECTIVES

The field of lipid oxidation has come a long way since it was first discovered, more than a century ago. From the early perception of it as undesirable accidental damage, it evolved to being understood as a process leading to a complex myriad of biological effects. The current state-of-the-art is that the effects can be either beneficial or detrimental, depending on the oxidized lipid species in question and the cellular conditions prevalent at the time, and that some lipid oxidation products even have actions that can be considered as pathophysiological signaling. The field is complicated by the fact that as well as the addition of oxygen atoms, lipids can be modified by enzyme-catalyzed and oxidant-induced reactions leading to introduction of heteroatoms (e.g., nitration and halogenation), and these products also show bioactivity. As different individual oxidized lipids can have rather diverse and even opposing actions, and even the same oxidized lipid compound can cause different effects in different cell or tissue types, a knowledge of the specific lipid oxidation products present in any particular condition is essential to future understanding of their role in cell biology and disease. Future research will need to use the advanced methodologies that are being developed to address this issue, and provide detailed information on the levels of these compounds *in vivo*. It is also important to leave behind the old-fashioned view that all lipid oxidation is bad, and keep an open mind toward the occurrence of protective effects and potential for certain oxidized lipids to be developed as therapeutics.

ABBREVIATIONS

CS1-fibronectin	Connecting segment 1-fibronectin
CYP450	Cytochrome P450

eNOS   endothelial nitric oxide synthase
HETEs   hydroxyeicosatetraenoic acids
4-HNE   4-hydroxynonenal
HPLC   high-performance liquid chromatography
Keap1   Kelch-like ECH-associated protein 1
MDA   malondialdehyde
MS   mass spectrometry
NFκB   nuclear factor-kappa B
NRF2   Nuclear factor (erythroid-derived 2)-like 2
oxPAPC   oxidized palmitoyl arachidonoyl phosphatidylcholine
PAF   platelet activating factor
PC   phosphatidylcholine
PE   phosphatidylethanolamine
PGPC   1-palmitoyl-2-glutaroyl-sn-glycero-3-phosphocholine
POVPC   1-palmitoyl-2-(5-oxovaleroyl)-sn-glycero-3-phosphocholine
PUFAs   polyunsaturated fatty acids; ROS, reactive oxygen species.

REFERENCES

Afonyushkin, T., Oskolkova, O. V., Philippova, M., Resink, T. J., Erne, P., Binder, B. R., and Bochkov, V. N. 2010. Oxidized phospholipids regulate expression of ATF4 and VEGF in endothelial cells via NRF2-dependent mechanism: Novel point of convergence between electrophilic and unfolded protein stress pathways. *Arterioscler Thromb Vasc Biol*, 30, 1007–13.

Andreyev, A. Y., Kushnareva, Y. E., and Starkov, A. A. 2005. Mitochondrial metabolism of reactive oxygen species. *Biochemistry (Mosc)*, 70, 200–14.

Arnhold, J., Osipov, A. N., Spalteholz, H., Panasenko, O. M., and Schiller, J. 2002. Formation of lysophospholipids from unsaturated phosphatidylcholines under the influence of hypochlorous acid. *Biochim Biophys Acta-General Subjects*, 1572, 91–100.

Back, M. 2009. Leukotriene signaling in atherosclerosis and ischemia. *Cardiovasc Drugs Ther*, 23, 41–8.

Baker, P. R., Lin, Y., Schopfer, F. J., Woodcock, S. R., Groeger, A. L., Batthyany, C., Sweeney, S. et al. 2005. Fatty acid transduction of nitric oxide signaling: Multiple nitrated unsaturated fatty acid derivatives exist in human blood and urine and serve as endogenous peroxisome proliferator-activated receptor ligands. *J Biol Chem*, 280, 42464–75.

Benzie, I. F. 1996. Lipid peroxidation: A review of causes, consequences, measurement and dietary influences. *Int J Food Sci Nutr*, 47, 233–61.

Birukov, K. G., Bochkov, V. N., Birukova, A. A., Kawkitinarong, K., Rios, A., Leitner, A., Verin, A. D., Bokoch, G. M., Leitinger, N., and Garcia, J. G. 2004. Epoxycyclopentenone-containing oxidized phospholipids restore endothelial barrier function via Cdc42 and Rac. *Circ Res*, 95, 892–901.

Birukova, A. A., Starosta, V., Tian, X., Higginbotham, K., Koroniak, L., Berliner, J. A., and Birukov, K. G. 2013. Fragmented oxidation products define barrier disruptive endothelial cell response to OxPAPC. *Transl Res*, 161, 495–504.

Bochkov, V. N., Kadl, A., Huber, J., Gruber, F., Binder, B. R., and Leitinger, N. 2002a. Protective role of phospholipid oxidation products in endotoxin-induced tissue damage. *Nature*, 419, 77–81.

Bochkov, V. N., Mechtcheriakova, D., Lucerna, M., Huber, J., Malli, R., Graier, W. F., Hofer, E., Binder, B. R., and Leitinger, N. 2002b. Oxidized phospholipids stimulate tissue factor expression in human endothelial cells via activation of ERK/EGR-1 and Ca(++)/NFAT. *Blood*, 99, 199–206.

Bochkov, V. N., Oskolkova, O. V., Birukov, K. G., Levonen, A. L., Binder, C. J., and Stockl, J. 2010. Generation and biological activities of oxidized phospholipids. *Antioxid Redox Signal*, 12, 1009–59.

Breusing, N. and Grune, T. 2008. Regulation of proteasome-mediated protein degradation during oxidative stress and aging. *Biol Chem*, 389, 203–9.

Burke, J. E. and Dennis, E. A. 2009. Phospholipase A2 structure/function, mechanism, and signaling. *J Lipid Res*, 50 Suppl, S237–42.

Calder, P. C. 2006. Polyunsaturated fatty acids and inflammation. *Prostaglandins Leukotrienes and Essential Fatty Acids*, 75, 197–202.

Catala, A. 2009. Lipid peroxidation of membrane phospholipids generates hydroxy-alkenals and oxidized phospholipids active in physiological and/or pathological conditions. *Chem Phys Lipids*, 157, 1–11.

Chatterjee, S., Berliner, J. A., Subbanagounder, G. G., Bhunia, A. K., and Koh, S. 2004. Identification of a biologically active component in minimally oxidized low density lipoprotein (MM-LDL) responsible for aortic smooth muscle cell proliferation. *Glycoconj J*, 20, 331–8.

Cui, R. R., Deng, L. Z., Shi, W. B., Yang, H. P., Sha, G. H., and Zhang, C. H. 2011. Liquid-liquid reaction of hydrogen peroxide and sodium hypochlorite for the production of singlet oxygen in a centrifugal flow singlet oxygen generator. *Quantum Electron*, 41, 139–144.

Dever, G., Stewart, L. J., Pitt, A. R., and Spickett, C. M. 2003. Phospholipid chlorohydrins cause ATP depletion and toxicity in human myeloid cells. *FEBS Lett*, 540, 245–50.

Dever, G., Wainwright, C. L., Kennedy, S., and Spickett, C. M. 2006. Fatty acid and phospholipid chlorohydrins cause cell stress and endothelial adhesion. *Acta Biochim Pol*, 53, 761–8.

Dever, G. J., Benson, R., Wainwright, C. L., Kennedy, S., and Spickett, C. M. 2008. Phospholipid chlorohydrin induces leukocyte adhesion to ApoE– /– mouse arteries via upregulation of P-selectin. *Free Radic Biol Med*, 44, 452–63.

Domingues, M. R., Reis, A., and Domingues, P. 2008. Mass spectrometry analysis of oxidized phospholipids. *Chem Phys Lipids,* 156, 1–12.

Domingues, R. M., Domingues, P., Melo, T., Perez-sala, D., Reis, A., and Spickett, C. M. 2013. Lipoxidation adducts with peptides and proteins: Deleterious modifications or signaling mechanisms? *J Proteomics*, 92, 110–31.

Duroudier, N. P., Tulah, A. S., and Sayers, I. 2009. Leukotriene pathway genetics and pharmacogenetics in allergy. *Allergy*, 64, 823–39.

Erridge, C., Kennedy, S., Spickett, C. M., and Webb, D. J. 2008. Oxidized phospholipid inhibition of toll-like receptor (TLR) signaling is restricted to TLR2 and TLR4: Roles for CD14, LPS-binding protein, and MD2 as targets for specificity of inhibition. *J Biol Chem*, 283, 24748–59.

Erridge, C., Webb, D. J., and Spickett, C. M. 2007. Toll-like receptor 4 signalling is neither sufficient nor required for oxidised phospholipid mediated induction of interleukin-8 expression. *Atherosclerosis*, 193, 77–85.

Ethen, C. M., Reilly, C., Feng, X., Olsen, T. W., and Ferrington, D. A. 2007. Age-related macular degeneration and retinal protein modification by 4-hydroxy-2-nonenal. *Invest Ophthalmol Vis Sci*, 48, 3469–79.

Ford, D. A. 2010. Lipid oxidation by hypochlorous acid: Chlorinated lipids in atherosclerosis and myocardial ischemia. *Clin Lipidol*, 5, 835–852.

Forman H. J. and Boveris, A. 1982. Superoxide radical and hydrogen peroxide in mitochondria. In: Pryor WA, editor. *Free Radical in Biology,* New York: Academic Press; pp. 65–90.

Forman, H. J. and Moore, K. E. 2013. Methods of lipid oxidation product identification and quantification. *Free Radic Biol Med*, 59, 1–108.

Forman, H. J., Ursini, F., and Maiorino, M. 2014. An overview of mechanisms of redox signaling. *J Mol Cell Cardiol*, 73, 2–9.

Fritz, K. S. and Petersen, D. R. 2013. An overview of the chemistry and biology of reactive aldehydes. *Free Radic Biol Med*, 59, 85–91.

Fu, P. F. and Birukov, K. G. 2009. Oxidized phospholipids in control of inflammation and endothelial barrier. *Transl Res*, 153, 166–176.

Girotti, A. W. 1998. Lipid hydroperoxide generation, turnover, and effector action in biological systems. *J Lipid Res*, 39, 1529–42.

Gracy, R. W., Talent, J. M., Kong, Y., and Conrad, C. C. 1999. Reactive oxygen species: The unavoidable environmental insult? *Mutat Res*, 428, 17–22.

Greenberg, M. E., Li, X. M., Gugiu, B. G., Gu, X., Qin, J., Salomon, R. G., and Hazen, S. L. 2008. The lipid whisker model of the structure of oxidized cell membranes. *J Biol Chem*, 283, 2385–96.

Greig, F. H., Kennedy, S., and Spickett, C. M. 2012. Physiological effects of oxidized phospholipids and their cellular signaling mechanisms in inflammation. *Free Radic Biol Med*, 52, 266–80.

Grintzalis, K., Zisimopoulos, D., Grune, T., Weber, D., and Georgiou, C. D. 2013. Method for the simultaneous determination of free/protein malondialdehyde and lipid/protein hydroperoxides. *Free Radic Biol Med*, 59, 27–35.

Gruber, F., Bicker, W., Oskolkova, O. V., Tschachler, E., and Bochkov, V. N. 2012. A simplified procedure for semi-targeted lipidomic analysis of oxidized phosphatidylcholines induced by UVA irradiation. *J Lipid Res*, 53, 1232–42.

Gutteridge, J. M. and Halliwell, B. 1990. The measurement and mechanism of lipid peroxidation in biological systems. *Trends Biochem Sci*, 15, 129–35.

Hampton, M. B., Kettle, A. J., and Winterbourn, C. C. 1998. Inside the neutrophil phagosome: Oxidants, myeloperoxidase, and bacterial killing. *Blood*, 92, 3007–17.

Imlay, J. A. 2003. Pathways of oxidative damage. *Annu Rev Microbiol*, 57, 395–418.

Jaskolla, T., Fuchs, B., Karas, M., and Schiller, J. 2009. The new matrix 4-chloro-alpha-cyanocinnamic acid allows the detection of phosphatidylethanolamine chloramines by MALDI-TOF mass spectrometry. *J Am Soc Mass Spectrom*, 20, 867–74.

Jiang, Z.-Y., Hunt, J. V., and Wolff, S. P. 1992. Ferrous ion oxidation in the presence of xylenol orange for detection of lipid hydroperoxide in low density lipoprotein. *Anal Biochem*, 202, 384–389.

Kanofsky, J. R., Hoogland, H., Wever, R., and Weiss, S. J. 1988a. Singlet oxygen production by human eosinophils. *J Biol Chem*, 263, 9692–6.

Kanofsky, J. R., Hoogland, H., Wever, R., and Weiss, S. J. 1988b. Singlet oxygen production by human eosinophils. *J Biol Chem*, 263, 9692–9696.

Kinnunen, P. K., Kaarniranta, K., and Mahalka, A. K. 2012. Protein-oxidized phospholipid interactions in cellular signaling for cell death: From biophysics to clinical correlations. *Biochim Biophys Acta*, 1818, 2446–55.

Konkel, A. and Schunck, W. H. 2011. Role of cytochrome P450 enzymes in the bioactivation of polyunsaturated fatty acids. *Biochim Biophys Acta*, 1814, 210–22.

Landar, A., Zmijewski, J. W., Dickinson, D. A., Le Goffe, C., Johnson, M. S., Milne, G. L., Zanoni, G., Vidari, G., Morrow, J. D., and Darley-Usmar, V. M. 2006. Interaction of electrophilic lipid oxidation products with mitochondria in endothelial cells and formation of reactive oxygen species. *Am J Physiol Heart Circ Physiol*, 290, H1777–87.

Lee, H., Shi, W., Tontonoz, P., Wang, S., Subbanagounder, G., Hedrick, C. C., Hama, S. et al. 2000. Role for peroxisome proliferator-activated receptor alpha in oxidized phospholipid-induced synthesis of monocyte chemotactic protein-1 and interleukin-8 by endothelial cells. *Circ Res*, 87, 516–21.

Leitinger, N., Tyner, T. R., Oslund, L., Rizza, C., Subbanagounder, G., Lee, H., Shih, P. T. et al. 1999. Structurally similar oxidized phospholipids differentially regulate endothelial binding of monocytes and neutrophils. *Proc Natl Acad Sci USA*, 96, 12010–5.

Leonarduzzi, G., Sevanian, A., Sottero, B., Arkan, M. C., Biasi, F., Chiarpotto, E., Basaga, H., and Poli, G. 2001. Up-regulation of the fibrogenic cytokine TGF-beta1 by oxysterols: A mechanistic link between cholesterol and atherosclerosis. *FASEB J*, 15, 1619–21.

Leonarduzzi, G., Sottero, B., Verde, V., Poli, G., Preedy, V., and Watson, R. 2005. Oxidized products of cholesterol: Toxic effects. *Rev Food Nutr Tox*, 3, 129–164.

Levonen, A. L., Landar, A., Ramachandran, A., Ceaser, E. K., Dickinson, D. A., Zanoni, G., Morrow, J. D., and Darley-Usmar, V. M. 2004. Cellular mechanisms of redox cell signalling: Role of cysteine modification in controlling antioxidant defences in response to electrophilic lipid oxidation products. *Biochem J*, 378, 373–382.

Lima, E. S., Bonini, M. G., Augusto, O., Barbeiro, H. V., Souza, H. P., and Abdalla, D. S. 2005. Nitrated lipids decompose to nitric oxide and lipid radicals and cause vasorelaxation. *Free Radic Biol Med*, 39, 532–9.

Lima, E. S., Di Mascio, P., and Abdalla, D. S. 2003. Cholesteryl nitrolinoleate, a nitrated lipid present in human blood plasma and lipoproteins. *J Lipid Res*, 44, 1660–6.

Lima, E. S., Di Mascio, P., Rubbo, H., and Abdalla, D. S. 2002. Characterization of linoleic acid nitration in human blood plasma by mass spectrometry. *Biochemistry*, 41, 10717–22.

Maciel, E., Da Silva, R. N., Simoes, C., Melo, T., Ferreira, R., Domingues, P., and Domingues, M. R. 2013a. Liquid chromatography-tandem mass spectrometry of phosphatidylserine advanced glycated end products. *Chem Phys Lipids*, 174, 1–7.

Maciel, E., Faria, R., Santinha, D., Domingues, M. R., and Domingues, P. 2013b. Evaluation of oxidation and glyco-oxidation of 1-palmitoyl-2-arachidonoyl-phosphatidylserine by LC-MS/MS. *J Chromatogr B Analyt Technol Biomed Life Sci*, 929, 76–83.

Maciel, E., Neves, B. M., Santinha, D., Reis, A., Domingues, P., Teresa Cruz, M., Pitt, A. R., Spickett, C. M., and Domingues, M. R. 2014. Detection of phosphatidylserine with a modified polar head group in human keratinocytes exposed to the radical generator AAPH. *Arch Biochem Biophys*, 548, 38–45.

Massey, K. A. and Nicolaou, A. 2011. Lipidomics of polyunsaturated-fatty-acid-derived oxygenated metabolites. *Biochem Soc Trans*, 39, 1240–6.

Mcintyre, T. M. 2012. Bioactive oxidatively truncated phospholipids in inflammation and apoptosis: Formation, targets, and inactivation. *Biochim Biophys Acta-Biomembr*, 1818, 2456–2464.

Mikkelsen, R. B. and Wardman, P. 2003. Biological chemistry of reactive oxygen and nitrogen and radiation-induced signal transduction mechanisms. *Oncogene*, 22, 5734–54.

Milne, G. L., Gao, B., Terry, E. S., Zackert, W. E., and Sanchez, S. C. 2013. Measurement of F2- isoprostanes and isofurans using gas chromatography-mass spectrometry. *Free Radic Biol Med*, 59, 36–44.

Moore, K. and Roberts, L. J., 2nd 1998. Measurement of lipid peroxidation. *Free Radic Res*, 28, 659–71.

Morgan, A. H., Hammond, V. J., Morgan, L., Thomas, C. P., Tallman, K. A., Garcia-Diaz, Y. R., Mcguigan, C. et al. 2010. Quantitative assays for esterified oxylipins generated by immune cells. *Nat Protoc*, 5, 1919–31.

Niki, E. 2009. Lipid peroxidation: Physiological levels and dual biological effects. *Free Radic Biol Med*, 47, 469–84.

O'Donnell, V. B. and Freeman, B. A. 2001. Interactions between nitric oxide and lipid oxidation pathways—Implications for vascular disease. *Circ Res*, 88, 12–21.

Orrenius, S., Gogvadze, V., and Zhivotovsky, B. 2007. Mitochondrial oxidative stress: Implications for cell death. *Annu Rev Pharmacol Toxicol*, 47, 143–83.

Ott, M., Gogvadze, V., Orrenius, S., and Zhivotovsky, B. 2007. Mitochondria, oxidative stress and cell death. *Apoptosis*, 12, 913–22.

Pavlovic, R. and Santaniello, E. 2007. Peroxynitrite and nitrosoperoxycarbonate, a tightly connected oxidizing-nitrating couple in the reactive nitrogen-oxygen species family: New perspectives for protection from radical-promoted injury by flavonoids. *J Pharm Pharmacol*, 59, 1687–95.

Podrez, E. A., Poliakov, E., Shen, Z., Zhang, R., Deng, Y., Sun, M., Finton, P. J. et al. 2002a. A novel family of atherogenic oxidized phospholipids promotes macrophage foam cell formation via the scavenger receptor CD36 and is enriched in atherosclerotic lesions. *J Biol Chem*, 277, 38517–23.

Podrez, E. A., Poliakov, E., Shen, Z., Zhang, R., Deng, Y., Sun, M., Finton, P. J. et al. 2002b. Identification of a novel family of oxidized phospholipids that serve as ligands for the macrophage scavenger receptor CD36. *J Biol Chem*, 277, 38503–16.

Premasekharan, G., Nguyen, K., Contreras, J., Ramon, V., Leppert, V. J., and Forman, H. J. 2011. Iron-mediated lipid peroxidation and lipid raft disruption in low-dose silica-induced macrophage cytokine production. *Free Radic Biol Med*, 51, 1184–1194.

Prokopowicz, Z., Marcinkiewicz, J., Katz, D. R., and Chain, B. M. 2012. Neutrophil myeloperoxidase: Soldier and statesman. *Arch Immunol Ther Exp (Warsz)*, 60, 43–54.

Pryor, W. A., Castle, L., and Church, D. F. 1985. Nitrosation of organic hydroperoxides by nitrogen dioxide/dinitrogen tetraoxide. *J Am Chem Soc*, 107, 211–217.

Qin, F., Simeone, M., and Patel, R. 2007. Inhibition of NADPH oxidase reduces myocardial oxidative stress and apoptosis and improves cardiac function in heart failure after myocardial infarction. *Free Radic Biol Med*, 43, 271–81.

Ravandi, A., Kuksis, A., Marai, L., Myher, J. J., Steiner, G., Lewisa, G., and Kamido, H. 1996. Isolation and identification of glycated aminophospholipids from red cells and plasma of diabetic blood. *FEBS Lett*, 381, 77–81.

Ravandi, A., Kuksis, A., and Shaikh, N. A. 2000. Glucosylated glycerophosphoethanolamines are the major LDL glycation products and increase LDL susceptibility to oxidation: Evidence of their presence in atherosclerotic lesions. *Arterioscler Thromb Vasc Biol*, 20, 467–77.

Reis, A. and Spickett, C. M. 2012. Chemistry of phospholipid oxidation. *Biochim Biophys Acta*, 1818, 2374–87.

Rouhanizadeh, M., Hwang, J., Clempus, R. E., Marcu, L., Lassegue, B., Sevanian, A., and Hsiai, T. K. 2005. Oxidized-1-palmitoyl-2-arachidonoyl-sn-glycero-3-phosphorylcholine induces vascular endothelial superoxide production: Implication of NADPH oxidase. *Free Radic Biol Med*, 39, 1512–22.

Rubbo, H. and Radi, R. 2008 Protein and lipid nitration: Role in redox signaling and injury. *Biochim Biophys Acta*, 1780, 1318–24.

Rubbo, H., Trostchansky, A., and O'Donnell, V. B. 2009. Peroxynitrite-mediated lipid oxidation and nitration: Mechanisms and consequences. *Arch Biochem Biophys*, 484, 167–72.

Salvayre, R., Auge, N., Benoist, H., and Negre-Salvayre, A. 2002. Oxidized low-density lipoprotein-induced apoptosis. *Biochim Biophys Acta*, 1585, 213–21.

Sayre, L. M., Perry, G., and Smith, M. A. 2008. Oxidative stress and neurotoxicity. *Chem Res Toxicol*, 21, 172–88.

Shoji, N., Nakagawa, K., Asai, A., Fujita, I., Hashiura, A., Nakajima, Y., Oikawa, S., and Miyazawa, T. 2010. LC-MS/MS analysis of carboxymethylated and carboxyethylated phosphatidylethanolamines in human erythrocytes and blood plasma. *J Lipid Res*, 51, 2445–53.

Simoes, C., Silva, A. C., Domingues, P., Laranjeira, P., Paiva, A., and Domingues, M. R. 2013. Phosphatidylethanolamines glycation, oxidation, and glycoxidation: Effects on monocyte and dendritic cell stimulation. *Cell Biochem Biophys*, 66, 477–87.

Smith, W. L., Dewitt, D. L., and Garavito, R. M. 2000a. Cyclooxygenases: Structural, cellular, and molecular biology. *Annu Rev Biochem*, 69, 145–82.

Smith, W. L., Rieke, C. J., Thuresson, E. D., Mulichak, A. M., and Garavito, R. M. 2000b. Fatty-acid substrate interactions with cyclo-oxygenases. *Ernst Schering Res Found Workshop*, 31, 53–64.

Spalteholz, H., Wenske, K., Panasenko, O. M., Schiller, J., and Arnhold, J. 2004. Evaluation of products upon the reaction of hypohalous acid with unsaturated phosphatidylcholines. *Chem Phys Lipids*, 129, 85–96.

Spickett, C. M. 2007. Chlorinated lipids and fatty acids: An emerging role in pathology. *Pharmacol Ther*, 115, 400–9.

Spickett, C. M. 2013 The lipid peroxidation product 4-hydroxy-2-nonenal: Advances in chemistry and analysis. *Redox Biol*, 1, 145–152.

Spickett, C. M., Pitt, A. R., and Brown, A. J. 1998. Direct observation of lipid hydroperoxides in phospholipid vesicles by electrospray mass spectrometry. *Free Radic Biol Med*, 25, 613–20.

Spickett, C. M., Reis, A., and Pitt, A. R. 2011. Identification of oxidized phospholipids by electrospray ionization mass spectrometry and LC-MS using a QQLIT instrument. *Free Radic Biol Med*, 51, 2133–49.

Spickett, C. M., Wiswedel, I., Siems, W., Zarkovic, K., and Zarkovic, N. 2010. Advances in methods for the determination of biologically relevant lipid peroxidation products. *Free Radic Res*, 44, 1172–202.

Spin, J. M., Maegdefessel, L., and Tsao, P. S. 2012. Vascular smooth muscle cell phenotypic plasticity: Focus on chromatin remodelling. *Cardiovasc Res*, 95, 147–55.

Subbanagounder, G., Leitinger, N., Schwenke, D. C., Wong, J. W., Lee, H., Rizza, C., Watson, A. D., Faull, K. F., Fogelman, A. M., and Berliner, J. A. 2000. Determinants of bioactivity of oxidized phospholipids. Specific oxidized fatty acyl groups at the sn-2 position. *Arterioscler Thromb Vasc Biol*, 20, 2248–54.

Thum, T., Fraccarollo, D., Schultheiss, M., Froese, S., Galuppo, P., Widder, J. D., Tsikas, D., Ertl, G., and Bauersachs, J. 2007. Endothelial nitric oxide synthase uncoupling impairs endothelial progenitor cell mobilization and function in diabetes. *Diabetes*, 56, 666–74.

Tsimikas, S., Lau, H. K., Han, K. R., Shortal, B., Miller, E. R., Segev, A., Curtiss, L. K., Witztum, J. L., and Strauss, B. H. 2004. Percutaneous coronary intervention results in acute increases in oxidized phospholipids and lipoprotein(a): Short-term and long-term immunologic responses to oxidized low-density lipoprotein. *Circulation*, 109, 3164–70.

Turrens, J. F. 2003. Mitochondrial formation of reactive oxygen species. *J Physiol*, 552, 335–44.

Vistoli, G., De maddis, D., Cipak, A., Zarkovic, N., Carini, M., and Aldini, G. 2013. Advanced glycoxidation and lipoxidation end products (AGEs and ALEs): An overview of their mechanisms of formation. *Free Radic Res*, 47 (Suppl 1), 3–27.

Vora, D. K., Fang, Z. T., Liva, S. M., Tyner, T. R., Parhami, F., Watson, A. D., Drake, T. A., Territo, M. C., and Berliner, J. A. 1997. Induction of P-selectin by oxidized lipoproteins. Separate effects on synthesis and surface expression. *Circ Res*, 80, 810–8.

Wacker, B. K., Albert, C. J., Ford, B. A., and Ford, D. A. 2013. Strategies for the analysis of chlorinated lipids in biological systems. *Free Radic Biol Med*, 59, 92–9.

Walton, K. A., Cole, A. L., Yeh, M., Subbanagounder, G., Krutzik, S. R., Modlin, R. L., Lucas, R. M. et al. 2003a. Specific phospholipid oxidation products inhibit ligand activation of toll-like receptors 4 and 2. *Arterioscler Thromb Vasc Biol*, 23, 1197–203.

Walton, K. A., Hsieh, X., Gharavi, N., Wang, S., Wang, G., Yeh, M., Cole, A. L., and Berliner, J. A. 2003b. Receptors involved in the oxidized 1-palmitoyl-2-arachidonoyl-sn-glycero-3-phosphorylcholine-mediated synthesis of interleukin-8. A role for Toll-like receptor 4 and a glycosylphosphatidylinositol-anchored protein. *J Biol Chem*, 278, 29661–6.

Watson, A. D., Leitinger, N., Navab, M., Faull, K. F., Horkko, S., Witztum, J. L., Palinski, W. et al. 1997. Structural identification by mass spectrometry of oxidized phospholipids in minimally oxidized low density lipoprotein that induce monocyte/endothelial interactions and evidence for their presence in vivo. *J Biol Chem*, 272, 13597–607.

Wenk, M. R. 2010. Lipidomics: New tools and applications. *Cell*, 143, 888–95.

Westphal, C., Konkel, A., and Schunck, W. H. 2011. CYP-eicosanoids—A new link between omega-3 fatty acids and cardiac disease? *Prostaglandins Other Lipid Mediat*, 96, 99–108.

Witztum, J. L. and Steinberg, D. 1991. Role of oxidized low density lipoprotein in atherogenesis. *J Clin Invest*, 88, 1785–92.

Wright, M. M., Kim, J., Hock, T. D., Leitinger, N., Freeman, B. A., and Agarwal, A. 2009. Human haem oxygenase-1 induction by nitro-linoleic acid is mediated by cAMP, AP-1 and E-box response element interactions. *Biochem J*, 422, 353–61.

Wu, D. C., Teismann, P., Tieu, K., Vila, M., Jackson-Lewis, V., Ischiropoulos, H., and Przedborski, S. 2003. NADPH oxidase mediates oxidative stress in the 1-methyl-4-phenyl-1,2,3,6-tetrahydropyridine model of Parkinson's disease. *Proc Natl Acad Sci USA*, 100, 6145–50.

Yeh, M., Cole, A. L., Choi, J., Liu, Y., Tulchinsky, D., Qiao, J. H., Fishbein, M. C. et al. 2004a. Role for sterol regulatory element-binding protein in activation of endothelial cells by phospholipid oxidation products. *Circ Res*, 95, 780–8.

Yeh, M., Gharavi, N. M., Choi, J., Hsieh, X., Reed, E., Mouillesseaux, K. P., Cole, A. L., Reddy, S. T., and Berliner, J. A. 2004b. Oxidized phospholipids increase interleukin 8 (IL-8) synthesis by activation of the c-src/signal transducers and activators of transcription (STAT)3 pathway. *J Biol Chem*, 279, 30175–81.

Zhang, H. and Forman, H. J. 2009. Signaling pathways involved in phase II gene induction by alpha, beta-unsaturated aldehydes. *Toxicol Ind Health*, 25, 269–78.

Zhang, X., Gregg, E. W., Williamson, D. F., Barker, L. E., Thomas, W., Bullard, K. M., Imperatore, G., Williams, D. E., and Albright, A. L. 2010. A1C level and future risk of diabetes: A systematic review. *Diabetes Care*, 33, 1665–73.

Section I

Chemistry and Biochemistry of Lipid Oxidation

2 Nonenzymatic Mechanisms of Lipid Oxidation

Claus Schneider and Sean S. Davies

CONTENTS

INTRODUCTION

Fatty acids with two or more double bonds, such as linoleic acid (LA; C18:2ω6) and arachidonic acid (AA; C20:4ω6), readily undergo autoxidation due to the ease by which a hydrogen atom can be abstracted from the molecule. Hydrogen abstraction occurs preferentially from the *bis*-allylic methylene and forms a carbon-centered radical that is delocalized over the pentadiene system. The fatty acid pentadienyl radical readily reacts with molecular oxygen to form a peroxyl radical. The peroxyl radical propagates the chain reaction by abstracting a *bis*-allylic hydrogen from an adjacent polyunsaturated fatty acid (PUFA). Thus, by cycling between carbon-centered and oxygen-centered radical forms, the initial radical generates multiple fatty acid hydroperoxides while fatty acid and O_2 are fed into the reaction chain. But the chain reaction does not go on indefinitely and termination occurs when two radical intermediates combine to form nonradical products.

Basis reaction steps of lipid peroxidation

A number of excellent reviews have been published over the years summarizing the fundamental reaction steps and products of lipid peroxidation [1–6]. For the purpose of this introductory chapter, we will first highlight the principal reactions involved, then describe how these reactions are involved in the formation of relevant secondary transformation products, and conclude with a section on recent discoveries in the field.

PRINCIPAL REACTION PATHWAYS FOR FATTY ACID PEROXIDES

Reaction of molecular oxygen, a diradical, with the fatty acid pentadienyl radical is diffusion limited as the activation energy of this reaction is essentially zero. In contrast, the resulting fatty acid peroxyl radical has a considerable lifetime, enabling it to undergo several possibilities for further reaction. The likelihood of an individual fatty acid peroxyl radical to form a particular product depends on the number of available double bonds in the molecule, the concentration of molecular oxygen, the number of alternative hydrogen donors such as vitamin E in the local environment, and the reaction rate of each specific pathway.

Four principal reactions of the peroxyl radical are important for the outcome of lipid peroxidation: formation of a fatty acid hydroperoxide, β-fragmentation, addition to a double bond, and formation of peroxide bridges.

FORMATION OF FATTY ACID HYDROPEROXIDES

The primary products of lipid peroxidation are fatty acid hydroperoxides with a conjugated E,Z-double bond system (e.g. from linoleic acid 13-hydroperoxy-9Z,11E-octadecadienoic acid, 13-HPODE). The peroxyl radical forms a hydroperoxide preferentially by abstracting a hydrogen atom from a hydrogen donor such as another PUFA. This propagates the reaction by handing the radical from one fatty acid molecule to the next. The newly formed fatty acyl instantly combines with oxygen to give a peroxyl radical. The chain continues.

In the presence of an antioxidant such as vitamin E (α-tocopherol), the peroxyl radical abstracts the hydrogen atom from the antioxidant, and the radical chain reaction is interrupted [7]. Hydrogen abstraction from the antioxidant itself generates a radical. The antioxidant radical, however, has greatly increased stability compared to the fatty acid peroxyl such that it fails to initiate another peroxidation chain reaction. The more efficient at transferring a hydrogen atom to the peroxyl radical, the better the antioxidant. Importantly, antioxidants do not intercept the carbon-centered fatty acyl radical but only the peroxyl radical.

β-Fragmentation

A second option for reaction of the peroxyl radical is to revert back to a carbon-centered radical and release O_2. Because the bond in β-position to the radical is cleaved in this process, the reaction is called β-fragmentation [8,9]. β-Fragmentation only temporarily retards the peroxidation process because the regenerated fatty acyl radical will eventually react with molecular oxygen in a way that leads to stable incorporation of O_2.

E,Z configuration   β-fragmentation   E,E configuration

Switching of an E,Z conjugated diene to a Z,Z diene as a result of β-fragmentation

β-Fragmentation accounts for the change of the double bond configuration of hydroperoxides that is characteristic of prolonged autoxidation reactions. While the initially formed hydroperoxides (the kinetic products) have E,Z configuration, with time the more stable E,E configuration hydroperoxides predominate (the thermodynamic products) [10]. The change in double bond configuration occurs as follows: Hydroperoxides can revert back to peroxyl radicals, for example, by exchange of the hydrogen atom with another peroxyl. The peroxyl radical allows free rotation at the position of the original 1,2 double bond, giving a transoid configuration. If β-fragmentation occurs at this point, the resulting pentadienyl radical will have the 1,2 carbon bond in the E configuration. Addition of oxygen at the opposite end of the pentadiene results in reformation of the conjugated diene, now as E,E. Thus, via this sequence of reactions a kinetic product such as 13(9Z,11E)-HPODE changes into a thermodynamic product, 9(10E,12E)-HPODE [10].

Formation of Endoperoxides

The presence of a double bond two carbons away from the peroxyl radical enables attack of the peroxyl radical on this double bond to form an endoperoxide and a carbon-centered radical [11–13]. The 1,2-dioxolanyl radical generated by 5-*exo* cyclization is a key intermediate in the formation of secondary products such as serial cyclic endoperoxides, isoprostanes, and diepoxides.

Peroxyl radical   1,2-Dioxolanyl radical

5-*Exo* cyclization of a peroxyl radical

The 1,2-dioxolanyl radical can react with a second molecule of oxygen. If an additional double bond is present in the appropriate position, the resulting peroxyl radical can once again react with this double bond giving rise to serial cyclic endoperoxides [14–16].

Serial cyclic endoperoxides

Formation of serial cyclic endoperoxides

Alternatively, the 1,2-dioxolanyl radical can react with the double bond adjacent to the endoperoxide to create a highly unstable cyclopentane ring [13,17]. The bicyclic endoperoxides created in this reaction are analogous to prostaglandin H_2 formed enzymatically by cyclooxygenases; the resulting products, therefore, have been given the trivial name isoprostanes [18].

1,2-Dioxolanyl radical   Bicyclic endoperoxide H_2-Isoprostane

5-*Exo* cyclization of the 1,2-dioxolanyl radical leads to bicyclic endoperoxides and isoprostanes

A fourth reaction option for the 1,2-dioxolanyl radical is the intramolecular radical substitution (S_Hi) of the peroxide by the carbon-centered radical [19,20]. This reaction can subsequently form diepoxyalcohols that also have been formed in the reaction of mutant cyclooxygenase-2 [21]. These diepoxyalcohols have been proposed as alternative precursors for isofuran and isopyran compounds [6,22].

1,2-Dioxolanyl radical   Diepoxyalcohol   Isofuran

Intramolecular radical substitution leads to an epoxide and alkoxyl radical; further transformations can give rise to diepoxyalcohols and isofurans

There is yet a fifth possible reaction for the 1,2-dioxolanyl radical. Following oxygen addition the 1,2-dioxolanyl-3-peroxyl radical can also react across the chain to give a bicyclic endoperoxide with a five- and a seven-membered ring, with the seven-membered ring containing a second peroxide bridge [23]. Such a diendoperoxide has not yet been described to be formed under conditions of autoxidation but has been isolated from the cyclooxygenase-2 catalyzed oxygenation of 5S-HETE. Similar to the PGH_2 endoperoxide, the diendoperoxide is unstable and rearranges nonenzymatically. Two hemiketal (HK) eicosanoids have been isolated as rearrangement products and were termed HKD_2 and HKE_2 [24].

Oxygenation of 5S-HETE by COX-2 gives a diendoperoxide intermediate that rearranges to the hemiketal eicosanoids HKD_2 and HKE_2

As an alternative to 5-*exo* cyclization the peroxyl radical may also undergo 4-*exo* cyclization to form 1,2-dioxetanyl carbinyl radicals. Such reactions appear to only be favored if there are no double bonds appropriately positioned for 5-*exo* cyclization. Dioxetanes have been postulated to serve as precursors to lipid aldehydes by fragmentation reactions.

Fatty acid peroxyl radical   1,2-Dioxetanyl radical

4-*Exo* cyclization of a peroxyl radical to a 1,2-dioxetanyl radical

FATTY ACID DIMERS AND OLIGOMERS

While formation of intramolecular peroxides is favored, the fatty acid peroxyl radical can also undergo addition to a double bond of another fatty acid, leading to dimerization and oligomerization. Due to the technical challenges in their isolation and structural identification, the products of acyl chain oligomerization are poorly characterized [25–28]. Some evidence suggests that such fatty acid oligomers provide a pool of radicals during the onset and early stages of lipid peroxidation [29].

Interestingly, polymerization of acyl chains is the basis for the drying and hardening of oil-based paint.

If two peroxyl radicals form simultaneously on adjacent fatty acids, a termination reaction is possible where the two peroxyl radicals combine to form a tetra-oxygen bridge between the two lipids (R_1OOOOR_2) [30]. Not too surprisingly, such compounds are highly unstable and rapidly rearrange to lose O_2 and recombine to peroxides R_1OOR_2 or to a keto and a hydroxy fatty acid. Whether oxygen is released as singlet oxygen has been debated [2].

RATE CONSTANTS AS DETERMINANTS OF THE OUTCOME OF LIPID PEROXIDATION

What factors determine which of the four possible reactions described above the peroxyl radical will undergo? Obviously, the type and concentration of available reaction partners is one key determinant. For instance, forming a bicyclic endoper-oxide requires at least three double bonds in the fatty acid. High concentrations of oxygen are likely to favor serial endoperoxides whereas low concentrations of oxygen favor elimination, scission, and addition reactions. For this reason, depletion of reaction partners over the course of autoxidation leads to a shift of products formed. The other key determinant of product outcome is the rate constant for each of the possible reactions. Reactions proceeding at a higher rate constant are preferred over slower reactions.

A major breakthrough in the understanding of the importance of rate constants on product formation came in the early 2000s when the long-elusive bis-allylic 11-hydroperoxide was isolated from an autoxidation reaction of methyl linoleate [31]. A bis-allylic hydroperoxide is expected from the delocalization of the pentadienyl radical, invoking radical character, and thus the possibility for oxygen addition, at carbons 1, 3, and 5, equivalent to C9, C11, and C13 of linoleate. The predicted radical character at C11 was also backed by ESR data [32]. Yet only the conjugated 9- and 13-HPODE had been observed as products. The apparent lack of formation of the bis-allylic 11-HPODE was attributed to an increase in stability when the radical is localized to the terminal carbons since this leads to conjugated double bonds.

Primary products from linoleic acid autoxidation: the conjugated 9- and 13-HPODE and the bis-allylic 11-HPODE

11-HPODE (methyl ester) was isolated from an autoxidation reaction of methyl linoleate conducted using a higher than usual amount of α-tocopherol [31]. α-Tocopherol was included in order to drive the reaction toward formation of the allylic 8- and 14-hydroperoxides, as had been described to occur earlier [33]. While these products were not found, the bis-allylic 11(9Z,12Z)-HPODE was obtained as one of the major products in addition to the expected 9- and 13-HPODEs. Importantly, 11-HPODE failed to form when α-tocopherol was absent.

A detailed study was conducted to clarify the mechanistic role of α-tocopherol: Acting as an antioxidant, α-tocopherol traps the 11-peroxyl radical as the hydroperoxide before it can β-fragment back to the pentadienyl radical [34]. It turned out that the rate constant for β-fragmentation of the *bis*-allylic peroxyl radical is orders of magnitude higher than for the conjugated peroxyls. As a consequence, the loss of O_2 from the *bis*-allylic peroxyl is much faster than the propagation reaction, that is, the abstraction of a hydrogen atom from a *bis*-allylic methylene [34]. In other words, the oxygen at the *bis*-allylic position is prone to come off before it is fixed as the hydroperoxide. The formation of the *bis*-allylic hydroperoxide can only happen if an efficient antioxidant (H-atom donor) is present during the reaction. In contrast, if only linoleate is present in an autoxidation reaction, the *bis*-allylic peroxyl radical β-fragments before it can abstract an H-atom. Under optimal conditions (>1.5 M α-tocopherol) the *bis*-allylic 11-HPODE becomes the most prevalent product, outcompeting the conjugated 9- and 13-HPODEs [34].

PROMINENT REACTIONS IN THE FORMATION OF SECONDARY PRODUCTS

It is an oversimplification to assume that there is an early stage of autoxidation when only fatty acid hydroperoxides accumulate and that these only give rise to the secondary products in a later phase. While simple fatty acid hydroperoxides are the primary and major isolated products of lipid peroxidation, secondary products begin to form almost immediately after initiation of reaction. In some cases, these products are highly unstable, quickly decomposing into what might be rightfully described as tertiary and quaternary products. Given the multiple reactions that are required for some of the most abundant products, and the lack of clarity on the precise order of steps that lead to these products, such designations would be highly cumbersome. Thus, all products other than fatty acid hydroperoxides are best grouped together as secondary products.

Some secondary products, such as serial cyclic endoperoxides, isoprostanes, and diepoxyalcohols have already been discussed above. Lipid peroxidation gives rise to many other secondary products including hydroxy fatty acids (e.g., HETEs and HODEs), keto-fatty acids (e.g., KETEs and KODEs), hydro(pero)xyalkenals and oxoalkenals (e.g., 4-hydroxy-2-nonenal and 4-oxo-2-nonenal), malondialdehyde, short-chained alkanes (e.g., ethane and pentane), and highly complex products like isolevuglandins. In many cases, these products form on fatty acids esterified to phospholipids, resulting in bioactive, oxidative fragmented phospholipids (e.g., butyryl-PAF and azelaoyl-PC). Reactions leading to the formation of these various products are briefly outlined below.

ALKOXYL RADICAL PRODUCTS

Several other key reactions besides those discussed above for the peroxyl radical must be considered to account for the formation of currently known secondary products.

One important reaction is the formation of alkoxyl radicals by iron or copper catalyzed scission at the oxygen–oxygen bond of the hydroperoxide or endoperoxides. The alkoxyl radical can then abstract a *bis*-allylic hydrogen from a nearby PUFA to form the stable hydroxyl fatty acid.

Cleavage of a hydroperoxide to an alkoxyl radical

The alkoxyl radical can also generate a carbonyl to form keto-fatty acids, for example, 13-KODE from linoleic acid hydroperoxide [35,36]. Unlike their hydroxyl counterparts, keto-fatty acids can undergo Michael addition which may account for their ability to activate the transcription factors PPARγ and Nrf2/Keap-1 [37–39].

Oxidation of an alkoxyl radical to a keto fatty acid

Alternatively, the alkoxyl radical can attack an adjacent double bond to form an epoxide and a carbon-centered radical [36]. Subsequent addition of O_2 generates an epoxyperoxyl radical that can form an epoxyhydroperoxide or undergo the reduction/oxidation steps above to give an epoxyalcohol or epoxyketo fatty acid. Hydrolysis of the epoxide gives triol compounds.

Oxygenation of the alkoxyl radical leading to epoxyketo, epoxyhydroxy, and trihydroxy fatty acids

Another important pathway by which alkoxyl radicals form secondary products is via β-scission reactions. Here, cleavage of the carbon–carbon bond on the side opposite from the conjugated double bonds results in an alkanyl radical fragment and an aldehydic fragment. Abstraction of hydrogen by the alkanyl radical fragment (from an adjacent PUFA) generates a stable alkane product. The aldehydic fragment can undergo two electron oxidation to yield a carboxylate.

Carbon chain cleavage through β-scission of an alkoxyl radical

β-Scission reactions account for the formation of pentane (from 15-HPETE or 13-HPODE) and ethane (from 17-HPEPA or 19-HPDHA) that have been used to measure lipid peroxidation in breath condensate [40,41]. β-Scission occurring on PUFA esterified as diacyl and alkylacyl phospholipids gives rise to butyryl-PC and butyryl-PAF (from phospholipids with AA) and propionyl-PC and propionyl-PAF (from phospholipids with DHA), respectively [42,43]. These phospholipids are ligands for the platelet-activating factor (PAF) receptor.

Hock Rearrangement

Carbon chain fragmentation of fatty acid hydroperoxides to aldehydes also occurs by "Hock rearrangement" or "Hock cleavage." Hock cleavage is catalyzed by Brønsted acids or Lewis acids in aprotic solvents [44]. One of the alkyl chains migrates to the peroxide moiety to generate a hemiacetal bond that readily undergoes hydrolysis to form two shorter aldehydes. The fragment retaining the double bond contains a bis-allylic hydrogen that can readily be abstracted to generate γ-hydro(pero)xy- and γ-oxo-alkenals.

Hock cleavage of a fatty acid hydroperoxide forming two aldehyde fragments

Formation and Fragmentation of Bicyclic Endoperoxides

Important variations of the reactions described for fatty acid hydroperoxides occur with the bicyclic endoperoxides. As described previously, bicyclic endoperoxides can be generated when PUFA with at least three double bonds form the 1,2-dioxolanyl radical. This allows the carbon-centered radical to react with the E double bond to form a pentane ring. The chemical structure of these bicyclic endoperoxides, given the trivial name of H_2-isoprostanes, are similar to prostaglandin H_2 (PGH_2) except that the predominant species of H_2-isoprostanes have their two side chains in the *cis* configuration rather than the *trans* configuration found in PGH_2 [18]. Furthermore, there are four regioisomers of H_2-isoprostanes formed during non-enzymatic formation of bicyclic endoperoxides from AA. These differences allow distinction of the products derived from H_2-isoprostanes from those of PGH_2 on various chromatographic systems.

The entropic strain caused by the formation of two five-membered rings makes fragmentation of at least one of these rings highly likely. The endoperoxide ring undergoes scission to generate two alkoxyl radicals that then abstract hydrogens from H-donors to form 1,3-diols. This reaction accounts for the formation of F_2-isoprostanes, one of the most commonly used biomarkers for lipid peroxidation *in vivo* [45]. When scission of the endoperoxide occurs under less reducing conditions, one oxygen gives rise to a ketone, generating D_2- and E_2-isoprostanes [46].

Fragmentation of a bicyclic endoperoxide into D_2, E_2, and F_2 ring isoprostanes

Fragmentation of both the cyclopentane ring and endoperoxide can also occur in a base catalyzed rearrangement to give γ-ketoaldehydes [47]. These γ-ketoaldehydes have been given the trivial names isolevuglandins or isoketals when formed from AA and neuroketals when formed from DHA. γ-Ketoaldehydes are among the most reactive secondary products of lipid peroxidation, rapidly adducting to cellular amines such as lysines, phosphatidylethanolamines, and nucleic acids [48–52].

Formation of γ-ketoaldehydes from a bicyclic endoperoxide

In the presence of iron, fragmentation of both the cyclopentane and endoperoxide rings can give rise to malondialdehyde (MDA), an ubiquitous product of lipid peroxidation [53–55]. MDA is the intended target of the thiobarbituric acid reactive substances (TBARS) assay commonly used to measure lipid peroxidation, although a broad spectrum of other lipid electrophiles are also detected by this assay.

Bicyclic
endoperoxide

Malon-
dialdehyde

Fragmentation of a bicyclic endoperoxide to
malondialdehyde and an ene-fatty acid

CARBON CHAIN CLEAVAGE TO ALKENALS

It is obvious that by concatenating the various reactions described above for every major PUFA, an overwhelming number of secondary products of lipid peroxidation can be expected. Depending on one's perspective, it is of most interest to focus either on the most abundant or the most bioactive of these compounds. By either criteria, hydro(pero)xy- and oxo-alkenals are clearly important secondary products of lipid peroxidation.

4-Hydroxy-2E-nonenal (4-HNE) was first described by Esterbauer and colleagues in 1980 [56], and the mechanisms of its formation and bioactivity have been of keen interest since [57]. There have been a number of mechanisms proposed for the formation of 4-HNE through the years, with various combinations of reactions invoked to rationalize experimental data [58]. The realization that 4-hydroperoxy-2E-nonenal (4-HPNE) is a precursor to 4-HNE further increased the mechanistic possibilities [59,60]. There can be little doubt that different pathways culminate in the formation of 4-HNE in vitro and in vivo. Which of the possible pathways contributes to 4-HNE formation in a particular biological system is difficult to determine.

One proposed pathway for the formation of 4-HNE was described under Hock rearrangement. This mechanism has been invoked for the formation of 4-H(P)NE from 9-HPODE [61]. In the same study, it was found that ω-6 hydroperoxides form 4-H(P)NE through a different mechanism. A crucial finding was that 9S-HPODE formed the aldehyde as a racemic mixture whereas 13S-HPODE gave rise to chiral 4S-H(P)NE.

9S-HPODE chain cleaves to give racemic 4-HPNE, 13S-HPODE forms
chiral 4S-HPNE

Retention of the chirality of 4S-H(P)NE from 13S-HPODE was attributed to Hock cleavage of a 10,13-diHPODE [61]. When 10,13-diHPODE was synthesized, however, it was found that this intermediate does not form 4-HPNE under autoxidation conditions thus ruling out this mechanistic proposal [62,63].

10,13-DiHPODE is not an intermediate in the formation of 4S-HPNE from 13

Another mechanism postulates the formation of fatty acid peroxide dimers and oligomers that give rise to vicinal peroxides [64].

Dimerization and oligomerization of fatty acid peroxides

Fragmentation (S_Hi) of the vicinal peroxides yields 4-H(P)NE and hydroperoxy epoxides in a reaction that is similar to the mechanism of styrene/oxygen co-polymerization and subsequent degradation to benzaldehyde and formaldehyde [57,65,66]. An equivalent cross-chain mechanism is likely to contribute to the formation of 4-H(P)NE during autoxidation of cardiolipin [67].

Intramolecular radical substitution of a peroxide-linked fatty acid oligomer resulting in carbon chain cleavage to aldehyde fragments

A different mechanism, proceeding via reversible diepoxycarbinyl radical and vinyl ether radical intermediates has recently been proposed for generation of γ-hydroxy- and γ-oxo-alkenals including 4-HNE [68].

A diepoxycarbinyl radical intermediate in the formation of 4-HNE from 13-HPODE

RECENT DEVELOPMENTS IN LIPID PEROXIDATION

Much recent work has focused on defining the bioactivity of the individual lipid peroxidation products and will be the topic of chapters following in this volume. Important recent mechanistic discoveries include the refinement of "radical clocks" that are used for determination of rate constants of peroxyl radical reactions, insight into the limitation of dietary antioxidants to prevent or reduce lipid peroxidation, and the development of "super-antioxidants." Each of these will be discussed briefly.

RADICAL CLOCKS

Discovery of the *bis*-allylic 11-HPODE and understanding how it is formed have renewed interest in reaction kinetics in lipid peroxidation. This work has led to the development and calibration of novel radical clocks to measure absolute rate constants for key transformation steps involved in lipid peroxidation [69].

Radical clocks are set up such that rate constants are determined in a reaction comparing a unimolecular reaction of known rate constant with a bimolecular reaction of which the rate constant is to be determined. The ratio of the unimolecular over the bimolecular reaction products is directly proportional to the ratio of the rate constants, thus allowing direct calculation of the unknown rate constant. Radical clocks have been developed using the distribution of E,Z versus E,E HPODEs and formation of the *bis*-allylic 11-HPODE from linoleate and the distribution of conjugated versus nonconjugated hydroperoxides from allylbenzene oxidation to cover a wide range of inhibition and propagation rate constants ranging from 10^0 to 10^7 $M^{-1}s^{-1}$ [69,70].

LIMITATION OF DIETARY ANTIOXIDANTS TO INHIBIT LIPID PEROXIDATION

The 1990s saw great excitement about the use of dietary antioxidants for the treatment of diseases associated with increased lipid peroxidation. Deficiency in vitamin E is associated with increased lipid peroxidation, and treatment with vitamins E and C showed clear efficacy in rodent trials, even in vitamin-sufficient animals. Unfortunately, large randomized clinical trials with these and other dietary antioxidants, including β-carotene, have generally been negative [71–76]. The reasons for this failure are discussed with much controversy. Of note, subsequent investigations with vitamin E found that it required extremely high doses (800 IU or higher) given for many months (minimum 16 weeks) in order to significantly lower levels of F2-isoprostanes, and this effect was only evident in women who began with highly elevated levels [77]. Thus, many prevention trials with vitamin E are unlikely to have directly tested the hypothesis that lowering lipid peroxidation products can prevent disease [78]. Even so, it is unclear whether increasing the dose of vitamin E or combinations of vitamins E and C beyond the recommended intake levels would be efficacious, especially since high doses of vitamin E have been associated with adverse effects.

"SUPER-ANTIOXIDANTS"

For a long time, α-tocopherol (the most active form of vitamin E) was the most efficient lipid-soluble antioxidant known. An understanding of the factors involved in enabling H-atom transfer by a phenolic hydroxyl versus its reaction with oxygen has led to the rational development of 5-pyrimidols and 6-amino-3-pyridinols (=tetra-hydro-1,8-naphthyridinol) antioxidants [79,80]. Antioxidant activity of the phenolic core structure was enhanced by placing electron-donating methyl groups in the *ortho* position to the hydroxyl (contributing to lowering the bond dissociation enthalpy) and a strongly donating dimethyl (alkyl) amino group in *para* position. The stability against direct reaction with oxygen was increased by incorporating nitrogen atoms at the 3- and/or 5-position of the ring. The resulting 6-amino-3-pyridinols with a fused aliphatic 6- or 5-membered ring exceeded the inhibition rates of α-tocopherol by about 25- and 90-fold, respectively, under *in vitro* conditions [80]. The insight gained from these studies was then used to prepare analogs of vitamin E that contain nitrogen atoms in the rings and a phythyl side chain [81]. The novel "N-tocopherol" (N-TOH) was as efficient as α-tocopherol in inhibiting cholesterol ester oxidation in human low-density lipoprotein and in binding to recombinant human tocopherol transfer protein [81]. The novel super-antioxidants might find application for testing in animal models with the goal to overcome the limitations observed in clinical trials with α-tocopherol.

6-amino-3-pyridinols   N-TOH

Novel rationally designed antioxidants

OPEN QUESTIONS

Despite nearly 40 years of intense study, a great deal remains unknown both in terms of mechanism and biological relevance of lipid peroxidation. While the area of bio-activity continues to be of great interest, and significant headway has been made, we feel it appropriate to note mechanistic areas that have been neglected and that might make significant contributions.

The question of oligomerization and polymerization of fatty acyl chains has been recognized since the earliest studies, but little headway seems to have been made. While polymerization is thought to contribute to pathology and biochemical hall-marks of aging such as lipofuscin, there is little detailed knowledge about reaction rates or structures formed during dimerization and polymerization of fatty acyl chains.

Similarly, while carbon chain cleavage has long been recognized as the basis of the smell associated with lipid-containing food going rancid, there is no clear under-standing how many pathways lead to carbon chain cleavage and what their relative contribution to product formation may be. Some of the reactions involved in these processes have been touched upon in this chapter but, by and large, neither process has found satisfactory explanation of all of the chemical transformations involved.

Another major future challenge will be to unravel how lipid peroxidation proceeds in complex systems encountered *in vivo*, where the phospholipid bilayer includes integral and accessory proteins, cholesterol, and other lipid-soluble small molecules. Understanding lipid peroxidation *in vivo* becomes especially important as it has become apparent that while reactive oxygen species drive lipid peroxidation, the ultimate source of the initiating reactive oxygen species is almost certainly via enzymes such as NADPH oxidase and myeloperoxidase. The interplay between these and other enzymes (including mitochondrial electron transport chain enzymes) and the bulk levels of lipid peroxidation products is only now beginning to be understood, but may be critical in designing interventions that lead to reduction in human diseases.

ACKNOWLEDGMENT

Work in the authors' labs has been funded by awards GM076592, AT006896, GM015431, HL111945, and AT007830 from the National Institutes of Health (NIH).

REFERENCES

1. Porter, N. A. 1986. Mechanism for the autoxidation of polyunsaturated lipids. *Acc. Chem. Res.* **19**: 262–268.
2. Gardner, H. W. 1989. Oxygen radical chemistry of polyunsaturated fatty acids. *Free Radical Biol. Med.* **7**: 65–86.
3. Porter, N. A., S. E. Caldwell, and K. A. Mills. 1995. Mechanisms of free radical oxidation of unsaturated lipids. *Lipids* **30**: 277–290.
4. Yin, H., L. Xu, and N. A. Porter. 2011. Free radical lipid peroxidation: Mechanisms and analysis. *Chem. Rev.* **111**: 5944–5972.
5. Porter, N. A. 2013. A perspective on free radical autoxidation: The physical organic chemistry of polyunsaturated fatty acid and sterol peroxidation. *J. Org. Chem.* **78**: 3511–3524.
6. Jahn, U., J. M. Galano, and T. Durand. 2008. Beyond prostaglandins—Chemistry and biology of cyclic oxygenated metabolites formed by free-radical pathways from polyunsaturated fatty acids. *Angew. Chem. Int. Ed. Engl.* **47**: 5894–5955.
7. Schneider, C. 2005. Chemistry and biology of vitamin E. *Mol. Nutr. Food Res.* **49**: 7–30.
8. Chan, H. W.-S., G. Levett, and J. A. Matthew. 1979. The mechanism of the rearrangement of linoleate hydroperoxides. *Chem. Phys. Lipids* **24**: 245–256.
9. Porter, N. A., B. A. Weber, H. Weenen, and J. A. Khan. 1980. Autoxidation of polyunsaturated lipids. Factors controlling the stereochemistry of product hydroperoxides. *J. Am. Chem. Soc.* **102**: 5597–5601.
10. Porter, N. A., and D. G. Wujek. 1984. Autoxidation of polyunsaturated fatty acids, an expanded mechanistic study. *J. Am. Chem. Soc.* **106**: 2626–2629.
11. Nugteren, D. H., H. Vonkeman, and D. A. Van Dorp. 1967. Non-enzymic conversion of all-cis 8, 11, 14-eicosatrienoic acid into prostaglandin E_1. *Recl. Trav. Chim. Pays-Bas* **86**: 1237–1245.
12. Pryor, W. A., and J. P. Stanley. 1975. Letter: A suggested mechanism for the production of malonaldehyde during the autoxidation of polyunsaturated fatty acids. Nonenzymatic production of prostaglandin endoperoxides during autoxidation. *J. Org. Chem.* **40**: 3615–3617.
13. Porter, N. A., and M. O. Funk. 1975. Peroxy radical cyclization as a model for prostaglandin biosynthesis. *J. Org. Chem.* **40**: 3614–3615.

14. Khan, J. A., and N. A. Porter. 1982. Serial cyclizations of an arachidonic-acid hydroperoxide. *Angew. Chem. Int. Ed. Engl.* **21**: 217–218.

15. Frankel, E. N., W. E. Neff, and D. Weisleder. 1982. Formation of hydroperoxy bisepidioxides in sensitized photo-oxidized methyl linolenate. *J. Chem. Soc. Chem. Commun.* 599–600.

16. Yin, H., and N. A. Porter. 2005. New insights regarding the autoxidation of polyunsaturated fatty acids. *Antioxid. Redox Signal* **7**: 170–184.

17. O'Connor, D. E., E. D. Mihelich, and M. C. Coleman. 1984. Stereochemical course of the autoxidative cyclization of lipid hydroperoxides to prostaglandin-like bicyclo endoperoxides. *J. Am. Chem. Soc.* **106**: 3577–3584.

18. Morrow, J. D., K. E. Hill, R. F. Burk, T. M. Nammour, K. F. Badr, and L. J. Roberts, 2nd. 1990. A series of prostaglandin F2-like compounds are produced *in vivo* in humans by a non-cyclooxygenase, free radical-catalyzed mechanism. *Proc. Natl. Acad. Sci. USA* **87**: 9383–9387.

19. Mihelich, E. D. 1980. Structure and stereochemistry of novel endoperoxides isolated from the sensitized photo-oxidation of methyl linoleate—Implications for prostaglandin biosynthesis. *J. Am. Chem. Soc.* **102**: 7141–7143.

20. Porter, N. A., P. J. Zuraw, and J. A. Sullivan. 1984. Peroxymercuration-demercuration of lipid hydroperoxides. *Tetrahed. Lett.* **25**: 807–810.

21. Schneider, C., W. E. Boeglin, and A. R. Brash. 2004. Identification of two cyclooxygenase active site residues, leucine-384 and glycine-526, that control carbon ring cyclization in prostaglandin biosynthesis. *J. Biol. Chem.* **279**: 4404–4414.

22. Fessel, J. P., N. A. Porter, K. P. Moore, J. R. Sheller, and L. J. Roberts, 2nd. 2002. Discovery of lipid peroxidation products formed *in vivo* with a substituted tetrahydrofuran ring (isofurans) that are favored by increased oxygen tension. *Proc. Natl. Acad. Sci. USA* **99**: 16713–16718.

23. Schneider, C., W. E. Boeglin, H. Yin, D. F. Stec, and M. Voehler. 2006. Convergent oxygenation of arachidonic acid by 5-lipoxygenase and cyclooxygenase-2. *J. Am. Chem. Soc.* **128**: 720–721.

24. Griesser, M., W. E. Boeglin, T. Suzuki, and C. Schneider. 2009. Convergence of the 5-LOX and COX-2 pathways. Heme-catalyzed cleavage of the 5S-HETE-derived di-endoperoxide into aldehyde fragments. *J. Lipid Res.* **50**: 2455–2462.

25. Witting, L. A., S. S. Chang, and F. A. Kummerow. 1957. The isolation and characterization of the polymers formed during autoxidation of ethyl linoleate. *J. Am. Oil Chem. Soc.* **34**: 470–473.

26. Morita, M., and M. Kujimaki. 1973. Minor peroxide components as catalysts and precursors to monocarbonyls in the authoxidation of methyl linoleate. *J. Agr. Food Chem.* **21**: 860–863.

27. Schieberle, P., and W. Grosch. 1981. Decomposition of linoleic-acid hydroperoxides. 2. Breakdown of methyl 13-hydroperoxy-*cis*-9-*trans*-11-octadecadienoate by radicals or copper-Ii ions. *Z. Lebensm. Unters. Forsch.* **173**: 192–198.

28. Miyashita, K., K. Fujimoto, and T. Kaneda. 1982. Formation of dimers during the initial-stage of autoxidation in methyl linoleate. *Agr. Biol. Chem. Tokyo* **46**: 751–755.

29. Miyashita, K., K. Fujimoto, and T. Kaneda. 1984. Structural studies of polar dimers in autoxidized methyl linoleate during the initial stages of autoxidation. *Agr. Biol. Chem. Tokyo* **48**: 2511–2515.

30. Russell, G. A. 1957. Deuterium-isotope effects in the autoxidation of aralkyl hydrocarbons. Mechanism of the interaction of peroxy radicals. *J. Am. Chem. Soc.* **79**: 3871–3877.

31. Brash, A. R. 2000. Autoxidation of methyl linoleate: Identification of the bis-allylic 11-hydroperoxide. *Lipids* **35**: 947–952.

32. Griller, D., K. U. Ingold, and J. C. Walton. 1979. 2 Conformations of the pentadienyl radical. *J. Am. Chem. Soc.* **101**: 758–759.

33. Haslbeck, F., W. Grosch, and J. Firl. 1983. Formation of hydroperoxides with unconjugated diene systems during autoxidation and enzymic oxygenation of linoleic acid. *Biochim. Biophys. Acta* **750**: 185–193.
34. Tallman, K. A., D. A. Pratt, and N. A. Porter. 2001. Kinetic products of linoleate peroxidation: Rapid β-fragmentation of nonconjugated peroxyls. *J. Am. Chem. Soc.* **123**: 11827–11828.
35. Landino, L. M., and L. J. Marnett. 1996. Mechanism of hydroperoxide reduction by mangano-prostaglandin endoperoxide synthase. *Biochemistry* **35**: 2637–2643.
36. Dix, T. A., and L. J. Marnett. 1985. Conversion of linoleic acid hydroperoxide to hydroxy, keto, epoxyhydroxy, and trihydroxy fatty acids by hematin. *J. Biol. Chem.* **260**: 5351–5357.
37. O'Flaherty, J. T., L. C. Rogers, C. M. Paumi, R. R. Hantgan, L. R. Thomas, C. E. Clay, K. High et al. 2005. 5-Oxo-ETE analogs and the proliferation of cancer cells. *Biochim. Biophys. Acta* **1736**: 228–236.
38. Hammond, V. J., A. H. Morgan, S. Lauder, C. P. Thomas, S. Brown, B. A. Freeman, C. M. Lloyd et al. 2012. Novel keto-phospholipids are generated by monocytes and macrophages, detected in cystic fibrosis, and activate peroxisome proliferator-activated receptor-gamma. *J. Biol. Chem.* **287**: 41651–41666.
39. Groeger, A. L., C. Cipollina, M. P. Cole, S. R. Woodcock, G. Bonacci, T. K. Rudolph, V. Rudolph, B. A. Freeman, and F. J. Schopfer. 2010. Cyclooxygenase-2 generates antiinflammatory mediators from omega-3 fatty acids. *Nat. Chem. Biol.* **6**: 433–441.
40. Dillard, C. J., E. E. Dumelin, and A. L. Tappel. 1977. Effect of dietary vitamin E on expiration of pentane and ethane by the rat. *Lipids* **12**: 109–114.
41. Knutson, M. D., G. J. Handelman, and F. E. Viteri. 2000. Methods for measuring ethane and pentane in expired air from rats and humans. *Free Radic. Biol. Med.* **28**: 514–519.
42. Marathe, G. K., S. S. Davies, K. A. Harrison, A. R. Silva, R. C. Murphy, H. Castro-Faria-Neto, S. M. Prescott, G. A. Zimmerman, and T. M. McIntyre. 1999. Inflammatory platelet-activating factor-like phospholipids in oxidized low density lipoproteins are fragmented alkyl phosphatidylcholines. *J. Biol. Chem.* **274**: 28395–28404.
43. Tanaka, T., H. Minamino, S. Unezaki, H. Tsukatani, and A. Tokumura. 1993. Formation of platelet-activating factor-like phospholipids by Fe2 + /ascorbate/EDTA-induced lipid peroxidation. *Biochim. Biophys. Acta* **1166**: 264–274.
44. Gardner, H. W., and R. D. Plattner. 1984. Linoleate hydroperoxides are cleaved heterolytically into aldehydes by a Lewis acid in aprotic solvent. *Lipids* **19**: 294–299.
45. Kadiiska, M. B., B. C. Gladen, D. D. Baird, D. Germolec, L. B. Graham, C. E. Parker, A. Nyska et al. 2005. Biomarkers of oxidative stress study II: Are oxidation products of lipids, proteins, and DNA markers of CCl4 poisoning? *Free Radic. Biol. Med.* **38**: 698–710.
46. Morrow, J. D., L. J. Roberts, V. C. Daniel, J. A. Awad, O. Mirochnitchenko, L. L. Swift, and R. F. Burk. 1998. Comparison of formation of D2/E2-isoprostanes and F2-isoprostanes *in vitro* and in vivo—Effects of oxygen tension and glutathione. *Arch. Biochem. Biophys.* **353**: 160–171.
47. Salomon, R. G., D. B. Miller, M. G. Zagorski, and D. J. Coughlin. 1984. Solvent induced fragmentation of prostaglandin endoperoxides. New aldehyde products from PGH2 and novel intramolecular 1,2-hydride shift during endoperoxide fragmentation in aqueous solution. *J. Am. Chem. Soc.* **106**: 6049–6060.
48. Brame, C. J., R. G. Salomon, J. D. Morrow, and L. J. Roberts, 2nd. 1999. Identification of extremely reactive gamma-ketoaldehydes (isolevuglandins) as products of the isoprostane pathway and characterization of their lysyl protein adducts. *J. Biol. Chem.* **274**: 13139–13146.
49. Bernoud-Hubac, N., L. B. Fay, V. Armarnath, M. Guichardant, S. Bacot, S. S. Davies, L. J. Roberts, II, and M. Lagarde. 2004. Covalent binding of isoketals to ethanolamine phospholipids. *Free Radic. Biol. Med.* **37**: 1604–1611.

50. Carrier, E. J., Amarnath, J. A. Oates, and O. Boutaud. 2009. Characterization of covalent adducts of nucleosides and DNA fomred by reaction with levuglandin. *Biochemistry* **48**: 10775–10781.

51. Murthi, K. K., L. R. Friedman, N. L. Oleinick, and R. G. Salomon. 1993. Formation of DNA-protein cross-links in mammalian cells by levuglandin E2. *Biochemistry* **32**: 4090–4097.

52. Iyer, R. S., S. Ghosh, and R. G. Salomon. 1989. Levuglandin E2 crosslinks proteins. *Prostaglandins* **37**: 471–480.

53. Mason, R. J., T. P. Stossel, and M. Vaughan. 1972. Lipids of alveolar macrophages, polymorphonuclear leukocytes, and their phagocytic vesicles. *J. Clin. Invest.* **51**: 2399–2407.

54. Fogelman, A. M., I. Shechter, J. Seager, M. Hokom, J. S. Child, and P. A. Edwards. 1980. Malondialdehyde alteration of low density lipoproteins leads to cholesteryl ester accumulation in human monocyte-macrophages. *Proc. Natl. Acad. Sci. USA* **77**: 2214–2218.

55. Esterbauer, H., R. J. Schaur, and H. Zollner. 1991. Chemistry and biochemistry of 4-hydroxynonenal, malonaldehyde and related aldehydes. *Free Radic. Biol. Med.* **11**: 81–128.

56. Benedetti, A., M. Comporti, and H. Esterbauer. 1980. Identification of 4-hydroxy-nonenal as a cytotoxic product originating from the peroxidation of liver microsomal lipids. *Biochim. Biophys. Acta* **620**: 281–296.

57. Schneider, C., N. A. Porter, and A. R. Brash. 2008. Routes to 4-hydroxynonenal: Fundamental issues in the mechanisms of lipid peroxidation. *J. Biol. Chem.* **283**: 15539–15543.

58. Porter, N. A., and W. A. Pryor. 1990. Suggested mechanisms for the production of 4-hydroxy-2-nonenal from the autoxidation of polyunsaturated fatty acids. *Free Radic. Biol. Med.* **8**: 541–543.

59. Gardner, H. W., and M. J. Grove. 1998. Soybean lipoxygenase-1 oxidizes 3Z-nonenal. A route to 4S-hydroperoxy-2E-nonenal and related products. *Plant Physiol.* **116**: 1359–1366.

60. Gardner, H. W., and M. Hamberg. 1993. Oxygenation of (3Z)-nonenal to (2E)-4-hydroxy-2-nonenal in the broad bean (*Vicia faba* L.). *J. Biol. Chem.* **268**: 6971–6977.

61. Schneider, C., K. A. Tallman, N. A. Porter, and A. R. Brash. 2001. Two distinct pathways of formation of 4-hydroxynonenal. Mechanisms of nonenzymatic transformation of the 9- and 13-hydroperoxides of linoleic acid to 4-hydroxyalkenals. *J. Biol. Chem.* **276**: 20831–20838.

62. Schneider, C., W. E. Boeglin, H. Yin, D. F. Stec, D. L. Hachey, N. A. Porter, and A. R. Brash. 2005. Synthesis of dihydroperoxides of linoleic and linolenic acids and studies on their transformation to 4-hydroperoxynonenal. *Lipids* **40**: 1155–1162.

63. Zhang, W., M. Sun, and R. G. Salomon. 2006. Preparative singlet oxygenation of linoleate provides doubly allylic dihydroperoxides: Putative intermediates in the generation of biologically active aldehydes in vivo. *J. Org. Chem.* **71**: 5607–5615.

64. Schneider, C., W. E. Boeglin, H. Yin, N. A. Porter, and A. R. Brash. 2008. Intermolecular peroxyl radical reactions during autoxidation of hydroxy and hydroperoxy arachidonic acids generate a novel series of epoxidized products. *Chem. Res. Toxicol.* **21**: 895–903.

65. Miller, A. A., and F. R. Mayo. 1956. Oxidation of unsaturated compounds. I. The oxidation of styrene. *J. Am. Chem. Soc.* **78**: 1017–1023.

66. Mayo, F. R., and A. A. Miller. 1956. Oxidation of unsaturated compounds. II. Reactions of styrene peroxide. *J. Am. Chem. Soc.* **78**: 1023–1034.

67. Liu, W., N. A. Porter, C. Schneider, A. R. Brash, and H. Yin. 2011. Formation of 4-hydroxynonenal from cardiolipin oxidation: Intramolecular peroxyl radical addition and decomposition. *Free Radic. Biol. Med.* **50**: 166–178.

68. Gu, X., and R. G. Salomon. 2012. Fragmentation of a linoleate-derived gamma-hydroperoxy-alpha,beta-unsaturated epoxide to gamma-hydroxy- and gamma-oxo-alkenals

involves a unique pseudo-symmetrical diepoxycarbinyl radical. *Free Radic. Biol. Med.* **52**: 601–606.

69. Roschek, B., Jr., K. A. Tallman, C. L. Rector, J. G. Gillmore, D. A. Pratt, C. Punta, and N. A. Porter. 2006. Peroxyl radical clocks. *J. Org. Chem.* **71**: 3527–3532.

70. Xu, L., T. A. Davis, and N. A. Porter. 2009. Rate constants for peroxidation of polyunsaturated fatty acids and sterols in solution and in liposomes. *J. Am. Chem. Soc.* **131**: 13037–13044.

71. Waters, D. D., E. L. Alderman, J. Hsia, B. V. Howard, F. R. Cobb, W. J. Rogers, P. Ouyang et al. 2002. Effects of hormone replacement therapy and antioxidant vitamin supplements on coronary atherosclerosis in postmenopausal women: A randomized controlled trial. *JAMA* **288**: 2432–2440.

72. Vivekananthan, D. P., M. S. Penn, S. K. Sapp, A. Hsu, and E. J. Topol. 2003. Use of antioxidant vitamins for the prevention of cardiovascular disease: Meta-analysis of randomised trials. *Lancet* **361**: 2017–2023.

73. Salonen, R. M., K. Nyyssonen, J. Kaikkonen, E. Porkkala-Sarataho, S. Voutilainen, T. H. Rissanen, T. P. Tuomainen et al. Antioxidant Supplementation in Atherosclerosis Prevention. 2003. Six-year effect of combined vitamin C and E supplementation on atherosclerotic progression: The Antioxidant Supplementation in Atherosclerosis Prevention (ASAP) Study. *Circulation* **107**: 947–953.

74. Hodis, H. N., W. J. Mack, L. LaBree, P. R. Mahrer, A. Sevanian, C. R. Liu, C. H. Liu et al. 2002. Alpha-tocopherol supplementation in healthy individuals reduces low-density lipoprotein oxidation but not atherosclerosis: The Vitamin E Atherosclerosis Prevention Study (VEAPS). *Circulation* **106**: 1453–1459.

75. Yusuf, S., P. Sleight, J. Pogue, J. Bosch, R. Davies, and G. Dagenais. 2000. Effects of an angiotensin-converting-enzyme inhibitor, ramipril, on cardiovascular events in high-risk patients. The Heart Outcomes Prevention Evaluation Study Investigators. *New England J. Med.* **342**: 145–153.

76. Redlich, C. A., J. S. Chung, M. R. Cullen, W. S. Blaner, A. M. Van Bennekum, and L. Berglund. 1999. Effect of long-term beta-carotene and vitamin A on serum cholesterol and triglyceride levels among participants in the Carotene and Retinol Efficacy Trial (CARET). *Atherosclerosis* **143**: 427–434.

77. Roberts, L. J., 2nd, J. A. Oates, M. F. Linton, S. Fazio, B. P. Meador, M. D. Gross, Y. Shyr, and J. D. Morrow. 2007. The relationship between dose of vitamin E and suppression of oxidative stress in humans. *Free Radic. Biol. Med.* **43**: 1388–1393.

78. Blumberg, J. B., and B. Frei. 2007. Why clinical trials of vitamin E and cardiovascular diseases may be fatally flawed. Commentary on "The relationship between dose of vitamin E and suppression of oxidative stress in humans". *Free Radic. Biol. Med.* **43**: 1374–1376.

79. Pratt, D. A., G. A. DiLabio, G. Brigati, G. F. Pedulli, and L. Valgimigli. 2001. 5-Pyrimidinols: Novel chain-breaking antioxidants more effective than phenols. *J. Am. Chem. Soc.* **123**: 4625–4626.

80. Wijtmans, M., D. A. Pratt, L. Valgimigli, G. A. DiLabio, G. F. Pedulli, and N. A. Porter. 2003. 6-Amino-3-pyridinols: Towards diffusion-controlled chain-breaking antioxidants. *Angewandte Chem.* **42**: 4847.

81. Nam, T. G., C. L. Rector, H. Y. Kim, A. F. P. Sonnen, R. Meyer, W. M. Nau, J. Atkinson, J. Rintoul, D. A. Pratt, and N. A. Porter. 2007. Tetrahydro-1,8-naphthyridinol analogues of alpha-tocopherol as antioxidants in lipid membranes and low-density lipoproteins. *J. Am. Chem. Soc.* **129**: 10211–10219.

3 Enzymatic Oxidation of Polyunsaturated Fatty Acids

Sharon A. Murphy, Naser M. I. Al-Aaswad, and Anna Nicolaou

CONTENTS

INTRODUCTION

Fatty acids are carboxylic acids with long saturated or unsaturated aliphatic chains. Fatty acids with more than one double bond are referred to as polyunsaturated fatty acids (PUFA), and those with chains of C12–C22 are the main species in mammals. PUFA can be esterified in phospholipids, acyl glycerols, and cholesterol esters found in cell membranes, storage cells, and organelles, and are crucial for a plethora of biological functions including membrane structure and function, synthesis of lipid

45

mediators, signaling events, and regulation of gene expression (Calder 2012). The essential fatty acids required in mammalian diets are linoleic acid (LA) and α-linoleic acid (ALA), precursors to the biologically important families of ω-6 and ω-3 PUFA that include the long-chain PUFA arachidonic acid (AA, 20:4n-6), eicosapentaenoic acid (EPA, 20:5n-3), and docosahexaenoic acid (DHA, 22:6n-3) (Simopoulos 2009).

Enzymatic oxidation of PUFA results in a large number of bioactive metabolites that are produced on demand and in response to various stressors and stimuli (Sala, Folco et al. 2010; Haeggstrom and Funk 2011; Stables and Gilroy 2011; Kendall and Nicolaou 2013). Important lipid metabolizing enzymes include the phospholipases (PL), which control the release of fatty acids from membrane phospholipids, cyclooxygenases (COX), which metabolize PUFA to generate prostaglandins, prostacyclin, and thromboxanes (collectively termed *prostanoids*), lipoxygenases (LOX), which generate a variety of hydroxy fatty acids, and, finally, cytochrome P450 (CYP) monooxygenases and epoxygenases that generate fatty acyl epoxides and an array of hydroxylated species. The oxygenated derivatives of 20-carbon PUFA, predominantly AA, are termed *eicosanoids* and are known for their involvement in various aspects of inflammation.

This chapter will focus on the families of enzymes that oxidize PUFA, their cellular location, function and substrate specificity, and introduce the role they and their enzymatic products play in cellular health and pathophysiological conditions.

PHOSPHOLIPASES

Phospholipases (PL) are a superfamily of enzymes which hydrolyze phospholipids to provide free fatty acids, and in this way are instrumental in the availability of PUFA for enzymatically mediated oxygenation. Of particular relevance to this are the enzymes PLA_2 that cleave phospholipids at the sn-2 position resulting in free PUFA and lysophospholipids, and PLC and PLD that hydrolyze the phosphate head group yielding IP_3 (from phosphatidylinositols) and 1,2-diacylglycerol (1,2-DAG) or PA, respectively (Jenkins and Frohman 2005; Fukami et al. 2010; Dennis et al. 2011; Jang et al. 2012; Kadamur and Ross 2013).

PLA_2 isoforms are grouped into six distinctive types: cytosolic PLA_2 ($cPLA_2$), secretory PLA_2 ($sPLA_2$), Ca^{2+} independent PLA_2 ($iPLA_2$), platelet-activating factor acetyl hydrolase (PAF-AH), adipose PLA_2 ($AdPLA_2$), and lysosomal PLA_2 ($LysoPLA_2$). $cPLA_2$ are some of the best-studied isoforms exhibiting PLA_2 and lysophospholipase activities (Murakami et al. 2011). They are Ca^{2+}-dependent enzymes (0.1–3 μM) with high molecular masses ranging from 60 to 115 kDa (Leslie 1997; Murakami et al. 2011). $cPLA_2$ shows substrate specificity toward AA-containing phospholipids, although it can also hydrolyze esterified EPA and has some low affinity for C18 fatty acids (Leslie and Channon 1990; Clark et al. 1991; Shikano et al. 1994). $cPLA_2$ isoforms are expressed in most mammalian cells, are localized in the cytosol and, following phosphorylation-mediated activation, translocate to the perinuclear membrane (Clark et al. 1995; Akiba and Sato 2004; Tucker et al. 2009).

$iPLA_2$ are Ca^{2+} independent isozymes located in the cytosol and membrane fractions (Akiba and Sato 2004). They tend to be high molecular mass proteins (84–90 kDa), though smaller sized isoforms have been found (Allyson et al. 2012).

iPLA$_2$ isoforms exhibit lysophospholipase, transacylase, as well as PLA$_1$ and PLA$_2$ activities (Dennis et al. 2011; Allyson et al. 2012). They do not exhibit specificity for any individual PUFA and are involved in lysophospholipid regulation and membrane phospholipid remodelling (Balsinde et al. 1995; Burke and Dennis 2009; Dennis et al. 2011; Cheon et al. 2012). Their activation and regulation is complex, mediated through different mechanisms including ATP binding, caspase cleavage, fatty acyl CoA, and calmodulin to date (Burke and Dennis 2009; Dennis et al. 2011).

sPLA$_2$ are extracellular secreted enzymes of small molecular weight (14–19, and 55 kDa) that require high Ca^{2+} concentrations for activity (mM) (Dennis et al. 2011). sPLA$_2$ enzymes show higher activity with anionic phospholipids, although they do not have any specific PUFA selectivity. They are involved in phospholipid digestion (pancreatic sPLA$_2$) as well as a host of inflammatory diseases such as arthritis, Crohn's disease, atherosclerosis, and cancer (Fenard et al. 1999; Haapamaki et al. 1999; Koduri et al. 2002; Bostrom et al. 2007; Mounier et al. 2008; Pucer et al. 2013).

LysoPLA$_2$ has low molecular weight (45 kDA) and exhibits Ca^{2+} independent PLA$_2$, transacylase and 1-o-acylceramide synthase activities, with specificity for PC- and PE-esterified unsaturated fatty acids (Abe and Shayman 1998; Dennis et al. 2011). Is it highly expressed in alveolar macrophages and has been implicated in foam cell formation, surfactant lipid accumulation, phospholipidosis and, potentially, atherogenesis (Taniyama et al. 2005; Hiraoka et al. 2006). Platelet-activating factor acetyl hydrolases (PAF-AH) are Ca^{2+} independent acylhydrolases targeting PAF and phospholipids with short acyl groups (Stafforini and McIntyre 2013). Adipose PLA$_2$ is a newly discovered PLA$_2$ enzyme highly expressed in adipose tissue that is Ca^{2+} independent and shows specificity for PC acting as PLA$_2$ and PLA$_1$ (Uyama et al. 2009). This enzyme has been shown to control adipose tissue lipogenesis through the production of AA-derived PGE$_2$ and its expression is upregulated by feeding and insulin (Duncan et al. 2008; Jaworski et al. 2009).

PLC is a family of phosphodiesterases that catalyze phosphatidylinositol 4,5-bisphosphate (PIP$_2$) hydrolysis to inositol-1,4,5-triphosphate (IP$_3$) and DAG, that can be further hydrolyzed to produce AA for eicosanoid biosynthesis. PLC is localized in the cytosol and translocated to the plasma membrane upon cell activation, while its activity is regulated by Ca^{2+}, heterotrimeric G proteins, protein tyrosine kinases, and small GTPases (Fukami et al. 2010; Kadamur and Ross 2013). PLD hydrolyze primarily PC to give PA and free choline, and the two mammalian isoforms PLD1 and PLD2 are regulated by a number of factors including PKCs, phosphoinositides, ARF and Rho GTPases (Jenkins and Frohman 2005; Jang et al. 2012). A PLD with specificity for N-arachidonyl phosphatidylethanolamines is involved with production of the endocannabinoid anandamide (AEA), and other PUFA-ethanolamides (Fonseca et al. 2013).

Interestingly, the activity of many phospholipases can be affected by reactive oxygen species (ROS), thus impacting upon the related signaling pathways (Korbecki et al. 2013). Ultraviolet radiation (UVR)-induced ROS formation has been shown to increase the activity and expression of cPLA$_2$ with concomitant stimulation of the AA cascade in skin, while the activity of NADPH oxidases are affected by AA, DAG, and PA, produced by PLA$_2$, PLC, and PLD, respectively (Chen et al. 1996; Cummings et al. 2002; Frey et al. 2002; Pendyala et al. 2009).

BIOSYNTHESIS OF PROSTANOIDS

CYCLOOXYGENASE-MEDIATED REACTIONS AND RELATED MEDIATORS

Prostaglandin-endoperoxide synthases or prostaglandin H synthases (PGHS) are commonly known as cyclooxygenases (COX). There are two isoforms of COX, the constitutive COX-1 and the inducible COX-2, and a splice variant of COX-1 has been reported as COX-3 (Smith et al. 2000; Chandrasekharan et al. 2002; Smith et al. 2011). COX are bi-functional enzymes exhibiting both cyclooxygenase and peroxidase activities: the first step of this multistep reaction (cyclooxygenase) adds molecular oxygen to AA forming PGG_2, followed by the peroxidase reaction that gives PGH_2 (Figure 3.1). This unstable prostaglandin is then converted into various prostanoids depending on the cellular prevalence of terminal prostanoid synthases. COX-1 is located in the membranes of the ER, whereas COX-2 is located on the ER and the nuclear membrane. The two isoforms are similar, exhibiting 63% sequence homology, and although identical in length, the active site is larger in COX-2 than in COX-1; giving COX-2 wider substrate specificity (Lecomte et al. 1994; Smith et al. 2000).

In addition to AA, which is a precursor to the biologically dominant prostanoids (series-2), both COX-1 and -2 can also metabolize the C20 fatty acids DGLA (20:3n-6) and EPA to form series-1 and series-3 prostanoids, respectively (Figure 3.2). Though both COX-1 and -2 have similar affinity toward the C20 PUFA AA, EPA, and DGLA, COX-2 can oxygenate a wider range of fatty acids, including DHA and LA, although it only exhibits peroxidase activity forming 13-HDHA and 9-HODE, respectively (Kaduce et al. 1989). The endocannabinoids AEA and 2-AG are also substrates for COX-2 enzymes and they are converted into prostamides (e.g., PGE_2-EA, $PGF_{2\alpha}$-EA) or glycerol esters (such as PGE_2-G, PGI_2-G), respectively (Figure 3.2) (Woodward et al. 2008).

Aspirin acetylates a serine residue positioned in the catalytic site of the enzyme (Ser529 of COX-1 and Ser516 of COX-2) resulting in irreversible inhibition of COX-1/2 (Loll et al. 1995). However, the acetylated-COX-2 (ac-COX-2) retains part of its catalytic activity and can still act as a peroxidase forming R monohydroxy fatty acids such as 15R-HETE from AA, 18R-HEPE from EPA and 17R-HDHA (Mancini et al. 1994; Serhan et al. 2008; Sharma et al. 2010).

Although the COX reaction requires catalytic levels of peroxide, the enzyme activity is not directly affected by ROS production (Smith 2008). However, ROS-mediated increase of the inducible enzymes involved in the arachidonic acid cascade, including phospholipases that provide free AA and other PUFA substrates for COX-mediated reactions, COX-2 and prostaglandin synthases (see Section 3.1), makes the formation of eicosanoids sensitive to the cellular redox status, and this can impact on disease states characterized by oxidative stress (Korbecki et al. 2013).

PROSTANOID SYNTHASES FORM PROSTAGLANDINS, PROSTACYCLIN, AND THROMBOXANES

Prostaglandin E synthase (PGES) catalyzes the isomerization of PGH to PGE. Three types of PGES have been characterized: the cytosolic PGES (cPGES) and

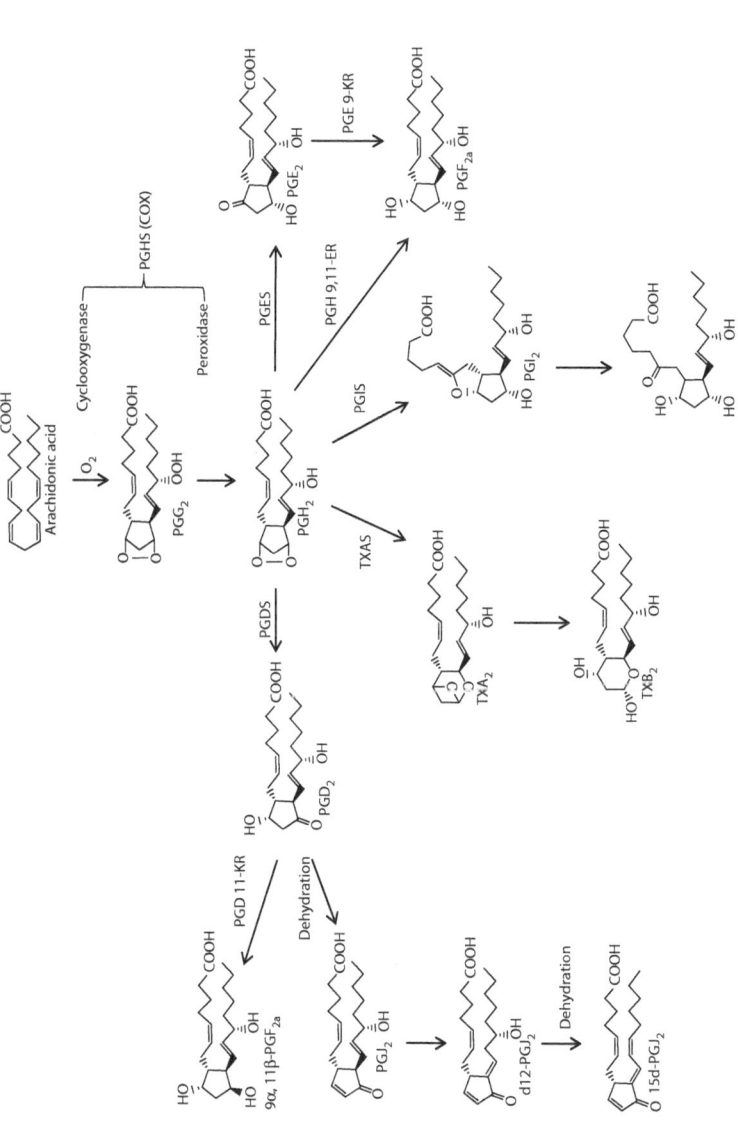

FIGURE 3.1 Schematic outline of the conversion of arachidonic acid to prostanoids via cyclooxygenase (COX) and prostanoid synthase (PGS) catalyzed reactions. PGDS, prostaglandin D synthase; PGES, prostaglandin E synthase; PGH 9,11-ER, prostaglandin 9,11-endoperoxide reductase; PGE 9-KR, prostaglandin E 9-ketoreductase; PGD 11-KR, prostaglandin D 11-ketoreductase; TXAS, thromboxane A synthase; PGIS, prostcyclin synthase.

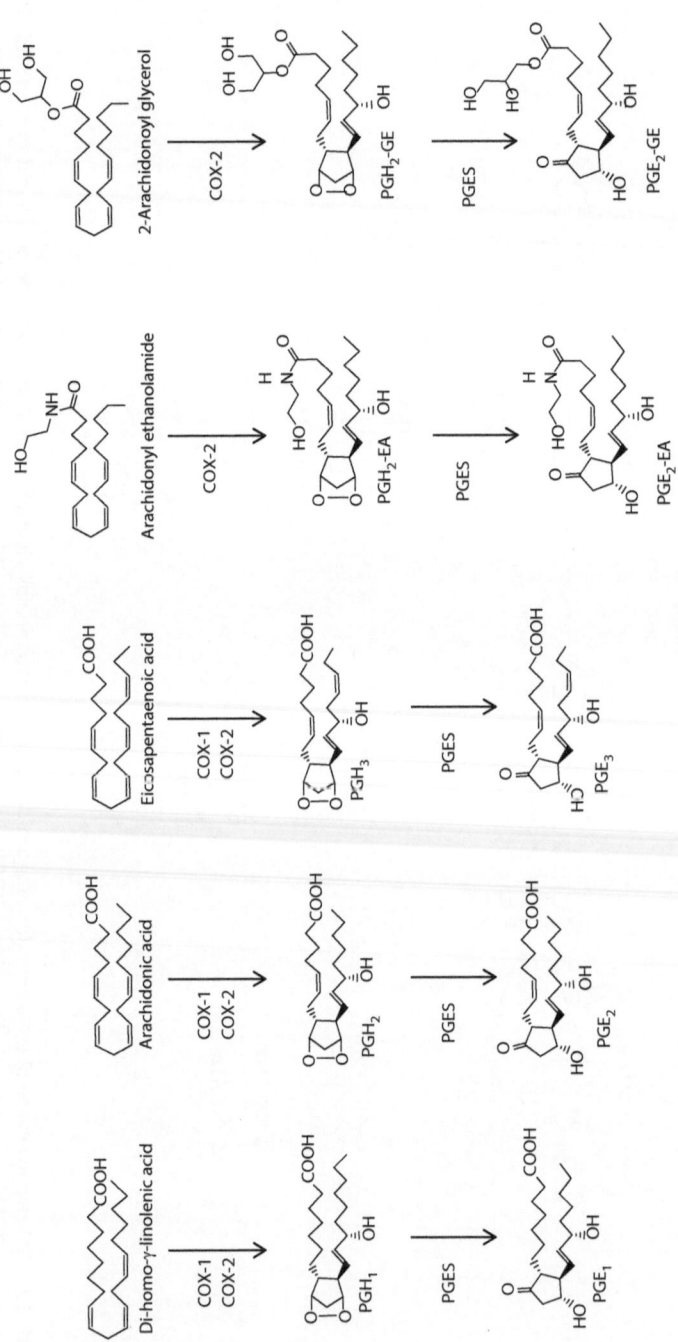

FIGURE 3.2 Examples of cyclooxygenase (COX)-catalyzed conversion of free fatty acids and endocannabinoids to prostaglandins, prostamides, and glycerol esters. PGES, prostaglandin E synthase; COX, PGH, prostaglandin H_1; PGH$_2$, prostaglandin H_2; PGH$_3$, prostaglandin H_3; PGH$_2$-EA, prostaglandin H_2 ethanolamide' PGH$_2$-GE, prostaglandin H_2 glycerol ester.

the membrane-bound mPGES-1 and mPGES-2 (Jakobsson et al. 1999; Tanioka et al. 2000; Tanikawa et al. 2002). The AA-derived PGE_2 is one of the best-studied prostaglandins, exhibiting a wide range of bioactivities: low concentrations PGE_2 are important for cellular homeostasis, while high concentrations generated in a response to inflammatory and other stimuli are found in many pathological conditions; in these cases coordinated upregulation and functional coupling of COX-2 and mPGES have been observed (Mancini et al. 2001; Stichtenoth et al. 2001; Park et al. 2006; Samuelsson et al. 2007). Overall, PGE_2 is considered to be a pro-inflammatory and immunosuppresive lipid mediator that promotes cellular proliferation (Chen and Smyth 2011; Kalinski 2011). PGE_1 and PGE_3 are considered less inflammatory and can tone down the effect of PGE_2; for this reason nutritional supplementation with PUFA including the n-3PUFA EPA, precursor of PGE_3, have been considered in order to address systemic inflammation and disorders related to it (Simopoulos 2009; Calder 2012).

Prostaglandin D synthases (PGDS) catalyzes the formation of PGD. Two types of PGDS have been identified: the glutathione-dependent hematopoietic-type PGDS (H-PGDS) and the lipocaline-type PGDS (L-PGDS). H-PGDS is found in antigen-presenting cells while L-PGDS is found in the brain and other cells of neural crest origin such as epidermal melanocytes (Urade and Eguchi 2002; Takeda et al. 2006; Yazaki et al. 2012). The cyclopentenone prostaglandin PGJ_2 is produced by spontaneous dehydration of PGD_2 (Urade and Eguchi 2002). Similarly, PGJ_2 isomerizes to form Δ^{12}-PGJ_2 and then undergoes a secondary dehydration to15-deoxy-$\Delta^{12,14}$-PGJ_2. PGD_2 is a powerful sleep-inducing endogenous lipid mediator with immunomodulatory properties (Urade and Hayaishi 2011).

Prostaglandin F synthases (PGFS) belong to the aldoketoreductase family of enzymes (AKR) catalyze the reduction of PGH_2 to $PGF_{2\alpha}$ though there are additional indirect synthetic routes to $PGF_{2\alpha}$ from both PGD_2 and PGE_2 (Figure 3.1). PGFS were originally isolated from bovine tissues, though human homologues were later discovered, with a total of three established human PGFS enzymes: AKR1C3, AKR1B1, and AKR1A1 (Watanabe 2011; Lacroix Pepin et al. 2013). PGH 9-, 11-endoperoxide reductase (PGH 9,11ER) catalyzes the formation of $PGF_{2\alpha}$ from PGH_2 while PGD 11-ketoreductase (PGD11-KR) catalyzes the formation of 9α, 11β-$PGF_{2\alpha}$ from PGD_2. AKR1C3 has both PGH 9,11ER and PGD11-KR activities and can convert PGD_2-EA into $PGF_{2\alpha}$-EA (Smith et al. 2011), and PGE 9-ketoreductase (PGE9-KR) catalyzes the formation of $PGF_{2\alpha}$ from PGE_2 (Suzuki-Yamamoto et al. 1999). $PGF_{2\alpha}$ is a potent vasoconstrictor and has pro-thrombotic activity similar to thromboxane (Lacroix Pepin et al. 2013).

Prostacyclin (PGI_2) is formed from PGH_2 by the enzyme prostacyclin synthase (PGIS). PGIS is a membrane-bound hemoprotein enzyme from the CYP450 family (CYP8A1), though it is not an oxygenase but an isomerase (Wu and Liou 2005; Smith et al. 2011). PGIS is widely expressed in human tissues particularly in the endothelial cells and smooth muscle cells. In endothelial cells, the expression and production of both PGIS-mRNA and PGIS-protein can by induced by TNFα and some inflammatory cytokines. PGIS is co-localized with COX-1 and COX-2 in the nuclear envelope and functionally couples with COX and $cPLA_2$ after their translocation to the nuclear membrane (Wu and Liou 2005). PGI_2 is a potent vasodilator and also inhibits platelet aggregation. Once formed, PGI_2 is unstable and hydrolyzed nonenzymatically to the inactive metabolite 6-keto-$F_{1\alpha}$, which is used to measure the prevalence of PGI_2 (Hara et al. 1994; Liou et al. 2000).

Thromboxane A synthase (TXAS) converts PGH_2 into thromboxane A_2 (TXA_2) or EPA-derived PGH_3 to TXA_3. TXAS belongs to the P450 epoxygenase family and is localized in the membrane of the endoplasmic reticulum (Ekambaram et al. 2011). Once formed, TXA_2 is very unstable and is nonenzymatically converted into TXB_2, a stable inactive product used to measure TXA_2 (Nakahata 2008). It has a half-life of 30s and functions as an autocrine or paracrine mediator in the body (Smith et al. 1991). Discovered initially in platelets, TXs are involved in platelet aggregation and vasoconstriction (Smith et al. 2011).

PROSTANOID RECEPTORS, TRANSPORTERS, AND DEACTIVATION REACTIONS

Prostanoids act via specific G-protein-coupled receptors described as follows: DP for PGD; EP_1, EP_2, EP_3 and EP_4 for PGE; FP (two isoforms: FPA and FPB) for PGF; TP (TP_α and TP_β) for TX and IP for PGI; moreover, PGJ_2 and 15-deoxy- ^{12,14}-PGJ_2 (15d-PGJ_2) activate the PPAR nuclear receptors. In addition, there are eight isoforms for EP_3 that only differ in their C-terminal and these are generated through alternative splicing (Breyer et al. 2001; Scher and Pillinger 2005; Woodward et al. 2011).

This diversity can explain the multitude of effects mediated by prostanoids. Depending on the subtype of EP it signals through, PGE_2 can mediate a plethora of effects. For example, PGE_2 has vaso-constrictive effects when acting via EP1 or EP3 receptors but it acts as vasodilator when binding to EP2 or EP4 (Tang et al. 2000). Furthermore, the effect of prostanoid receptor activation on cell function is mediated by different intracellular signaling pathways. EP2, EP4, and IP receptors activate adenyl ayclase, which increases production of intracellular cAMP, while via Gq EP1 and FP activate PI metabolism. Finally, TP acts through two types of G-protein coupled receptors, Gq and Gl3, which both activate phospholipase C. Although both TP isoforms are coupled to phospholipase C activation, $P\alpha$ stimulates adenylyl cyclase, whereas TP_β inhibits it (Hatae et al. 2002).

Members of the organic anion transporting polypeptide superfamily (OATPs) can transport prostamides, with OATP2A1 (previously termed PGT) capable of transporting multiple substrates, namely PGE_2, PGE_1, PGD_2, PGF_{2a}, and TXB_2 (Lu et al. 1996; Hagenbuch and Stieger 2013). OATP2A1 mRNA has also been found in multiple tissues analysis including brain, heart, prostate, and colon (Schuster 2002; Roth et al. 2012).

Inactivation of PGs and TXs is undertaken by 15-prostaglandin dehydrogenase (15-PGDH). There are two types of PGDH, type 1 and type 2, the second isoform also has a 9-ketoPG function. After initial oxidation by PGDH, further oxidation to 13,14-dihydro-PGs occurs by the enzyme 13-PGR, the same enzyme which deactivates LTB_4 to 12-keto-LTB_4 (Tai et al. 2002). 15-PGDH can oxidize a wide range of prostanoid substrates, including PGE_1, PGE_2, $PGF_{1\alpha}$, PGI_2, and 6-keto-$PGF_{1\alpha}$. Deactivation of TXs occurs via the enzyme 11-TXB_2DH to 11-dehydro-TXB_2 (Tai et al. 2002). Interestingly, PG transporters can also be involved in the process of inactivation, as exemplified by work in HeLa cell lines showing that oxidation to dihydro-keto PGs can only occur when PGT and PGDH are coexpressed (Nomura et al. 2004). Finally, prostanoids can be subjected to β-oxidation and transformed to dinor- and tetranor-prostanoids (species with 2- and 4-carbon shorter aliphatic chains, respectively). The products of these reactions are found in the urine and have been used as markers of systemic levels of n-3 or n-6 PUFA and their metabolism via COX (Song et al. 2008; Jabr et al. 2013; Kuklev et al. 2013).

LIPOXYGENASE-MEDIATED FORMATION OF HYDROXY FATTY ACIDS

LIPOXYGENASES

Lipoxygenases (LOX) are a family of dioxygenases catalyzing the insertion of molecular oxygen stereoselectively and regioselectively into PUFA with at least 1 cis,cis-1,4-pentadiene in their structure (Haeggstrom and Funk 2011; Joo and Oh 2012). They are commonly labeled according to the regiospecific carbon at which they oxidize AA (i.e., 5-LOX, 12-LOX, etc.). In terms of the phylogenetic classification, mammalian LOXs include 5-, 8-, 12-, and 15-LOX (with 8-LOX expressed only in rodents), plants include 9-, 13-LOXs, fungi: 11-, 13-, and 15-LOXs, and bacteria 9-, 10-, 11-, 13-, and 15-LOX. The mechanism for lipid oxidation by LOX is free radical based and is composed of three consecutive steps (Haeggstrom and Funk 2011). The first step is hydrogen abstraction of an allylic hydrogen from a PUFA substrate, followed by a radical rearrangement to an adjacent carbon (+1) and finally the stereospecific insertion of molecular oxygen (Kuhn and Thiele 1999).

5-LOX-MEDIATED REACTIONS AND METABOLITES

The enzyme 5-LOX is expressed predominantly in human leukocytes (eosinophils, neutrophils, macrophages, etc.) and can be found in both the cytoplasm and the nucleus depending on the cell type (Woods et al. 1995; Luo et al. 2003; Brock 2005).

5-LOX inserts molecular oxygen into AA to form the intermediate hydroperoxy compound 5-HpETE. This is readily reduced to 5-HETE or subsequently transformed into the unstable epoxide LTA_4 that is further hydrolyzed to LTB_4 by the ubiquitous enzyme LTA_4 hydrolase (Brock et al. 2005) (Figure 3.3). 5-LOX-activating protein (FLAP) is an 18 kDa protein located in the nuclear envelope and is essential for 5-LOX activity by transferring the substrate fatty acid and enhancing the sequential oxidation to 5-HpETE and dehydration to 5-HETE and LTA_4 (Aparoy et al. 2012). The initial translocation of 5-LOX to the nuclear membrane is FLAP independent but Ca^{2+} dependent (4–10 μM for full-activation) and is followed by FLAP-dependent activation of the enzyme (Percival et al. 1992; Woods et al. 1995; Luo et al. 2003). Interestingly both cPLA2, which provides free AA substrate, and 5-LOX are activated by phosphorylation (Werz et al. 2002; Werz et al. 2002; Luo et al. 2004). Finally, 5-LOX appears to have features of interfacially activated enzymes with preference for positively charged membrane phospholipids (Pande et al. 2004).

5-HETE is a 5-LOX-derived AA metabolite, considered to be pro-inflammatory. It was found to be elevated in patients with chronic kidney disease (Maaloe et al. 2011)

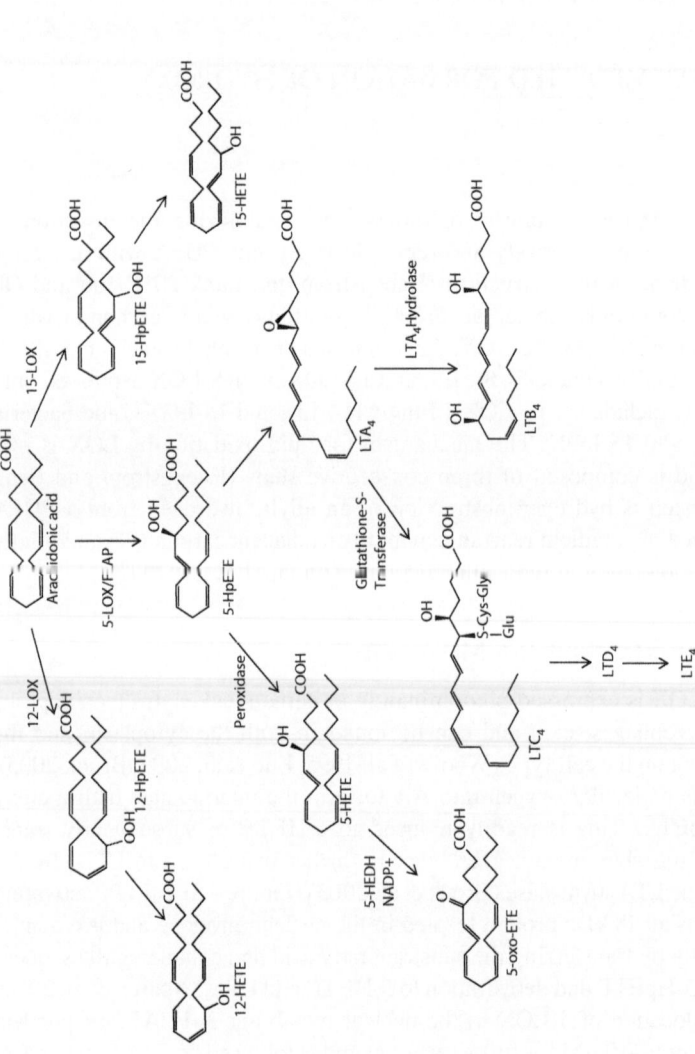

FIGURE 3.3 Conversion of arachidonic acid to hydroxy fatty acids and leukotrienes via the 5-, 12-, and 15-lipoxygenase (LOX)-catalyzed reactions. 5-HETE, 5-hydroxyeicosatetraenoic acid; 5-HpETE, 5-hydroperoxyeicosatetraenoic acid; 5-oxo-ETE, 5-oxoeicosatetranoic acid; 12-HETE, 12-hydroxyeicosatetraenoic acid; 15-HETE, 15-hydroxyeicosatetraenoic acid; LTA₄, leukotriene A₄; 5S,6S-epoxy-7E,9E,11Z,14Z-eicosatetraenoic acid; LTA₄ hydrolase, leukotriene A₄ hydrolase; LTB₄, leukotriene B₄; 5S,12R dihydroxy-6Z,8E,10E,14Z-eicosatetraenoic acid; LTC₄, leukotriene C₄, LTD₄, leukotriene D₄; LTE₄, leukotriene E₄; 5-LOX, 5-lipoxygenase; 12-LOX, 12-lipoxygenase; 15-LOX, 15-lipoxygenase.

and in a prostate adenocarcinoma mouse model (Pham et al. 2006). It was also shown to stimulate growth in a breast cancer cell line; this growth was reduced with the application of 5-LOX inhibitors (Avis et al. 2001). 4-HDHA and 7-HDHA are DHA-derived 5-LOX products and 4-HDHA exhibits antiangiogenic properties. In studies of mice fed omega -3 PUFA diets, the activity of 4-HDHA was found to be mediated by the PPARγ receptor and this antiangiogenic effect was absent in 5-LOX knockout mice (Sapieha et al. 2011).

5-Oxo-ETE is a 5-LOX-derived mediator produced by eosinophils and is a very powerful chemoattractant for neutrophils, monocytes, and basophils (Sozzani et al. 1996; Sturm et al. 2005) (Figure 3.3). After reaction with 5-LOX, 5-hydroxyeicosanoid dehydrogenase (5-HEDH) further oxidizes 5S-HETE to generate 5-oxo-ETE (Powell and Rokach 2013). 5-HEDH is highly selective for 5-S-HETE and its cofactor for activity is NADP$^+$ while it is also inhibited by NADPH (Erlemann et al. 2007). 5-Oxo-ETE generates its effects via a specific GPCR OXE-receptor that is closely related to the CysLT receptors. It is found in monocytes, neutrophils, eosinophils, and in tumor cells and it can also bind 5-HpETE and 5-HETE, though with a reduced selectivity (Powell and Rokach 2005; Sundaram and Ghosh 2006). Due to its location on inflammatory/tumor cells, this receptor has been implicated in chronic inflammatory diseases such as asthma and in tumor progression (Sundaram and Ghosh 2006).

Leukotrienes (LT) are potent pro-inflammatory mediators formed by 5-LOX expressed in different types of leukocytes (monocytes, granulocytes, mast cells, dendritic, macrophages) (Haeggstrom and Wetterholm 2002). When AA is metabolized by 5-LOX it gives rise to series-4 LTA$_4$ an LTB$_4$ and the cysteinyl leukotrienes (CysLTs) LTC$_4$, LTD$_4$, and LTE$_4$ (Figure 3.3). CysLTs are generated in cells expressing LTC$_4$ synthase, and then glutathione conjugation forms LTC$_4$ (Lam 2003; Singh et al. 2013). Post-synthesis, LTB$_4$ and LTC$_4$ are exported from the cells via transporters (Lam et al. 1989; Haeggstrom and Funk 2011; Singh et al. 2013), and LTC$_4$ is cleaved by a transpeptidase resulting in LTD$_4$, which can then undergo further reaction with a dipeptidase to cleave glycine yielding LTE$_4$ (Lee et al. 2009). LTB$_4$ can be inactivated via two routes: by ω-oxidation in human PMN with a CYP enzyme, or via LTB$_4$ dehydrogenase (LTB4DH) (Yokomizo et al. 1993; Tobin et al. 2013). It should be noted that LTs can also be generated by transcellular metabolism and this has been observed in co-culturing studies with pneumocytes and neutrophils (Grimminger et al. 1992). Finally, 5-LOX can also generate series-5 LT (e.g., LTB$_5$) and cysLTs from EPA (Sapieha et al. 2011; Joo and Oh 2012).

LTs have been implicated in a number of allergic and chronic inflammatory disease pathologies, including arthritis, asthma, atherosclerosis, and cancer, and LTB$_4$ is a very potent chemoattractant, recruiting inflammatory cells to site in these inflammatory diseases (Haeggstrom and Funk 2011; Singh et al. 2013). CysLTs elicit other responses such as vascular permeability, plasma leakage, pulmonary edema and smooth muscle contraction, and their role as potent bronchoconstrictors makes them the main mediators associated with the symptoms of asthma sufferers (Haeggstrom and Funk 2011; Singh et al. 2013).

Leukotrienes elicit their effects via GPCRs: BLT$_1$ and BLT$_2$ for LTB$_4$, and CysLT1 and CysLT2 for the CysLTs (Haeggstrom and Funk 2011). CysLT1 binds LTD$_4$ with the highest affinity, followed by LTC$_4$ and LTE$_4$. CysLT2 binds LTC$_4$ and LTD$_4$ but

LTE$_4$ shows low affinity. Therapeutic intervention strategies for asthma are focused around antagonists for CysLT and BLT receptors. Currently, drugs such as the asthma medication Montelukast are effective CysLT1 antagonists; since the CysLT2 receptors pathophysiological role is less well understood, there is no antagonist available (Singh et al. 2013). BLT1 is expressed in leukocytes whereas BLT$_2$ is expressed in multiple tissues (Haeggstrom and Funk 2011). Finally, LTE$_4$ is a ligand for the CYSLT2 receptor and may have a role in vascular permeability (Di Gennaro and Haeggstrom 2012).

5-LOX is associated with inflammatory allergic diseases (i.e., asthma, rhinitis), cardiovascular diseases (atherosclerosis), but also is associated with gastrointestinal diseases (i.e., Crohn's, colitis). 5-LOX has also been implicated in cancers of the colon, esophagus, lung, pancreas, and prostate. Specific roles of the 5-LOXs in cancer include inhibition of apoptosis, tumorigenesis, angiogenesis, genotoxicity, metastasis, and cell proliferation. Interestingly, 5-LOX inhibitors used in studies have indicated roles in the inhibition of many of these processes.

12- AND 15-LOX-MEDIATED REACTIONS

There are three distinct 12-LOX types: the platelet type (p12-LOX) that is expressed in platelets and megakaryocytes, the reticulocyte/leukocyte type (L12-LOX) and the epidermal type (e12-LOX and eLOX-3) (Boeglin et al. 1998; Ikei et al. 2012). 15-LOX reactions are catalyzed by two isozymes, the reticulocyte-type 15-LOX-1 and the epidermal-type 15-LOX-2. Interestingly, 15-LOX-1 has been found to prefer oxidation of LA to generate 13S-HODE, while 15-LOX-2 has been shown to prefer AA over LA as a substrate (Vincent et al. 2008; Wecksler et al. 2008; O'Flaherty et al. 2012). The enzymes L12-LOX and 15-LOX-1 are classified as 12/15LOX since they are able to produce both 12S-HpETE and 15S-HpETE from AA, albeit in varying ratios depending on the cell type (Dobrian et al. 2011; Krieg and Furstenberger 2013). Furthermore, 15-LOX-2 shares 78% sequence identity with the murine-exclusive 8-LOX, but both display different tissue distributions and products, namely 15S-HpETE and 8S-HpETE, respectively (Krieg and Furstenberger 2013).

12R-LOX is the only mammalian LOX that forms R derived products (12R-HpETE) and this enzyme has been found to be important in both epidermal barrier function and terminal differentiation of the skin (Epp et al. 2007; Agarwal et al. 2009). eLOX-3 is a hydroperoxide isomerase recently determined to be important in hepoxilin synthesis (Krieg and Furstenberger 2013; Munoz-Garcia et al. 2014). It has also been found to be important in epidermal barrier function along with 12R-LOX, with which it shares sequence similarities: 12-R-LOX (54%) and 15-LOX-2 (51%) (Krieg and Furstenberger 2013; Munoz-Garcia et al. 2014).

Overall, and based on their phylogenetic similarities, 12-LOX and 15-LOX are grouped into three types: 12-LOX, 12/15-LOX (comprising the isoforms: leukocyte 12S-LOX and 15-LOX-1) and epidermal LOX (comprising the isoforms: 12R-LOX, 15-LOX-2, 8-LOX, and eLOX-3) (Krieg and Furstenberger 2013). It appears that 12-LOX enzymes are generally associated with inflammation and hypertension as well as diabetes (Joo and Oh 2012), while 15-LOX enzymes have been linked to atherogenesis, carcinogenesis, asthma, inflammation, and cell differentiation (Joo and Oh 2012).

12-,15-LOX-Derived Mono Hydroxy Fatty Acids

Lipoxygenation of PUFA can result in a range of positional hydroxyl fatty acid iso-mers that have been shown to exhibit a range of bioactivities. When AA is the main PUFA substrate, a range of HETE are formed (Figure 3.3).

12-HETE is considered to be a potent leukocyte chemo-attractant that can also stimulate cell proliferation and enhance tumor cell survival (Honn et al. 1994; Dailey and Imming 1999). Binding sites for 12-HETE have been identified in human kerati-nocytes and Langerhans cells, suggesting its potential involvement in wound healing and cutaneous inflammation (Ruzicka 1992; Arenberger et al. 1993). Work on skin-derived lymphocytes involved in psoriasis has shown that 12(R)-HETE has modest chemotactic properties for T cells but is less potent than LTB_4 (Bacon and Camp 1990). Furthermore, it has been shown that inhibition of 12/15-LOX enhanced the production of Th2 cytokines and attenuated the development of allergic inflamma-tion in a mouse model of allergic lung disease, whilst delivery of 12(S)-HETE had the opposite effect (Cai et al. 2009). Increased levels of 12-HETE were also associ-ated with metabolic changes in T cells leading to the development of autoimmune disease (Kato et al. 1983). Some studies have shown that 12-HETE is involved in growth and survival of cancer cells, possibly through inhibition of cancer cell apop-tosis (Pidgeon et al. 2002). Interestingly, 12-LOX mRNA expression was elevated in one clinical study of prostate cancer patients, and was correlated with aggressive late-stage prostate cancer, which could possibly be applied in prostate cancer pro-gression/prognosis (Gao et al. 1995). In another study, 12-LOX was also shown to be upregulated in CRC tumors, with its inhibition limiting both growth and migration (Klampfl et al. 2012).

15-HETE has multiple anti-inflammatory effects that can counteract some of the pro-inflammatory properties of 12-HETE and PGE_2, may contribute to the resolution of cutaneous inflammation, inhibition of superoxide production and neutrophil migration across cytokine-activated endothelium, and modulation of monocyte adhesion mediated by β-integrins, in atherosclerosis (Vachier et al. 2002; Serhan et al. 2003; Cai et al. 2004; Wittwer and Hersberger 2007; Rhodes et al. 2009). 15-HETE regulates T cell division and displays antiproliferative effects on a leuke-mia T-cell line (Bailey et al. 1997; Kumar et al. 2009). Furthermore, 15-HETE is thought to be anticarcinogenic in the colon (Bhattacharya et al. 2009). Similar prop-erties have been attributed to 8-HETE, product of the murine 8-LOX, which has also been found to induce differentiation in the epidermis via a PPAR-mediated pathway (Muga et al. 2000; Schweiger et al. 2007). It has been suggested that 15-HETE may mediate some of its anticancer and protective effects through further metabolism to produce lipoxins LXA_4 and LXB_4 (Serhan et al. 2003).

The same anti-inflammatory profile is shared by the 15-LOX-derived LA and DGLA metabolites, 13-HODE and 15-HETrE (Xi et al. 2000; Ziboh et al. 2000; Nicolaou et al. 2012). 13-HODE is one of the most abundant hydroxy fatty acids in the skin and has been shown to reverse epidermal hyperproliferation (Cho and Ziboh 1994). Recent reports have shown a possible involvement of HODE and oxo-ODE in pain, although this has not yet been explored in skin (Ruparel et al. 2012; Sisignano et al. 2013). 13-HODE can also inhibit colonic cell proliferation and has been shown to induce cell apoptosis (Shureiqi et al. 1999). It may also act as an antiadhesive molecule and has been postulated as having a pro-tumorigenic role in prostate cancer cells via a mitogenic pathway by prolonging survival and the proliferation of prostate epithelial cells (Kelavkar and Cohen 2004).

15-HETrE is also considered to be anti-inflammatory and exhibits antiproliferative activity proposed to be propagated via the modulation of AP-1, a nuclear transcription factor associated with apoptosis (Xi et al. 2000). It has been shown to suppress COX-2 expression and therefore PGE_2 production in a prostate cancer cells (Pham et al. 2004). Finally, 17-HDHA, a 15-LOX product of DHA, has vasodilatory effects associated with Ca^{2+}-activated K^+ channels in coronary arterial smooth muscle cells (Li et al. 2011), while 14,21-dihydroxy-docosahexaenoic acid (HDHA) has recently been reported to mediate wound healing (Lu et al. 2010). The EPA-product 18-HEPE is an important precursor for RvE and has also been found to be anti-inflammatory in nature. 18-HEPE levels (along with 17-HDHA) were elevated in the liver tissue of a murine hepatic tumorigenesis model when subjected to diethylnitrosamine and this was accompanied by a decrease in $TNF\alpha$ in mice plasma (Weylandt et al. 2011). Parallel *in vitro* work on a murine macrophage cell line demonstrated reduced $TNF\alpha$ following an LPS challenge. Synthetic was found 18R-HEPE reduced neutrophil recruitment when tested on a murine zymosan induced peritonitis model (Krishnamurthy et al. 2011).

Finally, it should be noted also that the endocannabinoids AEA and 2-AG can be hydroxylated by 12- and 15-LOX, forming the equivalent 12-hydroxy fatty acid ethanolamides: 12-HETE-EA, 15-HETE-EA, and 12-HETE-G (Rouzer and Marnett 2011; Fonseca et al. 2013) These metabolites, like their AA analogues, have shown bioactivity though more work is needed to establish their significance (Fonseca et al. 2013).

LOX–LOX Reactions: Lipoxins, Resolvins, Protectins, and Maresins

A number of PUFA products are generated through sequential LOX reactions and transcellular metabolism involving multiple enzymes that may be native to different cell types (Sala et al. 2010). These reactions generate poly-hydroxy fatty acids such as lipoxins, resolvins, protectins, and maresins, products of AA, EPA, and DHA (Figure 3.4).

Lipoxins (LX) are trihydroxytetraene lipid mediators formed from arachidonic acid via multiple sequential LOX reactions (Serhan 2005). They were initially found to be important counter regulators of some of the pro-inflammatory lipid mediators (prostaglandins, leukotrienes), but have since been established as potent anti-inflammatory and pro-resolving lipid mediators, and low levels of these mediators have been implicated in multiple chronic inflammatory diseases such as arthritis, asthma, and colitis (Thomas et al. 1995; Bonnans et al. 2004; Fiorucci et al. 2004). LXs are formed from leukocyte 5-LOX activity on AA, which generates LTA_4; further enzymatic oxidation with 15-LOX or 12-LOX then generates LXB_4. 15-LOX activity on AA yields 15-S-HpETE, which can undergo further reaction with 5-LOX to give LXA_4. Epimers of lipoxins (epi-LX) have opposing stereochemistry at one stereocenter and can be generated from the Ac-COX-2 pathway. In this way, AA

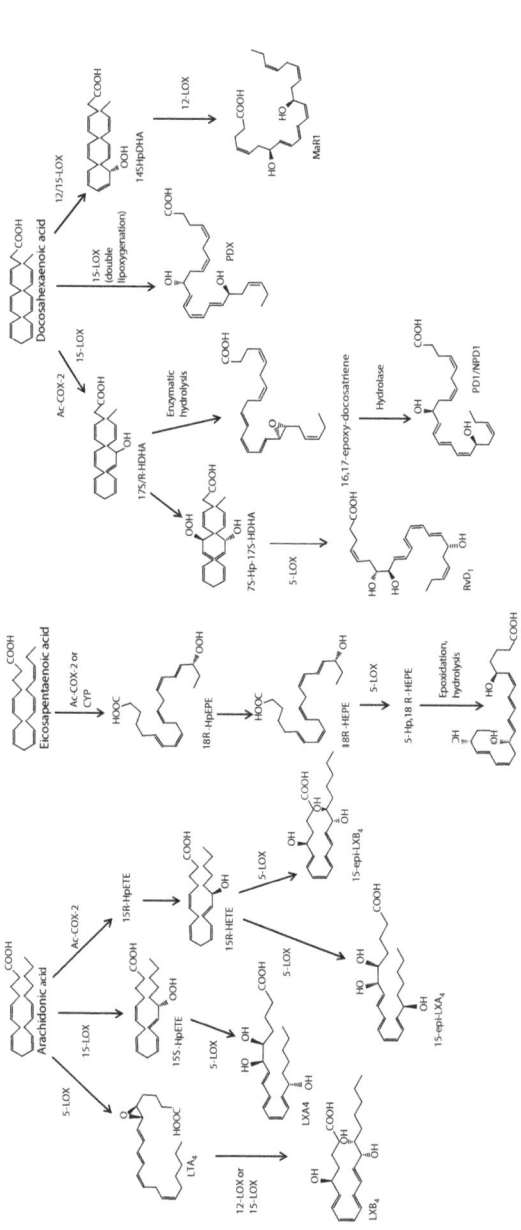

FIGURE 3.4 Examples of oxidation products of selected PUFA formed via sequential and transcellular biosynthetic oxygenations catalyzed by lipoxygenase (LOX) and acetylated-cyclooxygenase-2 (Ac-COX-2). LTA₄, leukotriene A₄; 5S,6S-epoxy-7E,9E,11Z,14Z-eicosatetraenoic acid; LXA₄, lipoxin A₄; 5S,6R,15S-trihydroxy-7E,9E,11Z,13E-eicosatetraenoic acid; LXB₄, lipoxin B₄; 5S,14R,15S-trihydroxy-6E,8Z,10E,12E-eicosatetraenoic acid; 15-epi-LXA₄, aspirin triggered 15R lipoxin A₄; 5S,6R,15R-trihydroxy-7E,9E,11Z,13E-eicosatetraenoic acid 15-epi-LXA₄, aspirin triggered 15R lipoxin B₄, 5S,14R,15S-trihydroxy-6E,8Z,10E,12E-eicosatetraenoic acid; 15R/S-HETE, 15R/S-hydroxyeicosatetranoic acid; 15-HpETE, 15-hydroperoxyeicosatetranoic acid; 18R-HEPE, 18-R-hydroxyeicosapentaenoic acid; 18R-HpEPE, 18-R-hydroperoxyeicosapentaenoic acid; RvE₁, resolvin E₁; 5S,12R,18R-trihydroxy-6Z,8E,10E,14Z,16E-eicosapentaenoic acid; 17-S/R-HDHA, 17-R/S- hydroxydocosahexaenoic acid; 7S-Hp-17S-HDHA, 7-S-hydroperoxy-17S-hydroxydocosahexaenoic acid; 14-S-Hp-DHA, 14S-hydroperoxydocosahexaenoic acid; 17-HDHA, 17S/R-hydroxydocosahexaenoic acid; RvD₁, resolvin D₁; 7S,8R,17S-trihydroxy-4Z,9E,11E,13Z,15E,19Z-docosahexaenoic acid; PD1/NPD1, protectin D1/neuroprotectin D1; 10R,17S-dihydroxy-4Z,7Z,11E,13E,15Z,19Z-docosahexaenoic acid; PDX, protectin DX; 10S,17S-dihydroxy-4Z,7Z,11E,13Z,15E,19Z-docosahexaenoic acid; MaR1, maresin 1; 17R,14S-dihydroxy-4Z,8E,10E,12Z,16Z,19Z-docosahexaenoic acid.

is metabolized to 15R-HEPE, which can further be metabolized by 5-LOX to give 15-epi-LXA$_4$ or 15-epi-LXB$_4$ (Serhan 2005; Das 2013). The receptor for lipoxins, ALX/FPR2, is a formyl peptide receptor type 2 GPCR that is expressed by a number of tissues and cells and is capable of binding LXA$_4$ as well as RvD1, epi-LXA4 and AT-RvD1 (Levy and Serhan 2014). Deactivation of lipoxins occurs by PG-dehydrogenase 15-PGDH found in lung, liver, kidney, placenta, spleen, and immune cells (Tai et al. 2002).

Resolvins (Rv) are derivatives of EPA and/or DHA and are now well-established as exhibiting anti-inflammatory or pro-resolving properties. They were first identified in a murine air pouch model by Serhan in 2000 (Serhan et al. 2000), and since then have been both applied and detected in various animal inflammatory disease models (i.e., colitis, ischemia–reperfusion injury, wound healing) and human nutritional intervention/disease studies (Kasuga et al. 2008; Bento et al. 2011; Norling et al. 2011; Giera et al. 2012; Mas et al. 2012).

E-series resolvins (RvE$_1$–RvE$_3$) are derived from EPA with 18R-HEPE generated from Ac-COX-2, that is subsequently metabolized by 5-LOX to 5-Hp-18R-HEPE that is followed by enzymatic epoxide hydrolysis to RvE$_1$ or eventually reduced to RvE$_2$ (Figure 3.4) (Uddin and Levy 2011). RvE$_3$ was more recently discovered in human and mouse eosinophils (Isobe et al. 2012, 2013). RvE$_3$ like RvE$_1$ is formed from Ac-COX-2, although it undergoes a further hydroxylation by 12/15-LOX to give 17R,18R/S dihydroxy-5Z,8Z,11Z,13E,15E-EPA (Isobe et al. 2012). RvE$_3$ was found to inhibit PMN chemotaxis in vitro and their numbers in a zymosan-induced peritonitis model in vivo. RvE$_1$ interacts with a number of receptors to elicit its anti-inflammatory actions, Chem R23 or CMKLR1 on monocytes and dendritic cells and it is also an antagonist for BLT1 on neutrophils (Arita et al. 2007). It is thought that RvE$_1$ receptors may be shared by RvE$_2$ (Levy and Serhan 2014). Metabolism of RvE$_1$ can occur by multiple pathways, though its deactivation is considered to be via conversion into 12-oxo-RvE$_1$ followed by conversion into 10,11-dihydro-RvE1 (Arita et al. 2006; Hong et al. 2008).

The D-series resolvins are tri- (RvD$_1$-RvD$_4$) or di- (RvD$_5$-RvD$_6$) hydroxylated stereoselective mediators derived from DHA via LOX/LOX enzymatic oxidations. 17S-HpDHA is generated from 15-LOX and subsequent reaction with 5-LOX generates RvD$_1$-RvD$_4$. The 17R series resolvins are generated from an Ac-COX-2 and follows the same pathway as before, yielding aspirin-triggered (AT)-RvD$_1$ - AT-RvD$_4$. Receptors for RvD$_1$, namely ALX and GPR32, were discovered on human phagocytes (Krishnamoorthy et al. 2010). Now named DRV1/GPR32, this receptor is expressed in peripheral blood leukocytes and vascular tissue. Binding of RvD1 at DRV1/GPR32 was found to be linked with homeostatic processes, whereas binding at ALX/FRP2 was associated with regulating the inflammatory response (Levy and Serhan 2014). 17-oxo-RvD1 has been established as the inactivation product of RvD1 through the action of eicosanoid oxidoreductases (EORs) (Sun et al. 2007).

Protectins (PD1) or Neuroprotectin D1 (NPD1) have been identified in the brain and other CNS model systems, and were first reported in 2003 (Hong et al. 2003). In other disease models not related to the brain, PD1 has been found to have both anti-inflammatory and protective properties in a wide range of pathophysiological models including ischemic stroke, obesity–insulin resistance and peritonitis amongst others

(White et al. 2010; Belayev et al. 2011; Serhan et al. 2011). PD1 is biosynthesized from 15-LOX-1 forming an epoxide intermediate, followed by an enzymatic hydrolysis resulting in the 10R, 17S-di-HDHA. An aspirin-triggered (AT) pathway through Ac-COX-2 produces 10R, 17R-diHDHA, termed AT-PD1 or AT-NDP1 (Serhan et al. 2006; Petasis et al. 2012). Recently, a conformational isomer of PD1 with the same 10S,17S stereocenters but a differing configuration of the double bonds (E,Z,E), termed PDX, has been reported. While PD1 is formed via a mono-lipoxygenation, PDX is generated via a double lipoxygenation (Chen et al. 2009; Balas et al. 2014). (Balas et al. 2014). PDX, unlike PD1, has shown potent antithrombotic activity, and inhibited platelet aggregation induced by collagen, AA or thromboxane, activity which appears central to its E,Z,E configuration (Chen et al. 2009). Receptors for PD1 are still to be established, though the actions of PD1 are cell-specific and there are thought to be multiple receptors (Kenchegowda et al. 2013; Levy and Serhan 2014).

Maresins were the most recent specialized pro-resolving mediators to be identified, and were isolated from exudates of resolution-phase murine peritonitis (Serhan et al. 2009). They are formed by 12-LOX action on DHA to give 14-HpDHA, which subsequently undergoes epoxidation/enzymatic hydrolysis to give 7R,14S-HDHA (Dalli et al. 2013). They have potent anti-inflammatory and pro-resolving activities and like Rvs and protectins (PDs), block PMN infiltration and stimulate macrophage phagocytosis. They have also been found to possess effective analgesic properties in both inflammatory and neuropathic pain models (Stables and Gilroy 2011), and more recently, reduce pro-inflammatory cytokine production in bronchial epithelial cells exposed to dust (Nordgren et al. 2013) and stimulate tissue regeneration in planaria (Serhan et al. 2012). Recently, the MaR1 precursor, 13,14-epoxy-DHA, was found to have anti-inflammatory bioactivity by blocking LTA$_4$ biosynthesis via LTA4H (Dalli et al. 2013).

Recently, this group of poly-hydroxy PUFA metabolites has been further extended with the discovery of n-6 docosapentaenoic acid (DPA)-derived Rvs (RvD1$_{DPA}$, RvD2$_{DPA}$, RvD5$_{DPA}$) MaRs (MaR1$_{DPA}$, MaR2$_{DPA}$) and PDs (PD1$_{DPA}$ PD2$_{DPA}$) (Dalli et al. 2013). These products, similar to the DHA-derived mediators, also exhibit anti-inflammatory and pro-resolving activities (Dangi et al. 2010; Chiu et al. 2012; Dalli et al. 2013).

CYTOCHROME P450-MEDIATED METABOLISM OF PUFA

CYP Monooxygenases

Microsomal cytochrome P450 (CYP) monooxygenases are better known for their role in xenobiotic metabolism, a process which is crucial in the therapeutic and toxic effects associated with pharmaceuticals. However, they are also involved in the biosynthetic transformation of PUFA to epoxy-, mono-hydroxylated- and di-hydroxylated-metabolites. When AA is the substrate, CYP-reactions produce a range of epoxyeicosatrienoic acids (EET), hydroxyeicosatrienoic acids (HETE), and dihydroxyeicosatrienoic acids (DHET) (Figure 3.5).

CYPs are NADPH dependent and cleave molecular oxygen before its insertion into fatty acids (Capdevila et al. 2000). In AA, olefinic oxidation is possible at all

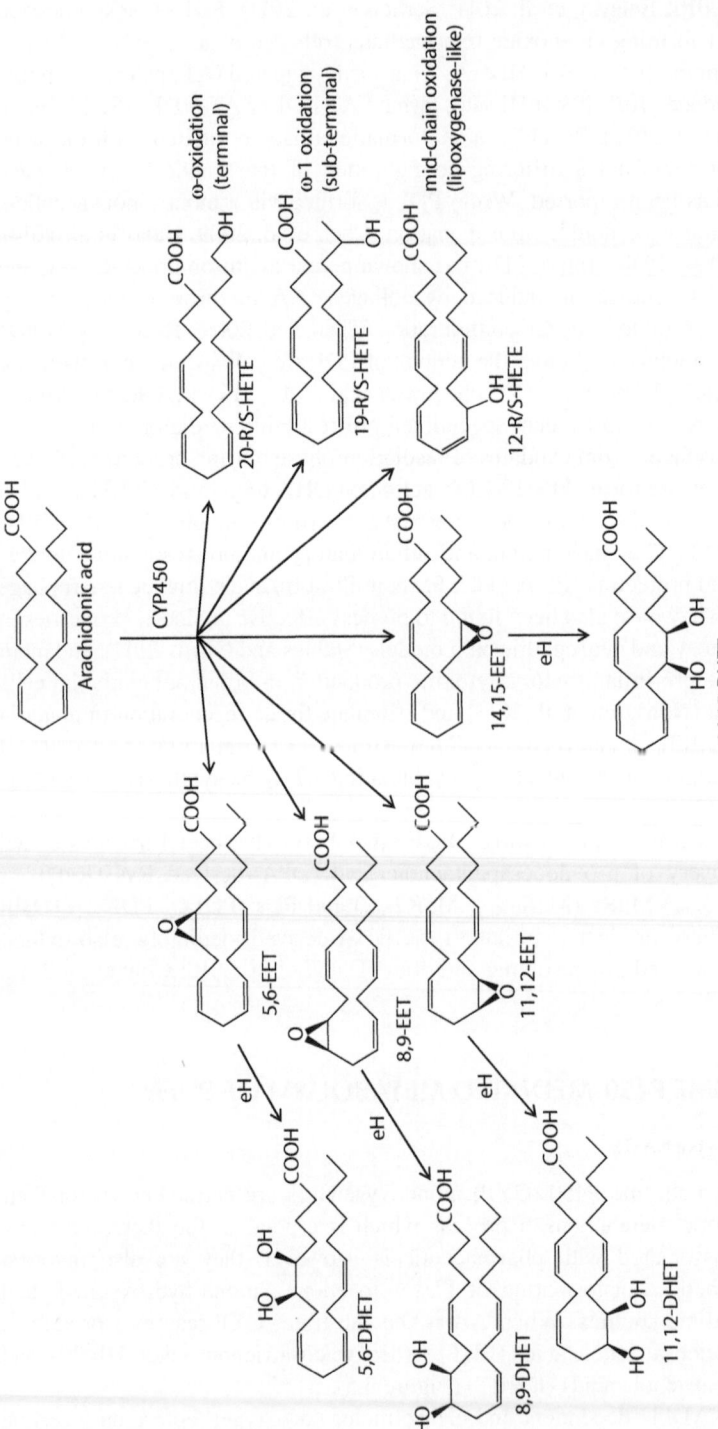

FIGURE 3.5 Conversion of arachidonic acid to epoxy- and hydroxy-eicosatrienes (EET) via CYP monooxygenase enzymes. CYP, CYP450 monooxygenases; eH, epoxide hydrolases; 5,6-EET, 5,6-epoxyeicosatrienes; 8,9-EET, 8,9-epoxyeicosatrienes; 11,12-epoxyeicosatrienes; 14,15-EET, 14,15-epoxyeicosatrienes; 5,6-DHET, 5,6-dihydroxyeicosatriene; 8,9-DHET, 8,9-dihydroxyeicosatriene; 11,12-DHET, 11,12-dihydroxyeicosatriene; 14,15-DHET, 14,15-dihydroxyeicosatriene; 19-R/S-HETE, 19-hydroxyeicosatriene; 20-R/S-HETE, 20-hydroxyeicosatriene.

four double bonds resulting in 4-regioisomeric *cis*-EET products with either *R* or *S* chirality; these can be further hydrolyzed by epoxide hydrolases to give dihydroxyl-ated PUFA species such as the DHETs, derivatives of AA (Figure 3.5). Oxidation can also occur in an allylic fashion producing an array of mid-chain HETEs (5-, 8-, 9-, 11-, 12-, 15-HETE) while ω-hydroxylations can generate end-chain hydroxylated PUFA species: 16-HETE up to 20-HETE. There is a wide range of fatty acid sub-strates for CYP-mediated reactions including LA, EPA, and DHA (Van Rollins et al. 1988; Roman 2002; Fer et al. 2008).

CYPs are membrane-bound enzymes and are found not only in the kidney and liver, but also the brain, heart, cardiovascular system and the lung, therefore their bioactive products have a myriad of functions in these tissues (Roman 2002). They are categorized according to the clan/family, subfamily, followed by a number. There are 18 mammalian CYP enzyme families, with 41 protein coding subfamilies, and 57 genes that are encoded in the human genome (Roman 2002; Nebert et al. 2013).

EPOXYEICOSATRIENOIC ACIDS AND DIHYDROXYEICOSATRIENOIC ACIDS

EETs can be formed from oxidation at any of the four double bonds found in AA, resulting in four possible regioisomers: 5,6-EET, 8,9-EET, 11,12-EET, or 14,15-EET. These compounds are potent anti-inflammatory mediators with vasodilatory prop-erties making them important for vascular function, renal function and cardiovas-cular homeostasis (Arnold et al. 2010; Bellien and Joannides 2013). They have also recently been shown to play a role in tissue growth, regeneration, and wound healing in a number of animal models (Panigrahy et al. 2013). EETs may also be involved in tumorigenesis as well as multi-organ metastasis (Panigrahy et al. 2012). Further metabolism of EETs via epoxide hydrolases results in DHETs, which exhibit much reduced bioactivities.

Recent work on the epoxidation of EPA and DHA using recombinant human CYP enzymes has shown that the predominant products were 17,18-EETeTr and 19,20-EDP, respectively. Interestingly, a concomitant reduction in AA-derived epoxidase derivatives was observed, suggesting that some of the EPA and DHA observed effects may be attributed to a shift in the generation of such metabolites (Van Rollins et al. 1988; Fer et al. 2008).

Epoxidation of LA generates epoxides such as 9,10-EpOME, 9HpODE, and 12,13-EpOME, while further metabolism by epoxide hydrolases generates the correspond-ing 9,10-DHOME and 12,13-DHOME. 9,10-epoxy-OME was termed "Leukotoxin" after its discovery in leukocytes and subsequent implication in cardiac failure and acute respiratory distress syndrome in severely burned patients (Hayakawa et al. 1986; Moghaddam et al. 1997). However, since then, it was discovered that the toxic activity was actually due to 9,10-DHOME. The epoxy has since been labeled as "protoxin," and the diol versions were termed "leukotoxin diol" (9,10-DHOME) and "isoleukotoxin" (12,13-DHOME). Oleic acid (OA) can also undergo epoxidation by CYP enzymes generating *cis*-9,10-epoxyoctadecanoic acid. This metabolite has been proposed as a potential biomarker of liver function, with low levels being detected in patients with severe chronic liver disease and cirrhosis (Ulsaker and Teien 1983; Thum et al. 2010).

OTHER CYP-DERIVED HYDROXY FATTY ACIDS

In addition to LOX-mediated reactions, AA-derived HETEs and LA-derived HODEs can also be biosynthesized by CYP pathways (Bylund et al. 1998). HODEs which have been determined as CYP products include 9-, 13-, and 18-HODE (Bylund et al. 1998), while some work on rodents and primates found the additional products 8-HODE, 11-HODE (acid-labile converts into 9/13-HODE), and 14-HODE (Oliw et al. 1993). 8-HODE, 11-HODE, and 14-HODE were all formed via an allylic hydroxylation, whereas the 9R-HODE and 13R-HODE were stereoselectively biosynthesized via hydroxylation with double bond migration (Oliw et al. 1993). AA-derived HETEs that can be generated by CYP include 7-, 10-, and 13-HETE (acid-labile) and 5,6-diHETrE. CYP enzymes can also catalyze the formation of terminal and subterminal hydroxylated species such as 16-, 18-, 19-, 20-HETE, all of which have been shown to exhibit vasodilatory properties (Carroll et al. 1996). 19-HETE has been reported to affect vascular tone and ion transport. 20-HETE is a CYP4 family AA oxidation product, which has been found to be a potent vasoconstrictor but it also has been identified in some studies as having a pro-carcinogenic role contributing to cell proliferation and tumor growth (Guo et al. 2008; Alexanian et al. 2009; Liu et al. 2010).

CONCLUDING REMARKS

We have given a comprehensive overview of the enzymatic pathways involved in PUFA oxidation, with particular emphasis on those enzymes that PUFA from phospholipid membranes (PLases), cyclize/peroxidize (COX), and stereoselectively oxygenate (LOX and CYP). We have also discussed some key downstream enzymes such as terminal prostanoid synthases, prostaglandin dehydrogenses/reductases, epoxide hydrolases, and so on, their products and related bioactivities, as well as the main receptors facilitating the multitude of activities mediated by enzymatically oxygenated PUFA. This is an expanding field and we have no doubt that there will be further discoveries supported by the advancement of the bioanalytical platforms used in metabolomics and lipidomics (Murphy and Nicolaou 2013). These developments are bound to enhancing our understanding of the importance of these mediators in health and disease, support the discovery of diagnostic biomarkers and inform new therapeutic strategies.

ACKNOWLEDGMENTS

We gratefully acknowledge research funding from The Wellcome Trust (WT094028).

ABBREVIATIONS

AA	Arachidonic acid
Ac-COX	Acetylated COX
AdPLA$_2$	Adipose phospholipase A$_2$
AEA	Arachidonyl ethanolamide or anandamide

2-AG	2-arachidonylglycerol
AKR	Aldoketoreductases
CaLB	Calcium-dependent lipid binding domain
COX	Cyclooxygenase
cPLA$_2$	Cytosolic phospholipase
CYP	Cytochrome P450
CysLT	Cysteinyl leukotrienes
DAG	Diacylglycerol
DHA	Docosahexaenoic acid
EOR	Eicosanoid oxidoreductase
EPA	Eicosapentaenoic acid
ER	Endoplasmic reticulum
FLAP	5-lipoxygenase activating protein
HDHA	Hydroxydocosahexaenoic acid
HDL	High-density lipoprotein
HEDH	Hydroxyeicosanoid dehydrogenase
HEPE	Hydroxyeicosapentaenoic acid
HETE	Hydroxyeicosatetraenoic acid
HODE	Hydroxyoctadecadienoic acid
HpDHA	Hydroperoxydocosahexaenoic acid
HPGDS	Hematopoietic-type PGDS
IP$_3$	Inositol-1,4,5-triphosphate
iPLA$_2$	Calcium_independent phospholipase A$_2$
LDL	Low-density lipoprotein
LOX	Lipoxygenase
LPGDS	Lipocaline-type PGDS
LT	Leukotriene
LysoPLA$_2$	Lysosomal phospholipase A$_2$
MaR	Maresin
NAPE	*N*-arachidonyl phosphatidylethanolamine
OATP	Organic anion transport polypeptides
OCTs	Organic cation transporters
PA	Phosphatidic acid
PAF-AH	Platelet activating factor acetyl hydrolase
PC	Phopsphatidylcholine
PD	Protectin
PE	Phosphatidylethanolamine
PG	Prostaglandin
PGDH	Prostaglandin dehydrogenase
PGE 9-KR	PGE 9-ketoreductase
PGS	Prostaglandin synthase
PI	Phosphatidylinositol
PIP$_2$	Phosphatidylinositol 4,5-bisphosphate
PL	Phospholipase
PMN	Polymorphonuclear leukocytes
PS	Phosphatidylserine

PUFA   Polyunsaturated fatty acid
ROS   Reactive oxygen species
Rv   Resolvin
sPLA$_2$ Secretory phospholipase A$_2$
TX   Thromboxane

REFERENCES

Abe, A. and J. A. Shayman. 1998. Purification and characterization of 1-O-acylceramide synthase, a novel phospholipase A2 with transacylase activity. *J Biol Chem* **273**(14): 8467–8474.

Agarwal, S., C. Achari et al. 2009. Inhibition of 12-LOX and COX-2 reduces the proliferation of human epidermoid carcinoma cells (A431) by modulating the ERK and PI3K-Akt signalling pathways. *Exp Dermatol* **18**(11): 939–946.

Akiba, S. and T. Sato. 2004. Cellular function of calcium-independent phospholipase A2. *Biol Pharm Bull* **27**(8): 1174–1178.

Alexanian, A., V. A. Rufanova et al. 2009. Down-regulation of 20-HETE synthesis and signaling inhibits renal adenocarcinoma cell proliferation and tumor growth. *Anticancer Res* **29**(10): 3819–3824.

Allyson, J., X. Bi et al. 2012. Maintenance of synaptic stability requires calcium-independent phospholipase A(2) activity. *Neural Plast* **2012**: 569149.

Aparoy, P., K. K. Reddy et al. 2012. Structure and ligand based drug design strategies in the development of novel 5- LOX inhibitors. *Curr Med Chem* **19**(22): 3763–3778.

Arenberger, P., L. Kemeny et al. 1993. Characterization of high-affinity 12(S)-hydroxyeicosatetraenoic acid (12(S)-HETE) binding sites on normal human keratinocytes. *Epithelial Cell Biol* **2**(1): 1–6.

Arita, M., S. F. Oh et al. 2006. Metabolic inactivation of resolvin E1 and stabilization of its anti-inflammatory actions. *J Biol Chem* **281**(32): 22847–22854.

Arita, M., T. Ohira et al. 2007. Resolvin E1 selectively interacts with leukotriene B4 receptor BLT1 and ChemR23 to regulate inflammation. *J Immunol* **178**(6): 3912–3917.

Arnold, C., A. Konkel et al. 2010. Cytochrome P450-dependent metabolism of omega-6 and omega-3 long-chain polyunsaturated fatty acids. *Pharmacol Rep* **62**(3): 536–547.

Avis, I., S. H. Hong et al. 2001. Five-lipoxygenase inhibitors can mediate apoptosis in human breast cancer cell lines through complex eicosanoid interactions. *FASEB J* **15**(11): 2007–2009.

Bacon, K. B. and R. D. Camp 1990. Lipid lymphocyte chemoattractants in psoriasis. *Prostaglandins* **40**(6): 603–614.

Bailey, J. M., M. Fletcher et al. 1997. Regulation of human T-lymphocyte proliferative responses by the lipoxygenase product 15-HETE. *Biochem Soc Trans* **25**(2): 247S.

Balas, L., M. Guichardant et al. 2014. Confusion between protectin D1 (PD1) and its isomer protectin DX (PDX). An overview on the dihydroxy-docosatrienes described to date. *Biochimie* **99**: 1–7.

Balsinde, J., I. D. Bianco et al. 1995. Inhibition of calcium-independent phospholipase A2 prevents arachidonic acid incorporation and phospholipid remodeling in P388D1 macrophages. *Proc Natl Acad Sci USA* **92**(18): 8527–8531.

Belayev, L., L. Khoutorova et al. 2011. Docosahexaenoic acid therapy of experimental ischemic stroke. *Transl Stroke Res* **2**(1): 33–41.

Bellien, J. and R. Joannides 2013. Epoxyeicosatrienoic acid pathway in human health and diseases. *J Cardiovasc Pharmacol* **61**(3): 188–196.

Bento, A. F., R. F. Claudino et al. 2011. Omega-3 fatty acid-derived mediators 17(R)-hydroxy docosahexaenoic acid, aspirin-triggered resolvin D1 and resolvin D2 prevent experimental colitis in mice. *J Immunol* **187**(4): 1957–1969.

Bhattacharya, S., G. Mathew et al. 2009. 15-Lipoxygenase-1 in colorectal cancer: A review. *Tumour Biol* **30**(4): 185–199.

Boeglin, W. E., R. B. Kim et al. 1998. A 12R-lipoxygenase in human skin: Mechanistic evidence, molecular cloning, and expression. *Proc Natl Acad Sci USA* **95**(12): 6744–6749.

Bonnans, C., P. Chanez et al. 2004. Lipoxins in asthma: Potential therapeutic mediators on bronchial inflammation? *Allergy* **59**(10): 1027–1041.

Bostrom, M. A., B. B. Boyanovsky et al. 2007. Group v secretory phospholipase A2 promotes atherosclerosis: Evidence from genetically altered mice. *Arterioscler Thromb Vasc Biol* **27**(3): 600–606.

Breyer, R. M., C. K. Bagdassarian et al. 2001. Prostanoid receptors: Subtypes and signaling. *Annu Rev Pharmacol Toxicol* **41**: 661–690.

Brock, T. G. 2005. Regulating leukotriene synthesis: The role of nuclear 5-lipoxygenase. *J Cell Biochem* **96**(6): 1203–1211.

Brock, T. G., Y. J. Lee et al. 2005. Nuclear localization of leukotriene A4 hydrolase in type II alveolar epithelial cells in normal and fibrotic lung. *Am J Physiol Lung Cell Mol Physiol* **289**(2): L224–232.

Burke, J. E. and E. A. Dennis 2009. Phospholipase A2 biochemistry. *Cardiovasc Drugs Ther* **23**(1): 49–59.

Bylund, J., J. Ericsson et al. 1998. Analysis of cytochrome P450 metabolites of arachidonic and linoleic acids by liquid chromatography-mass spectrometry with ion trap MS. *Anal Biochem* **265**(1): 55–68.

Cai, Q., L. Lanting et al. 2004. Growth factors induce monocyte binding to vascular smooth muscle cells: Implications for monocyte retention in atherosclerosis. *Am J Physiol Cell Physiol* **287**(3): C707–714.

Cai, Y., R. K. Kumar et al. 2009. Ym1/2 promotes Th2 cytokine expression by inhibiting 12/15(S)-lipoxygenase: Identification of a novel pathway for regulating allergic inflammation. *J Immunol* **182**(9): 5393–5399.

Calder, P. C. 2012. Mechanisms of action of (n-3) fatty acids. *J Nutr* **142**(3): 592S–599S.

Capdevila, J. H., J. R. Falck et al. 2000. Cytochrome P450 and arachidonic acid bioactivation. Molecular and functional properties of the arachidonate monooxygenase. *J Lipid Res* **41**(2): 163–181.

Carroll, M. A., M. Balazy et al. 1996. Cytochrome P-450-dependent HETEs: Profile of biological activity and stimulation by vasoactive peptides. *Am J Physiol* **271**(4 Pt 2): R863–869.

Chandrasekharan, N. V., H. Dai et al. 2002. COX-3, a cyclooxygenase-1 variant inhibited by acetaminophen and other analgesic/antipyretic drugs: Cloning, structure, and expression. *Proc Natl Acad Sci USA* **99**(21): 13926–13931.

Chen, E. P. and E. M. Smyth 2011. COX-2 and PGE2-dependent immunomodulation in breast cancer. *Prostaglandins Other Lipid Mediat* **96**(1–4): 14–20.

Chen, P., B. Fenet et al. 2009. Full characterization of PDX, a neuroprotectin/protectin D1 isomer, which inhibits blood platelet aggregation. *FEBS Lett* **583**(21): 3478–3484.

Chen, X., A. Gresham et al. 1996. Oxidative stress mediates synthesis of cytosolic phospholipase A2 after UVB injury. *Biochim Biophys Acta* **1299**(1): 23–33.

Cheon, Y., H. W. Kim et al. 2012. Disturbed brain phospholipid and docosahexaenoic acid metabolism in calcium-independent phospholipase A(2)-VIA (iPLA(2)beta)-knockout mice. *Biochim Biophys Acta* **1821**(9): 1278–1286.

Chiu, C. Y., B. Gomolka et al. 2012. Omega-6 docosapentaenoic acid-derived resolvins and 17-hydroxydocosahexaenoic acid modulate macrophage function and alleviate experimental colitis. *Inflamm Res* **61**(9): 967–976.

Cho, Y. and V. A. Ziboh 1994. 13-Hydroxyoctadecadienoic acid reverses epidermal hyperproliferation via selective inhibition of protein kinase C-beta activity. *Biochem Biophys Res Commun* **201**(1): 257–265.

Clark, J. D., L. L. Lin et al. 1991. A novel arachidonic acid-selective cytosolic PLA2 contains a Ca(2+)-dependent translocation domain with homology to PKC and GAP. *Cell* **65**(6): 1043–1051.

Clark, J. D., A. R. Schievella et al. 1995. Cytosolic phospholipase A2. *J Lipid Mediat Cell Signal* **12**(2–3): 83–117.

Cummings, R., N. Parinandi et al. 2002. Phospholipase D/phosphatidic acid signal transduction: Role and physiological significance in lung. *Mol Cell Biochem* **234–235**(1–2): 99–109.

Dailey, L. A. and P. Imming 1999. 12-Lipoxygenase: Classification, possible therapeutic benefits from inhibition, and inhibitors. *Curr Med Chem* **6**(5): 389–398.

Dalli, J., R. A. Colas et al. 2013. Novel n-3 immunoresolvents: Structures and actions. *Sci Rep* **3**: 1940.

Dalli, J., M. Zhu et al. 2013. The novel 13S,14S-epoxy-maresin is converted by human macrophages to maresin 1 (MaR1), inhibits leukotriene A4 hydrolase (LTA4H), and shifts macrophage phenotype. *FASEB J* **27**(7): 2573–2583.

Dangi, B., M. Obeng et al. 2010. Metabolism and biological production of resolvins derived from docosapentaenoic acid (DPAn-6). *Biochem Pharmacol* **79**(2): 251–260.

Das, U. N. 2013. Arachidonic acid and lipoxin A4 as possible endogenous anti-diabetic molecules. *Prostaglandins Leukot Essent Fatty Acids* **88**(3): 201–210.

Dennis, E. A., J. Cao et al. 2011. Phospholipase A2 enzymes: Physical structure, biological function, disease implication, chemical inhibition, and therapeutic intervention. *Chem Rev* **111**(10): 6130–6185.

Di Gennaro, A. and J. Z. Haeggstrom 2012. The leukotrienes: Immune-modulating lipid mediators of disease. *Adv Immunol* **116**: 51–92.

Dobrian, A. D., D. C. Lieb et al. 2011. Functional and pathological roles of the 12- and 15-lipoxygenases. *Prog Lipid Res* **50**(1): 115–131.

Duncan, R. E., E. Sarkadi-Nagy et al. 2008. Identification and functional characterization of adipose specific phospholipase A2 (AdPLA). *J Biol Chem* **283**(37): 25428–25436.

Ekambaram, P., W. Lambiv et al. 2011. The thromboxane synthase and receptor signaling pathway in cancer: An emerging paradigm in cancer progression and metastasis. *Cancer Metastasis Rev* **30**(3–4): 397–408.

Epp, N., G. Furstenberger et al. 2007. 12R-lipoxygenase deficiency disrupts epidermal barrier function. *J Cell Biol* **177**(1): 173–182.

Erlemann, K. R., C. Cossette et al. 2007. Regulation of 5-hydroxyeicosanoid dehydrogenase activity in monocytic cells. *Biochem J* **403**(1): 157–165.

Fenard, D., G. Lambeau et al. 1999. Secreted phospholipases A(2), a new class of HIV inhibitors that block virus entry into host cells. *J Clin Invest* **104**(5): 611–618.

Fer, M., Y. Dreano et al. 2008. Metabolism of eicosapentaenoic and docosahexaenoic acids by recombinant human cytochromes P450. *Arch Biochem Biophys* **471**(2): 116–125.

Fiorucci, S., J. L. Wallace et al. 2004. A beta-oxidation-resistant lipoxin A4 analog treats hapten-induced colitis by attenuating inflammation and immune dysfunction. *Proc Natl Acad Sci USA* **101**(44): 15736–15741.

Fonseca, B. M., M. A. Costa et al. 2013. Endogenous cannabinoids revisited: A biochemistry perspective. *Prostaglandins Other Lipid Mediat* **102–103**: 13–30.

Frey, R. S., A. Rahman et al. 2002. PKCzeta regulates TNF-alpha-induced activation of NADPH oxidase in endothelial cells. *Circ Res* **90**(9): 1012–1019.

Fukami, K., S. Inanobe et al. 2010. Phospholipase C is a key enzyme regulating intracellular calcium and modulating the phosphoinositide balance. *Prog Lipid Res* **49**(4): 429–437.

Gao, X., D. J. Grignon et al. 1995. Elevated 12-lipoxygenase mRNA expression correlates with advanced stage and poor differentiation of human prostate cancer. *Urology* **46**(2): 227–237.

Giera, M., A. Ioan-Facsinay et al. 2012. Lipid and lipid mediator profiling of human synovial fluid in rheumatoid arthritis patients by means of LC-MS/MS. *Biochim Biophys Acta* **1821**(11): 1415–1424.

Grimminger, F., I. von Kurten et al. 1992. Type II alveolar epithelial eicosanoid metabolism: Predominance of cyclooxygenase pathways and transcellular lipoxygenase metabolism in co-culture with neutrophils. *Am J Respir Cell Mol Biol* **6**(1): 9–16.

Guo, A. M., J. Sheng et al. 2008. Expression of CYP4A1 in U251 human glioma cell induces hyperproliferative phenotype *in vitro* and rapidly growing tumors in vivo. *J Pharmacol Exp Ther* **327**(1): 10–19.

Haapamaki, M. M., J. M. Gronroos et al. 1999. Gene expression of group II phospholipase A2 in intestine in Crohn's disease. *Am J Gastroenterol* **94**(3): 713–720.

Haeggstrom, J. Z. and C. D. Funk 2011. Lipoxygenase and leukotriene pathways: Biochemistry, biology, and roles in disease. *Chem Rev* **111**(10): 5866–5898.

Haeggstrom, J. Z. and A. Wetterholm 2002. Enzymes and receptors in the leukotriene cascade. *Cell Mol Life Sci* **59**(5): 742–753.

Hagenbuch, B. and B. Stieger 2013. The SLCO (former SLC21) superfamily of transporters. *Mol Aspects Med* **34**(2–3): 396–412.

Hara, S., A. Miyata et al. 1994. Isolation and molecular cloning of prostacyclin synthase from bovine endothelial cells. *J Biol Chem* **269**(31): 19897–19903.

Hatae, N., Y. Sugimoto et al. 2002. Prostaglandin receptors: Advances in the study of EP3 receptor signaling. *J Biochem* **131**(6): 781–784.

Hayakawa, M., S. Sugiyama et al. 1986. Neutrophils biosynthesize leukotoxin, 9, 10-epoxy-12-octadecenoate. *Biochem Biophys Res Commun* **137**(1): 424–430.

Hiraoka, M., A. Abe et al. 2006. Lysosomal phospholipase A2 and phospholipidosis. *Mol Cell Biol* **26**(16): 6139–6148.

Hong, S., K. Gronert et al. 2003. Novel docosatrienes and 17S-resolvins generated from docosahexaenoic acid in murine brain, human blood, and glial cells. Autacoids in anti-inflammation. *J Biol Chem* **278**(17): 14677–14687.

Hong, S., T. F. Porter et al. 2008. Resolvin E1 metabolome in local inactivation during inflammation-resolution. *J Immunol* **180**(5): 3512–3519.

Honn, K. V., D. G. Tang et al. 1994. 12-lipoxygenases and 12(S)-HETE: Role in cancer metastasis. *Cancer Metastasis Rev* **13**(3–4): 365–396.

Ikei, K. N., J. Yeung et al. 2012. Investigations of human platelet-type 12-lipoxygenase: Role of lipoxygenase products in platelet activation. *J Lipid Res* **53**(12): 2546–2559.

Isobe, Y., M. Arita et al. 2012. Identification and structure determination of novel anti-inflammatory mediator resolvin E3, 17,18-dihydroxyeicosapentaenoic acid. *J Biol Chem* **287**(13): 10525–10534.

Isobe, Y., M. Arita et al. 2013. Stereochemical assignment and anti-inflammatory properties of the omega-3 lipid mediator resolvin E3. *J Biochem* **153**(4): 355–360.

Jabr, S., S. Gartner et al. 2013. Quantification of major urinary metabolites of PGE2 and PGD2 in cystic fibrosis: Correlation with disease severity. *Prostaglandins, Leukotrienes, and Essential Fatty Acids* **89**(2–3): 121–126.

Jakobsson, P. J., S. Thoren et al. 1999. Identification of human prostaglandin E synthase: A microsomal, glutathione-dependent, inducible enzyme, constituting a potential novel drug target. *Proc Natl Acad Sci USA* **96**(13): 7220–7225.

Jang, J. H., C. S. Lee et al. 2012. Understanding of the roles of phospholipase D and phosphatidic acid through their binding partners. *Prog Lipid Res* **51**(2): 71–81.

Jaworski, K., M. Ahmadian et al. 2009. AdPLA ablation increases lipolysis and prevents obesity induced by high-fat feeding or leptin deficiency. *Nat Med* **15**(2): 159–168.

Jenkins, G. M. and M. A. Frohman 2005. Phospholipase D: A lipid centric review. *Cell Mol Life Sci* **62**(19–20): 2305–2316.

Joo, Y. C. and D. K. Oh 2012. Lipoxygenases: Potential starting biocatalysts for the synthesis of signaling compounds. *Biotechnol Adv* **30**(6): 1524–1532.

Kadamur, G. and E. M. Ross 2013. Mammalian phospholipase C. *Annu Rev Physiol* **75**: 127–154.

Kaduce, T. L., P. H. Figard et al. 1989. Formation of 9-hydroxyoctadecadienoic acid from linoleic acid in endothelial cells. *J Biol Chem* **264**(12): 6823–6830.

Kalinski, P. 2011. Regulation of immune responses by prostaglandin E2. *J Immunol* **188**(1): 21–28.

Kasuga, K., R. Yang et al. 2008. Rapid appearance of resolvin precursors in inflammatory exudates: Novel mechanisms in resolution. *J Immunol* **181**(12): 8677–8687.

Kato, K., Y. Koshihara et al. 1983. Augmentation of 12-lipoxygenase activity of lymph node and spleen T cells in autoimmune mice MRL/1. *Prostaglandins Leukot Med* **12**(3): 273–280.

Kelavkar, U. P. and C. Cohen 2004. 15-Lipoxygenase-1 expression upregulates and activates insulin-like growth factor-1 receptor in prostate cancer cells. *Neoplasia* **6**(1): 41–52.

Kenchegowda, S., J. He et al. 2013. Involvement of pigment epithelium-derived factor, docosahexaenoic acid and neuroprotectin D1 in corneal inflammation and nerve integrity after refractive surgery. *Prostaglandins Leukot Essent Fatty Acids* **88**(1): 27–31.

Kendall, A. C. and A. Nicolaou 2013. Bioactive lipid mediators in skin inflammation and immunity. *Prog Lipid Res* **52**(1): 141–164.

Klampfl, T., E. Bogner et al. 2012. Up-regulation of 12(S)-lipoxygenase induces a migratory phenotype in colorectal cancer cells. *Exp Cell Res* **318**(6): 768–778.

Koduri, R. S., J. O. Gronroos et al. 2002. Bactericidal properties of human and murine groups I, II, V, X, and XII secreted phospholipases A(2). *J Biol Chem* **277**(8): 5849–5857.

Korbecki, J., I. Baranowska-Bosiacka et al. 2013. The effect of reactive oxygen species on the synthesis of prostanoids from arachidonic acid. *J Physiol Pharmacol* **64**(4): 409–421.

Krieg, P. and G. Furstenberger 2013. The role of lipoxygenases in epidermis. *Biochim Biophys Acta*.

Krishnamoorthy, S., A. Recchiuti et al. 2010. Resolvin D1 binds human phagocytes with evidence for proresolving receptors. *Proc Natl Acad Sci USA* **107**(4): 1660–1665.

Krishnamurthy, V. R., A. Dougherty et al. 2011. Total synthesis and bioactivity of 18(R)-hydroxyeicosapentaenoic acid. *J Org Chem* **76**(13): 5433–5437.

Kuhn, H. and B. J. Thiele 1999. The diversity of the lipoxygenase family: Many sequence data but little information on biological significance. *FEBS Lett* **449**(1): 7–11.

Kuklev, D. V., J. A. Hankin et al. 2013. Major urinary metabolites of 6-keto-prostaglandin F2alpha in mice. *Journal of Lipid Research* **54**(7): 1906–1914.

Kumar, K. A., K. M. Arunasree et al. 2009. Effects of (15S)-hydroperoxyeicosatetraenoic acid and (15S)-hydroxyeicosatetraenoic acid on the acute- lymphoblastic-leukaemia cell line Jurkat: Activation of the Fas-mediated death pathway. *Biotechnol Appl Biochem* **52**(Pt 2): 121–133.

Lacroix Pepin, N., P. Chapdelaine et al. 2013. Evaluation of the prostaglandin F synthase activity of human and bovine aldo-keto reductases: AKR1A1s complement AKR1B1s as potent PGF synthases. *Prostaglandins Other Lipid Mediat* **106**: 124–132.

Lam, B. K. 2003. Leukotriene C(4) synthase. *Prostaglandins Leukot Essent Fatty Acids* **69** (2–3): 111–116.

Lam, B. K., W. F. Owen, Jr. et al. 1989. The identification of a distinct export step following the biosynthesis of leukotriene C4 by human eosinophils. *J Biol Chem* **264**(22): 12885–12889.

Lecomte, M., O. Laneuville et al. 1994. Acetylation of human prostaglandin endoperoxide synthase-2 (cyclooxygenase-2) by aspirin. *J Biol Chem* **269**(18): 13207–13215.

Lee, T. H., G. Woszczek et al. 2009. Leukotriene E4: Perspective on the forgotten mediator. *J Allergy Clin Immunol* **124**(3): 417–421.

Leslie, C. C. 1997. Properties and regulation of cytosolic phospholipase A2. *J Biol Chem* **272**(27): 16709–16712.

Leslie, C. C. and J. Y. Channon 1990. Anionic phospholipids stimulate an arachidonoyl-hydrolyzing phospholipase A2 from macrophages and reduce the calcium requirement for activity. *Biochim Biophys Acta* **1045**(3): 261–270.

Levy, B. D. and C. N. Serhan 2014. Resolution of acute inflammation in the lung. *Annu Rev Physiol* **76**: 467–492.

Li, X., S. Hong et al. 2011. Docosahexanoic acid-induced coronary arterial dilation: Actions of 17S-hydroxy docosahexanoic acid on K+ channel activity. *J Pharmacol Exp Ther* **336**(3): 891–899.

Liou, J. Y., S. K. Shyue et al. 2000. Colocalization of prostacyclin synthase with prostaglandin H synthase-1 (PGHS-1) but not phorbol ester-induced PGHS-2 in cultured endothelial cells. *J Biol Chem* **275**(20): 15314–15320.

Liu, J. Y., N. Li et al. 2010. Metabolic profiling of murine plasma reveals an unexpected bio-marker in rofecoxib-mediated cardiovascular events. *Proc Natl Acad Sci USA* **107**(39): 17017–17022.

Loll, P. J., D. Picot et al. 1995. The structural basis of aspirin activity inferred from the crystal structure of inactivated prostaglandin H2 synthase. *Nat Struct Biol* **2**(8): 637–643.

Lu, R., N. Kanai et al. 1996. Cloning, *in vitro* expression, and tissue distribution of a human prostaglandin transporter cDNA(hPGT). *J Clin Invest* **98**(5): 1142–1149.

Lu, Y., H. Tian et al. 2010. Novel 14,21-dihydroxy-docosahexaenoic acids: Structures, forma-tion pathways, and enhancement of wound healing. *J Lipid Res* **51**(5): 923–932.

Luo, M., S. M. Jones et al. 2003. Nuclear localization of 5-lipoxygenase as a determinant of leukotriene B4 synthetic capacity. *Proc Natl Acad Sci USA* **100**(21): 12165–12170.

Luo, M., S. M. Jones et al. 2004. Protein kinase A inhibits leukotriene synthesis by phosphory-lation of 5-lipoxygenase on serine 523. *J Biol Chem* **279**(40): 41512–41520.

Maaloe, T., E. B. Schmidt et al. 2011. The effect of n-3 polyunsaturated fatty acids on leukot-riene B(4) and leukotriene B(5) production from stimulated neutrophil granulocytes in patients with chronic kidney disease. *Prostaglandins Leukot Essent Fatty Acids* **85**(1): 37–41.

Mancini, J. A., K. Blood et al. 2001. Cloning, expression, and up-regulation of inducible rat prostaglandin e synthase during lipopolysaccharide-induced pyresis and adjuvant-induced arthritis. *J Biol Chem* **276**(6): 4469–4475.

Mancini, J. A., G. P. O'Neill et al. 1994. Mutation of serine-516 in human prostaglandin G/H synthase-2 to methionine or aspirin acetylation of this residue stimulates 15-R-HETE synthesis. *FEBS Lett* **342**(1): 33–37.

Mas, E., K. D. Croft et al. 2012. Resolvins D1, D2, and other mediators of self-limited reso-lution of inflammation in human blood following n-3 fatty acid supplementation. *Clin Chem* **58**(10): 1476–1484.

Moghaddam, M. F., D. F. Grant et al. 1997. Bioactivation of leukotoxins to their toxic diols by epoxide hydrolase. *Nat Med* **3**(5): 562–566.

Mounier, C. M., D. Wendum et al. 2008. Distinct expression pattern of the full set of secreted phospholipases A2 in human colorectal adenocarcinomas: sPLA2-III as a biomarker candidate. *Br J Cancer* **98**(3): 587–595.

Muga, S. J., P. Thuillier et al. 2000. 8S-lipoxygenase products activate peroxisome proliferator-activated receptor alpha and induce differentiation in murine keratinocytes. *Cell Growth Differ* **11**(8): 447–454.

Munoz-Garcia, A., C. P. Thomas et al. 2014. The importance of the lipoxygenase-hepoxilin pathway in the mammalian epidermal barrier. *Biochim Biophys Acta* **1841**(3): 401–408.

Murakami, M., Y. Taketomi et al. 2011. Recent progress in phospholipase A(2) research: From cells to animals to humans. *Prog Lipid Res* **50**(2): 152–192.

Murphy, S. A. and A. Nicolaou 2013. Lipidomics applications in health, disease and nutrition research. *Mol Nutr Food Res* **57**(8): 1336–1346.

Nakahata, N. 2008. Thromboxane A2: Physiology/pathophysiology, cellular signal transduc-tion and pharmacology. *Pharmacol Ther* **118**(1): 18–35.

Nebert, D. W., K. Wikvall et al. 2013. Human cytochromes P450 in health and disease. *Philos Trans R Soc Lond B Biol Sci* **368**(1612): 20120431.

Nicolaou, A., M. Masoodi et al. 2012. The eicosanoid response to high dose UVR exposure of individuals prone and resistant to sunburn. *Photochem Photobiol Sci* **11**(2): 371–380.

Nomura, T., R. Lu et al. 2004. The two-step model of prostaglandin signal termination: *In vitro* reconstitution with the prostaglandin transporter and prostaglandin 15 dehydrogenase. *Mol Pharmacol* **65**(4): 973–978.

Nordgren, T. M., A. J. Heires et al. 2013. Maresin-1 reduces the pro-inflammatory response of bronchial epithelial cells to organic dust. *Respir Res* **14**(1): 51.

Norling, L. V., M. Spite et al. 2011. Cutting edge: Humanized nano-proresolving medicines mimic inflammation-resolution and enhance wound healing. *J Immunol* **186**(10): 5543–5547.

O'Flaherty, J. T., Y. Hu et al. 2012. 15-Lipoxygenase metabolites of docosahexaenoic acid inhibit prostate cancer cell proliferation and survival. *PLoS One* **7**(9): e45480.

Oliw, E. H., I. D. Brodowsky et al. 1993. Bis-allylic hydroxylation of polyunsaturated fatty acids by hepatic monooxygenases and its relation to the enzymatic and nonenzymatic formation of conjugated hydroxy fatty acids. *Arch Biochem Biophys* **300**(1): 434–439.

Pande, A. H., D. Moe et al. 2004. Modulation of human 5-lipoxygenase activity by membrane lipids. *Biochemistry* **43**(46): 14653–14666.

Panigrahy, D., M. L. Edin et al. 2012. Epoxyeicosanoids stimulate multiorgan metastasis and tumor dormancy escape in mice. *J Clin Invest* **122**(1): 178–191.

Panigrahy, D., B. T. Kalish et al. 2013. Epoxyeicosanoids promote organ and tissue regeneration. *Proc Natl Acad Sci USA* **110**(33): 13528–13533.

Park, J. Y., M. H. Pillinger et al. 2006. Prostaglandin E2 synthesis and secretion: The role of PGE2 synthases. *Clin Immunol* **119**(3): 229–240.

Pendyala, S., P. V. Usatyuk et al. 2009. Regulation of NADPH oxidase in vascular endothelium: The role of phospholipases, protein kinases, and cytoskeletal proteins. *Antioxid Redox Signal* **11**(4): 841–860.

Percival, M. D., D. Denis et al. 1992. Investigation of the mechanism of non-turnover-dependent inactivation of purified human 5-lipoxygenase. Inactivation by H_2O_2 and inhibition by metal ions. *Eur J Biochem* **210**(1): 109–117.

Petasis, N. A., R. Yang et al. 2012. Stereocontrolled total synthesis of neuroprotectin D1/protectin D1 and its aspirin-triggered stereoisomer. *Tetrahedron Lett* **53**(14): 1695–1698.

Pham, H., T. Banerjee et al. 2004. Suppression of cyclooxygenase-2 overexpression by 15S-hydroxyeicosatrienoic acid in androgen-dependent prostatic adenocarcinoma cells. *Int J Cancer* **111**(2): 192–197.

Pham, H., K. Vang et al. 2006. Dietary gamma-linolenate attenuates tumor growth in a rodent model of prostatic adenocarcinoma via suppression of elevated generation of PGE(2) and 5S-HETE. *Prostaglandins Leukot Essent Fatty Acids* **74**(4): 271–282.

Pidgeon, G. P., M. Kandouz et al. 2002. Mechanisms controlling cell cycle arrest and induction of apoptosis after 12-lipoxygenase inhibition in prostate cancer cells. *Cancer Res* **62**(9): 2721–2727.

Powell, W. S. and J. Rokach 2005. Biochemistry, biology and chemistry of the 5-lipoxygenase product 5-oxo-ETE. *Progr Lipid Res* **44**(2–3): 154–183.

Powell, W. S. and J. Rokach 2013. The eosinophil chemoattractant 5-oxo-ETE and the OXE receptor. *Progr Lipid Res* **52**(4): 651–665.

Pucer, A., V. Brglez et al. 2013. Group X secreted phospholipase A2 induces lipid droplet formation and prolongs breast cancer cell survival. *Mol Cancer* **12**(1): 111.

Rhodes, L. E., K. Gledhill et al. 2009. The sunburn response in human skin is characterized by sequential eicosanoid profiles that may mediate its early and late phases. *FASEB J* **23**(11): 3947–3956.

Roman, R. J. 2002. P-450 metabolites of arachidonic acid in the control of cardiovascular function. *Physiol Rev* **82**(1): 131–185.

Roth, M., A. Obaidat et al. 2012. OATPs, OATs and OCTs: The organic anion and cation transporters of the SLCO and SLC22A gene superfamilies. *Br J Pharmacol* **165**(5): 1260–1287.

Rouzer, C. A. and L. J. Marnett 2011. Endocannabinoid oxygenation by cyclooxygenases, lipoxygenases, and cytochromes P450: Cross-talk between the eicosanoid and endocannabinoid signaling pathways. *Chem Rev* **111**(10): 5899–5921.

Ruparel, S., D. Green et al. 2012. The cytochrome P450 inhibitor, ketoconazole, inhibits oxidized linoleic acid metabolite-mediated peripheral inflammatory pain. *Mol Pain* **8**: 73.

Ruzicka, T. 1992. The role of the epidermal 12-hydroxyeicosatetraenoic acid receptor in the skin. *Eicosanoids* **5** Suppl: S63–65.

Sala, A., G. Folco et al. 2010. Transcellular biosynthesis of eicosanoids. *Pharmacol Rep* **62**(3): 503–510.

Samuelsson, B., R. Morgenstern et al. 2007. Membrane prostaglandin E synthase-1: A novel therapeutic target. *Pharmacol Rev* **59**(3): 207–224.

Sapieha, P., A. Stahl et al. 2011. 5-Lipoxygenase metabolite 4-HDHA is a mediator of the antiangiogenic effect of omega-3 polyunsaturated fatty acids. *Sci Transl Med* **3**(69): 69ra12.

Scher, J. U. and M. H. Pillinger 2005. 15d-PGJ2: The anti-inflammatory prostaglandin? *Clin Immunol (Orlando, FL)* **114**(2): 100–109.

Schuster, V. L. 2002. Prostaglandin transport. *Prostaglandins Other Lipid Mediat* **68–69**: 633–647.

Schweiger, D., G. Furstenberger et al. 2007. Inducible expression of 15-lipoxygenase-2 and 8-lipoxygenase inhibits cell growth via common signaling pathways. *J Lipid Res* **48**(3): 553–564.

Serhan, C. N. 2005. Lipoxins and aspirin-triggered 15-epi-lipoxins are the first lipid mediators of endogenous anti-inflammation and resolution. *Prostaglandins Leukot Essent Fatty Acids* **73**(3–4): 141–162.

Serhan, C. N., C. B. Clish et al. 2000. Novel functional sets of lipid-derived mediators with antiinflammatory actions generated from omega-3 fatty acids via cyclooxygenase 2-nonsteroidal antiinflammatory drugs and transcellular processing. *J Exp Med* **192**(8): 1197–1204.

Serhan, C. N., J. Dalli et al. 2012. Macrophage proresolving mediator maresin 1 stimulates tissue regeneration and controls pain. *FASEB J* **26**(4): 1755–1765.

Serhan, C. N., K. Gotlinger et al. 2006. Anti-inflammatory actions of neuroprotectin D1/protectin D1 and its natural stereoisomers: Assignments of dihydroxy-containing docosatrienes. *J Immunol* **176**(3): 1848–1859.

Serhan, C. N., A. Jain et al. 2003. Reduced inflammation and tissue damage in transgenic rabbits overexpressing 15-lipoxygenase and endogenous anti-inflammatory lipid mediators. *J Immunol* **171**(12): 6856–6865.

Serhan, C. N., S. Krishnamoorthy et al. 2011. Novel anti-inflammatory— pro-resolving mediators and their receptors. *Curr Top Med Chem* **11**(6): 629–647.

Serhan, C. N., S. Yacoubian et al. 2008. Anti-inflammatory and proresolving lipid mediators. *Annu Rev Pathol* **3**: 279–312.

Serhan, C. N., R. Yang et al. 2009. Maresins: Novel macrophage mediators with potent antiinflammatory and proresolving actions. *J Exp Med* **206**(1): 15–23.

Sharma, N. P., L. Dong et al. 2010. Asymmetric acetylation of the cyclooxygenase-2 homodimer by aspirin and its effects on the oxygenation of arachidonic, eicosapentaenoic, and docosahexaenoic acids. *Mol Pharmacol* **77**(6): 979–986.

Shikano, M., Y. Masuzawa et al. 1994. Complete discrimination of docosahexaenoate from arachidonate by 85 kDa cytosolic phospholipase A2 during the hydrolysis of diacyl- and alkenylacylglycerophosphoethanolamine. *Biochim Biophys Acta* **1212**(2): 211–216.

Shureiqi, I., K. J. Wojno et al. 1999. Decreased 13-S-hydroxyoctadecadienoic acid levels and 15-lipoxygenase-1 expression in human colon cancers. *Carcinogenesis* **20**(10): 1985–1995.

Simopoulos, A. P. 2009. Omega-6/omega-3 essential fatty acids: Biological effects. *World Rev Nutr Diet* **99**: 1–16.

Singh, R. K., R. Tandon et al. 2013. A review on leukotrienes and their receptors with reference to asthma. *J Asthma* **50**(9): 922–931.

Sisignano, M., C. Angioni et al. 2013. Synthesis of lipid mediators during UVB-induced inflammatory hyperalgesia in rats and mice. *PLoS One* **8**(12): e81228.

Smith, W. L. 2008. Nutritionally essential fatty acids and biologically indispensable cyclooxygenases. *Trends Biochem Sci* **33**(1): 27–37.

Smith, W. L., D. L. DeWitt et al. 2000. Cyclooxygenases: Structural, cellular, and molecular biology. *Annu Rev Biochem* **69**: 145–182.

Smith, W. L., L. J. Marnett et al. 1991. Prostaglandin and thromboxane biosynthesis. *Pharmacol Ther* **49**(3): 153–179.

Smith, W. L., Y. Urade et al. 2011. Enzymes of the cyclooxygenase pathways of prostanoid biosynthesis. *Chem Rev* **111**(10): 5821–5865.

Song, W. L., M. Wang et al. 2008. Tetranor PGDM, an abundant urinary metabolite reflects biosynthesis of prostaglandin D2 in mice and humans. *J Biol Chem* **283**(2): 1179–1188.

Sozzani, S., D. Zhou et al. 1996. Stimulating properties of 5-oxo-eicosanoids for human monocytes: Synergism with monocyte chemotactic protein-1 and -3. *J Immunol* **157**(10): 4664–4671.

Stables, M. J. and D. W. Gilroy 2011. Old and new generation lipid mediators in acute inflammation and resolution. *Prog Lipid Res* **50**(1): 35–51.

Stafforini, D. M. and T. M. McIntyre 2013. Determination of phospholipase activity of PAF acetylhydrolase. *Free Radic Biol Med* **59**: 100–107.

Stichtenoth, D. O., S. Thoren et al. 2001. Microsomal prostaglandin E synthase is regulated by proinflammatory cytokines and glucocorticoids in primary rheumatoid synovial cells. *J Immunol* **167**(1): 469–474.

Sturm, G. J., R. Schuligoi et al. 2005. 5-Oxo-6,8,11,14-eicosatetraenoic acid is a potent chemoattractant for human basophils. *J Allergy Clin Immunol* **116**(5): 1014–1019.

Sun, Y. P., S. F. Oh et al. 2007. Resolvin D1 and its aspirin-triggered 17R epimer. Stereochemical assignments, anti-inflammatory properties, and enzymatic inactivation. *J Biol Chem* **282**(13): 9323–9334.

Sundaram, S. and J. Ghosh 2006. Expression of 5-oxoETE receptor in prostate cancer cells: Critical role in survival. *Biochem Biophys Res Commun* **339**(1): 93–98.

Suzuki-Yamamoto, T., M. Nishizawa et al. 1999. cDNA cloning, expression and characterization of human prostaglandin F synthase. *FEBS Lett* **462**(3): 335–340.

Tai, H. H., C. M. Ensor et al. 2002. Prostaglandin catabolizing enzymes. *Prostaglandins Other Lipid Mediat* **68–69**: 483–493.

Takeda, K., S. Yokoyama et al. 2006. Lipocalin-type prostaglandin D synthase as a melanocyte marker regulated by MITF. *Biochem Biophys Res Commun* **339**(4): 1098–1106.

Tang, L., K. Loutzenhiser et al. 2000. Biphasic actions of prostaglandin E(2) on the renal afferent arteriole: Role of EP(3) and EP(4) receptors. *Circ Res* **86**(6): 663–670.

Tanikawa, N., Y. Ohmiya et al. 2002. Identification and characterization of a novel type of membrane-associated prostaglandin E synthase. *Biochem Biophys Res Commun* **291**(4): 884–889.

Tanioka, T., Y. Nakatani et al. 2000. Molecular identification of cytosolic prostaglandin E2 synthase that is functionally coupled with cyclooxygenase-1 in immediate prostaglandin E2 biosynthesis. *J Biol Chem* **275**(42): 32775–32782.

Taniyama, Y., H. Fuse et al. 2005. Loss of lysophospholipase 3 increases atherosclerosis in apolipoprotein E-deficient mice. *Biochem Biophys Res Commun* **330**(1): 104–110.

Thomas, E., J. L. Leroux et al. 1995. Conversion of endogenous arachidonic acid to 5,15-diHETE and lipoxins by polymorphonuclear cells from patients with rheumatoid arthritis. *Inflamm Res* **44**(3): 121–124.

Thum, T., S. Batkai et al. 2010. Measurement and diagnostic use of hepatic cytochrome P450 metabolism of oleic acid in liver disease. *Liver Int* **30**(8): 1181–1188.

Tobin, D. M., F. J. Roca et al. 2013. An enzyme that inactivates the inflammatory mediator leukotriene b4 restricts mycobacterial infection. *PLoS One* **8**(7): e67828.

Tucker, D. E., M. Ghosh et al. 2009. Role of phosphorylation and basic residues in the catalytic domain of cytosolic phospholipase A2alpha in regulating interfacial kinetics and binding and cellular function. *J Biol Chem* **284**(14): 9596–9611.

Uddin, M. and B. D. Levy 2011. Resolvins: Natural agonists for resolution of pulmonary inflammation. *Prog Lipid Res* **50**(1): 75–88.

Ulsaker, G. A. and G. Teien 1983. Gas chromatographic–mass spectrometric identification of 9,10-epoxystearate in human blood. *Analyst* **108**(1285): 521–524.

Urade, Y. and N. Eguchi 2002. Lipocalin-type and hematopoietic prostaglandin D synthases as a novel example of functional convergence. *Prostaglandins Other Lipid Mediat* **68–69**: 375–382.

Urade, Y. and O. Hayaishi 2011. Prostaglandin D2 and sleep/wake regulation. *Sleep Med Rev* **15**(6): 411–418.

Uyama, T., J. Morishita et al. 2009. The tumor suppressor gene H-Rev107 functions as a novel Ca2 + -independent cytosolic phospholipase A1/2 of the thiol hydrolase type. *J Lipid Res* **50**(4): 685–693.

Vachier, I., P. Chanez et al. 2002. Endogenous anti-inflammatory mediators from arachidonate in human neutrophils. *Biochem Biophys Res Commun* **290**(1): 219–224.

Van Rollins, M., P. D. Frade et al. 1988. Oxidation of 5,8,11,14,17-eicosapentaenoic acid by hepatic and renal microsomes. *Biochim Biophys Acta* **966**(1): 133–149.

Vincent, C., R. Fiancette et al. 2008. 5-LOX, 12-LOX and 15-LOX in immature forms of human leukemic blasts. *Leuk Res* **32**(11): 1756–1762.

Watanabe, K. 2011. Recent reports about enzymes related to the synthesis of prostaglandin (PG) F(2) (PGF(2alpha) and 9alpha, 11beta-PGF(2)). *J Biochem* **150**(6): 593–596.

Wecksler, A. T., V. Kenyon et al. 2008. Substrate specificity changes for human reticulocyte and epithelial 15-lipoxygenases reveal allosteric product regulation. *Biochemistry* **47**(28): 7364–7375.

Werz, O., E. Burkert et al. 2002. Extracellular signal-regulated kinases phosphorylate 5-lipoxygenase and stimulate 5-lipoxygenase product formation in leukocytes. *FASEB J* **16**(11): 1441–1443.

Werz, O., D. Szellas et al. 2002. Arachidonic acid promotes phosphorylation of 5-lipoxygenase at Ser-271 by MAPK-activated protein kinase 2 (MK2). *J Biol Chem* **277**(17): 14793–14800.

Weylandt, K. H., L. F. Krause et al. 2011. Suppressed liver tumorigenesis in fat-1 mice with elevated omega-3 fatty acids is associated with increased omega-3 derived lipid mediators and reduced TNF-alpha. *Carcinogenesis* **32**(6): 897–903.

White, P. J., M. Arita et al. 2010. Transgenic restoration of long-chain n-3 fatty acids in insulin target tissues improves resolution capacity and alleviates obesity-linked inflammation and insulin resistance in high-fat-fed mice. *Diabetes* **59**(12): 3066–3073.

Wittwer, J. and M. Hersberger 2007. The two faces of the 15-lipoxygenase in atherosclerosis. *Prostaglandins Leukot Essent Fatty Acids* **77**(2): 67–77.

Woods, J. W., M. J. Coffey et al. 1995. 5-Lipoxygenase is located in the euchromatin of the nucleus in resting human alveolar macrophages and translocates to the nuclear envelope upon cell activation. *J Clin Invest* **95**(5): 2035–2046.

Woodward, D. F., R. L. Jones et al. 2011. International Union of Basic and Clinical Pharmacology. LXXXIII: Classification of prostanoid receptors, updating 15 years of progress. *Pharmacol Rev* **63**(3): 471–538.

Woodward, D. F., Y. Liang et al. 2008. Prostamides (prostaglandin-ethanolamides) and their pharmacology. *Br J Pharmacol* **153**(3): 410–419.

Wu, K. K. and J. Y. Liou 2005. Cellular and molecular biology of prostacyclin synthase. *Biochem Biophys Res Commun* **338**(1): 45–52.

Xi, S., H. Pham et al. 2000. Suppression of proto-oncogene (AP-1) in a model of skin epidermal hyperproliferation is reversed by topical application of 13-hydroxyoctadecadienoic acid and 15-hydroxyeicosatrienoic acid. *Prostaglandins Leukot Essent Fatty Acids* **62**(1): 13–19.

Xi, S., H. Pham et al. 2000. 15-Hydroxyeicosatrienoic acid (15-HETrE) suppresses epidermal hyperproliferation via the modulation of nuclear transcription factor (AP-1) and apoptosis. *Arch Dermatol Res* **292**(8): 397–403.

Yazaki, M., K. Kashiwagi et al. 2012. Rapid degradation of cyclooxygenase-1 and hematopoietic prostaglandin D synthase through ubiquitin-proteasome system in response to intracellular calcium level. *Mol Biol Cell* **23**(1): 12–21.

Yokomizo, T., T. Izumi et al. 1993. Enzymatic inactivation of leukotriene B4 by a novel enzyme found in the porcine kidney. Purification and properties of leukotriene B4 12-hydroxydehydrogenase. *J Biol Chem* **268**(24): 18128–18135.

Ziboh, V. A., C. C. Miller et al. 2000. Significance of lipoxygenase-derived monohydroxy fatty acids in cutaneous biology. *Prostaglandins Other Lipid Mediat* **63**(1–2): 3–13.

4 Halogenated Lipids
Products of Peroxidase-Derived Reactive Halogenating Species Targeting Conventional Lipids

David A. Ford

CONTENTS

INTRODUCTION: PEROXIDASE PRODUCTS TARGETING THE PHOSPHOLIPID MILIEU OF MEMBRANES AND THE LIBERATION OF HALOGENATED LIPIDS

Peroxidases are important enzymes that are present in leukocytes and some tissues that have important physiological roles in the innate immune system and have been implicated in a broad spectrum of tissue pathologies. Three peroxidases that produce reactive halogenating species have been shown to target conventional lipids producing halogenated lipids. These peroxidases include myeloperoxidase (MPO), eosinophil peroxidase (EPO), and thyroid peroxidase (TPO). These peroxidases are important in both physiological and pathophysiological repair mechanisms, innate immunity and thyroid hormone production. The reactive halogenating species produced by these peroxidases include HOCl, Cl$_2$, HOBr, Br$_2$, BrCl, and either I$^+$ or I^0. Figure 4.1 shows the structural motifs of plasmenylethanolamine that are targeted by reactive halogenating species, which include the vinyl ether bond of plasmalogens, double bonds of alkenes, and primary amines of polar lipids. Although the *in vivo* oxidation of conventional lipids by reactive halogenating systems is considered to be off-site targeting of host tissue, identifying the halogenated lipids that are produced as well as understanding the role of the loss of their precursors may provide important insights into physiological and pathophysiological functions mediated by this system.

TPO produces reactive iodinating species that target plasmalogens resulting in the production of 2-iodohexadecanal (Panneels et al. 1994, 1996). These studies

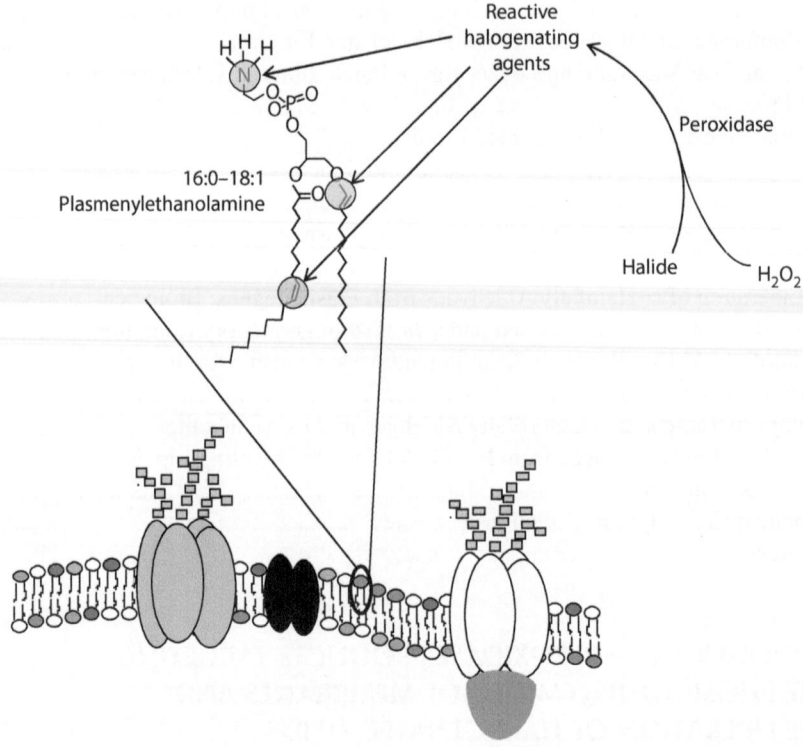

FIGURE 4.1 Lipidic targets of reactive halogenating species produced by peroxidases. Membrane-associated plasmenylethanolamine (16:0–18:1 [# of carbons: # of double bonds in the *sn*-1-*sn*-2 positions of the glycerol backbone]) contains three reactive sites for reactive halogenating species to target as shaded in gray.

suggested that the active oxidizing agent was either I^+ or I^0. Based on these earlier findings it seemed likely that plasmalogens would also be targeted by reactive chlorinating species produced by MPO as well as by reactive brominating species produced by EPO (Albert et al. 2003; Thukkani et al. 2002). To date these halogenated aldehydes liberated by reactive halogenating agents appear to be one of the major mechanisms for the endogenous production of halogenated lipids in mammalian systems.

Leukocytes, containing MPO, produce the reactive chlorinating species, HOCl. Human neutrophils, monocytes, and some macrophages are enriched with MPO (Daugherty et al. 1994; Harrison and Schultz 1976; Lampert and Weiss 1983; Sugiyama et al. 2001; Weiss et al. 1982). Notably however, C57Bl mouse macrophages and monocytes do not contain MPO (Brennan et al. 2001), while their neutrophils do contain MPO (van Leeuwen et al. 2008). HOCl is produced by MPO-catalyzed oxidation of Cl^- using H_2O_2 as co-substrate. MPO also produces to a lesser extent HOBr, which is limited in its synthesis by the nearly three orders of magnitude greater physiological concentration of Cl^- compared to Br^- (Henderson et al. 2001; Teitz 1999). Additionally, Hazen and Heinecke showed that under acidic conditions some chlorination is mediated following the conversion of HOCl into Cl_2 (Hazen et al. 1996). Furthermore, others have suggested that the mixed halide, BrCl, may mediate some halogenation reactions (Henderson et al. 2001). The other major peroxidase of phagocytes is EPO, which is found in the eosinophil and in the presence of physiological levels of bromide selectively produces HOBr (Weiss et al. 1986). This mechanism is important in the antiviral role of eosinophils and is important in asthma (Mitra et al. 2000; Wu et al. 2000). Although this review will focus on halogenated lipid products resulting from peroxidase activities, it should be appreciated that numerous oxidation products mediated by MPO have been described that do not include halogenated lipid products. These products include the oxidation products from MPO utilization of nitric oxide leading to hydroxyeicosatetraenoic acid and hydroperoxyoctadecadienoic acid products from arachidonic and linolenic acid (Podrez et al. 1999, 2002b; Zhang et al. 2002).

This chapter will highlight the targeting of membrane lipids by peroxidase-derived reactive halogenating species and how this leads to the generation of new lipid species that are oxidized and, in the context of this review, are halogenated. In particular this chapter will focus on the targeting of the plasmalogen vinyl ether bond (Figure 4.1), which gives rise to α-halofatty aldehydes. These molecules have great potential to provide new information in our understanding of inflammation and the associated diseases. Also one must consider that the destruction of key lipids can disrupt the membrane milieu that solvates critical peripheral and transmembrane proteins. From our experience less than 0.5% of plasmalogens are degraded by HOCl targeting during neutrophil activation (Thukkani et al. 2002). This loss of plasmalogen apparently does not disrupt membrane integrity, but changes in plasmalogens in specific membrane domains may have profound effects on neighboring protein function. Also the localized liberation of a chlorinated aldehyde from plasmalogen potentially could be highly reactive with local proteins. To date, significant studies have focused on the types of halogenated lipids produced from the reactive motifs on specific lipids, followed by examining their biological roles and metabolic clearance.

In this chapter these individual lipid targets, their halogenated products and the biochemical and biological consequences of the liberated halogenated products will be reviewed.

MPO- AND EPO-CONTAINING LEUKOCYTES IN INNATE IMMUNITY, INFLAMMATION, AND DISEASE

MPO is important in the generation of leukocyte-derived oxidants that combat invading pathogens (Klebanoff et al. 1984; Nathan 2006). In addition to hypochlorous acid and its conjugate base (OCl$^-$), HOCl is in equilibrium with chlorine gas and secondary chlorinating intermediates, such as N-monochloramines (Harrison and Schultz 1976; Hazen et al. 1996; Lampert and Weiss 1983). The microbicidal and cytotoxic properties of HOCl are attributed to its chemical reactivity with amines and unsaturated lipids, and oxidation of heme groups and iron sulfur centers (Heinecke et al. 1994; Nathan 2006; Thomas et al. 1982; Weiss 1989; Winterbourn et al. 1992). Like MPO, EPO amplifies the oxidant potential of hydrogen peroxide by generating HOBr. HOBr attacks viral and microbial proteins and membrane components. The targeting of biomolecules by HOCl and HOBr by these cells of the innate immune system, while potent in killing many invading organisms, also results in the production of halogenated biomolecules that may have important biological effects and remain in the host until metabolically cleared.

The role of neutrophils, monocytes, and macrophages in cardiovascular disease, including atherosclerosis and ischemia/reperfusion injury, is of critical importance, and will be the underlying focus of this chapter. However, it should be appreciated that the role of MPO in inflammation and wound healing is not limited to cardiovascular disease. Pancreatitis leads to neutrophil and macrophage infiltrates that have been implicated in the pathophysiology of this disease and recently lipid oxidation mediated by MPO has also been demonstrated (Franco-Pons et al. 2013; Frossard et al. 2008). Microglial cell and macrophage MPO has also been associated with neurogeneration and demyelination (Chen et al. 2008; Gray et al. 2008; Marik et al. 2007; Nagra et al. 1997). Monocyte-derived MPO has been implicated as a mediator of protein modification associated with glomerular renal disease (Johnson et al. 1987). It should also be noted that a major proportion of the human body's neutrophils are sequestered in the lung, and acute lung injury is associated with neutrophil activation (Grommes and Soehnlein 2011; Kolaczkowska and Kubes 2013; Kreisel et al. 2010). In addition to phagocytes containing MPO, eosinophils containing EPO are important mediators of asthma, allergic reactions, and tumor biology. It is likely that EPO-derived oxidants and oxidation products are important in these eosinophil-mediated diseases. Taken together, MPO- and EPO-containing phagocytes mediate their roles in pathology, at least in part, through their peroxidases that yield strong oxidant hypohalous acids that target lipids generating halogenated lipids. These halogenated lipids represent an important research area to investigate in order to delineate their role in pathology.

Atherosclerosis is a chronic inflammatory disease of the vascular wall. During the initiation of atherosclerotic disease, pro-inflammatory monocytes are recruited to the subintimal layer of the vascular wall, differentiate to macrophages, and further promote inflammation in the vascular wall. In conjunction to the inflammatory milieu provided by activated leukocyte is the uptake of oxidized LDL by macrophages leading to foam cell production which is a hallmark of the early stages of atherosclerosis (Berliner and Heinecke 1996; Marathe et al. 2001; Witztum and Steinberg 1991). Considerable attention has been given to delineating the oxidants and oxidized products that mediate atherosclerotic lesion formation and propagation. Multiple independent studies have supported a role for MPO in human atherosclerosis (Daugherty et al. 1994; Hazell et al. 1996; Hazen and Heinecke 1997; Malle et al. 2000). MPO likely has a role in the early and late stages of atherosclerosis (Podrez et al. 1999, 2002a,b; Sugiyama et al. 2001). MPO localizes in the shoulder regions of mature human atherosclerotic plaques (Sugiyama et al. 2001). Additionally, MPO may have a role in plaque rupture and acute coronary syndromes since hypochlorous acid activates the metalloproteinase, matrilysin, *in vitro* (Fu et al. 2001). Furthermore, MPO co-localizes with matrilysin in tissue sections from vulnerable plaques suggesting that MPO may have an important role in plaque rupture and acute coronary syndromes (Fu et al. 2001; Sugiyama et al. 2001). Multiple MPO-derived oxidation products have been found in human atherosclerotic lesions including nitrotyrosine and chlorotyrosine (Hazen and Heinecke 1997; Pennathur et al. 2004; Zheng et al. 2004). Hazen and coworkers have shown that MPO uses nitrite to form reactive nitrogen species that oxidize low-density lipoprotein (LDL), leading to the modified polyunsaturated aliphatic residues associated with choline glycerophospholipids present in LDL (Podrez et al. 1999, 2002a,b). This modified LDL is a high uptake form for macrophages leading to foam cell formation. Additionally, the MPO-derived oxidation product of plasmalogen, α-chlorofatty aldehyde, accumulates 1400-fold in human atherosclerotic lesions compared to normal vascular tissue (Thukkani et al. 2003b). The level of α-chlorofatty aldehyde in atherosclerotic lesions has been estimated at ~10 μM (Thukkani et al. 2003b).

The involvement of MPO in mouse models of atherosclerosis is complicated since murine monocytes lack MPO. Studies using MPO$^{-/-}$ mice crossed with apolipoprotein E$^{-/-}$ mice did not show a reduction in mouse atherosclerosis in the absence of MPO (Brennan et al. 2001). However, wild-type mouse atherosclerotic lesions revealed that invading monocytes and macrophages do not possess MPO (Brennan et al. 2001). However, transgenic mice expressing human MPO in their macrophages in the LDL receptor$^{-/-}$ mouse background subjected to a Western-type (pro-atherogenic) diet have a twofold increase in atherosclerotic lesion area compared to mice that do not express human MPO in their macrophages (McMillen et al. 2005). Taken together, multiple lines of evidence suggest a role for MPO in human atherosclerosis, and enhancing the MPO content in the monocytes and macrophages of mice with human MPO results in the appearance of MPO-containing cells in atherosclerotic lesions that accelerate the pathophysiological sequelae of mouse atherosclerosis.

Plasma MPO levels correlate with the risk of coronary artery disease. While MPO deficiency in humans is associated with increased susceptibility to fungal and yeast infections, it is also associated with reduced incidence of cardiovascular disease (Kutter et al. 2000). Functional polymorphisms in the promoter region of MPO resulting in decreased MPO expression are associated with decreased risk for cardiovascular disease (Kutter et al. 2000; Nikpoor et al. 2001; Pecoits-Filho et al. 2003). Plasma levels of chlorotyrosine and nitrotyrosine are elevated in the plasma and serum of individuals with established cardiovascular disease (Bergt et al. 2004; Pennathur et al. 2004; Zheng et al. 2004). However, unlike plasma levels of MPO, plasma levels of chlorotyrosine and nitrotyrosine do not predict mortality in patients following myocardial infarction and they are not prognostic indicators of future adverse coronary events (Mocatta et al. 2007).

Additionally in cardiovascular disease, neutrophils alter post-ischemic contractile function (Engler and Covell 1987; Jordan et al. 1999; Pabla et al. 1996; Sheridan et al. 1991). Activated neutrophils mediate endothelial dysfunction via secretion of proteolytic enzymes, such as elastase, and oxygen radicals, which can have short-term and long-term effects on endothelial function (Jordan et al. 1999). Coupling the activation of NADPH oxidase during the respiratory burst in neutrophils with MPO release in ischemic myocardium, the activated neutrophil amplifies the oxidizing potential of hydrogen peroxide with the production of HOCl (Harrison and Schultz 1976). Neutrophil infiltration into previously ischemic zones mediates, in part, myocardial reperfusion injury (Hayasaki et al. 2006; Lucchesi 1990; Mehta et al. 1989; Ockaili et al. 2005). Myocardial damage in response to ischemia–reperfusion is reduced in animals that are either rendered neutropenic, pretreated with antibodies that block neutrophil–endothelial cell binding, or genetically modified to reduce neutrophil infiltration (Hayasaki et al. 2006; Litt et al. 1989; Palazzo et al. 1998a,b). Undoubtedly, some degree of myocardial damage during reperfusion following ischemia is mediated by neutrophil-derived free radicals (Chen et al. 2006; Duilio et al. 2001; Hansen 1995; Shandelya et al. 1993).

While neutrophils have a role in post-ischemic myocardial dysfunction, the role of MPO-derived products is only partially understood. MPO$^{-/-}$ mice have an increased risk of ventricular rupture following left anterior descending coronary artery (LAD) ligation compared to wild-type mice, which may involve decreased inhibition of plasminogen activator inhibitor 1 (Askari et al. 2003). Other studies have shown that MPO-derived glycine and threonine oxidation products do not impact on post-ischemic function, but have a role in ventricular remodeling (Vasilyev et al. 2005). On the other hand, the chlorinated lipid, 2-chlorohexadecanal (Thukkani et al. 2005), decreases ventricular function. Studies have also shown that hypochlorous acid added to isolated heart perfusates elicits contractile dysfunction and arrhythmias, possibly by targeting adrenergic receptors and sodium–potassium ATPase (Kato et al. 1998; Okabe et al. 1993; Persad et al. 1999).

While previous studies have focused on MPO in cardiovascular disease, it is important to consider the role of monocytes, macrophages, eosinophils, and neutrophils as mediators of innate immunity, and as important responsive cells in inflammation. The role of MPO and EPO and their hypohalous acids targeting lipids to generate novel halogenated lipids is relatively unexplored, and should be thoroughly examined, in diseases including pancreatitis, asthma, renal glomerular, and tubular disease.

PLASMALOGEN-DERIVED HALOGENATED LIPIDS

Plasmalogens are phospholipid molecular subclasses characterized by the unique structural motif of a vinyl ether-linked aliphatic chain attached to the sn-1 position of the glycerol backbone. The vinyl ether is a masked aldehyde that is liberated as an aldehyde in the presence of acid. This vinyl ether bond is a target for HOCl, HOBr, Cl₂, and either I⁺ or I⁰, which are the reactive halogenating species produced by MPO, EPO, and TPO. The liberated α-halogenated fatty aldehydes derived from the vinyl ether bond have been identified under physiological conditions and likely represent the major halogenated lipids and precursors of other halogenated lipids.

PLASMALOGENS: AN ABUNDANT PHOSPHOLIPID SUBCLASS FOUND IN MAMMALIAN TISSUES

Plasmalogens were first characterized as the aldehyde-like lipid of the plasma, which gives rise to their name. The vinyl ether linkage is a masked aldehyde that is acid labile (pH ≤ 2). The molecular structure of the ethanolamine plasmalogen, plasmenylethanolamine is shown in Figure 4.1. Biophysical studies measuring two-dimensional nuclear Overhauser effects by nuclear magnetic resonance have shown that plasmalogens pack more tightly in membranes and the orientation of the polar head group of the plasmalogen is more perpendicular with the plane of the bilayer compared to diacyl molecular species (Han and Gross 1990). This orientation was originally envisaged to suggest greater accessibility for phospholipase A₂ targeting plasmalogens for hydrolysis (Han and Gross 1990), but should also be considered as this structure could explain increased accessibility of reactive halogenating species targeting the vinyl ether bond that is near the hydrophilic/hydrophobic interface of the membrane (Figure 4.1).

Plasmalogens are the predominant ethanolamine and choline glycerophospholipids in human myocardium (Hazen et al. 1993), and plasmenylethanolamine is a major phospholipid found in many tissues. In particular, endothelial cells, smooth muscle cells, neutrophils, monocytes, lipoprotein, and lung have abundant levels of plasmalogens (Chilton and Connell 1988; Chilton and Murphy 1986; Ford and Gross 1989; Gross 1985, 1984; Hazen et al. 1993; Malavolta et al. 2004; Post et al. 1988; Vance 1990; Zoeller et al. 2002). Myocardium is somewhat unique in that it contains robust amounts of plasmenylcholine, but it should be appreciated that this enrichment is not as great in rats and mice compared to human, dog, rabbit, pig, and cow (Schulz 1996). Plasmalogens are found in intracellular membrane pools, plasma membrane and are highly abundant in lipid rafts (Gross 1984; Han and Gross 1994; Pike et al. 2002). Considerable interest has been directed toward the biological role of plasmalogens due to: (1) their enrichment with arachidonic acid residues esterified to the sn-2 position of the glycerol backbone; (2) their potential to generate other bioactive lipids; and (3) their antioxidant capacity (via the vinyl ether bond).

The masked aldehyde of plasmalogens is not readily metabolized, but plasmalogenase activity has been reported (McMaster et al. 1992). The turnover of the vinyl ether aliphatic group is 300-times slower compared to that of turnover of the sn-2 esterified fatty acid and the sn-3 polar head group (Ford and Gross 1994). However, the vinyl ether reactivity to chemicals as well as its localization in the hydrophilic domain of membranes suggest that this is a structural motif for oxidant targeting. Indeed, plasmalogens protect cells from free radical damage through their antioxidant properties (Morand et al. 1988; Zoeller et al. 1988, 2002) resulting in aldehyde production without further free radical production (in contrast to that occurring from free radical attack of arachidonic acid) (Morand et al. 1988; Scherrer and Gross 1989; Vance 1990; Zoeller et al. 1988, 2002). In fact, plasmalogens that are complexed to lipoproteins have been hypothesized to serve as endogenous plasma antioxidants (Vance 1990).

LIBERATION OF α-HALOFATTY ALDEHYDES FROM PLASMALOGENS: CHEMICAL IDENTITY

In the 1990s, α-iodofatty aldehyde was shown to be the halogenated product of plasmalogens targeted by the reactive iodinating species generated by TPO. Based on these studies and the vulnerable nature of the vinyl ether bond of plasmalogens to oxidation, it was predicted that this bond is a general target for reactive halogenating species including hypochlorous acid and hypobromous acid. Indeed, the plasmalogens, lysoplasmenylcholine, and plasmenylcholine, are targets of either hypochlorous acid or chlorine gas resulting in the release of the α-chlorofatty aldehyde, 2-chlorohexadecanal (Albert et al. 2001). These studies also showed that only the masked aldehyde is targeted by hypochlorous acid compared to the free aldehyde (Albert et al. 2001). Oxidation products from HOCl targeting the plasmalogen, 1-O-hexadec-1'-enyl-2-octadec-9'-enoyl-sn-glycero-3-phosphocholine, are 2-chlorohexadecanal and the lysophosphatidylcholine, 2-octadec-9'-enoyl-sn-glycero-3-phosphocholine. Furthermore, as shown in Figure 4.2, the unsaturated lysophosphatidylcholine can be targeted by secondary hypochlorous attack leading to the production of chlorohydrins of unsaturated lysophosphatidylcholines (Messner et al. 2006). Figure 4.2 shows the sequential oxidation of the plasmalogen vinyl ether bond, then the oxidation of the sn-2 aliphatic alkene of the reaction product, lysophosphatidylcholine. The preferred target is the plasmalogen vinyl ether bond, but excess hypochlorous acid compared to plasmalogen vinyl ether bond targets leads to the complete ablation of vinyl ether containing plasmalogens and subsequent consumption of remaining hypochlorous acid through the oxidation of sn-2 aliphatic alkenes. Others have also shown that excess HOCl leads to complete degradation of plasmalogens with glycerophosphorylcholine product (LeBig et al. 2007). Studies with purified plasmenylcholine and HOBr also revealed similar degradation products from the oxidation of the vinyl ether bond (e.g., both the α-bromofatty aldehyde, 2-bromohexadecanal and lysophosphatidylcholine) (Albert et al. 2002). It is also likely that sequential conversion of unsaturated molecular species of lysophosphatidylcholine to bromohydrin molecular species occurs following the initial targeting of the plasmalogen vinyl ether bond.

The characterization of the product of RCS targeting of plasmalogens led to several methods that were developed to both purify and quantify α-halofatty aldehyde. For example, thin-layer chromatography with silica gel G as a solid phase and petroleum ether/ethyl ether/acetic acid (90/10/1) separates α-chlorofatty aldehydes ($R_f = 0.46$) and α-bromofatty aldehydes ($R_f = 0.58$) (Albert et al. 2001, 2002). The confirmation of the structure of 2-chlorohexadecanal and 2-bromohexadecanal using GC-MS following derivatization to its pentafluorobenzyl oxime also proved to be an extremely

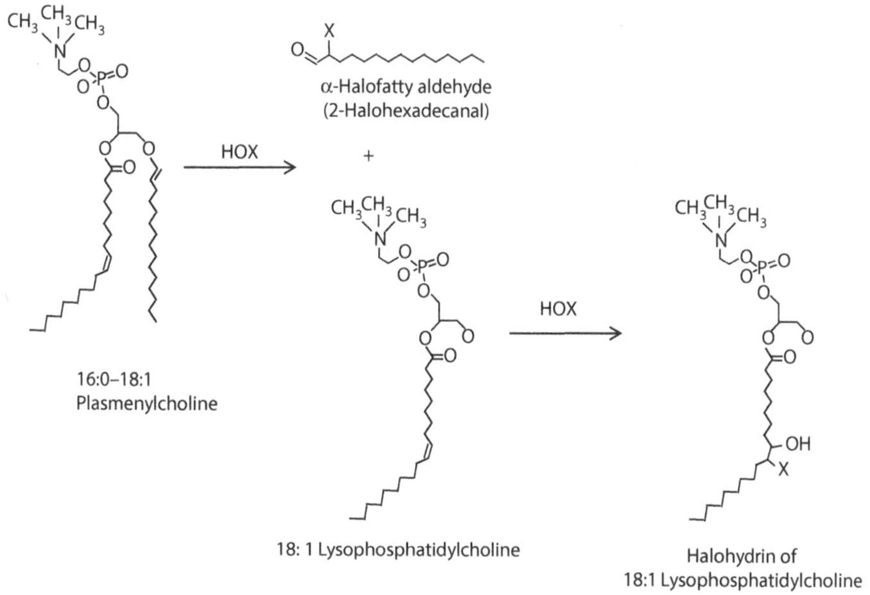

FIGURE 4.2 Sequential targeting of plasmalogen reactive sites by hypohalous acids. Hypohalous acids (HOX) first target the vinyl ether bond of plasmalogens and then the alkene of the *sn*-2 aliphatic group of unsaturated molecular species of lysophosphatidylcholine.

sensitive analytical tool to detect α-chlorofatty aldehydes and α-bromofatty aldehydes (Albert et al. 2001; Thukkani et al. 2002, 2003a,b, 2005). Utilizing a deuterated internal standard analog (e.g., 2-chloro-[7,7,8,8-d_4]-hexadecanal and 2-bromo-[7,7,8,8-d_4]-hexadecanal), these molecules could be readily quantified in biological samples employing selected ion monitoring and negative ion-chemical ionization mass spectrometry (e.g., Albert et al. 2003; Thukkani et al. 2002, 2003b).

Other groups also showed the production of chlorinated fatty aldehydes from plasmalogens. For example, Malle and coworkers showed lipoprotein-associated plasmalogens are targeted by hypochlorous acid resulting in 2-chlorohexadecanal production (Marsche et al. 2004). Additionally, important kinetic studies by Davies and coworkers (Skaff et al. 2008) demonstrated that the rate constant for HOCl targeting the vinyl ether bond of plasmalogens is 2 orders of magnitude faster than the targeting of alkenes. These elegant kinetic studies also supported the finding that plasmalogen oxidation appears to first involve oxidation of the vinyl ether bond and then chlorohydrins are produced from the unsaturated lysophosphatidylcholine molecular species (Messner et al. 2006).

Liberation of α-Halofatty Aldehydes from Plasmalogens: Biological Production in Leukocytes and under *In Vivo* Inflammatory Conditions

Human leukocytes are not only enriched with either MPO or EPO (Albrich et al. 1981; Lampert and Weiss 1983; Weiss et al. 1986), but also contain robust plasmalogen pools (Albert et al. 2003; Chilton and Connell 1988; Hsu et al. 2003; Thukkani et al. 2003a). Thus, it was not surprising that phorbol ester- and formyl-methione–leucine–phenylalanine peptide-activated human neutrophils accumulate the α-chlorofatty aldehyde, 2-chlorohexadecanal (Thukkani et al. 2002) and phorbol ester-activated human eosinophils accumulate the α-bromofatty aldehyde, 2-bromohexadecanal (Albert et al. 2003). Furthermore, at least in the case of neutrophil activation, it was shown that neutrophil-derived HOCl target endothelial cell plasmalogens, and thus neutrophil-derived HOCl has the potential to impact neighboring cells, which in the case of the vascular wall may be critical for endothelial function. Both 2-chlorohexadecanal and 2-bromohexadecanal production by neutrophils and eosinophils, respectively, are blocked by heme enzyme inhibitors (azide, cyanide, and 3-aminotriazole) that have previously been used to inhibit HOCl production by MPO (Albert et al. 2003; Heim et al. 1956; Kukreja et al. 1989; Nauseef et al. 1983). Additionally, the dependence of 2-chlorohexadecanal production on MPO activity is supported by the attenuation of 2-chlorohexadecanal accumulation in zymosan-activated, thioglycolate-recruited peritoneal lavage cells from MPO-knockout mice compared to wild-type mice (Thukkani et al. 2003b). Monocyte-derived hypochlorous acid has also been shown to target lipoprotein-associated plasmalogens in addition to monocyte-associated plasmalogens, resulting in increased production of both 2-chlorooctadecanal and 2-chlorohexadecanal (Thukkani et al. 2003a). Comparisons have also been made between human monocytes and mouse monocytes, which have confirmed that α-chlorofatty aldehydes are produced in the human but not in the mouse monocyte, which is devoid of MPO activity (Thukkani et al. 2003a)

Studies by Boeynaems and coworkers showed that plasmalogens are targeted by reactive iodinating species produced in cultured dog thyroid cells (Panneels et al. 1996; Pereira et al. 1990). Furthermore, 2-iodohexadecanal is the major iodinated lipid found in horse thyroid gland (Pereira et al. 1990). The role of this lipid in the thyroid has yet to be fully understood; however, its metabolism to 2-iodohexadecanol has been shown (Panneels et al. 1996) and seems similar to that described for 2-chlorohexadecanal (Wildsmith et al. 2006). Interestingly, one study has shown that 2-iodohexadecanal inhibits thyroid adenylyl cyclase, which has been hypothesized to mediate the effect of iodide treatment leading to thyroid tissue adenylyl cyclase inhibition (Panneels et al. 1994).

As previously mentioned, monocyte-derived HOCl targets low-density lipoprotein-associated plasmalogen leading to the production of 2-chlorohexadecanal and unsaturated molecular species of lysophosphatidylcholine (Thukkani et al. 2003a). The implication of this finding is that HOCl produced by MPO present in atherosclerotic lesions could be a mechanism that could lead to the accumulation of α-chlorofatty aldehydes and lysophospholipids in atherosclerotic vascular wall. Indeed, human atherosclerotic lesions have a 1400- and 34-fold increase in the content of 2-chlorohexadecanal and unsaturated molecular species of lysophosphatidylcholine, respectively, in comparison to nonlesioned vascular tissue (Thukkani et al. 2003b). Both of these products of plasmalogen oxidation by HOCl are present in atherosclerotic tissue at levels ≥10 μM. These plasmalogen oxidation products may augment inflammatory mechanisms since 2-chlorohexadecanal is a chemoattractant for phagocytes (Thukkani et al. 2002) and unsaturated molecular species of lyso-phosphatidylcholine stimulate membrane surface expression of phagocyte-tethering proteins (e.g., P-selectin) on endothelial cells (Thukkani et al. 2003b). Taken together, these *in vitro* experiments suggest that through the attack of the vinyl ether bond of plasmenylcholine, two products are made that concertedly attract phagocytes (2-chlorohexadecanal) and facilitate the tethering of phagocytes to endothelium (unsaturated molecular species of lysophosphatidylcholine).

Studies using the rat LAD occlusion model of myocardial ischemia have shown that neutrophil infiltration into myocardial infarct zones leads to the accumulation of 2-chlorohexadecanal in the myocardium (Thukkani et al. 2005). In these studies, heart tissue from rats subjected to surgical infarction had elevated levels of 2-chlorohexadecanal and neutrophil infiltration levels, compared with heart tissue from rats subjected to sham surgery. To show the role of neutrophils in 2-chlorohexadecanal accumulation in infarcted hearts, rats rendered neutropenic and then subjected to LAD occlusion had reduced levels of myocardial 2-chlorohexadecanal and reduced myocardial neutrophil infiltration compared with that from rats that had normal levels of circulating neutrophils (Thukkani et al. 2005). Furthermore, 2-chlorohexadecanal elicits cardiac injury and reduced contractile performance in isolated perfused rat hearts treated with 2-chlorohexadecanal concentrations found in infarcted myocardium (Thukkani et al. 2005).

Studies by Wolfgang Sattler and coworkers have shown that brain plasmalogens are susceptible to HOCl targeting, leading to 2-chlorohexadecanal (Ullen et al. 2010). Furthermore, endotoxin injection into mice leads to an increase in brain MPO levels, brain inflammation and the accumulation of brain 2-chlorohexadecanal (Ullen et al. 2010). These investigators postulate that this neuroinflammatory process compromises synaptic transmission. Additional recent studies have shown that 2-chlorohexadecanal has profound effects on brain microvascular endothelial cell function, resulting in loss of barrier function and increased apoptosis and related mechanisms (Ullen et al. 2012). These studies also showed the polyphenol, phloretin, was able to reduce these deleterious effects of 2-chlorohexadecanal on brain microvascular endothelial cells.

Taken together, multiple studies have shown the production of α-halofatty aldehydes as a result of the activities of the peroxidases of neutrophils, monocytes, macrophages, eosinophils, and thyroid cells. These aldehydes are produced in response to inflammation and have been shown to alter important physiological processes including vascular integrity and cardiac contractile performance.

α-HALOFATTY ALDEHYDE METABOLISM: POTENTIAL MECHANISMS FOR THE ACTION OF HALOGENATED LIPIDS

One potential mechanism by which α-halofatty aldehydes could elicit functional changes in targeted cells would be through Schiff-base adduct formation with primary amines of proteins and lipids. This is particularly important since Schiff-base adduct formation could potentially alter membrane dynamics and protein function. 2-Chlorohexadecanal Schiff-base adducts with ethanolamine glycerophospholipids and lysine have been identified (Wildsmith et al. 2006). In these studies, Schiff-base stabilization by reduction with cyanoborohydride resulted in an adduct containing an unsaturated carbon–carbon bond from the elimination of HCl (Wildsmith et al. 2006).

Other studies using radiolabeled, as well as stable isotope labeled, 2-chlorohexadecanal showed that neutrophils and endothelial cells can metabolize 2-chlorohexadecanal to 2-chlorohexadecanol (an α-chlorofatty alcohol) and 2-chlorohexadecanoic acid (an α-chlorofatty acid) (Wildsmith et al. 2006) (Figure 4.3). This metabolism is reminiscent to the earlier studies that demonstrated that iodohexadecanal can be reduced to iodohexadecanol in cultured dog thyroid cells (Panneels et al. 1996). Interestingly, it appears that endothelial cells and neutrophils actively secrete the metabolites of 2-chlorohexadecanal rather than accumulate them within the cell (Wildsmith et al. 2006). Some radiolabeled 2-chlorohexadecanal is also incorporated into triglycerides and phospholipids, and it is likely that this represents the esterification of 2-chlorohexadecanoic acid into these pools. These intermediates of α-chlorofatty aldehyde metabolism make sense because fatty aldehyde is an intermediate of the fatty acid–fatty alcohol cycle (Rizzo et al. 1987). In fact, using a cell culture model of fatty aldehyde dehydrogenase (FALDH) (one of the enzymes of the fatty acid–fatty alcohol cycle), deficiency, FAA.K1A cells, it was shown that 2-chlorohexadecanal oxidation to 2-chlorohexadecanoic acid was dependent on FALDH activity (Anbukumar et al. 2010). Endogenously produced 2-chlorohexadecanal in activated neutrophils is metabolized to 2-chlorohexadecanoic acid and 2-chlorohexadecanol and these metabolites are released from neutrophils (Anbukumar et al. 2010). Brain microvascular endothelial cells have also been shown to take up 2-chlorohexadecanal and metabolize it to a great extent to 2-chlorohexadecanol compared to 2-chlorohexadecanoic acid, and both of these metabolites are readily released from microvascular

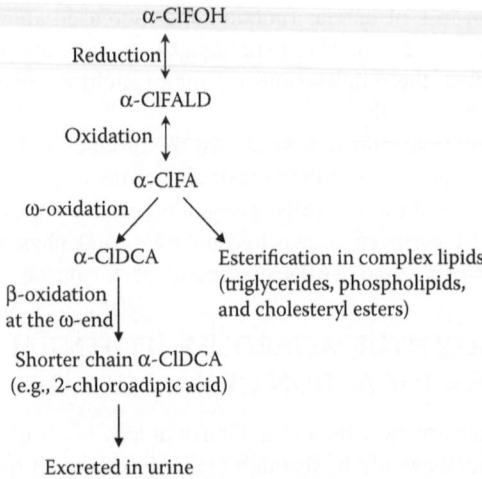

FIGURE 4.3 Metabolism of α-chlorofatty aldehyde. α-Chlorofatty aldehyde (α-ClFALD); α-chlorofatty alcohol (α-ClFOH); α-chlorofatty acid (α-ClFA); α-chlorodicarboxylic acid (α-ClDCA).

endothelial cells (Ullen et al. 2012). Taken together, cell studies have shown that plasmalogen targeting by HOCl leads to α-chlorofatty aldehyde, which is the precursor of a family of chlorinated lipids that are present *in vivo*.

Endotoxin treatment of rodents leading to inflammation has been shown to lead to the production of chlorinated lipids. As previously described, Sattler and coworkers showed that endotoxin treatment led to neuroinflammation with concomitant increased levels of both MPO mRNA and 2-chlorohexadecanal in the brain (Ullen et al. 2010). Our studies have also shown that endotoxin leads to chlorinated lipid production in various systems *in vivo* (Brahmbhatt et al. 2010). In fact, lipopolysaccharide treatment led to elevated 2-chlorohexadecanoic acid levels in the plasma. These studies also delineated the metabolic clearance pathway for 2-chlorohexadecanoic acid, which is mediated by sequential hepatic ω-oxidation, β-oxidation from the ω-end, and then subsequent excretion of 2-chloroadipic acid in the urine (Figure 4.3). Furthermore, lipopolysaccharide treatment of rats led to not only the elevation of plasma 2-chlorohexadecanoic acid, but also increased excretion of 2-chloroadipic acid in the urine.

Similar studies for the metabolic products of α-iodofatty aldehyde and α-bromofatty aldehyde are limited. Although 2-iodohexadecanol has been demonstrated in the thyroid gland, its biological function has not been determined (Panneels et al. 1996), and it has not been identified *in vivo*. Interestingly, these investigations were not able to detect 2-iodohexadecanoic acid (Panneels et al. 1996), which is different from those assessing the parallel metabolism of α-chlorofatty aldehyde in other cell types (Wildsmith et al. 2006). For the bromolipids, only α-bromofatty aldehyde has been identified, and the delineation of its metabolites remains to be described.

While 2-chlorohexadecanal is produced in both atherosclerotic lesions and infarcted myocardium, it remains to be seen whether this aldehyde gives rise to chlorinated lipid metabolites as well as the putative incorporation of chlorinated aliphatic groups into complex lipids including triglycerides, phospholipids, and cholesteryl esters. Furthermore, these metabolites need to be assessed in experiments designed to demonstrate biological effects of α-chlorofatty aldehyde.

BIOLOGICAL TARGETS OF α-CHLOROFATTY ALDEHYDE AND ITS METABOLITES

The biological targets of chlorinated lipids derived from plasmalogen oxidation by HOCl are shown in Figure 4.4. Several studies have investigated the biological role of α-chlorofatty aldehyde. For example, Marsche et al. showed that LDL-associated 2-chlorohexadecanal reduces nitric oxide production in endothelial cells through inhibition of endothelial nitric oxide synthase (eNOS) activity via decreasing eNOS mRNA stability (Marsche et al. 2004). Additionally, studies with human coronary artery endothelial cells revealed that both 2-chlorohexadecanal and 2-chlorohexadecanoic acid increase COX-2 expression and prostacyclin production through increased NFκB signaling (Messner et al. 2008). These two studies examining eNOS and COX-2 are interesting in that they both show targeting of vasodilators by 2-chlorohexadecanal with opposing effects (i.e., decreasing nitric oxide, while increasing prostacyclin). Also, since 2-chlorohexadecanal increases NFκB signaling (Messner et al. 2008), it is likely that other pro-inflammatory pathways may be activated.

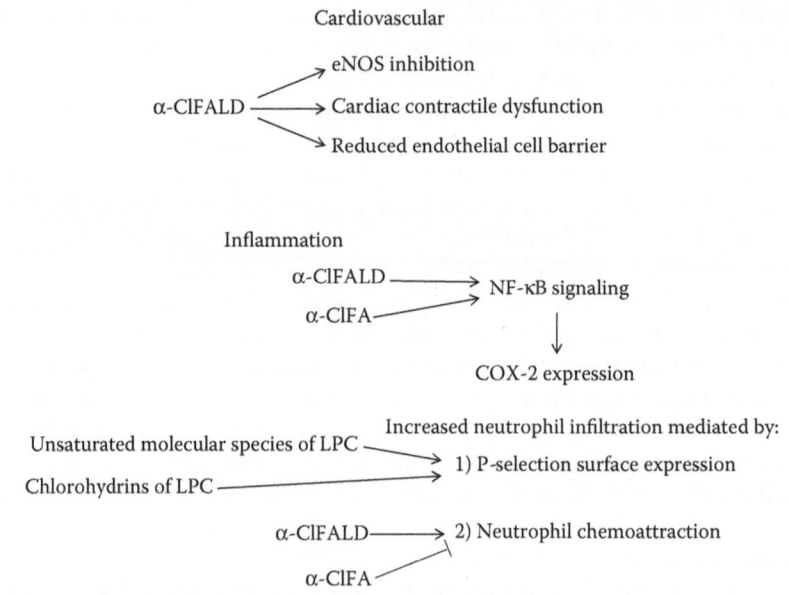

FIGURE 4.4 Biological targets of chlorinated lipids. Cyclooxygenase-2 (COX-2); endothelial nitric oxide synthase (eNOS); lysophosphatidylcholine (LPC); and nuclear factor kappa-B (NF-κB).

2-Chlorohexadecanal and 2-chlorohexadecanoic acid are a chemoattractant and chemorepellant, respectively, for neutrophils (Anbukumar et al. 2010; Thukkani et al. 2002). Thus, 2-chlorohexadecanal may be produced as a result of infiltration and activation of neutrophils, and initially participate in the recruitment of additional neutrophils. However, the conversion of 2-chlorohexadecanal into 2-chlorohexadecanoic acid, may have an important role in decelerating further neutrophil recruitment. Also, it should be appreciated that the tandem production of 2-chlorohexadecanal and unsaturated molecular species of lysophosphatidylcholine (and perhaps their chlorohydrins) may enhance neutrophil infiltration by not only chemoattraction, but also stimulating the surface expression of P-selectin (Messner et al. 2008; Thukkani et al. 2003b).

The recent studies on MPO-derived α-chlorofatty aldehyde in the brain during neuroinflammation are exciting. In contrast to other studies they found not only a significant increase in α-chlorofatty aldehyde in response to endotoxin treatment, but also the loss of several plasmalogen molecular species (Ullen et al. 2010). Decreased brain plasmalogen content has been shown in Alzheimer's disease and Gaucher's disease, and has been associated with neurological dysfunction (Ginsberg et al. 1995; Goodenowe et al. 2007; Moraitou et al. 2008). Accordingly changes in brain plasmalogens mediated by HOCl targeting has been proposed to be one mechanism that may lead to disease associated alterations in synaptic transmission (Ullen et al. 2010). In additional studies by this group they have shown that 2-chlorohexadecanal has significant effects on brain microvascular endothelial cells causing apoptosis with associated caspase 3 activation, ATP depletion, and decreased cell–cell barrier function (Ullen et al. 2012). Mechanistically, 2-chlorohexadecanal appears to be eliciting intracellular ROS production and cell death, and the antioxidant polyphenol, phloretin, is able to reduce the injury elicited by 2-chlorohexadecanal.

While significant studies have focused on the biological roles of α-chlorofatty aldehyde, the roles of its metabolites need to be assessed, and the potential roles assigned to α-chlorofatty aldehyde that are actually mediated by its metabolites should be considered. Since both 2-chlorohexadecanal and 2-chlorohexadecanoic acid increase endothelial COX-2 levels, it is possible that 2-chlorohexadecanal does not directly elicit COX-2 expression, but rather its fatty acid metabolite may be the mediator of this pathway (Anbukumar et al. 2010; Messner et al. 2008). It should be noted that the role of metabolites of both α-bromofatty aldehyde and α-iodofatty aldehyde remain to be examined.

OTHER HALOLIPIDS PRODUCED FROM HYPOHALOUS ACIDS TARGETING LIPIDS OTHER THAN PLASMALOGENS

As shown in Figure 4.1, there are additional reactive sites in phospholipids that potentially may be halogenated in the presence of peroxidase-derived hypohalous acids. In 1998, studies by Winterbourne et al. showed both hypochlorous acid and hypobromous acids react with both unsaturated aliphatic residues and the ethanolamine of phosphatidylethanolamine (Carr et al. 1998). These studies showed that the amine was more reactive than alkenes and that the alkenes formed halohydrins only after the amine was consumed by the hypohalous acid. This was similar to the findings with plasmalogens which showed the chlorohydrins were only produced once the vinyl ether bond was consumed. Additionally these earlier studies demonstrated that bromoamine was reactive in producing bromohydrins, and this activity was greater than that observed with chloramines (Carr et al. 1998). Others have also shown the reactivity of hypohalous acids with ethanolamine (Jaskolla et al. 2009), but it is unclear whether these compounds are stable in cells or *in vivo*.

Several studies have also examined the targeting of sphingolipids by HOCl. The N-deacylated sphingolipid, sphingosylphosphorylcholine, reacts with HOCl to produce 2-hexadecenal and 1-cyano methano phosphocholine (Brahmbhatt et al. 2007). Similarly other N-deacylated sphingolipids such as sphingosine are degraded to 2-hexadecenal when treated with HOCl. Acylation at the amine of sphingolipids protects these lipids from degradation (Brahmbhatt et al. 2007). Additionally, the alkene of the sphingosine backbone of sphingomyelin can be chlorinated to yield chlorohydrins (Nusshold et al. 2010). These sphingomyelin oxidation products have also been shown to cause apoptosis and dissipation of mitochondrial membrane potential in PC12 cells.

Considerable work has been directed at characterizing halohydrins of the alkenes of lipids. The treatment of tissues with hypochlorous acid and subsequent lipid chlorohydrin production has been associated with biological consequences, including erythocyte and endothelial cell lysis, as well as decreased contractile tension in cardiac papillary muscle preparations (Carr et al. 1997; Iwase et al. 1997). Early studies by Winterbourne showed monounsaturated and polyunsaturated fatty acids, either as free fatty acids or esterified in phosphatidylcholine, could be converted into chlorohydrin species (Winterbourn et al. 1992). While monounsaturated fatty acids (e.g., oleic acid) formed monochlorohydrins, polyunsaturated fatty acids (e.g., linoleic acid) formed multiple bischlorohydrins at higher relative molar concentrations of HOCl. At lower relative concentrations of HOCl predominantly monochlorohydrins are observed with linoleic acid as the precursor. Several investigators have shown HOCl targeting of cholesterol resulting in cholesterol chlorohydrin production (Carr et al. 1996, 1997; Heinecke et al. 1994; van den Berg et al. 1993). Studies by Hazen et al. also showed that dichlorinated cholesterol species could be produced by the MPO–HOCl generating system which suggested that Cl_2, which is in equilibrium with HOCl, is the responsible agent for dichlorination (Hazen et al. 1996).

It is important to understand the targets and products of HOCl modified LDL since it is known that HOCl modified LDL is a high uptake form of LDL that can lead to foam cell production, which is an early event in atherogenesis (Hazell and Stocker 1993). Although protein chlorination is considered as the primary mediator of converting LDL into a high uptake form by HOCl (Hazell and Stocker 1993), vinyl ether bonds and alkenes of phospholipids may also be targeted (Jerlich et al. 1998; Thukkani et al. 2003a). One confounding issue with MPO-catalyzed oxidation of LDL lipids is that both chlorohydrins and peroxidation products can be formed. However, using a liquid chromatography–mass spectrometry method Spickett and coworkers found chlorohydrins to be the major products of LDL oxidation by HOCl (Jerlich et al. 2000).

Chlorohydrin production occurs following consumption of HOCl by sulfhydryl, amine, and vinyl ether consumption, and chlorohydrins convert into epoxides. Accordingly, the presence of chlorohydrins *in vivo* is limited, but a couple of studies have suggested that these compounds may be very significant in inflammation. Chlorohydrins of unsaturated lysophosphatidylcholines accumulate in human atherosclerotic tissue (Messner et al. 2008), and the chlorohydrins of unsaturated lysophosphatidylcholine may enhance the tethering of pro-inflammatory leukocytes to regions of inflammation since they promote P-selectin surface expression on endothelial cells (Messner et al. 2008). Similar results have been reported for the chlorohydrin of stearoyloleoyl phosphatidylcholine (Dever et al. 2008). Other recent evidence has shown that fatty acid chlorohydrins are released from adipose tissue in an acute pancreatitis model elicited by intraductal infusion of taurocholate, which also is associated with increases in MPO levels in the pancreas, adipose tissue, and the lung (Franco-Pons et al. 2013). This release led to the appearance of micromolar levels of fatty acid chlorohydrins in the plasma and levels up to 250 µM in the ascites fluid. These chlorohydrins are from oleic and linoleic acid and they do appear to be very robust in this model. At these concentrations many of the studies examining the toxicity of chlorohydrins become physiologically important.

CONCLUSIONS AND FUTURE DIRECTIONS

While the production of hypohalous acids by leukocytes is an important mechanism in innate immunity to destroy microorganisms, targeting host tissue biomolecules may be important in disease processes. Among reactive targets within lipids, the plasmalogen vinyl ether bond is one of the most reactive leading to the production of α-halofatty aldehydes and their metabolites. Our understanding of the role of halogenated lipids in physiological and pathophysiological conditions is at the early stages, and these molecules may prove to be extremely important in the pathophysiological sequelae of diseases, including neurodegenerative diseases, asthma, atherosclerosis, and ischemic heart disease. Recent studies have shown that these halogenated lipids accumulate in some of these diseases, yet the family of halogenated lipid metabolites needs to be fully characterized and the role of these metabolites as mediators of disease remains to be understood. The targets of the different chlorinated lipids that are produced *in vivo* need to be defined. While signaling pathways for the halogenated lipids have been partially determined, it will be important to identify specific receptors for these lipids and what role they might have in physiological responses.

REFERENCES

Albert, C. J., J. R. Crowley, F. F. Hsu, A. K. Thukkani, and D. A. Ford. 2001. Reactive chlorinating species produced by myeloperoxidase target the vinyl ether bond of plasmalogens: Identification of 2-chlorohexadecanal. *J Biol Chem* 276(26):23733–41.

Albert, C. J., J. R. Crowley, F. F. Hsu, A. K. Thukkani, and D. A. Ford. 2002. Reactive brominating species produced by myeloperoxidase target the vinyl ether bond of plasmalogens: Disparate utilization of sodium halides in the production of alpha-halo fatty aldehydes. *J Biol Chem* 277(7):4694–703.

Albert, C. J., A. K. Thukkani, R. M. Heuertz, A. Slungaard, S. L. Hazen, and D. A. Ford. 2003. Eosinophil peroxidase-derived reactive brominating species target the vinyl ether bond of plasmalogens generating a novel chemoattractant, alpha-bromo fatty aldehyde. *J Biol Chem* 278(11):8942–50.

Albrich, J. M., C. A. McCarthy, and J. K. Hurst. 1981. Biological reactivity of hypochlorous acid: Implications for microbicidal mechanisms of leukocyte myeloperoxidase. *Proc Natl Acad Sci USA* 78(1):210–4.

Anbukumar, D. S., L. P. Shornick, C. J. Albert, M. M. Steward, R. A. Zoeller, W. L. Neumann, and D. A. Ford. 2010. Chlorinated lipid species in activated human neutrophils: Lipid metabolites of 2-chlorohexadecanal. *J. Lipid Res.* 51:1085–1092.

Askari, A. T., M. L. Brennan, X. Zhou, M. Geraci, J. Drinko, A. Morehead, J. D. Thomas, E. J. Topol, S. L. Hazen, and M. S. Penn. 2003. Myeloperoxidase and plasminogen activator inhibitor-1 play a central role in ventricular modeling after myocardial infarction. *J. Exp. Med.* 197:615–24.

Bergt, C., S. Pennathur, X. Fu, J. Byun, K. O'Brien, T. O. McDonald, P. Singh et al. 2004. The myeloperoxidase product hypochlorous acid oxidizes HDL in the human artery wall and impairs ABCA1-dependent cholesterol transport. *Proc Natl Acad Sci USA* 101(35):13032–7.

Berliner, J. A., and J. W. Heinecke. 1996. The role of oxidized lipoproteins in atherogenesis. *Free Radic Biol Med* 20(5):707–27.

Brahmbhatt, V. V., C. J. Albert, D. S. Anbukumar, B. A. Cunningham, W. L. Neumann, and D. A. Ford. 2010. {Omega}-oxidation of {alpha}-chlorinated fatty acids: Identification of {alpha}-chlorinated dicarboxylic acids. *J Biol Chem* 285(53):41255–69.

Brahmbhatt, V. V., F. F. Hsu, J. L. Kao, E. C. Frank, and D. A. Ford. 2007. Novel carbonyl and nitrile products from reactive chlorinating species attack of lysosphingolipid. *Chem Phys Lipids* 145(2):72–84.

Brennan, M. L., M. M. Anderson, D. M. Shih, X. D. Qu, X. Wang, A. C. Mehta, L. L. Lim et al. 2001. Increased atherosclerosis in myeloperoxidase-deficient mice. *J Clin Invest* 107(4):419–30.

Carr, A. C., J. J. van den Berg, and C. C. Winterbourn. 1996. Chlorination of cholesterol in cell membranes by hypochlorous acid. *Arch Biochem Biophys* 332(1):63–9.

Carr, A. C., J. J. van den Berg, and C. C. Winterbourn. 1998. Differential reactivities of hypochlorous and hypobromous acids with purified *Escherichia coli* phospholipid: Formation of haloamines and halohydrins. *Biochim Biophys Acta* 1392(2–3):254–64.

Carr, A. C., M. C. Vissers, N. M. Domigan, and C. C. Winterbourn. 1997. Modification of red cell membrane lipids by hypochlorous acid and haemolysis by preformed lipid chlorohydrins. *Redox Report* 3(5–6):263–71.

Carr, A. C., C. C. Winterbourn, J. W. Blunt, A. J. Phillips, and A. D. Abell. 1997. Nuclear magnetic resonance characterization of 6 alpha-chloro-5 beta-cholestane-3 beta, 5-diol formed from the reaction of hypochlorous acid with cholesterol. *Lipids* 32(4):363–7.

Chen, J. W., M. O. Breckwoldt, E. Aikawa, G. Chiang, and R. Weissleder. 2008. Myeloperoxidase-targeted imaging of active inflammatory lesions in murine experimental autoimmune encephalomyelitis. *Brain* 131(Pt 4):1123–33.

Chen, S. Y., G. Hsiao, H. R. Hwang, P. Y. Cheng, and Y. M. Lee. 2006. Tetramethylpyrazine induces heme oxygenase-1 expression and attenuates myocardial ischemia/reperfusion injury in rats. *J Biomed Sci* 13(5):731–40.

Chilton, F. H., and T. R. Connell. 1988. 1-ether-linked phosphoglycerides. Major endogenous sources of arachidonate in the human neutrophil. *J Biol Chem* 263(11):5260–5.

Chilton, F. H., and R. C. Murphy. 1986. Remodeling of arachidonate-containing phosphoglycerides within the human neutrophil. *J Biol Chem* 261(17):7771–7.

Daugherty, A., J. L. Dunn, D. L. Rateri, and J. W. Heinecke. 1994. Myeloperoxidase, a catalyst for lipoprotein oxidation, is expressed in human atherosclerotic lesions. *J Clin Invest* 94(1):437–44.

Dever, G. J., R. Benson, C. L. Wainwright, S. Kennedy and C. M. Spickett. 2008. Phospholipid chlorohydrin induces leukocyte adhesion to ApoE –/– mouse arteries via upregulation of P-selectin. *Free Radic Biol Med* 44:452–463.

Duilio, C., G. Ambrosio, P. Kuppusamy, A. DiPaula, L. C. Becker, and J. L. Zweier. 2001. Neutrophils are primary source of O2 radicals during reperfusion after prolonged myocardial ischemia. *Am J Physiol—Heart Circ Physiol.* 280(6):H2649–57.

Engler, R, and J. W. Covell. 1987. Granulocytes cause reperfusion ventricular dysfunction after 15-minute ischemia in the dog. *Circ Res* 61(1):20–28.

Ford, D. A., and R. W. Gross. 1989. Plasmenylethanolamine is the major storage depot for arachidonic acid in rabbit vascular smooth muscle and is rapidly hydrolyzed after angiotensin II stimulation. *Proc Natl Acad Sci USA* 86(10):3479–83.

Ford, D. A., and R. W. Gross. 1994. The discordant rates of sn-1 aliphatic chain and polar head group incorporation into plasmalogen molecular species demonstrate the fundamental importance of polar head group remodeling in plasmalogen metabolism in rabbit myocardium. *Biochemistry* 33(5):1216–22.

Franco-Pons, N., J. Casas, G. Fabrias, S. Gea-Sorli, E. de-Madaria, E. Gelpi, and D. Closa. 2013. Fat necrosis generates proinflammatory halogenated lipids during acute pancreatitis. *Ann Surg* 257(5):943–51.

Frossard, J. L., M. L. Steer, and C. M. Pastor. 2008. Acute pancreatitis. *Lancet* 371(9607): 143–52.

Fu, X., S. Y. Kassim, W. C. Parks, and J. W. Heinecke. 2001. Hypochlorous acid oxygenates the cysteine switch domain of pro-matrilysin (MMP-7). A mechanism for matrix metalloproteinase activation and atherosclerotic plaque rupture by myeloperoxidase. *J Biol Chem* 276(44):41279–87.

Ginsberg, L., S. Rafique, J. H. Xuereb, S. I. Rapoport, and N. L. Gershfeld. 1995. Disease and anatomic specificity of ethanolamine plasmalogen deficiency in Alzheimer's disease brain. *Brain Res* 698(1–2):223–6.

Goodenowe, D. B., L. L. Cook, J. Liu, Y. Lu, D. A. Jayasinghe, P. W. Ahiahonu, D. Heath et al. 2007. Peripheral ethanolamine plasmalogen deficiency: A logical causative factor in Alzheimer's disease and dementia. *J Lipid Res* 48(11):2485–98.

Gray, E., T. L. Thomas, S. Betmouni, N. Scolding, and S. Love. 2008. Elevated activity and microglial expression of myeloperoxidase in demyelinated cerebral cortex in multiple sclerosis. *Brain Pathol* 18(1):86–95.

Grommes, J., and O. Soehnlein. 2011. Contribution of neutrophils to acute lung injury. *Mol Med* 17(3–4):293–307.

Gross, R. W. 1984. High plasmalogen and arachidonic acid content of canine myocardial sarcolemma: A fast atom bombardment mass spectroscopic and gas chromatography-mass spectroscopic characterization. *Biochemistry* 23(1):158–65.

Gross, R. W. 1985. Identification of plasmalogen as the major phospholipid constituent of cardiac sarcoplasmic reticulum. *Biochemistry* 24(7):1662–8.

Han, X., and R. W. Gross. 1994. Electrospray ionization mass spectroscopic analysis of human erythrocyte plasma membrane phospholipids. *Proc Natl Acad Sci USA* 91(22):10635–9.

Han, X. L., and R. W. Gross. 1990. Plasmenylcholine and phosphatidylcholine membrane bilayers possess distinct conformational motifs. *Biochemistry* 29(20):4992–6.

Hansen, P. R. 1995. Role of neutrophils in myocardial ischemia and reperfusion. *Circulation* 91(6):1872–85.

Harrison, J. E., and J. Schultz. 1976. Studies on the chlorinating activity of myeloperoxidase. *J Biol Chem* 251(5):1371–4.

Hayasaki, T., K. Kaikita, T. Okuma, E. Yamamoto, W. A. Kuziel, H. Ogawa, and M. Takeya. 2006. CC chemokine receptor-2 deficiency attenuates oxidative stress and infarct size caused by myocardial ischemia–reperfusion in mice. *Circulation J* 70(3):342–51.

Hazell, L. J., L. Arnold, D. Flowers, G. Waeg, E. Malle, and R. Stocker. 1996. Presence of hypochlorite-modified proteins in human atherosclerotic lesions. *J Clin Invest* 97(6): 1535–44.

Hazell, L. J., and R. Stocker. 1993. Oxidation of low-density lipoprotein with hypochlorite causes transformation of the lipoprotein into a high-uptake form for macrophages. *Biochem J* 290(Pt 1):165–72.

Hazen, S. L., C. R. Hall, D. A. Ford, and R. W. Gross. 1993. Isolation of a human myocardial cytosolic phospholipase A₂ isoform. Fast atom bombardment mass spectroscopic and reverse-phase high pressure liquid chromatography identification of choline and ethanolamine glycerophospholipid substrates. *J Clin Invest* 91(6):2513–22.

Hazen, S. L., and J. W. Heinecke. 1997. 3-Chlorotyrosine, a specific marker of myeloperoxidase-catalyzed oxidation, is markedly elevated in low density lipoprotein isolated from human atherosclerotic intima. *J Clin Invest* 99(9):2075–81.

Hazen, S. L., F. F. Hsu, K. Duffin, and J. W. Heinecke. 1996. Molecular chlorine generated by the myeloperoxidase-hydrogen peroxide–chloride system of phagocytes converts low density lipoprotein cholesterol into a family of chlorinated sterols. *J Biol Chem* 271(38):23080–8.

Heim, Werner G., David Appleman, and H. T. Pyfrom. 1956. Effects of 3-amino-1,2,4-triazole (AT) on catalase and other compounds. *Am J Physiol* 186(1):19–23.

Heinecke, J. W., W. Li, D. M. Mueller, A. Bohrer, and J. Turk. 1994. Cholesterol chlorohydrin synthesis by the myeloperoxidase-hydrogen peroxide-chloride system: Potential markers for lipoproteins oxidatively damaged by phagocytes. *Biochemistry* 33(33):10127–36.

Henderson, J. P., J. Byun, M. V. Williams, D. M. Mueller, M. L. McCormick, and J. W. Heinecke. 2001. Production of brominating intermediates by myeloperoxidase. A transhalogenation pathway for generating mutagenic nucleobases during inflammation. *J Biol Chem* 276(11):7867–75.

Hsu, F. F., J. Turk, A. K. Thukkani, M. C. Messner, K. R. Wildsmith, and D. A. Ford. 2003. Characterization of alkylacyl, alk-1-enylacyl and lyso subclasses of glycerophosphocholine by tandem quadrupole mass spectrometry with electrospray ionization. *J Mass Spectrom* 38(7):752–63.

Iwase, H., Y. Yamada, H. Uemura, H. Nakaya, T. Takatori, M. Nagao, and K. Iwadate. 1997. Effect of monochlorohydrins of linoleic acid on guinea-pig cardiac papillary muscles. *Biochem Biophys Res Commun* 231(2):295–8.

Jaskolla, T., B. Fuchs, M. Karas, and J. Schiller. 2009. The new matrix 4-chloro-alpha-cyanocin-namic acid allows the detection of phosphatidylethanolamine chloramines by MALDI-TOF mass spectrometry. *J Am Soc Mass Spectrom* 20(5):867–74.

Jerlich, A., J. S. Fabjan, S. Tschabuschnig, A. V. Smirnova, L. Horakova, M. Hayn, H. Auer et al. 1998. Human low density lipoprotein as a target of hypochlorite generated by myeloperoxidase. *Free Radic Biol Med* 24(7–8):1139–48.

Jerlich, A., A. R. Pitt, R. J. Schaur, and C. M. Spickett. 2000. Pathways of phospholipid oxidation by HOCl in human LDL detected by LC-MS. *Free Radic Biol Med* 28(5):673–82.

Johnson, R. J., W. G. Couser, E. Y. Chi, S. Adler, and S. J. Klebanoff. 1987. New mechanism for glomerular injury. Myeloperoxidase-hydrogen peroxide-halide system. *J Clin Invest* 79(5):1379–87.

Jordan, J. E., Z.-Q. Zhao, and J. Vinten-Johansen. 1999. The role of neutrophils in myocardial ischemia–reperfusion injury. *Cardiovasc Res* 43(4):860–878.

Kato, K., Q. Shao, V. Elimban, A. Lukas, and N. S. Dhalla. 1998. Mechanism of depression in cardiac sarcolemmal Na + -K + -ATPase by hypochlorous acid. *Am J Physiol.* 275 (3 Pt 1):C826–31.

Klebanoff, S. J., A. M. Waltersdorph, and H. Rosen. 1984. Antimicrobial activity of myeloper-oxidase. *Meth Enzymol.* 105:399–403.

Kolaczkowska, E., and P. Kubes. 2013. Neutrophil recruitment and function in health and inflammation. *Nat Rev Immunol* 13(3):159–75.

Kreisel, D., R. G. Nava, W. Li, B. H. Zinselmeyer, B. Wang, J. Lai, R. Pless, A. E. Gelman, A. S. Krupnick, and M. J. Miller. 2010. *In vivo* two-photon imaging reveals monocyte-dependent neutrophil extravasation during pulmonary inflammation. *Proc Natl Acad Sci USA* 107(42):18073–8.

Kukreja, R. C., A. B. Weaver, and M. L. Hess. 1989. Stimulated human neutrophils damage cardiac sarcoplasmic reticulum function by generation of oxidants. *Biochim Biophys Acta* 990(2):198–205.

Kutter, D., P. Devaquet, G. Vanderstocken, J. M. Paulus, V. Marchal, and A. Gothot. 2000. Consequences of total and subtotal myeloperoxidase deficiency: Risk or benefit? *Acta Haematol* 104(1):10–5.

Lampert, M. B., and S. J. Weiss. 1983. The chlorinating potential of the human monocyte. *Blood* 62(3):645–51.

LeBig, J., J. Schiller, J. Arnhold, and B. Fuchs. 2007. Hypochlorous acid-mediated generation of glycerophosphocholine from unsaturated plasmalogen glycerophosphocholine lipids. *J Lipid Res* 48:1316–1324.

Litt, M. R., R. W. Jeremy, H. F. Weisman, J. A. Winkelstein, and L. C. Becker. 1989. Neutrophil depletion limited to reperfusion reduces myocardial infarct size after 90 minutes of ischemia. Evidence for neutrophil-mediated reperfusion injury. *Circulation* 80(6):1816–27.

Lucchesi, B. R. 1990. Modulation of leukocyte-mediated myocardial reperfusion injury. *Ann Rev Physiol* 52:561–76.

Malavolta, M., F. Bocci, E. Boselli, and N. G. Frega. 2004. Normal phase liquid chromatography-electrospray ionization tandem mass spectrometry analysis of phos-pholipid molecular species in blood mononuclear cells: Application to cystic fibrosis. *J Chromatogr B Analyt Technol Biomed Life Sci* 810(2):173–86.

Malle, E., G. Waeg, R. Schreiber, E. F. Grone, W. Sattler, and H. J. Grone. 2000. Immunohistochemical evidence for the myeloperoxidase/H$_2$O$_2$/halide system in human atherosclerotic lesions: Colocalization of myeloperoxidase and hypochlorite-modified proteins. *Eur J Biochem* 267(14):4495–503.

Marathe, G. K., S. M. Prescott, G. A. Zimmerman, and T. M. McIntyre. 2001. Oxidized LDL contains inflammatory PAF-like phospholipids. *Trends Cardiovasc Med.* 11(3–4):139–42.

Marik, C., P. A. Felts, J. Bauer, H. Lassmann, and K. J. Smith. 2007. Lesion genesis in a subset of patients with multiple sclerosis: A role for innate immunity? *Brain* 130(Pt 11): 2800–15.

Marsche, G., R. Heller, G. Fauler, A. Kovacevic, A. Nuszkowski, W. Graier, W. Sattler, and E. Malle. 2004. 2-Chlorohexadecanal derived from hypochlorite-modified high-density lipoprotein-associated plasmalogen is a natural inhibitor of endothelial nitric oxide biosynthesis. *Arterioscler Thromb Vasc Biol* 24(12):2302–6.

McMaster, C. R., C. Q. Lu, and P. C. Choy. 1992. The existence of a soluble plasmalogenase in guinea pig tissues. *Lipids* 27(12):945–9.

McMillen, T. S., J. W. Heinecke, and R. C. LeBoeuf. 2005. Expression of human myeloperoxidase by macrophages promotes atherosclerosis in mice. *Circulation* 111(21): 2798–804.

Mehta, J., J. Dinerman, P. Mehta, T. G. Saldeen, D. Lawson, W. H. Donnelly, and R. Wallin. 1989. Neutrophil function in ischemic heart disease. *Circulation* 79(3):549–56.

Messner, M. C., C. J. Albert, and D. A. Ford. 2008. 2-Chlorohexadecanal and 2-chlorohexadecanoic acid induce COX-2 expression in human coronary artery endothelial cells. *Lipids* 43(7):581–8.

Messner, M. C., C. J. Albert, F. F. Hsu, and D. A. Ford. 2006. Selective plasmenylcholine oxidation by hypochlorous acid: Formation of lysophosphatidylcholine chlorohydrins. *Chem Phys Lipids* 144(1):34–44.

Messner, M. C., C. J. Albert, J. McHowat, and D. A. Ford. 2008. Identification of lysophosphatidylcholine–chlorohydrin in human atherosclerotic lesions. *Lipids* 43:243–249.

Mitra, S. N., A. Slungaard, and S. L. Hazen. 2000. Role of eosinophil peroxidase in the origins of protein oxidation in asthma. *Redox Report.* 5(4):215–24.

Mocatta, T. J., A. P. Pilbrow, V. A. Cameron, R. Senthilmohan, C. M. Frampton, A. M. Richards, and C. C. Winterbourn. 2007. Plasma concentrations of myeloperoxidase predict mortality after myocardial infarction. *J Am Coll Cardiol.* 49:1993–2000.

Moraitou, M., E. Dimitriou, D. Zafeiriou, C. Reppa, T. Marinakis, J. Sarafidou, and H. Michelakakis. 2008. Plasmalogen levels in Gaucher disease. *Blood Cells Mol Dis* 41(2):196–9.

Morand, O. H., R. A. Zoeller, and C. R. Raetz. 1988. Disappearance of plasmalogens from membranes of animal cells subjected to photosensitized oxidation. *J Biol Chem* 263(23):11597–606.

Nagra, R. M., B. Becher, W. W. Tourtellotte, J. P. Antel, D. Gold, T. Paladino, R. A. Smith, J. R. Nelson, and W. F. Reynolds. 1997. Immunohistochemical and genetic evidence of myeloperoxidase involvement in multiple sclerosis. *J Neuroimmunol* 78(1–2):97–107.

Nathan, C. 2006. Neutrophils and immunity: Challenges and opportunities. *Nature Rev Immunology.* 6(3):173–82.

Nauseef, W. M., J. A. Metcalf, and R. K. Root. 1983. Role of myeloperoxidase in the respiratory burst of human neutrophils. *Blood* 61(3):483–92.

Nikpoor, B., G. Turecki, C. Fournier, P. Theroux, and G. A. Rouleau. 2001. A functional myeloperoxidase polymorphic variant is associated with coronary artery disease in French-Canadians. *Am Heart J* 142(2):336–9.

Nusshold, C., M. Kollroser, H. Kofeler, G. Rechberger, H. Reicher, A. Ullen, E. Bernhart et al. 2010. Hypochlorite modification of sphingomyelin generates chlorinated lipid species that induce apoptosis and proteome alterations in dopaminergic PC12 neurons in vitro. *Free Radic Biol Med* 48(12):1588–600.

Ockaili, R., R. Natarajan, F. Salloum, B. J. Fisher, D. Jones, A. A. Fowler, 3rd, and R. C. Kukreja. 2005. HIF-1 activation attenuates postischemic myocardial injury: Role for heme oxygenase-1 in modulating microvascular chemokine generation. *Am J Physiol-Heart Circ Physiol* 289(2):H542–8.

Okabe, E., S. Takahashi, M. Norisue, N. H. Manson, R. C. Kukreja, M. L. Hess, and H. Ito. 1993. The effect of hypochlorous acid and hydrogen peroxide on coronary flow and arrhythmogenesis in myocardial ischemia and reperfusion. *Eur J Pharmacol.* 248(1):33–9.

Pabla, R., A. J. Buda, D. M. Flynn, S. A. Blesse, A. M. Shin, M. J. Curtis, and D. J. Lefer. 1996. Nitric oxide attenuates neutrophil-mediated myocardial contractile dysfunction after ischemia and reperfusion. *Circ Res* 78(1):65–72.

Palazzo, A. J., S. P. Jones, D. C. Anderson, D. N. Granger, and D. J. Lefer. 1998a. Coronary endothelial P-selectin in pathogenesis of myocardial ischemia-reperfusion injury. *Am J Physiol-Heart Circ Physiol* 275(5 Pt 2):H1865–72.

Palazzo, A. J., S. P. Jones, W. G. Girod, D. C. Anderson, D. N. Granger, and D. J. Lefer. 1998b. Myocardial ischemia-reperfusion injury in CD18- and ICAM-1-deficient mice. *Am J Physiol-Heart Circ Physiol* 275(6 Pt 2):H2300–7.

Panneels, V., P. Macours, H. Van den Bergen, J. C. Braekman, J. Van Sande, and J. M. Boeynaems. 1996. Biosynthesis and metabolism of 2-iodohexadecanal in cultured dog thyroid cells. *J Biol Chem* 271(38):23006–14.

Panneels, V., J. Van Sande, H. Van den Bergen, C. Jacoby, J. C. Braekman, J. E. Dumont, and J. M. Boeynaems. 1994. Inhibition of human thyroid adenylyl cyclase by 2-iodoaldehydes. *Mol Cell Endocrinol* 106(1–2):41–50.

Pecoits-Filho, R., P. Stenvinkel, A. Marchlewska, O. Heimburger, P. Barany, C. M. Hoff, C. J. Holmes, M. Suliman, B. Lindholm, M. Schalling, and L. Nordfors. 2003. A functional variant of the myeloperoxidase gene is associated with cardiovascular disease in end-stage renal disease patients. *Kidney Intl Supplement* (84):S172–6.

Pennathur, S., C. Bergt, B. Shao, J. Byun, S. Y. Kassim, P. Singh, P. S. Green et al. 2004. Human atherosclerotic intima and blood of patients with established coronary artery disease contain high density lipoprotein damaged by reactive nitrogen species. *J Biol Chem* 279(41):42977–83.

Pereira, A., J. C. Braekman, J. E. Dumont, and J. M. Boeynaems. 1990. Identification of a major iodolipid from the horse thyroid gland as 2-iodohexadecanal. *J Biol Chem* 265(28):17018–25.

Persad, S., V. Elimban, F. Siddiqui, and N. S. Dhalla. 1999. Alterations in cardiac membrane beta-adrenoceptors and adenylyl cyclase due to hypochlorous acid. *J Molec Cell Cardiol* 31(1):101–11.

Pike, L. J., X. Han, K. N. Chung, and R. W. Gross. 2002. Lipid rafts are enriched in arachidonic acid and plasmenylethanolamine and their composition is independent of caveolin-1 expression: A quantitative electrospray ionization/mass spectrometric analysis. *Biochemistry* 41:2075–88.

Podrez, E. A., E. Poliakov, Z. Shen, R. Zhang, Y. Deng, M. Sun, P. J. Finton et al. 2002a. A novel family of atherogenic oxidized phospholipids promotes macrophage foam cell formation via the scavenger receptor CD36 and is enriched in atherosclerotic lesions. *J Biol Chem* 277(41):38517–23.

Podrez, E. A., E. Poliakov, Z. Shen, R. Zhang, Y. Deng, M. Sun, P. J. Finton et al. 2002b. Identification of a novel family of oxidized phospholipids that serve as ligands for the macrophage scavenger receptor CD36. *J Biol Chem* 277(41):38503–16.

Podrez, E. A., D. Schmitt, H. F. Hoff, and S. L. Hazen. 1999. Myeloperoxidase-generated reactive nitrogen species convert LDL into an atherogenic form in vitro. *J Clin Invest* 103(11):1547–60.

Post, J. A., A. J. Verkleij, B. Roelofsen, and J. A. Op de Kamp. 1988. Plasmalogen content and distribution in the sarcolemma of cultured neonatal rat myocytes. *FEBS Letters* 240(1–2):78–82.

Rizzo, W. B., D. A. Craft, A. L. Dammann, and M. W. Phillips. 1987. Fatty alcohol metabolism in cultured human fibroblasts. Evidence for a fatty alcohol cycle. *J Biol Chem* 262(36):17412–9.

Scherrer, L. A., and R. W. Gross. 1989. Subcellular distribution, molecular dynamics and catabolism of plasmalogens in myocardium. *Molec Cell Biochem* 88(1–2):97–105.

Schulz, R. 1996. Plasmalogens, NO, and the ischemic-reperfused heart. In *Advances in Lipobiology*, edited by W. G. Richard: JAI Press.

Shandelya, S. M., P. Kuppusamy, M. L. Weisfeldt, and J. L. Zweier. 1993. Evaluation of the role of polymorphonuclear leukocytes on contractile function in myocardial reperfusion injury. Evidence for plasma-mediated leukocyte activation. *Circulation* 87(2):536–46.

Sheridan, F. M., I. M. Dauber, I. F. McMurtry, E. J. Lesnefsky, and L. D. Horwitz. 1991. Role of leukocytes in coronary vascular endothelial injury due to ischemia and reperfusion. *Circ Res* 69(6):1566–1574.

Skaff, O., D. I. Pattison, and M. J. Davies. 2008. The vinyl ether linkages of plasmalogens are favored targets for myeloperoxidase-derived oxidants: A kinetic study. *Biochemistry* 47:8237–8245.

Sugiyama, S., Y. Okada, G. K. Sukhova, R. Virmani, J. W. Heinecke, and P. Libby. 2001. Macrophage myeloperoxidase regulation by granulocyte macrophage colony-stimulating factor in human atherosclerosis and implications in acute coronary syndromes. *Am J Pathol* 158(3):879–91.

Teitz, N. W. 1999. Edited by C. A. Burtis and E. R. Ashwood, In: *Teitz Textbook of Clinical Chemistry*. Philadelphia: W. B. Saunders Co. p. 2216.

Thomas, E. L., M. M. Jefferson, and M. B. Grisham. 1982. Myeloperoxidase-catalyzed incorporation of amines into proteins: Role of hypochlorous acid and dichloramines. *Biochemistry* 21(24):6299–308.

Thukkani, A. K., C. J. Albert, K. R. Wildsmith, M. C. Messner, B. D. Martinson, F. F. Hsu, and D. A. Ford. 2003a. Myeloperoxidase-derived reactive chlorinating species from human monocytes target plasmalogens in low density lipoprotein. *J Biol Chem* 278(38):36365–72.

Thukkani, A. K., F. F. Hsu, J. R. Crowley, R. B. Wysolmerski, C. J. Albert, and D. A. Ford. 2002. Reactive chlorinating species produced during neutrophil activation target tissue plasmalogens: Production of the chemoattractant, 2-chlorohexadecanal. *J Biol Chem* 277(6):3842–9.

Thukkani, A. K., B. D. Martinson, C. J. Albert, G.A. Vogler, and D. A. Ford. 2005. Neutrophil-mediated accumulation of 2-ClHDA during myocardial infarction: 2-ClHDA-mediated myocardial injury. *Am J Physiol-Heart Circ Physiol* 288:H2955–64.

Thukkani, A. K., J. McHowat, F. F. Hsu, M. L. Brennan, S. L. Hazen, and D. A. Ford. 2003b. Identification of alpha-chloro fatty aldehydes and unsaturated lysophosphatidylcholine molecular species in human atherosclerotic lesions. *Circulation* 108(25):3128–33.

Ullen, A., G. Fauler, E. Bernhart, C. Nusshold, H. Reicher, H. J. Leis, E. Malle, and W. Sattler. 2012. Phloretin ameliorates 2-chlorohexadecanal-mediated brain microvascular endothelial cell dysfunction in vitro. *Free Radic Biol Med* 53(9):1770–81.

Ullen, A., G. Fauler, H. Kofeler, S. Waltl, C. Nusshold, E. Bernhart, H. Reicher et al. 2010. Mouse brain plasmalogens are targets for hypochlorous acid-mediated modification *in vitro* and in vivo. *Free Radic Biol Med* 49(11):1655–65.

van den Berg, J. J., C. C. Winterbourn, and F. A. Kuypers. 1993. Hypochlorous acid-mediated modification of cholesterol and phospholipid: Analysis of reaction products by gas chromatography-mass spectrometry. *J Lipid Res* 34(11):2005–12.

van Leeuwen, M., M. J. Gijbels, A. Duijvestijn, M. Smook, M. J. van de Gaar, P. Heeringa, M. P. de Winther, and J. W. Tervaert. 2008. Accumulation of myeloperoxidase-positive neutrophils in atherosclerotic lesions in LDLR −/− mice. *Arterioscler Thromb Vasc Biol* 28(1):84–9.

Vance, J. E. 1990. Lipoproteins secreted by cultured rat hepatocytes contain the antioxidant 1-alk-1-enyl-2-acylglycerophosphoethanolamine. *Biochim Biophys Acta* 1045(2): 128–34.

Vasilyev, N., T. Williams, M. L. Brennan, S. Unzek, X. Zhou, J. W. Heinecke, D. R. Spitz, E. J. Topol, S. L. Hazen, and M. S. Penn. 2005. Myeloperoxidase-generated oxidants modulate left ventricular remodeling but not infarct size after myocardial infarction. *Circulation* 112(18):2812–20.

Weiss, S. J. 1989. Tissue destruction by neutrophils. *N Engl J Med* 320(6):365–76.

Weiss, S. J., R. Klein, A. Slivka, and M. Wei. 1982. Chlorination of taurine by human neutrophils. Evidence for hypochlorous acid generation. *J Clin Invest* 70(3):598–607.

Weiss, S. J., S. T. Test, C. M. Eckmann, D. Roos, and S. Regiani. 1986. Brominating oxidants generated by human eosinophils. *Science* 234(4773):200–3.

Wildsmith, K. R., C. J. Albert, D.S. Anbukumar, and D. A. Ford. 2006. Metabolism of myeloperoxidase-derived 2-chlorohexadecanal. *J Biol Chem* 281:16849–16860.

Wildsmith, K. R., C. J. Albert, F. F. Hsu, J. L.-F. Kao, and D. A. Ford. 2006. Myeloperoxidase-derived 2-chlorohexadecanal forms Schiff bases with primary amines of ethanolamine glycerophospholipids and lysine. *Chem Phys Lipids* 139:157–170.

Winterbourn, C. C., J. J. van den Berg, E. Roitman, and F. A. Kuypers. 1992. Chlorohydrin formation from unsaturated fatty acids reacted with hypochlorous acid. *Arch Biochem Biophys* 296(2):547–55.

Witztum, J. L., and D. Steinberg. 1991. Role of oxidized low density lipoprotein in atherogenesis. *J Clin Invest* 88(6):1785–92.

Wu, W., M. K. Samoszuk, S. A. Comhair, M. J. Thomassen, C. F. Farver, R. A. Dweik, M. S. Kavuru, S. C. Erzurum, and S. L. Hazen. 2000. Eosinophils generate brominating oxidants in allergen-induced asthma. *J Clin Invest* 105(10):1455–63.

Zhang, R., M. L. Brennan, Z. Shen, J. C. MacPherson, D. Schmitt, C. E. Molenda, and S. L. Hazen. 2002. Myeloperoxidase functions as a major enzymatic catalyst for initiation of lipid peroxidation at sites of inflammation. *J Biol Chem* 277(48):46116–22.

Zheng, L., B. Nukuna, M. L. Brennan, M. Sun, M. Goormastic, M. Settle, D. Schmitt et al. 2004. Apolipoprotein A-I is a selective target for myeloperoxidase-catalyzed oxidation and functional impairment in subjects with cardiovascular disease. *J Clin Invest* 114(4):529–41.

Zoeller, R. A., T. J. Grazia, P. LaCamera, J. Park, D. P. Gaposchkin, and H. W. Farber. 2002. Increasing plasmalogen levels protects human endothelial cells during hypoxia. *Am J Physiol-Heart Circ Physiol* 283(2):H671–9.

Zoeller, R. A., O. H. Morand, and C. R. Raetz. 1988. A possible role for plasmalogens in protecting animal cells against photosensitized killing. *J Biol Chem* 263(23):11590–6.

5 Formation of Nitrated Lipids and Their Biological Relevance

Andrés Trostchansky and Homero Rubbo

CONTENTS

INTRODUCTION

Nitration of biomolecules represents a key biologically relevant redox signaling and injury event. There is a wide array of chemical reactions that might lead to nitration of targets, including proteins, lipids, and DNA, linked to overproduction of reactive nitrogen species (RNS) derived from the interaction of nitric oxide ('NO) with reactive oxygen species (ROS). Since production of RNS is intimately connected with oxidative stress, this condition is referred as nitro-oxidative stress. Reactions involving synthesis of RNS are very relevant to pathophysiology since they lead to removal of the endothelial-derived relaxing factor (EDRF) from the vasculature, critical for regulation of vascular tone. In the last decade, biological nitration has also been reported to modulate lipid activity in a physiologically relevant way. In fact, nitration of aliphatic compounds has been a subject of interest in organic chemistry since the early nineteenth century. Nonetheless, it was only recently linked to the pathobiological conditions associated with inflammation and nitro-oxidative stress; alongside with the breakthrough of ˙NO identification as the EDRF came the acknowledgment that RNS participate in physiological processes involving modification of cellular targets. Lipid nitration gained biological relevance in the 1990s as a potential pathway for nitro-oxidative modification of biological molecules. In 2002, for the first time, nitrated derivatives of linoleic acid were detected in the plasma of healthy individuals (Lima et al. 2002). Moreover, increased concentrations of these nitro-derivatives were found in hyperlipidemic donors, thus reinforcing their biological relevance and positioning nitro-fatty acids (nitroalkenes, NO_2-FA) as potential footprints for nitro-oxidative damage in human disease. At this time, the current hypothesis promotes that NO_2-FA may represent redox signaling mediators that are able to modulate a variety of cell signaling pathways.

LIPID NITRATION

Nitration is classically defined as the reaction between an organic compound and a nitrating agent to introduce a nitro group ($-NO_2$) on to a carbon atom (C-nitration) or to produce nitrates (O-nitration) or nitramines (N-nitration). The nitro group most frequently substitutes a hydrogen atom. A good candidate as nitrating agent *in vivo* is nitrogen dioxide (˙NO_2), which reacts with several biological targets such as thiols, ascorbate, and urate ($k \sim 10^7$ $M^{-1}s^{-1}$), as well as unsaturated fatty acids, tryptophan and tyrosine residues ($k \sim 10^5$ $M^{-1}s^{-1}$). Sources for ˙NO_2 formation *in vivo* include (a) ˙NO autoxidation in cellular hydrophobic compartments, (b) peroxynitrite, (c) dinitrogen trioxide (N_2O_3), (d) nitrite reduction catalyzed by peroxidases, and (e) nitrite (NO_2^-) reduction at acidic pH (Radi et al. 2000).

Nitration has been studied in a great detail in proteins, since protein tyrosine nitration represents a well characterized biomarker of cell injury *in vivo*. This post-translational modification is a by-product of cell metabolism that could render proteins more susceptible to proteolytic degradation. Nitration of tyrosine involves an initial one-electron oxidation of a tyrosine residue to form tyrosyl radical—an aromatic radical cation—followed by a diffusion-controlled radical–radical termination reaction with ˙NO_2 to yield 3-nitrotyrosine (Radi 2013). Different nitration pathways can contribute to *in vivo* nitration of tyrosine. Radicals arising from the homolysis of peroxynitrite were originally postulated as the major oxidizing species leading to the formation of the tyrosyl radical in hydrophilic compartments. More recently, alternative routes such as the nitrite/ H_2O_2/heme-peroxidase and transition metal-dependent mechanisms or the reaction of tyrosine with lipid peroxyl radicals in hydrophobic compartments have been proposed (Radi 2013). Since protein nitration has been studied in detail, this chapter focuses on lipid nitration, reviewing the major chemical and biological aspects of NO_2-FA, highlighting their roles as novel endogenously produced anti-inflammatory cell signaling mediators.

MECHANISMS OF FATTY ACID NITRATION: CHEMICAL AND CELL SYNTHESIS OF NITROALKENES

RNS react with unsaturated fatty acids (UFAs) yielding NO_2-FA in human plasma, cell membranes, and tissues. These novel species represent redox anti-inflammatory signaling mediators that are able to modulate a variety of cell signaling pathways. For the last 10 years many different experimental approaches have been used to chemically synthesize NO_2-FA (Gorczynski et al. 2006; Lim et al. 2002; Napolitano et al. 2000; O'Donnell et al. 1999; Woodcock et al. 2006). Mechanisms for lipid nitration involve (a) $\cdot NO_2$ reaction with UFAs and lipid radicals: $\cdot NO_2$ can be formed from $\cdot NO$ oxidation (Radi et al. 2000) or acidic nitrite (NO_2^- Pannala et al. 2003) reacting with UFAs and lipid radicals or homolytically attacking double bonds to yield a β-nitroalkyl radical, which in turn combines with another molecule of $\cdot NO_2$ generating nitro/nitrite intermediates. Nitroalkenes are formed from these intermediates due to the loss of nitrous acid HNO_2, (Gallon and Pryor 1994; O'Donnell and Freeman 2001). Since $\cdot NO_2$ can also initiate lipid oxidation reactions, the yields of nitrated derivatives will depend on oxygen levels, predominantly NO_2-FA formation at low O_2 concentrations (Gallon and Pryor 1994; O'Donnell and Freeman 2001). (b) Chemical synthesis based on the Henry reaction (nitroaldol addition) has been used for the formation of regioisomeric derivatives of NO_2-FA (Gorczynski et al. 2006; Woodcock et al. 2006). (c) An additional lipid nitration mechanism involves peroxynitrite ($ONOO^-$ plus $ONOOH$) (Radi et al. 2000). Peroxynitrite *per se* does not react with the C–H bonds in UFAs; rather, the main route for peroxynitrite-dependent fatty acid nitration is the generation of $\cdot NO_2$ following $ONOOH$ homolysis (Botti et al. 2005; Khairutdinov et al. 2000; Radi et al. 1991; Rubbo et al. 1994, 1998). The biological chemistry of peroxynitrite is fairly complicated. In particular, pH and CO_2 redirect peroxynitrite reactivity, enhancing lipid nitration and limiting oxidation reactions (Radi et al. 2000). Since both peroxynitrite and $\cdot NO_2$ readily diffuse through the membrane bilayers, reactions leading to $\cdot NO_2$ generation may take place in the aqueous environment in proximity to the membrane or inside the lipid bilayer. In the absence of CO_2, the proton-catalyzed homolysis of peroxynitrite at the O–O bond yields $\cdot NO_2$ and hydroxyl radical ($\cdot OH$). The latter is an extremely potent oxidant capable of initiating lipid peroxidation, while $\cdot NO_2$ is capable of both oxidation and nitration of UFAs. In the presence of CO_2, peroxynitrite rapidly decomposes to $\cdot NO_2$ and carbonate radical ($CO_3^{\cdot-}$); $CO_3^{\cdot-}$ radical is also a strong oxidant but its oxidative damage is constrained to surface-exposed proteins (Botti et al. 2004), because it is unable to penetrate the biological membranes due to its anionic nature.

Nitroalkenes are present endogenously as free, esterified and nucleophilic-adducted species (Ferreira et al. 2009; Nadtochiy et al. 2009; Schopfer et al. 2009). We have demonstrated that during macrophages activation cholesteryl linoleate (CL) is nitrated at the fatty acid moiety in a process prevented by nitric oxide synthase (NOS) inhibitors, demonstrating that lipid nitration is in concert with increased inducible NOS (NOS2) expression and activity (Ferreira et al. 2009). Thus, overproduction of $\cdot NO$ could drive lipid nitration generating endogenous, secondary signaling mediators as will be discussed later. More recently, site-specific formation of NO_2-FA has been demonstrated in mitochondria from cardiac ischemia/reperfusion (Rudolph et al. 2010) or ischemic preconditioned (Nadtochiy et al. 2009) hearts.

CHARACTERIZATION AND IDENTIFICATION OF NITRO-FATTY ACIDS BY MASS SPECTROMETRY

Owing to its rapid consumption by β-oxidation or saturation of double bonds, identification and characterization of NO_2-FA by liquid chromatography-mass spectrometry (LC-MS) methods has been chosen over other methods that lose material and/ or allow alteration of structure (Woodcock et al. 2013). Analysis of chemical synthesized products and those obtained from biological samples differ in sample preparation before LC-MS analysis. In fact, for biological samples one needs to take care of the matrix used, for example, cell media, plasma or tissue, with extraction strategies which involve protein precipitation and solid-phase extraction (Salvatore et al. 2013; Woodcock et al. 2013). Precipitation with acetonitrile allows the recovery of different NO_2-FA derivatives and metabolites, which can be lost with classical organic extraction methods, although the latter method exhibits higher yields of recovery (Woodcock et al. 2013). Many strategies have been used by different groups to identify and chemically characterize NO_2-FA by LC-MS: nitro-linoleic acid (NO_2-LA), a two double bond nitroalkene, can be detected in the negative ion having a mass/charge (*m/z*) ratio of 324 (Baker et al. 2004; Lim et al. 2002; Lima et al. 2002). However, different positional isomers exhibit similar retention times under these experimental conditions. The presence of a –NO_2 group attached to the carbon chain of the fatty acid can be detected by following the neutral loss of HNO_2 (*m/z* 47) and the final structure is confirmed by infrared and nuclear magnetic resonance analysis (Baker et al. 2005; Ferreira et al. 2009; Lim et al. 2002; Trostchansky et al. 2007). A recent report of structural elucidation of the positional isomers of NO_2-LA is based on the collision-induced dissociation fragmentation of the NO_2-FA (Bonacci 2011). This method showed a complex cyclization and fragmentation of nitroalkenes to nitrile and aldehyde products characteristic of each positional isomer (Bonacci 2011). LC-MS/ MS characterization of the nitration derivatives of the conjugated linoleic acid (cLA) compared to the non-conjugated isomers has also been performed (Bonacci et al. 2012). Although both products in the negative ion mode exhibit *m/z* of 324 and neutral loss of HNO_2 (*m/z* 324/277), the chromatographic behavior of both isomers is different (Bonacci et al. 2012). In fact, reverse-phase chromatography of a mixture of these products showed that NO_2-LA elutes earlier than conjugated nitrolinoleic acid (NO_2-cLA, Bonacci et al. 2012). Experiments in the positive ion mode using lithium, previously reported for the characterization of nitroarachidonic acid (NO_2-AA) isomers (Trostchansky et al. 2007), confirmed the structure of both NO_2-LA and NO_2-cLA (Bonacci et al. 2012). The presence of lithium locks the charge at the carboxy moiety and allows the cleavage directly across the nitroalkene double bond, depending on the position of the NO_2 group, generating two specific fragments that can be detected by MS/MS in the positive mode (Figure 5.1). The fragments carrying the charge, allowing its detection by MS, corresponded to those having the carbon closer to the carboxyl group in the nitroalkene double bond (Figure 5.1) (Trostchansky et al. 2007).

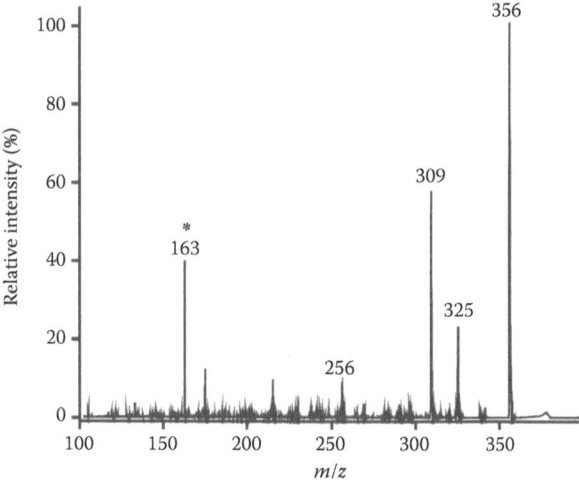

Isomer	m/z of (M + Li)+	m/z of lithium derived fragment
5-NO₂AA	356	136
6-NO₂AA	356	123
8-NO₂AA	356	176
9-NO₂AA	356	163
11-NO₂AA	356	216
12-NO₂AA	356	203
14-NO₂AA	356	256
15-NO₂AA	356	243

FIGURE 5.1 Analysis of NO₂-AA isomers by mass spectrometry using lithium. NO₂-AA isomers were incubated with lithium which blocks the positive charge at the carboxyl group allowing a specific fragmentation at the double bond where the nitro group is bonded. This yields specific ions as well as MS/MS spectrum exemplified for the 9-NO₂-AA, that can be used for isomers identification in a reaction mixture. The observed m/z in the spectrum correspond to the lithium adduct ([M + Li] +, m/z 356), the neutral loss of the nitro group (M + Li-HNO₂ +, m/z 309) and the specific fragment for 9-NO₂-AA (m/z 163). (Adapted from Trostchansky, A. et al. 2007. *Biochemistry* 46 (15):4645–53.)

ELECTROPHILICITY OF NITRO-FATTY ACIDS: CELL SIGNALING ANTI-INFLAMMATORY PROPERTIES

The highly electronegative nitro functional group facilitates reaction of the carbon β adjacent to the nitro group with nucleophilic cellular targets *via* Michael addition reversible reactions (Alexander et al. 2006; Baker et al. 2007; Batthyany et al. 2006). This rapid and reversible addition reaction with thiols and to a lesser extent primary and secondary amines (i.e., Cys or His residues) is of biological importance for cell signaling (Alexander et al. 2006; Batthyany et al. 2006; Schopfer et al. 2009). Nitroalkenes react with thiolates (i.e., GS−), and the reversion of this reaction is due to the presence of other thiols that result in transalkylation reactions of these nitroalkenes yielding the end of the signaling action (Baker et al. 2007; Schopfer et al. 2009).

Nitroalkylation of different transcriptional regulatory proteins at specific thiols residues affect downstream gene expression and the metabolic and inflammatory responses under their regulation. One of the most studied transcriptional factors activated by NO_2-FA is the peroxisome proliferator activated receptor (PPAR). In the vasculature, PPARγ expressed in monocytes/macrophages, smooth muscle cells and endothelium regulates energy balance and adipogenesis (Asakawa et al. 2002; Baker et al. 2005; Rosen and Spiegelman 2001; Schopfer et al. 2005; Tontonoz et al. 1998). For example, nitroalkenes modulate the expression of multiple PPAR target genes in human aortic smooth muscle cells (Li et al. 2008; Schopfer et al. 2005; Villacorta et al. 2009) exhibiting greater reactivity and effects on PPARγ than PPARα and PPARβ/δ (Cui et al. 2006). Activation of this receptor by NO_2-FA regulates the expression of several pro-inflammatory cytokines and chemokines in activated macrophages. Cys^{285} in PPARγ is nitroalkylated by NO_2-FA and crystallization of this adduct has been reported previously (Li et al. 2008). The biological relevance of the regulation of PPARγ was determined when nitrooleic acid (NO_2-OA) was administrated to *ob/ob* mice modulating insulin and glucose levels without inducing adverse side effects (Schopfer et al. 2010).

Other mechanisms, apart from PPARγ modulation, have been reported and proposed to explain the observed protective effects exerted by nitroalkenes. An important pro-inflammatory pathway regulated by NO_2-FA is the translocation to the nucleus of the nuclear factor-kappa B (NF-κB) transcription factor (Cui et al. 2006; Rudolph et al. 2010). NF-κB remains inactive in the cytosol bounded to the IκB repressor; both NO_2-LA and NO_2-OA modulate the release of the transcription factor by nitroalkylating the NF-κB-p65 protein subunit leading to a decrease on lipopolysaccharide-induced secretion of pro-inflammatory cytokines in macrophages (e.g. IL-6, TNFα, MCP-1). Similar mechanisms were reported for the inhibition of the expression of the vascular cell adhesion molecule 1 (VCAM-1) as well as monocyte rolling and adhesion (Cui et al. 2006).

Nitroalkenes not only act by downregulating the expression of pro-inflammatory mediators; of highly biological importance, expression of phase II antioxidants enzymes are activated by NO_2-FA (Ferreira et al. 2009; Kansanen et al. 2009, 2011; Wright et al. 2006, 2009) (Figure 5.2). Hemeoxygenase-1 (HO-1), responsible for the oxidative degradation of heme thus exerting antioxidant and anti-inflammatory

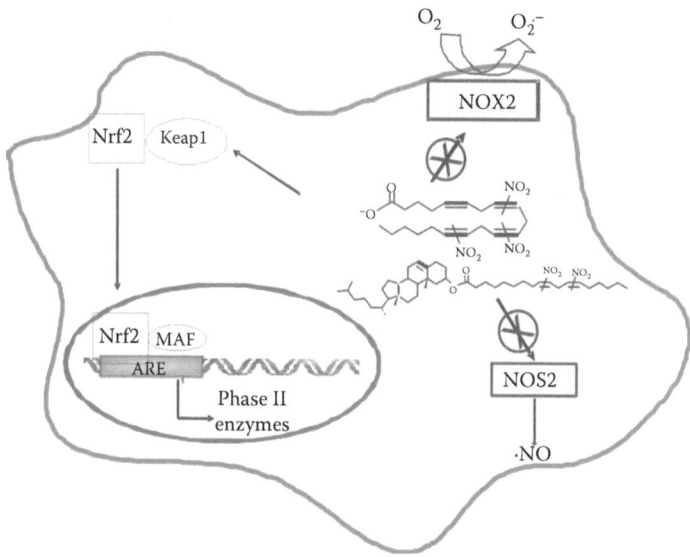

FIGURE 5.2 Anti-inflammatory effects of nitro-fatty acids. Free that is, nitroarachi-donic acid (NO$_2$-AA) as well as esterified, that is, cholesteryl-nitrolinoleic acid (NO$_2$-CL) nitroalkenes downregulate NOS2 expression and activate Nrf2 pathway and phase II enzymes expression in activated macrophages. In addition, NO$_2$-AA decreases superoxide formation by NOX2 preventing the correct membrane assembly of the active enzyme.

actions, is an endogenous cytoprotective pathway triggered by a variety of stress-related signals and electrophilic species (Foresti and Motterlini 1999; Sarady et al. 2002). In a wide variety of cell models (Cui et al. 2006; Ferreira et al. 2009; Wright et al. 2006), nitroalkenes induce HO-1 expression by PPARγ-independent and both ˙NO-dependent and -independent mechanisms (Wright et al. 2006). Phase II enzymes expression such as HO-1 are regulated by the transcription factor nuclear factor E2-related factor 2 (Nrf2) (Dinkova-Kostova et al. 2002; Gao et al. 2007; Hong et al. 2005). Inactive Nrf2 turns over rapidly in the cytosol due to its interaction with Keap1/Cul3, which promotes Nrf2 proteosomal degradation. Electrophiles, including NO$_2$-FA, alkylates critical Cys residues in Keap1 allowing Nrf2 to escape degradation and migrate to the nucleus where it binds as an heterodimer to the antioxidant response element (ARE) in DNA, activating the expression of phase 2 enzymes (Dinkova-Kostova et al. 2002; Gao et al. 2007; Hong et al. 2005; Jyrkkanen et al. 2008; Kansanen et al. 2009, 2011; Levonen et al. 2007) (Figure 5.2). Although diverse studies show that Cys[151] in Keap1 is an electrophile sensor residue (Eggler et al. 2009; Li et al. 2008; Yamamoto et al. 2008), NO$_2$-OA is a Cys[151]-independent Nrf2 activator (Kansanen et al. 2011). Moreover, nitroalkenes increase the intracellular antioxidant capacity by induction of glutamate–cysteine ligase *via* the Nrf2/ARE pathway, thus increasing glutathione (GSH) levels (unpublished results). In addition, NO$_2$-OA is induced to a great extent by many heat-shock transcription factor-regulated heat-shock genes (Kansanen et al. 2009). This mechanism can be viewed as a novel anti-inflammatory and cytoprotective action of NO_2-FA in addition to the other protective cell signaling functions reported for nitroalkenes.

The serine/threonine kinase, AMP-activated protein kinase (AMPK) acts as a cellular energy sensor by both stimulating catabolic and inhibiting anabolic pathways, respectively (Kahn et al. 2005). Phosphorylation of a threonine residue (Thr172) is followed by enzyme activation; in addition to its reported role in the regulation of metabolism it has been proposed that AMPK maintains endothelial function through regulation of ˙NO bioavailability, decreasing ROS generation as well as modulating vascular tone (Freeman et al. 2008; Zou and Wu 2008). Recently, it was demonstrated that NO_2-FA, in particular NO_2-OA, activate AMPK through an HO-1/hypoxia-inducible factor-1α/Ca^{+2} pathway promoting the phosphorylation of the endothelial NOS (Wu et al. 2012).

MODULATION OF THE PRODUCTION OF REACTIVE SPECIES: NITRIC OXIDE SYNTHASE AND NADPH OXIDASE

Arachidonic acid (AA) and ˙NO signaling pathways are intrinsically related (Blanchard-Fillion et al. 2001; Balazy et al. 2001; Di Rosa et al. 1996; Goodwin et al. 1999; Marnett 2002; Rouzer and Marnett 2003; Salvemini et al. 1995, 1996). We propose that nitration of AA could divert the fatty acid from its normal metabolic pathway with unknown cell signaling consequences. One of the demonstrated anti-inflammatory actions of NO_2-FA in activated macrophages is the downregulation of the expression of nitric oxide synthase 2 (NOS2, Figure 5.2). This effect was also shown for NO_2-CL (Ferreira et al. 2009). The studies using cycloheximide show that the effect exerted by NO_2-FA was at the transcription level without any effect on the enzyme activity (Ferreira et al. 2009; Trostchansky et al. 2007). We have recently reported that NO_2-AA is an almost irreversible inhibitor of the inducible isoform of the prostaglandin endoperoxide H synthase (PGHS-2) (Trostchansky et al. 2011) (Figure 5.3), which in addition to the downregulation of NOS2 under inflammatory stimulus, should contribute to the physiological shut down of inflammatory responses in macrophages. In endothelial cells, NO_2-OA activation of AMPK leads to the phosphorylation of eNOS-Ser1177 thus increasing the production of ˙NO (Wu et al. 2012). This effect is of biological relevance giving NO_2-OA anti-inflammatory and antiatherogenic vascular effects. However, proangiogenic effects reported for NO_2-FA are independent of endothelial NOS activation (Rudnicki et al. 2011).

How do NO_2-FA affect intracellular ROS production? Macrophages act as the first line of defense by detecting infectious agents through membrane receptors, initiating signaling pathways leading to a wide variety of cellular responses (Alvarez et al. 2002, 2011). Superoxide anion ($O_2^{\cdot-}$) formation due to the NADPH-dependent univalent reduction of oxygen by the phagocytic NADPH oxidase isoform (NOX2) is highly relevant (Alvarez et al. 2002, 2011; Cathcart 2004). Superoxide production is accomplished after cell activation, due to the correct assembly in the membrane of the membrane-bound flavocytochrome b_{558} (gp91[phox] and p22[phox]), three cytosolic subunits (p47[phox], p40[phox], and p67[phox]) and GTPase Rac2 (Dinauer et al. 1990; Juliet et al. 2003; Sheppard et al. 2005; Vignais 2002; Wientjes et al. 1997). Activation promotes phosphorylation by protein kinase C (PKC) (Park et al. 2001;

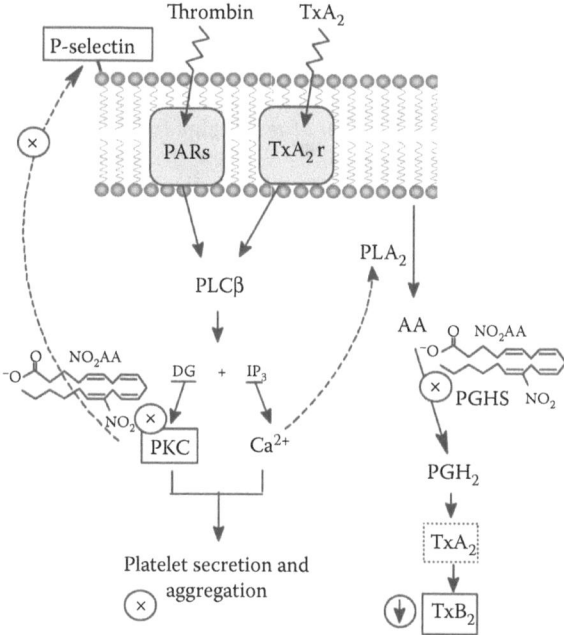

FIGURE 5.3 Modulation of human platelet signaling by NO$_2$-AA. When added exogenously, platelet aggregation as well as integrin translocation to the membrane is modulated by NO$_2$-AA. Mechanisms involved include inhibition of PGHS-1 and PKCα transient translocation to the membrane associated with PKCα activation.

Waite et al. 1997) and migration to the membrane of the cytosolic subunits, which then associate with gp91^{phox} and p22^{phox} forming the active enzyme complex able to generate O$_2^{\cdot-}$ (Ambruso et al. 2004; Azumi et al. 1999; Doussiere and Vignais 1992; Gorlach et al. 2000; Juliet et al. 2003). It has been suggested that there is a strong relationship between AA pathway and NOX2 assembly/activation (Chang et al. 2009; Shiose and Sumimoto 2000). This led us to investigate if NO$_2$-AA has any effect on O$_2^{\cdot-}$ formation in activated macrophages by inhibiting NOX2 (Gonzalez-Perilli et al. 2013). In this way, we have demonstrated the inhibition of phagocytic NOX2 activity by NO$_2$-AA both in activated macrophages and *in vivo* (Gonzalez-Perilli et al. 2013) (Figure 5.2). NOX2 activity was analyzed by direct formation of O$_2^{\cdot-}$ or its dismutation product hydrogen peroxide (H$_2$O$_2$). Importantly, the inhibitory effects of NO$_2$-AA were not observed in macrophages from gp91^{phox} −/− mice confirming that the effects were due to the involvement of NOX2. Inhibition of O$_2^{\cdot-}$ formation by NO$_2$-AA was observed both in macrophage cell lines and in primary macrophages (Gonzalez-Perilli et al. 2013). The capacity of NO$_2$-AA to regulate NOX2 was also studied in mice injected with thioglycolate, which elicits an inflammatory response including macrophage activation. As expected, macrophages obtained from NO$_2$-AA treated mice exhibited lower amounts of O$_2^{\cdot-}$ compared to nontreated animals further supporting a strong anti-inflammatory role (Gonzalez-Perilli et al. 2013).

NITRIC OXIDE AND ARACHIDONIC ACID SIGNALING ARE LINKED THROUGH
NITRO-FATTY ACIDS: MODULATION OF PROSTAGLANDIN ENDOPEROXIDE H SYNTHASE

Prostaglandin endoperoxide H synthase (PGHS) is a heme-key enzyme of AA
metabolism, catalyzing the formation of prostaglandin H_2 (PGH_2) whose final des-
tination depends on cell type (Marnett et al. 1999; Rouzer and Marnett 2003; Smith
et al. 2000; van der Donk et al. 2002). Two isoforms of PGHS (PGHS-1 and -2) are
found in mammalian tissues being targets for non-steroidal anti-inflammatory drugs
(Malkowski et al. 2000). The enzyme has two separate active sites, one of them
catalyzing the oxidation of AA to the PGG_2 hydroperoxide (cyclooxygenase [COX]
reaction) whereas the other reduces the hydroperoxyl group at C15 by the peroxidase
(POX) reaction yielding the final product PGH_2 (Rouzer and Marnett 2003; Smith
and Song 2002; van der Donk et al. 2002). The nitrated derivative of AA is not a
substrate for PGHS but is able to exert POX inhibition in a time- and concentration-
dependent manner in both PGHS-1 and 2 (Trostchansky et al. 2011). Kinetic analysis
suggests a two-step mechanism of inactivation: an initial reversible binding followed
by a practically irreversible event leading to an inactivated enzyme. Importantly,
a disruption of heme moiety interaction with the protein was observed concomitant to inactivation. These inhibitory effects were selective for NO_2-AA since other
nitroalkenes had no effect (Trostchansky et al. 2011). When analyzed in a cell model
where PGHS activity is highly relevant, for example, activated human platelets,
NO_2-AA significantly decreased PGHS-1-dependent thromboxane-B_2 formation
(Trostchansky et al. 2011), supporting a novel vascular protective role (Figure 5.3).

IN VIVO DETECTION AND THERAPEUTIC POTENTIAL
OF NITRATED LIPIDS

While the biochemical mechanisms described for lipid nitration are still under discussion, different labs showed unambiguous evidence of NO_2-FA formation *in vivo*
as well as their increase during inflammatory conditions (Balazy et al. 2001; Bonacci
et al. 2012; Ferreira et al. 2009; Lima et al. 2002; Salvatore et al. 2013; Tsikas et al.
2009). In the last 10 years, the reported concentrations fluctuated between the nanomolar (Tsikas et al. 2009) to the low micromolar range (Baker et al. 2005) all of
them capable of exerting biological actions, including inhibition of platelet and macrophage activation (Coles et al. 2002; Trostchansky et al. 2007), pro-inflammatory
cytokine secretion (Schopfer et al. 2005) and vascular smooth muscle cell proliferation (Villacorta et al. 2007).

There are emerging data on the therapeutic potential of NO_2-FA. We recently
explored a model of inflammation based on thioglycolate injection into mice peri-
toneum (Zhang et al. 2008), evaluating the effects of NO_2-AA administration on
macrophage activation. Subcutaneous injection of low-micromolar concentrations
of NO_2-AA decreased macrophage activation, indicating that NO_2-AA is capable
to exert anti-inflammatory effects *in vivo*. There are several reports using NO_2-FA
as pharmacological modulators of inflammatory-related diseases in animal models (Nadtochiy et al. 2009; Rudolph et al. 2010; Zhang et al. 2010). Of interest,
NO_2-OA subcutaneous administration to Ang-II-treated mice significantly lowered

the increase in blood pressure as well as the contractile response to Ang-II in mesenteric arteries (Zhang et al. 2010). Nitrooleic acid binds to the AT1R modulating intracellular signaling cascades (inositol-1,4,5-trisphosphate and calcium mobilization) (Zhang et al. 2010). Overall, the reported results show that NO_2-OA diminishes the pressor response to Ang II, inhibiting AT1R-dependent vasoconstriction and suggesting that NO_2-OA can be a pharmacological relevant modulator of Ang II-induced hypertension (Zhang et al. 2010). Nitroalkenes were also tested in C57/BL6 mice subjected to coronary artery ligation followed by 30 min reperfusion (I/R). Under these experimental conditions, both NO_2-OA and NO_2-LA were formed. When administered exogenously during the ischemic episode, NO_2-OA was able to exert protection against I/R injury reducing the infarct size as well as preserving the left ventricular function (Rudolph et al. 2010). The proposed mechanisms involve NO_2-OA-mediated inactivation of the p65 subunit of NFκB in I/R tissue as well as to suppression of downstream intercellular adhesion molecule 1 (ICAM-1), monocyte chemotactic protein 1 (MCP-1), neutrophil infiltration and myocyte apoptosis (Rudolph et al. 2010). The same group of researchers evaluated the capacity of NO_2-FA to modulate atherosclerosis, a chronic inflammatory disease (Rudolph et al. 2010). Subcutaneous administration of NO_2-OA potently reduced atherosclerotic lesion formation in apolipoprotein E-deficient mice (Rudolph et al. 2010). Atherosclerotic lesions of NO_2-OA-treated animals showed an increased content of collagen and smooth muscle actin, suggesting conferral of higher plaque stability. Overall, the results reveal the antiatherogenic potential of electrophilic NO_2-FA.

Reactive lipid species including NO_2-FA participate in several physiological pathways due to their electrophilic capacity through modification of specific signaling proteins (Higdon et al. 2012). The formation of covalent adducts with proteins give particular characteristics to the signaling capacity of NO_2-FA, leading the alteration of a signal over time. Moreover, the reversibility of the covalent reaction can transform NO_2-FA into important intracellular mediators (Schopfer et al. 2009). In fact, low electrophile concentrations may accumulate over time (Higdon et al. 2012) leading to persistent signaling while higher concentrations should activate a wide array of signaling cascades. The signaling activity of NO_2-FA is different than other electrophiles (e.g., 4 hydroxy-nonenal, HNE), as it has not yet been demonstrated that nitroalkylation involves protein ubiquitination and degradation as HNE does (Forman et al. 2008). Although beneficial effects of NO_2-FA in different *in vivo* models are clearly demonstrated (Cui et al. 2006; Nadtochiy et al. 2009; Rudolph et al. 2010; Schopfer et al. 2009; Zhang et al. 2010) there are still no reports evaluating the potential toxicity that these compounds could exert when administered for longer periods of time. The manner of NO_2-FA administration should have also effects in the levels reached *in vivo* (i.e., i/p. versus subcutaneous) with the concomitant signaling and biological effects that may vary.

POTENTIAL PITFALLS

The main pitfall that has been described for the formation, detection, and quantitation of NO_2-FA *in vivo* is related to the use of particular nitroalkene isoforms as standards. Thus, other structural possibilities for NO_2-FA as well as different positional isomers that should be present may be underestimated. Recently, β-oxidation as well as keto-derivatives of different NO_2-FA were reported (Rudolph and Freeman 2009; Rudolph et al. 2009; Schopfer et al. 2009, 2012) that can be underestimated. Another issue is that the capacity to bind thiol residues in proteins as well as being in their free and esterified forms concomitant to the release and stability of NO_2-FA in different compartments could affect their quantification (Schopfer et al. 2005). Thus, detection and quantitation of nitrated lipids *in vivo* is complex, being determined by the biological environment and abundance of target molecules. Even with the presence of these limitations, we can consider NO_2-FA at nanomolar concentration able to mediate potent signaling transduction cascades. Thus, it is possible that during inflammatory conditions, nitrated lipids may serve not only as biomarkers of pathophysiological processes but also as protective agents diminishing the pro-inflammatory effects of oxidant exposure.

ABBREVIATIONS

AA	Arachidonic acid
AMPK	AMP-activated protein kinase
Ang-II	Angiotensin-II
AT_1	Angiotensin-II type I receptor
CL	Cholesteryl linoleate
cLA	Conjugated linoleic acid
CO_3^{-}	Carbonate radical
COX	Cyclooxygenase
EDRF	Endothelial-derived relaxing factor
GSH	Glutathione
HNE	4 hydroxy-nonenal
HO-1	Hemoxygenase 1
H_2O_2	Hydrogen peroxide
ICAM-1	Intercellular adhesion molecule 1
LA	Linoleic acid
LC-MS	Liquid chromatography-mass spectrometry
MCP-1	Monocyte chemotactic protein 1
NF-κB	Nuclear factor-kappa B
˙NO	Nitric oxide
NO_2^{-}	Nitrite
˙NO_2	Nitrogen dioxide
NO_2-AA	Nitro-arachidonic acid
NO_2-CL	Cholesteryl nitro-linoleate
NO_2-cLA	Conjugated nitrolinoleic acid
NO_2-FA	Nitro fatty acids
NO_2-LA	Nitro-linoleic acid
NO_2-OA	Nitro-oleic acid
NOS	Nitric oxide synthase
NOS2	Inducible nitric oxide synthase

NOX2   Phagocytic NADPH oxidase
Nrf2   Nuclear factor E2-related factor 2
$O_2^{\cdot-}$ Superoxide anion
OA   Oleic acid
$\cdot$OH   Hydroxyl radical
$ONOO^-$ Peroxynitrite
ONOOH   Peroxynitrous acid
PGHS   Prostaglandin endoperoxide H synthase
PKC   Protein Kinase C
POX   Peroxidase activity
PPAR   Peroxisome proliferator activated receptor
PPARγ   Peroxisome-proliferator activated receptor gamma
RNS   Reactive nitrogen species
ROS   Reactive oxygen species
UFA   Unsaturated fatty acid
VCAM-1   Vascular cell adhesion molecule 1

ACKNOWLEDGMENTS

This work was supported by grants from Fondo Clemente Estable-ANII (FCE_6353-Uruguay) and Fundación Mapfre (Spain) to AT, ICGEB (Italy) and CSIC-Uruguay to HR.

AUTHOR DISCLOSURE STATEMENT

None of the authors has any conflict of interest.

REFERENCES

Alexander, R. L., D. J. Bates, M. W. Wright, S. B. King, and C. S. Morrow. 2006. Modulation of nitrated lipid signaling by multidrug resistance protein 1 (MRP1): Glutathione conjugation and MRP1-mediated efflux inhibit nitrolinoleic acid-induced, PPARgamma-dependent transcription activation. *Biochemistry* 45 (25):7889–96.

Alvarez, M. N., G. Peluffo, L. Piacenza, and R. Radi. 2011. Intraphagosomal peroxynitrite as a macrophage-derived cytotoxin against internalized *Trypanosoma cruzi*: Consequences for oxidative killing and role of microbial peroxiredoxins in infectivity. *J Biol Chem* 286 (8):6627–40.

Alvarez, M. N., M. Trujillo, and R. Radi. 2002. Peroxynitrite formation from biochemical and cellular fluxes of nitric oxide and superoxide. *Methods Enzymol* 359:353–66.

Ambruso, D. R., N. Cusack, and G. Thurman. 2004. NADPH oxidase activity of neutrophil specific granules: Requirements for cytosolic components and evidence of assembly during cell activation. *Mol Genet Metab* 81 (4):313–21.

Asakawa, M., H. Takano, T. Nagai et al. 2002. Peroxisome proliferator-activated receptor {gamma} plays a critical role in inhibition of cardiac hypertrophy *in vitro* and in vivo. *Circulation* 105 (10):1240–46.

Azumi, H., N. Inoue, S. Takeshita et al. 1999. Expression of NADH/NADPH oxidase p22phox in human coronary arteries. *Circulation* 100 (14):1494–8.

Baker, L. M., P. R. Baker, F. Golin-Bisello et al. 2007. Nitro-fatty acid reaction with glutathione and cysteine. Kinetic analysis of thiol alkylation by a Michael addition reaction. *J Biol Chem* 282 (42):31085–93.

Baker, P. R., Y. Lin, F. J. Schopfer et al. 2005. Fatty acid transduction of nitric oxide signaling: Multiple nitrated unsaturated fatty acid derivatives exist in human blood and urine and serve as endogenous peroxisome proliferator-activated receptor ligands. *J Biol Chem* 280 (51):42464–75.

Baker, P. R., F. J. Schopfer, S. Sweeney, and B. A. Freeman. 2004. Red cell membrane and plasma linoleic acid nitration products: Synthesis, clinical identification, and quantitation. *Proc Natl Acad Sci USA* 101 (32):11577–82.

Balazy, M., T. Iesaki, J. L. Park et al. 2001. Vicinal nitrohydroxyeicosatrienoic acids: Vasodilator lipids formed by reaction of nitrogen dioxide with arachidonic acid. *J Pharmacol Exp Ther* 299 (2):611–9.

Batthyany, C., F. J. Schopfer, P. R. Baker et al. 2006. Reversible post-translational modification of proteins by nitrated fatty acids in vivo. *J Biol Chem* 281 (29):20450–63.

Blanchard-Fillion, B., J. M. Souza, T. Friel et al. 2001. Nitration and inactivation of tyrosine hydroxylase by peroxynitrite. *J Biol Chem* 276 (49):46017–23.

Bonacci, G., Asciutto E.K., Woodcock S.R., Salvatore S.R., Freeman B.A., and Schopfer F.J. 2011. Gas-phase fragmentation analysis of nitro fatty acids. *J. Am. Soc. Mass Spectrom.* 22 (9):1534–51.

Bonacci, G., P. R. Baker, S. R. Salvatore et al. 2012. Conjugated linoleic acid is a preferential substrate for fatty acid nitration. *J Biol Chem* 287 (53):44071–82.

Botti, H., C. Batthyany, A. Trostchansky et al. 2004. Peroxynitrite-mediated alpha-tocopherol oxidation in low-density lipoprotein: A mechanistic approach. *Free Radic Biol Med* 36 (2):152–62.

Botti, H., A. Trostchansky, C. Batthyany, and H. Rubbo. 2005. Reactivity of peroxynitrite and nitric oxide with LDL. *IUBMB Life* 57 (6):407–12.

Cathcart, M. K. 2004. Regulation of superoxide anion production by NADPH oxidase in monocytes/macrophages: contributions to atherosclerosis. *Arterioscler Thromb Vasc Biol* 24 (1):23–8.

Coles, B., Allison B., Stephen R. C. et al. 2002. Nitrolinoleate inhibits superoxide generation, degranulation, and integrin expression by human neutrophils: Novel anti-inflammatory properties of nitric oxide-derived reactive species in vascular cells. *Circ Res* 91 (5):375–381.

Cui, T., F. J. Schopfer, J. Zhang et al. 2006. Nitrated fatty acids: Endogenous anti-inflammatory signaling mediators. *J Biol Chem* 281 (47):35686–98.

Chang, L. C., R. H. Lin, L. J. Huang et al. 2009. Inhibition of superoxide anion generation by CHS-111 via blockade of the p21-activated kinase, protein kinase B/Akt and protein kinase C signaling pathways in rat neutrophils. *Eur J Pharmacol* 615 (1–3):207–17.

Di Rosa, M., A. Ialenti, A. Ianaro, and L. Sautebin. 1996. Interaction between nitric oxide and cyclooxygenase pathways. *Prostaglandins Leukot Essent Fatty Acids* 54 (4):229–38.

Dinauer, M. C., E. A. Pierce, G. A. Bruns, J. T. Curnutte, and S. H. Orkin. 1990. Human neutrophil cytochrome b light chain (p22-phox). Gene structure, chromosomal location, and mutations in cytochrome-negative autosomal recessive chronic granulomatous disease. *J Clin Invest* 86 (5):1729–37.

Dinkova-Kostova, A. T., W. D. Holtzclaw, R. N. Cole et al. 2002. Direct evidence that sulfhydryl groups of Keap1 are the sensors regulating induction of phase 2 enzymes that protect against carcinogens and oxidants. *Proc Natl Acad Sci USA* 99 (18):11908–13.

Doussiere, J., and P. V. Vignais. 1992. Diphenylene iodonium as an inhibitor of the NADPH oxidase complex of bovine neutrophils. Factors controlling the inhibitory potency of diphenylene iodonium in a cell-free system of oxidase activation. *Eur J Biochem* 208 (1):61–71.

Eggler, A. L., E. Small, M. Hannink, and A. D. Mesecar. 2009. Cul3-mediated Nrf2 ubiquitination and antioxidant response element (ARE) activation are dependent on the partial molar volume at position 151 of Keap1. *Biochem J* 422 (1):171–80.

Ferreira, A. M., M. I. Ferrari, A. Trostchansky et al. 2009. Macrophage activation induces formation of the anti-inflammatory lipid cholesteryl-nitrolinoleate. *Biochem J* 417 (1):223–34.

Foresti, R., and R. Motterlini. 1999. The heme oxygenase pathway and its interaction with nitric oxide in the control of cellular homeostasis. *Free Radic Res* 31 (6):459–75.

Forman, H. J., J. M. Fukuto, T. Miller et al. 2008. The chemistry of cell signaling by reactive oxygen and nitrogen species and 4-hydroxynonenal. *Arch Biochem Biophys* 477 (2):183–95.

Freeman, B. A., P. R. Baker, F. J. Schopfer et al. 2008. Nitro-fatty acid formation and signaling. *J Biol Chem* 283 (23):15515–9.

Gallon, A. A., and W. A. Pryor. 1994. The reaction of low levels of nitrogen dioxide with methyl linoleate in the presence and absence of oxygen. *Lipids* 29 (3):171–6.

Gao, L., J. Wang, K. R. Sekhar et al. 2007. Novel n-3 fatty acid oxidation products activate Nrf2 by destabilizing the association between Keap1 and Cullin3. *J Biol Chem* 282 (4):2529–37.

Gonzalez-Perilli, L., M. N. Alvarez, C. Prolo et al. 2013. Nitroarachidonic acid prevents NADPH oxidase assembly and superoxide radical production in activated macrophages. *Free Radic Biol Med* 58:126–33.

Goodwin, D. C., L. M. Landino, and L. J. Marnett. 1999. Effects of nitric oxide and nitric oxide-derived species on prostaglandin endoperoxide synthase and prostaglandin biosynthesis. *Faseb J* 13 (10):1121–36. 3: Goodwin DC et al. Reactions of prostaglandin en ... [PMID: 10065376] Related Articles, Links.

Gorczynski, M. J., J. Huang, and S. B. King. 2006. Regio- and stereospecific syntheses and nitric oxide donor properties of (E)-9- and (E)-10-nitrooctadec-9-enoic acids. *Org Lett* 8 (11):2305–8.

Gorlach, A., R. P. Brandes, K. Nguyen et al. 2000. A gp91phox containing NADPH oxidase selectively expressed in endothelial cells is a major source of oxygen radical generation in the arterial wall. *Circ Res* 87 (1):26–32.

Higdon, A., A. R. Diers, J. Y. Oh, A. Landar, and V. M. Darley-Usmar. 2012. Cell signalling by reactive lipid species: New concepts and molecular mechanisms. *Biochem J* 442 (3):453–64.

Hong, F., K. R. Sekhar, M. L. Freeman, and D. C. Liebler. 2005. Specific patterns of electrophile adduction trigger Keap1 ubiquitination and Nrf2 activation. *J Biol Chem* 280 (36):31768–75.

Juliet, P. A., T. Hayashi, A. Iguchi, and L. J. Ignarro. 2003. Concomitant production of nitric oxide and superoxide in human macrophages. *Biochem Biophys Res Commun* 310 (2):367–70.

Jyrkkanen, H. K., E. Kansanen, M. Inkala et al. 2008. Nrf2 regulates antioxidant gene expression evoked by oxidized phospholipids in endothelial cells and murine arteries in vivo. *Circ Res* 103 (1):e1–9.

Kahn, B. B., T. Alquier, D. Carling, and D. G. Hardie. 2005. AMP-activated protein kinase: Ancient energy gauge provides clues to modern understanding of metabolism. *Cell Metab* 1 (1):15–25.

Kansanen, E., G. Bonacci, F. J. Schopfer et al. 2011. Electrophilic nitro-fatty acids activate NRF2 by a KEAP1 cysteine 151-independent mechanism. *J Biol Chem* 286 (16):14019–27.

Kansanen, E., H. K. Jyrkkanen, O. L. Volger et al. 2009. Nrf2-dependent and -independent responses to nitro-fatty acids in human endothelial cells: Identification of heat shock response as the major pathway activated by nitro-oleic acid. *J Biol Chem* 284 (48):33233–41.

Khairutdinov, R. F., J. W. Coddington, and J. K. Hurst. 2000. Permeation of phospholipid membranes by peroxynitrite. *Biochemistry* 39 (46):14238–49.

Levonen, A. L., Matias, I., Tommi, H. et al. 2007. Nrf2 gene transfer induces antioxidant enzymes and suppresses smooth muscle cell growth *in vitro* and reduces oxidative stress in rabbit aorta *in vivo*. *Arteriosclerosis, Thrombosis, and Vascular Biology* 27 (4):741–747.

Li, L., M. Kobayashi, H. Kaneko et al. 2008. Molecular evolution of Keap1. Two Keap1 molecules with distinctive intervening region structures are conserved among fish. *J Biol Chem* 283 (6):3248–55.

Li, Y., J. Zhang, F. J. Schopfer et al. 2008. Molecular recognition of nitrated fatty acids by PPAR gamma. *Nat Struct Mol Biol* 15 (8):865–7.

Lim, D. G., S. Sweeney, A. Bloodsworth et al. 2002. Nitrolinoleate, a nitric oxide-derived mediator of cell function: Synthesis, characterization, and vasomotor activity. *Proc Natl Acad Sci USA* 99 (25):15941–6.

Lima, E. S., P. Di Mascio, H. Rubbo, and D. S. Abdalla. 2002. Characterization of linoleic acid nitration in human blood plasma by mass spectrometry. *Biochemistry* 41 (34):10717–22.

Malkowski, M. G., S. L. Ginell, W. L. Smith, and R. M. Garavito. 2000. The productive conformation of arachidonic acid bound to prostaglandin synthase. *Science* 289 (5486):1933–7.

Marnett, L. J. 2002. Recent developments in cyclooxygenase inhibition. *Prostaglandins Other Lipid Mediat* 68–69:153–64.

Marnett, L. J., S. W. Rowlinson, D. C. Goodwin, A. S. Kalgutkar, and C. A. Lanzo. 1999. Arachidonic acid oxygenation by COX-1 and COX-2. Mechanisms of catalysis and inhibition. *J Biol Chem* 274 (33):22903–6.

Nadtochiy, S. M., P. R. Baker, B. A. Freeman, and P. S. Brookes. 2009. Mitochondrial nitroalkene formation and mild uncoupling in ischaemic preconditioning: Implications for cardioprotection. *Cardiovasc Res* 82 (2):333–40.

Napolitano, A., E. Camera, M. Picardo, and M. d'Ischia. 2000. Acid-promoted reactions of ethyl linoleate with nitrite ions: Formation and structural characterization of isomeric nitroalkene, nitrohydroxy, and novel 3-nitro-1,5-hexadiene and 1,5-dinitro-1, 3-pentadiene products. *J Org Chem* 65 (16):4853–60.

O'Donnell, V. B., J. P. Eiserich, P. H. Chumley et al. 1999. Nitration of unsaturated fatty acids by nitric oxide-derived reactive nitrogen species peroxynitrite, nitrous acid, nitrogen dioxide, and nitronium ion. *Chem Res Toxicol* 12 (1):83–92.

O'Donnell, V. B., and B. A. Freeman. 2001. Interactions between nitric oxide and lipid oxidation pathways: implications for vascular disease. *Circ Res* 88 (1):12–21

Pannala, A. S., A. R. Mani, J. P. Spencer et al. 2003. The effect of dietary nitrate on salivary, plasma, and urinary nitrate metabolism in humans. *Free Radic Biol Med* 34 (5):576–84.

Park, H. S., S. M. Lee, J. H. Lee et al. 2001. Phosphorylation of the leucocyte NADPH oxidase subunit p47(phox) by casein kinase 2: Conformation-dependent phosphorylation and modulation of oxidase activity. *Biochem J* 358 (Pt 3):783–90.

Radi, R. 2013. Peroxynitrite, a stealthy biological oxidant. *J Biol Chem* 288 (37):26464–72.

Radi, R., J. S. Beckman, K. M. Bush, and B. A. Freeman. 1991. Peroxynitrite-induced membrane lipid peroxidation: The cytotoxic potential of superoxide and nitric oxide. *Arch Biochem Biophys* 288 (2):481–7.

Radi, R., A. Denicola, B. Alvarez, G. Ferrer, and H. Rubbo. 2000. The biological chemistry of peroxynitrite. In *Nitric Oxide Biology and Pathobiology*, edited by L. J. Ignarro. San Diego: Academic Press.

Rosen, E. D., and B. M. Spiegelman. 2001. PPARgamma: A nuclear regulator of metabolism, differentiation, and cell growth. *J Biol Chem* 276 (41):37731–4.

Rouzer, C. A., and L. J. Marnett. 2003. Mechanism of free radical oxygenation of polyunsaturated fatty acids by cyclooxygenases. *Chem Rev* 103 (6):2239–304.

Rubbo, H., C. Batthyány, B. A. Freeman, R. Radi, and A. Denicola. 1998. Nitric oxide diffusion across low density lipoprotein and inhibition of lipid oxidation-dependent chemiluminescence. *Nitric Oxide* 2 (supp. 2):117.

Rubbo, H., R. Radi, M. Trujillo et al. 1994. Nitric oxide regulation of superoxide and peroxynitrite-dependent lipid peroxidation. Formation of novel nitrogen-containing oxidized lipid derivatives. *J Biol Chem* 269 (42):26066–75.

Rudnicki, M., L. A. Faine, N. Dehne et al. 2011. Hypoxia inducible factor-dependent regulation of angiogenesis by nitro-fatty acids. *Arterioscler Thromb Vasc Biol* 31 (6):1360–7.

Rudolph, T. K., and B. A. Freeman. 2009. Transduction of redox signaling by electrophile–protein reactions. *Sci Signal* 2 (90):re7.

Rudolph, T. K., V. Rudolph, M. M. Edreira et al. 2010. Nitro-fatty acids reduce atherosclerosis in apolipoprotein E-deficient mice. *Arterioscler Thromb Vasc Biol* 30 (5):938–45.

Rudolph, V., T. K. Rudolph, F. J. Schopfer et al. 2010. Endogenous generation and protective effects of nitro-fatty acids in a murine model of focal cardiac ischaemia and reperfusion. *Cardiovasc Res* 85 (1):155–66.

Rudolph, V., F. J. Schopfer, N. K. Khoo et al. 2009. Nitro-fatty acid metabolome: Saturation, desaturation, {beta}-oxidation, and protein adduction. *J Biol Chem* 284 (3):1461–73.

Salvatore, S. R., D. A. Vitturi, P. R. Baker et al. 2013. Characterization and quantification of endogenous fatty acid nitroalkene metabolites in human urine. *J Lipid Res* 54 (7):1998–2009.

Salvemini, D., M. G. Currie, and V. Mollace. 1996. Nitric oxide-mediated cyclooxygenase activation. A key event in the antiplatelet effects of nitrovasodilators. *J Clin Invest* 97 (11):2562–8.

Salvemini, D., K. Seibert, J. L. Masferrer et al. 1995. Nitric oxide activates the cyclooxygenase pathway in inflammation. *Am J Ther* 2 (9):616–619.

Sarady, J. K., S. L. Otterbein, F. Liu, L. E. Otterbein, and A. M. Choi. 2002. Carbon monoxide modulates endotoxin-induced production of granulocyte macrophage colony-stimulating factor in macrophages. *Am J Respir Cell Mol Biol* 27 (6):739–45.

Schopfer, F. J., P. R. Baker, G. Giles et al. 2005. Fatty acid transduction of nitric oxide signaling. Nitrolinoleic acid is a hydrophobically stabilized nitric oxide donor. *J Biol Chem* 280 (19):19289–97.

Schopfer, F. J., C. Batthyany, P. R. Baker et al. 2009. Detection and quantification of protein adduction by electrophilic fatty acids: Mitochondrial generation of fatty acid nitroalkene derivatives. *Free Radic Biol Med* 46 (9):1250–9.

Schopfer, F. J., C. Cipollina, and B. A. Freeman. 2012. Formation and signaling actions of electrophilic lipids. *Chem Rev* 111 (10):5997–6021.

Schopfer, F. J., M. P. Cole, A. L. Groeger et al. 2010. Covalent peroxisome proliferator-activated receptor gamma adduction by nitro-fatty acids: Selective ligand activity and anti-diabetic signaling actions. *J Biol Chem* 285 (16):12321–33.

Schopfer, F. J., Y. Lin, P. R. Baker et al. 2005. Nitrolinoleic acid: An endogenous peroxisome proliferator-activated receptor gamma ligand. *Proc Natl Acad Sci USA* 102 (7):2340–5.

Sheppard, F. R., M. R. Kelher, E. E. Moore et al. 2005. Structural organization of the neutrophil NADPH oxidase: Phosphorylation and translocation during priming and activation. *J Leukoc Biol* 78 (5):1025–42.

Shiose, A., and H. Sumimoto. 2000. Arachidonic acid and phosphorylation synergistically induce a conformational change of p47phox to activate the phagocyte NADPH oxidase. *J Biol Chem* 275 (18):13793–801.

Smith, W. L., D. L. DeWitt, and R. M. Garavito. 2000. Cyclooxygenases: Structural, cellular, and molecular biology. *Annu Rev Biochem* 69:145–82.

Smith, W. L., and I. Song. 2002. The enzymology of prostaglandin endoperoxide H synthases-1 and -2. *Prostaglandins Other Lipid Mediat* 68–69:115–28.

Tontonoz, P., L. Nagy, J. G. Alvarez, V. A. Thomazy, and R. M. Evans. 1998. PPARgamma promotes monocyte/macrophage differentiation and uptake of oxidized LDL. *Cell* 93 (2):241–52.

Trostchansky, A., L. Bonilla, C. P. Thomas et al. 2011. Nitroarachidonic acid, a novel peroxidase inhibitor of prostaglandin endoperoxide H synthases 1 and 2. *J Biol Chem* 286 (15):12891–900.

Trostchansky, A., J. M. Souza, A. Ferreira et al. 2007. Synthesis, isomer characterization, and anti-inflammatory properties of nitroarachidonate. *Biochemistry* 46 (15):4645–53.

Tsikas, D., A. A. Zoerner, A. Mitschke, and F. M. Gutzki. 2009. Nitro-fatty acids occur in human plasma in the picomolar range: A targeted nitro-lipidomics GC-MS/MS study. *Lipids* 44 (9):855–65.

van der Donk, W. A., A. L. Tsai, and R. J. Kulmacz. 2002. The cyclooxygenase reaction mechanism. *Biochemistry* 41 (52):15451–8.

Vignais, P. V. 2002. The superoxide-generating NADPH oxidase: Structural aspects and activation mechanism. *Cell Mol Life Sci* 59 (9):1428–59.

Villacorta, L., F. J. Schopfer, J. Zhang, B. A. Freeman, and Y. E. Chen. 2009. PPARgamma and its ligands: Therapeutic implications in cardiovascular disease. *Clin Sci (Lond)* 116 (3):205–18.

Villacorta, L., J. Zhang, M. T. Garcia-Barrio et al. 2007. Nitro-linoleic acid inhibits vascular smooth muscle cell proliferation via the Keap1/Nrf2 signaling pathway. *Am J Physiol Heart Circ Physiol* 293 (1):H770–6.

Waite, K. A., R. Wallin, D. Qualliotine-Mann, and L. C. McPhail. 1997. Phosphatidic acid-mediated phosphorylation of the NADPH oxidase component p47-phox. Evidence that phosphatidic acid may activate a novel protein kinase. *J Biol Chem* 272 (24):15569–78.

Wientjes, F. B., A. W. Segal, and J. H. Hartwig. 1997. Immunoelectron microscopy shows a clustered distribution of NADPH oxidase components in the human neutrophil plasma membrane. *J Leukoc Biol* 61 (3):303–12.

Woodcock, S. R., G. Bonacci, S. L. Gelhaus, and F. J. Schopfer. 2013. Nitrated fatty acids: synthesis and measurement. *Free Radic Biol Med* 59:14–26.

Woodcock, S. R., A. J. Marwitz, P. Bruno, and B. P. Branchaud. 2006. Synthesis of nitrolipids. All four possible diastereomers of nitrooleic acids: (E)- and (Z)-, 9- and 10-nitro-octadec-9-enoic acids. *Org Lett* 8 (18):3931–4.

Wright, M. M., J. Kim, T. D. Hock et al. 2009. Human haem oxygenase-1 induction by nitro-linoleic acid is mediated by cAMP, AP-1 and E-box response element interactions. *Biochem J* 422 (2):353–61.

Wright, M. M., F. J. Schopfer, P. R. Baker et al. 2006. Fatty acid transduction of nitric oxide signaling: Nitrolinoleic acid potently activates endothelial heme oxygenase 1 expression. *Proc Natl Acad Sci USA* 103 (11):4299–304.

Wu, Y., Y. Dong, P. Song, and M. H. Zou. 2012. Activation of the AMP-activated protein kinase (AMPK) by nitrated lipids in endothelial cells. *PLoS One* 7 (2):e31056.

Yamamoto, T., T. Suzuki, A. Kobayashi et al. 2008. Physiological significance of reactive cysteine residues of Keap1 in determining Nrf2 activity. *Mol Cell Biol* 28 (8):2758–70.

Zhang, J., L. Villacorta, L. Chang et al. 2010. Nitro-oleic acid inhibits angiotensin II-induced hypertension. *Circ Res* 107 (4):540–8.

Zhang, X., R. Goncalves, and D. M. Mosser. 2008. The isolation and characterization of murine macrophages. *Curr Protoc Immunol* Chapter 14:Unit 14 1.

Zou, M. H., and Y. Wu. 2008. AMP-activated protein kinase activation as a strategy for protecting vascular endothelial function. *Clin Exp Pharmacol Physiol* 35 (5–6):535–45.

6 Protein Lipoxidation

Koji Uchida

CONTENTS

INTRODUCTION

Lipid peroxidation has been implicated in the pathogenesis of numerous diseases, including atherosclerosis, diabetes, cancer, and rheumatoid arthritis, as well as in drug-associated toxicity, postischemic reoxygenation injury, and aging. Lipid peroxidation proceeds by a free radical chain reaction mechanism and yields lipid hydroperoxides as the major initial reaction products (Figure 6.1). The lipid hydroperoxides further undergo carbon–carbon bond cleavage via alkoxyl radicals in the presence of transition metals giving rise to the formation of short-chain, unesterified aldehydes of 3–9 carbons in length, and a second class of aldehydes still esterified to the parent lipid (Esterbauer et al., 1991). The important agents that give rise to the modification of a protein may be represented by reactive aldehydic intermediates, such as 2-alkenals, and 4-hydroxy-2-alkenals (Esterbauer et al., 1991; Uchida,

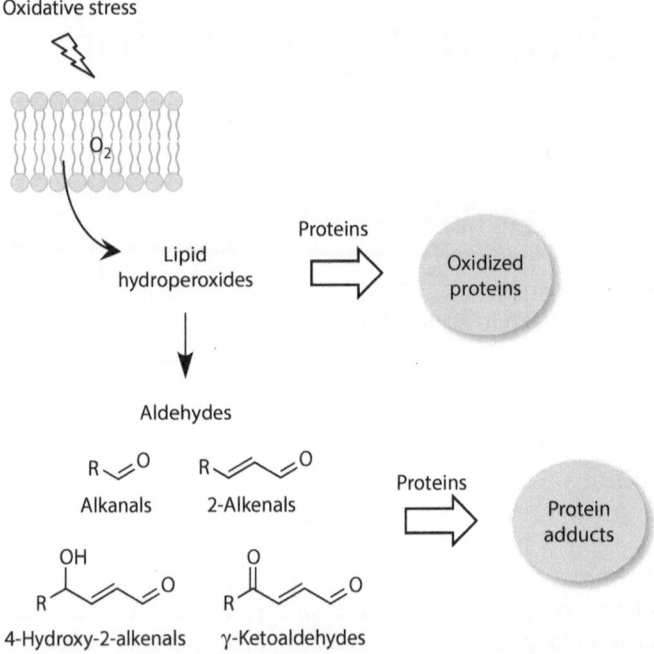

FIGURE 6.1 A mechanism for the formation of protein lipoxidation products. Oxidative stress induces lipid peroxidation, leading to the production of reactive molecules, such as lipid hydroperoxides and aldehydes.

2003). These reactive aldehydes are considered important mediators of cell damage due to their ability to covalently modify biomolecules, which can disrupt important cellular functions and can cause mutations (Esterbauer et al., 1991). The most sensitive cellular target of the lipid peroxidation products is protein, generating a variety of covalent adducts ("protein lipoxidation"). The pattern of protein lipoxidation tends to be broad and incompletely resolved upon SDS-gel electrophoresis, due probably to the varying degrees of oxidative cleavage of the polypeptide chain and the modification of amino acid side chains. The protein lipoxidation is also associated with the introduction of carbonyl groups into amino acid residues, which is also mediated by a variety of modification pathways *in vivo* and *in vitro*, such as metal-catalyzed oxidation of specific amino acid residues and glycation.

This chapter summarizes the state of knowledge about protein lipoxidation, focusing on the chemistry of lipid peroxidation-derived aldehydes and their adducts with amino acid residues. In addition, the recent findings on ligand function of lipoxidation products are also reviewed.

REACTIVE SPECIES INVOLVED IN PROTEIN LIPOXIDATION

Reactive species related to protein lipoxidation are classified into two major types, lipid hydroperoxides and their decomposition products, aldehydes.

Lipid Hydroperoxides

Lipid hydroperoxides are essentially an oxidant like H_2O_2, mediating oxidative modification of protein, and do not directly form covalent adducts. However, as described later, the lipid hydroperoxides in addition to H_2O_2 have been shown to mediate the binding of aldehydes (alkanals) to the lysine residues of protein to generate N-acylation adducts (Ishino et al., 2008).

Aldehydes

Representative aldehydes involved in protein lipoxidation are shown in Figure 6.2.

2-Alkenals

2-Alkenals represent a group of highly reactive aldehydes containing two electrophilic reaction centers. A partially positive carbon 1 or 3 in such molecules can

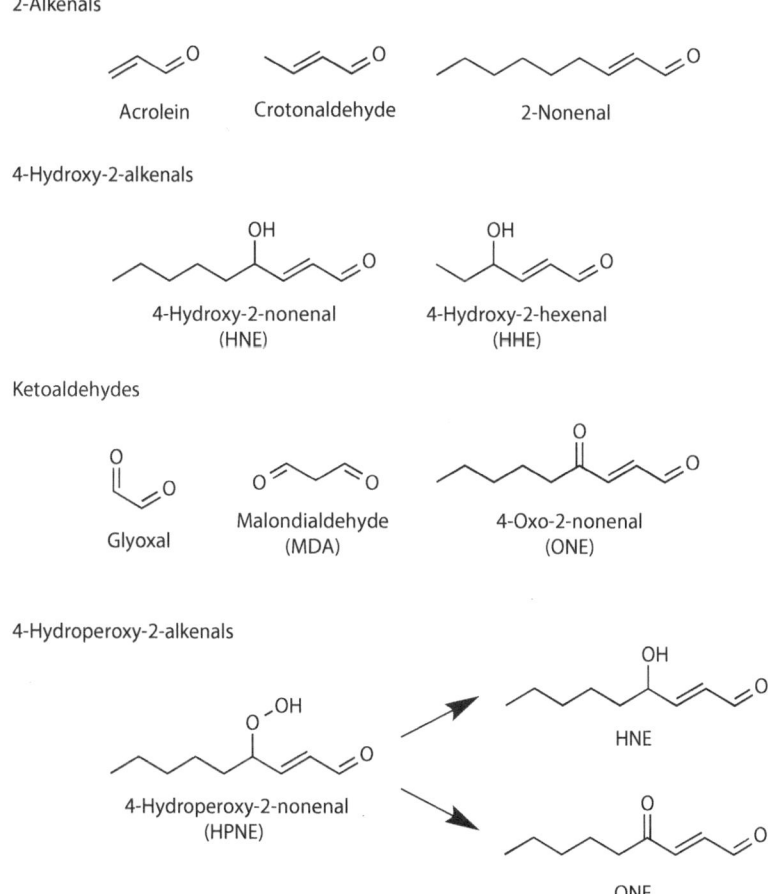

FIGURE 6.2   Major lipid peroxidation-derived aldehydes.

attack nucleophiles, such as proteins. 2-Hexenal is one of the most well established 2-alkenals generated from lipid peroxidation. Acrolein and its methyl derivative, crotonaldehyde, represent the most potent electrophilic 2-alkenals commonly detected in mobile source emissions, cigarette smoke, and other products of thermal degradation (International Agency for Research on Cancer, 1995). They have been considered as the "unnatural" environmental pollutants; however, recent studies revealed that these aldehydes were endogenously produced under oxidative stress (Uchida et al., 1998a, b; Ichihashi et al., 2001; Kondo et al., 2001). 2-Nonenal is also generated through the peroxidation of fatty acids (Haze et al., 2001; Ishino et al., 2010).

4-Hydroxy-2-Alkenals

4-Hydroxy-2-alkenals, which contain two electrophilic reaction centers like 2-alkenals, represent the most prominent aldehyde substances generated during the peroxidation of ω6 polyunsaturated fatty acids (Esterbauer et al., 1991). Of these, 4-hydroxy-2-nonenal (HNE) has achieved the status of one of the best recognized and most studied of the lipid peroxidation-derived aldehydes (Esterbauer et al., 1991). The peroxidation of LDL indeed generates HNE as one of the major aldehydes (Esterbauer et al., 1990). It has been suggested that HNE can be causally involved in many of the pathophysiological effects associated with oxidative stress in cells and tissues. In addition to studies on its bioactivity, HNE is commonly used as a biomarker for the occurrence and/or the extent of lipid peroxidation. The peroxidation of ω3 polyunsaturated fatty acids, such as docosahexaenoic and eicosapentaenoic acids, generates a closely related compound, 4-hydroxy-2-hexenal (HHE).

Ketoaldehydes

Other important reactive aldehydes that originate from lipid peroxidation include ketoaldehydes, such as glyoxal, malondialdehyde (MDA), and 4-oxo-2-nonenal (ONE). Glyoxal, a well-established α-ketoaldehyde intermediate in the glycation reaction, was demonstrated to be the product of the lipid peroxidation reaction by Thorpe and her colleagues (Fu et al., 1996). A β-ketoaldehyde MDA is considered to be the most abundant individual aldehyde resulting from the lipid peroxidation, and its determination by 2-thiobarbituric acid is one of the most common assays in lipid peroxidation studies. ONE is a 4-keto cousin of HNE and a particularly potent lipid hydroperoxide-derived genotoxin, which can readily react with nucleophilic biomacromolecules, such as protein and DNA (Rindgen et al., 1999).

4-Hydroperoxy-2-Alkenals

4-Hydroperoxy-2-nonenal (HPNE), the 4-hydroperoxy analog of HNE, was originally recognized as an enzymatic product of the ω6 polyunsaturated fatty acid metabolism in plants. However, Brash and his colleagues previously established that the formation of HPNE is also generated during nonenzymatic decomposition of a polyunsaturated fatty acid hydroperoxide (Schneider et al., 2001). Besides HNE, HPNE has been reported to generate ONE. Thus, HPNE has been recognized as a direct precursor of the most representative α,β-unsaturated aldehydes, HNE and ONE.

PROTEIN LIPOXIDATION

PROTEIN OXIDATION

The oxidation of proteins can result in cleavage of the polypeptide backbone, cross-linking, and modification of the side chains of amino acids. It is generally believed that lipid hydroperoxides are involved in the oxidative modification of protein through metal-catalyzed free radical production. The free radical species generated at the metal-binding sites on the proteins mediate the conversion of histidine into 2-oxo-histidine and asparagine; of lysine into α-aminoadipic semialdehyde; of proline into γ-glutamic semialdehyde, 2-pyrrolidone, and pyroglutamic or glutamic acid; and of arginine to γ-glutamic semialdehyde (Stadtman, 1992; Berlett and Stadtman, 1997). Among a wide variety of protein modifications, introduction of carbonyl groups into amino acid residues is a hallmark for oxidative damage to proteins (Figure 6.3). Both α-aminoadipic semialdehyde and γ-glutamic semialdehyde are identified as the major carbonyl products formed through protein oxidation.

PROTEIN ADDUCTION

2-Alkenal Adducts

The representative 2-alkenal adducts are summarized in Figure 6.4. The reactions of lysine with 2-alkenals have been mainly studied with acrolein, crotonaldehyde, and 2-nonenal. Similar to other α,β-unsaturated aldehydes, acrolein selectively reacts with the cysteine, histidine, and lysine residues of proteins. The primary products are their β-substituted propanals (**1**) (Figure 6.4a). These β-substituted propanals or Schiff's base crosslinks had been suggested as the predominant acrolein–lysine adducts; however, the major product formed upon the reaction of acrolein with a protein was identified to be a novel lysine product, N^{ε}-(3-formyl-3,4-dehydropiperidino) lysine (FDP-lysine) (**2**), which requires attachment of two acrolein molecules to one lysine side chain (Uchida et al., 1998b). This and the fact that crotonaldehyde also forms a similar FDP-type adduct, N^{ε}-(2,5-dimethyl-3-formyl-3,4-dehydropiperidino) lysine (dimethyl-FDP-lysine) (Ichihashi et al., 2001), suggest that this type of condensation reaction is characteristic of the reaction of 2-alkenals with primary amines. Indeed, upon reaction with a lysine derivative, other 2-alkenals, such as

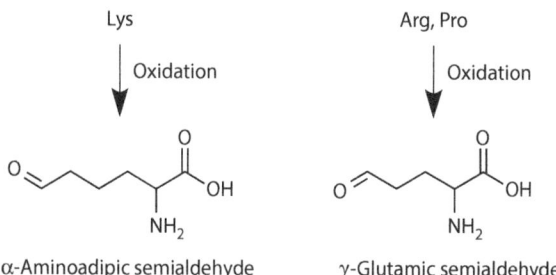

α-Aminoadipic semialdehyde   γ-Glutamic semialdehyde

FIGURE 6.3 Introduction of carbonyl groups into amino acid residues.

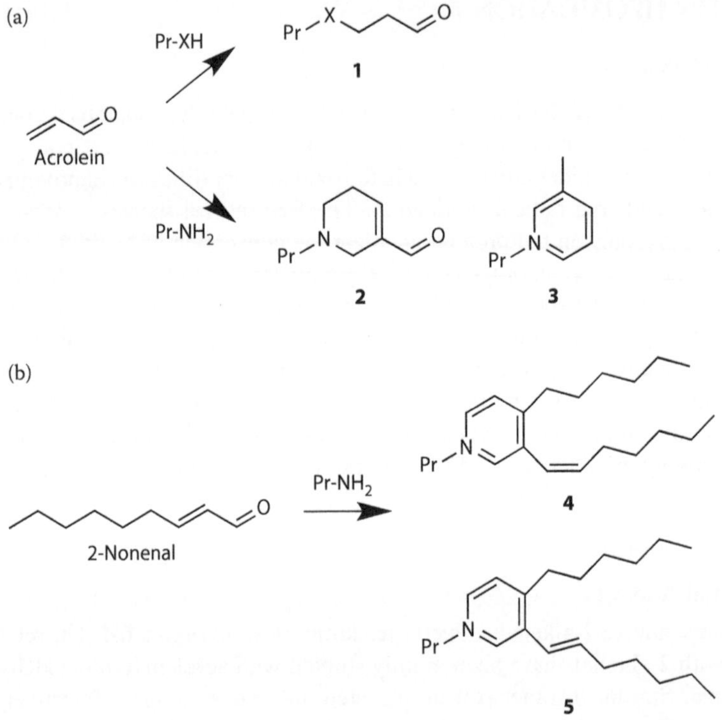

　　　　　　　　　　　　　　　　　Lipid Oxidation in Health and Disease

FIGURE 6.4 The structures of 2-alkenal adducts. (a) The structures of acrolein adducts (1~3). (b) The structures of 2-nonenal adducts (*cis-* and *trans-*HHP-lysine) (4 and 5).

2-pentenal and 2-hexenal, generate diethyl-FDP-lysine and dipropyl-FDP-lysine, respectively, while the formation of the FDP adducts is inversely proportional to the length of the alkyl groups of the 2-alkenals (Ichihashi et al., 2001). In a later study, Furuhata et al. (2002) reported that the FDP adducts are not the end products but the electrophilic intermediates that potently react with thiol compounds (Furuhata et al., 2002). On the other hand, lysine adduction with 2-alkenals generates another class of condensation adducts possessing a pyridinium ring. The pyridinium adducts, N^ε-(2-methylpyridinium)lysine (MP-lysine) (3) and N^ε-(5-ethyl-2-methylpyridinium) lysine (EMP-lysine), were identified as the products upon the reaction of lysine with acrolein and crotonaldehyde, respectively (Ichihashi et al., 2001; Furuhata et al., 2003). Ishino et al. recently identified novel lysine-pyridinium adducts, *cis-* and *trans-*N^ε- 3-[(hept-1-enyl)-4-hexylpyridinium]lysine (HHP-lysine) (4 and 5) upon the investigation of antigenic adducts recognized by the antibody against the 2-nonenal-modified proteins (Figure 6.4b) (Ishino et al., 2010). The formation of the 3,4-substituted pyridinium adducts has also been reported to be a dominant pathway for modification of the primary amine with 2-alkenals, such as 2-hexenal and 2-octenal (Alaiz and Barragan, 1995; Baker et al., 1998). Of interest, the monoclonal antibodies raised against the protein-bound 2-alkenals, such as acrolein, crotonaldehyde, and 2-nonenal, commonly recognize pyridinium-containing adducts as the major epitopes (Ichihashi et al., 2001; Furuhata et al., 2003; Ishino et al., 2010). Therefore, it is likely that, due to the placement of a fixed, positive charge on the ε-amino group, the pyridinium-containing adducts could be an important immunological epitope generated in 2-alkenal-modified proteins.

4-Hydroxy-2-Alkenal Adducts

Figure 6.5 summarizes the representative HNE adducts (**6~8**). The formation of thiol-derived Michael adducts (**6**) was initially considered to constitute the main reactivity of the 4-hydroxy-2-alkenals (Esterbauer et al., 1991; Uchida and Stadtman, 1992a). However, other studies concluded that these aldehydes could also form Michael adducts with the imidazole moiety of the histidine residues. The modification of histidine by 2-alkenals and 4-hydroxy-2-alkenals primarily consists of a Michael-type addition of the imidazole nitrogen atom of histidine to the α,β-unsaturated bond, while the primary 4-hydroxy-2-alkenal-histidine Michael adducts, possessing a free aldehyde group, undergo further cyclization to form cyclic hemiacetal derivatives (Uchida and Stadtman, 1992b, 1993). Because the oxo-cyclo equilibrium favors hemiacetal formation, the free aldehyde moiety of the primary product may react with the 4-hydroxyl group to form the hemiacetal derivative (Esterbauer et al., 1991). It has also been suggested that the 4-hydroxy-2-alkenal-histidine Michael adduct may be stabilized toward a retro-Michael reaction, because of the poorer leaving group ability of the imidazole under neutral conditions. The structures of the histidine adduct with 4-hydroxy-2-alkenals have been well characterized with HNE. It was first speculated that the HNE-histidine adduct was a mixture of the isomeric form of the N^{π}- and N^{τ}-substituted adducts of the imidazole ring; however, based on an NMR spectral analysis of the adducts, it appeared that the reaction exclusively occurs at one position (N^{τ}-alkylation) (Nadkarni and Sayre, 1995). Because HNE generated during the lipid peroxidation is a racemic mixture of the 4R- and 4S-isomers, the HNE cyclic hemiacetal

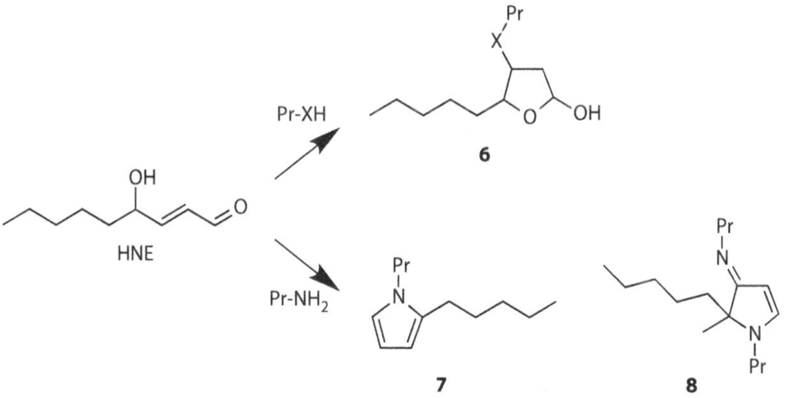

FIGURE 6.5 The structures of HNE adducts. The structures of the HNE Michael addition adducts (**6**) and other pyrrole-type (**7**) and cross-linking-type (**8**) adducts. (b) The structures of the HPNE-lysine adducts (**14** and **15**).

adducts contain chiral centers at C-1, C-3, and C-4 of the HNE moiety. Accordingly, the cyclic hemiacetal adducts are composed of at least eight isomers (Hashimoto et al., 2003; Wakita et al., 2009). With regard to the modification of the protein-based lysine ε-amino groups by the 4-hydroxy-2-alkenals, most of the data has been obtained from the reaction with HNE. In a manner similar to the modification of histidine, HNE forms a Michael-type lysine adduct. Sayre et al. (1993) later reported a novel HNE–lysine adduct (7) possessing a pyrrole structure (Sayre et al., 1993). On the other hand, Esterbauer and his colleagues demonstrated that the treatment of LDL with HNE generates the same lipofuscin-like fluorescence properties as seen in the Cu^{2+}-oxidized LDL (Esterbauer et al., 1986). This finding suggested that HNE could be the major contributor to the fluorescence generated in the oxidized LDL. The chemical nature of the fluorophore arising from the HNE protein modification had remained elusive; however, Itakura et al. (1998) identified for the first time the major lipofuscin-like fluorophore derived from HNE and lysine to be the 3-hydroxy-3-imino-1,2-dihydropyrrole derivative (8) and found that the fluorescent properties of this pigment are similar to those of the oxidized LDL (Itakura et al., 1998). The same adduct was later reported by Xu and Sayre (1998) and Tsai et al. (1998) (Tsai et al., 1998; Xu and Sayre, 1998). A pathway for the formation of the fluorophore that has been proposed is that the ε-amino group of lysine readily reacts with the C-1 and C-3 of HNE via Schiff base formation and Michael addition, respectively, to form the initial 1:2 HNE-amine intermediate, which is subsequently converted into the fluorophore via two oxidation steps and intermolecular cyclization. Mechanistic studies of the HNE-derived fluorophore formation have proposed an alternative mechanism, involving two 2e oxidations following the initial Schiff base formation (Xu et al., 1999).

Ketoaldehyde Adducts

The representative ketoaldehyde adducts are summarized in Figure 6.6. Glyoxal is a common intermediate in the formation of carboxymethyllysine (CML) (9) (Fu et al., 1996) and a glyoxal-derived lysine–lysine dimer (Wells-Knecht et al., 1995) during the oxidation of both carbohydrates and lipids followed by reaction with lysine residues. Glyoxal also reacts with arginine residues to form an imidazolone adduct (Schwarzenbolz et al., 1997). MDA specifically modifies the lysine residues of proteins. The major reaction of MDA comprises the addition to primary amines, generating N^ε-(2-propenal)lysine (10) (Chio and Tappel, 1969a, b). This adduct has been detected as the major form in which endogenous MDA is excreted in rat and human urine (McGirr et al., 1985; Draper et al., 1988). MDA also forms fluorescent products with primary amino compounds, and it had long been considered that the MDA-derived fluorophores might be responsible for the fluorescence of lipofuscin. In earlier studies, Chio and Tappel proposed the aminoenimines (R-NH-CH=CH-CH=N-R) (11) as the MDA-derived fluorophores and suggested that the fluorescence of lipofuscin was due to these cross-links (Chio and Tappel, 1969a, b). In both cases, however, the structural confirmations were conducted after reduction of the aminoenimines to the nonfluorescent R-NH-CH$_2$-CH$_2$-CH$_2$-NH-R forms with NaBH$_4$. In a later study, Itakura and Uchida (2001) succeeded in isolating the aminoenimine formed from MDA and lysine residues and found that this

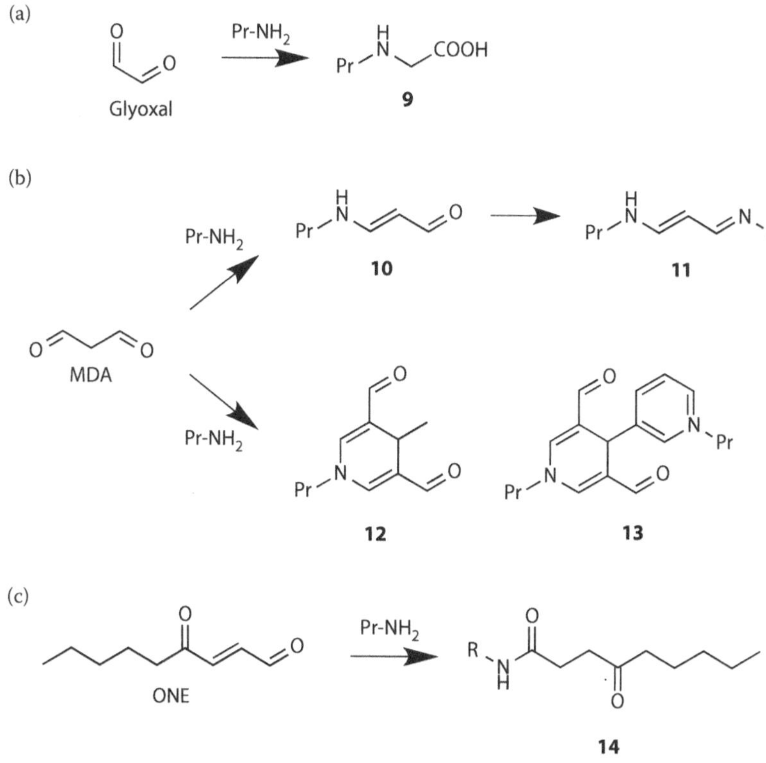

FIGURE 6.6 The structures of the ketoaldehyde adducts. (a) The structures of the glyoxal adduct (CML) (9). (b) The structures of the MDA–lysine adducts (10–13). (c) The structures of the ONE-lysine adduct (14).

cross-link did not contribute to the fluorescence formation of lipofuscin. On the other hand, the 1,4-dihydropyridine-3,5-dicarboxaldehydes were identified as the major fluorophore generated upon the reaction of MDA with primary amines (12) (Kikugawa et al., 1981; Kikugawa and Ido, 1984). Moreover, Itakura et al. (1996) also reported another fluorescent MDA–lysine adduct (13) containing the dihydro-pyridine and pyridinium rings (Itakura et al., 1996). ONE, upon reaction with pro-teins, selectively modifies the nucleophilic side chains of lysine, histidine, cysteine, and arginine (Doorn and Petersen, 2002). The predominant initial reaction appears to involve the Michael addition to the central ONE double bond, more at C3 than at C2, to produce substituted 4-oxononanals. These adducts are relatively unstable and could be further converted to stable long-lived products, which include dihydro-furan, dihydropyrrole, and isomeric 4-ketoamide derivatives (14) originating from the reaction of ONE with lysine (Zhu and Sayre, 2007) and a substituted imidazole derivative with arginine (Oe et al., 2003). ONE also forms furan derivatives upon its reaction with cysteine and histidine derivatives (Zhang et al., 2003; Yocum et al., 2005). Shimozu et al. (2009) recently characterized a series of products originating from the ONE-cysteine Michael adducts.

HYDROPEROXIDE-DEPENDENT PROTEIN ADDUCTION

PROTEIN LIPOXIDATION BY 2-ALKANALS IN THE PRESENCE OF HYDROPEROXIDES

Alkanals are most abundantly formed during lipid peroxidation (Esterbauer and Zollner, 1989; Spiteller et al., 2001). Upon the reaction with proteins, these aldehydes react with lysine residues to form an imine or Schiff base adduct (Fenaille et al., 2001). Due to the reversible nature of such unconjugated Schiff bases, these aldehydes have received relatively little attention as the causative agent for the modification of nucleophilic biomolecules. Ishino et al. (2008) discovered that H_2O_2, and to a lesser extent alkyl hydroperoxides, are capable of mediating covalent modification of proteins by saturated aldehydes (Figure 6.7a). This finding suggests the possibility that saturated aldehydes, in combination with hydroperoxides, may contribute to the modification of nucleophilic biomolecules and the development of tissue damage under oxidative stress. A probable mechanism for the reaction has been suggested to be the imine analog of the Baeyer–Villiger reaction of ketones with peroxides to give esters, which also pertains to the mechanism of oxidation of aldehydes to carboxylic acids by the hydroperoxides. It has been proposed that the reaction would proceed by addition of hydroperoxides to the Schiff base, followed by 1,2-migration of hydride and expulsion of H_2O or alkyl-OH, respectively.

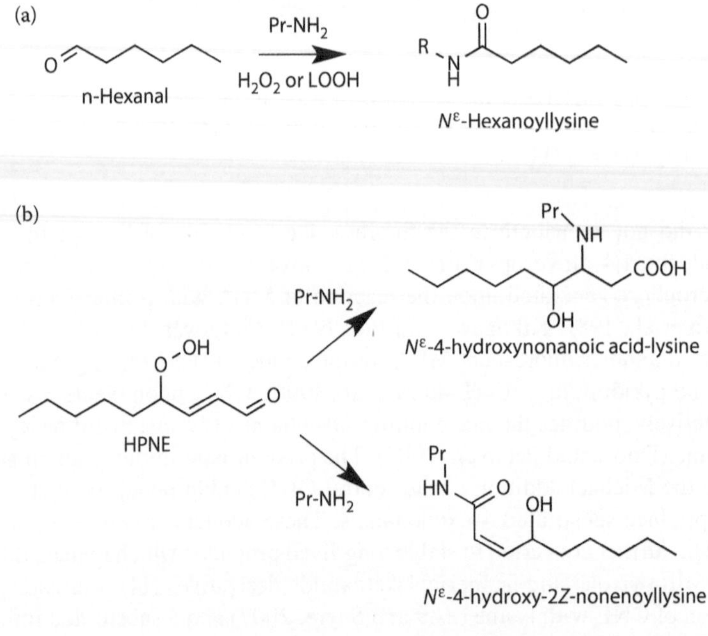

FIGURE 6.7 Hydroperoxide-dependent protein lipoxidation. (a) Hydroperoxide-mediated covalent modification of protein by lipid peroxidation-derived saturated aldehydes. (b) HPNE is a reactive molecule that covalently modifies proteins to generate unique intramolecular oxidation products.

PROTEIN LIPOXIDATION BY 4-HYDROPEROXY-2-ALKENALS

Owing to the unstable nature of the hydroperoxy group, HPNE has received relatively little attention as the causative agent for modification of the nucleophilic biomolecules. Because HPNE is a direct precursor of HNE and ONE, it may be inevitable to speculate that HPNE generates the same lysine adducts as those obtained from the reaction with these end products. However, Shimozu et al. characterized the HPNE modification of lysine residues and identified two HPNE-specific lysine adducts, N^ε-4-hydroxynonanoic acid-lysine and N^ε-4-hydroxy-2Z-nonenoyllysine (Figure 6.7b) (Shimozu et al., 2011), both of which are suggested to be formed through mechanisms in which the initial HPNE–lysine adducts undergo Baeyer–Villiger-like reactions proceeding through an intramolecular oxidation catalyzed by the hydroperoxy group. Thus, HPNE is not just a precursor of HNE and ONE, but also a reactive molecule that could directly modify proteins in biological systems.

LIGAND FUNCTION OF LIPOXIDATION PRODUCTS

The lipoxidation products function as danger signals called damage-associated molecular patterns (DAMPs), which represent endogenous danger molecules as a group that is separated from pathogen-derived pathogen-associated molecular patterns. In the extracellular space, DAMPs can bind to pattern recognition receptors (PRRs), which recognize conserved molecular patterns that distinguish foreign organisms, or to specialized receptors to elicit an immune response by promoting the release of pro-inflammatory mediators and recruiting immune cells to infiltrate the tissue (Janeway and Medzhitov, 2002). DAMPs, possessing an exposed epitope, are also accessible for recognition by the soluble PRRs, such as natural antibodies and regulatory proteins (Miller et al., 2011; Weismann and Binder, 2012). DAMPs stimulate the adaptive immunity and participate in autoimmune responses and tissue repair. It has been suggested that the DAMPs-mediated activation of the innate immune system has an important role in the pathogenesis of various immune and inflammatory diseases. Some of the lipoxidation products, such as ω-(2-carboxyethyl)pyrrole and other related pyrroles, HNE-histidine, ONE-lysine, and 2-nonenal-lysine adducts (Kumano-Kuramochi et al., 2012; West et al., 2010; Shibata et al., 2011; Ishino et al., 2010), have been identified as candidate ligands of PRRs, leading to downstream inflammation. On the other hand, the plasma complement factor H has been identified as a soluble PRR, which binds MDA-modified proteins and block both the uptake of the modified proteins by macrophages and MDA-induced pro-inflammatory effects *in vivo* (Weismann et al., 2011). These findings suggest that some of the lipoxidation products could function as ligands of PRRs and play a pivotal role in triggering innate immune responses.

CONCLUSION

This chapter summarized protein lipoxidation, mainly focusing on protein adduction chemistry with reactive aldehydes generated from the peroxidation of polyunsaturated fatty acids. On the basis of a large number of reports concerning the chemical and immunochemical detection of lipoxidation products *in vivo*, there may be no doubt that their steady-state levels increase under pathophysiological states associated with oxidative stress. However, only a limited number of lipoxidation products are presently ascribed to the chemically characterized and assayable products. A large fraction of amino acid residues modified during lipid peroxidation therefore remains uncharacterized. On the contrary, considerable progress has recently been made toward understanding the mechanisms of action of lipoxidation products. The most striking finding is that some of the protein lipoxidation products could function as DAMPs that could be recognized by PRRs, suggesting that the protein lipoxidation may be a major contributing factor in the innate immune response. Because of the biological effects exerted by lipoxidation products, identification and quantification of lipoxidation products may be essential for further clarifying the role of protein lipoxidation in health and diseases.

ABBREVIATIONS

apo B	apolipoprotein B-100
CML	carboxymethyllysine
FDP-lysine	N^{ε}-(3-formyl-3,4-dehydropiperidino)lysine
HHP-lysine	N^{ε}-3-[(hept-1-enyl)-4-hexylpyridinium]lysine
HNE	4-hydroxy-2-nonenal
LDL	low-density lipoproteins
MDA	malondialdehyde
MP-lysine	N^{ε}-(2-methylpyridinium)lysine
ONE	4-oxo-2-nonenal.

REFERENCES

Alaiz, M. and Barragan, S. Changes induced in bovine serum albumin following interactions with the lipid peroxidation product E-2-octenal. *Chem. Phys. Lipids* 77; 1995: 217–223.

Baker, A., Zidek, L., Wiesler, D., Chmelik, J., and Pagel, M., and Novotny, M. V. Reaction of N-acetylglycyllysine methyl ester with 2-alkenals: An alternative model for covalent modification of proteins. *Chem. Res. Toxicol.* 11; 1998: 730–40.

Berlett, B. S. and Stadtman, E. R. Protein oxidation in aging, disease, and oxidative stress. *J. Biol. Chem.* 272; 1997: 20313–6.

Chio, K. S. and Tappel, A. L. Synthesis and characterization of the fluorescent products derived from malonaldehyde and amino acids. *Biochemistry* 8; 1969a: 2821–2827.

Chio, K. S. and Tappel, A. L. Inactivation of ribonuclease and other enzymes by peroxidizing lipids and by malonaldehyde. *Biochemistry* 8; 1969b: 2827–2832.

Doorn, J. A. and Petersen, D. R. Covalent modification of amino acid nucleophiles by the lipid peroxidation products 4-hydroxy-2-nonenal and 4-oxo-2-nonenal. *Chem. Res. Toxicol.* 15; 2002: 1445–1450.

Draper, H. H., Hadley, M., Lissemore, L., Laing, N. M., and Cole, P. D. Identification of Ne-(2-propenal)lysine as a major urinary metabolite of malondialdehyde. *Lipids* 23; 1988: 626–628.

Esterbauer, H., Dieber-Rotheneder, M., Waeg, G., Striegl, G., and Jürgens, G. Biochemical, structural, and functional properties of oxidized low-density lipoprotein. *Chem. Res. Toxicol.* 3; 1990: 77–92.

Esterbauer, H., Jurgens, G., Quehenberger, O., and Koller, E. Autoxidation of human low density lipoprotein: Loss of polyunsaturated fatty acids and vitamin E and generation of aldehydes. *J. Lipid Res.* 28; 1987: 495–509.

Esterbauer, H., Koller, E., Slee, R. G., and Koster, J. F. Possible involvement of the lipid-peroxidation product 4-hydroxynonenal in the formation of fluorescent chromolipids. *Biochem J.* 239; 1986: 405–409.

Esterbauer, H., Schaur, R. J., and Zollner, H. Chemistry and biochemistry of 4-hydroxynonenal, malondialdehyde and related aldehydes. *Free Radic. Biol. Med.* 11; 1991: 81–128.

Esterbauer, H. and Zollner, H. Methods for determination of aldehydic lipid peroxidation products. *Free Radic. Biol. Med.* 7; 1989: 197–203.

Fenaille, F., Guy, P. A., and Tabet, J. C. Study of protein modification by 4-hydroxy-2-nonenal and other short chain aldehydes analyzed by electrospray ionization tandem mass spectrometry. *J. Am. Soc. Mass Spectrom.* 14; 2003: 215–226.

Fu, M. X., Requena, J. R., Jenkins, A. J., Lyons, T. J., Baynes, J. W., and Thorpe, S. R. The advanced glycation end product, Nε-(carboxymethyl)lysine, is a product of both lipid peroxidation and glycoxidation reactions. *J. Biol. Chem.* 271; 1996: 9982–9986.

Furuhata, A., Ishii, T., Kumazawa, S., Yamada, T., Nakayama, T., and Uchida, K. Nε-(3-methylpyridinium)lysine, a major antigenic adduct generated in acrolein-modified protein. *J. Biol. Chem.* 278; 2003: 48658–48665.

Furuhata, A., Nakamura, M., Osawa, T., and Uchida, K. Thiolation of protein-bound carcinogenic aldehyde: An electrophilic acrolein-lysine adduct that covalently binds to thiols. *J. Biol. Chem.* 277; 2002: 27919–27926.

Hashimoto, M., Shibata, T., Wasada, H., Toyokuni, S., and Uchida, K. Structural basis of protein-bound endogenous aldehydes: Chemical and immunochemical characterizations of configurational isomers of a 4-hydroxy-2-nonenal-histidine adduct. *J. Biol. Chem.* 278; 2003: 5044–5051.

Haze, S., Gozu, Y., Nakamura, S., Kohno, Y., Sawano, K., Ohta, H., and Yamazaki, K. 2-Nonenal newly found in human body odor tends to increase with aging. *J. Invest. Dermatol.* 116; 2001: 520–524.

Ichihashi, K., Osawa, T., Toyokuni, S., and Uchida, K. Endogenous formation of protein adducts with carcinogenic aldehydes: Implication for oxidative stress. *J. Biol. Chem.* 276; 2001: 23903–23913.

International Agency for Research on Cancer 1995. IARC Monographs on the Evaluation of Carcinogenic Risks to Humans: Dry Cleaning, *Some Chlorinated Solvents and other Industrial Chemicals*, Vol. 63, pp 373–391, International Agency for Research on Cancer, Lyon, France.

Ishino, K., Shibata, T., Ishii, T., Liu, Y. T., Toyokuni, S., Zhu, X., Sayre, L. M., and Uchida, K. Protein N-acylation: H_2O_2-mediated covalent modification of protein by lipid peroxidation-derived saturated aldehydes. *Chem. Res. Toxicol.* 21; 2008: 1261–1270.

Ishino, K., Wakita, C., Shibata, T., Toyokuni, S., Machida, S., Matsuda, S., Matsuda, T., and Uchida, K. Lipid peroxidation generates a body odor component *trans*-2-nonenal covalently bound to protein in vivo. *J. Biol. Chem.* 285; 2010: 15302–15313.

Itakura, K., Osawa, T., and Uchida, K. Structure of a fluorescent compound formed from 4-hydroxy-2-nonenal and Na-hippuryllysine: A model for fluorophores derived from protein modifications by lipid peroxidation. *J. Org. Chem.* 63; 1998: 185–187.

Itakura, K. and Uchida, K. Evidence that malondialdehyde-derived aminoenimine is not a fluorescent age pigment. *Chem. Res. Toxicol.* 14; 2001: 473–475.

Itakura, K., Uchida, K., and Osawa, T. A novel fluorescent malondialdehyde-lysine adduct. *Chem. Phys. Lipids* 84; 1996: 75–79.

Janeway Jr., C. A. and Medzhitov, R. Innate immune recognition. *Ann. Rev. Immunol.* 20; 2002: 197–216.

Kikugawa, K. and Ido, Y. Studies on peroxidized lipids. V. Formation and characterization of 1,4-dihydropyridine-3,5-dicarbaldehydes as model of fluorescent components in lipofuscin. *Lipids* 19; 1984: 600–608.

Kikugawa, K., Machida, Y., Kida, M., and Kurachi, T. Studies on peroxidized lipids. III. Fluorescent pigments derived from the reaction of malondialdehyde and amino acids. *Chem. Pharm. Bull.* 29; 1981: 3003–3011.

Kondo, M., Oya-Ito, T., Kumagai, T., Osawa, T., and Uchida, K. Cyclopentenone prostaglandins as potential inducers of intracellular oxidative stress. *J. Biol. Chem.* 276; 2001: 12076–12083.

Kumano-Kuramochi, M., Shimozu, Y., Wakita, C., Ohnishi-Kameyama, M., Shibata, T., Matsunaga, S., Takano-Ishikawa, Y. et al. Identification of 4-hydroxy-2-nonenal- histidine adducts that serve as ligands for human lectin-like oxidized LDL receptor-1. *Biochem. J.* 442; 2012: 171–180.

Maeshima, T., Honda, K., Chikazawa, M., Shibata, T., Kawai, Y., Akagawa, M., and Uchida, K. Quantitative analysis of acrolein-specific adducts generated during lipid peroxidation modification of proteins in vitro: Identification of Nt-formylethylhistidine as the major adduct. *Chem. Res. Toxicol.* 25; 2012: 1384–1392.

McGirr, L. G., Hadley, M., and Draper, H. H. Identification of Na-acetyl-e-(2-propenal)lysine as a urinary metabolite of malondialdehyde. *J. Biol. Chem.* 260; 1985: 15427–15431.

Miller, Y. I., Choi, S. H., Wiesner, P., Fang, L., Harkewicz, R., Hartvigsen, K., Boullier, A. et al. Oxidation-specific epitopes are danger-associated molecular patterns recognized by pattern recognition receptors of innate immunity. *Circ. Res.* 108; 2011: 2352–2348.

Nadkarni, D. V. and Sayre, L. M. Structural definition of early lysine and histidine adduction chemistry of 4-hydroxynonenal. *Chem. Res. Toxicol.* 8; 1995: 284–291.

Obama, T., Kato, R., Masuda, Y., Takahashi, K., Aiuchi, T., and Itabe, H. Analysis of modified apolipoprotein B-100 structures formed in oxidized low-density lipoprotein using LC-MS/MS. *Proteomics* 7; 2007: 2132–2141.

Oe, T., Lee, S. H., Silva Elipe, M. V., Arison, B. H., and Blair, I. A. A novel lipid hydroperoxide-derived modification to arginine. *Chem. Res. Toxicol.* 16; 2003: 1598–1605.

Rindgen, D., Nakajima, M., Wehrli, S., and Xu, K., and Blair I. A. *Chem. Res. Toxicol.* 12; 1999: 1195–1204.

Sayre, L. M., Arora, P. K., Iyer, R. S., and Salomon, R. G. Pyrrole formation from 4-hydroxynonenal and primary amines. *Chem. Res. Toxicol.* 6; 1993: 19–22.

Schneider, C., Tallman, K. A., Porter, N. A., and Brash, A. R. Two distinct pathways of formation of 4-hydroxynonenal. Mechanisms of nonenzymatic transformation of the 9- and 13-hydroperoxides of linoleic acid to 4-hydroxyalkenals. *J. Biol. Chem.* 276; 2001: 20831–20838.

Schwarzenbolz, U., Henle, T., Haeßner, R., and Klostermeyer, H. Z. On the reaction of glyoxal with proteins. *Lebensm Unters Forsch A* 205; 1997: 121–124.

Shibata, T., Shimozu, Y., Wakita, C., Shibata, N., Kobayashi, M., Machida, S., Kato, R. et al. Lipid peroxidation modification of protein generates Nε-(4-oxononanoyl)lysine as a pro-inflammatory ligand. *J. Biol. Chem.* 286; 2011: 9943–9957.

Shimozu, Y., Hirano, K., Shibata, T., Shibata, N., and Uchida, K. 4-Hydroperoxy-2-nonenal is not just an intermediate, but a reactive molecule that covalently modifies proteins to generate unique intramolecular oxidation products. *J. Biol. Chem.* 286; 2011: 29313–29324.

Shimozu, Y., Shibata, T., Ojika, M., and Uchida, K. Identification of advanced reaction products originating from the initial 4-oxo-2-nonenal-cysteine Michael adducts. *Chem. Res. Toxicol.* 22; 2009: 957–964.

Spiteller, P., Kern, W., Reiner, J., and Spiteller, G. Measurement of n-alkanals and hydroxyalkenals in biological samples. *Biochim. Biophys. Acta* 1531; 2001: 188–208.

Stadtman, E. R. Protein oxidation and aging. *Science* 257; 1992: 1220–4.

Tsai, L., Szweda, P. A., Vinogradova, O., and Szweda, L. I. Structural characterization and immunochemical detection of a fluorophore derived from 4-hydroxy-2-nonenal and lysine. *Proc. Natl. Acad. Sci. USA* 95; 1998: 7975–7980.

Uchida, K. 4-Hydroxy-2-nonenal: A product and mediator of oxidative stress. *Prog. Lipid Res.* 42; 2003: 318–343.

Uchida, K. and Stadtman, E. R. Selective cleavage of thioether linkage in protein modified with 4-hydroxynonenal. *Proc. Natl. Acad. Sci. USA* 89; 1992a: 5611–5615.

Uchida, K. and Stadtman, E. R. Modification of histidine residues in proteins by reaction with 4-hydroxynonenal. *Proc. Natl. Acad. Sci. USA* 89; 1992b: 4544–4548.

Uchida, K. and Stadtman, E. R. Covalent attachment of 4-hydroxynonenal to glyceraldehyde-3-phosphate dehydrogenase: A possible involvement of intramolecular and intermolecular cross-linking reactions. *J. Biol. Chem.* 268; 1993: 6388–6393.

Uchida, K., Kanematsu, M., Sakai, K., Matsuda, M., Hattori, N., Mizuno, Y., Suzuki, D. et al. Protein-bound acrolein: Potential markers for oxidative stress. *Proc. Natl. Acad. Sci. USA* 95; 1998a: 4882–4887.

Uchida, K., Kanematsu, M., Morimitsu, Y., Osawa, T., Noguchi, N., and Niki, E. Acrolein is a product of lipid peroxidation reaction: formation of acrolein and its conjugate with lysine residues in oxidized low-density lipoprotein. *J. Biol. Chem.* 273; 1998b: 16058–16066.

Wakita, C., Honda, K., Shibata, T., Akagawa, M., and Uchida, K. A method for detection of 4-hydroxy-2-nonenal adducts in proteins. *Free Radic. Biol. Med.* 51; 2011: 1–4.

Wakita, C., Maeshima, T., Yamazaki, A., Shibata, T., Ito, S., Akagawa, M., Ojika, M., Yodoi, J., and Uchida, K. Stereochemical configuration of 4-hydroxy-2-nonenal-cysteine adducts and their stereoselective formation in a redox-regulated protein. *J. Biol. Chem.* 284; 2009: 28810–28822.

Weismann, D. and Binder, C. J. The innate immune response to products of phospholipid peroxidation. *Biochim. Biophys. Acta* 1818; 2012: 2465–2475.

Weismann, D., Hartvigsen, K., Lauer, K. N., Bennett, K. L., Scholl, H. P., Charbel Issa, P., Cano, M. et al. Complement factor H binds malondialdehyde epitopes and protects from oxidative stress. *Nature* 478; 2011: 76–81.

Wells-Knecht, K. J., Brinkman, E., and Baynes, J. W. Characterization of an imidazolium salt formed from glyoxal and Na-hippuryllysine: A model for Maillard reaction crosslinks in proteins. *J. Org. Chem.* 60; 1995:6246–6247.

West, X. Z., Malinin, N. L., Merkulova, A. A., Tischenko, M., Kerr, B. A., Borde, E. C., Podrez, E. A., Salomon, R. G., and Byzova, T. V. Oxidative stress induces angiogenesis by activating TLR2 with novel endogenous ligands. *Nature* 467; 2010: 972–976.

Xu, G. and Sayre L. M. Structural characterization of a 4-hydroxy-2-alkenal-derived fluorophor that contributes to lipoperoxidation-dependent protein cross-linking in aging and degenerative disease. *Chem. Res. Toxicol.* 11; 1998: 247–251.

Xu, G., Liu Y., and Sayre L. M. Independent synthesis, solution behavior, and studies on the mechanism of formation of a primary amine-derived fluorophore representing cross-linking of proteins by (E)-4-hydroxy-2-nonenal. *Chem. Res. Toxicol.* 64; 1999: 5732–5745.

Yocum, A. K., Oe, T., Yergey, A. L., and Blair, I. A. Novel lipid hydroperoxide-derived hemoglobin histidine adducts as biomarkers of oxidative stress. *J. Mass Spectrom.* 40; 2005: 754–764.

Zhang, W. H., Liu, J., Xu, G., Yuan, Q., and Sayre, L. M. Model studies on protein side chain modification by 4-oxo-2-nonenal. *Chem. Res. Toxicol.* 16; 2003: 512–523.

Zhu, X. and Sayre, L. M. Long-lived 4-oxo-2-enal-derived apparent lysine michael adducts are actually the isomeric 4-ketoamides. *Chem. Res. Toxicol.* 2; 2007:165–170.

7 Analysis of Lipid Peroxidation Products in Health and Disease

Michael J. Thomas

CONTENTS

INTRODUCTION: LIPID PEROXIDATION

Lipid peroxidation implies that one or more oxygen atoms has been added to a lipid backbone, usually accompanied by changes in the number and/or the position of double bonds. Before discussing methods for quantifying oxidized lipids, the process of product formation will be discussed. Two factors that are essential for understanding peroxidation are the "strength" of the radical and "lability" of the attacked species.

The first step of the oxidation process is radical formation by removal of a hydrogen atom from the susceptible lipid. In biological systems, like cells or tissues, the source of initiating radicals is often unknown. However, a consequence of living in an oxygen-rich environment and using oxygen as a terminal electron acceptor, is that the process of respiration generates free radicals. Despite a battery of antioxidant molecules and enzymes that trap radicals or reduce potential oxidants, a very small number of these radicals escape to initiate oxidation. In some circumstances, the respiratory process can become uncoupled and release more free radicals that cause significant oxidant stress in the cell. Free radicals may also be generated through light-induced photolysis, ionizing radiation or by molecules that can cycle between oxidized and reduced states, for example, quinones, NADPH, oxidases, and so on. In the presence of dioxygen an electron can be transferred to form superoxide in a redox cycle. Transition metals can readily cycle between oxidized and reduced states catalyzing radical oxidations, but low concentrations of unbound transition metals suggest that this mode of oxidation is not a primary contributor to *in vivo* lipid oxidation. Organic molecules, like quinones, can cycle between quinone–hydroquinone releasing electrons to oxygen. Redox cycling of quinone–hydroquinone may be a significant component of toxicity of cigarette smoke [1–3].

The ability to abstract hydrogen atoms (H$^\bullet$) has been extensively studied and an approximation of strengths for various elements is, for example, RO$^\bullet$ > RHN$^\bullet$ > RH$_2$C$^\bullet$ > RS$^\bullet$. Radical strength depends on the degree of stabilization of the radical center from atoms or groups of atoms attached to the radical center, with the ability to abstract H$^\bullet$ reduced with greater stabilization. Abstraction of H$^\bullet$ is the primary mode for starting or initiating lipid oxidation, but radicals can abstract other atoms, like halogens, and add to double bonds, a process used for the commercial preparation of plastics. The most powerful radical in biological systems is the hydroxyl radical (HO$^\bullet$). This radical is extremely reactive. Once formed it can move only a few molecular diameters before abstracting a susceptible hydrogen.

A second important facet of radical-induced oxidation relates to the strength of the carbon–hydrogen (C–H) bond under attack by a free radical. Hydrocarbons like hexane have the highest C–H bond strength. Adjacent substituents that stabilize a developing radical center, even alkyl groups can do this, promoting H$^\bullet$ abstraction. For most hydrocarbons susceptibility to H$^\bullet$ abstraction has the order $CH_2=CH_2$ < CH_4 < RCH_3 < R_2CH_2 < $R(CH_2=CHR)CH_2$ < $(RCH=CH)(R'CH=CH)CH_2$ a series that goes from unreactive hydrocarbons to the substantially more reactive bis-allylic moiety found in polyunsaturated fatty acids (PUFA).

Addition of dioxygen, O_2, to a newly formed radical center is the second step in the autoxidation process as is shown for the PUFA, linoleic acid in Figure 7.1. The stable ground state for dioxygen is a diradical that facilitates the reaction with a carbon-centered free radical to generate a peroxyradical. Because the oxygen-centered electron of the peroxyradical is stabilized by an interaction with the adjacent oxygen it is much less energetic than the hydroxyl radical or the alkoxyl radical and, therefore, more selective in its attack on C–H bonds. In a lipophilic environment, the hydroperoxyl radical will abstract H$^\bullet$ from a bis-allylic moiety in preference to H$^\bullet$ from an allylic moiety, for example, it will prefer linoleic acid over oleic acid or cholesterol [4–7], Figure 7.2. A summary of free radical autoxidation was recently published by Porter [8].

In a lipid membrane the hydrogen atom most susceptible to abstraction is a bis-allylic moiety, like those present in PUFA. As shown in Figure 7.3, the process will continue to propagate in a chain reaction that will oxidize many PUFA from a single initiating event. The longer the chain length, the greater will be the number of PUFA molecules that are oxidized to hydroperoxides. Only reactions with the hydroperoxyl radical that destroy the radical center will stop the chain. This process of chain termination usually involves reaction with an antioxidant like α-tocopherol, vitamin E, that donates H$^\bullet$ to the hydroperoxyl radical forming a stable tocopheroxyl radical, or by a self-reaction like disproportionation, which will generate oxygen and nonradical products.

Many of the mechanistic aspects of free radical chemistry, including the rate constants for reaction, have been well studied in homogeneous solution. Products from lipid oxidation in membranes were the same as products formed by oxidation in

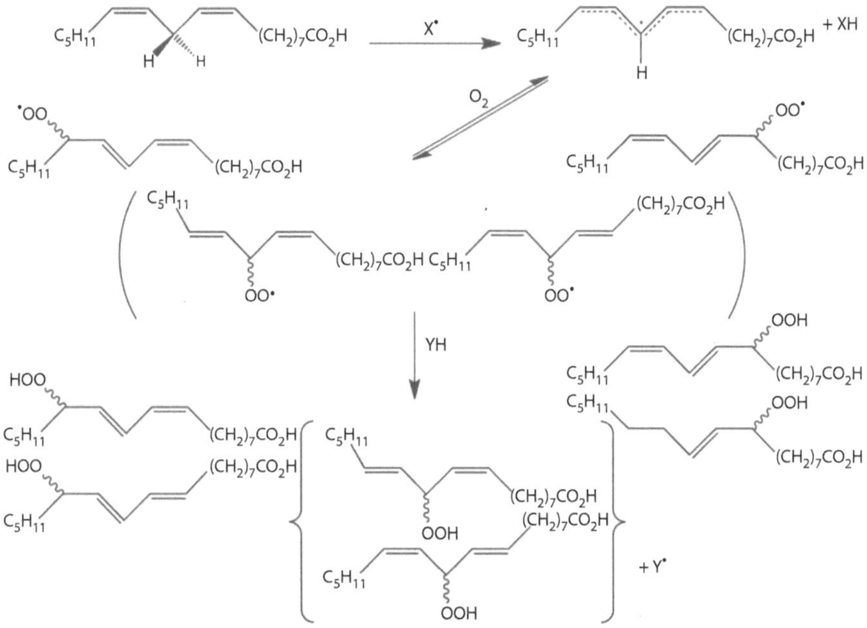

FIGURE 7.1 Radical catalyzed oxidation of linoleic acid. X* is used to represent a generalized radical sufficiently reactive to remove a bis-allylic hydrogen atom. YH is a generalized hydrogen source that can transfer H* to -OO*. The products in brackets are only observed when the intermediate products are rapidly trapped by H* transfer.

homogeneous solution, but in cases where multiple products are formed, the distribution of products was changed. A second difference between oxidation in homogeneous solution and a membrane is that in the membrane the process is taking place in a two-dimensional environment. A significant change between membranes and homogeneous solution was that the rates of propagation were reported to be slower in membranes compared to homogeneous solution [9,10].

PRODUCTS OF AUTOXIDATION

Typically, the first formed and primary product of autoxidation is a lipid hydroperoxide. Often the most abundant products generated in biological systems are hydroxylated lipids formed after hydroperoxide reduction. In many instances these alcohols are further oxidized to ketones. Decompositions of peroxidized PUFA having more than three double bonds leads to the formation of malonaldehyde [11,12], Figure 7.4a, which is also formed in a step that generates thromboxane A_2, by a pathway that also gives hydroxyheptadecatrienoic acid [13,14]. Therefore, in many instances measuring malonaldehyde may not be the best indicator of lipid autoxidation [15,16]. The most commonly used test for malonaldehyde, the thiobarbituric acid test, was also reported to give a positive result with cyclic peroxides and with products derived from carbohydrates rather than lipids [17]. Another well-studied group of breakdown products from PUFA-hydroperoxides are the hydroxy alkenals,

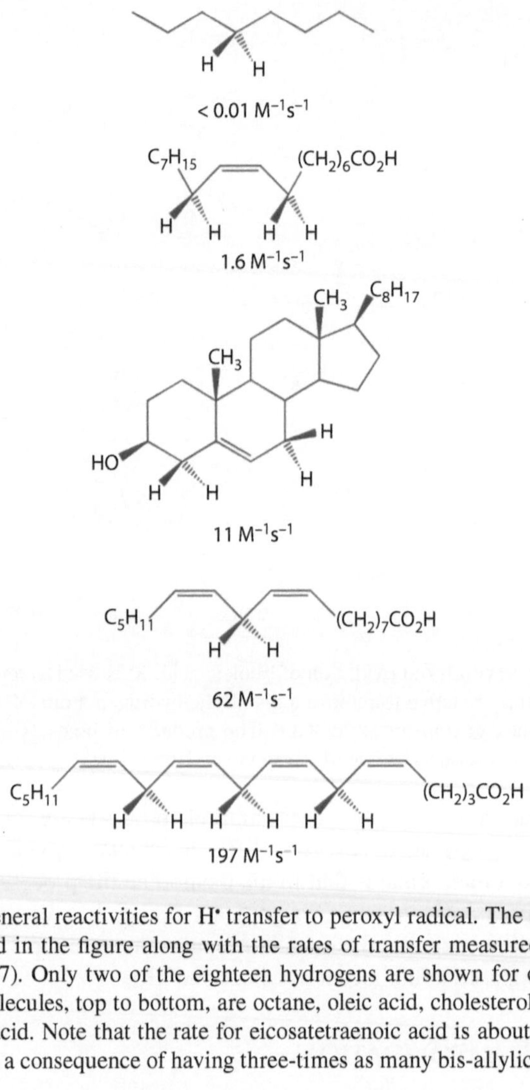

FIGURE 7.2 General reactivities for H• transfer to peroxyl radical. The susceptible hydrogens are indicated in the figure along with the rates of transfer measured or estimated for each molecule (4,7). Only two of the eighteen hydrogens are shown for octane. Names for the individual molecules, top to bottom, are octane, oleic acid, cholesterol, linoleic acid and eicosatetraenoic acid. Note that the rate for eicosatetraenoic acid is about three fold greater than for linoleate, a consequence of having three-times as many bis-allylic hydrogen atoms.

like 4-hydroxy-2-nonenal (4-HNE) [18]. Pryor and Porter described a mechanism for generation of 4-HNE by transition metal catalysis [19]. An outline of this process is shown for eicosatetraenoic acid in Figure 7.4b.

A particularly novel reaction that was used to explain the enzymic formation of prostaglandins is cyclization of the first formed hydroperoxyl radical yielding an endoperoxide [11,20,21]. Later free radical derived products, called F_2-isoprostanes, were identified that had a skeleton similar to enzymically generated prostaglandins [22,23]. Because there is no stereochemical control of this process as many as 64 different isomers, 32 racemic pairs, can be formed [24], Figure 7.4c. Autoxidation with cyclization gives low yields of prostaglandin $F_{2\alpha}$ ($PGF_{2\alpha}$) that is formed in high yield by enzymic oxidation of eicosatetraenoic acid.

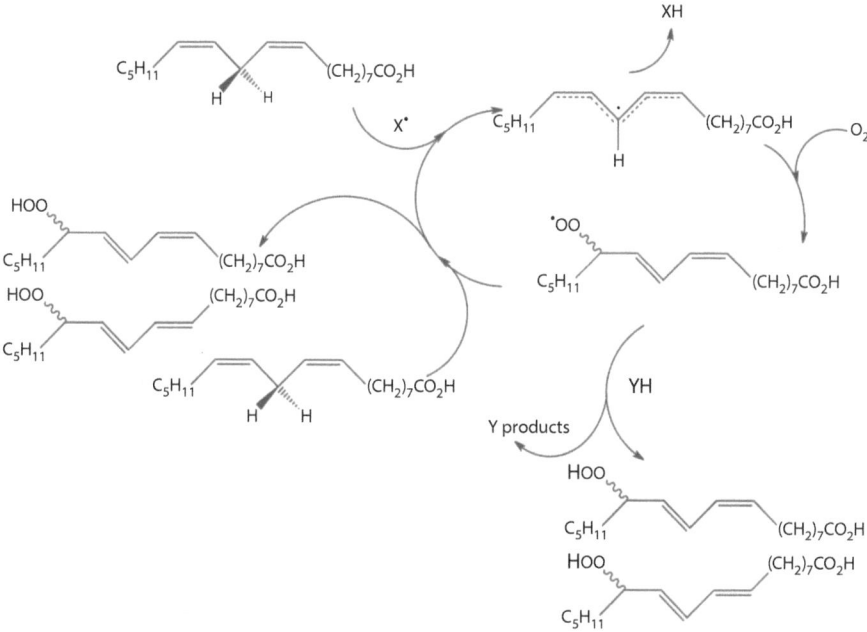

FIGURE 7.3 Free radical chain autoxidation demonstrated using linoleic acid. Autoxidation is initiated by a radical species X• generating a bis-allylic radical. In an oxygenated environment rich in linoleate or other PUFA the H• donor to the PUFA hydroperoxyl radical is a doubly allylic hydrogen from PUFA that yields more bis-allylic radical. This is called the propagation phase of autoxidation. In PUFA-rich environments each event initiated by X• can generate hundreds of PUFA hydroperoxides. When PUFA levels are low the chain reaction is terminated by self-reaction of the PUFA hydroperoxyl radical or by reaction with an inhibitor (YH) that will break the chain reaction. α-Tocopherol is an excellent chain-breaking inhibitor present in many biological systems.

Other oxidation products include epoxides that can arise from hydroperoxide rearrangement [25–27], but are also formed by enzymic processes [28,29]. Products of lipid oxidation may be quantified after formation from simple lipids, for example, oxidation of cholesterol or PUFA. These lipids may be components of more complex lipids like cholesteryl esters [30] or PUFA-containing phospholipids [31]. An analysis of the oxidized products can be performed on the intact lipid or after the individual lipid components are separated by hydrolysis.

Many free radical-derived oxidation products are isobaric with products generated by enzymes such as the lipoxygenases and the cyclooxygenases. Enzymic hydroxylations or peroxidations of PUFAs yield a single set of chiral centers through stereospecific addition of oxygen and shifting a double bond to give E-stereochemistry. For example, 5-lipoxygenase will oxidize (5Z,8Z,11Z,14Z)-5,8,11,14-eicosatetraenoic acid (eicosatetraenoic acid or arachidonic acid) to a single enantiomer (5S,6E,8Z,11Z,14Z)-5-hydroxyicosa-6,8,11,14-tetraenoic acid. Other lipoxygenases show similar degree of stereospecificity. Lipoxygenase regiospecificity determines which of the six possible oxygenated products are generated from eicosatetraenoic acid, for example, the

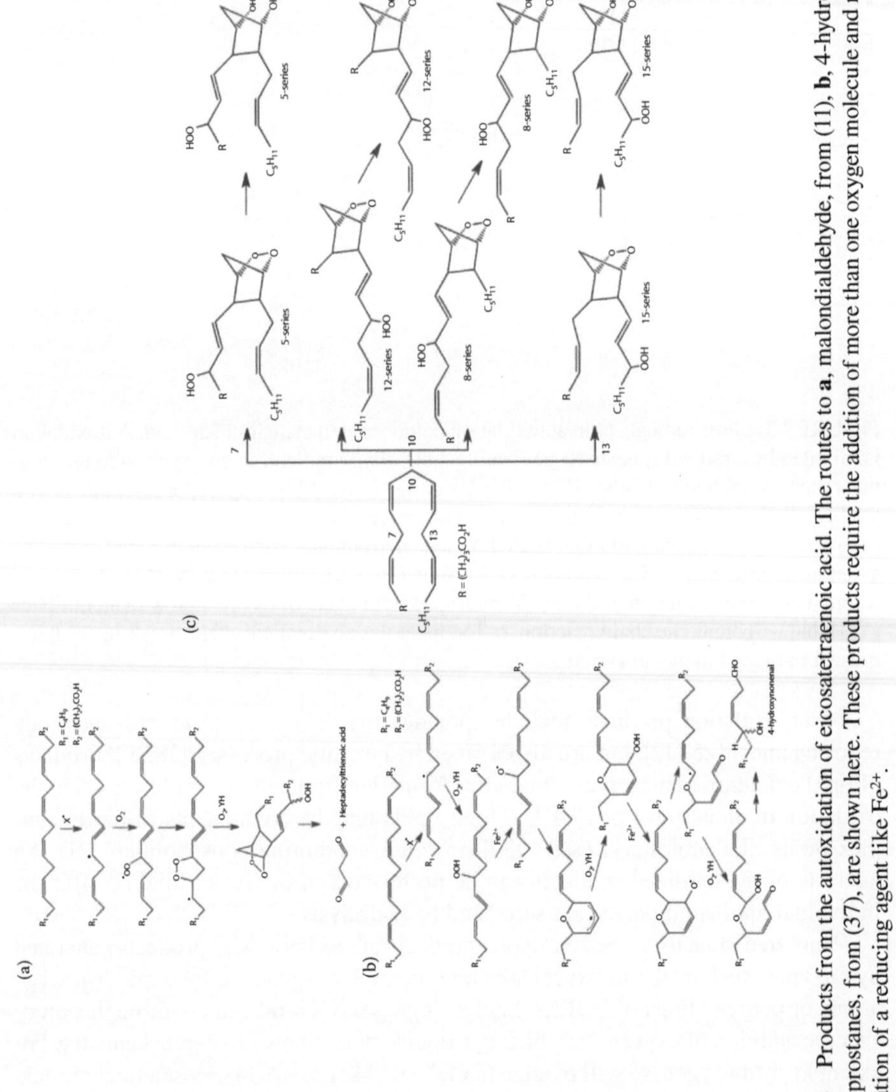

FIGURE 7.4 Products from the autoxidation of eicosatetraenoic acid. The routes to **a**, malondialdehyde, from (11), **b**, 4-hydroxy-2-nonenal, from (24), and **c**, F$_2$-isoprostanes, from (37), are shown here. These products require the addition of more than one oxygen molecule and malondialdehyde requires the participation of a reducing agent like Fe^{2+}.

oxidation of carbon centers 5, 8, 9, 11, 12, or 15. Likewise, cyclooxygenase converts eicosatetraenoic acid into prostaglandin H_2 (PGH_2), an unstable endoperoxide, that is subsequently converted by various synthases to prostaglandins, for example, prostaglandin E synthase converts PGH_2 into prostaglandin E_2 having four chiral centers. Enzymes exert stereochemical control through highly selective binding of a substrate and steric control of $H^{\bullet}$ removal and/or oxygen addition.

Free-radical oxidations usually yield both enantiomers at each site oxidized because of the lack of stereochemical control provided by an enzyme. In some cases these oxidations give both the E- along with some Z-double bond stereochemistry, although the E-stereochemistry is thermodynamically preferred. In contrast to the selective generation of a single enantiomer by a lipoxygenase, free radical autoxidation of eicosatetraenoic acid will produce all six possible hydroperoxide products, in varying yields, giving both enantiomers at each oxidized carbon center. In most instances free radical chain oxidation of PUFA yields a thermodynamic distribution of products although in some circumstances kinetic intermediates can be trapped.

Because hydroperoxy-PUFA, hydroxy-PUFA, and F_2-isoprostanes are often the first-formed products of lipid oxidation, these species are excellent candidates for assessing oxidant stress. How do we distinguish products of free radical oxidation from those generated by enzymes? This is best done by identifying the number and types of products. For example, lipoxygenase oxidation of eicosatetraenoic acid will usually produce only 1 product and this product will be a single enantiomer compared with the free radical autoxidation that generates all six possible regioisomeric hydroxyicosatetraenoic acids (HETE) with each oxidized center having both enantiomers. Extending this observation to reactions that generate the prostaglandin skeleton the F_2-isoprostanes mixtures can have as many as 32 racemic products with the thermodynamic favored products predominating. Gas chromatographic tandem mass spectrometric analysis of autoxidized eicosanoids would identify $PGF_{2\alpha}$ show many, partially resolved isobaric products [22,23,32], the F_2-isoprostanes, with $PGF_{2\alpha}$ a minor product from free radical autoxidation [33,34]. However, some samples from biological sources will show a large peak that coelutes with $PGF_{2\alpha}$ suggesting that the enzymic synthesis of $PGF_{2\alpha}$ may have contributed to the formation of prostaglandin species [35]. This conclusion would have to be verified by MS/MS methods that can discriminate $PGF_{2\alpha}$ from other F_2-isoprostanes.

Therefore, to determine if oxidized lipids were formed by enzymic processes or by free radical autoxidation, a first step is to visualize the distribution of products. This step requires previous knowledge of the maximum number of oxidized products, their chromatographic behavior and ions associated with mass spectrometric detection of each product. Quantitative analyses almost always require the use of appropriate, pure standards. For samples from more complex sources where the lipids of interest are present at low concentration there may be many interfering ions. In these instances, tandem mass spectrometry can be used to select pairs of precursor ions and product ions formed by collision-induced dissociation in a procedure called selected reaction monitoring (SRM). This type of analysis usually provides a significant improvement in signal to noise so that the product can be accurately quantified. With modern instruments many, up to hundreds, of these transitions can be measured in a single analysis. In conjunction with retention time from chromatography individual products can be identified with a reasonable degree of certainty. SRM pairs that are used for measuring oxidized eicosatetraenoic acids are listed in a recent review [36,37]. Because of the large number of F_2-isoprostanes generated by free-radical oxidation it is relatively easy to conclude whether or not autoxidation was the primary mechanism of lipid oxidation. Within a set of similar products it is usually not significant whether a single isomer or a group of isomers is monitored, unless it is important for the study to quantify only those autoxidation products that are bioactive, such as 8-iso-PGF$_{2\alpha}$ (15-F$_{2t}$-IsoP), 8-iso-PGE2 (15-E$_{2t}$-IsoP), or others [38–40]. If a limited distribution of HETEs was formed there can be questions as to the extent to which free-radical autoxidation participated in the oxidation. Chiral analysis can be used to resolve the issue [41].

METHODS OF ANALYSIS

In this chapter analysis by liquid chromatography-tandem mass spectrometry (LC/MS/MS) will be discussed as the primary tool of analysis. Liquid chromatography provides a gentle, yet highly efficient method to separate molecules. Mass spectrometry provides high sensitivity along with the capability to provide molecular fragmentation data in real time that adds confidence to the results.

EXTRACTION

The first hurtle is to reproducibly extract lipids from a matrix. The most common lipids extraction methods are those of Bligh and Dyer [42] and Folch [43]. Recent analysis of these two methods has shown that the Folch method tends to have a greater total recovery of lipid [44]. A variety of other solvent mixtures has been compared and may offer fewer hazards with similar recoveries [45]. These extraction methods are designed to recover the principal lipid classes, but may not be as useful for recovery of lipids that have unique charge characteristics. For example, fatty acids, phosphatidic acids, and lyso-phosphatidic acids usually require acidic solvents to facilitate recovery from an aqueous solution while neutral lipids may not be sufficiently soluble in an organic solvent [46]. Other complexities include solvent manipulations required to extract more polar lipids like the phosphatidylinositol phosphates.

SEPARATION

Because oxidized lipids constitute a minor fraction of the lipid extract and other lipids predominate, a prepurification may be required to identify and quantify peroxidation products. A similar problem is associated with quantifying lipid mediators that influence a variety of physiological functions [47–52]. Depending on the method of separation minor components can be separated if their polarity is significantly different from the bulk lipids. Prepurification can be accomplished using solid-phase separation columns and general methods have been published for rapidly separating lipid classes [53,54]. A complication that often arises in electrospray mass spectrometry coupled with liquid chromatography is a process called ion suppression, where a molecule or molecules coeluting with the molecule of interest are more easily ionized within the electrospray droplets. Complication due to ion suppression has been recognized for some time and the best means to eliminate or at least reduce the effects are to purify the sample and to use a electrospray source that generates very small droplets [55–59], for example, a nanospray source.

OXIDIZED LIPIDS: FATTY ACIDS

In many instances, oxidized products are considerably more stable when formed in complex lipids, like phospholipids or cholesteryl esters, and alkaline hydrolysis is required to release the oxidized lipid for analysis. Separation and quantitation of mono-hydroxyicosanoids by HPLC has been regularly used for the last 30 years. The procedures have become routine and there are now published procedures for automating the analysis [60]. A recent review by Yin et al. [36] describes in detail how to quantify both mono-hydroxyeicosatetraenoates and F_2-isoprostanes. Similar procedures can be used to separate and identify products derived from (5Z,8Z,11Z,14Z,17Z)-eicosa-5,8,11,14,17-pentenoic acid, (7Z,10Z,13Z,16Z,19Z)-docosa-7,10,13,16,19-pentaenoic acid, and (4Z,7Z,10Z,13Z,16Z,19Z)-docosa-4,7,10,13,16,19-hexaenoic acid.

CHOLESTEROL

Although cholesterol is about 10-fold less reactive in autoxidation than PUFA there is still interest in the products that have been associated with oxidant stress. Cholesterol, free and bound to ester, is measured after separation into free and ester cholesterol by selective extraction and hydrolysis using gas chromatography-tandem mass spectrometry (GC/MS/MS) [61]. Oxidized cholesterol, oxysterols, can be analyzed using LC-MS/MS with a minimum of manipulation or by using GC/MS/MS after derivatization [62–66].

SUMMARY

Oxygen is a necessary component for the life of large organisms. Many organisms employ oxygen to enzymically modify selected lipids, then use these oxidized lipids in complex signaling processes. Examples of oxidized lipids used include, hydroxylated eicosanoids, leukotrienes, prostaglandins, oxysterols, and so on. However, oxidation of essential biological components like the polyunsaturated fatty acids present in lipid membranes is also a consequence of an oxygen-rich environment. Uncontrolled oxidation due to various forms of oxidant stress can severely debilitate cells. To combat random oxidation aerobic tissues contain enzymes and a variety of small molecules that, under normal conditions, substantially retard or inhibit uncontrolled lipid oxidation. However, even lesser degrees of oxidation can lead to the formation of lipid oxidation products that interfere with normal signaling processes. This chapter summarizes the modes and mechanisms for free radical chain oxidation of lipids and summarizes methods for recovering and quantifying the most abundant products of lipid oxidation.

REFERENCES

1. Church, D. F., and Pryor, W. A. 1985. Free-radical chemistry of cigarette smoke and its toxicological implications, *Environ Health Perspect 64*, 111–126.
2. Schmeltz, I., Tosk, J., Jacobs, G., and Hoffmann, D. 1977. Redox potential and quinone content of cigarette smoke, *Anal Chem 49*, 1924–1929.
3. Chouchane, S., Wooten, J. B., Tewes, F. J., Wittig, A., Muller, B. P., Veltel, D., and Diekmann, J. 2006. Involvement of semiquinone radicals in the *in vitro* cytotoxicity of cigarette mainstream smoke, *Chem Res Toxicol 19*, 1602–1610.
4. Yakupova, L. R., and Safiullin, R. L. 2011. Kinetics of the initiated and inhibited oxidation of methyl oleate in homogeneous and aqueous emulsion media, *Kinet Katal 52*, 785–792.
5. Wong, W.-S. D., and Hammond, E. G. 1977. Analysis of oleate and linoleate hydroperoxides in oxidized ester mixtures, *Lipids 12*, 475–479.
6. Cosgrove, J. P., Church, D. F., and Pryor, W. A. 1987. The kinetics of the autoxidation of polyunsaturated fatty acids, *Lipids 22*, 299–304.
7. Xu, L., Davis, T. A., and Porter, N. A. 2009. Rate constants for peroxidation of polyunsaturated fatty acids and sterols in solution and in liposomes, *J Am Chem Soc 131*, 13,037–13,044.
8. Porter, N. A. 2013. A perspective on free radical autoxidation: The physical organic chemistry of polyunsaturated fatty acid and sterol peroxidation, *J Org Chem 78*, 3511–3524.
9. Barclay, L. R. C., Baskin, K. A., Dakin, K. A., Locke, S. J., and Vinqvist, M. R. 1990. The antioxidant activities of phenolic antioxidants in free radical peroxidation of phospholipid membranes, *Can J Chem 68*, 2258–2269.
10. Barclay, L. R. C., Baskin, K. A., Locke, S. J., and Vinqvist, M. R. 1989. Absolute rate constants for lipid peroxidation and inhibition in model biomembranes, *Can J Chem 67*, 1366–1369.
11. Pryor, W. A., and Stanley, J. P. 1975. Letter: A suggested mechanism for the production of malonaldehyde during the autoxidation of polyunsaturated fatty acids. Nonenzymatic production of prostaglandin endoperoxides during autoxidation, *J Org Chem 40*, 3615–3617.
12. Dahle, L. K., Hill, E. G., and Holman, R. T. 1962. The thiobarbituric acid reaction and the autoxidations of polyunsaturated fatty acid methyl esters, *Arch Biochem Biophys 98*, 253–261.
13. McMillan, R. M., MacIntyre, D. E., Booth, A., and Gordon, J. L. 1978. Malonaldehyde formation in intact platelets is catalysed by thromboxane synthase, *Biochem J 176*, 595–598.
14. Minkes, M., Stanford, N., Chi, M. M., Roth, G. J., Raz, A., Needleman, P., and Majerus, P. W. 1977. Cyclic adenosine 3′,5′-monophosphate inhibits the availability of arachidonate to prostaglandin synthetase in human platelet suspensions, *J Clin Invest 59*, 449–454.
15. Strauss, R. G. 1981. Malonaldehyde formation is not a suitable screening test to detect oxidation in human neutrophils, *J Clin Pathol 34*, 800–802.
16. Smith, J. B., Ingerman, C. M., and Silver, M. J. 1976. Malondialdehyde formation as an indicator of prostaglandin production by human platelets, *J Lab Clin Med 88*, 167–172.
17. Porter, N. A., Nixon, J., and Isaac, R. 1976. Cyclic peroxides and the thiobarbituric assay, *Biochim Biophys Acta 441*, 506–512.
18. Esterbauer, H., Benedetti, A., Lang, J., Fulceri, R., Fauler, G., and Comporti, M. 1986. Studies on the mechanism of formation of 4-hydroxynonenal during microsomal lipid peroxidation, *Biochim Biophys Acta 876*, 154–166.

19. Pryor, W. A., and Porter, N. A. 1990. Suggested mechanisms for the production of 4-hydroxy-2-nonenal from the autoxidation of polyunsaturated fatty acids, *Free Radic Biol Med 8*, 541–543.
20. Porter, N. A., and Funk, M. O. 1975. Letter: Peroxy radical cyclization as a model for prostaglandin biosynthesis, *J Org Chem 40*, 3614–3615.
21. Funk, M. O., Isaac, R., and Porter, N. A. 1975. Letter: Free radical cyclization of unsaturated hydroperoxides, *J Am Chem Soc 97*, 1281–1282.
22. Morrow, J. D., Harris, T. M., and Roberts, L. J., 2nd. 1990. Noncyclooxygenase oxidative formation of a series of novel prostaglandins: Analytical ramifications for measurement of eicosanoids, *Anal Biochem 184*, 1–10.
23. Morrow, J. D., Hill, K. E., Burk, R. F., Nammour, T. M., Badr, K. F., and Roberts, L. J., 2nd. 1990. A series of prostaglandin F2-like compounds are produced *in vivo* in humans by a non-cyclooxygenase, free radical-catalyzed mechanism, *Proc Natl Acad Sci USA 87*, 9383–9387.
24. Morrow, J. D., and Roberts, L. J., 2nd. 1996. The isoprostanes. Current knowledge and directions for future research, *Biochem Pharmacol 51*, 1–9.
25. Sevanian, A., Mead, J. F., and Stein, R. A. 1979. Epoxides as products of lipid autoxidation in rat lungs, *Lipids 14*, 634–643.
26. Sevanian, A., and Peterson, A. R. 1986. The cytotoxic and mutagenic properties of cholesterol oxidation products, *Food Chem Toxicol 24*, 1103–1110.
27. Smith, L. L. 1987. Cholesterol autoxidation 1981–1986, *Chem Phys Lipids 44*, 87–125.
28. Morisseau, C. 2013. Role of epoxide hydrolases in lipid metabolism, *Biochimie 95*, 91–95.
29. Morisseau, C., Inceoglu, B., Schmelzer, K., Tsai, H. J., Jinks, S. L., Hegedus, C. M., and Hammock, B. D. 2010. Naturally occurring monoepoxides of eicosapentaenoic acid and docosahexaenoic acid are bioactive antihyperalgesic lipids, *J Lipid Res 51*, 3481–3490.
30. Yin, H., Havrilla, C. M., Morrow, J. D., and Porter, N. A. 2002. Formation of isoprostane bicyclic endoperoxides from the autoxidation of cholesteryl arachidonate, *J Am Chem Soc 124*, 7745–7754.
31. Morrow, J. D., Awad, J. A., Boss, H. J., Blair, I. A., and Roberts, L. J., 2nd. 1992. Non-cyclooxygenase-derived prostanoids (F2-isoprostanes) are formed *in situ* on phospholipids, *Proc Natl Acad Sci USA 89*, 10,721–10,725.
32. Thomas, M. J., Chen, Q., Sorci-Thomas, M. G., and Rudel, L. L. 2001. Isoprostane levels in lipids extracted from atherosclerotic arteries of nonhuman primates, *Free Radic Biol Med 30*, 1337–1346.
33. Waugh, R. J., Morrow, J. D., Roberts, L. J., 2nd, and Murphy, R. C. 1997. Identification and relative quantitation of F2-isoprostane regioisomers formed *in vivo* in the rat, *Free Radic Biol Med 23*, 943–954.
34. Waugh, R. J., and Murphy, R. C. 1996. Mass spectrometric analysis of four regioisomers of F2-Isoprostanes formed by free radical oxidation of arachidonic acid, *J Am Soc Mass Spectrom 7*, 490–499.
35. Wagner, J. D., Thomas, M. J., Williams, J. K., Zhang, L., Greaves, K. A., and Cefalu, W. T. 1998. Insulin sensitivity and cardiovascular risk factors in ovariectomized monkeys with estradiol alone or combined with nomegestrol acetate, *J Clin Endocrinol Metab 83*, 896–901.
36. Yin, H., Davis, T., and Porter, N. A. 2010. Simultaneous analysis of multiple lipid oxidation products *in vivo* by liquid chromatographic-mass spectrometry (LC-MS), *Methods Mol Biol 610*, 375–386.
37. Yin, H., Porter, N. A., and Morrow, J. D. 2005. Separation and identification of F2-isoprostane regioisomers and diastereomers by novel liquid chromatographic/mass spectrometric methods, *J Chromatogr B Analyt Technol Biomed Life Sci 827*, 157–164.

38. Morrow, J. D., Minton, T. A., Badr, K. F., and Roberts, L. J., 2nd. 1994. Evidence that the F2-isoprostane, 8-epi-prostaglandin F2 alpha, is formed in vivo, *Biochim Biophys Acta 1210*, 244–248.

39. Roberts, L. J., 2nd, Moore, K. P., Zackert, W. E., Oates, J. A., and Morrow, J. D. 1996. Identification of the major urinary metabolite of the F2-isoprostane 8-iso-prostaglandin F2alpha in humans, *J Biol Chem 271*, 20,617–20,620.

40. Cracowski, J. L., and Ormezzano, O. 2004. Isoprostanes, emerging biomarkers and potential mediators in cardiovascular diseases, *Eur Heart J 25*, 1675–1678.

41. Mesaros, C., and Blair, I. A. 2012. Target chiral analysis of bioactive arachidonic acid metabolites using liquid-chromagography-mass spectrometry, *Metabolites 2*, 337–365.

42. Bligh, E. G., and Dyer, W. J. 1959. A rapid method of total lipid extraction and purification, *Can J Biochem Physiol 37*, 911–917.

43. Folch, J., Lees, M., and Sloane Stanley, G. H. 1957. A simple method for the isolation and purification of total lipides from animal tissues, *J Biol Chem 226*, 497–509.

44. Iverson, S. J., Lang, S. L., and Cooper, M. H. 2001. Comparison of the Bligh and Dyer and Folch methods for total lipid determination in a broad range of marine tissue, *Lipids 36*, 1283–1287.

45. Matyash, V., Liebisch, G., Kurzchalia, T. V., Shevchenko, A., and Schwudke, D. 2008. Lipid extraction by methyl-tert-butyl ether for high-throughput lipidomics, *J Lipid Res 49*, 1137–1146.

46. Bollinger, J. G., Ii, H., Sadilek, M., and Gelb, M. H. 2010. Improved method for the quantification of lysophospholipids including enol ether species by liquid chromatography-tandem mass spectrometry, *J Lipid Res 51*, 440–447.

47. Serhan, C. N., Chiang, N., and Van Dyke, T. E. 2008. Resolving inflammation: Dual anti-inflammatory and pro-resolution lipid mediators, *Nat Rev Immunol 8*, 349–361.

48. Buckley, C. D., Gilroy, D. W., Serhan, C. N., Stockinger, B., and Tak, P. P. 2013. The resolution of inflammation, *Nat Rev Immunol 13*, 59–66.

49. Wymann, M. P., and Schneiter, R. 2008. Lipid signalling in disease, *Nat Rev Mol Cell Biol 9*, 162–176.

50. Hannun, Y. A., and Obeid, L. M. 2008. Principles of bioactive lipid signalling: Lessons from sphingolipids, *Nat Rev Mol Cell Biol 9*, 139–150.

51. Spiegel, S., and Milstien, S. 2003. Sphingosine-1-phosphate: An enigmatic signalling lipid, *Nat Rev Mol Cell Biol 4*, 397–407.

52. Maceyka, M., Harikumar, K. B., Milstien, S., and Spiegel, S. 2012. Sphingosine-1-phosphate signaling and its role in disease, *Trends Cell Biol 22*, 50–60.

53. Bodennec, J., Brichon, G., Zwingelstein, G., and Portoukalian, J. 2000. Purification of sphingolipid classes by solid-phase extraction with aminopropyl and weak cation exchanger cartridges, *Methods Enzymol 312*, 101–114.

54. Kaluzny, M. A., Duncan, L. A., Merritt, M. V., and Epps, D. E. 1985. Rapid separation of lipid classes in high yield and purity using bonded phase columns, *J Lipid Res 26*, 135–140.

55. Mallet, C. R., Lu, Z., and Mazzeo, J. R. 2004. A study of ion suppression effects in electrospray ionization from mobile phase additives and solid-phase extracts, *Rapid Commun Mass Spectrom 18*, 49–58.

56. Annesley, T. M. 2003. Ion suppression in mass spectrometry, *Clin Chem 49*, 1041–1044.

57. King, R., Bonfiglio, R., Fernandez-Metzler, C., Miller-Stein, C., and Olah, T. 2000. Mechanistic investigation of ionization suppression in electrospray ionization, *J Am Soc Mass Spectrom 11*, 942–950.

58. Köfeler, H. C., Fauland, A., Rechberger, G. N., and Trötzmüller, M. 2012. Mass spectrometry based lipidomics: An overview of metabolites technological platforms, *Metabolites 2*, 19–38.

59. Schmidt, A., Karas, M., and Dulcks, T. 2003. Effect of different solution flow rates on analyte ion signals in nano-ESI MS, or: When does ESI turn into nano-ESI? *J Am Soc Mass Spectrom 14*, 492–500.

60. Kita, Y., Takahashi, T., Uozumi, N., and Shimizu, T. 2005. A multiplex quantitation method for eicosanoids and platelet-activating factor using column-switching reversed-phase liquid chromatography-tandem mass spectrometry, *Anal Biochem 342*, 134–143.

61. Sorci-Thomas, M. G., Owen, J. S., Fulp, B., Bhat, S., Zhu, X., Parks, J. S., Shah, D. et al. 2012. Nascent high density lipoproteins formed by ABCA1 resemble lipid rafts and are structurally organized by three ApoA-I monomers, *J Lipid Res 53*, 1890–1909.

62. Liu, W., Xu, L., Lamberson, C. R., Merkens, L. S., Steiner, R. D., Elias, E. R., Haas, D., and Porter, N. A. 2013. Assays of plasma dehydrocholesteryl esters and oxysterols from Smith–Lemli–Opitz syndrome patients, *J Lipid Res 54*, 244–253.

63. Hannedouche, S., Zhang, J., Yi, T., Shen, W., Nguyen, D., Pereira, J. P., Guerini, D. et al. 2011. Oxysterols direct immune cell migration via EBI2, *Nature 475*, 524–527.

64. Griffiths, W. J., Crick, P. J., and Wang, Y. 2013. Methods for oxysterol analysis: Past, present and future, *Biochem Pharmacol 86*, 3–14.

65. Griffiths, W. J., Crick, P. J., Wang, Y., Ogundare, M., Tuschl, K., Morris, A. A., Bigger, B. W., and Clayton, P. T. 2013. Analytical strategies for characterization of oxysterol lipidomes: Liver X receptor ligands in plasma, *Free Radic Biol Med 59*, 69–84.

66. Griffiths, W. J., and Wang, Y. 2011. Analysis of oxysterol metabolomes, *Biochim Biophys Acta 1811*, 784–799.

Section II

Sites of Biological Actions
of Oxidized Lipids

8 Oxidized Lipids as Damage-Associated Molecular Patterns in Inflammatory Responses

Maria Fedorova and Ralf Hoffmann

CONTENTS

Reactive oxygen and nitrogen species (ROS/RNS) have the ability to modify, either directly or indirectly, the majority of biomolecules, including proteins, lipids, DNA, and carbohydrates.[1,2] More than 200 clinical disorders have been suggested to depend on ROS/RNS, which are considered important initiators and mediators of cancer, heart failure, endothelial dysfunction, atherosclerosis and other cardiovascular disorders, brain degenerative impairments, diabetes, and ischemic pathologies.[3] Recent experimental evidence linked ROS/RNS overproduction to inflammatory processes.[4] Many human disorders, such as atherosclerosis, cardiovascular diseases, rheumatoid arthritis, diabetes, and obesity are accompanied by inflammatory response and innate immune system activation. Inflammation, as a defense mechanism against invading pathogenic microorganisms, usually leads to increased production rates of ROS via activation of neutrophils, which produce superoxide anions in high quantities to fight the pathogens. However, recent data indicate that oxidative stress and oxidized biomolecules are involved in sterile inflammation (i.e., inflammation in the absence of pathogens) that accompanies many human disorders.[5] Oxidized biomolecules were shown to trigger chronic inflammation, which determines the onset of many diseases and favors their progression. The underlying mechanisms by which ROS/RNS modified molecules initiate and participate in inflammatory reactions are mostly unknown and need to be resolved.

OXIDATION AND IMMUNE RESPONSE: DAMAGE-ASSOCIATED MOLECULAR PATTERNS

A new hypothesis links oxidized biomolecules to immune reactions via oxidation of "self" biomolecules generating new damage-associated molecular patterns (DAMPs) or oxidation-specific epitopes (OSE).[6] Oxidation-derived DAMPs (oxDAMPs) are related to certain oxidative stress-associated disorders and can be recognized by pattern recognition receptors (PRRs) of the innate immune system, and thus activate innate or adaptive immune responses.

The relevance of oxidized "self" biomolecules to trigger DAMP production was recognized first in atherosclerosis research linking LDL oxidation to the etiology of the disease. Steinberg and coauthors showed that oxidation converts LDL to the versions recognized by receptors on the surface of macrophages.[7] Two receptors present on macrophages, namely SR-A1 and CD36, bind more than 90% of all oxLDL species. The molecular patterns of oxLDL specifically recognized by CD36 consisted of oxidized phospholipids (oxPL) and oxidized phosphatidylcholines (oxPC) in particular. Oxidation of polyunsaturated fatty acyl moieties in the sn-2 position of PL is a prerequisite of oxidation-specific epitopes or oxDAMPs, but the PC quaternary ammonium head group is essential for ligand recognition. Furthermore, oxPC-modified proteins (oxPC-ApoB, oxPC-BSA) bind well to these receptors and thus can initiate pro-inflammatory cytokine production[8] leading to chronic sterile inflammation.

The "danger model" introduced by Polly Matzinger altered the previously favored "self-nonself" hypothesis of immune response, by postulating that "self" biomolecules are potential DAMPs that can be recognized by germline encoded PRRs, which previously were defined as receptors recognizing only pathogen-associated molecular patterns (PAMPs).[9]

PATTERN RECOGNITION RECEPTORS

DAMP/PAMP recognition and response is an essential part of innate immunity, which is not specific for certain antigens but rather recognizes groups of macromolecules to signal danger. PAMPs or DAMPs signal potential damage via a wide set of PRRs, such as toll-like receptors (TLRs), non-TLRs (e.g., NODs), scavenger receptors (SR; e.g., CD36 and SR-A1), RAGEs, and many others.[10,11] PRRs are often cell associated and can be found on antigen-presenting cells (APCs), cell membranes, intracellular membrane organelles or in the cytosol. A number of secreted PRRs were described including secreted forms of cell-associated PRRs, components of the complement system, lectins, and pentaxins. Most PRRs recognizing PAMPs were recently shown to interact also with endogenous DAMPs, that is, endogenous tissue damage signals produced during necrosis and apoptosis.[11,12] Thus, chaperones, Ca^{2+}-binding proteins, and chromatin-associated proteins released by cell damage can all act as DAMPs and trigger systematic pro-inflammatory responses through activation of immune and endothelial cells.[11] Studies of oxLDL in combination with the initiation of inflammation and activation of an immune response showed the significance of "self" biomolecule oxidative modifications in the formation of oxDAMPs, as most PRRs are able to recognize oxDAMPs and initiate the inflammation and/or immune response.[7]

Scavenger Receptors

Scavenger receptors (SRs) were discovered after the observation that certain molecular patterns on acetyl-LDL can be recognized by receptors of the innate immune system. Steinberg and coworkers showed that oxidation of LDL produces molecular patterns recognized by macrophage receptors triggering internalization of oxLDL followed by the formation of lipid-laden macrophage foam cells.[13,14] This process can be inhibited by antioxidants confirming the oxidation-derived ligand structure. Further SR-AI and SR-AII were purified from bovine macrophages and SR-AI was identified in human atherosclerotic lesions. SR-AI is an important PRRs for multiple pathogen-derived ligands (PAMPs), self-altered (lipo)proteins, and lipids. Further, well-characterized SRs are CD36, CD68, CD163, MACRO, SR-B1, SR-PSOX, LOX-1, SREC-1, CL-P1, FEEL-1/-2, and SCARA5.[13] All bind modified self-molecules, apoptotic cells, cell debris, and different PAMPs, and transduce intracellular signals via pro-inflammatory cytokines in innate immunity cells. One of the main characteristics of SRs and other PRRs is their capability to recognize and bind multiple, structurally diverse ligands. CD36 was identified as a major PRR of macrophages responsible for binding oxLDL. Oxidized PLs are specific ligands for CD36. In particular, oxidized PC species provide at least two different recognition motifs. First, the PC head group of the oxidized but not the native lipid represents an essential epitope.[15] It was also demonstrated that phagocytosis of apoptotic bodies by macrophages is driven by CD36 recognition of oxPC moieties exposed on the surface of dying but not viable cells. Additionally, several pathogens, such as *Staphylococcus aureus*, carry PC groups on their surface that are recognized by SRs. Second, Podrez et al. demonstrated that PC oxidatively cleaved at the *sn-2* position are recognized by the CD36 receptor and this recognition is determined by the structure of the truncated *sn-2* part.[16] The presence of γ-hydroxy(oxo)-α,β-unsaturated groups was shown to be crucial for CD36 recognition. Thus, CD36 is characterized as a multi-ligand receptor with different sites of recognition for different compounds.[17] This confirmed the capability of PRRs to recognize simultaneously multiple epitopes resulting in cooperative high-affinity binding of DAMPs and PAMPs.

Toll-Like Receptors

TLRs were originally described as classical innate immunity PRRs responsible for recognition of various PAMPs. TLR activation often relies on co-receptor proteins. For example, a ligand is recognized by CD14 glycoprotein, the complex binds to TLR4 together with adaptor protein MD-2, and the receptor dimerization initiates a signaling cascade leading to NF-κB activation and pro-inflammatory cytokine expression.[18] CD14 also serves as co-receptor for TLR2. Interestingly, CD36 also presents ligands for TLR2 and TLR6 activation. It has been confirmed that TLRs can bind not only PAMPs but recognize multiple DAMPs including oxidized PL and cholesterol esters. Thus, oxPAPC species can activate TLR4 via CD14 recognition on macrophages and endothelial cells. TLR4 participates in macrophage response to minimally modified LDL[19] and in HeLa production of IL-8 in response to oxPAPC treatment mediated by TLR4 activation.[20] LDL oxidized by incubation with 15-LOX-expressing cells was recognized by CD14 and activated cytokine production via the TLR4/MD-2 complex.[21] The molecular pattern responsible for ligand recognition was a 15-LOX-oxidized CE species and reduction of hydroperoxide diminishes receptor binding.

Recent studies by West et al. revealed that TLR2 recognizes proteins modified by carboxyethylpyrrole (CEP) resulting from docosahexaenoic acid (DHA) oxidation.[22] Oxidation of DHA-containing PL yields 4-hydroxy-7-oxohept-5-enoates, which can modify primary amino groups of lysine residues due to their electrophilicity. Binding of CEP-modified proteins by TLR2 plays a significant role in wound healing and angiogenesis. In a mouse model (wild types versus TLR2$^{-/-}$ mice) of carbon tetrachloride treatment and by coincubation of bone marrow-derived macrophages with *in vitro* oxidized PAPC, Kadl et al.[23] demonstrated induction of TLR2-mediated inflammatory response via JNK and p38 downstream signaling pathway. Interestingly, it was shown that reduction of short-chain products of PAPC oxidation by NaBH$_4$ diminished TLR2 activation,[23] which is in agreement with data from West et al., and might highlight the significance of hydroxy(oxo)-α,β-unsaturated groups in TLR2 recognition of truncated oxPL products.

Overall, a wide panel of oxDAMPs can interact with TLRs[22,24] thereby inducing inflammation even for minimally oxidized LDL, mixtures of oxPL and oxPC, long-chain oxPC containing reactive carbonyls and protein-bound pyrroles resulting from LPP–protein interactions.[22] However, different cells exhibit different responses to oxPL mediated by TLRs. HeLa cells treated with oxPL activate TLR-mediated NF-κB response and increase TNF-α production, whereas other cell lines do not respond via NF-κB/TNF-α axis.[25,26] Thus, oxPAPC is capable to induce mouse lung injury and promote IL-6 expression in macrophages via TLR4 activation but downstream signaling was correlated to TRIF, TRAF6, and IKKε activation.[27] The diversity of TLRs-mediated signaling upon oxPL activation was also reviewed in respect of oxPL interference of TLRs pathogen-derived responses.[28] OxPL were proposed as weak agonists but strong antagonists of TLR-mediated inflammatory responses and thus dampen the acute inflammation derived from classical PAMPs.[23] Number of studies indicated antagonistic or competing effects of oxPL, in particular oxPAPC, in TLRs activation by bacterial PAMPs.[28-30] Thus, oxPAPC was shown to compete for CD14 and MD-2 binding with LPS and inhibiting TLR4 activation and NF-κB downstream signaling.[26] Similarly, TLR2 activation by bacterial PAMPs was inhibited in the presence of oxPL.[26] Antagonistic effects of oxPL were also demonstrated for TLR3 and TLR9 PAMPs-mediated activation.[31,32] It is important to note that the majority of studies regarding oxPL-mediated activation of TLRs used oxidized PC species, which are indeed the most abundant phospholipids *in vivo*. However, effects of other oxidized PLs should be evaluated in details for deeper understanding of oxPL signaling via recognition by TLRs.

Receptors for Advanced Glycation End-Products

An interesting example of PRRs originally identified as oxDAMP binding ligands represent RAGEs. RAGEs are members of the immunoglobulin superfamily that bind a wide range of endogenous ligands. Originally they were identified by their recognition of advanced glycation end-products (AGEs). Afterwards it was shown that advanced oxidation protein products are specific RAGE ligands confirming the role of RAGE in perception of oxDAMPs.[33] RAGEs can also recognize endogenous ligands associated to tissue damage and inflammation, such as S100/calgranulin family polypeptides overexpressed in rheumatoid arthritis. RAGEs are expressed on a variety of cells, including vascular, inflammatory cells, neurons, cardiomyocytes, and epithelial cells. RAGE-mediated signaling pathways were proved to be connected with different MAPKs, such as ERK1/2, p38, and JNK. RAGE expression increased upon inflammatory conditions and activation of RAGE by oxDAMPs and other endogenous ligands triggered inflammatory and CD4+ T cell influx.[34] The key role of RAGE signaling in T cells during adaptive immune response was demonstrated as well. Additionally, soluble forms of RAGE (sRAGE) were identified and their serum levels were shown to correlate negatively with inflammatory disease progression (e.g., diabetes, rheumatoid arthritis, and septic shock). Based on these findings it was suggested that sRAGE can act as a decoy for oxDAMPs, lowering the risk of vascular damage and inflammation.[6] Soluble circulating forms are characteristic for many PRRs, including CD36 and CD14. Other soluble PRRs binding oxDAMPs are lectins, LBP, MD-2, LOX-1, and pentaxins, such as the C-reactive protein (CRP) binding oxPC and oxLDL. Despite some evidence indicating that AGE-modified PL are formed *in vivo*, little is known about its interactions with RAGE.

Currently there is intensive debate about when PRRs appeared during evolution. The current paradigm might be altered toward the concept that PRRs developed first to recognize various DAMPs formed during the development of organisms (e.g., apoptotic cells and oxDAMPs) and later on adapted to recognize also PAMPs. The fact that apoptotic cells are removed via CD36 and SR-A already in embryonic stages when the fetus is still protected from pathogens supports this hypothesis. This is also true for TLR4 and natural antibodies.

C-Reactive Protein

C-reactive protein (CRP) is a widely known acute-phase protein and valuable marker of inflammation. This pentameric protein is produced mostly by hepatocytes and its plasma concentration significantly increases during acute and chronic inflammation.[35] As a general pattern recognition molecule, CRP is able to recognize PCs on the surface of pathogenic microorganisms.[36] CRP was shown to recognize specifically the PC head group in oxPC species in oxLDL and apoptotic cells, but not in native PC lipids. Binding of CRP to oxPC initiates the classical complement pathway as well as opsonization of apoptotic cells and microorganisms to clear them via macrophage uptake. It was also demonstrated that CRP co-localizes in atherosclerotic lesions with PC-containing oxPL.[37,38]

Complement Factor H

Complement Factor H (CFH) or simply Factor H is one of the main inhibitors of the alternative complement pathway and acts both in plasma and on cell surfaces by binding C3b. Binding of CHF to malondialdehyde (MDA) epitopes was demonstrated for necrotic and apoptotic cells as well as for apoptotic blebs. CFH–MDA–adduct complexes bind to C3b and generate locally the anti-inflammatory iC3b fragment. CHF binding of MDA–adducts can inhibit IL-8 and TNFα production.[39]

Natural Antibodies

Another mechanism responsible for immune-mediated clearance of oxDAMPs is connected to atherosclerosis and oxidation of LDL, which was revealed by oxDAMP-specific natural antibodies (nAbs). Generally, nAbs are part of the innate immune system providing a rapid response to invading pathogens as well as eliminating self-specific structures exposed during necrotic and apoptotic cell death. This provides an important housekeeping mechanism by removing damaged cells and cell debris. The role of natural antibodies against oxidation-specific epitopes was demonstrated for oxLDL, where most auto-antibodies produced against Cu(I)-oxidized or MDA-modified LDL belong to the IgM subclass.[40] Antibodies recognized mostly the ApoB protein modified by phosphatidylcholine peroxidation products. The specific IgM antibody clone EO6 isolated from the spleen of nonimmunized atherosclerotic mice recognizes the head group of oxidized but not of native PC. The same nAb bound to the surface of apoptotic cells containing elevated amounts of oxPC lipids. nAbs recognizing oxidized but not natural cardiolipin can distinguish apoptotic from viable cells. The same is true for MDA- and HNE-modified proteins. Thus, nAb can eliminate oxidative DAMPs and thus prevent inflammatory reactions induced by lipid and protein oxidation.[15]

Auto-Antibodies (autoAb) and Immune Complexes

The role of the adaptive immune response triggered by the presence of oxLDL was confirmed by identifying immune complexes of autoAbs directed against oxLDL. Both oxLDL-specific Abs and T cells have been shown to be common in humans. LPP produced from oxLDL can modify extracellular matrix proteins binding to

LDL, such as fibronectin, collagen, and laminin. MDA-modified fibronectin is present in atherosclerotic plaques. Elevated levels of autoAbs directed against ApoB peptides, aldehyde-modified fibronectin and laminin reduce the risk of cardiovascular diseases.[41]

Overall, a vast amount of evidence supports the high significance of oxDAMPs in the development and progression of inflammatory and immune system-mediated disorders, such as atherosclerosis, rheumatoid arthritis, multiple sclerosis, and type 1 diabetes. However, most data concerning the influence of oxDAMPs on the innate and adaptive immune cells are random and poorly systematized. The exact molecular patterns that trigger PRRs recognition have yet to be revealed.

MODIFIED LIPIDS AS oxDAMPs

PHOSPHATIDYLCHOLINES

The relevance of oxidized lipids as oxDAMPs has been mostly studied for PC lipids due to their natural abundance in cell membranes and lipoproteins. Most effects of minimally modified LDL (mmLDL) and oxLDL can be attributed to non-enzymatically oxidized PC lipids. The oxidation of PC containing highly unsaturated PUFA, especially PAPC, by free radicals yields a mixture of diverse products that induce a complex biological response. In HUVEC cells, for example, the expression levels of more than 1000 proteins and the phosphorylation degree of several hundred proteins changed. Recent advances in analytics and the unambiguous chemical synthesis of standards allow the identification of specific pathways trigged or influenced by certain oxPC species.[15,42,43]

Truncated oxPC

Free radical-driven oxidation of PC often produces truncated oxPC species of which most are biologically active, as demonstrated for 1-palmitoyl-2-(5-oxovaleryl)-glycerophosphatidylcholine (POVPC), an aldehyde derived from PAPC oxidation, in proliferation of aortic smooth muscle cells. POVPC formed within or internalized by the cells can activate UDP galactose:glucosylceramide ($\beta1 \rightarrow 4$) galactosyltransferase leading to the production of lactosylceramide (LacCer), which can stimulate NADPH oxidase activity. NADPH oxidase produces superoxide anions that activate p44 MAPK kinase increasing the expression of c-fos and proliferating nuclear antigen (PCNA) leading to cell proliferation.[44] Additionally, POVPC was shown to induce adhesion of monocytes but not neutrophiles to endothelial cells (ECs) by increased expression of the connecting segment 1 (CS-1) domain of fibronectin on EC surfaces.[45] Similar activities were reported for 1-palmitoyl-2-(5-hydroxy-8-oxooct-6-enoyl)glyceroPC (HOOA-PC).

In contrast, 1-palmitoyl-2-glutaryl-glycerophosphatidylcholine (PGPC) stimulates the adhesion of monocytes and neutrophils via increased expression levels of E-selectin and vascular cell adhesion molecule 1 (VCAM-1).[45] PGPC can additionally activate peroxisome proliferator-activated receptor γ (PPARγ) inducing maturation of monocytes to macrophages, CD36 expression and oxLDL uptake.[46] The oxLDL uptake via CD36 activation was demonstrated for several oxPCs, which is initiated by the interaction of oxPC with CD36 followed by Lyn-kinase-dependent phosphorylation of JNK, finally leading through still unknown signaling pathways to membrane invagination and internalization of oxLDL.[47] Other PC species can also bind to CD36, such as oxidized PAPC (e.g., POVPC, oxo-octenedioic acid, keto-octendioyl, and keto-oxo-octenoyl-containing truncated PC) and PLPC (hydroxy(or keto)-dodecenoic acid and hydroxy-oxododecenoyl-containing PC) species. Binding of oxPC to the platelet CD36 receptor increased P-selectin expression at the cellular surface and activated platelet fibronectin receptor integrin αIIβ3, which was shown to sensitize platelet cell activation and trigger the development of a prothrombic phenotype. Additionally, PGPC increases the intracellular Ca^{2+}-concentration, which triggers the expression of the tissue factor on ECs with pro-coagulant activity resulting in enhanced blood clotting.[42]

Both POVPC and PGPC can induce expression of monocyte chemotactic protein-1 (MCP-1) and IL-8 in ECs, which are well-known markers of inflammation. The signaling pathway responsible for both gene expressions was linked with activation of PPARα and transcription of genes of the PPAR response element.

Oxygenated oxPC

Another class of bioactive oxPC are formed by nonenzymatic hydroxylation at the double bounds of PC-esterifed PUFA of which a few were identified in oxLDL, such as 1-palmitoyl-2-(epoxyisoprostane)-glycerophosphatidylcholine (PEIPC) and its dehydrated form 1-palmitoyl-2-(epoxycyclopentenone)-glycerophosphatidylcholine (PECPC). PEIPCs can induce monocyte adhesion to ECs by stimulating the expression of CS-1 fibronectin via the activation of E-type prostaglandin receptor (EP2) on the EC surface.[48] The activation of EP2 generates cAMP as a second messenger, which turns on a kinase cascade with PKA, R-Ras, and PI3K leading to the activation of α5β1 integrin and the deposition of CS-1 fibronectin on the EC surface. Additionally, both PEIPC and PECPC can induce MCP1 and IL-8 synthesis in activated ECs, which is mediated for PEIPC via interaction with the 37 kDa GPI-anchored protein that interacts with TLR4 to initiate a PPARα-mediated signaling cascade finally leading to the expression of pro-inflammatory cytokines.[46]

Interestingly, increased production of cAMP resulting from PEIPC-EP2 binding, as described above, results in the phosphorylation of transcriptional factor CREB (cAMP response binding protein) and transcription of genes regulated by the cAMP response element leading to the expression of anti-inflammatory genes, such as hemeoxygenase 1 (HO-1). Additionally, PEIPC was shown to participate in resolution of acute inflammation via cAMP-mediated decreased TNFα and IL-10 expression. These mechanisms confirm the dual role of oxygenated oxPCs with respect to pro- and anti-inflammatory effects.[42]

oxPC in Mitochondria Dysfunction and Apoptosis

Several oxPC species, such as PGPC, PONPC, and PAzPC, were shown to induce mitochondria dysfunction and promote apopotic cell death. Endogenously produced or internalized via the TMEM30a transporter, truncated oxPC can directly intercalate into the mitochondria membrane and thus induce mitochondria membrane depolarization, swelling, and mitochondria permeability transition (MPT) pore opening.[49] The resulting leakage of cytochrome c into the cytoplasm leads to the activation of the caspase cascade including effector caspase 3.[50] Additionally, opening of MPT pores results in cytoplasmic localization of apoptosis-inducing factor (AIF), which translocates to the nucleus and induces DNA fragmentation. It was demonstrated that ether species of truncated PC are two-fold more effective in mitochondria disruption and apoptosis induction.[51] Additionally, lysoPC can interact with the Bid mitochondrial protein, which possesses lipid binding and transfer activity.[52] The binding alters tertiary structure of Bid stimulating cytochrome c release to the cytoplasm and initiation of caspase-dependent apoptosis.

Platelet-Activating Factor-Like Oxidized PC

1-O-alkyl-2-acetyl-glycerophosphatidylcholine, also known as PAF, is a PC lipid with an ether bond in *sn*-1 position and an acetyl residue in *sn*-2. PAF has a wide range of inflammatory and non-inflammatory actions including the induction of chemotaxis in monocytes and their adhesion to ECs, changes in vascular permeability, and platelet aggregation.[53] PAF acts through highly specific G-protein-coupled receptors, located on the platelet and many other cells involved in inflammatory response, which recognize the ether bond at *sn*-1, short acyl group at *sn*-2 and phosphatidylcholine head group at *sn*-3. PAF synthesis is tightly regulated and mostly performed following the stimulation of inflammatory cells via ether PC remodeling when *sn*-2 FA from the ether PC is removed by PLA_2 and acetylated by acetyl-coenzyme A. However, several oxPC species with truncated PUFA can mimic the PAF activities.[49] These PAF-like oxPC can be recognized by the PAF receptor resulting in platelet activation via increased Ca^{2+} concentration, P-selectin surface deposition and aggregate spreading. The structures of some oxidized lipids with PAF-like activities were determined and confirmed by chemically synthesized standards. Thus, 2-lyso-PAF esters of oxidatively truncated LA, AA, and DHA were shown to induce platelet activation. PAF-like oxPC with short-chain aldehydes and acids at *sn*-2 position were more active agonists then long-chain derivatives.[49]

PHOSPHATIDYLETHANOLAMINES

Phosphatidylethanolamine (PE) lipids are present in all mammalian cells and body fluids, but at lower amounts than PC. Different types of modifications were reported for PE. The oxidation of *sn*-2 esterified PUFA occurs by similar reactions as described for PC species above (Figure 8.1). However, the nucleophilic primary amino group of the head group moiety provides an additional modification site. The biological activities of modified PE have been much less studied than modified PC, although several studies showed that modified PE contributes to inflammatory processes.

PE-Esterified Eicosanoids

Until recently, it was generally accepted that enzymatic oxygenation and formation of various eicosanoids can occur only for free arachidonic acid. However, in the late 1990s it was demonstrated that 15-HETE was formed by alkaline hydrolysis of PL. Later, PE-esterified prostaglandins were identified in stimulated RAW and HCA-7 cells. So far, several families of PE-esterified eicosanoids have been identified, such

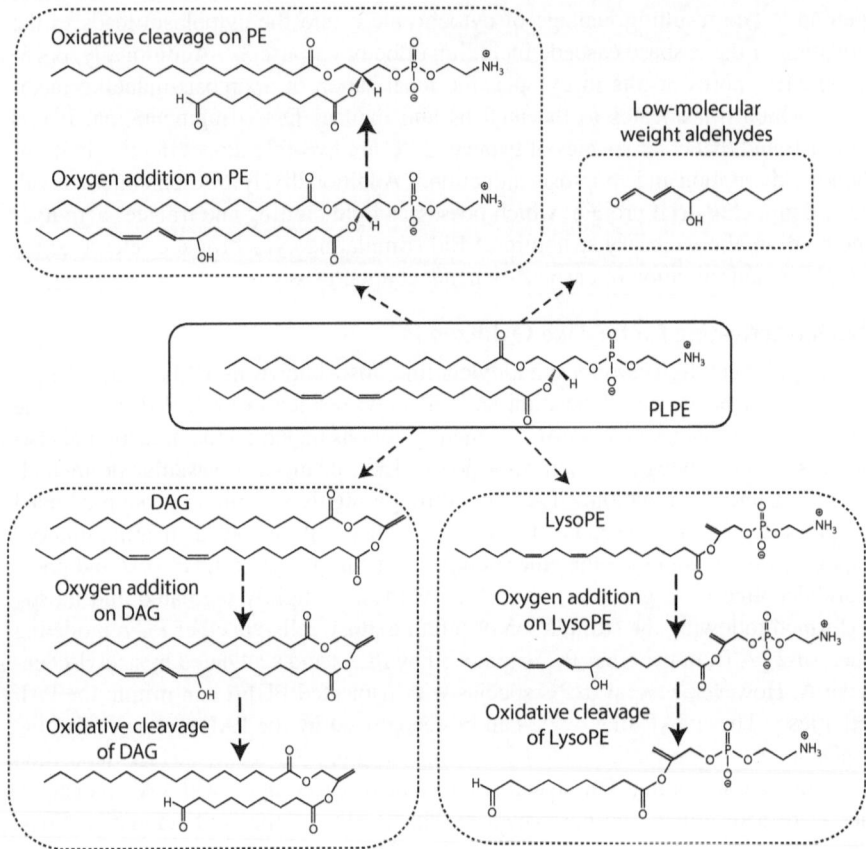

FIGURE 8.1 Oxidation products formed from phosphatidylethanolamine (PLPE) lipid, including low-molecular weight aldehydes, oxygen addition, and oxidative cleavage products of PLPE, corresponding DAG and lysoPE forms.

as 15-HETE-, 12/15-HpETE-, and 12/15-KETE-PE in macrophages, 5-HETE-PE in neutrophils, and 12-HETE-PE in platelets.[54-56] Unlike free eicosanoids, these lipids are usually not secreted and act only locally. 15-HETE-PE generated in macrophages can slightly activate PPARγ and inhibit a LPS-driven inflammatory response. Although the exact mechanisms are still unclear it was assumed that 15-HETE-PE can compete with LPS for CD14 binding and thus modulate the TLR4-mediated response. Similar activities were reported for some oxPC. Overall, 15-HETE-PE mostly acts as an anti-inflammatory agent. Neutrophil generated 5-HETE-PE was shown to be formed via esterification of 5-HETE, generated from free arachidonic acid by enzymatic activity of 5-LOX, and induced superoxide anion production and IL-8 release. Additionally, 5-HETE-PE can inhibit the release of neutrophil extracellular traps. But again, the exact mechanisms of these biological activities of PE-eicosanoids remain to be disclosed.[57]

Head Group Modifications

The amino functions of both PE and PS lipids can be glycated *in vivo* (Figure 8.2). Glycated PE are dominantly formed in glycated LDL and induce production of pro-inflammatory cytokines, such as IL-6, TNFα, and MCP1 in stimulated macrophages.[58] The glycation degree of PE is significantly increased in diabetes patients and appears to be proportional to glycated hemoglobin.[59] Thus it was hypothesized that some hyperglycemia complications might be attributed to glycated PE. Amadori-modified PE, for example, induces proliferation, migration, and secretion of matrix metalloproteinase 2 (MMP2) in human umbilical vein endothelial cells (HUVECs), which play an important role in angiogenesis.[60] Moreover, glycated PE significantly enhances lipid peroxidation even in the absence of transition metal ions. In monocytes and dendritic cells glycated and glycoxidized PE (e.g., PLPE) increase secretion of IL-1β, IL-6, IL-8, MIP1β, and TNFα.[61] It has been noticed that PE with more double bonds in *sn*-2 PUFA showed more glycation (on head group) and oxidation products (on PUFA), which stimulated monocyte and dendritic cells most efficiently.[62]

The primary amino groups present in PE and PS lipids represent major nucleophilic targets for electrophilic compounds produced by lipid oxidation. PE modified by highly reactive isolevuglandins (isoLG or γ-ketoaldehydes formed during oxidation of arachidonic acid via the isoprostane pathway), induced monocyte binding to ECs via higher expression levels of ICAM1, VCAM1, and E-selectin as well as secretion of pro-inflammatory cytokines, such as MCP1 and IL-8.[63,64] Pro-inflammatory effects of PE modified by 4-hydroxy-nonenal (HNE) and malondialdehyde (MDA), another major electrophile produced by lipid oxidation, were demonstrated as well.[65,66] It should be noted that modifications of primary amino groups of PE and PS lipids are poorly characterized and their biological effects have not been much investigated.

LYSOPHOSPHOLIPIDS

PL can be oxidized to lysoforms by nonenzymatic reactions or enzymatic hydrolysis by phospholipase A_2.[67,68] LysoPL accumulate in oxLDL and appear to be associated with the development of atherosclerosis and its complications.[69] Numerous immunomodulator effects were demonstrated for lysoPL, though the receptors responsible for lysoPL action were identified only recently.[70] Similar to PAF, lysoPL is recognized by immune cells via G protein-coupled receptors. Lysophosphatidic acid (LPA) was shown to induce IL-2 expression by T cells, if recognized by endothelial differentiation gene-encoded receptor EDG2, and thus can promote their proliferation. However, it was noticed that expression of EDG2 on T cells is mediated by mitogen activation, whereas unstimulated T cells mostly express the EDR4 form of the receptor. Recognition of LPA by ERG4 results in leukocyte migration and upregulation of matrix metalloproteinases, which are necessary for penetration of T cells into the tissue. This pattern of receptor expression can shift T-cell behavior from proliferation (EDG2-mediated response) to migration (EDG4-mediated response) upon LPA stimulation.[71]

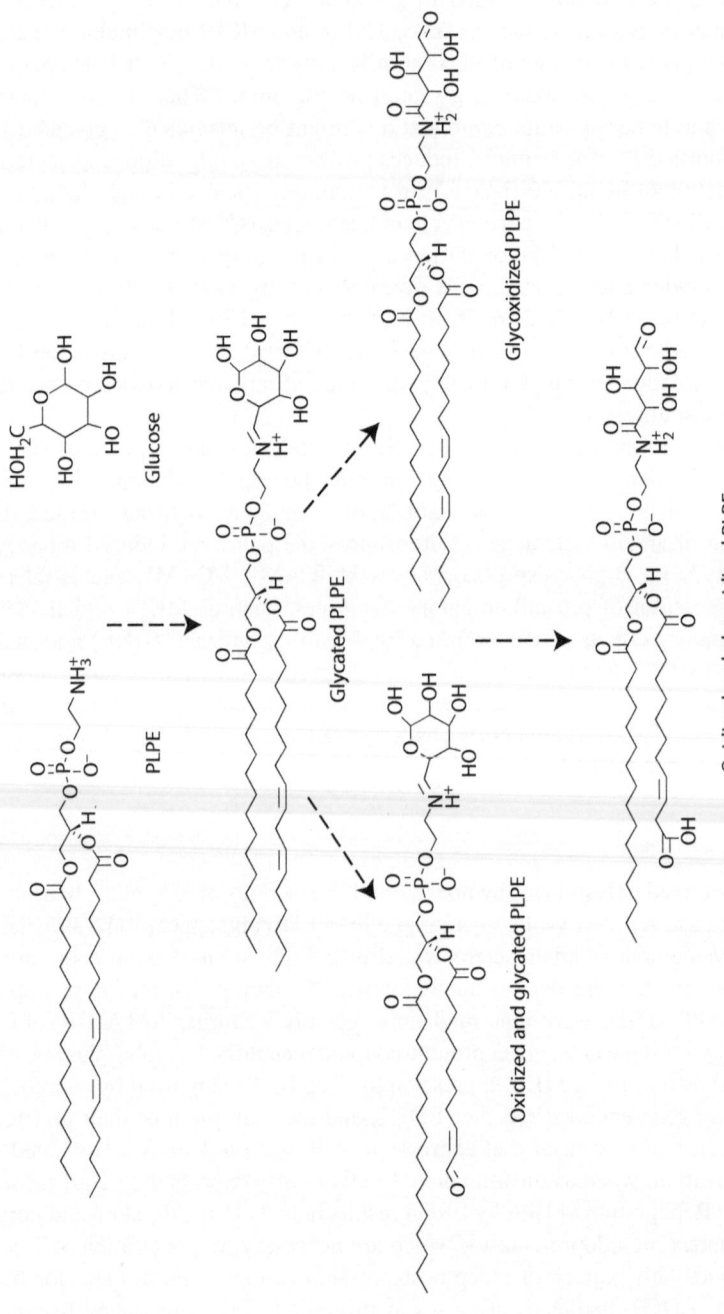

FIGURE 8.2 Glycation and glycoxidation products formed from phosphatidylethanolamine (PLPE) lipid.

Lysophosphatidylcholine (LPC) is one of the most studied lysoPL, for which many biological activities have been reported. LPC is typically present at the 1–5% level in LDL, but represents 40–50% of the whole PC pool in oxLDL.[72] LPC can be generated nonenzymatically or by lecithin–cholesterol acyltransferase in its secreted or lipoprotein-associated form. LPC acts mainly as atherogenic and pro-inflammatory via the G2A receptor. Thus, LPC results in increased expression of main ECs adhesion molecules, such as VCAM-1, intracellular adhesion molecule 1 (ICAM-1) and MCP1. Additionally, LPC was shown to increase the expression of COX1 and eNOS. It can further induce expression of CD36 on monocytes promoting its maturation and transformation into foam cells. A wide range of LPC inflammatory responses on T cells were shown including synthesis of inflammatory mediators (e.g., INFγ and IL-12), increased chemotaxis, and apoptosis.[72]

OXYSTEROLS AND CHOLESTEROL ESTERS

Oxidative modifications of cholesterol and cholesterol esters (CE) have been intensively studied in the context of atherosclerosis and its complications.[73] There are many experimental evidences indicating significant pro-inflammatory and pro-apoptotic responses for this class of compounds.[74,75] The role of oxidized CE in inflammatory responses was shown for polyoxygenated CE hydroperoxides produced by 12/15LOX-catalyzed oxidation, which are strikingly similar to some effects observed for mmLDL on macrophages. CE hydroperoxides can initiate the TLR4-mediated signaling cascade via activation of the TLR4/MD-2 complex and Syk kinase recruitment.[76–78] Syk phosphorylation triggers a cascade of events by activating the ERK1/2 pathway, which produces the N-WASP/Arp2 complex necessary to induce of actin-dependent membrane ruffling, which stimulates phagocytosis of large volumes of extracellular material (including modified and unmodified lipoproteins) resulting in intracellular cholesterol accumulation and foam cell formation. In another study, a mixture of oxysterols resembling the *in vivo* conditions, stimulated expression and synthesis of CD36 PRR on macrophages. Induction of CD36 expression was mediated via PKC delta, ERK1/2 and PPARγ signaling pathways and resulted in an uptake of oxLDL by oxysterol-stimulated macrophages and foam cell formation.[79] Another report revealed the signaling cascade of CD36 expression upon oxysterol treatment, which involved G proteins, c-Src and phospholipase C upstream of PKC-delta signaling.[80]

In addition to the few elucidated mechanisms of oxysterol and CE action, multiple effects on inflammatory cells, ECs, and vascular muscle cells have been demonstrated. 7β-Hydroxycholesterol identified in oxLDL was capable to induce secretion of IL-1β and to increase expression of adhesion molecules on ECs. 7-Keto-cholesterol can polarize both M1 and M2 macrophages toward pro-inflammatory, pro-invasive, and pro-angiogenic phenotypes.[81] A mixture of biologically relevant oxysterols induced expression of β1-integrin on macrophages and thus promoted their adhesion to ECs.[82] Overall, oxysterols can induce the expression of a wide range of cytokines and chemoattractants, such as MCP1, TGF-β1, MIP-1β, TNFα, IL-1β, and IL-8. Additionally, the level of autoantibodies against oxLDL correlates with increasing amounts of hydroxycholesterols in blood plasma of patients with coronary artery disease, which is one of the main atherosclerotic complications.[73]

Several oxysterol classes present in oxLDL appear to be cytotoxic toward fibro-blasts, ECs, and vascular smooth muscle cells, especially 7-hydroperoxycholes-terol (7-OOH-chol), 7β- and 7α-hydroxycholesterol (7-OH-chol), 7-ketocholesterol (7-keto-chol), and cholesterol epoxides (epoxy-chol). 7β-OOH-chol, a precur-sor of hydroxyl- and keto-oxysterols, was reported to be the most toxic.[83] During LDL oxidation 7β-OOH-chol was produced in three to five times higher quanti-ties than 7α-OOH-chol, other oxysterols and even hydroxy-nonenal, which is one of the most abundant lipid oxidation products.[83] Cytotoxicity of oxysterols was con-nected to increased cellular oxidative stress. Some studies suggest that oxysterols are even involved in oxidative stress induction. Animal models indicate that dietary oxysterols can significantly decrease glutathione levels and increase expression of glutathione peroxidase and superoxide dismutase. In apolipoprotein-deficient mice, the NADPH-oxidase activity was induced by 7-keto-chol, 7β-OH-chol, and 5β,6β-epoxy-chol.[84] The increased activity of NADPH oxidase yields more superoxide anions that amplify oxidative stress.

PEPTIDE–LIPID ADDUCTS

Lipids and FA can be oxidatively cleaved to yield carbonylated low-molecular weight lipid peroxidation products (oxoLPP), such as hydroxy/oxoalkenals, epoxy-alkenals, and γ-ketoaldehydes.[85] The fragments contain reactive carbonyl groups, which represent very strong electrophiles, and thus can readily react with nucleo-philic side chains of lysine, cysteine, and histidine residues to predominantly form 1,4-Michael or Schiff base adducts.[86,87] In recent decades, α,β-unsaturated aldehydes and ketones have attracted much attention due to their high reactivity toward nucleophilic protein residues. Multiple inflammatory responses from innate and adaptive immune systems have been described for hydroxy- and oxo-alkenals, which covalently modify extra- and intracellular proteins in many inflammation-related disorders.[88,89] 4-Hydroxy-2-nonenal (HNE), one of the major oxoLPP formed during oxidation of free or PL esterified LA, forms covalent adducts with ApoB100 protein in oxLDL.[90] This adduct was recognized by lectin-like oxidized LDL receptor1 (LOX1), that is, PPRs present on ECs, and connected to the develop-ment of endothelial dysfunction. Additionally, protein adducts with oxoLPP, such as HNE and malondialdehyde (MDA), were shown to activate a Th17 cell response, leading to the production of auto-antibodies directed to HNE- and MDA-modified host molecules.[41] MDA-decorated proteins were also recognized by complement factor H, which blocks their uptake by macrophages.[39] An adduct of ApoB100 with oxidized electrophilic PL was also detected in oxLDL using natural Ab EO6, which recognizes oxPC.[91] The exact biological activities of oxPL-protein adducts still remain to be discovered.

NITRATED FATTY ACIDS

Phospholipids containing unsaturated fatty acids can be nitrated to form nitro-fatty acids or nitroalkenes, as shown in Figure 8.3 for PLPE. Nitroalkenes affect cell signaling in different ways, as shown for cGMP-dependent and -independent

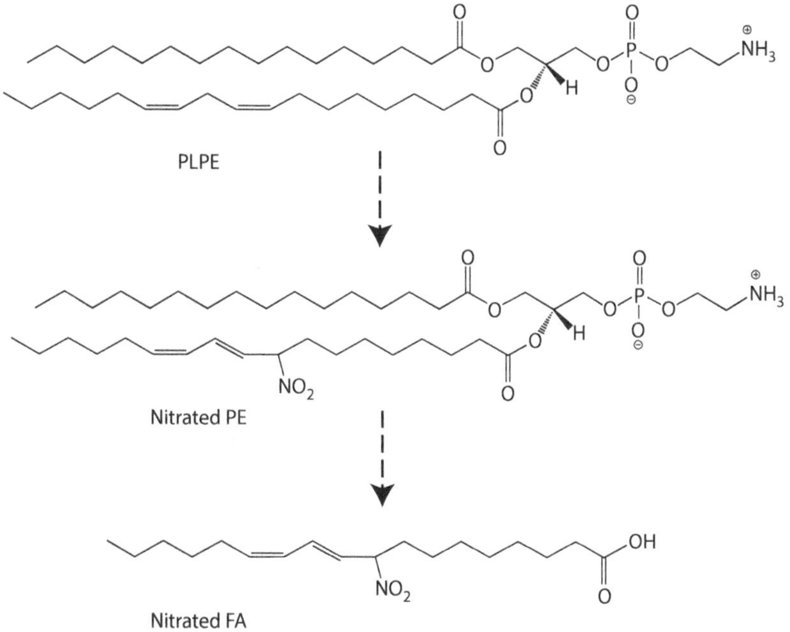

FIGURE 8.3 Nitrated phosphatidylethanolamine (PLPE) lipid and the corresponding nitrated fatty acids.

pathways.[92] Importantly, PUFA-NO$_2$ can decompose via the Nef reaction by releasing HNO and thus represent physiologically relevant NO-donors and NO-deposers. Nitration of PUFA could modulate lipid peroxidation by nitration-mediated antioxidant reactions cleaving the carbon chain.[93] PUFA-NO$_2$ have anti-inflammatory effects and can also modulate activation of the PPAR and inhibit human peripheral blood neutrophil activation and thrombin-induced platelet aggregation.[92,94–96] They can induce apoptosis, as shown for rat aortic smooth muscle cells[97] and regulate the expression for several inflammatory, cell proliferation, and differentiation proteins.[95]

The underlying mechanisms and physiological aspects of PUFA-NO$_2$, however, are poorly understood. It was hypothesized that the regulatory and mediator effects of nitrated fatty acids are mainly connected to their reactivity toward proteins.[98] The electrophilic nature of the β-carbon conjugated to the nitrated carbon atom promotes the interaction with nucleophilic amino acid residues, such as cysteine, lysine, and histidine via Michael addition.[93,98,99] Nitroalkylation of nucleophilic amino acid residues can alter protein structures and protein functions (e.g., enzyme activity, protein interaction, and cofactor binding) as well as the protein subcellular localization by shifting it to more hydrophobic regions.[99] When glyceraldehyde-3-phosphate dehydrogenase (GAPDH) was treated with OA-NO$_2$, overall five modified histidine and cysteine residues were identified. The decreased enzyme activity can be explained by modification of a cysteine residue in the catalytic center, which can be reversed by reduction with glutathione. Nitroalkylation of GAPDH was also confirmed *in vivo* for red blood cells obtained from healthy individuals.[99] Transcriptional factor NF-κB signaling was inhibited by L-NO$_2$ and OA-NO$_2$ via nitroalkylation of the DNA binding site of the p65 subunit.[95]

A proteome wide profiling of PUFA-NO$_2$ targets, which has not been reported so far, could reveal the extent and the cellular distribution of nitroalkylation providing details about the mechanisms underlying NO-mediated fatty acid signaling and regulation. Such a set of modified proteins, especially proteins with reactive cysteine residues, will also advance our understanding of general redox signaling events and their functional link to cysteine-mediated loss of protein functions described for many proteins during oxidative stress.

CONCLUSION

Lipid modifications, induced by various ROS/RNS and driven by enzymatic reactions, significantly enhance the complexity of the lipidome. At the current state of research it is already clear that many modified lipids, if not all, play an important role in inflammatory reactions via activation of innate and adaptive immune responses. Modified endogenous lipids represent powerful DAMPs that are recognized by an array of cellular and soluble PRRs. Recognition by PRRs, such as CD36 scavenger receptor, and TLRs, usually results in increased secretion of pro-inflammatory cytokines and/or expression of ECs adhesion molecules as a primary trigger of inflammation. In most cases lipid-derived DAMPs do not induce acute inflammatory response but rather promote low-level chronic inflammation, which is known to accompany many human disorders such as atherosclerosis, rheumatoid arthritis, diabetes, obesity, and metabolic syndrome. Additionally, some modified lipids, such as PC- and PE-esterified eicosanoids and nitrated FA, promote anti-inflammatory or mixed mode activities. Undoubtedly, advances in analytics, especially high-resolution mass spectrometry, will allow detection and identification of new bioactive lipids. In the future, close interdisciplinary studies combining different analytical techniques, cell biology, and immunology will clarify the role of lipid-derived DAMPs in disease etiology.

REFERENCES

1. Kohen, R.; Nyska, A., Oxidation of biological systems: Oxidative stress phenomena, antioxidants, redox reactions, and methods for their quantification. *Toxicol Pathol* 2002, 30(6), 620–50.
2. Halliwell, B., Biochemistry of oxidative stress. *Biochem Soc Trans* 2007, 35(Pt 5), 1147–50.
3. Garcia, A. G.; Rodriguez-Rocha, H.; Madayiputhiya, N.; Pappa, A.; Panayiotidis, M. I.; Franco, R., Biomarkers of protein oxidation in human disease. *Curr Mol Med* 2012, 12(6), 681–97.
4. Khansari, N.; Shakiba, Y.; Mahmoudi, M., Chronic inflammation and oxidative stress as a major cause of age-related diseases and cancer. *Recent Patents on Inflammation & Allergy Drug Discovery* 2009, 3(1), 73–80.
5. Bartsch, H.; Nair, J., Chronic inflammation and oxidative stress in the genesis and perpetuation of cancer: Role of lipid peroxidation, DNA damage, and repair. *Langenbecks Arch Surg* 2006, 391(5), 499–510.

6. Miller, Y. I.; Choi, S. H.; Wiesner, P.; Fang, L.; Harkewicz, R.; Hartvigsen, K.; Boullier, A. et al. Oxidation-specific epitopes are danger-associated molecular patterns recognized by pattern recognition receptors of innate immunity. *Circ Res* 2011, 108(2), 235–48.

7. Steinberg, D.; Witztum, J. L., Oxidized low-density lipoprotein and atherosclerosis. *Arteriosclerosis, Thrombosis, and Vascular Biology* 2010, 30(12), 2311–6.

8. Levitan, I.; Volkov, S.; Subbaiah, P. V., Oxidized LDL: Diversity, patterns of recognition, and pathophysiology. *Antioxid Redox Signal* 2010, 13(1), 39–75.

9. Matzinger, P., The danger model: A renewed sense of self. *Science* 2002, 296(5566), 301–5.

10. Sirisinha, S., Insight into the mechanisms regulating immune homeostasis in health and disease. *Asian Pac J Allergy Immunol* 2011, 29(1), 1–14.

11. Foell, D.; Wittkowski, H.; Roth, J., Mechanisms of disease: A "DAMP" view of inflammatory arthritis. *Nature Clinical Practice. Rheumatology* 2007, 3(7), 382–90.

12. Foell, D.; Wittkowski, H.; Vogl, T.; Roth, J., S100 proteins expressed in phagocytes: A novel group of damage-associated molecular pattern molecules. *Journal of Leukocyte Biology* 2007, 81(1), 28–37.

13. Greaves, D. R.; Gordon, S., The macrophage scavenger receptor at 30 years of age: Current knowledge and future challenges. *Journal of Lipid Research* 2009 (50 Suppl), S282–6.

14. Krieger, M.; Herz, J., Structures and functions of multiligand lipoprotein receptors: Macrophage scavenger receptors and LDL receptor-related protein (LRP). *Annu Rev Biochem* 1994, 63, 601–37.

15. Weismann, D.; Binder, C. J., The innate immune response to products of phospholipid peroxidation. *Biochim Biophys Acta* 2012, 1818(10), 2465–75.

16. Podrez, E. A.; Poliakov, E.; Shen, Z.; Zhang, R.; Deng, Y.; Sun, M.; Finton, P. J. et al. Identification of a novel family of oxidized phospholipids that serve as ligands for the macrophage scavenger receptor CD36. *J Biol Chem* 2002, 277(41), 38503–16.

17. Nergiz-Unal, R.; Rademakers, T.; Cosemans, J. M.; Heemskerk, J. W., CD36 as a multiple-ligand signaling receptor in atherothrombosis. *Cardiovasc Hematol Agents Med Chem* 2011, 9(1), 42–55.

18. Werts, C.; Tapping, R. I.; Mathison, J. C.; Chuang, T. H.; Kravchenko, V.; Saint Girons, I.; Haake, D. A. et al. Leptospiral lipopolysaccharide activates cells through a TLR2-dependent mechanism. *Nat Immunol* 2001, 2(4), 346–52.

19. Miller, Y. I.; Viriyakosol, S.; Worrall, D. S.; Boullier, A.; Butler, S.; Witztum, J. L., Toll-like receptor 4-dependent and -independent cytokine secretion induced by minimally oxidized low-density lipoprotein in macrophages. *Arterioscler, Thrombosis, Vasc Biol* 2005, 25(6), 1213–9.

20. Walton, K. A.; Hsieh, X.; Gharavi, N.; Wang, S.; Wang, G.; Yeh, M.; Cole, A. L.; Berliner, J. A., Receptors involved in the oxidized 1-palmitoyl-2-arachidonoyl-*sn*-glycero-3-phosphorylcholine-mediated synthesis of interleukin-8. A role for Toll-like receptor 4 and a glycosylphosphatidylinositol-anchored protein. *J Biol Chem* 2003, 278(32), 29661–6.

21. Miller, Y. I.; Viriyakosol, S.; Binder, C. J.; Feramisco, J. R.; Kirkland, T. N.; Witztum, J. L., Minimally modified LDL binds to CD14, induces macrophage spreading via TLR4/MD-2, and inhibits phagocytosis of apoptotic cells. *J Biol Chem* 2003, 278(3), 1561–8.

22. West, X. Z.; Malinin, N. L.; Merkulova, A. A.; Tischenko, M.; Kerr, B. A.; Borden, E. C.; Podrez, E. A.; Salomon, R. G.; Byzova, T. V., Oxidative stress induces angiogenesis by activating TLR2 with novel endogenous ligands. *Nature* 2010, 467(7318), 972–6.

23. Kadl, A.; Sharma, P. R.; Chen, W.; Agrawal, R.; Meher, A. K.; Rudraiah, S.; Grubbs, N.; Sharma, R.; Leitinger, N., Oxidized phospholipid-induced inflammation is mediated by Toll-like receptor 2. *Free Radic Biol Med* 2011, 51(10), 1903–9.

24. Podrez, E. A.; Byzova, T. V.; Febbraio, M.; Salomon, R. G.; Ma, Y.; Valiyaveettil, M.; Poliakov, E. et al. Platelet CD36 links hyperlipidemia, oxidant stress and a prothrombotic phenotype. *Nat Med* 2007, 13(9), 1086–95.

25. Erridge, C.; Webb, D. J.; Spickett, C. M., Toll-like receptor 4 signalling is neither sufficient nor required for oxidised phospholipid mediated induction of interleukin-8 expression. *Atherosclerosis* 2007, 193(1), 77–85.

26. Erridge, C.; Kennedy, S.; Spickett, C. M.; Webb, D. J., Oxidized phospholipid inhibition of toll-like receptor (TLR) signaling is restricted to TLR2 and TLR4: Roles for CD14, LPS-binding protein, and MD2 as targets for specificity of inhibition. *J Biol Chem* 2008, 283(36), 24748–59.

27. Imai, Y.; Kuba, K.; Neely, G. G.; Yaghubian-Malhami, R.; Perkmann, T.; van Loo, G.; Ermolaeva, M. et al. Identification of oxidative stress and Toll-like receptor 4 signaling as a key pathway of acute lung injury. *Cell* 2008, 133(2), 235–49.

28. Greig, F. H.; Kennedy, S.; Spickett, C. M., Physiological effects of oxidized phospholipids and their cellular signaling mechanisms in inflammation. *Free Radic Biol Med* 2012, 52(2), 266–80.

29. Bochkov, V. N.; Kadl, A.; Huber, J.; Gruber, F.; Binder, B. R.; Leitinger, N., Protective role of phospholipid oxidation products in endotoxin-induced tissue damage. *Nature* 2002, 419(6902), 77–81.

30. Subbanagounder, G.; Deng, Y.; Borromeo, C.; Dooley, A. N.; Berliner, J. A.; Salomon, R. G., Hydroxy alkenal phospholipids regulate inflammatory functions of endothelial cells. *Vasc Pharmacol* 2002, 38(4), 201–9.

31. Bluml, S.; Kirchberger, S.; Bochkov, V. N.; Kronke, G.; Stuhlmeier, K.; Majdic, O.; Zlabinger, G. J. et al. Oxidized phospholipids negatively regulate dendritic cell maturation induced by TLRs and CD40. *J Immunol* 2005, 175(1), 501–8.

32. Ma, Z.; Li, J.; Yang, L.; Mu, Y.; Xie, W.; Pitt, B.; Li, S., Inhibition of LPS- and CpG DNA-induced TNF-alpha response by oxidized phospholipids. *Am J Physiol Lung Cell Mol Physiol* 2004, 286(4), L808–16.

33. Yan, S. F.; Ramasamy, R.; Schmidt, A. M., The RAGE axis: A fundamental mechanism signaling danger to the vulnerable vasculature. *Circ Res* 2010, 106(5), 842–53.

34. Ramasamy, R.; Yan, S. F.; Schmidt, A. M., RAGE: Therapeutic target and biomarker of the inflammatory response—The evidence mounts. *J Leukoc Biol* 2009, 86(3), 505–12.

35. Verma, S.; Szmitko, P. E.; Ridker, P. M., C-reactive protein comes of age. *Nature Clinical Practice. Cardiovascular Medicine* 2005, 2(1), 29–36; quiz 58.

36. Lysenko, E.; Richards, J. C.; Cox, A. D.; Stewart, A.; Martin, A.; Kapoor, M.; Weiser, J. N., The position of phosphorylcholine on the lipopolysaccharide of *Haemophilus influenzae* affects binding and sensitivity to C-reactive protein-mediated killing. *Mol Microbiol* 2000, 35(1), 234–45.

37. Singh, U.; Dasu, M. R.; Yancey, P. G.; Afify, A.; Devaraj, S.; Jialal, I., Human C-reactive protein promotes oxidized low density lipoprotein uptake and matrix metalloproteinase-9 release in Wistar rats. *J Lipid Res* 2008, 49(5), 1015–23.

38. van Tits, L.; de Graaf, J.; Toenhake, H.; van Heerde, W.; Stalenhoef, A., C-reactive protein and annexin A5 bind to distinct sites of negatively charged phospholipids present in oxidized low-density lipoprotein. *Arterioscler, Thrombosis, Vasc Biol* 2005, 25(4), 717–22.

39. Weismann, D.; Hartvigsen, K.; Lauer, N.; Bennett, K. L.; Scholl, H. P.; Charbel Issa, P.; Cano, M. et al. Complement factor H binds malondialdehyde epitopes and protects from oxidative stress. *Nature* 2011, 478(7367), 76–81.

40. Binder, C. J., Natural IgM antibodies against oxidation-specific epitopes. *J Clin Immunol* 2010 (30 Suppl 1), S56–60.

41. Duner, P.; To, F.; Berg, K.; Alm, R.; Bjorkbacka, H.; Engelbertsen, D.; Fredrikson, G. N.; Nilsson, J.; Bengtsson, E., Immune responses against aldehyde-modified laminin accelerate atherosclerosis in Apoe-/- mice. *Atherosclerosis* 2010, 212(2), 457–65.
42. Salomon, R. G., Structural identification and cardiovascular activities of oxidized phospholipids. *Circulation Research* 2012, 111(7), 930–46.
43. Bochkov, V. N.; Oskolkova, O. V.; Birukov, K. G.; Levonen, A. L.; Binder, C. J.; Stockl, J., Generation and biological activities of oxidized phospholipids. *Antioxid Redox Signal* 2010, 12(8), 1009–59.
44. Chatterjee, S.; Berliner, J. A.; Subbanagounder, G. G.; Bhunia, A. K.; Koh, S., Identification of a biologically active component in minimally oxidized low density lipoprotein (MM-LDL) responsible for aortic smooth muscle cell proliferation. *Glycoconj J* 2004, 20(5), 331–8.
45. Leitinger, N.; Tyner, T. R.; Oslund, L.; Rizza, C.; Subbanagounder, G.; Lee, H.; Shih, P. T. et al. Structurally similar oxidized phospholipids differentially regulate endothelial binding of monocytes and neutrophils. *Proc Natl Acad Sci USA* 1999, 96(21), 12010–5.
46. Lee, H.; Shi, W.; Tontonoz, P.; Wang, S.; Subbanagounder, G.; Hedrick, C. C.; Hama, S. et al. Role for peroxisome proliferator-activated receptor alpha in oxidized phospholipid-induced synthesis of monocyte chemotactic protein-1 and interleukin-8 by endothelial cells. *Circ Res* 2000, 87(6), 516–21.
47. Rahaman, S. O.; Lennon, D. J.; Febbraio, M.; Podrez, E. A.; Hazen, S. L.; Silverstein, R. L., A CD36-dependent signaling cascade is necessary for macrophage foam cell formation. *Cell Metab* 2006, 4(3), 211–21.
48. Li, R.; Mouillesseaux, K. P.; Montoya, D.; Cruz, D.; Gharavi, N.; Dun, M.; Koroniak, L.; Berliner, J. A., Identification of prostaglandin E2 receptor subtype 2 as a receptor activated by OxPAPC. *Circ Res* 2006, 98(5), 642–50.
49. McIntyre, T. M., Bioactive oxidatively truncated phospholipids in inflammation and apoptosis: Formation, targets, and inactivation. *Biochim Biophys Acta* 2012, 1818(10), 2456–64.
50. Chen, R.; Yang, L.; McIntyre, T. M., Cytotoxic phospholipid oxidation products. Cell death from mitochondrial damage and the intrinsic caspase cascade. *J Biol Chem* 2007, 282(34), 24842–50.
51. Chen, R.; Feldstein, A. E.; McIntyre, T. M., Suppression of mitochondrial function by oxidatively truncated phospholipids is reversible, aided by bid, and suppressed by Bcl-XL. *J Biol Chem* 2009, 284(39), 26297–308.
52. Crimi, M.; Astegno, A.; Zoccatelli, G.; Esposti, M. D., Pro-apoptotic effect of maize lipid transfer protein on mammalian mitochondria. *Arch Biochem Biophys* 2006, 445(1), 65–71.
53. Snyder, F.; Lee, T. C.; Blank, M. L., Platelet-activating factor and related ether lipid mediators. Biological activities, metabolism, and regulation. *Ann N Y Acad Sci* 1989, 568, 35–43.
54. Hammond, V. J.; Morgan, A. H.; Lauder, S.; Thomas, C. P.; Brown, S.; Freeman, B. A.; Lloyd, C. M. et al. Novel keto-phospholipids are generated by monocytes and macrophages, detected in cystic fibrosis, and activate peroxisome proliferator-activated receptor-gamma. *J Biol Chem* 2012, 287(50), 41651–66.
55. Hammond, V. J.; O'Donnell, V. B., Esterified eicosanoids: Generation, characterization and function. *Biochim Biophys Acta* 2012, 1818(10), 2403–12.
56. O'Donnell, V. B., Mass spectrometry analysis of oxidized phosphatidylcholine and phosphatidylethanolamine. *Biochim Biophys Acta* 2011, 1811(11), 818–26.
57. O'Donnell, V. B.; Murphy, R. C., New families of bioactive oxidized phospholipids generated by immune cells: Identification and signaling actions. *Blood* 2012, 120(10), 1985–92.

58. Ravandi, A.; Kuksis, A.; Shaikh, N. A., Glucosylated glycerophosphoethanolamines are the major LDL glycation products and increase LDL susceptibility to oxidation: Evidence of their presence in atherosclerotic lesions. *Arterioscler, Thrombosis, Vasc Biol* 2000, 20(2), 467–77.

59. Breitling-Utzmann, C. M.; Unger, A.; Friedl, D. A.; Lederer, M. O., Identification and quantification of phosphatidylethanolamine-derived glucosylamines and aminoketoses from human erythrocytes—Influence of glycation products on lipid peroxidation. *Arch Biochem Biophys* 2001, 391(2), 245–54.

60. Oak, J. H.; Nakagawa, K.; Oikawa, S.; Miyazawa, T., Amadori-glycated phosphatidylethanolamine induces angiogenic differentiations in cultured human umbilical vein endothelial cells. *FEBS Lett* 2003, 555(2), 419–23.

61. Simoes, C.; Silva, A. C.; Domingues, P.; Laranjeira, P.; Paiva, A.; Domingues, M. R., Modified phosphatidylethanolamines induce different levels of cytokine expression in monocytes and dendritic cells. *Chem Phys Lipids* 2013, 175–176, 57–64.

62. Simoes, C.; Silva, A. C.; Domingues, P.; Laranjeira, P.; Paiva, A.; Domingues, M. R., Phosphatidylethanolamines glycation, oxidation, and glycoxidation: Effects on monocyte and dendritic cell stimulation. *Cell Biochem Biophys* 2013, 66(3), 477–87.

63. Guo, L.; Chen, Z.; Amarnath, V.; Davies, S. S., Identification of novel bioactive aldehyde-modified phosphatidylethanolamines formed by lipid peroxidation. *Free Radic Biol Med* 2012, 53(6), 1226–38.

64. Guo, L.; Chen, Z.; Cox, B. E.; Amarnath, V.; Epand, R. F.; Epand, R. M.; Davies, S. S., Phosphatidylethanolamines modified by gamma-ketoaldehyde (gammaKA) induce endoplasmic reticulum stress and endothelial activation. *J Biol Chem* 2011, 286(20), 18170–80.

65. Bhuyan, K. C.; Master, R. W.; Coles, R. S.; Bhuyan, D. K., Molecular mechanisms of cataractogenesis: IV. Evidence of phospholipid. malondialdehyde adduct in human senile cataract. *Mech Ageing Dev* 1986, 34(3), 289–96.

66. Hagemann, H.; Marcillat, O.; Buchet, R.; Vial, C., Magnesium-adenosine diphosphate binding sites in wild-type creatine kinase and in mutants: Role of aromatic residues probed by Raman and infrared spectroscopies. *Biochemistry* 2000, 39(31), 9251–6.

67. McHowat, J.; Corr, P. B., Thrombin-induced release of lysophosphatidylcholine from endothelial cells. *J Biol Chem* 1993, 268(21), 15605–10.

68. Kougias, P.; Chai, H.; Lin, P. H.; Lumsden, A. B.; Yao, Q.; Chen, C., Lysophosphatidylcholine and secretory phospholipase A2 in vascular disease: Mediators of endothelial dysfunction and atherosclerosis. *Med Sci Monit: Int Med J Exp Clin Res* 2006, 12(1), RA5–16.

69. Tselepis, A. D.; John Chapman, M., Inflammation, bioactive lipids and atherosclerosis: Potential roles of a lipoprotein-associated phospholipase A2, platelet activating factor-acetylhydrolase. *Atherosclerosis. Supplements* 2002, 3(4), 57–68.

70. Graler, M. H.; Goetzl, E. J., Lysophospholipids and their G protein-coupled receptors in inflammation and immunity. *Biochim Biophys Acta* 2002, 1582(1–3), 168–74.

71. Zheng, Y.; Voice, J. K.; Kong, Y.; Goetzl, E. J., Altered expression and functional profile of lysophosphatidic acid receptors in mitogen-activated human blood T lymphocytes. *FASEB J* 2000, 14(15), 2387–9.

72. Matsumoto, T.; Kobayashi, T.; Kamata, K., Role of lysophosphatidylcholine (LPC) in atherosclerosis. *Curr Med Chem* 2007, 14(30), 3209–20.

73. Yasunobu, Y.; Hayashi, K.; Shingu, T.; Yamagata, T.; Kajiyama, G.; Kambe, M., Coronary atherosclerosis and oxidative stress as reflected by autoantibodies against oxidized low-density lipoprotein and oxysterols. *Atherosclerosis* 2001, 155(2), 445–53.

74. Vejux, A.; Malvitte, L.; Lizard, G., Side effects of oxysterols: Cytotoxicity, oxidation, inflammation, and phospholipidosis. *Brazil J Med Biol Res = Rev Brasil Pesquisas Med e Biol/Soc Brasil Biofis... [et al.]* 2008, 41(7), 545–56.

75. Spann, N. J.; Glass, C. K., Sterols and oxysterols in immune cell function. *Nat Immunol* 2013, 14(9), 893–900.
76. Choi, S. H.; Harkewicz, R.; Lee, J. H.; Boullier, A.; Almazan, F.; Li, A. C.; Witztum, J. L.; Bae, Y. S.; Miller, Y. I., Lipoprotein accumulation in macrophages via toll-like receptor-4-dependent fluid phase uptake. *Circ Res* 2009, 104(12), 1355–63.
77. Miller, Y. I.; Choi, S. H.; Fang, L.; Tsimikas, S., Lipoprotein modification and macrophage uptake: Role of pathologic cholesterol transport in atherogenesis. *Sub-Cell Biochem* 2010, 51, 229–51.
78. Murdolo, G.; Bartolini, D.; Tortoioli, C.; Piroddi, M.; Iuliano, L.; Galli, F., Lipokines and oxysterols: Novel adipose-derived lipid hormones linking adipose dysfunction and insulin resistance. *Free Radic Biol Med* 2013, 65C, 811–820.
79. Leonarduzzi, G.; Gamba, P.; Gargiulo, S.; Sottero, B.; Kadl, A.; Biasi, F.; Chiarpotto, E. et al. Oxidation as a crucial reaction for cholesterol to induce tissue degeneration: CD36 overexpression in human promonocytic cells treated with a biologically relevant oxysterol mixture. *Aging Cell* 2008, 7(3), 375–82.
80. Leonarduzzi, G.; Gargiulo, S.; Gamba, P.; Perrelli, M. G.; Castellano, I.; Sapino, A.; Sottero, B.; Poli, G., Molecular signaling operated by a diet-compatible mixture of oxy-sterols in up-regulating CD36 receptor in CD68 positive cells. *Mol Nutr Food Res* 2010, 54 (Suppl 1), S31–41.
81. Buttari, B.; Segoni, L.; Profumo, E.; D'Arcangelo, D.; Rossi, S.; Facchiano, F.; Businaro, R.; Iuliano, L.; Rigano, R., 7-Oxo-cholesterol potentiates pro-inflammatory signaling in human M1 and M2 macrophages. *Biochem Pharmacol* 2013, 86(1), 130–7.
82. Gargiulo, S.; Gamba, P.; Testa, G.; Sottero, B.; Maina, M.; Guina, T.; Biasi, F.; Poli, G.; Leonarduzzi, G., Molecular signaling involved in oxysterol-induced beta1-integrin over-expression in human macrophages. *Int J Mol Sci* 2012, 13(11), 14278–93.
83. Colles, S. M.; Irwin, K. C.; Chisolm, G. M., Roles of multiple oxidized LDL lipids in cellular injury: Dominance of 7 beta-hydroperoxycholesterol. *J Lipid Res* 1996, 37(9), 2018–28.
84. Rosenblat, M.; Aviram, M., Oxysterol-induced activation of macrophage NADPH-oxidase enhances cell-mediated oxidation of LDL in the atherosclerotic apolipopro-tein E deficient mouse: Inhibitory role for vitamin E. *Atherosclerosis* 2002, 160(1), 69–80.
85. Domingues, R. M.; Domingues, P.; Melo, T.; Perez-Sala, D.; Reis, A.; Spickett, C. M., Lipoxidation adducts with peptides and proteins: Deleterious modifications or signaling mechanisms? *J Proteomics* 2013, 92, 110–31.
86. Zhu, X.; Tang, X.; Anderson, V. E.; Sayre, L. M., Mass spectrometric characterization of protein modification by the products of nonenzymatic oxidation of linoleic acid. *Chem Res Toxicol* 2009, 22(8), 1386–97.
87. Doorn, J. A.; Petersen, D. R., Covalent adduction of nucleophilic amino acids by 4-hydroxynonenal and 4-oxononenal. *Chem Biol Interact* 2003, 143–144, 93–100.
88. Zarkovic, N.; Cipak, A.; Jaganjac, M.; Borovic, S.; Zarkovic, K., Pathophysiological relevance of aldehydic protein modifications. *J Proteomics* 2013, 92, 239–47.
89. Uchida, K., Redox-derived damage-associated molecular patterns: Ligand function of lipid peroxidation adducts. *Redox Biol* 2013, 1(1), 94–96.
90. Kumano-Kuramochi, M.; Shimozu, Y.; Wakita, C.; Ohnishi-Kameyama, M.; Shibata, T.; Matsunaga, S.; Takano-Ishikawa, Y. et al. Identification of 4-hydroxy-2-nonenal-histidine adducts that serve as ligands for human lectin-like oxidized LDL receptor-1. *Biochem J* 2012, 442(1), 171–80.
91. Frostegard, J.; Svenungsson, E.; Wu, R.; Gunnarsson, I.; Lundberg, I. E.; Klareskog, L.; Horkko, S.; Witztum, J. L., Lipid peroxidation is enhanced in patients with systemic lupus erythematosus and is associated with arterial and renal disease manifestations. *Arthritis Rheumat* 2005, 52(1), 192–200.

92. Baker, P. R.; Schopfer, F. J.; O'Donnell, V. B.; Freeman, B. A., Convergence of nitric oxide and lipid signaling: Anti-inflammatory nitro-fatty acids. *Free Radic Biol Med* 2009, 46(8), 989–1003.

93. Trostchansky, A.; Rubbo, H., Nitrated fatty acids: Mechanisms of formation, chemical characterization, and biological properties. *Free Radic Biol Med* 2008, 44(11), 1887–96.

94. Baker, P. R.; Lin, Y.; Schopfer, F. J.; Woodcock, S. R.; Groeger, A. L.; Batthyany, C.; Sweeney, S. et al. Fatty acid transduction of nitric oxide signaling: Multiple nitrated unsaturated fatty acid derivatives exist in human blood and urine and serve as endogenous peroxisome proliferator-activated receptor ligands. *J Biol Chem* 2005, 280(51), 42464–75.

95. Cui, T.; Schopfer, F. J.; Zhang, J.; Chen, K.; Ichikawa, T.; Baker, P. R.; Batthyany, C. et al. Nitrated fatty acids: Endogenous anti-inflammatory signaling mediators. *J Biol Chem* 2006, 281(47), 35686–98.

96. Trostchansky, A.; Souza, J. M.; Ferreira, A.; Ferrari, M.; Blanco, F.; Trujillo, M.; Castro, D. et al. Synthesis, isomer characterization, and anti-inflammatory properties of nitroarachidonate. *Biochemistry* 2007, 46(15), 4645–53.

97. Tang, X.; Guo, Y.; Nakamura, K.; Huang, H.; Hamblin, M.; Chang, L.; Villacorta, L.; Yin, K.; Ouyang, H.; Zhang, J., Nitroalkenes induce rat aortic smooth muscle cell apoptosis via activation of caspase-dependent pathways. *Biochem Biophys Res Commun* 2010, 397(2), 239–44.

98. Rudolph, V.; Schopfer, F. J.; Khoo, N. K.; Rudolph, T. K.; Cole, M. P.; Woodcock, S. R.; Bonacci, G. et al. Nitro-fatty acid metabolome: Saturation, desaturation, beta-oxidation, and protein adduction. *J Biol Chem* 2009, 284(3), 1461–73.

99. Batthyany, C.; Schopfer, F. J.; Baker, P. R.; Duran, R.; Baker, L. M.; Huang, Y.; Cervenansky, C.; Branchaud, B. P.; Freeman, B. A., Reversible post-translational modification of proteins by nitrated fatty acids in vivo. *J Biol Chem* 2006, 281(29), 20450–63.

9 Formation and Beneficial Roles of Polyunsaturated Lipid Mediators

Lipoxins, Resolvins, Protectins, and Maresins

Nicos A. Petasis

CONTENTS

INTRODUCTION

Inflammation begins as a protective response to injury or infection and involves the mobilization of a series of local mediators and inflammatory cells. Among the molecules involved in the first line of the inflammatory response are polyunsaturated fatty acids (PUFA) released from local membrane phospholipids. As a result of their high degree of unsaturation, PUFA undergo facile free-radical oxidation with local reactive oxygen species to form lipid oxidation products. More importantly, PUFA also serve as endogenous substrates to oxygenases and associated enzymes that transform them into potent lipid mediators that are key players in the initiation, progression, and resolution of inflammation.

During the initial acute inflammation phase, which is protective in nature, PUFA provide a steady supply of pro-inflammatory chemotactic lipid mediators such as prostaglandins and leukotrienes that help to recruit polymorphonuclear neutrophils

(PMN) and other leukocytes to the local site. If left uncontrolled, however, persistent and chronic inflammation becomes detrimental to the host, and results in a wide range of inflammatory diseases that affect large numbers of people. Among these are some of the greatest health challenges of our time, including cardiovascular disease, rheumatoid arthritis, asthma, diabetes, cancer, periodontal disease, stroke and other neurological disorders, as well as neurodegenerative diseases such as Alzheimer's disease and age-related macular degeneration.

This chapter summarizes recent findings involving the formation and biological roles of several new types of polyunsaturated anti-inflammatory and pro-resolving lipid mediators [1], such as lipoxins, resolvins, protectins, and maresins (MaR). These molecules are able to dampen the inflammatory response, promote the resolution of inflammation, and exhibit other beneficial actions at inflammatory sites [2–4].

LIPID MEDIATORS FROM THE ENZYMATIC OXIDATION OF PUFAs

The many health benefits of PUFA, especially the omega-3 fatty acids, have long been recognized [5], resulting in the growing popularity of omega-3 nutritional supplements and related diets. Despite extensive epidemiological and nutritional evidence for the ability of dietary omega-3 PUFA to improve health and prevent disease, the detailed molecular mechanisms responsible for these benefits has long remained poorly understood. Recent discoveries involving new enzymatic pathways of arachidonic acid (ARA), eicosapentaenoic acid (EPA), and docosahexaenoic acid (DHA) led to the identification of a series of new potent lipid mediators that have beneficial roles in inflammation. These efforts unraveled new enzymatic processes of ARA, EPA, and DHA (Figure 9.1), and significantly expanded the biosynthetic cascades of these fatty acids with the addition of multiple new metabolites with novel structures and biological actions (Figure 9.2) [1].

Arachidonic acid (ARA or AA), a C20:4 omega-6 (ω-6) fatty acid, is the most extensively studied PUFA, which was shown to form numerous oxygenated metabolites (eicosanoids) that were found to play major roles in health and disease. The early discoveries that ARA is converted by cyclooxygenase enzymes (COX1, COX2) to prostaglandins (PG) and by 5-lipoxygenase (5-LO) to leukotrienes (LT), established ARA as a key precursor to these pro-inflammatory lipid mediators. Although the transcellular biosynthesis of the lipoxins (LX) from ARA was also discovered early [6,7], the detailed biological roles of LX remained elusive for nearly a decade, due in part to their limited chemical and biological stability. The study of the metabolic inactivation of LX and the initial design, synthesis and characterization of biostable LX analogs [8–11] led to multiple discoveries that established the lipoxins as major anti-inflammatory and pro-resolving ARA-derived lipid mediators.

EPA, a C20:5 omega-3 (ω-3) fatty acid, is a common ingredient of fish oil and is similar to ARA with an additional double bond. Despite this seemingly small difference from ARA, the enzymatic oxygenation of EPA affords new series of anti-inflammatory metabolites [12], termed E-resolvins (RvE), which were the first omega-3 lipid mediators to be identified for their ability to resolve inflammation via receptor-specific actions [13]. These findings provided the first molecular-level mechanistic evidence for the beneficial roles of omega-3 PUFA in health and disease.

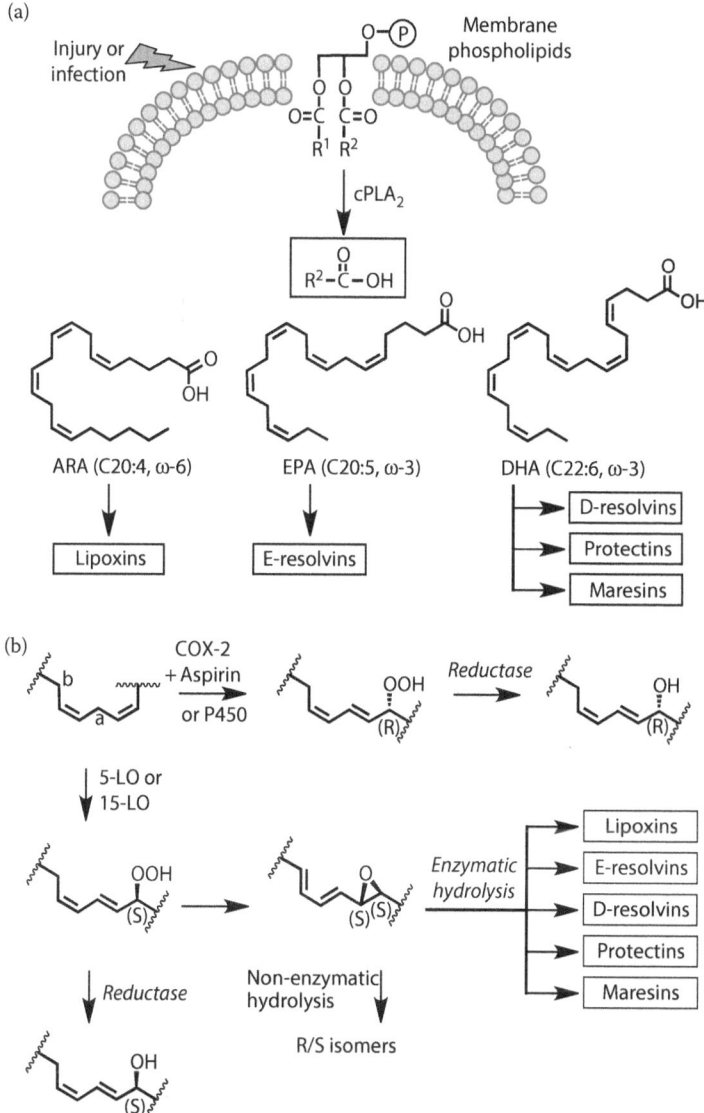

FIGURE 9.1 Common enzymatic processes involved in the formation of anti-inflammatory and pro-resolving lipid mediators from PUFA. (a) Selective release of PUFA catalyzed by cytosolic phospholipase A2; (b) Stereocontrolled oxygenation of PUFA mediated by oxygenase enzymes (COX2/aspirin, P450, 5-LO, 15-LO), followed by peroxidases and epoxide hydrolases.

DHA, a C22:6 omega-3 (ω-3) fatty acid, is a major lipid component in the brain and eye [14], and a common ingredient of fish oil. DHA is an extensively studied PUFA, and its enzymatic oxygenation leads to several series of beneficial lipid mediators (docosanoids), including D-resolvins (RvD) [15], protectins/neuroprotectins (PD/NPD) [16–19], and maresins (MaR) [20–22].

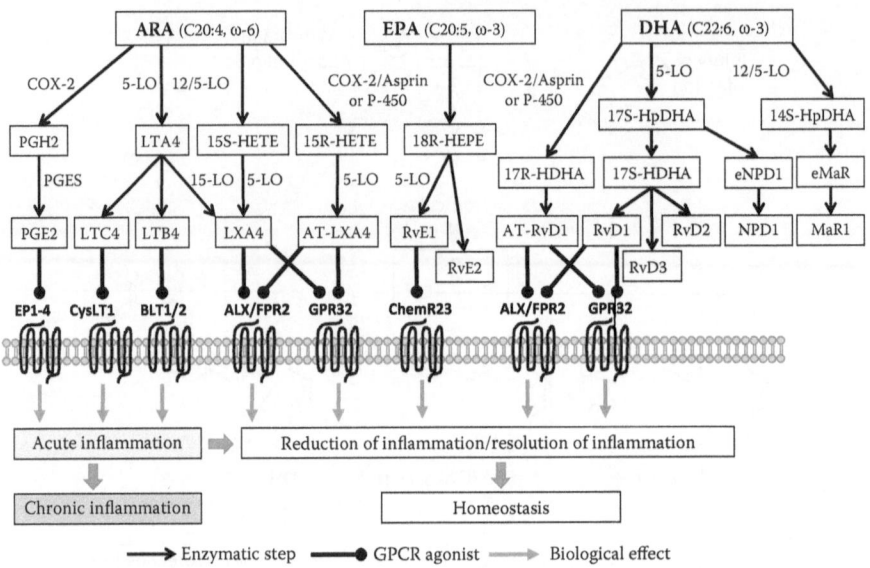

FIGURE 9.2 Biosynthetic pathways and actions of selected lipid mediators derived from ARA, EPA, and DHA.

The biosynthetic formation of the various structural types of PUFA-derived anti-inflammatory and pro-resolving lipid mediators involve common enzymatic processes with certain enzymes and similar biochemical transformations (Figure 9.1). As a result of the inflammatory response, triggered by tissue injury or infection, cytosolic phospholipase A2 (cPLA2) selectively acts at position 2 of phospholipids and catalyzes the release of PUFA, including ARA, EPA, and DHA (Figure 9.1a). The free PUFA and their hydroxylated metabolites serve as substrates of local oxygenating enzymes such as lipoxygenases (LO), as well as aspirin-modified cyclooxygenase-2 (COX2) or cytochrome P450 enzymes.

Of particular significance are 5- and/or 12- or 15-lipoxygenases (5-LO, 12-LO, 15-LO), which catalyze stereocontrolled oxygenation at specific PUFA sites (Figure 9.1b, labeled a, b). Lipoxygenation of PUFA gives predominately (S)-hydroperoxides, which are either reduced to (S)-hydroxy derivatives or are transformed to specific epoxide intermediates. These labile epoxides undergo another stereocontrolled transformation catalyzed by specific epoxide hydrolases, resulting in the formation of potent lipid mediators, including lipoxins, E-resolvins, D-resolvins, protectins/neuroprotectins or maresins, depending on the substrate and oxygenation sites. Nonenzymatic hydrolysis of these epoxides forms mixtures of R/S and Z/E stereoisomers that have much lower bioactivity or are generally inactive.

The two main cyclooxygenase enzymes, the constitutive COX1 and inflammation-induced COX2, are primarily involved in the conversion of ARA to prostaglandins and related eicosanoids. However, in the presence of aspirin these enzymes behave differently. Although aspirin acetylates the active site serine of both COX1 and COX2, only COX1 becomes completely inhibited by aspirin. Despite being unable to produce prostaglandins, acetylated COX2 continues to function as (R)-selective lipoxygenase enzyme with PUFA substrates, including ARA, EPA, and DHA. The resulting (R)-hydroperoxides are converted into (R)-hydroxy derivatives (Figure 9.1b) that can be further transformed into the corresponding lipid mediators, such as aspirin-triggered lipoxins (ATL) and aspirin-triggered resolvins (ATRv).

The most potent and biologically relevant hydroxylated metabolites derived from ARA, EPA, and DHA are formed via a sequence of enzymatic transformations and have stereochemically defined structures with distinct R/S and Z/E configurations. Since these molecules are typically formed in very small quantities, the elucidation of their complete structure and stereochemistry is typically determined via a direct comparison and matching with stereochemically pure materials and related stereoisomers prepared unambiguously via total synthesis [1]. These synthetic molecules also enable the investigation of the biological roles and mechanism of action of these lipid mediators.

The discovery, structural and stereochemical elucidation, and biological investigation of several new series of lipid mediators derived from ARA, EPA, and DHA has expanded significantly the biosynthetic pathways and signaling cascades involved in the inflammatory response (Figure 9.2). The initiation of inflammation involves the enzymatic conversion of ARA into pro-inflammatory prostaglandins (e.g., PGE_2, PGD_2) and leukotrienes (e.g., LTB_4, LTC_4), which upon binding to their perspective G-protein-coupled receptors (GPCR) trigger the expression of chemokines, cytokines, and pro-inflammatory enzymes (COX2, 5-LO). These actions lead quickly to the onset of acute inflammation that characterizes the early phase of host defense. If left uncontrolled, however, this inflammatory response can lead to chronic inflammation, which is implicated in the pathogenesis of numerous diseases [1,3,4].

Initially, the resolution of acute inflammation was thought to be a passive process, that eventually takes place without the active involvement of particular pathways or mediators. Recent investigations, however, revealed that resolution is an active process mediated by certain molecular mediators that activate pro-resolving biochemical pathways and cellular signaling. Notably, several lipid mediators derived from ARA, EPA, and DHA were shown to play major roles in promoting resolution [1–4].

While ARA initially forms pro-inflammatory PG and LT, it eventually participates in a lipid-mediator class switching [23] that involves the formation of lipoxins (LX) and aspirin-triggered lipoxins (ATL), which are now established as anti-inflammatory and pro-resolving lipid mediators acting on newly identified GPCRs (Figure 9.2). Moreover, the resolution of inflammation is also actively promoted by several specialized anti-inflammatory and pro-resolving lipid mediators derived from EPA (e.g., RvE1, RvE2) and DHA (e.g., RvD1, RvD2, RvD3, PD1/NPD1, MaR1).

These new findings have led to a new paradigm for the nature of the inflammatory response (Figure 9.2). Initially, the pathways that generate pro-inflammatory mediators result in acute inflammation, which is followed by pathways that generate anti-inflammatory and pro-resolving mediators that reduce inflammation and lead to resolution of inflammation. The failure of resolution leads to chronic inflammation and the associated diseases.

This chapter is focused on the biosynthesis and actions of lipoxins, resolvins, protectins, and maresins. Each group of PUFA-derived lipid mediators has its own distinct biosynthetic pathways and features a specialized biological activity profile.

LIPOXINS

The transcellular formation of lipoxins (LX) and aspirin-triggered lipoxins (ATL) from ARA [9,24,25] involve the combined actions of two lipoxygenase enzymes (Figure 9.3a). The name *"lipoxin"* was introduced to indicate that they are derived via *lipox*ygenase *in*teractions. Pathway **1** involves the initial lipoxygenation by 5-LO to produce 5S-HpETE, which is transformed into the leukotriene epoxide LTA$_4$. Initially, LTA$_4$ is enzymatically converted into pro-inflammatory leukotrienes (LTB$_4$, LTC$_4$). In the presence of another lipoxygenase (2-LO or 15-LO) LTA$_4$ is also converted into a hydroxylated epoxide intermediate (15S-H, 5S,6S-Epoxide), which is enzymatically converted into lipoxin A$_4$ (LXA$_4$) and the isomeric lipoxin B$_4$ (LXB$_4$). Pathway **2** begins with lipoxygenation by 15-LO to form 15S-HpETE, which is reduced to 15S-HETE by a peroxidase. A second lipoxygenation mediated by 5-LO leads to the same hydroxylated epoxide and subsequently to LXA$_4$ and LXB$_4$. Pathway **3** is initiated by COX2 in the presence of aspirin, or by cytochrome P450. The resulting 15R-HpETE is enzymatically reduced to 15R-HETE, which undergoes further lipoxygenation by 5-LO leading to the 15R-hydroxylated epoxide (15R-H, 5S,6S-epoxide) and subsequently to the aspirin-triggered lipoxins AT-LXA$_4$ and AT-LXB$_4$.

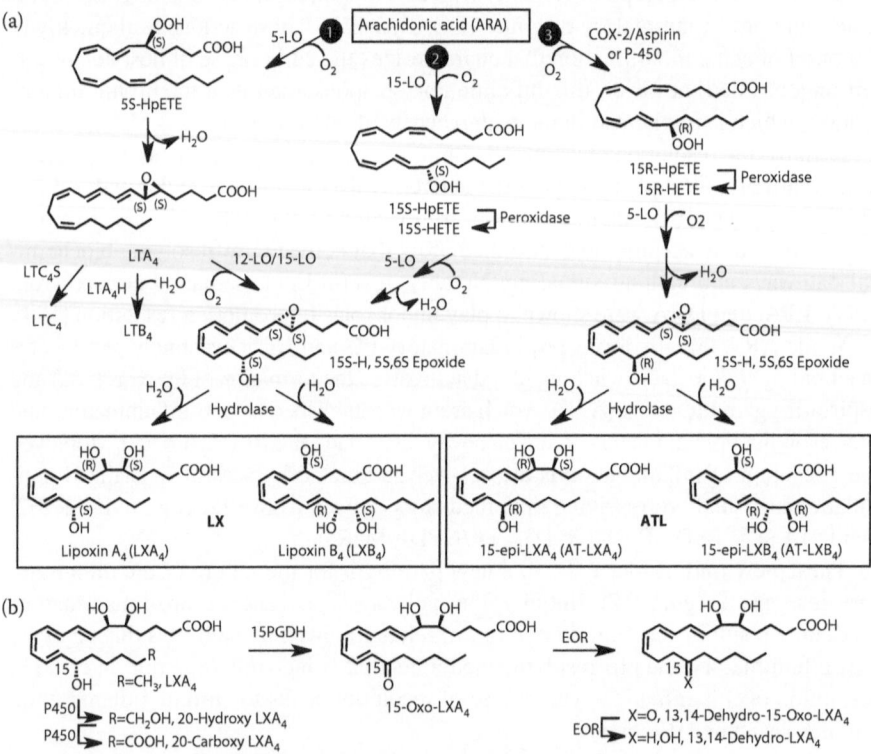

FIGURE 9.3   (a) Biosynthesis of LX and ATL from AA; (b) metabolic inactivation of lipoxin A4.

The involvement of aspirin in ATL biosynthesis revealed a novel therapeutic role of aspirin that does not involve the reduced production of pro-inflammatory lipid mediators (e.g., PGE_2, LTB_4, LTC_4), resulting from the inhibition of COX1 and COX2. Aspirin-mediated acetylation does not completely inactivate COX2, which continues to function as R-selective lipoxygenase that initiates the production of ATL that were found to exert potent anti-inflammatory and pro-resolving actions.

Although the lipoxins were discovered earlier [6,7], the investigation of their endogenous biological roles has been hampered by their relatively short half-life, which is about half a minute in some cell systems. A detailed study [26] showed that the conversion of LXA_4 to its inactive metabolites involves the same dehydrogenase and oxidoreductase enzymes that are involved in the metabolic inactivation of PG and LT [10]. These include 15-prostaglandin dehydrogenase (15-PGDH), the enzyme that inactivates prostaglandin E2 (PGE_2) and eicosanoid oxidoreductase (EOR), the enzyme that inactivates leukotriene B_4 (LTB_4) (Figure 9.3b). Selective dehydrogenation by 15-PGDH of the 15-hydroxyl group of LXA_4 and LXB_4 mimics the same process in PGE_2, forming the corresponding unsaturated ketones. Subsequent reduction by EOR of the double bond adjacent to the ketone carbonyl forms 13,14-dehydro-15-oxo-LXA_4, which is further reduced to the corresponding alcohol 13,14-dehydro-LXA_4. A similar metabolic inactivation of LXB_4 was shown to take place similarly at the C5 position, forming analogous products. Another common metabolic process to PUFA-derived lipid mediators involves omega-oxidation by cytochrome P450. Both LXA_4 and LXB_4 are oxidized by cytochrome P450 at the omega position (C20) to form the corresponding 20-hydroxy and 20-carboxy metabolites (Figure 9.3b). Overall, through sequential redox processes these very potent lipid mediators are rapidly converted into inactive products.

To overcome the oxidative inactivation of LX, the design and synthesis of biostable LX analogs was pursued, resulting in several synthetic molecules with significantly enhanced half-life while retaining all of the properties of the native lipoxins [9]. The first synthetic stable analogs of both LXA_4 [8,10] and LXB_4 [27] served as key molecular probes that enable the first detailed investigation of the bioactivity of these lipid mediators in multiple cell types and disease models. These studies also helped to identify the related molecular pathways and overall endogenous biological roles of these molecules. These LX analogs also served as prototypes for developing analogs and new drug candidates that were advanced toward clinical applications [9,28].

Using these synthetic LX compounds and their stable analogs it became possible to identify key molecular pathways associated with the bioactions of these lipid mediators. It was found that LXA_4 and AT-LXA_4 act as potent and selective agonists of ALX/FPR2 [29], a GPCR involved in the inflammatory response [30,31]. More recently, LXA_4 was also found to act as agonist of GPR32, a GPCR orphan receptor [32]. Additionally, LXA_4 was identified recently as an endogenous allosteric enhancer of the CB1 cannabinoid receptor [33].

Investigations on the actions of lipoxins in various cell types and *in vivo*, helped to establish these lipid mediators as immunomodulators that exhibit potent anti-inflammatory and pro-resolving actions. In addition to reducing excessive and persistent inflammation, LXA_4 and its synthetic stable analogs were the first molecules

TABLE 9.1

Selected Biological Actions of Anti-Inflammatory and Pro-Resolving Lipid Mediators in Disease Models

Lipid Mediator	Disease Models
LXA$_4$, AT-LXA$_4$	Microvasculature [37]; peritonitis [38,53]; dermal inflammation [28,39,40,53]; intestinal inflammation [41,45,54]; acute inflammation [42]; blood [43]; periodontal disease [44,51]; airway inflammation [48]; glomerulonephritis [35]; renal ischemia reperfusion [49,61]; bone loss [51]; cystic fibrosis [55,56]; acute lung injury [57,62]; asthma [58]; corneal wound healing [59]; vascular inflammation [60]; rheumatoid arthritis [63].
RvE1	Colitis [71,72]; periodontitis [73]; asthma [74]; dry eye disease [76]; inflammatory pain [77]; ischemia–reperfusion injury [75]; allergic airway inflammation [78].
RvD1	Vascular inflammation [60]; inflammatory pain [77]; colitis [84]; infection [85]; acute lung injury [86]
RvD2	Sepsis [80]; colitis [84]; obesity [87]; burn wounds [104]
PD1/NPD1, AT-PD1/ AT-NPD1	Retinal diseases [89]; asthma [58]; vascular inflammation [60]; renal ischemia reperfusion [61]; Alzheimer's disease [93]; stroke [94]; infection [85]; neurodegenerative disease [95].
MaR1	Peritonitis [20]; tissue regeneration [21]; pain [21]; rheumatoid arthritis [63]; colitis [96].

of this type to be recognized as potent pro-resolving lipid mediators. At picomolar concentrations, these molecules were able to stimulate nonphlogistic phagocytosis of apoptotic neutrophils (efferocytosis) by monocyte-derived macrophages [34]. Similar effects were also observed with LXB$_4$ [35]. The ability of these lipids to promote clearance of damaged cells and cellular debris resulting from the inflammatory response is now well recognized as an endogenous process that leads to the resolution of inflammation [25,36] and ultimately to homeostasis.

A large number of studies on the biological actions of the lipoxins and their stable analogs revealed the multi-faceted biological roles of this class of lipid mediators [8,10,11,27,28,30,31,35,37–63]. Selected examples of applications in various disease models are summarized in the section "Beneficial Roles of Lipoxins, Resolvins, Protectins, and Maresins" (Table 9.1).

E-RESOLVINS

The resolvins are lipid mediators derived from eicosapentaenoic acid (E-resolvins) or docosahexaenoic acid (D-resolvins), which are generated during the resolution phase of inflammation and act to dampen and *resolve* inflammation.

The formation of E-resolvins (RvE) from EPA was first identified [12] in the presence of COX2 and aspirin or via P450 (Figure 9.4a). The resulting peroxide 18R-HpEPE is reduced to 18R-HEPE, which serves as a substrate of 5-LO to initially form another peroxide intermediate (5S-Hp, 18R-HEPE) that is transformed into the

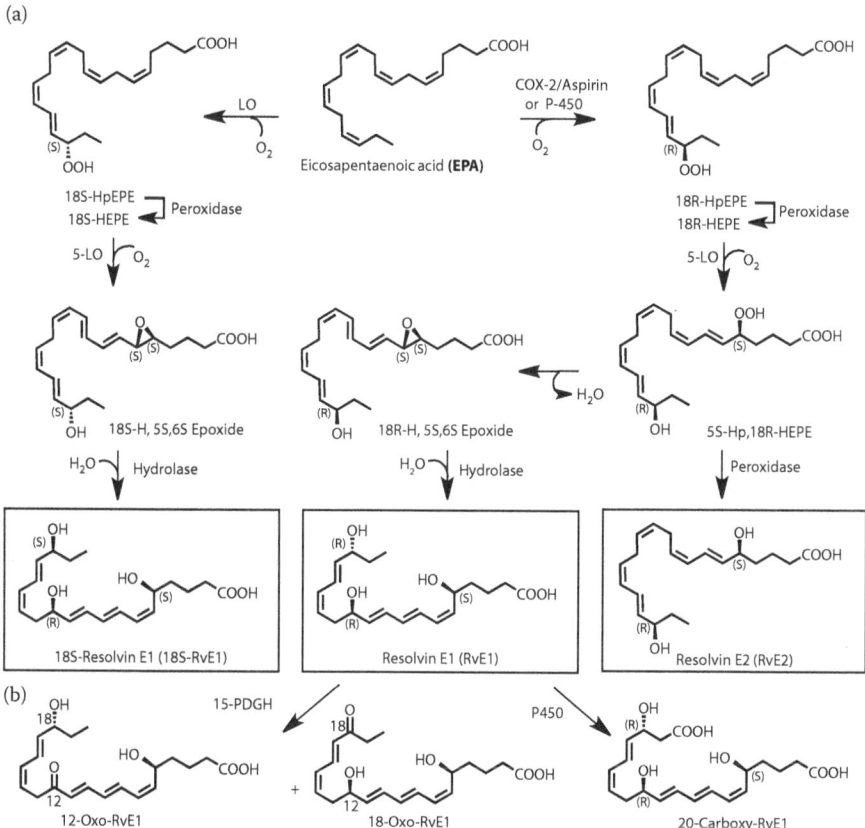

(a)

Eicosapentaenoic acid (EPA)

18S-HpEPE
18S-HEPE

18R-HpEPE
18R-HEPE

18S-H, 5S,6S Epoxide

18R-H, 5S,6S Epoxide

5S-Hp,18R-HEPE

18S-Resolvin E1 (18S-RvE1)

Resolvin E1 (RvE1)

Resolvin E2 (RvE2)

(b)

12-Oxo-RvE1

18-Oxo-RvE1

20-Carboxy-RvE1

FIGURE 9.4   (a) Biosynthesis of E-resolvins (RvE) from EPA; (b) metabolic inactivation of resolvin E1.

18R-H, 5S,6S epoxide, which undergoes enzymatic hydrolysis to form RvE1. The peroxide 5S-Hp, 18R-HEPE is also reduced via a peroxidase to form RvE2 [64,65]. RvE1 was detected in the blood of healthy human donors taking EPA and aspirin, along with the isomeric 18S-RvE1 [66], which is presumably formed via the initial lipoxygenation of EPA to form 18S-HpEPE, followed by similar biosynthetic steps.

The complete stereochemical assignment of RvE1 was established using synthetic stereochemically pure materials [13]. Investigation of the metabolism of RvE1 (Figure 9.4b) [67,68], showed the involvement of dehydrogenase and oxidoreductase enzymes to produce the 12-oxo-RvE1 and 18-oxo-RvE1, similarly to PG, LT, and LX. Other inactive metabolites, such as 20-carboxy-RvE1, are generated from omega-oxidation by cytochrome P450. Synthetic RvE1 analogs were shown to resist metabolic inactivation [67].

The investigation of the biological role of RvE1 revealed potent anti-inflammatory and pro-resolving actions, which are mediated via the GPCR receptor ChemR23 [13] and involves receptor-dependent phosphorylation [69]. Additional actions of RvE1 were also identified with the LTB$_4$ receptor BLT1, where RvE1 serves as a partial agonist dampening the signaling effects of this receptor [70]. Considering that omega-6 fatty acids, such as ARA, cannot produce hydroxylated molecules analogous to RvE1, these receptor-specific actions provided the first clear mechanistic evidence for the biological differences among omega-6 and omega-3 PUFA.

Several studies have investigated the biological activity of RvE1 in various cells and disease models. At nanomolar levels, RvE1 dramatically reduced dermal inflammation, peritonitis, dendritic cell migration, and IL-12 expression [13]. RvE1 also induced the expression of intestinal alkaline phosphatase, an enzyme involved in the detoxification of bacterial LPS by intestinal epithelia [71]. The biological role and activities of RvE1 have been investigated extensively [4,13,66–68,70–78]. Some representative examples involving disease models are summarized in the section "Beneficial Roles of Lipoxins, Resolvins, Protectins, and Maresins" (Table 9.1).

D-RESOLVINS

Similarly to the biosynthesis of lipoxins and E-resolvins, the postulated formation of D-resolvins [15] takes place from DHA and involves similar dual lipoxygenations (Figure 9.5). The action of 15-LO on DHA forms 17S-HpDHA followed

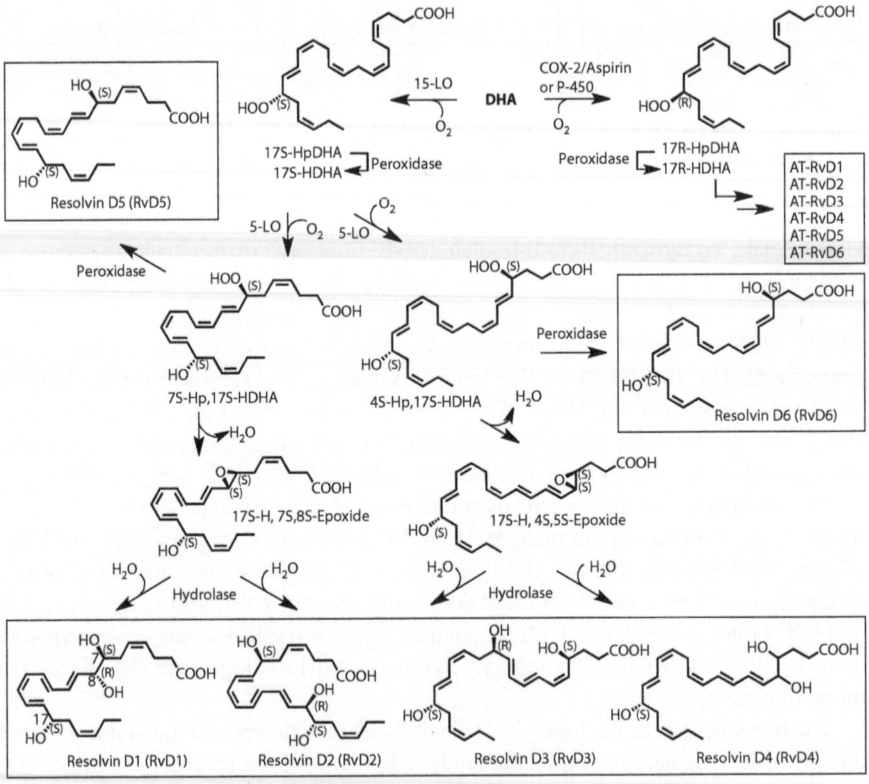

FIGURE 9.5　Biosynthesis of D-resolvins (RvD) from DHA.

by the reduction to 17S-HDHA, which is further oxygenated by 5-LO to form 7S-Hp, 17S-HDHA (7S-hydroperoxy, 17S-hydroxy-DHA) and 4S-Hp, 17S-HDHA (4S-hydroperoxy, 17S-hydroxy-DHA). Both of these hydroperoxide intermediates are then transformed to their perspective epoxides, namely 17S-H, 7S,8S-epoxide, and 17S-H, 4S,5S-epoxide, respectively. Enzymatic hydrolysis of the 7S,8S-epoxide forms resolvin D1 (RvD1) [32,79] and resolvin D2 (RvD2) [80]. Similarly, the enzymatic hydrolysis of the 4S,5S-epoxide forms resolvin D3 (RvD3) [81,82] and resolvin D4 (RvD4). Peroxidase-mediated reduction of 7S-Hp, 17S-HDHA leads to resolvin D5 (RvD5), while 4S-Hp, 17S-HDHA is similarly converted into resolvin D6 (RvD6).

An alternative oxygenation pathway of DHA involving COX2 in the presence of aspirin or via P450 leads to 17R-HDHA, which is converted into the aspirin-triggered D-resolvins (AT-RvD1, AT-RvD2, AT-RvD3, AT-RvD4, AT-RvD5, and AT-RvD6) having the 17R stereochemistry.

The complete stereochemical configuration of RvD1 [79], RvD2 [80], and RvD3 [81,82] have been elucidated by matching with stereochemically pure synthetic materials [1]. An investigation of the metabolism of RvD1 established the involvement of similar pathways with other lipid mediators, and identified several RvD1 metabolites, including 8-oxo-RvD1 and 17-oxo-RvD1, as well C20 oxygenated products derived via omega-oxidation by cytochrome P450 [79].

Several distinct molecular interactions and biological roles for D-resolvins have been identified. In particular, RvD1 was shown to bind to specific sites on human leukocytes and to interact specifically with two GPCRs, namely ALX/FPR2 and GPR32, both of which were also identified as lipoxin receptors [32]. A variety of biological activities in different systems have been reported for RvD1, [32,60,77,79,83–87], including several disease models summarized in the section "Beneficial Roles of Lipoxins, Resolvins, Protectins, and Maresins" (Table 9.1).

PROTECTINS

Initial investigations of DHA metabolites from various cells and tissues has identified a 10,17-dihydroxylated derivative [16,88], which was subsequently termed neuroprotectin D1 (NPD1) [17] due to its protective role in the brain and retina [89].

The complete R/S and Z/E stereochemistry of this docosanoid was established by direct comparison and matching of its chemical and biological properties using several synthetic stereochemically pure stereoisomers [18]. The synthetic materials were also used to investigate in detail the anti-inflammatory properties and other actions of this lipid mediator, which was also termed protectin D1 (PD1) to indicate its broader scope [18].

The elucidation of the stereochemistry of PD1/NPD1 led to its postulated biosynthesis from DHA (Figure 9.6). A single lipoxygenation of DHA by 15-LO forms 17S-HpDHA that is converted to the 16S,17S epoxide (eNPD1), which undergoes enzymatic hydrolysis to form PD1/NPD1 having the 10R,17S configuration. Another investigation using different synthetic diastereomers [90] established the stereochemistry, biosynthesis, and actions of the corresponding aspirin-triggered isomer [19]. This, initial lipoxygenation of DHA by COX2 in the presence of aspirin, or by P450, leads to the formation of the aspirin-triggered stereoisomer AT-PD1/

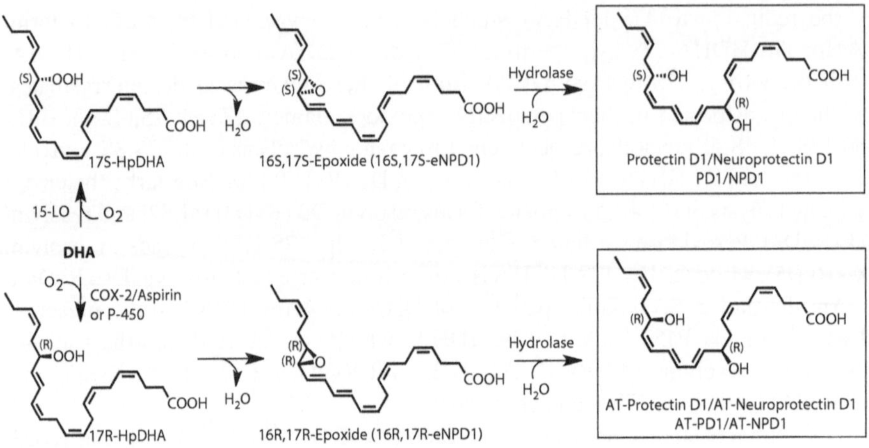

FIGURE 9.6 Biosynthesis of protectins from DHA.

AT-NPD1, having the 10R,17R configuration [19]. Alternative stereoisomers of protectins that are derived via double lipoxygenation of DHA have completely different stereochemical configuration as well as reduced potency and a different bioactivity profile [18].

Neuroprotectin D1/protectin D1 was shown to bind specifically to neutrophils and to human retinal pigment epithelial cells, in a stereoselective manner in comparison with other stereoisomers [91]. Although the detailed molecular mechanisms involving the protectins are still under investigation, several beneficial biological actions have been identified [16–19,58,61,85,88,91–95], including multiple studies involving disease models, which are summarized in the section "Beneficial Roles of Lipoxins, Resolvins, Protectins, and Maresins" (Table 9.1).

MARESINS

The maresins (MaR) are lipid mediators generated from DHA in macrophages and have beneficial actions by enhancing resolution. The name *"maresin"* was coined to indicate that they are *ma*crophage mediators that *res*olve *in*flammation [20].

The formation of maresins from DHA (Figure 9.7) involves a single lipoxygenation by 12-LO or 15-LO to produce the hydroperoxides 14S-HpDHA, which is converted into the 13S,14S epoxide that undergoes enzymatic hydrolysis to form maresin 1 (MaR1). The complete stereochemical configuration of MaR1 was fully determined by comparison with synthetic materials and was shown to have the designated Z/E geometry and the 7R,14S configuration [21].

The postulated epoxide intermediates in the biosynthesis of the above-described lipid mediators is generally difficult to confirm experimentally due to synthetic challenges and the labile nature of these epoxides that are very readily hydrolyzed. Despite these difficulties, the involvement of the 13S,14S epoxide precursor in the biosynthesis of MaR1 (Figure 9.7) was recently demonstrated [22]. Stereochemically pure 13S,14S epoxide, prepared via stereocontrolled total synthesis, upon incubation

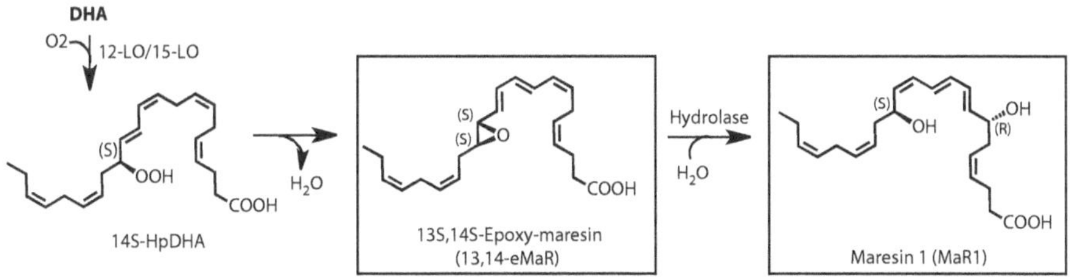

FIGURE 9.7   Biosynthesis of maresins from DHA.

with macrophages was converted into MaR1, confirming its key role in this pathway. This epoxide was termed 13,14-epoxy-maresin (13,14-eMaR), and exhibited its own bioactions, suppressing the production of LTB_4 via LTA_4 hydrolase, and preventing the lipoxygenation of ARA by 12-LO.

While the biological roles of the maresins are still under investigation, some distinct activities have been described [20–22,63,96], including several disease models (Table 9.1).

BENEFICIAL ROLES OF LIPOXINS, RESOLVINS, PROTECTINS, AND MARESINS

The lipid mediators derived from ARA, EPA, and DHA discussed above (i.e., lipoxins, resolvins, protectins, and maresins) have different biosynthetic origins and variable overall bioactivity profiles. These molecules have been collectively termed specialized pro-resolving lipid mediators (SPM), due to their common biological role, namely their ability to reduce the inflammatory response, and to promote the resolution of inflammation and return to homeostasis.

Anti-inflammatory activities involve reduction of pro-inflammatory mediators and limiting trafficking of neutrophils and other pro-inflammatory cells. Pro-resolving actions, however, are distinct from anti-inflammatory actions. Being pro-resolving means the recruitment of cells and the activation of pathways that actively promote the clearance of molecular and cellular debris, as well as damaged or apoptotic cells resulting from the initial injury or infection, or during the inflammatory response. These actions include macrophage-mediated efferocytosis (i.e., removal of intact apoptotic cells), as well as clearance of damage-associated molecular patterns (DAMPs) or pathogen-associated molecular patterns (PAMPs). Ultimately, these pro-resolving actions lead to the dampening of acute inflammation, the prevention of chronic inflammation, the acceleration of healing, and the return to homeostasis.

Several SPM were found to act as endogenous receptor agonists of certain GPCRs that are involved in signaling associated with inflammation and related processes (Figure 9.2). The first of these receptors was ALX/FPR2, which was originally identified as the LXA_4 receptor [29–31] and subsequently as a receptor for RvD1 [32]. The orphan receptor GPR32 was initially identified as the RvD1 receptor, which is also activated by LXA4 [32]. The first receptor identified for a lipid mediator derived from omega-3 fatty acid was ChemR23, which was found to be the RvE1 receptor [13]. Notably, these receptors have multiple agonist ligands, including lipid mediators and peptidic molecules, which exhibit variable potency and distinct cellular responses [29]. The role of microRNAs in regulating the actions of SPM and their receptors has been reported recently, and the existence of miRNA circuits for a given lipid mediator (e.g., RvD1) has been demonstrated [97].

The most well-studied SPM are the lipoxins, particularly LXA_4, $AT-LXA_4$, and their stable analogs [9,25]. The activation of ALX/FPR2 by its endogenous LX or RvD ligands has been investigated in various cell types and some key aspects of the observed anti-inflammatory and pro-resolving activities have been elucidated. In human neutrophils LXA4 dampens the LTB_4-mediated inflammatory response. Upon binding to its neutrophil receptors, LTB_4 promotes rapid remodeling of presqualene diphosphate (PSDP) to presqualene monophosphate (PSMP), that results in the activation of both phospholipase D (PLD) and PI3 kinase (PI3K/Akt), resulting in the activation of NADPH oxidase that produces reactive oxygen species (ROS). It was shown that LXA4 reverses this LTB_4-mediated signaling and inhibits the activation of phospholipase D (PLD) mediated by presqualene diphosphate (PSDP), which was identified as an endogenous regulator of PI3K/Akt [98–100]. The activation of the ALX/FPR2 receptor by LX leads to anti-inflammatory and pro-resolving effects, including a reduction in ROS formation, and a reduction in the gene expression of inflammatory chemokines (e.g., IL-8) and cytokines (e.g., IL-1, TNF-α) mediated by the transcription factor NF-κB, which regulates the immune response.

LXA4 was shown to inhibit peroxynitrite formation, NF-κB and AP-1 activation and IL-8 gene expression in human leukocytes [47]. LXA_4, $AT-LXA_4$ and their stable analogs inhibit transmigration of human neutrophils across epithelial cells [52]. $AT-LXA_4$ reverses the effects of serum amyloid A (SAA), a protein whose levels are elevated during acute inflammation [101]. While SAA activates neutrophils and stimulates both the MAPK kinase/ERK and PI3K/Akt signaling pathways, both of these SAA-mediated effects are reversed upon treatment with $AT-LXA_4$. This lipid mediator also inhibits myeloperoxidase (MPO) signaling (even though it does not inhibit the catalytic activity of MPO), and overrides the suppression apoptosis of human neutrophils, resulting in accelerated resolution of neutrophil-dependent inflammation [62]. LXA_4 stimulates calcium mobilization in monocytes that promotes chemotaxis and adherence, but not in a pro-inflammatory manner. In monocytes and macrophages (but not in neutrophils), LXA_4 was also found to induce actin reorganization mediated by Rho GTPases, resulting in cell mobility [46]. In human T-cells, LXA_4, LXB_4, and their analogs regulate T-cell response in inflammation by inhibiting ERK-dependent secretion of TNF-α [50]. Moreover, it was reported that LXA_4 inhibits the apoptosis of macrophages by activating the PI3K/Akt and ERK/Nrf2 pathways [102]. The combined actions of lipoxins to mobilize noninflammatory phagocytic cells and to also suppress their apoptosis are indicative of their unique ability to promote the resolution of inflammation and prevent tissue damage. If these LX-mediated functions are not operating properly, the resulting sustained inflammatory response can lead to irreversible tissue damage and disease pathogenesis. For example, in patients with cystic fibrosis (CF), a genetic disorder where progressive lung disease leads to recurrent bacterial infections, it was shown that the amount of LXA4 in their airway fluids was substantially decreased [55]. Using a CF mouse model it was shown that treatment with a stable LX analog can restore the normal inflammatory response and prevent uncontrolled infections and airway damage [55,56,103].

The ability of certain lipoxins, resolvins, and protectins to prevent tissue injury and promote wound healing has been shown in a number of disease models. These cytoprotective effects have been characterized in several tissues, including the cornea [59] and the kidney [49,61], and were attributed in part to the amplification of expression of heme oxygenase 1 (HO-1). Important neuroprotective actions were identified for NPD1, which was shown to protect the brain [16,93–95] and the retina [17,91,92].

Depending on the cells or tissues involved in the formation and site of action of these lipid mediators, a number of beneficial roles have been demonstrated *in vivo*. The various SPM acting on specific GPCRs and/or via other mechanisms are associated with particular cell types or tissues, resulting in additional localized actions that are relevant to specific diseases. A wide range of beneficial actions by SPM that are relevant to human disease have been identified in cell studies, disease models, as well as human studies [1,3,4]. Some representative examples are listed in Table 9.1.

CONCLUSIONS

The discovery and investigation of endogenous polyunsaturated lipid mediators, such as lipoxins, resolvins, protectins, and maresins, established several new biomedical paradigms. New modes of enzymatic oxidation of ARA, EPA, and DHA have been identified that produce stereochemically defined oxygenated derivatives with potent receptor-mediated actions. These lipid mediators have unique anti-inflammatory, pro-resolving, and cytoprotective activities, which result in multiple beneficial roles in health and disease. The ability of these lipids to reduce excessive inflammation and promote homeostasis provide new insights for the pathogenesis and therapy of inflammatory diseases. Further investigations of these molecules and their respective molecular and cellular pathways, promise to identify new approaches for managing and treating a wide range of major diseases.

ACKNOWLEDGMENTS

The author thanks his collaborators and the current and past members of his research group for their numerous contributions to this research, exemplified in the cited references. The author also thanks the National Institutes of Health for support of his work in this area under grants DE013499, DE016191, HL079312, AT005909, and GM095467.

REFERENCES

1. Serhan C. N., Petasis N. A. 2011. Resolvins and protectins in inflammation resolution. *Chemical Reviews*. 111(10):5922–5943.
2. Lawrence T., Willoughby D. A., Gilroy D. W. 2002. Anti-inflammatory lipid mediators and insights into the resolution of inflammation. *Nature Reviews Immunology*. 2(10):787–795.

3. Serhan C. N. 2007. Resolution phase of inflammation: Novel endogenous anti-inflammatory and proresolving lipid mediators and pathways. *Annual Review of Immunology.* 25(1):101–137.

4. Serhan C. N., Yacoubian S., Yang R. 2008. Anti-inflammatory and proresolving lipid mediators. *Annual Review of Pathology: Mechanisms of Disease.* 3(1):279–312.

5. Calder P. C. 2012. Mechanisms of action of (n-3) fatty acids. *Journal of Nutrition.* 142(3):592S–599S.

6. Serhan C. N., Hamberg M., Samuelsson B. 1984. Lipoxins: Novel series of biologically active compounds formed from arachidonic acid in human leukocytes. *Proceedings of the National Academy of Sciences.* 81(17):5335–5339.

7. Samuelsson B., Dahlen S. E., Lindgren J. A., Rouzer C. A., Serhan C. N. 1987. Leukotrienes and lipoxins: Structures, biosynthesis, and biological effects. *Science.* 237(4819):1171–1176.

8. Serhan C. N., Maddox J. F., Petasis N. A., Akritopoulou-Zanze I., Papayianni A., Brady H. R., Colgan S. P., Madara J. L. 1995. Design of lipoxin A4 stable analogs that block transmigration and adhesion of human neutrophils. *Biochemistry.* 34(44):14, 609–146, 15.

9. Petasis N. A., Akritopoulou-Zanze I., Fokin V. V., Bernasconi G., Keledjian R., Yang R., Uddin J., Nagulapalli K. C., Serhan C. N. 2005. Design, synthesis and bioactions of novel stable mimetics of lipoxins and aspirin-triggered lipoxins. *Prostaglandins, Leukotrienes and Essential Fatty Acids.* 73(3–4):301–321.

10. Petasis N. A., Keledjian R., Sun Y.-P., Nagulapalli K. C., Tjonahen E., Yang R., Serhan C. N. 2008. Design and synthesis of benzo-lipoxin A4 analogs with enhanced stability and potent anti-inflammatory properties. *Bioorganic & Medicinal Chemistry Letters.* 18(4):1382–1387.

11. Sun Y.-P., Tjonahen E., Keledjian R., Zhu M., Yang R., Recchiuti A., Pillai P. S., Petasis N. A., Serhan C. N. 2009. Anti-inflammatory and pro-resolving properties of benzo-lipoxin A4 analogs. *Prostaglandins, Leukotrienes and Essential Fatty Acids.* 81(5–6):357–366.

12. Serhan C. N., Clish C. B., Brannon J., Colgan S. P., Chiang N., Gronert K. 2000. Novel functional sets of lipid-derived mediators with anti-inflammatory actions generated from omega-3 fatty acids via cyclooxygenase 2-nonsteroidal anti-inflammatory drugs and transcellular processing. *Journal of Experimental Medicine.* 192(8):1197–1204.

13. Arita M., Bianchini F., Aliberti J., Sher A., Chiang N., Hong S., Yang R., Petasis N. A., Serhan C. N. 2005. Stereochemical assignment, anti-inflammatory properties, and receptor for the omega-3 lipid mediator resolvin E1. *Journal of Experimental Medicine.* 201(5):713–722.

14. Bazan N. G., Molina M. F., Gordon W. C. 2011. Docosahexaenoic acid signalolipidomics in nutrition: Significance in aging, neuroinflammation, macular degeneration, Alzheimer's, and other neurodegenerative diseases. *Annual Review of Nutrition.* 31(1):321–351.

15. Serhan C. N., Hong S., Gronert K., Colgan S. P., Devchand P. R., Mirick G., Moussignac R.-L. 2002. Resolvins: A family of bioactive products of omega-3 fatty acid transformation circuits initiated by aspirin treatment that counter proinflammation signals. *Journal of Experimental Medicine.* 196(8):1025–1037.

16. Marcheselli V. L., Hong S., Lukiw W. J., Tian X. H., Gronert K., Musto A., Hardy M. et al. 2003. Novel docosanoids inhibit brain ischemia–reperfusion-mediated leukocyte infiltration and pro-inflammatory gene expression. *Journal of Biological Chemistry.* 278(44):43, 807–43, 817.

17. Mukherjee P. K., Marcheselli V. L., Serhan C. N., Bazan N. G. 2004. Neuroprotectin D1: A docosahexaenoic acid-derived docosatriene protects human retinal pigment epithelial cells from oxidative stress. *Proceedings of the National Academy of Sciences.* 101(22):8491–8496.

18. Serhan C. N., Gotlinger K., Hong S., Lu Y., Siegelman J., Baer T., Yang R., Colgan S. P., Petasis N. A. 2006. Anti-inflammatory actions of neuroprotectin D1/protectin D1 and its natural stereoisomers: Assignments of dihydroxy-containing docosatrienes. *Journal of Immunology.* 176(3):1848–1859.

19. Serhan C. N., Fredman G., Yang R., Karamnov S., Belayev L. S., Bazan N. G., Zhu M., Winkler J. W., Petasis N. A. 2011. Novel proresolving aspirin-triggered DHA pathway. *Chemistry & Biology (Cell).* 18(8):976–987.

20. Serhan C. N., Yang R., Martinod K., Kasuga K., Pillai P. S., Porter T. F., Oh S. F., Spite M. 2009. Maresins: Novel macrophage mediators with potent antiinflammatory and proresolving actions. *Journal of Experimental Medicine.* 206(1):15–23.

21. Serhan C. N., Dalli J., Karamnov S., Choi A., Park C.-K., Xu Z.-Z., Ji R.-R., Zhu M., Petasis N. A. 2012. Macrophage proresolving mediator maresin 1 stimulates tissue regeneration and controls pain. *FASEB Journal.* 26(4):1755–1765.

22. Dalli J., Zhu M., Vlasenko N. A., Deng B., Haeggström J. Z., Petasis N. A., Serhan C. N. 2013. The novel 13S,14S-epoxy-maresin is converted by human macrophages to maresin 1 (MaR1), inhibits leukotriene A4 hydrolase (LTA4H), and shifts macrophage phenotype. *FASEB Journal.* 27(7):2573–2583.

23. Levy B. D., Clish C. B., Schmidt B., Gronert K., Serhan C. N. 2001. Lipid mediator class switching during acute inflammation: Signals in resolution. *Nature Immunology.* 2(7):612–619.

24. Serhan C. N. 2002. Lipoxins and aspirin-triggered 15-epi-lipoxin biosynthesis: An update and role in anti-inflammation and pro-resolution. *Prostaglandins & Other Lipid Mediators.* 68–69:433–455.

25. Serhan C. N. 2005. Lipoxins and aspirin-triggered 15-epi-lipoxins are the first lipid mediators of endogenous anti-inflammation and resolution. *Prostaglandins, Leukotrienes and Essential Fatty Acids.* 73(3–4):141–162.

26. Serhan C. N., Fiore S., Brezinski D. A., Lynch S. 1993. Lipoxin A4 metabolism by differentiated HL-60 cells and human monocytes: Conversion to novel 15-oxo and dihydro products. *Biochemistry.* 32(25):6313–6319.

27. Maddox J. F., Colgan S. P., Clish C. B., Petasis N. A., Fokin V. V., Serhan C. N. 1998. Lipoxin B4 regulates human monocyte/neutrophil adherence and motility: Design of stable lipoxin B4 analogs with increased biologic activity. *FASEB Journal.* 12(6):487–494.

28. Guilford W. J., Bauman J. G., Skuballa W., Bauer S., Wei G. P., Davey D., Schaefer C. et al. 2004. Novel 3-oxa lipoxin A4 analogues with enhanced chemical and metabolic stability have anti-inflammatory activity in vivo. *Journal of Medicinal Chemistry.* 47(8):2157–2165.

29. Chiang N., Serhan C. N., Dahlen S.-E., Drazen J. M., Hay D. W. P., Rovati G. E., Shimizu T., Yokomizo T., Brink C. 2006. The lipoxin receptor ALX: Potent ligand-specific and stereoselective actions in vivo. *Pharmacological Reviews.* 58(3):463–487.

30. Maddox J. F., Hachicha M., Takano T., Petasis N. A., Fokin V. V., Serhan C. N. 1997. Lipoxin A4 stable analogs are potent mimetics that stimulate human monocytes and THP-1 cells via a G-protein-linked lipoxin A4 receptor. *Journal of Biological Chemistry.* 272(11):6972–6978.

31. Takano T., Fiore S., Maddox J. F., Brady H. R., Petasis N. A., Serhan C. N. 1997. Aspirin-triggered 15-epi-lipoxin A4 (LXA4) and LXA4 stable analogues are potent inhibitors of acute inflammation: Evidence for anti-inflammatory receptors. *Journal of Experimental Medicine.* 185(9):1693–1704.

32. Krishnamoorthy S., Recchiuti A., Chiang N., Yacoubian S., Lee C.-H., Yang R., Petasis N. A., Serhan C. N. 2010. Resolvin D1 binds human phagocytes with evidence for proresolving receptors. *Proceedings of the National Academy of Sciences.* 107(4):1660–1665.

33. Pamplona F. A., Ferreira J., Menezes de Lima O., Duarte F. S., Bento A. F., Forner S., Villarinho J. G. et al. 2012. Anti-inflammatory lipoxin A4 is an endogenous allosteric

enhancer of CB1 cannabinoid receptor. *Proceedings of the National Academy of Sciences.* 109(51):21,134–21,139.

34. Godson C., Mitchell S., Harvey K., Petasis N. A., Hogg N., Brady H. R. 2000. Cutting edge: Lipoxins rapidly stimulate nonphlogistic phagocytosis of apoptotic neutrophils by monocyte-derived macrophages. *Journal of Immunology.* 164(4):1663–1667.

35. Mitchell S., Thomas G., Harvey K., Cottell D., Reville K., Bernasconi G., Petasis N. A. et al. 2002. Lipoxins, aspirin-triggered epi-lipoxins, lipoxin stable analogues, and the resolution of inflammation: Stimulation of macrophage phagocytosis of apoptotic neutrophils in vivo. *Journal of the American Society of Nephrology.* 13(10):2497–2507.

36. Schwab J. M., Serhan C. N. 2006. Lipoxins and new lipid mediators in the resolution of inflammation. *Current Opinion in Pharmacology.* 6(4):414–420.

37. Scalia R., Gefen J., Petasis N. A., Serhan C. N., Lefer A. M. 1997. Lipoxin A4 stable analogs inhibit leukocyte rolling and adherence in the rat mesenteric microvasculature: Role of P-selectin. *Proceedings of the National Academy of Sciences.* 94(18):9967–9972.

38. Chiang N., Takano T., Clish C. B., Petasis N. A., Tai H.-H., Serhan C. N. 1998. Aspirin-triggered 15-epi-lipoxin A4 (ATL) generation by human leukocytes and murine peritonitis exudates: Development of a specific 15-epi-LXA4 ELISA. *Journal of Pharmacology and Experimental Therapeutics.* 287(2):779–790.

39. Takano T., Clish C. B., Gronert K., Petasis N., Serhan C. N. 1998. Neutrophil-mediated changes in vascular permeability are inhibited by topical application of aspirin-triggered 15-epi-lipoxin A4 and novel lipoxin B4 stable analogues. *Journal of Clinical Investigation.* 101(4):819–826.

40. Clish C. B., O'Brien J. A., Gronert K., Stahl G. L., Petasis N. A., Serhan C. N. 1999. Local and systemic delivery of a stable aspirin-triggered lipoxin prevents neutrophil recruitment in vivo. *Proceedings of the National Academy of Sciences.* 96(14):8247–8252.

41. Gewirtz A. T., McCormick B., Neish A. S., Petasis N. Á., Gronert K., Serhan C. N., Madara J. L. 1998. Pathogen-induced chemokine secretion from model intestinal epithelium is inhibited by lipoxin A_4 analogs. *Journal of Clinical Investigation.* 101(9):1860–1869.

42. Hachicha M., Pouliot M., Petasis N. A., Serhan C. N. 1999. Lipoxin (LX)A4 and aspirin-triggered 15-epi-LXA4 inhibit tumor necrosis factor 1alpha -initiated neutrophil responses and trafficking: Regulators of a cytokine–chemokine axis. *Journal of Experimental Medicine.* 189(12):1923–1930.

43. Filep J. G., Zouki C., Petasis N. A., Hachicha M., Serhan C. N. 1999. Anti-inflammatory actions of lipoxin A_4 stable analogs are demonstrable in human whole blood: Modulation of leukocyte adhesion molecules and inhibition of neutrophil-endothelial interactions. *Blood.* 94(12):4132–4142.

44. Pouliot M., Clish C. B., Petasis N. A., Van Dyke T. E., Serhan C. N. 2000. Lipoxin A4 analogues inhibit leukocyte recruitment to porphyromonas gingivalis: A role for cyclooxygenase-2 and lipoxins in periodontal disease. *Biochemistry.* 39(16):4761–4768.

45. Goh J., Baird A. W., O'Keane C., Watson R. W. G., Cottell D., Bernasconi G., Petasis N. A., Godson C., Brady H. R., MacMathuna P. 2001. Lipoxin A_4 and aspirin-triggered 15-epi-lipoxin A_4 antagonize TNFa-stimulated neutrophil–enterocyte interactions *in vitro* and attenuate TNFa-induced chemokine release and colonocyte apoptosis in human intestinal mucosa ex vivo. *Journal of Immunology.* 167(5):2772–2780.

46. Maderna P., Cottell D. C., Bernasconi G., Petasis N. A., Brady H. R., Godson C. 2002. Lipoxins induce actin reorganization in monocytes and macrophages but not in neutrophils: Differential involvement of Rho GTPases. *American Journal of Pathology.* 160(6):2275–2283.

47. József L., Zouki C., Petasis N. A., Serhan C. N., Filep J. G. 2002. Lipoxin A4 and aspirin-triggered 15-epi-lipoxin A4 inhibit peroxynitrite formation, NF-kB and AP-1

activation, and IL-8 gene expression in human leukocytes. *Proceedings of the National Academy of Sciences*. 99(20):13, 266–13, 271.

48. Levy B. D., De Sanctis G. T., Devchand P. R., Kim E., Ackerman K., Schmidt B. A., Szczeklik W., Drazen J. M., Serhan C. N. 2002. Multi-pronged inhibition of airway hyper-responsiveness and inflammation by lipoxin A4. *Nature Medicine*. 8(9):1018–1023.

49. Leonard M. O., Hannan K., Burne M. J., Lappin D. W. P., Doran P., Coleman P., Stenson C. et al. 2002. 15-Epi-16-(para-fluorophenoxy)-Lipoxin A4-methyl ester, a synthetic analogue of 15-epi-Lipoxin A4, is protective in experimental ischemic acute renal failure. *Journal of the American Society of Nephrology*. 13(6):1657–1662.

50. Ariel A., Chiang N., Arita M., Petasis N. A., Serhan C. N. 2003. Aspirin-triggered lipoxin A_4 and B_4 analogs block extracellular signal-regulated kinase-dependent TNF-a secretion from human T cells. *Journal of Immunology*. 170(12):6266–6272.

51. Serhan C. N., Jain A., Marleau S., Clish C., Kantarci A., Behbehani B., Colgan S. P. et al. 2003. Reduced inflammation and tissue damage in transgenic rabbits overexpress-ing 15-lipoxygenase and endogenous anti-inflammatory lipid mediators. *Journal of Immunology*. 171(12):6856–6865.

52. Fierro I. M., Colgan S. P., Bernasconi G., Petasis N. A., Clish C. B., Arita M., Serhan C. N. 2003. Lipoxin A4 and aspirin-triggered 15-epi-lipoxin A4 inhibit human neutro-phil migration: Comparisons between synthetic 15 epimers in chemotaxis and transmi-gration with microvessel endothelial cells and epithelial cells. *Journal of Immunology*. 170(5):2688–2694.

53. Bannenberg G., Moussignac R.-L., Gronert K., Devchand P. R., Schmidt B. A., Guilford W. J., Bauman J. G. et al. 2004. Lipoxins and novel 15-epi-lipoxin analogs display potent anti-inflammatory actions after oral administration. *British Journal of Pharmacology*. 143(1):43–52.

54. Fiorucci S., Wallace J. L., Mencarelli A., Distrutti E., Rizzo G., Farneti S., Morelli A. et al. 2004. A β-oxidation-resistant lipoxin A4 analog treats hapten-induced colitis by attenuating inflammation and immune dysfunction. *Proceedings of the National Academy of Sciences*. 101(44):15, 736–15, 741.

55. Karp C. L., Flick L. M., Park K. W., Softic S., Greer T. M., Keledjian R., Yang R. et al. 2004. Defective lipoxin-mediated anti-inflammatory activity in the cystic fibrosis air-way. *Nature Immunology*. 5(4):388–392.

56. Karp C. L., Flick L. M., Yang R., Uddin J., Petasis N. A. 2005. Cystic fibrosis and lipox-ins. *Prostaglandins, Leukotrienes and Essential Fatty Acids*. 73(3–4):263–270.

57. Bonnans C., Fukunaga K., Levy M. A., Levy B. D. 2006. Lipoxin A4 regulates bronchial epithelial cell responses to acid injury. *American Journal of Pathology*. 168(4):1064–1072.

58. Levy B. D., Kohli P., Gotlinger K., Haworth O., Hong S., Kazani S., Israel E., Haley K. J., Serhan C. N. 2007. Protectin D1 is generated in asthma and dampens airway inflam-mation and hyperresponsiveness. *Journal of Immunology*. 178(1):496–502.

59. Biteman B., Hassan I. R., Walker E., Leedom A. J., Dunn M., Seta F., Laniado-Schwartzman M., Gronert K. 2007. Interdependence of lipoxin A4 and heme-oxygen-ase in counter-regulating inflammation during corneal wound healing. *FASEB Journal*. 21(9):2257–2266.

60. Merched A. J., Ko K., Gotlinger K. H., Serhan C. N., Chan L. 2008. Atherosclerosis: Evidence for impairment of resolution of vascular inflammation governed by specific lipid mediators. *FASEB Journal*. 22(10):3595–3606.

61. Hassan I. R., Gronert K. 2009. Acute changes in dietary ω-3 and ω-6 polyunsaturated fatty acids have a pronounced impact on survival following ischemic renal injury and formation of renoprotective docosahexaenoic acid-derived protectin D1. *Journal of Immunology*. 182(5):3223–3232.

62. El Kebir D., Jozsef L., Pan W., Wang L., Petasis N. A., Serhan C. N., Filep J. G. 2009. 15-Epi-lipoxin A4 inhibits myeloperoxidase signaling and enhances resolution of acute lung injury. *American Journal of Respiratory and Critical Care Medicine* 180(4):311–319.

63. Giera M., Ioan-Facsinay A., Toes R., Gao F., Dalli J., Deelder A. M., Serhan C. N., Mayboroda O. A. 2012. Lipid and lipid mediator profiling of human synovial fluid in rheumatoid arthritis patients by means of LC–MS/MS. *Biochimica et Biophysica Acta (BBA)—Molecular and Cell Biology of Lipids*. 1821(11):1415–1424.

64. Oh S. F., Dona M., Fredman G., Krishnamoorthy S., Irimia D., Serhan C. N. 2012. Resolvin E2 formation and impact in inflammation resolution. *Journal of Immunology*. 188(9):4527–4534.

65. Tjonahen E., Oh S. F., Siegelman J., Elangovan S., Percarpio K. B., Hong S., Arita M., Serhan C. N. 2006. Resolvin E2: Identification and anti-inflammatory actions: Pivotal role of human 5-lipoxygenase in resolvin E series biosynthesis. *Chemistry & Biology (Cell)*. 13(11):1193–1202.

66. Oh S. F., Pillai P. S., Recchiuti A., Yang R., Serhan C. N. 2011. Pro-resolving actions and stereoselective biosynthesis of 18S E-series resolvins in human leukocytes and murine inflammation. *Journal of Clinical Investigation*. 121(2):569–581.

67. Arita M., Oh S. F., Chonan T., Hong S., Elangovan S., Sun Y.-P., Uddin J., Petasis N. A., Serhan C. N. 2006. Metabolic inactivation of resolvin E1 and stabilization of its anti-inflammatory actions. *Journal of Biological Chemistry*. 281(32):22, 847–22, 854.

68. Hong S., Porter T. F., Lu Y., Oh S. F., Pillai P. S., Serhan C. N. 2008. Resolvin E1 metabolome in local inactivation during inflammation-resolution. *Journal of Immunology*. 180(5):3512–3519.

69. Ohira T., Arita M., Omori K., Recchiuti A., Van Dyke T. E., Serhan C. N. 2010. Resolvin E1 receptor activation signals phosphorylation and phagocytosis. *Journal of Biological Chemistry*. 285(5):3451–3461.

70. Arita M., Ohira T., Sun Y.-P., Elangovan S., Chiang N., Serhan C. N. 2007. Resolvin E1 selectively interacts with leukotriene B4 receptor BLT1 and ChemR23 to regulate inflammation. *Journal of Immunology*. 178(6):3912–3917.

71. Campbell E. L., MacManus C. F., Kominsky D. J., Keely S., Glover L. E., Bowers B. E., Scully M., Bruyninckx W. J., Colgan S. P. 2010. Resolvin E1-induced intestinal alkaline phosphatase promotes resolution of inflammation through LPS detoxification. *Proceedings of the National Academy of Sciences*. 107(32):14, 298–14, 303.

72. Arita M., Yoshida M., Hong S., Tjonahen E., Glickman J. N., Petasis N. A., Blumberg R. S., Serhan C. N. 2005. Resolvin E1, an endogenous lipid mediator derived from omega-3 eicosapentaenoic acid, protects against 2,4,6-trinitrobenzene sulfonic acid-induced colitis. *Proceedings of the National Academy of Sciences*. 102(21):7671–7676.

73. Hasturk H., Kantarci A., Ohira T., Arita M., Ebrahimi N., Chiang N., Petasis N. A., Levy B. D., Serhan C. N., Van Dyke T. E. 2006. RvE1 protects from local inflammation and osteoclast-mediated bone destruction in periodontitis. *FASEB Journal*. 20(2):401–403.

74. Aoki H., Hisada T., Ishizuka T., Utsugi M., Kawata T., Shimizu Y., Okajima F., Dobashi K., Mori M. 2008. Resolvin E1 dampens airway inflammation and hyperresponsiveness in a murine model of asthma. *Biochemical and Biophysical Research Communications*. 367(2):509–515.

75. Keyes K. T., Ye Y., Lin Y., Zhang C., Perez-Polo J. R., Gjorstrup P., Birnbaum Y. 2010. Resolvin E1 protects the rat heart against reperfusion injury. *American Journal of Physiology—Heart and Circulatory Physiology*. 299(1):H153–H164.

76. Li N., He J., Schwartz C. E., Gjorstrup P., Bazan H. E. P. 2010. Resolvin E1 improves tear production and decreases inflammation in a dry eye mouse model *Journal of Ocular Pharmacology and Therapeutics*. 26(5):431–439.

77. Xu Z.-Z., Zhang L., Liu T., Park J. Y., Berta T., Yang R., Serhan C. N., Ji R.-R. 2010. Resolvins RvE1 and RvD1 attenuate inflammatory pain via central and peripheral actions. *Nature Medicine*. 16(5):592–597.
78. Levy B. D. 2012. Resolvin D1 and resolvin E1 promote the resolution of allergic airway inflammation via shared and distinct molecular counter-regulatory pathways. *Frontiers in Immunology*. 3:1–10.
79. Sun Y.-P., Oh S. F., Uddin J., Yang R., Gotlinger K., Campbell E., Colgan S. P., Petasis N. A., Serhan C. N. 2007. Resolvin D1 and its aspirin-triggered 17R epimer: Stereochemical assignments, anti-inflammatory properties, and enzymatic inactivation. *Journal of Biological Chemistry*. 282(13):9323–9334.
80. Spite M., Norling L. V., Summers L., Yang R., Cooper D., Petasis N. A., Flower R. J., Perretti M., Serhan C. N. 2009. Resolvin D2 is a potent regulator of leukocytes and controls microbial sepsis. *Nature*. 461(7268):1287–1291.
81. Dalli J., Winkler Jeremy W., Colas Romain A., Arnardottir H., Cheng C.-Yee C., Chiang N., Petasis Nicos A., Serhan Charles N. 2013. Resolvin D3 and aspirin-triggered resolvin D3 are potent immunoresolvents. *Chemistry & Biology (Cell)*. 20(2):188–201.
82. Winkler J. W., Uddin J., Serhan C. N., Petasis N. A. 2013. Stereocontrolled total synthesis of the potent anti-inflammatory and pro-resolving lipid mediator resolvin D3 and its aspirin-triggered 17R-epimer. *Organic Letters*. 15(7):1424–1427.
83. Kasuga K., Yang R., Porter T. F., Agrawal N., Petasis N. A., Irimia D., Toner M., Serhan C. N. 2008. Rapid appearance of resolvin precursors in inflammatory exudates: Novel mechanisms in resolution. *Journal of Immunology*. 181(12):8677–8687.
84. Bento A. F., Claudino R. F., Dutra R. C., Marcon R., Calixto J. o. B. 2011. Omega-3 fatty acid-derived mediators 17(R)-hydroxy docosahexaenoic acid, aspirin-triggered resolvin D1 and resolvin D2 prevent experimental colitis in mice. *Journal of Immunology*. 187(4):1957–1969.
85. Chiang N., Fredman G., Backhed F., Oh S. F., Vickery T., Schmidt B. A., Serhan C. N. 2012. Infection regulates pro-resolving mediators that lower antibiotic requirements. *Nature*. 484(7395):524–528.
86. Eickmeier O., Seki H., Haworth O., Hilberath J. N., Gao F., Uddin M., Croze R. H., Carlo T., Pfeffer M. A., Levy B. D. 2013. Aspirin-triggered resolvin D1 reduces mucosal inflammation and promotes resolution in a murine model of acute lung injury. *Mucosal Immunology*. 6(2):256–266.
87. Clària J., Dalli J., Yacoubian S., Gao F., Serhan C. N. 2012. Resolvin D1 and Resolvin D2 govern local inflammatory tone in obese fat. *Journal of Immunology*. 189(5):2597–2605.
88. Hong S., Gronert K., Devchand P. R., Moussignac R.-L., Serhan C. N. 2003. Novel docosatrienes and 17S-resolvins generated from docosahexaenoic acid in murine brain, human blood, and glial cells: Autacoids in anti-inflammation. *Journal of Biological Chemistry*. 278(17):14, 677–14, 687.
89. Bazan N. G. 2006. Cell survival matters: Docosahexaenoic acid signaling, neuroprotection and photoreceptors. *Trends in Neurosciences*. 29(5):263–271.
90. Petasis N. A., Yang R., Winkler J. W., Zhu M., Uddin J., Bazan N. G., Serhan C. N. 2012. Stereocontrolled total synthesis of Neuroprotectin D1/Protectin D1 and its aspirin-triggered stereoisomer. *Tetrahedron Letters*. 53(14):1695–1698.
91. Marcheselli V. L., Mukherjee P. K., Arita M., Hong S., Antony R., Sheets K., Winkler J. W., Petasis N. A., Serhan C. N., Bazan N. G. 2010. Neuroprotectin D1/protectin D1 stereoselective and specific binding with human retinal pigment epithelial cells and neutrophils. *Prostaglandins, Leukotrienes and Essential Fatty Acids*. 82(1):27–34.
92. Calandria J. M., Marcheselli V. L., Mukherjee P. K., Uddin J., Winkler J. W., Petasis N. A., Bazan N. G. 2009. Selective survival rescue in 15-lipoxygenase-1-deficient retinal pigment epithelial cells by the novel docosahexaenoic acid-derived mediator, neuroprotectin D1. *Journal of Biological Chemistry*. 284(26):17, 877–17, 882.

93. Zhao Y., Calon F., Julien C., Winkler J. W., Petasis N. A., Lukiw W. J., Bazan N. G. 2011. Docosahexaenoic acid-derived neuroprotectin D1 induces neuronal survival via secretase- and PPARγ-mediated mechanisms in Alzheimer's disease models. *PLoS ONE.* 6(1):e15, 816.
94. Bazan N. G., Eady T. N., Khoutorova L., Atkins K. D., Hong S., Lu Y., Zhang C. et al. 2012. Novel aspirin-triggered neuroprotectin D1 attenuates cerebral ischemic injury after experimental stroke. *Experimental Neurology.* 236(1):122–130.
95. Calandria J. M., Mukherjee P. K., de Rivero Vaccari J. C., Zhu M., Petasis N. A., Bazan N. G. 2012. Ataxin-1 poly(Q)-induced proteotoxic stress and apoptosis are attenuated in neural cells by docosahexaenoic acid-derived neuroprotectin D1. *Journal of Biological Chemistry.* 287(28):23, 726–23, 739.
96. Marcon R., Bento A. F., Dutra R. C., Bicca M. A., Leite D. F. P., Calixto J. B. 2013. Maresin 1, a proresolving lipid mediator derived from omega-3 polyunsaturated fatty acids, exerts protective actions in murine models of colitis. *Journal of Immunology.* 191(8):4288–4298.
97. Recchiuti A., Krishnamoorthy S., Fredman G., Chiang N., Serhan C. N. 2011. MicroRNAs in resolution of acute inflammation: Identification of novel resolvin D1-miRNA circuits. *FASEB Journal.* 25(2):544–560.
98. Levy B. D., Petasis N. A., Serhan C. N. 1997. Polyisoprenyl phosphates in intracellular signalling. *Nature.* 389(6654):985–990.
99. Levy B. D., Fokin V. V., Clark J. M., Wakelam M. J. O., Petasis N. A., Serhan C. N. 1999. Polyisoprenyl phosphate (PIPP) signaling regulates phospholipase D activity: A 'stop' signaling switch for aspirin-triggered lipoxin A4. *FASEB Journal.* 13(8):903–911.
100. Bonnans C., Fukunaga K., Keledjian R., Petasis N. A., Levy B. D. 2006. Regulation of phosphatidylinositol 3-kinase by polyisoprenyl phosphates in neutrophil-mediated tissue injury. *Journal of Experimental Medicine.* 203(4):857–863.
101. El Kebir D., József L., Khreiss T., Pan W., Petasis N. A., Serhan C. N., Filep J. G. 2007. Aspirin-triggered lipoxins override the apoptosis-delaying action of serum amyloid A in human neutrophils: A novel mechanism for resolution of inflammation. *Journal of Immunology.* 179(1):616–622.
102. Prieto P., Cuenca J., Traves P. G., Fernandez-Velasco M., Martin-Sanz P., Bosca L. 2010. Lipoxin A4 impairment of apoptotic signaling in macrophages: Implication of the PI3K/Akt and the ERK/Nrf-2 defense pathways. *Cell Death and Differentiation.* 17(7):1179–1188.
103. Takai D., Nagase T., Shimizu T. 2004. New therapeutic key for cystic fibrosis: A role for lipoxins. *Nature Immunology.* 5(4):357–358.
104. Kurihara T., Jones C. N., Yu Y.-M., Fischman A. J., Watada S., Tompkins R. G., Fagan S. P., Irimia D. 2013. Resolvin D2 restores neutrophil directionality and improves survival after burns. *FASEB Journal.* 27(6):2270–2281.

10 Pro- and Anti-Inflammatory Action of Oxidized Phospholipids

Maria Philippova, Olga Oskolkova, and Valery N. Bochkov

CONTENTS

INTRODUCTION

Oxidized phospholipids (OxPLs) have been identified as an active component in minimally modified low-density lipoproteins (LDLs), which is a proinflammatory form of LDL generated by mild oxidation of lipoproteins in the presence of lipoxygenase-expressing cells [1]. *In vivo*, OxPLs are present in circulating lipoproteins [2], membrane vesicles, and apoptotic blebs [3] and accumulate locally in various disease states, as will be described in this chapter. OxPLs exert profound actions on intracellular signaling and transcription systems. For example, in endothelial cells, OxPLs regulate expression of about 1000 genes [4] and phosphorylation levels of hundreds of phosphoproteins [5]. OxPLs induce a wide variety of biological effects that sometimes may be functionally opposite, as illustrated by pro- and anti-inflammatory action of OxPLs (discussed in the next section). This chapter focuses on biological effects of OxPLs that have a straightforward link to innate immunity and inflammation, for example, regulation of leukocyte traffic, production of inflammatory cell adhesion molecules and cytokines, oxidative burst, phagocytosis, and so on. Related subjects such as modulation of permeability of lung vessels, regulation of adaptive immune responses, or the role of OxPLs in prothrombotic shift during inflammation, as well as other biological activities of OxPLs have been extensively reviewed elsewhere [2,6–22].

GENERATION OF OxPLs

Esterified mono- and polyunsaturated fatty acids represent the major target of oxidation within the phospholipid molecule. Oxidation can be initiated either by certain isoforms of lipid-oxidizing enzymes, for example, human 15-lipoxygenase, or nonenzymatically by free radicals. The mechanisms of oxidation of fatty acid residues, as well as the chemical structures of oxidized products, are relatively well understood. It is generally accepted that nonenzymatic oxidation of esterified PUFAs follows the same general mechanisms as peroxidation of free (unesterified) fatty acids. Lipid peroxidation is a stochastic process generating multiple structurally different oxidized molecular species from each molecular species of nonoxidized precursor. These can be roughly classified basing on the number of carbon atoms in the chain, namely, into a group having the same number of carbons as the precursor plus new oxy functions (nonfragmented or full-length OxPLs) and another group of oxidatively fragmented species, which may or may not contain new oxygen atoms (Figure 10.1). There is no straightforward correlation between the length of oxidized residue and its biological activity. For example, inflammatory cytokine IL-8 is induced both by fragmented and nonfragmented species. In contrast, the effects of long and short OxPLs on endothelial permeability are opposite: certain full-length species enhance endothelial barrier function, while fragmented OxPLs demonstrate strong barrier-disruptive action [23]. There is also no clear correlation between chemical reactivity of molecular species and their biological activity: chemically inert fragmented species having ω-terminal carboxy group in some cases induce similar effects on OxPLs having reactive α,β-unsaturated aldehyde functions. In other words, structure–activity relationship of OxPLs is not simple and requires further investigation. These points will be discussed further in this chapter.

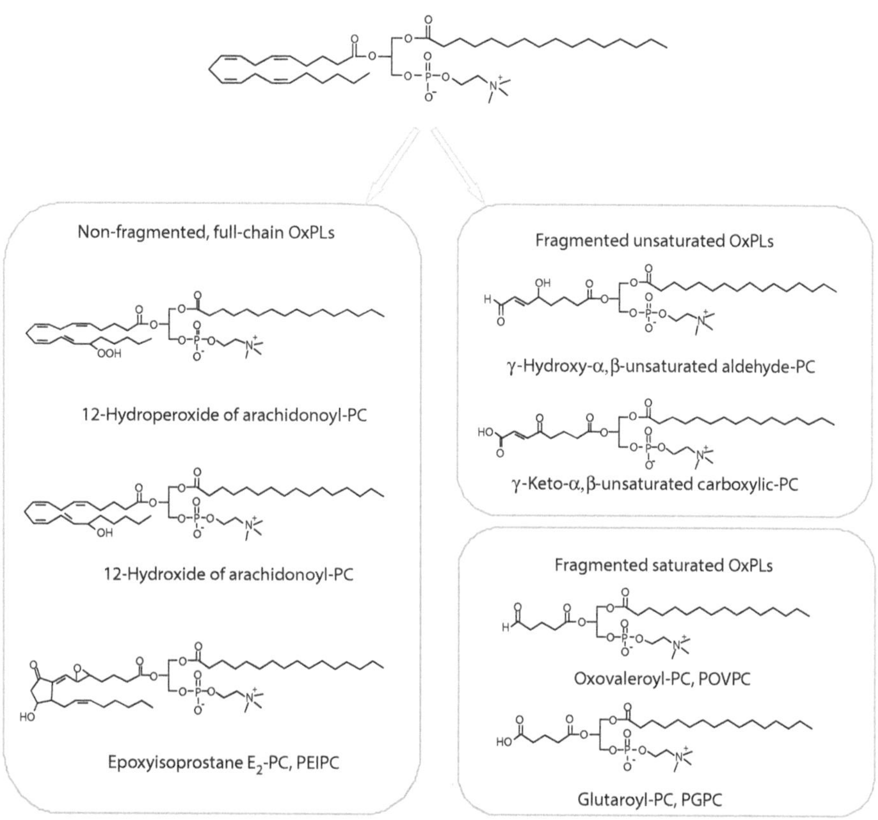

FIGURE 10.1 Chemical structures of typical OxPCs generated by peroxidation of palmitoyl-arachidonoyl-phosphatidylcholine. The figure presents a few most common molecular species selected from dozens of molecular species produced by nonenzymatic oxidation of a single precursor phospholipid. Note that oxidation can either add oxy functions to the full-size carbon chain or induce chain fragmentation. Both types of modifications produce phospholipids with abnormal properties and biological activities that were not characteristic of unoxidized phospholipid precursors.

BIOLOGICAL ACTIVITIES OF OxPLs POTENTIALLY RELEVANT TO INFLAMMATION

INDUCTION OF INFLAMMATORY CELL ADHESION MOLECULES AND MONOCYTE–ENDOTHELIAL INTERACTIONS

One of the best-documented activities of OxPLs related to inflammation is their ability to stimulate leukocyte binding to endothelial cells. Pretreatment of cultured endothelial cells with a mixture of oxidized species (OxPAPC) leads to selective adhesion of monocytes, but not granulocytes [1]. This effect is reproduced by fragmented OxPC species containing ω-terminal aldehyde group such as POVPC [24], fragmented α,β-unsaturated ω-terminal aldehyde HOOA-PC [25], as well as by full-chain OxPCs such as isoprostane-containing PEIPC [26]. OxPLs that were injected into an artificial cavity induced by subcutaneous injection of sterile air in mice (air pouch model) induced the accumulation of monocytes/macrophages in the pouch wall but not in the lavage. That was in strong contrast to the action of LPS, which induced accumulation of both monocytes and granulocytes not only in the wall, but also in the pouch lumen [27].

Increased adhesion of monocytes to OxPL-treated endothelial cells is mediated by connecting segment-1 (CS-1) fibronectin, which cross-links activated α5β1integrin on endothelial cells with integrin α4β1 on monocytes [24,28].

Induction of Inflammatory Cytokines

In human endothelial cells, OxPLs stimulate the production of inflammatory cyto- and chemokines including interleukin-8 (IL-8, [29]), MCP-1 [29], IL-6 [30], GROα [30], and GROγ [30]. The application of OxPAPC to the mouse carotid artery induced the expression of MCP-1, GROα, MIP-1α, and MIP-1β [31]. Furthermore, OxPLs induce VEGF [32] which, in addition to angiogenic properties, can stimulate recruitment of monocytes characteristic of atherosclerosis [33]. The majority of chemokines induced by OxPLs are specific for mononuclear cells, but IL-8 is also well known as a potent chemoattractant for granulocytes (apart from its ability to activate adhesion of monocytes [34]). However, a mixture of individual oxidized species such as OxPAPC does not stimulate endothelial cells to bind neutrophils [1]. The mechanisms underlying cell selectivity are not clear but may be due to the lack of concomitant upregulation of adhesion molecules necessary for binding of granulocytes. As a result and consistently with these *in vitro* data, application of OxPAPC *in vivo* leads to selective recruitment of monocytes [27,31].

In contrast to endothelial cells, OxPL effects on expression of inflammatory molecules in monocytes/macrophages have been investigated less. Cultured macrophages produced chemotactic compounds upon treatment with OxPLs; however, the effect was much lower than in cells treated with IL-4 or IFNγ/LPS [35]. OxPLs induced weakly COX-2 and IL-1β, but not other inflammatory molecules upregulated by the inducers of M1 or M2 macrophage phenotypes (IFNγ/LPS or IL-4, respectively) [35]. It has been hypothesized that OxPLs induce differentiation of macrophages into a new phenotype that was named Mox [35].

IL-8 expression represents the most common readout used in studies on proinflammatory action of OxPLs. Several species of OxPLs were shown to induce IL-8 in endothelial cells. IL-8 is upregulated in endothelial cells treated with fragmented OxPC species such as POVPC or PGPC [29], α,β-unsaturated ω-terminal aldehyde HOOA-PC [25] or full-chain isoprostane-containing species such as PEIPC, PECPC [29], and hydroperoxides of PC [36]. The activity does not significantly depend on the type of polar head group; different classes of PLs such as phosphatidylcholine, phosphatidylserine, phosphatidylglycerol, and phosphatidic acid all are capable of inducing IL-8 in endothelial cells [36]. Furthermore, oxidation of different PUFAs esterified in PLs, for example, linoleic (2 double bonds), arachidonic (4 double bonds), or docosahexaenoic (6 double bonds) produces OxPL species capable of inducing IL-8 [36]. Altogether, the data demonstrate the lack of strict structural specificity of OxPLs for the induction of this cytokine.

Mechanisms of IL-8 induction by OxPLs are complex and differ from the action of inflammatory cytokines such as TNFα [37]. Several signaling pathways and transcription factors were implicated in the upregulation of IL-8 in cells treated with OxPLs, including c-SRC/JAK2/STAT3 [38,39], eNOS/SREBP [40,41], PPARα [29], and ATF4 and XBP1 branches of unfolded protein response [30].

RECEPTORS MEDIATING INFLAMMATORY ACTION OF OxPLs

The complexity of signaling pathways mediating proinflammatory effects of OxPLs suggests that they may be initiated by several mechanisms, which however are only partially identified. Putative receptors mediating effects of OxPLs include PAF receptor, EP2 prostaglandin receptor, Toll-like receptors, as well as nonreceptor mechanisms described below.

EP2 prostaglandin receptor, which is expressed in ECs, monocytes, and macrophages, was shown to be activated by OxPLs. As a result, activation of integrins and enhanced binding of monocytes were observed [42]. The effect was specific for molecular species containing isoprostane moiety, which is structurally similar to prostaglandins and thus potentially can interact with prostaglandin receptors.

PAF-like phospholipids represent a group of phospholipid oxidation products that activate inflammatory cells through a characterized molecular mechanism. These are oxidatively truncated species of alkyl–acyl-phosphatidylcholines, which have *sn*-1 residue attached via an ether (alkyl) bond. Alkyl–acyl precursor species are very abundant in inflammatory cells [43]. Oxidative fragmentation of *sn*-2 residues produces short fragments without ω-terminal oxy function, mainly butanoyl or butenoyl residues, which can stimulate PAF receptor due to their structural similarity to platelet-activating factor (PAF). PAF receptor recognizes a combination of the choline head group, *sn*-1 alkyl residue and a short *sn*-2 residue [44]. This receptor is broadly presented on many cell types, including all cells of innate immunity. PAF-like lipids accumulate in OxLDL [45] and *in vivo*, for example, during UV irradiation of skin [46] and in chronic alcohol exposure accompanied by steatohepatitis [47]. PAF-like lipids induce various effects potentially relevant to inflammation, and in particular are capable of activating polymorphonuclear lymphocytes and monocytes through PAF receptor [48].

The evidence for the role of Toll-like receptors 4 or 2 as receptors for OxPLs is based on impaired responses to OxPLs or mmLDL in knockout animal models or after knockdown of TLRs in cultured cells [49–54]. OxPLs do not demonstrate canonical activation of TLRs in cells naturally expressing these receptors or in reporter cell lines [55–57]. It can be hypothesized that this discrepancy is explained by the need of additional soluble or membrane-associated proteins that present OxPLs to TLRs [51,52]. The role of TLRs as candidate receptors mediating proinflammatory action of OxPLs is discussed in more detail in Chapter 11 in this book.

It has been hypothesized that signaling effects of OxPLs are at least partially mediated by nonreceptor mechanisms such as membrane cholesterol depletion induced by

OxPLs in endothelial cells via an unidentified mechanism [41]. Furthermore, induction of IL-8 by OxPLs depends on the unfolded protein response [30]. This signaling pathway is activated by the accumulation of misfolded proteins in endoplasmic reticulum. The proteins are recognized by BiP/GRP78 chaperone which initiates signaling events. Further studies are necessary to clarify if OxPLs impair protein folding or directly bind to BiP/GRP78.

INHIBITION OF PHAGOCYTOSIS

OxPLs can influence the natural course of inflammation by interfering with the process of phagocytosis. OxPLs nonspecifically inhibit endo-, phago- and fluid pinocytosis *in vitro* and make mice more susceptible to bacterial inflammation *in vivo* [58]. The impaired phagocytosis is due to OxPL-induced activation of PKA and A-kinase anchoring protein WAVE1 [59]. In addition to oxidative burst induced by inflammation, smoking can also be among the mechanisms generating OxPLs in lungs thus suppressing clearance of bacteria by alveolar macrophages [60].

INHIBITION OF OXIDATIVE BURST

Generation of ROS represents an important mechanism of antibacterial defense. OxPLs inhibit oxidative burst induced by fMLP or phorbol ester in isolated neutrophil granulocytes [61]. Different classes of OxPLs suppressed assembly of phagocyte oxidase; the effect was not due to cytotoxicity because cell viability was not changed [61]. The inhibitory action on NADPH oxidase may represent a negative feedback preventing excessive production of ROS and damage to host tissues.

INHIBITION OF TLR ACTIVATION BY BACTERIAL PRODUCTS

Experimental evidence obtained in knockout animals (see above) suggests that proinflammatory effects of OxPLs are dependent on and probably initiated through TLRs. However, an opposite effect of OxPLs, namely inhibition of TLRs activation by bacterial products, was reported by several research groups. OxPLs suppress induction of various inflammatory molecules by bacterial lipopolysaccharide (LPS), among them are cell adhesion molecules such as E-selectin, VCAM-1, and ICAM-1 in endothelial cells [36,62]. OxPLs inhibit LPS-induced expression of inflammatory cytokines IL-8 and MCP-1 in endothelial cells and macrophages [36,52,63], suppress the induction of TNFα and IL6 induced by LPS in human whole blood and isolated monocytes [55,64] and prevent upregulation of TNFα, IL-1α, and IL-1β in human dermal fibroblasts treated with LPS [55]. Furthermore, OxPLs inhibit activation by LPS of NFκB-driven reporter constructs [55,57,65] and inhibit inflammation induced by LPS in animal models of sepsis [36,62,66]. In addition to TLR4, OxPLs also inhibit the activation of TLR2 *in vitro* and *in vivo* [57,63].

There is no strict lipid specificity for inhibition of TLRs by OxPLs. Different classes of OxPLs including oxidized phosphatidylcholine, phosphatidylethanolamine, phosphatidylserine, phosphatidylglycerol, and phosphatidic acid are capable of inhibiting effects of LPS [36,67]. Furthermore, both fragmented OxPLs as well as full-length OxPLs bearing hydroxide or isoprostane [25,36] residues can inhibit action of LPS. The most active are fragmented molecular species containing α,β-unsaturated carboxylic acids [67].

Mechanisms of TLR inhibition by OxPLs are complex and only partially understood. It is likely that OxPLs inhibit both extra- and intra-cellular steps in LPS recognition cascade [55]. One postulated mechanism is inhibition of recognition of bacterial products by TLRs. OxPLs bind to the LPS-binding protein (LBP), CD14 [55,62] and MD-2 [57], which are critically important for activation of TLR4. The binding prevents interaction of these proteins with LPS and its presentation to TLR4. Similarly, OxPLs inhibit the action of another CD14-dependent receptor, TLR2 [57]. In contrast, OxPLs did not influence the activation of CD14, MD-2-independent TLRs such as TLR3, 5, 7, 8, and 9 that were artificially expressed in NFκB-driven reporter cell line [57]. However, in cells that express TLRs naturally, inhibition of TLR3 [68] has been reported. A possible explanation reconciling contradictory findings is that additional currently unknown proteins are required to present OxPLs to various TLRs [51,52]. In addition to receptor antagonism, OxPLs were postulated to influence membrane- and intracellular mechanisms of TLR activation, for example, by impairing caveolar localization of TLR4 that is necessary for the activation of downstream signaling [63,67].

ACCUMULATION AND ROLE OF OxPLs IN INFLAMMATORY PATHOLOGIES

ACUTE INFLAMMATION

Acute inflammation is accompanied by oxidative burst, which kills bacteria but also can lead to oxidation of host molecules. For example, levels of circulating OxLDL are elevated in animal models of acute inflammation induced by zymozan, LPS, or turpentine [69] or in human infections such as dengue fever [70]. Elevation of OxPLs under inflammatory conditions was observed in different cell types, for example, in isolated monocytes and neutrophils stimulated with phorbol ester [71], or in macrophages infected with mycobacteria [72]. The treatment of human neutrophils with opsonized bacteria, phorbol ester, or fMLP resulted in fast (within 2 min) elevation of several molecular species of OxPCs and OxPEs containing esterified 5-hydroxide of arachidonic acid, which were formed by re-esterification of free (unesterified) 5-HETE produced by 5-LOX [73]. The same OxPL species containing 5-HETE were elevated in live bacterial peritonitis in mice and in human bacterial peritonitis [73]. Nonleukocyte cells such as IL-1-stimulated endothelial cells [74] or virally infected alveolar type II cells [75] also generate different forms of OxPLs. The accumulation of OxPLs in lungs during acute septic and aseptic inflammation is well documented and will be described below. Thus, the available data show that the accumulation of OxPLs is a typical feature of acute inflammation. Paradoxically, under certain conditions acute inflammation can lead to a decrease in OxPL levels; this may be due to variable abundance of cells producing OxPLs during acute phase and resolution. It was found that PE-hydroperoxides showed U-shaped time kinetics in mouse peritoneal inflammation induced by live *Staphylococcus epidermidis*: peritoneal lavage levels of phospholipid species containing 12- and 15-hydroxides of arachidonic acid were significantly reduced within the first hours and days of inflammation, and returned to basal values only one week later [64]. A similar reversible decrease in OxPEs was observed after the application of cell-free extract of *S. epidermidis* [64]. The authors postulated that the decrease in OxPEs resulted from clearance of the macrophage sub-population expressing 12/15 lipoxygenase during the acute phase and restoration of the number of such cells during the resolution phase [76].

ATHEROSCLEROSIS

Atherosclerotic vessels contain elevated concentrations of different classes and molecular species of OxPLs including oxidatively fragmented saturated and α,β-unsaturated species [1,77–79], PL-hydroperoxides and hydroxides [80] and PL-esterified isoprostanes and isolevuglandins [79,81–84]. There is no significant difference in the relative contents of different species in the course of plaque development, that is, in early and advanced lesions, thus suggesting that OxPLs are continuously formed and degraded [85].

It is likely that OxPLs are not merely markers of atherosclerosis accumulating in diseased vessels, but also play an active pathogenetic role. This notion is supported by the ability of OxPLs to induce proinflammatory reactions characteristic for atherosclerosis such as monocyte–endothelial interactions (see above) and production of cytokines and chemokines known to be upregulated in lesions, for example, IL-8 [86] or MCP-1 [87] that are known to play a causative role in atherogenesis [88,89]. Furthermore, OxPLs are high-affinity ligands for scavenger receptor CD36 and promote the key events in progression of atherogenesis such as formation of foam cells [78], phenotypic modulation and migration of smooth muscle cells [90], production of extracellular matrix [91], and vascular calcification [92]. Importantly, OxPLs demonstrate activities known to decrease plaque stability and promote acute atherothrombotic events, for example, angiogenesis [32], secretion of metalloproteinases [93], thrombogenic shift in the endothelium [94,95], and activation of platelets [96].

LEPROSY

Another type of chronic inflammatory condition where OxPLs are formed and can play a pathogenetic role is leprosy. Immunohistochemical staining and mass spectrometric analysis showed the accumulation of OxPCs in human macrophages infected with mycobacteria, as well as in lepromatous (disseminated) leprosy lesions where OxPCs accumulated in macrophage-derived foam cells [72]. The accumulation of OxPCs in leprosy can have functional consequences, as OxPCs impaired activation of dendritic and *M. leprae*-reactive CD1b-restricted T cells [72]. Furthermore, OxPCs modulated activation of TLR2/1 recognizing triacylated lipoproteins from mycobacteria: inhibited production of antimicrobial peptide cathelicidin and proinflammatory IL-12, but stimulated formation of anti-inflammatory IL-10 [72]. The data suggest that OxPLs play a negative role in leprosy by suppressing immune response to the pathogen.

Multiple Sclerosis

Inflammation represents an important pathogenic mechanism contributing to progression of neurodegenerative diseases such as multiple sclerosis (MS) [97]. This autoimmune disease is characterized by enhanced levels of ROS generated by macrophages and microglia [98]. Extracts of MS lesions contain high levels of OxPCs [99]. Immunohistochemical staining shows that OxPLs are present in different cell types, but mainly in acutely injured axons [100]. Furthermore, the frequency of OxPC-positive axonal spheroids correlated with the activity of lesions, that is, was minimal in normal white matter and progressively increased from inactive lesions to slowly expanding lesions reaching maximum in active lesions [100]. Since OxPLs can induce cellular death [14], it is tempting to speculate that the accumulation of OxPLs can trigger neuronal death.

Age-Related Macular Degeneration

Age-related macular degeneration (AMD) is a major cause of blindness in elderly people. Oxidative stress in the retina is regarded as an important pathogenic mechanism of this multifactorial disease [101]. OxPLs are known to accumulate in AMD retinas, where they are present in the photoreceptors and retinal pigment epithelium of the normal human macular area [102]. Eyes with AMD showed more intense immunoreactivity for OxPLs than age-matched normal eyes [102]. It is possible that OxPLs represent a mechanistic link between the oxidative stress and pathological changes in the retina characteristic of AMD, such as inflammation and neovascularization. In support of this notion, it has been shown that OxPLs induce in retinal pigmented cells expression of inflammatory cytokines such as MCP-1 [103] and the major angiogenic factor VEGF [103,104]. VEGF was induced by different classes and molecular species of OxPLs [104] showing the lack of strict lipid specificity for the effect. These data are in agreement with the involvement of the ATF4 branch of unfolded protein response in induction of VEGF by OxPLs [104] because previous studies demonstrated broad lipid specificity of UPR activation by OxPLs [105]. Functional importance of OxPL-induced MCP-1 and VEGF was illustrated by the ability of OxPLs to induce choroidal neovascularization upon subretinal application to mice [102].

The experimental *in vitro* and *in vivo* data suggesting that OxPLs may play a pathogenic role in AMD are further supported by the results of genome-wide association studies showing strong association between AMD and a SNP variant *rs1061170* in complement factor H (CFH), which confers a significant risk for AMD [106]. It is likely that this association results from the ability of CFH to inhibit proinflammatory and proangiogenic effects induced by lipid oxidation products including covalent adducts of OxPLs with proteins [107,108]. The mutant variant of CFH has impaired ability to bind OxLDL and POVPC-BSA adducts, and therefore may have a lower capacity to inhibit induction of inflammatory cytokines, scavenger receptors, and VEGF in RPE cells and macrophages [108]. As a result, homozygous carriers of the mutant allele have higher chances of developing AMD.

LUNG INJURY

Pulmonary surfactant mainly consists of phospholipids including oxidation-prone mono- and polyunsaturated molecular species. Contact of a surfactant with air oxidants such as oxygen and ozone leads to the formation of OxPLs and other oxidation products that can impair function of a surfactant [109] and influence epithelial and endothelial cells [23,110]. Furthermore, infections or other insults induce oxidative burst in lung cells, thus increasing concentrations of OxPLs [75,111]. The accumulation of OxPLs was observed in different types of acute lung injury including patients infected by H5N1 avian flu, anthrax-infected rabbits, and anthrax-, pox- and *Yersinia pestis*-infected monkeys [50], as well as in lung tissue of mice sensitized with ovalbumin upon application of an aerosol containing this protein [64].

OxPLs seem to play a proinflammatory role in acute lung pathology. OxPLs, as well as bronchoalveolar lavage fluid containing endogenous OxPLs, stimulated cytokine production in lung macrophages, including IL-6, which was a key cytokine determining the severity of disease [50]. Production of IL-6 was inhibited by E06 antibody selectively binding OxPCs. Furthermore, OxPLs impaired lung function after intratracheal application in mice [50]. Taken together, the data point to an important role of OxPLs in acute lung pathology, and suggest that at least partially the effects can be due to the induction of proinflammatory IL-6.

Other mechanisms of OxPL-induced lung injury include impairment of bacterial phagocytosis and clearance that was observed in mice subjected to cigarette smoke exposure ([60] and references above).

CHRONIC EXPOSURE TO FINE PARTICULATE MATTER

Exposure to particles of <2.5 μm that are present in polluted air represents a risk factor of cardiovascular disease [112]. OxPLs may play a role in the pathogenesis of inflammation induced by air pollution. In mice, chronic exposure to fine particulate matter induced inflammatory changes both systemically and in the lungs [53] and increased levels of OxPLs in bronchoalveolar lavage fluid [53]. The treatment of bone marrow-derived macrophages with OxPAPC mimicked effects of particulate matter exposure, for example, upregulated expression of MCP-1 and TNFα, that is, cytokines that were systemically elevated in particular matter-treated mice, and stimulated activation of NADPH oxidase. Systemic effects of particulate matter, as well as *in vitro* effects of OxPLs were lower in animals or isolated cells deficient in TLR4 [53]. These data show that OxPLs play a mechanistic role in deleterious action of air pollution by stimulating production of inflammatory cytokines and ROS.

SUMMARY: POTENTIAL ROLE OF OxPLs IN INFLAMMATORY CONDITIONS

BIPHASIC EFFECTS OF OxPLs IN LPS-INDUCED PROINFLAMMATORY REACTIONS

A large body of data described above demonstrates proinflammatory action of OxPLs on endothelial cells. On the other hand, OxPLs inhibit proinflammatory effects of

LPS. Analysis of concentration dependences using the same experimental setup for both pro- and anti-inflammatory action shows that these opposite effects develop at different concentration ranges (Figure 10.2). Low concentrations of OxPLs inhibit LPS-induced production of IL-8 [36]. When certain concentration of OxPLs is reached, LPS-induced production of IL-8 is completely inhibited. However, further elevation of OxPLs concentration restores IL-8 production apparently via an LPS- or TLR4-independent mechanism. In other words, the concentration dependence of OxPLs action on IL-8 production in LPS-stimulated cells is U-shaped [36]. For several classes and molecular species of OxPLs it was shown that half-maximal inhibition of LPS-induced IL-8 production is observed at concentrations of OxPLs that are 10-fold lower than those inducing IL-8 directly [36]. Thus, OxPLs inhibit inflammatory action of LPS at significantly lower concentrations than those inducing inflammation. These observations are matched with the data *in vivo* showing that LPS-induced inflammation in mice can be suppressed by doses of OxPLs that do not induce inflammation when injected alone. It has been shown that OxPLs inhibit elevation of inflammatory cytokines KC and IL-6 in mouse model of LPS-induced sepsis without inducing cytokine elevation when injected alone [36].

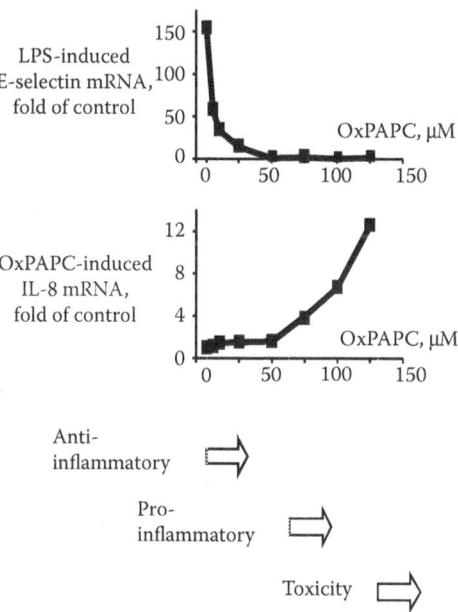

FIGURE 10.2 Biphasic pro- and anti-inflammatory action of OxPLs *in vitro*. The ability of OxPLs either to stimulate or inhibit inflammatory reactions is illustrated using as readouts LPS-stimulated production of E-selectin (upper panel) or OxPAPC-induced expression of IL-8 (lower panel) by endothelial cells. Low concentrations of OxPAPC inhibit production of E-selectin induced by LPS, while at higher concentrations OxPAPC stimulates synthesis of IL-8. In other words, low levels of OxPAPC are anti-inflammatory, while at higher concentrations the lipid demonstrates proinflammatory effects. Very high concentrations of OxPLs are toxic for cells.

Dual Role of OxPLs in Inflammation

The example above illustrates a well-established fact supported by a large body of published data, namely that OxPLs demonstrate both pro- and anti-inflammatory activities, which are mediated through different mechanisms. OxPLs are known to accumulate in a variety of acute and chronic inflammatory diseases. It is tempting to speculate that in acute bacterial inflammation low concentrations of OxPLs play a beneficial role by preventing undue or excessive activation of TLRs. In contrast, uncontrolled accumulation of high levels of OxPLs (which is more likely at sites of advanced and chronic pathology) can lead to proinflammatory shift through stimulated production of inflammatory cytokines, recruitment of monocytes, and impairment of antibacterial mechanisms such as oxidative burst and phagocytosis. Further work is necessary to establish how these opposite mechanisms are integrated during acute and chronic inflammatory conditions *in vivo*.

ABBREVIATIONS

AMD	age-related macular degeneration
ATF4	activating transcription factor 4
BiP/GRP78	binding immunoglobulin protein/78 kDa glucose-regulated protein
CFH	complement factor H
CS-1	connecting segment-1
EP2	prostaglandin E2 receptor 2
fMLP	N-formyl-methionyl-leucyl-phenylalanine
GRO	growth-regulated protein
HETE	hydroxyeicosatetraenoic acid
HOOA-PC	1-palmitoyl-2-(5-hydroxy-8-oxo-6-octenoyl)-*sn*-glycero-3-phosphocholine
ICAM-1	intercellular adhesion molecule-1
IFN	interferon
IL	interleukin
LDL	low-density lipoprotein
LOX	lipoxygenase
LPS	lipopolysaccharide
MCP-1	monocyte chemotactic protein-1
MIP	macrophage inflammatory protein
MS	multiple sclerosis
OxPAPC	oxidized 1-palmitoyl-2-arachidonoyl-*sn*-glycero-3-phosphocholine
OxPC	oxidized phosphatidylcholine
OxPE	oxidized phosphatidylethanolamine
OxPL	oxidized phospholipid
PAF	platelet activating factor
PE	phosphatidylethanolamine
PECPC	1-palmitoyl-2-(epoxycyclopentenone isoprostane)-*sn*-glycero-3-phosphocholine
PEIPC	1-palmitoyl-2-(epoxyisoprostane E2)-*sn*-glycero-3-phosphocholine

PGPC	1-palmitoyl-2-glutaroyl-*sn*-glycero-3-phosphocholine
PKA	protein kinase A
POVPC	1-palmitoyl-2-oxovaleroyl-*sn*-glycero-3-phosphocholine
ROS	reactive oxygen species
RPE	retinal pigment epithelial cells
PUFA	polyunsaturated fatty acid
TLR	Toll-like receptor
TNFα	tumor necrosis factor alfa
UPR	unfolded protein response
VCAM-1	vascular cell adhesion molecule-1
VEGF	vascular endothelial growth factor

REFERENCES

1. Watson, A.D., Leitinger, N., Navab, M., Faull, K.F., Horkko, S., Witztum, J.L., Palinski, W. et al. 1997. Structural identification by mass spectrometry of oxidized phospholipids in minimally oxidized low density lipoprotein that induce monocyte/endothelial interactions and evidence for their presence in vivo. *J. Biol. Chem.*, 272, 13597–13607.
2. Taleb, A., Witztum, J.L., and Tsimikas, S. 2011. Oxidized phospholipids on apoB-100-containing lipoproteins: A biomarker predicting cardiovascular disease and cardiovascular events. *Biomark. Med.*, 5, 673–694.
3. Huber, J., Vales, A., Mitulovic, G., Blumer, M., Schmid, R., Witztum, J.L., Binder, B.R., and Leitinger, N. 2002. Oxidized membrane vesicles and blebs from apoptotic cells contain biologically active oxidized phospholipids that induce monocyte–endothelial interactions. *Arterioscler. Thromb. Vasc. Biol.*, 22, 101–107.
4. Gargalovic, P.S., Imura, M., Zhang, B., Gharavi, N.M., Clark, M.J., Pagnon, J., Yang, W.P. et al. 2006. Identification of inflammatory gene modules based on variations of human endothelial cell responses to oxidized lipids. *Proc. Natl. Acad. Sci. USA*, 103, 12741–12746.
5. Zimman, A., Chen, S.S., Komisopoulou, E., Titz, B., Martinez-Pinna, R., Kafi, A., Berliner, J.A., and Graeber, T.G. 2010. Activation of aortic endothelial cells by oxidized phospholipids: A phosphoproteomic analysis. *J. Proteome. Res.*, 9, 2812–2824.
6. O'Donnell, V.B. and Murphy, R.C. 2012. New families of bioactive oxidized phospholipids generated by immune cells: Identification and signaling actions. *Blood*, 120, 1985–1992.
7. Aldrovandi, M. and O'Donnell, V.B. 2013. Oxidized PLs and vascular inflammation. *Curr. Atheroscler. Rep.*, 15, 323.
8. Salomon, R.G. 2012. Structural identification and cardiovascular activities of oxidized phospholipids. *Circ. Res.*, 111, 930–946.
9. Volinsky, R. and Kinnunen, P.K. 2013. Oxidized phosphatidylcholines in membrane-level cellular signaling: From biophysics to physiology and molecular pathology. *FEBS J.*, 280, 2806–2816.
10. Lee, S., Birukov, K.G., Romanoski, C.E., Springstead, J.R., Lusis, A.J., and Berliner, J.A. 2012. Role of phospholipid oxidation products in atherosclerosis. *Circ. Res.*, 111, 778–799.
11. Ullery, J.C. and Marnett, L.J. 2012. Protein modification by oxidized phospholipids and hydrolytically released lipid electrophiles: Investigating cellular responses. *Biochim. Biophys. Acta*, 1818, 2424–2435.
12. Kinnunen, P.K., Kaarniranta, K., and Mahalka, A.K. 2012. Protein-oxidized phospholipid interactions in cellular signaling for cell death: From biophysics to clinical correlations. *Biochim. Biophys. Acta*, 1818, 2446–2455.

13. Stemmer, U. and Hermetter, A. 2012. Protein modification by aldehydophospholipids and its functional consequences. *Biochim. Biophys. Acta*, 1818, 2436–2445.
14. McIntyre, T.M. 2012. Bioactive oxidatively truncated phospholipids in inflammation and apoptosis: Formation, targets, and inactivation. *Biochim. Biophys. Acta*, 1818, 2456–2464.
15. Thomas, C.P. and O'Donnell, V.B. 2012. Oxidized phospholipid signaling in immune cells. *Curr. Opin. Pharmacol.*, 12, 471–477.
16. Reis, A. and Spickett, C.M. 2012. Chemistry of phospholipid oxidation. *Biochim. Biophys. Acta*, 1818, 2374–2387.
17. Weismann, D. and Binder, C.J. 2012. The innate immune response to products of phospholipid peroxidation. *Biochim. Biophys. Acta*, 1818, 2465–2475.
18. Hammond, V.J. and O'Donnell, V.B. 2012. Esterified eicosanoids: Generation, characterization and function. *Biochim. Biophys. Acta*, 1818, 2403–2412.
19. Greig, F.H., Kennedy, S., and Spickett, C.M. 2012. Physiological effects of oxidized phospholipids and their cellular signaling mechanisms in inflammation. *Free Radic. Biol. Med.*, 52, 266–280.
20. Adamson, S. and Leitinger, N. 2011. Phenotypic modulation of macrophages in response to plaque lipids. *Curr. Opin. Lipidol.*, 22, 335–342.
21. O'Donnell, V.B. 2011. Mass spectrometry analysis of oxidized phosphatidylcholine and phosphatidylethanolamine. *Biochim. Biophys. Acta*, 1811, 818–826.
22. Miller, Y.I., Choi, S.H., Wiesner, P., Fang, L., Harkewicz, R., Hartvigsen, K., Boullier, A. et al. 2011. Oxidation-specific epitopes are danger-associated molecular patterns recognized by pattern recognition receptors of innate immunity. *Circ. Res.*, 108, 235–248.
23. Fu, P. and Birukov, K.G. 2009. Oxidized phospholipids in control of inflammation and endothelial barrier. *Transl. Res.*, 153, 166–176.
24. Leitinger, N., Tyner, T.R., Oslund, L., Rizza, C., Subbanagounder, G., Lee, H., Shih, P.T. et al. 1999. Structurally similar oxidized phospholipids differentially regulate endothelial binding of monocytes and neutrophils. *Proc. Natl. Acad. Sci. USA*, 96, 12010–12015.
25. Subbanagounder, G., Deng, Y., Borromeo, C., Dooley, A.N., Berliner, J.A., and Salomon, R.G. 2002. Hydroxy alkenal phospholipids regulate inflammatory functions of endothelial cells. *Vascul. Pharmacol.*, 38, 201–209.
26. Watson, A.D., Subbanagounder, G., Welsbie, D.S., Faull, K.F., Navab, M., Jung, M.E., Fogelman, A.M., and Berliner, J.A. 1999. Structural identification of a novel pro-inflammatory epoxyisoprostane phospholipid in mildly oxidized low density lipoprotein. *J. Biol. Chem.*, 274, 24787–24798.
27. Kadl, A., Galkina, E., and Leitinger, N. 2009. Induction of CCR2-dependent macrophage accumulation by oxidized phospholipids in the air-pouch model of inflammation. *Arthritis Rheum.*, 60, 1362–1371.
28. Cole, A.L., Subbanagounder, G., Mukhopadhyay, S., Berliner, J.A., and Vora, D.K. 2003. Oxidized phospholipid-induced endothelial cell/monocyte interaction is mediated by a cAMP-dependent R-Ras/PI3-kinase pathway. *Arterioscler. Thromb. Vasc. Biol.*, 23, 1384–1390.
29. Lee, H., Shi, W., Tontonoz, P., Wang, S., Subbanagounder, G., Hedrick, C.C., Hama, S. et al. 2000. Role for peroxisome proliferator-activated receptor alpha in oxidized phospholipid-induced synthesis of monocyte chemotactic protein-1 and interleukin-8 by endothelial cells. *Circ. Res.*, 87, 516–521.
30. Gargalovic, P.S., Gharavi, N.M., Clark, M.J., Pagnon, J., Yang, W.P., He, A., Truong, A. et al. 2006. The unfolded protein response is an important regulator of inflammatory genes in endothelial cells. *Arterioscler. Thromb. Vasc. Biol.*, 26, 2490–2496.
31. Furnkranz, A., Schober, A., Bochkov, V.N., Bashtrykov, P., Kronke, G., Kadl, A., Binder, B.R., Weber, C., and Leitinger, N. 2005. Oxidized phospholipids trigger atherogenic inflammation in murine arteries. *Arterioscler. Thromb. Vasc. Biol.*, 25, 633–638.

32. Bochkov, V.N., Philippova, M., Oskolkova, O., Kadl, A., Furnkranz, A., Karabeg, E., Afonyushkin, T. et al. 2006. Oxidized phospholipids stimulate angiogenesis via autocrine mechanisms, implicating a novel role for lipid oxidation in the evolution of atherosclerotic lesions. *Circ. Res.*, 99, 900–908.

33. Lucerna, M., Zernecke, A., de, N.R., de Jager, S.C., Bot, I., van der Lans, C., Kholova, I. et al. 2007. Vascular endothelial growth factor-A induces plaque expansion in ApoE knock-out mice by promoting de novo leukocyte recruitment. *Blood*, 109, 122–129.

34. Gerszten, R.E., Garcia-Zepeda, E.A., Lim, Y.C., Yoshida, M., Ding, H.A., Gimbrone, M.A., Jr., Luster, A.D., Luscinskas, F.W., and Rosenzweig, A. 1999. MCP-1 and IL-8 trigger firm adhesion of monocytes to vascular endothelium under flow conditions. *Nature*, 398, 718–723.

35. Kadl, A., Meher, A.K., Sharma, P.R., Lee, M.Y., Doran, A.C., Johnstone, S.R., Elliott, M.R. et al. 2010. Identification of a novel macrophage phenotype that develops in response to atherogenic phospholipids via Nrf2. *Circ. Res.*, 107, 737–746.

36. Oskolkova, O.V., Afonyushkin, T., Preinerstorfer, B., Bicker, W., von Schlieffen, E., Hainzl, E., Demyanets, S. et al. 2010. Oxidized phospholipids are more potent antagonists of lipopolysaccharide than inducers of inflammation. *J. Immunol.*, 185, 7706–7712.

37. Yeh, M., Leitinger, N., de, M.R., Onai, N., Matsushima, K., Vora, D.K., Berliner, J.A., and Reddy, S.T. 2001. Increased transcription of IL-8 in endothelial cells is differentially regulated by TNF-alpha and oxidized phospholipids. *Arterioscler. Thromb. Vasc. Biol.*, 21, 1585–1591.

38. Yeh, M., Gharavi, N.M., Choi, J., Hsieh, X., Reed, E., Mouillesseaux, K.P., Cole, A.L. et al. 2004. Oxidized phospholipids increase interleukin 8 (IL-8) synthesis by activation of the c-src/signal transducers and activators of transcription (STAT)3 pathway. *J. Biol. Chem.*, 279, 30175–30181.

39. Gharavi, N.M., Alva, J.A., Mouillesseaux, K.P., Lai, C., Yeh, M., Yeung, W., Johnson, J. et al. 2007. Role of the Jak/STAT pathway in the regulation of interleukin-8 transcription by oxidized phospholipids *in vitro* and in atherosclerosis in vivo. *J. Biol. Chem.*, 282, 31460–31468.

40. Gharavi, N.M., Baker, N.A., Mouillesseaux, K.P., Yeung, W., Honda, H.M., Hsieh, X., Yeh, M., Smart, E.J., and Berliner, J.A. 2006. Role of endothelial nitric oxide synthase in the regulation of SREBP activation by oxidized phospholipids. *Circ. Res.*, 98, 768–776.

41. Yeh, M., Cole, A.L., Choi, J., Liu, Y., Tulchinsky, D., Qiao, J.H., Fishbein, M.C. et al. 2004. Role for sterol regulatory element-binding protein in activation of endothelial cells by phospholipid oxidation products. *Circ. Res.*, 95, 780–788.

42. Li, R., Mouillesseaux, K.P., Montoya, D., Cruz, D., Gharavi, N., Dun, M., Koroniak, L., and Berliner, J.A. 2006. Identification of prostaglandin E2 receptor subtype 2 as a receptor activated by OxPAPC. *Circ. Res.*, 98, 642–650.

43. Mueller, H.W., O'Flaherty, J.T., and Wykle, R.L. 1982. Ether lipid content and fatty acid distribution in rabbit polymorphonuclear neutrophil phospholipids. *Lipids*, 17, 72–77.

44. Marathe, G.K., Harrison, K.A., Murphy, R.C., Prescott, S.M., Zimmerman, G.A., and McIntyre, T.M. 2000. Bioactive phospholipid oxidation products. *Free Radic. Biol. Med.*, 28, 1762–1770.

45. Marathe, G.K., Prescott, S.M., Zimmerman, G.A., and McIntyre, T.M. 2001. Oxidized LDL contains inflammatory PAF-like phospholipids. *Trends Cardiovasc. Med.*, 11, 139–142.

46. Marathe, G.K., Johnson, C., Billings, S.D., Southall, M.D., Pei, Y., Spandau, D., Murphy, R.C., Zimmerman, G.A., McIntyre, T.M., and Travers, J.B. 2005. Ultraviolet B radiation generates platelet-activating factor-like phospholipids underlying cutaneous damage. *J. Biol. Chem.*, 280, 35448–35457.

47. Yang, L., Latchoumycandane, C., McMullen, M.R., Pratt, B.T., Zhang, R., Papouchado, B.G., Nagy, L.E., Feldstein, A.E., and McIntyre, T.M. 2010. Chronic alcohol exposure increases circulating bioactive oxidized phospholipids. *J. Biol. Chem.*, 285, 22211–22220.

48. Marathe, G.K., Zimmerman, G.A., Prescott, S.M., and McIntyre, T.M. 2002. Activation of vascular cells by PAF-like lipids in oxidized LDL. *Vascul. Pharmacol.*, 38, 193–200.

49. Miller, Y.I., Viriyakosol, S., Worrall, D.S., Boullier, A., Butler, S., and Witztum, J.L. 2005. Toll-like receptor 4-dependent and -independent cytokine secretion induced by minimally oxidized low-density lipoprotein in macrophages. *Arterioscler. Thromb. Vasc. Biol.*, 25, 1213–1219.

50. Imai, Y., Kuba, K., Neely, G.G., Yaghubian-Malhami, R., Perkmann, T., van, L.G., Ermolaeva, M. et al. 2008. Identification of oxidative stress and Toll-like receptor 4 signaling as a key pathway of acute lung injury. *Cell*, 133, 235–249.

51. Kadl, A., Sharma, P.R., Chen, W., Agrawal, R., Meher, A.K., Rudraiah, S., Grubbs, N., Sharma, R., and Leitinger, N. 2011. Oxidized phospholipid-induced inflammation is mediated by Toll-like receptor 2. *Free Radic. Biol. Med.*, 51, 1903–1909.

52. Walton, K.A., Hsieh, X., Gharavi, N., Wang, S., Wang, G., Yeh, M., Cole, A.L., and Berliner, J.A. 2003. Receptors involved in the oxidized 1-palmitoyl-2-arachidonoyl-sn-glycero-3-phosphorylcholine-mediated synthesis of interleukin-8. A role for Toll-like receptor 4 and a glycosylphosphatidylinositol-anchored protein. *J. Biol. Chem.*, 278, 29661–29666.

53. Kampfrath, T., Maiseyeu, A., Ying, Z., Shah, Z., Deiuliis, J.A., Xu, X., Kherada, N. et al. 2011. Chronic fine particulate matter exposure induces systemic vascular dysfunction via NADPH oxidase and TLR4 pathways. *Circ. Res.*, 108, 716–726.

54. Shirey, K.A., Lai, W., Scott, A.J., Lipsky, M., Mistry, P., Pletneva, L.M., Karp, C.L. et al. 2013. The TLR4 antagonist Eritoran protects mice from lethal influenza infection. *Nature*, 497, 498–502.

55. von Schlieffen, E., Oskolkova, O.V., Schabbauer, G., Gruber, F., Bluml, S., Genest, M., Kadl, A. et al. 2009. Multi-hit inhibition of circulating and cell-associated components of the Toll-like receptor 4 pathway by oxidized phospholipids. *Arterioscler. Thromb. Vasc. Biol.*, 29, 356–362.

56. Erridge, C., Webb, D.J., and Spickett, C.M. 2007. Toll-like receptor 4 signalling is neither sufficient nor required for oxidised phospholipid mediated induction of interleukin-8 expression. *Atherosclerosis*, 193, 77–85.

57. Erridge, C., Kennedy, S., Spickett, C.M., and Webb, D.J. 2008. Oxidized phospholipid inhibition of toll-like receptor (TLR) signaling is restricted to TLR2 and TLR4: Roles for CD14, LPS-binding protein, and MD2 as targets for specificity of inhibition. *J. Biol. Chem.*, 283, 24748–24759.

58. Knapp, S., Matt, U., Leitinger, N., and van der Poll, T. 2007. Oxidized phospholipids inhibit phagocytosis and impair outcome in Gram-negative sepsis in vivo. *J. Immunol.*, 178, 993–1001.

59. Matt, U., Sharif, O., Martins, R., Furtner, T., Langeberg, L., Gawish, R., Elbau, I. et al. 2013. WAVE1 mediates suppression of phagocytosis by phospholipid-derived DAMPs. *J. Clin. Invest*, 123, 3014–3024.

60. Thimmulappa, R.K., Gang, X., Kim, J.H., Sussan, T.E., Witztum, J.L., and Biswal, S. 2012. Oxidized phospholipids impair pulmonary antibacterial defenses: Evidence in mice exposed to cigarette smoke. *Biochem. Biophys. Res. Commun.*, 426, 253–259.

61. Bluml, S., Rosc, B., Lorincz, A., Seyerl, M., Kirchberger, S., Oskolkova, O., Bochkov, V.N., Majdic, O., Ligeti, E., and Stockl, J. 2008. The oxidation state of phospholipids controls the oxidative burst in neutrophil granulocytes. *J. Immunol.*, 181, 4347–4353.

62. Bochkov, V.N., Kadl, A., Huber, J., Gruber, F., Binder, B.R., and Leitinger, N. 2002. Protective role of phospholipid oxidation products in endotoxin-induced tissue damage. *Nature*, 419, 77–81.

63. Walton, K.A., Cole, A.L., Yeh, M., Subbanagounder, G., Krutzik, S.R., Modlin, R.L., Lucas, R.M. et al. 2003. Specific phospholipid oxidation products inhibit ligand activation of toll-like receptors 4 and 2. *Arterioscler. Thromb. Vasc. Biol.*, 23, 1197–1203.

64. Morgan, A.H., Dioszeghy, V., Maskrey, B.H., Thomas, C.P., Clark, S.R., Mathie, S.A., Lloyd, C.M. et al. 2009. Phosphatidylethanolamine-esterified eicosanoids in the mouse: Tissue localization and inflammation-dependent formation in Th-2 disease. *J. Biol. Chem.*, 284, 21185–21191.
65. Kim, M.J., Choi, N.Y., Koo, J.E., Kim, S.Y., Joung, S.M., Jeong, E., and Lee, J.Y. 2013. Suppression of Toll-like receptor 4 activation by endogenous oxidized phosphatidylcholine, KOdiA-PC by inhibiting LPS binding to MD2. *Inflamm. Res.*, 62, 571–580.
66. Nonas, S., Birukova, A.A., Fu, P., Xing, J., Chatchavalvanich, S., Bochkov, V.N., Leitinger, N., Garcia, J.G., and Birukov, K.G. 2008. Oxidized phospholipids reduce ventilator-induced vascular leak and inflammation in vivo. *Crit Care*, 12, R27.
67. Walton, K.A., Gugiu, B.G., Thomas, M., Basseri, R.J., Eliav, D.R., Salomon, R.G., and Berliner, J.A. 2006. A role for neutral sphingomyelinase activation in the inhibition of LPS action by phospholipid oxidation products. *J. Lipid Res.*, 47, 1967–1974.
68. Bluml, S., Kirchberger, S., Bochkov, V.N., Kronke, G., Stuhlmeier, K., Majdic, O., Zlabinger, G.J. et al. 2005. Oxidized phospholipids negatively regulate dendritic cell maturation induced by TLRs and CD40. *J. Immunol.*, 175, 501–508.
69. Memon, R.A., Staprans, I., Noor, M., Holleran, W.M., Uchida, Y., Moser, A.H., Feingold, K.R., and Grunfeld, C. 2000. Infection and inflammation induce LDL oxidation in vivo. *Arterioscler. Thromb. Vasc. Biol.*, 20, 1536–1542.
70. Lee, C.Y., Seet, R.C., Huang, S.H., Long, L.H., and Halliwell, B. 2009. Different patterns of oxidized lipid products in plasma and urine of dengue fever, stroke, and Parkinson's disease patients: Cautions in the use of biomarkers of oxidative stress. *Antioxid. Redox. Signal.*, 11, 407–420.
71. Jerlich, A., Schaur, R.J., Pitt, A.R., and Spickett, C.M. 2003. The formation of phosphatidylcholine oxidation products by stimulated phagocytes. *Free Radic. Res.*, 37, 645–653.
72. Cruz, D., Watson, A.D., Miller, C.S., Montoya, D., Ochoa, M.T., Sieling, P.A., Gutierrez, M.A. et al. 2008. Host-derived oxidized phospholipids and HDL regulate innate immunity in human leprosy. *J. Clin. Invest.*, 118, 2917–2928.
73. Clark, S.R., Guy, C.J., Scurr, M.J., Taylor, P.R., Kift-Morgan, A.P., Hammond, V.J., Thomas, C.P. et al. 2011. Esterified eicosanoids are acutely generated by 5-lipoxygenase in primary human neutrophils and in human and murine infection. *Blood*, 117, 2033–2043.
74. Subbanagounder, G., Wong, J.W., Lee, H., Faull, K.F., Miller, E., Witztum, J.L., and Berliner, J.A. 2002. Epoxyisoprostane and epoxycyclopentenone phospholipids regulate monocyte chemotactic protein-1 and interleukin-8 synthesis. Formation of these oxidized phospholipids in response to interleukin-1beta. *J. Biol. Chem.*, 277, 7271–7281.
75. Van Lenten, B.J., Wagner, A.C., Navab, M., Anantharamaiah, G.M., Hui, E.K., Nayak, D.P., and Fogelman, A.M. 2004. D-4F, an apolipoprotein A-I mimetic peptide, inhibits the inflammatory response induced by influenza A infection of human type II pneumocytes. *Circulation*, 110, 3252–3258.
76. Dioszeghy, V., Rosas, M., Maskrey, B.H., Colmont, C., Topley, N., Chaitidis, P., Kuhn, H., Jones, S.A., Taylor, P.R., and O'Donnell, V.B. 2008. 12/15-Lipoxygenase regulates the inflammatory response to bacterial products in vivo. *J. Immunol.*, 181, 6514–6524.
77. Hoff, H.F., O'Neil, J., Wu, Z., Hoppe, G., and Salomon, R.L. 2003. Phospholipid hydroxyalkenals: Biological and chemical properties of specific oxidized lipids present in atherosclerotic lesions. *Arterioscler. Thromb. Vasc. Biol.*, 23, 275–282.
78. Podrez, E.A., Poliakov, E., Shen, Z., Zhang, R., Deng, Y., Sun, M., Finton, P.J. et al. 2002. A novel family of atherogenic oxidized phospholipids promotes macrophage foam cell formation via the scavenger receptor CD36 and is enriched in atherosclerotic lesions. *J. Biol. Chem.*, 277, 38517–38523.
79. Subbanagounder, G., Leitinger, N., Schwenke, D.C., Wong, J.W., Lee, H., Rizza, C., Watson, A.D., Faull, K.F., Fogelman, A.M., and Berliner, J.A. 2000. Determinants of

bioactivity of oxidized phospholipids. Specific oxidized fatty acyl groups at the sn-2 position. *Arterioscler. Thromb. Vasc. Biol.*, 20, 2248–2254.

80. Waddington, E., Sienuarine, K., Puddey, I., and Croft, K. 2001. Identification and quantitation of unique fatty acid oxidation products in human atherosclerotic plaque using high-performance liquid chromatography. *Anal. Biochem.*, 292, 234–244.

81. Gniwotta, C., Morrow, J.D., Roberts, L.J., and Kuhn, H. 1997. Prostaglandin F2-like compounds, F2-isoprostanes, are present in increased amounts in human atherosclerotic lesions. *Arterioscler. Thromb. Vasc. Biol.*, 17, 3236–3241.

82. Pratico, D., Iuliano, L., Mauriello, A., Spagnoli, L., Lawson, J.A., Rokach, J., Maclouf, J., Violi, F., and FitzGerald, G.A. 1997. Localization of distinct F2-isoprostanes in human atherosclerotic lesions. *J. Clin. Invest.*, 100, 2028–2034.

83. Poliakov, E., Meer, S.G., Roy, S.C., Mesaros, C., and Salomon, R.G. 2004. Iso[7]LGD2-protein adducts are abundant *in vivo* and free radical-induced oxidation of an arachidonyl phospholipid generates this D series isolevuglandin in vitro. *Chem. Res. Toxicol.*, 17, 613–622.

84. Salomon, R.G., Batyreva, E., Kaur, K., Sprecher, D.L., Schreiber, M.J., Crabb, J.W., Penn, M.S. et al. 2000. Isolevuglandin-protein adducts in humans: Products of free radical-induced lipid oxidation through the isoprostane pathway. *Biochim. Biophys. Acta*, 1485, 225–235.

85. Ravandi, A., Babaei, S., Leung, R., Monge, J.C., Hoppe, G., Hoff, H., Kamido, H., and Kuksis, A. 2004. Phospholipids and oxophospholipids in atherosclerotic plaques at different stages of plaque development. *Lipids*, 39, 97–109.

86. Cheng, C., Noordeloos, A.M., Jeney, V., Soares, M.P., Moll, F., Pasterkamp, G., Serruys, P.W., and Duckers, H.J. 2009. Heme oxygenase 1 determines atherosclerotic lesion progression into a vulnerable plaque. *Circulation*, 119, 3017–3027.

87. Ma, Y., Malbon, C.C., Williams, D.L., and Thorngate, F.E. 2008. Altered gene expression in early atherosclerosis is blocked by low level apolipoprotein E. *PLoS One*, 3, e2503.

88. Apostolakis, S., Vogiatzi, K., Amanatidou, V., and Spandidos, D.A. 2009. Interleukin 8 and cardiovascular disease. *Cardiovasc. Res.*, 84, 353–360.

89. Zernecke, A. and Weber, C. 2010. Chemokines in the vascular inflammatory response of atherosclerosis. *Cardiovasc. Res.*, 86, 192–201.

90. Pidkovka, N.A., Cherepanova, O.A., Yoshida, T., Alexander, M.R., Deaton, R.A., Thomas, J.A., Leitinger, N., and Owens, G.K. 2007. Oxidized phospholipids induce phenotypic switching of vascular smooth muscle cells *in vivo* and in vitro. *Circ. Res.*, 101, 792–801.

91. Cherepanova, O.A., Pidkovka, N.A., Sarmento, O.F., Yoshida, T., Gan, Q., Adiguzel, E., Bendeck, M.P., Berliner, J., Leitinger, N., and Owens, G.K. 2009. Oxidized phospholipids induce type VIII collagen expression and vascular smooth muscle cell migration. *Circ. Res.*, 104, 609–618.

92. Parhami, F., Morrow, A.D., Balucan, J., Leitinger, N., Watson, A.D., Tintut, Y., Berliner, J.A., and Demer, L.L. 1997. Lipid oxidation products have opposite effects on calcifying vascular cell and bone cell differentiation. A possible explanation for the paradox of arterial calcification in osteoporotic patients. *Arterioscler. Thromb. Vasc. Biol.*, 17, 680–687.

93. Lee, S., Springstead, J.R., Parks, B.W., Romanoski, C.E., Palvolgyi, R., Ho, T., Nguyen, P. et al. 2012. Metalloproteinase processing of HBEGF is a proximal event in the response of human aortic endothelial cells to oxidized phospholipids. *Arterioscler. Thromb. Vasc. Biol.*, 32, 1246–1254.

94. Ishii, H., Tezuka, T., Ishikawa, H., Takada, K., Oida, K., and Horie, S. 2003. Oxidized phospholipids in oxidized low-density lipoprotein down-regulate thrombomodulin transcription in vascular endothelial cells through a decrease in the binding of RARbeta-RXRalpha heterodimers and Sp1 and Sp3 to their binding sequences in the TM promoter. *Blood*, 101, 4765–4774.

95. Bochkov, V.N., Mechtcheriakova, D., Lucerna, M., Huber, J., Malli, R., Graier, W.F., Hofer, E., Binder, B.R., and Leitinger, N. 2002. Oxidized phospholipids stimulate tissue

factor expression in human endothelial cells via activation of ERK/EGR-1 and Ca(++)/ NFAT. *Blood*, 99, 199–206.

96. Zimman, A. and Podrez, E.A. 2010. Regulation of platelet function by class B scavenger receptors in hyperlipidemia. *Arterioscler. Thromb. Vasc. Biol.*, 30, 2350–2356.

97. Frohman, E.M., Racke, M.K., and Raine, C.S. 2006. Multiple sclerosis—The plaque and its pathogenesis. *N. Engl. J. Med.*, 354, 942–955.

98. van Horssen, J., Schreibelt, G., Drexhage, J., Hazes, T., Dijkstra, C.D., van, d., V and de Vries, H.E. 2008. Severe oxidative damage in multiple sclerosis lesions coincides with enhanced antioxidant enzyme expression. *Free Radic. Biol. Med.*, 45, 1729–1737.

99. Qin, J., Goswami, R., Balabanov, R., and Dawson, G. 2007. Oxidized phosphatidylcholine is a marker for neuroinflammation in multiple sclerosis brain. *J. Neurosci. Res.*, 85, 977–984.

100. Haider, L., Fischer, M.T., Frischer, J.M., Bauer, J., Hoftberger, R., Botond, G., Esterbauer, H., Binder, C.J., Witztum, J.L., and Lassmann, H. 2011. Oxidative damage in multiple sclerosis lesions. *Brain*, 134, 1914–1924.

101. Beatty, S., Koh, H., Phil, M., Henson, D., and Boulton, M. 2000. The role of oxidative stress in the pathogenesis of age-related macular degeneration. *Surv. Ophthalmol.*, 45, 115–134.

102. Suzuki, M., Kamei, M., Itabe, H., Yoneda, K., Bando, H., Kume, N., and Tano, Y. 2007. Oxidized phospholipids in the macula increase with age and in eyes with age-related macular degeneration. *Mol. Vis.*, 13, 772–778.

103. Suzuki, M., Tsujikawa, M., Itabe, H., Du, Z.J., Xie, P., Matsumura, N., Fu, X. et al. 2012. Chronic photo-oxidative stress and subsequent MCP-1 activation as causative factors for age-related macular degeneration. *J. Cell Sci.*, 125, 2407–2415.

104. Pollreisz, A., Afonyushkin, T., Oskolkova, O.V., Gruber, F., Bochkov, V.N., and Schmidt-Erfurth, U. 2013. Retinal pigment epithelium cells produce VEGF in response to oxidized phospholipids through mechanisms involving ATF4 and protein kinase CK2. *Exp. Eye Res.*, 116C, 177–184.

105. Oskolkova, O.V., Afonyushkin, T., Leitner, A., von Schlieffen, E., Gargalovic, P.S., Lusis, A.J., Binder, B.R., and Bochkov, V.N. 2008. ATF4-dependent transcription is a key mechanism in VEGF up-regulation by oxidized phospholipids: Critical role of oxidized sn-2 residues in activation of unfolded protein response. *Blood*, 112, 330–339.

106. Klein, R.J., Zeiss, C., Chew, E.Y., Tsai, J.Y., Sackler, R.S., Haynes, C., Henning, A.K. et al. 2005. Complement factor H polymorphism in age-related macular degeneration. *Science*, 308, 385–389.

107. Weismann, D., Hartvigsen, K., Lauer, N., Bennett, K.L., Scholl, H.P., Charbel, I.P., Cano, M. et al. 2011. Complement factor H binds malondialdehyde epitopes and protects from oxidative stress. *Nature*, 478, 76–81.

108. Shaw, P.X., Zhang, L., Zhang, M., Du, H., Zhao, L., Lee, C., Grob, S. et al. 2012 Complement factor H genotypes impact risk of age-related macular degeneration by interaction with oxidized phospholipids. *Proc. Natl. Acad. Sci. USA*, 109, 13757–13762.

109. Lang, J.D., McArdle, P.J., O'Reilly, P.J., and Matalon, S. 2002. Oxidant-antioxidant balance in acute lung injury. *Chest*, 122, 314S–320S.

110. Uhlson, C., Harrison, K., Allen, C.B., Ahmad, S., White, C.W., and Murphy, R.C. 2002. Oxidized phospholipids derived from ozone-treated lung surfactant extract reduce macrophage and epithelial cell viability. *Chem. Res. Toxicol.*, 15, 896–906.

111. Yoshimi, N., Ikura, Y., Sugama, Y., Kayo, S., Ohsawa, M., Yamamoto, S., Inoue, Y. et al. 2005. Oxidized phosphatidylcholine in alveolar macrophages in idiopathic interstitial pneumonias. *Lung*, 183, 109–121.

112. Brook, R.D., Rajagopalan, S., Pope, C.A., III, Brook, J.R., Bhatnagar, A., Diez-Roux, A.V., Holguin, F. et al. 2010. Particulate matter air pollution and cardiovascular disease: An update to the scientific statement from the American Heart Association. *Circulation*, 121, 2331–2378.

11 Modulation of Toll-Like Receptor Signaling by Oxidized Phospholipids

Vlad Serbulea and Norbert Leitinger

CONTENTS

INTRODUCTION

Oxidative tissue damage is a hallmark of many chronic inflammatory diseases, including atherosclerosis, fatty liver disease, type 2 diabetes, and autoimmune diseases, such as rheumatoid arthritis and multiple sclerosis.[1] The oxidatively modified molecules that accumulate in affected tissues include oxidized lipids, proteins, and nucleic acids, all of which have been suggested to contribute to the ongoing inflammatory reaction.[2] The importance of these molecules is highlighted by the fact that they are recognized by the immune system mainly in two ways: first, there are germline encoded antibodies that recognize oxidized phospholipids;[3] second, various immune and tissue cells respond to these molecules, partly by upregulating protective, antioxidant mechanisms,[4] but also by inducing pro-inflammatory responses, reminiscent of host-defense reactions.[5] These reactions are considerably weaker than those in response to exogenous danger signals, such as those derived from invading microorganisms of bacterial or viral origin. Nevertheless, the similarity of the reactions evoked the thought that similar recognition mechanisms—in other words, the same receptors—might be involved.

215

The first evidence that oxidized phospholipids (OxPL) can be recognized by TLRs (Toll-like receptors) comes from our study demonstrating that OxPL inhibit recognition of LPS by TLR4.[6] OxPL bind to TLR4 accessory proteins such as LBP, CD14, and MD2, thereby preventing binding of LPS and activation of TLR4 signaling.[7] Interestingly, this inhibitory activity was reported to be stronger than the pro-inflammatory, TLR-activating, activity of OxPL. In conclusion, this would make OxPL strong TLR4 antagonists, while being relatively weak agonists in general. Inhibition by OxPL also occurs at TLR2, and while some conventional TLR2 ligands have been shown to require CD14, others have been reported to involve CD36, which by itself is a receptor for OxPL.[8] Recently, we have shown that OxPL-induced inflammatory gene expression in macrophages requires TLR2.[5] A role for OxPL in TLR activated *in vivo* was suggested by studies using knockout mice, which demonstrated that mice deficient of TLR4 or TLR2 were protected against the development of atherosclerotic lesions—a process that is believed to require endogenously formed DAMPs, presumably derived from oxidative modification of phospholipids.

TLRs AND THE IMMUNE RESPONSE

TLRs belong to the class of pattern-recognition receptors (PRR). Activation of the Toll-pathway in *Drosophila* leads to NFκB-dependent expression of cytokines and antifungal/antibacterial products during the immune response and loss-of-function mutations in Toll receptors in *Drosophila* resulted in a higher rate of fungal infection.[9] Mammalian TLRs are similar in structure to the *Drosophila* Toll homologs, indicating that TLRs are highly conserved proteins involved in the eukaryotic immune response.[10] It was later determined that TLRs are expressed in immune cells such as dendritic cells, T lymphocytes, B lymphocytes, macrophages, and many other cell types.[11–13] To date, there are 10 identified TLRs in humans and 12 in mice, providing for an immune system that can recognize a plethora of structures (patterns), ranging from proteins to lipids and nucleic acids.[11,14]

In the context of infection, invading bacteria and viruses present unique molecular structures (pathogen-associated molecular patterns, PAMPs) to the host. Classical examples of these PAMPs are the bacterial cell wall constituents lipopolysaccharide (LPS) from Gram-negative and lipoteichoic acid (LTA) from Gram-positive bacteria. LPS is a potent activator of TLR4 while LTA is an activator of TLR2.[15–17] Other PAMPs include peptidoglycan[16] of Gram-positive bacteria and double-stranded RNA of viral origin, which stimulate signaling pathways through TLR2 and TLR3, respectively.[18] TLR5 is activated by bacterial flagellin of both Gram-negative and Gram-positive origins.[19] TLR7 and TLR8 are activated by viral, single-stranded RNA.[20,21] It is important to note that the TLRs capable of sensing nucleic acids (TLR3,[18] TLR7,[20] and TLR8,[21] TLR9[22]) are primarily restricted to cytoplasmic partitions.

It is thought that prior to signal activation, TLRs dimerize, as seen by TLR2, TLR3, and TLR4 homodimers,[18,23] or TLR2/6 or TLR2/1 heterodimers.[24,25] Ligand binding and dimerization induce the pairing of the intracellular Toll/Interleukin-1 receptor (TIR) domains, allowing a series of accessory proteins to bind and perpetuate the signal.[23,26,27] Activation of TLR signaling results in the induction of NFκB[27] and MAP-kinase pathways,[11] and PKC(protein kinase C)/Syk-dependent signaling.[28,29] In addition to NFκB, the final transcription factors that are activated by TLR-dependent signaling include AP-1, CREB, ATF-2, and STAT1.[30]

ROLE OF TLRs IN STERILE INFLAMMATION

DANGER MODEL AND THE HYPPO THEORY

It is thought that the recognition of PAMPs by PRRs evolved to serve a specific need to react to invading pathogens, as per the self–nonself (SNS) model of immunology.[31] This model seems to be intuitive at first glance, but there are a series of exceptions to this rule. First, there is the issue of autoimmunity—as Dr. Polly Matzinger and others have questioned—which does not seem to fit with the SNS model.[32] Why would autoimmunity arise if the immune system evolved to recognize only foreign components? On the same note, many have asked why and how a mother's immune system is able to tolerate the fetus, since the fetus does have a different, "foreign," genetic makeup. In order to address these concerns, the "Danger Model" was proposed, which led to a series of interesting interpretations of recent data.[32] According to this model, it is not only PAMPs that are crucial but also danger-associated molecular patterns (DAMPs), as they encompass both endogenous and foreign constituents. Particularly, the hydrophobic portions of cellular structures can act as DAMPs when removed from their natural setting.[33] Given that life evolved in a water-based environment, it was proposed that it was evolutionarily necessary for organisms to be able to recognize hydrophobic portions of molecules (hyppos) that were out of place—which would imply that something was amiss. This would be imperative in an organism's survival, while foreign components may or may not indicate real danger, damaged host components do. In accordance with this model, altered or damaged hydrophobic molecules may be recognized as DAMPs, eliciting an immune response.

Endogenous Structures Recognized by TLRs

In concordance with the danger model, a growing body of evidence supports the recognition of endogenous molecules by TLRs.[34] Among the first endogenous ligands to be identified for TLRs were heat-shock proteins (HSP), which are highly expressed in response to various forms of cellular stress including heat.[35,36] For instance, Vabulas et al. presented evidence for the activation of the TLR2 and TLR4 signaling pathways through HSP60[35] as well as HSP70,[36] indicating the importance of the immunological recognition of cell and tissue damage. Other examples of recognition of damaged or misplaced host components by TLRs are hyaluronan, one of the major components of extracellular matrix, and its fragments, which are known to activate TLR4- and TLR2-dependent signaling in dendritic cells,[37,38] and HMGB1, which is released by damaged cells, and recognized by TLR2 and TLR4.[39]

These studies support a concept by which parts released by the damaged host cells induce an immune response via TLR activation. These endogenous TLR ligands provide an infection-independent response of the immune system, arguing for the importance of danger recognition, primarily in the context of sterile tissue damage.

Another form of tissue damage is induced by increased oxidative stress. Since oxidative tissue damage is a primary feature of chronic inflammation, we propose that oxidatively modified lipids (such as oxidized phospholipids) represent DAMPs that elicit specific immune responses via PRRs including TLRs.

OxPL ARE DAMPs THAT MODULATE TLR ACTIVATION

FORMATION OF OxPL

Oxidized lipid formation occurs through nonenzymatic mechanisms involving free radical-induced peroxidation of polyunsaturated fatty acids (PUFAs), comprised of an initiation step by reactive species followed by multiple propagation steps.[40] Free radical species (ROS) may be derived from most cellular processes that involve the metabolism of oxygen—including the transport of electrons in mitochondria[41]—or from activated phagocytes[42] that release oxidizing species meant to damage pathogens prior to engulfment. Oxidized lipids may also form with the help of oxidative enzymes such as myeloperoxidase (MPO), lipoxygenases, and NADPH oxidase.[43] Both NADPH oxidase and MPO are key enzymes used by neutrophils to create superoxide and hydrogen peroxide, respectively, which in the context of inflammation results in a microenvironment enriched in these oxidizing agents.[43]

Lipid hydroperoxides can be further metabolized to form both esterified and unesterified aldehydes,[44] and fragmentation products such as 4-hydroxynonenal or malondialdehyde.[44] All these mechanisms result in the peroxidation of lipoproteins and membrane phospholipids, including phosphatidylcholines (PC), phosphatidylserines (PS), phosphatidylethanolamines (PE), and phosphatidylinositides (PI).[45] Phospholipid oxidation is also known to occur in apoptotic cells through the activation of NADPH oxidase.[46] The presence of oxidized phosphatidylserines[47] and, using the antibody E06[3], oxidized PC[48] have been confirmed on the surface of apoptotic cells. Moreover, membrane vesicles that are released from activated or apoptotic cells contain oxidized phosphatidylcholines that elicit pro-inflammatory effects.[48,49] A new model of membrane structural composition, known as the "lipid whisker model," suggests that oxidized lipids protrude into the extra-membranous space, allowing for receptor and antibody interaction.[50]

Experimentally, oxidized 1-palmitoyl-2-arachidonoyl-*sn*-glycero-3-phosphocholine (OxPAPC) has been widely used as a representative OxPL. It is important to note that OxPAPC consists of a mixture of PAPC derivatives, each having a distinct structure and possibly different biological activity.[43] Watson et al. identified and assessed the biological function of three different OxPAPC derivatives: 1-palmitoyl-2-oxovaleroyl-*sn*-glycero-3-phosphocholine (POVPC), 1-palmitoyl-2-glutaroyl-*sn*-glycero-3-phosphocholine (PGPC), and epoxyisoprostane-phosphocholine (PEIPC),[51,52] which laid the groundwork for many subsequent studies. Meanwhile, many additional structures of OxPL have been described and a variety of biological functions have emerged.[7] Of clinical relevance, E06-reactive OxPL are increased in patients with atherosclerosis and can be used as biomarkers and predictors of future cardiovascular events.[53]

OxPL Inhibit TLR Signaling: Evidence for the Involvement of Accessory Proteins

The first study to emphasize the importance of the individual lipid species in OxPAPC in the context of TLR signaling showed that POVPC inhibited the LPS-induced binding of neutrophils to endothelial cells while selectively stimulating monocyte adhesion to EC.[54] On the contrary, the same study revealed that PGPC activated the binding of neutrophils to endothelial cells by upregulating E-selectin (an endothelial cell-specific, cytokine stimulated, cell adhesion molecule) expression.[54] In addition to EC, it was shown that OxPL inhibited LPS-induced NFκB and consequent cyclooxygenase 2 expression in macrophages.[55] Bochkov et al. provided evidence of OxPAPC inhibition of LPS-induced TLR4 signaling in the context of murine LPS challenge.[6] The inhibition was so strong that injection of OxPAPC rescued mice from lethal doses of LPS. Von Schlieffen et al. showed, using human umbilical vein endothelial cells (HUVEC) and HEK293 cells, that OxPAPC prevents LPS-induced TLR4 signaling by binding to serum proteins CD14 and LPS-binding protein, both of which are necessary for LPS to bind to TLR4.[7] Further evidence for TLR interaction with OxPL comes from studies carried out by Erridge et al. showing that the inhibition of TLR-induced NFκB by OxPL can be associated with TLR2 and TLR4 in macrophages, endothelial cells, and smooth muscle cells.[56] Erridge et al. also conducted a series of binding assays (primarily using FLAG-tagged MD2 and ELISAs) showing that, *in vitro*, OxPAPC inhibits the binding of LPS to MD2—an accessory protein in the TLR4 signaling complex.[56] Recent evidence has shown that KOdiA-PC also inhibits LPS-induced TLR4 signaling by preventing the binding of LPS to MD2[57] (Table 11.1). Based on these results, it is concluded that OxPL inhibit recognition of LPS by TLR4 via blocking its interaction with accessory proteins CD14, LBP, or MD2 (Figure 11.1a). The identification of the binding sites for OxPL on these accessory proteins will reveal a mechanistic basis for this important inhibition.

As an alternative mechanism by which OxPL inhibit TLR signaling, Walton et al. showed that OxPAPC interferes with the formation of a TLR4-receptor lipid raft complex, presumably by causing the sequestration of caveolin-1 to the inner membrane.[58] Moreover, the inhibitory effect of OxPL on TLR4 is not restricted to oxidized phosphatidylcholines; a study by Subbanagounder et al. showed that both OxSAPC and OxSAPE inhibit LPS-induced binding of monocytes to endothelial cells.[59] Although some studies show that OxPL selectively inhibit TLR2 or TLR4,[56] evidence has been gathered that OxPL inhibit activation of other TLRs as well. Ma et al. confirmed the connection between TLR4 and OxPAPC while discovering that the CpG-DNA-induced TNFα-response (TLR9 dependent) was also inhibited.[60] Moreover, Blüml et al. showed that DC maturation induced by TLR4, TLR3, or TLR2 ligands was inhibited by OxPL.[61] These data point to a variety of mechanisms by which OxPL inhibit pro-inflammatory signaling in addition to blocking TLR–ligand interaction[62] (Table 11.1).

Evidence for OxPL Activating TLR-Dependent Signaling

Several studies have shown that TLR4 and/or TLR2 are involved in the recognition of oxLDL and minimally modified LDL resulting in promotion of inflammatory

TABLE 11.1
OxPL as TLR Ligands

Lipid	Receptor	Activate	Inhibit	Cell Type	In Vivo	References
OxPAPC	TLR2	+	+	THP1 macrophages; J774A.1 macrophages; HEK-293; isolated human monocytes; BMDM	Sepsis; neuroinflammation	5,56,61,84–86
	TLR3		+	Human peripheral blood monocytes		61
	TLR4	+	+	THP1 macrophages; J774A.1 macrophages; HEK-293; HUVEC; human peripheral blood monocytes; Isolated human monocytes; RAW264.7, BMDM; HeLa	Lung injury; influenza; sepsis; neuroinflammation	6,7,56,60,61,65,77,84–90
	TLR6		+	THP1 macrophages		88
	TLR9		+	RAW264.7		60
POVPC	TLR2		+	HEK-293		56
	TLR4	+	+	HAEC; MAEC; HUVEC		6,7,65,87
PGPC	TLR2		+	HEK-293		56
	TLR4	+	+	RAW264.7 macrophage		7,57,87
PEIPC	TLR4	+	+	HAEC; MAEC; HUVEC		65,87
KOdiA-PC	TLR4		+	RAW264.7 macrophage		57
7-ketoC; 25-OHC	TLR4	+		Placental trophoblasts		90
ω-2-carboxyethyl pyrrole (CEP)	TLR2	+		Murine lung endothelial cells; HUVEC	Hindlimb ischemia; tumor implantation	78
15LO-CE	TLR4	+		J774	High-fat diet	29

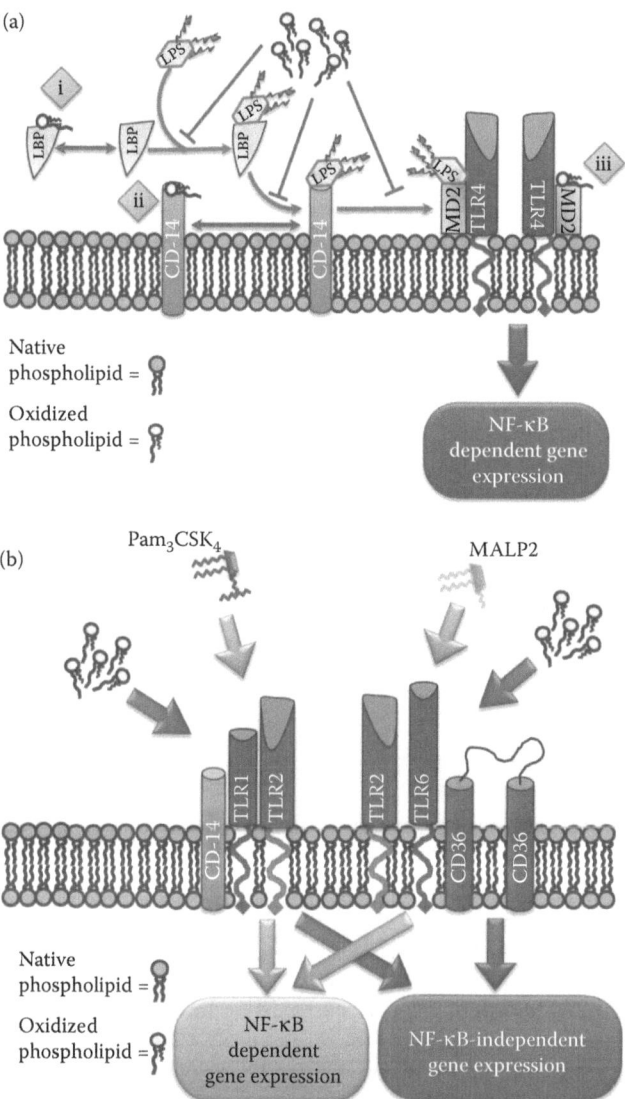

FIGURE 11.1 (a) OxPLs inhibit multiple steps of LPS-induced TLR4 signaling. (i) OxPLs are able to bind LPS-binding protein (LBP) directly, preventing LPS binding. (ii) OxPLs may bind CD-14, which exists in both membrane bound and soluble forms, inhibiting LPS shuttling to MD2-TLR4 complex. (iii) OxPLs can bind to MD2, preventing LPS binding and TLR4-dependent signal transduction. (b) TLR2 is necessary for OxPL-induced inflammatory gene expression. OxPLs, compared to classical ligands of Toll-like receptors, do not signal through NF-κB. CD-36 itself and in conjunction with TLR2 has been found to allow OxPL signaling. CD-14 is known to couple with TLR2/1 providing OxPL signaling. Pam$_3$CSK$_4$ and MALP2 are TLR2/1 and TLR2/6 agonists, respectively, which, unlike OxPL, induce NF-κB dependent signaling.

signaling.[63,64] Further evidence supporting the hypothesis that OxPL are recognized by PPRs comes from studies by Walton et al. who, using IL-8 expression as a readout, provided evidence for TLR4 as a putative receptor for OxPAPC, while at the same time indicating the involvement of a GPI-anchored protein for OxPAPC-induced signal transduction.[65] The activation of TLR4 by LPS takes place through a series of steps: LPS has been shown to bind directly to LBP, which then interacts and transfers the LPS molecule to the CD14 accessory protein and, working in conjunction with the newly bound LPS, activates the TLR4–MD2 complex and induces NFκB signaling.[66] Whether OxPL-induced TLR-dependent signaling requires membrane-bound and/or soluble accessory proteins is the focus of current investigations.

Of all the TLRs, TLR2 appears to have the most dissimilar sets of ligands that it recognizes. The reason behind this apparent nonspecificity lies with the ability of TLR2 to form heterodimers (Figure 11.1b). A series of studies revealed that TLR2 can form heterodimers with either TLR1 or TLR6. These studies show that TLR2/1 heterodimers recognize triacylated lipoproteins while TLR2/6 dimers recognize diacylated lipoproteins.[24] CD36 is a receptor for oxidized phospholipids,[8,67] and a role for CD36/TLR2-dependent mechanisms in lipid-induced macrophage apoptosis and development of atherosclerosis was proposed.[68] In the case of TLR2/6 heterodimers, knockout studies revealed that CD36 was necessary for TLR2/6 recognition of diacylated lipoproteins.[25,69] On the other hand, assembly of CD36 with a TLR4/TLR6 heterodimer was shown to mediate recognition of oxLDL and promote sterile inflammation.[70] Whether these homodimers, heterodimers, and associated protein complexes form before or because of the presence of the ligand is a different question entirely. Identification of different heterodimer pairs associated with accessory receptors such as CD14 or CD36 will help explain the complex recognition patterns of TLRs and may reveal possible new ligands.

A recent study by Kadl et al. provides evidence that the OxPAPC-induced expression of COX2, MIP2, and KC in macrophages is TLR2 dependent.[5] In the same study, it was discovered that the absence of TLR2 protected mice from carbon tetrachloride-induced liver damage, presumably by ablating inflammatory signaling. These results imply that TLR2 acts as a sensor for oxidative tissue damage via recognizing oxidized phospholipid-derived DAMPs. Interestingly, activation of TLR2 by OxPAPC in macrophages did not result in NFκB-dependent gene expression, but induced ERK and Jnk phosphorylation. The reasons for the absence of NFκB signaling are not clear. Oxidized lipids can act on a number of different tiers of TLR signaling. For instance, it was shown that TLR4-dependent signaling induced by oxidized cholesterol esters involves the activation of spleen tyrosine kinase (Syk) resulting in AP-1-dependent gene expression[28] (Table 11.1). Further studies are required to elucidate the exact signaling mechanisms and gene-expression patterns, as well as *in vivo* relevance of these mechanisms for disease initiation and progression.

RELEVANCE OF MODULATION OF TLR SIGNALING BY OxPL FOR PATHOLOGY

Here, we will discuss examples for the implications of endogenously formed OxPL activating TLR4 or TLR2 in major human or murine disease models.

A number of studies link TLRs and TLR-dependent signaling to the development of atherosclerosis.[71,72] Endogenous "danger signals," identified as TLR ligands, have been associated with lesion expansion and growth.[34,71] Vink et al. were among the first to draw an *in vivo* connection between atherosclerosis and TLR4, using immunohistochemistry and measurement of cytokine production.[73] Further evidence of TLR4's role in atherogenesis came from a study screening patients for Asp299Gly polymorphism, which concluded that patients bearing the mutation were more susceptible to bacterial infection, but less susceptible to atherosclerosis.[74] The C3H/HeJ mouse strain, containing a missense mutation in the Tlr4 gene, was characterized with a greater resistance to atherosclerosis but increased vulnerability to Gram-negative infection—similar to the Asp299Gly polymorphism in human patients.[15,75] Further support for this claim came from a study by which concluded, using aortic explants, that endothelial cells harvested from C3H/HeJ mice have a reduced response to both LPS and oxidized lipoproteins.[76] OxPL may also activate TLR4 in the lung. A study by Imai et al. showed that activation of TLR4 by OxPAPC was involved in causing lung tissue damage in various models of influenza. Using the E06 antibody, the authors could prevent TLR4 activation, which in this case caused TRIF-dependent signaling involving TRAF6 and IKKε, and IL6 production.[77]

Mullick et al. were able to link TLR2 and atherosclerosis when they generated LDLr$^{-/-}$/TLR2$^{-/-}$ mice, concluding that the TLR2 depletion resulted in much smaller lesion formation, while the treatment of the LDLr$^{-/-}$ mouse with TLR2 ligands resulted in larger lesion formation.[71] These studies strongly support the concept that endogenously formed TLR2 ligands exacerbate atherosclerotic lesion formation. A different class of oxidized phospholipids, the carboxyalkylpyrroles (CAPs) (including ω-(2-carboxyethyl) pyrrole, or CEP), were shown to activate TLR2 on endothelial cells to induce angiogenesis.[78] This finding turned out to be relevant for ischemia and general wound-healing processes, in which neovascularization is essential. Levels of CEP were increased during wound healing, in melanomas, and CEP-induced angiogenesis independently of VEGF. These lipids were also found to induce age-related macular degeneration (AMD) by triggering an immune response in the retina.[79]

Our findings that OxPL potently suppress TLR-dependent maturation of dendritic cells may have important implications for the function of the adaptive immune system. Especially in the elderly or in patients suffering from chronic inflammatory conditions, the immune system is compromised—a condition also known as "immunosenescence." The accumulation of OxPL may contribute to a compromised immune system in these conditions. Furthermore, evidence for compromising the host's immune response by endogenously formed OxPL comes from a study by Robert Modlin's group that showed that in human leprosy lesions, the OxPL 1-palmitoyl-2-(5,6-epoxyisoprostane E2)-*sn*-glycero-3-phosphorylcholine (PEIPC) accumulated—especially in macrophages. They went on to show that host-derived OxPL inhibited the immune response which, interestingly, was reversed by normal HDL but not HDL derived from leprosy patients.[80]

Finally, effective clearance of dead cells (efferocytosis) is essential for general tissue homeostasis, resolution of inflammation, and tissue repair. Defective efferocytosis may contribute to chronic inflammatory responses.[81] Several studies have shown that phagocytotic capacity of macrophages is decreased by oxidized phospholipids.[4,82,83] In this context, TLRs have been associated with the regulation of efferocytosis;[81] however, whether the inhibitory effect of OxPL on phagocytosis is mediated via TLRs remains to be shown.

CONCLUSIONS, OPEN QUESTIONS, AND FUTURE DIRECTIONS

On the basis of accumulating evidence, the modulation of TLR signaling by endogenously formed, oxidatively modified phospholipids seems to affect many pathological conditions ranging from host responses to infections to sterile chronic inflammatory settings. Clearly more research is needed to elucidate structure–function relationships as well as to understand biological consequences of such interactions. While inhibition of TLR signaling by OxPL may compromise the host's immune response, it may also provide a means to prevent overshooting inflammatory bursts or even contribute to the resolution of inflammation during infections. On the other hand, activation of TLRs by OxPL in settings of chronic inflammatory diseases—including atherosclerosis and diabetes—may explain the "low-grade" inflammatory status observed in these conditions.

The following questions remain: (1) Which structures are recognized by the TLR signaling complex? (2) Are the structural motifs for inhibition different from those for activation? (3) Which TLRs (or specific heterodimeric complexes) do recognize OxPL? and (4) Can the recognition of OxPL by TLRs be exploited for the design of novel drugs and treatment strategies against chronic inflammatory and autoimmune diseases? Future analyses of oxidized lipid formation by lipidomic approaches will lead to the discovery of novel oxidized phospholipids and through mass spectrometry we are able to identify chemical structures and to quantitatively assess the formation of individual oxidized phospholipid species. Characterizing more precisely the structure–function relationships that exist between these lipids and TLRs will lead to the design of novel compounds that modulate or mimic that interaction and thus open new avenues for therapeutic intervention.

ABBREVIATIONS

AMD	Age-related macular degeneration
CAP	Carboxyalkylpyrroles
CEP	ω-(2-carboxyethyl) pyrrole
DAMP	Danger-associated molecular patterns
HSP	Heat shock protein
HUVEC	Human umbilical vein endothelial cells
Hyppo	Hydrophobic portions
LPS	Lipopolysaccharide
LTA	Lipoteichoic acid
MAP-kinase	Mitogen-activated protein kinase
MPO	Myeloperoxidase
OxPAPC	Oxidized 1-palmitoyl-2-arachidonoyl-*sn*-glycero-3-phosphocholine
OxPL	Oxidized phospholipid
PAMP	Pathogen-associated molecular pattern
PAPC	1-palmitoyl-2-arachidonoyl-*sn*-glycero-3-phosphocholine

PC   Phosphatidylcholine
PE   Phosphatidylethanolamine
PEIPC   1-palmitoyl-2-(5,6-epoxyisoprostane E2)-*sn*-glycero-3-phosphorylcholine
PGPC   1-palmitoyl-2-glutaroyl-*sn*-glycero-3-phosphocholine
PI   Phosphatidylinositol
PKC   Protein kinase C
POVPC   1-palmitoyl-2-oxovaleroyl-*sn*-glycero-3-phosphocholine
PRR   Pattern recognition receptor
PS   Phosphatidylserine
PUFA   Polyunsaturated fatty acid
ROS   Reactive oxygen species
SNS   Self-non-self
SYK   Spleen tyrosine kinase
TIR   Toll/interleukin-1 receptor
TLR   Toll-like receptor

REFERENCES

1. Alves, J. D. and Ames, P. R. J. Atherosclerosis, oxidative stress and auto-antibodies in systemic lupus erythematosus and primary antiphospholipid syndrome. *Immunobiology* **207**, 23–8, 2003.
2. Halliwell, B. B. and Poulsen, H. E. (eds.) Oxidative stress. In *Cigarette Smoke Oxidative Stress* 1–4, 2006.
3. Shaw, P. X. et al. Natural antibodies with the T15 idiotype may act in atherosclerosis, apoptotic clearance, and protective immunity. *J. Clin. Invest.* **105**, 1731–40, 2000.
4. Kadl, A., Meher, A. and Sharma, P. Identification of a novel macrophage phenotype that develops in response to atherogenic phospholipids via Nrf2. *Circ. Res.* **107**, 737–46, 2010.
5. Kadl, A. et al. Oxidized phospholipid-induced inflammation is mediated by Toll-like receptor 2. *Free Radic. Biol. Med.* **51**, 1903–9, 2011.
6. Bochkov, V., Kadl, A., Huber, J. and Gruber, F. Protective role of phospholipid oxidation products in endotoxin-induced tissue damage. *Nature* **4**, 77–81, 2002.
7. Von Schlieffen, E. et al. Multi-hit inhibition of circulating and cell-associated components of the toll-like receptor 4 pathway by oxidized phospholipids. *Arterioscler. Thromb. Vasc. Biol.* **29**, 356–62, 2009.
8. Podrez, E. et al. Identification of a novel family of oxidized phospholipids that serve as ligands for the macrophage scavenger receptor CD36. *J. Biol. Chem.* **277**, 38503–16, 2002.
9. Lemaitre, B., Nicolas, E., Michaut, L., Reichhart, J. M. and Hoffmann, J. A. The dorsoventral regulatory gene cassette spätzle/Toll/cactus controls the potent antifungal response in Drosophila adults. *Cell* **86**, 973–83, 1996.
10. Rock, F., Hardiman, G., Timans, J., Kastlelein, R. and Bazan, J. F. A family of human receptors structurally related to Drosophila Toll. *Proc. Natl. Acad. Sci.* **95**, 588–593, 1998.
11. Kawai, T. and Akira, S. TLR signaling. *Cell Death Differ.* **13**, 816–25, 2006.
12. Beutler, B. Microbe sensing, positive feedback loops, and the pathogenesis of inflammatory diseases. *Immunol. Rev.* **227**, 248–63, 2009.
13. Pasare, C. and Medzhitov, R. Toll-like receptors: Linking innate and adaptive immunity. *Adv. Exp. Med. Biol.* **15**, 1382–7, 2005.
14. Lee, C. C., Avalos, A. M. and Ploegh, H. L. Accessory molecules for Toll-like receptors and their function. *Nat. Rev. Immunol.* **12**, 168–79, 2012.

15. Poltorak, a., He, X., Smirnova, I. and Liu, M. Defective LPS signaling in C3H/HeJ and C57BL/10ScCr mice: Mutations in Tlr4 gene. *Science* **282**, 2085–2088, 1998.

16. Schwandner, R., Dziarski, R. and Wesche, H. Peptidoglycan-and lipoteichoic acid-induced cell activation is mediated by toll-like receptor 2. *J. Biol. Chem.* **274**, 17406–17409, 1999.

17. Schröder, N. W. J. et al. Lipoteichoic acid (LTA) of *Streptococcus pneumoniae* and *Staphylococcus aureus* activates immune cells via Toll-like receptor (TLR)-2, lipopolysaccharide-binding protein (LBP), and CD14, whereas TLR-4 and MD-2 are not involved. *J. Biol. Chem.* **278**, 15587–94, 2003.

18. Wang, Y., Liu, L., Davies, D. and Segal, D. Dimerization of Toll-like receptor 3 (TLR3) is required for ligand binding. *J. Biol. Chem.* **285**, 36836–41, 2010.

19. Hayashi, F., Smith, K., Ozinsky, A. and Hawn, T. The innate immune response to bacterial flagellin is mediated by Toll-like receptor 5. *Nature* **410**, 1099–103, 2001.

20. Diebold, S., Kaisho, T., Hemmi, H., Akira, S. and Sousa, C. E. Innate antiviral responses by means of TLR7-mediated recognition of single-stranded RNA. *Science* **303**, 1529–31, 2004.

21. Heil, F. and Hemmi, H. Species-specific recognition of single-stranded RNA via toll-like receptor 7 and 8. *Science* **303**, 1526–9, 2004.

22. Lamphier, M. S., Sirois, C. M., Verma, A., Golenbock, D. T. and Latz, E. TLR9 and the recognition of self and non-self nucleic acids. *Ann. N. Y. Acad. Sci.* **1082**, 31–43, 2006.

23. O'Neill, L. and Bowie, A. The family of five: TIR-domain-containing adaptors in Toll-like receptor signalling. *Nat. Rev. Immunol.* **7**, 353–64, 2007.

24. Farhat, K. and Riekenberg, S. Heterodimerization of TLR2 with TLR1 or TLR6 expands the ligand spectrum but does not lead to differential signaling. *J. Leukoc. Biol.* **83**, 692–701, 2008.

25. Triantafilou, M. et al. Membrane sorting of toll-like receptor (TLR)-2/6 and TLR2/1 heterodimers at the cell surface determines heterotypic associations with CD36 and intracellular targeting. *J. Biol. Chem.* **281**, 31002–11, 2006.

26. Kawai, T., Adachi, O., Ogawa, T., Takeda, K. and Akira, S. Unresponsiveness of MyD88-deficient mice to endotoxin. *Immunity* **11**, 115–22, 1999.

27. Fitzgerald, K. and Rowe, D. LPS-TLR4 signaling to IRF-3/7 and NF-κB involves the Toll adapters TRAM and TRIF. *J. Exp. Med.* **198**, 1043–55, 2003.

28. Miller, Y. and Choi, S. The SYK side of TLR4: Signalling mechanisms in response to LPS and minimally oxidized LDL. *Br. J. Pharmacol.* **167**, 990–999, 2012.

29. Choi, S.-H. et al. Lipoprotein accumulation in macrophages via toll-like receptor-4-dependent fluid phase uptake. *Circ. Res.* **104**, 1355–63, 2009.

30. O'Neill, L. A J., Golenbock, D. and Bowie, A. G. The history of Toll-like receptors—Redefining innate immunity. *Nat. Rev. Immunol.* **13**, 453–60, 2013.

31. Janeway, C. A. and Medzhitov, R. Innate immune recognition. *Annu. Rev. Immunol.* **20**, 197–216, 2002.

32. Matzinger, P. Tolerance, danger, and the extended family. *Annu. Rev. Immunol.* **12**, 991–1045, 1994.

33. Seong, S. and Matzinger, P. Hydrophobicity: An ancient damage-associated molecular pattern that initiates innate immune responses. *Nat. Rev. Immunol.* **4**, 469–478, 2004.

34. Beg, A. Endogenous ligands of Toll-like receptors: Implications for regulating inflammatory and immune responses. *Trends Immunol.* **23**, 509–12, 2002.

35. Vabulas, R. and Ahmad-Nejad, P. Endocytosed HSP60s use toll-like receptor 2 (TLR2) and TLR4 to activate the toll/interleukin-1 receptor signaling pathway in innate immune cells. *J. Biol. Chem.* **276**, 31332–9, 2001.

36. Vabulas, R. and Ahmad-Nejad, P. HSP70 as endogenous stimulus of the Toll/interleukin-1 receptor signal pathway. *J. Biol. Chem.* **277**, 15107–12, 2002.

37. Termeer, C., Benedix, F. and Sleeman, J. Oligosaccharides of hyaluronan activate dendritic cells via toll-like receptor 4. *J. Exp. Med.* **195**, 99–111, 2002.
38. Scheibner, K. and Lutz, M. Hyaluronan fragments act as an endogenous danger signal by engaging TLR2. *J. Immunol.* **177**, 1272–81, 2006.
39. Yu, M., Wang, H., Ding, A. and Golenbock, D. HMGB1 signals through toll-like receptor (TLR) 4 and TLR2. *Shock* **26**, 174–9, 2006.
40. Smith, W. and Murphy, R. Oxidized lipids formed non-enzymatically by reactive oxygen species. *J. Biol. Chem.* **283**, 15513–4, 2008.
41. Richter, C. and Gogvadze, V. Oxidants in mitochondria: From physiology to diseases. *Biochim. Biophys. Acta* **1271**, 67–74, 1995.
42. Bae, Y. S. et al. Macrophages generate reactive oxygen species in response to minimally oxidized low-density lipoprotein: Toll-like receptor 4- and spleen tyrosine kinase-dependent activation of NADPH oxidase 2. *Circ. Res.* **104**, 210–8, 2009.
43. Bochkov, V. N. et al. Generation and biological activities of oxidized phospholipids. *Antioxid. Redox Signal.* **12**, 1009–59, 2010.
44. Esterbauer, H., Schaur, R. and Zollner, H. Chemistry and biochemistry of 4-hydroxynonenal, malonaldehyde and related aldehydes. *Free Radic. Biol. Med.* **11**, 81–128, 1991.
45. Field, C., Toyomizu, M. and Clandinin, M. Relationship between dietary fat, adipocyte membrane composition and insulin binding in the rat. *J. Nutr.* **89**, 1483–1489, 1989.
46. Arroyo, A. et al. NADPH oxidase-dependent oxidation and externalization of phosphatidylserine during apoptosis in Me$_2$SO-differentiated HL-60 cells. Role in phagocytic clearance. *J. Biol. Chem.* **277**, 49965–75, 2002.
47. Greenberg, M. E. et al. Oxidized phosphatidylserine-CD36 interactions play an essential role in macrophage-dependent phagocytosis of apoptotic cells. *J. Exp. Med.* **203**, 2613–25, 2006.
48. Chang, M.-K. et al. Apoptotic cells with oxidation-specific epitopes are immunogenic and proinflammatory. *J. Exp. Med.* **200**, 1359–70, 2004.
49. Huber, J., Vales, A. and Mitulovic, G. Oxidized membrane vesicles and blebs from apoptotic cells contain biologically active oxidized phospholipids that induce monocyte–endothelial interactions. *Arterioscler. Thromb. Vasc. Biol.* **22**, 101–107, 2002.
50. Greenberg, M., Li, X., Gugiu, B. and Gu, X. The lipid whisker model of the structure of oxidized cell membranes. *J. Biol. Chem.* **283**, 2385–96, 2008.
51. Watson, a D. et al. Structural identification by mass spectrometry of oxidized phospholipids in minimally oxidized low density lipoprotein that induce monocyte/endothelial interactions and evidence for their presence in vivo. *J. Biol. Chem.* **272**, 13597–607, 1997.
52. Watson, a D. et al. Structural identification of a novel pro-inflammatory epoxyisoprostane phospholipid in mildly oxidized low density lipoprotein. *J. Biol. Chem.* **274**, 24787–98, 1999.
53. Tsimikas, S. and Witztum, J. L. The role of oxidized phospholipids in mediating lipoprotein(a) atherogenicity. *Curr. Opin. Lipidol.* **19**, 369–77, 2008.
54. Leitinger, N. and Tyner, T. Structurally similar oxidized phospholipids differentially regulate endothelial binding of monocytes and neutrophils. *Proc. Natl. Acad. Sci.* **96**, 12010–5, 1999.
55. Eligini, S. et al. Oxidized phospholipids inhibit cyclooxygenase-2 in human macrophages via nuclear factor-kappaB/IkappaB- and ERK2-dependent mechanisms. *Cardiovasc. Res.* **55**, 406–15, 2002.
56. Erridge, C., Kennedy, S., Spickett, C. M. C. and Webb, D. D. J. Oxidized phospholipid inhibition of toll-like receptor (TLR) signaling is restricted to TLR2 and TLR4: Roles for CD14, LPS-binding protein, and MD2 as targets for specificity of inhibition. *J. Biol. Chem.* **283**, 24748–59, 2008.

57. Kim, M. J. et al. Suppression of Toll-like receptor 4 activation by endogenous oxidized phosphatidylcholine, KOdiA-PC by inhibiting LPS binding to MD2. *Inflamm. Res.* **62**, 571–80, 2013.

58. Walton, K. A. et al. Specific phospholipid oxidation products inhibit ligand activation of toll-like receptors 4 and 2. *Arterioscler. Thromb. Vasc. Biol.* **23**, 1197–203, 2003.

59. Subbanagounder, G. Determinants of bioactivity of oxidized phospholipids specific oxidized fatty acyl groups at the sn-2 position. *Arterioscler. Thromb. Vasc. Biol.* **20**, 2248–2254, 2000.

60. Ma, Z. et al. Inhibition of LPS- and CpG DNA-induced TNF-alpha response by oxidized phospholipids. *Am. J. Physiol. Lung Cell. Mol. Physiol.* **286**, L808–16, 2004.

61. Blüml, S. et al. Oxidized phospholipids negatively regulate dendritic cell maturation induced by TLRs and CD40. *J. Immunol.* **175**, 501–8, 2005.

62. Bochkov, V. and Leitinger, N. Anti-inflammatory properties of lipid oxidation products. *J. Mol. Med.* **81**, 613–26, 2003.

63. Yao, S. et al. Minimally modified low-density lipoprotein induces macrophage endoplasmic reticulum stress via toll-like receptor 4. *Biochim. Biophys. Acta* **1821**, 954–63, 2012.

64. Chávez-Sánchez, L. et al. Activation of TLR2 and TLR4 by minimally modified low-density lipoprotein in human macrophages and monocytes triggers the inflammatory response. *Hum. Immunol.* **71**, 737–44, 2010.

65. Walton, K. a K. et al. Receptors Involved in the Oxidized 1-Palmitoyl-2-arachidonoyl-sn-glycero-3-phosphorylcholine-mediated Synthesis of Interleukin-8 a role for Toll-like receptor 4 and a glycosylphosphatidylinositol-anchored protein. *J. Biol. Chem.* **278**, 29661–6, 2003.

66. Jiang, Z., Georgel, P., Du, X. and Shamel, L. CD14 is required for MyD88-independent LPS signaling. *Nat. Immunol.* **6**, 565–570, 2005.

67. Podrez, E. a et al. A novel family of atherogenic oxidized phospholipids promotes macrophage foam cell formation via the scavenger receptor CD36 and is enriched in atherosclerotic lesions. *J. Biol. Chem.* **277**, 38517–23, 2002.

68. Seimon, T., Nadolski, M. and Liao, X. Atherogenic lipids and lipoproteins trigger CD36-TLR2-dependent apoptosis in macrophages undergoing endoplasmic reticulum stress. *Cell Metab.* **12**, 467–482, 2010.

69. Hoebe, K., Georgel, P., Rutschmann, S. and Du, X. CD36 is a sensor of diacylglycerides. *Nature* **433**, 523–7, 2005.

70. Stewart, C. R. et al. CD36 ligands promote sterile inflammation through assembly of a Toll-like receptor 4 and 6 heterodimer. *Nat. Immunol.* **11**, 155–61, 2010.

71. Mullick, A., Tobias, P. and Curtiss, L. Modulation of atherosclerosis in mice by Toll-like receptor 2. *J. Clin. Immunol.* **115**, 3149–3156, 2005.

72. Curtiss, L. K. and Tobias, P. S. Emerging role of Toll-like receptors in atherosclerosis. *J. Lipid Res.* **50 Suppl**, S340–5, 2009.

73. Vink, A. and Schoneveld, A. *in vivo* evidence for a role of toll-like receptor 4 in the development of intimal lesions. *Circulation* **106**, 1985–1990, 2002.

74. Kiechl, S. and Lorenz, E. Toll-like receptor 4 polymorphisms and atherogenesis. *N. Engl. J. Med.* **347**, 1978–80, 2002.

75. Shi, W., Wang, N., Shih, D. and Sun, V. Determinants of atherosclerosis susceptibility in the C3H and C57BL/6 mouse model evidence for involvement of endothelial cells but not blood cells or cholesterol. *Circ. Res.* **86**, 1078–1084, 2000.

76. Shi, W., Haberland, M., Jien, M.-L., Shih, D. and Lusis, A. Endothelial responses to oxidized lipoproteins determine genetic susceptibility to atherosclerosis in mice. *Circulation* **102**, 75–81, 2000.

77. Imai, Y., Kuba, K. and Neely, G. Identification of oxidative stress and Toll-like receptor 4 signaling as a key pathway of acute lung injury. *Cell* **133**, 235–49, 2008.

78. West, X. Z. et al. Oxidative stress induces angiogenesis by activating TLR2 with novel endogenous ligands. *Nature* **467**, 972–6, 2010.

79. Perez, V. Infiltration of proinflammatory m1 macrophages into the outer retina precedes damage in a mouse model of age-related macular degeneration. *Int. J. Inflam.* **2013**, 503725, 2013.

80. Cruz, D., Watson, A. and Miller, C. Host-derived oxidized phospholipids and HDL regulate innate immunity in human leprosy. *J. Clin. Invest.* **118**, 2917–28, 2008.

81. Thorp, E. and Tabas, I. Mechanisms and consequences of efferocytosis in advanced atherosclerosis. *J. Leukoc. Biol.* **86**, 1089–95, 2009.

82. Knapp, S., Matt, U., Leitinger, N. and van der Poll, T. Oxidized phospholipids inhibit phagocytosis and impair outcome in gram-negative sepsis in vivo. *J. Immunol.* **178**, 993–1001, 2007.

83. Matt, U., Sharif, O. and Martins, R. WAVE1 mediates suppression of phagocytosis by phospholipid-derived DAMPs. *J. Clin. Invest.* **123**, 3014–24, 2013.

84. Weber, M. D., Frank, M. G., Sobesky, J. L., Watkins, L. R. and Maier, S. F. Blocking toll-like receptor 2 and 4 signaling during a stressor prevents stress-induced priming of neuroinflammatory responses to a subsequent immune challenge. *Brain. Behav. Immun.* **32**, 112–21, 2013.

85. Dueñas, A. I. et al. Selective attenuation of Toll-like receptor 2 signalling may explain the atheroprotective effect of sphingosine 1-phosphate. *Cardiovasc. Res.* **79**, 537–44, 2008.

86. Lee, S.-A. et al. Peptidoglycan enhances secretion of monocyte chemoattractants via multiple signaling pathways. *Biochem. Biophys. Res. Commun.* **408**, 132–8, 2011.

87. Oskolkova, O. V et al. Oxidized phospholipids are more potent antagonists of lipopolysaccharide than inducers of inflammation. *J. Immunol.* **185**, 7706–12, 2010.

88. Won, K., Kim, S. and Lee, S. Multiple signaling molecules are involved in expression of CCL2 and IL-1β in response to FSL-1, a Toll-like receptor 6 agonist, in macrophages. *Korean J. Physiol. Pharmacol.* **16**, 447–453, 2012.

89. Kampfrath, T. et al. Chronic fine particulate matter exposure induces systemic vascular dysfunction via NADPH oxidase and TLR4 pathways. *Circ. Res.* **108**, 716–26, 2011.

90. Aye, I. L. M. H., Waddell, B. J., Mark, P. J. and Keelan, J. A. Oxysterols exert proinflammatory effects in placental trophoblasts via TLR4-dependent, cholesterol-sensitive activation of NF-κB. *Mol. Hum. Reprod.* **18**, 341–53, 2012.

12 Dietary Oxidized Lipids as Regulators of Intracellular Signaling Pathways
PPAR and NF-κB

Robert Ringseis, Denise K. Gessner, and Klaus Eder

CONTENTS

INTRODUCTION

Consumption of deep-fried products is very popular in Western countries due to their desirable flavor, color, and crispy texture. In a Spanish cohort from the EPIC study the percentage of energy intake from fried food was estimated to be approximately 25% in the highest quintile of consumption (Guallar-Castillón et al. 2007). During deep frying a heterogeneous mixture of chemically distinct lipid oxidation products is generated in the frying fat (Choe and Min 2007), which is together with the lipid oxidation products absorbed into the fried food and thus ingested when the fried food is consumed. Thus, fried food is an important source for the consumption of oxidized lipids. Based on the fat content of typical fried foods, like potato crisps,

231

doughnuts, and French fries (10–40%; Moreira et al. 1999), and their consumption frequency in Western countries, the amounts of oxidized lipids taken up from fried foods can easily reach more than 50 g/day. Apart from the fact that consumption of high amounts of fried foods leads to a high fat and energy intake and is therefore associated with obesity (Guallar-Castillón et al. 2007; Sayon-Orea et al. 2013), a great number of studies with rats, mice, and pigs showed that administration of oxidized lipids compared with fresh fats induces versatile biological effects (reviewed by Ringseis and Eder 2011), from which some, like blood lipid-lowering effects, are surprisingly beneficial to health. Recent evidence from studies *in vivo* and *in vitro* suggests that these effects are mediated by specific lipid oxidation products contained in oxidized lipids through interfering with intracellular signaling pathways. Likely candidates of such lipid oxidation products are not only hydroxy and hydroperoxy fatty acids but also cyclic fatty acid monomers (CFAM). While hydroxy and hydroperoxy fatty acids are formed during heating of frying fats even at moderate temperatures of below 100°C (Choe and Min 2007; Toschi et al. 1997), the CFAM are only significantly formed from unsaturated 18-carbon fatty acids at temperatures above 200°C (Sebedio and Grandgirard 1989). Both hydroxy and hydroperoxy fatty acids and CFAM come into question as modulators of intracellular signaling pathways *in vivo* because absorption of these lipid oxidation products from the intestine and delivery as part of lipoproteins via the blood to tissues has been demonstrated (Martin et al. 1997; Wilson et al. 2002). This chapter will focus on two signaling pathways, peroxisome proliferator-activated receptor (PPAR) signaling and nuclear factor-kappa B (NF-κB) signaling, which are regulated by lipid oxidation products and are likely targets to mediate specific biological effects of dietary oxidized lipids.

LIPID OXIDATION PRODUCTS AS REGULATORS OF INTRACELLULAR SIGNALING PATHWAYS

LIPID OXIDATION PRODUCTS AND PPAR SIGNALING

A number of lipid oxidation products have been shown to be activators of the PPAR signaling pathway. PPAR are ligand-activated transcription factors, which act as important regulators of lipid and energy metabolism and inflammation (Desvergne and Wahli 1999). Upon binding of a ligand to the ligand-binding domain the PPAR forms a heterodimer with the retinoid X receptor (RXR), which causes binding of transcriptional coactivators and release of transcriptional corepressors. The activated PPAR/RXR heterodimer then binds to specific DNA sequences, called peroxisome proliferator response elements (PPREs), in the promoter, the intronic or the 5′-untranslated region of target genes, thereby stimulating transcription of these genes. Typical genes upregulated by PPARs, particularly PPARα, are genes involved in most aspects of lipid catabolism such as cellular fatty acid uptake, intracellular fatty acid transport, mitochondrial fatty acid uptake, fatty acid oxidation, carnitine uptake, and carnitine synthesis (Mandard et al. 2004; Ringseis and Eder 2012; Figure 12.1). This explains why activation of PPARα in the liver results in increased fatty acid catabolism, elevated liver carnitine concentrations and decreased triacylglycerol (TAG) concentrations in liver and plasma. However, PPARs can also negatively

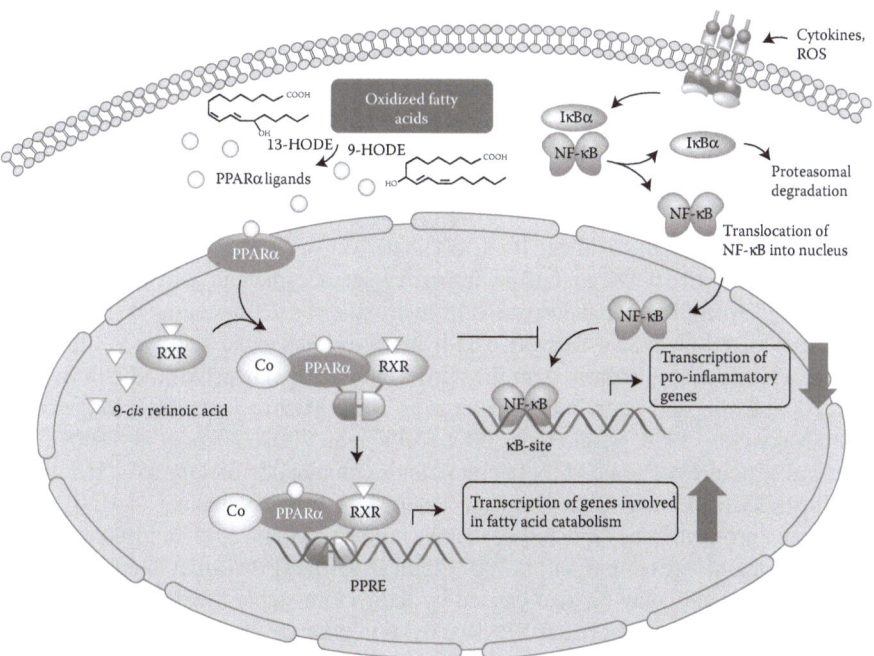

FIGURE 12.1 Activation of PPARα by dietary oxidized lipids is mediated by binding of hydroxy or hydroperoxy fatty acids, such as 9-HODE, 13-HODE, and 13-HPODE contained in the oxidized lipids to the PPARα protein which subsequently forms a complex with RXR. The PPARα/RXR complex binds to specific DNA sequences, called PPRE, present in and around the promoter of PPAR target genes, and, thereby, stimulates their transcription. PPARα can also negatively regulate transcription of genes, which are under the control of stress-sensitive transcription factors such as NF-κB and encode proteins involved in the stress and inflammation response. This is the molecular basis for the well-documented anti-inflammatory effects of PPARα.

regulate transcription of genes that are under the control of stress-sensitive transcription factors such as NF-κB and encode proteins involved in the stress and inflammation response (Figure 12.1). This is the molecular basis for the well-documented anti-inflammatory effects of PPARs. The PPARs exist in three isotypes, PPARα, PPARβ/δ, and PPARγ, all of which share a high degree of structural homology, particularly in the DNA-binding and ligand- and cofactor-binding domain. The relatively high structural homology in the ligand-binding domain explains that the different PPAR isotypes have at least some common ligands. Apart from synthetic ligands, such as the fibrate class of lipid-lowering drugs and the antidiabetic thiazolidinediones, fatty acids and their derivatives (e.g., eicosanoids) have been described as PPAR ligands (Forman et al. 1997; Göttlicher et al. 1992). In general, fatty acids bind best to PPARα, followed by PPARγ and PPARβ/δ (Krey et al. 1997). Polyunsaturated fatty acids such as linoleic acid, α-linolenic acid, arachidonic acid, eicosapentaenoic acid (EPA), and docosahexaenoic acid (DHA) were shown to be the strongest activators of PPARα (Forman et al. 1997; Krey et al. 1997), whereas saturated fatty acids with more than 10 carbon atoms are weak activators and saturated fatty acids with less than 10 carbon atoms are poor activators of PPARα (Forman et al. 1997). Monounsaturated fatty acids such as oleic acid and trans-fatty acids such as elaidic acid showed a similar potency to polyunsaturated fatty acids for binding and activating PPARα, PPARγ, and PPARβ/δ (Krey et al. 1997). Besides native (unoxidized) fatty acids, their oxidized counterparts are also ligands and activators of PPARs. Interestingly, the oxidized counterparts of EPA and DHA, respectively, 5-hydroxy-EPA and 4-hydroxy-DHA, the oxidized linoleic acid derivatives 9-hydroxy-octadecadienoic acid (9-HODE), 13-HODE and 13-hydroperoxy-octadecadienoic acid (13-HPODE), the oxidized arachidonic acid derivative 15-hydroxy-eicosatetraenoic acid (15-HETE) and the oxidized oleic acid derivative 11-hydroxy-octadecenoic acid were shown to activate PPARs more potently than the native fatty acids (Delerive et al. 2000; Itoh et al. 2008; König and Eder, 2006; Limor et al. 2008; Mishra et al. 2004; Muga et al. 2000; Nagy et al. 1998; Sethi et al. 2002; Waku et al. 2009; Yokoi et al. 2010). The chemical structures of some of these compounds contained in dietary oxidized lipids are illustrated in Figure 12.2. Even the aldehyde 4-hydroxynonenal (4-HNE), the breakdown product from oxidized derivatives of arachidonic acid and linoleic acid, is a PPAR agonist (Coleman et al. 2007) indicating that PPAR-mediated effects of oxidized fatty acids may be also caused by their more stable breakdown products. As an explanation for the greater PPAR activation potency of oxidized compared to nonoxidized fatty acids, it has been suggested that the hydroxy group of the oxidized

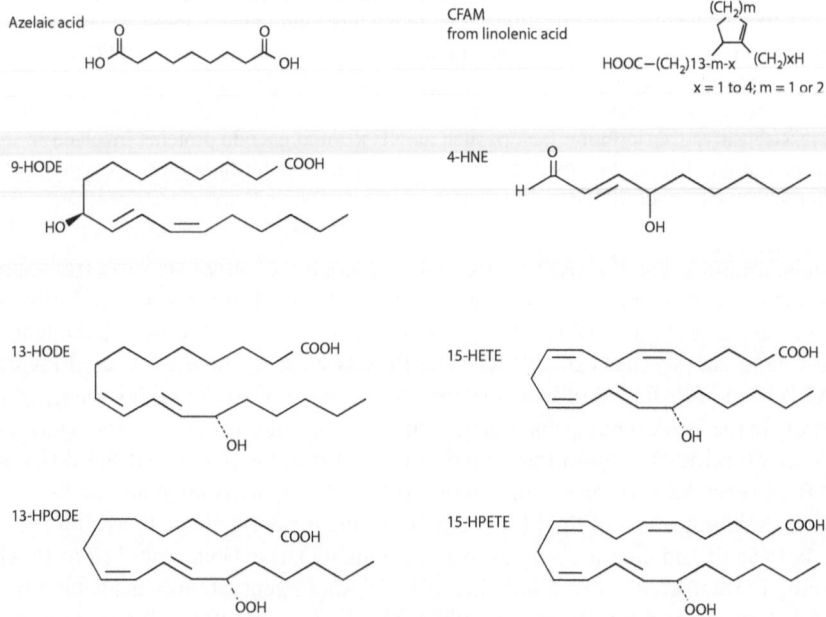

FIGURE 12.2   Chemical structures of specific lipid oxidation products contained in dietary oxidized lipids that are supposed to mediate several biological effects of oxidized lipids through interfering with intracellular signaling pathways, such as PPAR, NF-κB, and Nrf2 signaling.

fatty acids contributes a hydrogen bond to an amino acid in the ligand-binding pocket to stabilize the PPAR–ligand complex, resulting in a more potent activation of the PPARs (Yokoi et al. 2010). In addition, CFAM (the structure of a linolenic acid-derived CFAM is shown in Figure 12.2) were reported to cause activation of PPARα (Bretillon et al. 2003; Martin et al. 2000).

LIPID OXIDATION PRODUCTS AND NF-κB SIGNALING

NF-κB plays a central role in inflammatory processes through regulating the transcription of genes involved in inflammation, such as adhesion molecules, chemokines, and cytokines (Jones et al. 2005). In the inactive state, NF-κB exists in the cytosol of cells bound to the inhibitory proteins IκBs. Upon activation by cytokines, mitogens, or reactive oxygen species, the IκBs become phosphorylated and are degraded via the proteasome, resulting in the release of the active NF-κB, which translocates into the nucleus where it binds to specific DNA sequences, called NF-κB response elements. These response elements are present in the 5′-flanking regulatory region of target genes, and binding of NF-κB causes induction of their transcription (Jones et al. 2005). Activation of NF-κB in certain cell types or tissues has been associated with the etiology of different pathologies. For instance, activation of NF-κB in vascular cells significantly contributes to the development of vascular disorders such as atherosclerosis, while inhibition of NF-κB is accompanied by a reduced development of atherosclerosis (Jones et al. 2005).

Although a great number of studies demonstrated that feeding oxidized lipids causes oxidative stress, a known stimulus of NF-κB activation, in different tissues (Eder et al. 2003b; Izaki et al. 1984; Liu and Huang 1995), it has been shown that different oxidation products of fatty acids, like oxidized n-3 fatty acids (oxidized EPA, oxidized DHA) (Chaudhary et al. 2004; Majkova et al. 2011; Mishra et al. 2004; Musiek et al. 2008), oxidized derivatives from arachidonic acid (15-HETE, 15-HPETE) (Huang et al. 1997), but also 4-HNE (Marantos et al. 2008; Page et al. 1999), inhibit stimulus-induced NF-κB activation and NF-κB target gene expression in different cell culture models (umbilical vein and glomerular endothelial cells, monocytes/macrophages). In line with the above-mentioned negative regulation of NF-κB activation by PPARs (section "Lipid Oxidation Products and PPAR Signaling") it could be shown that oxidized EPA inhibits NF-κB activation via a PPARα-dependent pathway (Mishra et al. 2004). It is likely that the same mechanism also explains inhibition of NF-κB activation by oxidized DHA, 4-HNE, 15-HETE, and 15-HPETE, because these oxidized fatty acids and oxidation products, respectively, are also known to activate PPARs (Coleman et al. 2007; Limor et al. 2008; Sethi et al. 2002). However, since 15-HPETE breaks down to 4-HNE it is possible that 15-HPETE may exert some of its effects through its more stable breakdown product, 4-HNE. In contrast to these studies showing inhibitory effects of oxidized fatty acids on NF-κB activation, three studies using vascular smooth muscle cells (SMCs) reported that 4-HNE and the oxidized linoleic acid derivative 13-HPODE stimulated NF-κB activation (Dwarakanath et al. 2004; Natarajan et al. 2001; Ruef et al. 2001). Whether the stimulatory effect of oxidized fatty acids on NF-κB activation in SMCs is a cell type-specific phenomenon or simply explained by differences in the treatment conditions (e.g., time of incubation, fatty acid concentration) cannot be resolved from these studies because the treatment conditions differed in all studies.

BIOLOGICAL EFFECTS OF DIETARY OXIDIZED LIPIDS MEDIATED BY ACTIVATION OF INTRACELLULAR SIGNALING PATHWAYS

LOWERING OF PLASMA AND LIVER LIPIDS

Considering that oxidized lipids are widely thought to be detrimental on human health, it is quite surprising that dietary oxidized lipids were reported in a large number of feeding experiments with different animal species (rats, mice, guinea pigs, pigs) to decrease the concentrations of TAG and cholesterol in liver and plasma (Ammouche et al. 2002; Brandsch and Eder 2004; Chao et al. 2001, 2004; Eder 1999; Eder et al. 2003a,b, 2004; Eder and Kirchgessner, 1998; Huang et al. 1988; Kämmerer et al. 2011a; Liu and Lee 1998; Luci et al. 2007b; Ringseis et al. 2007a,b,c; Sülzle et al. 2004). The plasma lipid-lowering effect of oxidized lipids has to be considered as beneficial because increased lipid concentrations in plasma are well-known risk factors for the development of atherosclerosis. In a recent review from the authors' group the data from these studies in different species have been summarized (Ringseis and Eder, 2011). It is evident from this review that the TAG-lowering effect of oxidized lipids, which was consistently found in all these studies, was stronger than its cholesterol-lowering effect, which was somewhat less consistent. In addition, this review clearly shows that the TAG-lowering effects of oxidized lipids observed were more pronounced in rats and mice than in guinea pigs and pigs, indicating a species-dependence of this effect. The fat source and the heating regime for preparation of the oxidized lipids used in the above-mentioned studies are shown in Table 12.1. Since marked TAG-lowering effects were observed for oxidized lipids prepared from different fat sources and under quite different heating conditions which leads to a divergent spectrum of primary and secondary lipid peroxidation products, it can be concluded that both primary and secondary lipid peroxidation products contribute to this effect of dietary oxidized lipids.

In earlier studies describing the lipid-lowering effects of dietary oxidized lipids (Corcos Benedetti et al. 1990; Liu and Huang 1995; Nolen et al. 1967; Poling et al. 1970; Yoshida and Kajimoto 1989), mainly strongly oxidized lipids (e.g., hydrogenated soybean oil heated at 182°C with frying of potatoes, breaded scallops, and onion rings for 8 h per day over 1 to over 5 wk; soybean oil heated at 205°C with frying of potato sticks for 6 h per day on four consecutive days) were fed to the animals, which were not suitable for reuse as frying fat in restaurants due to their high levels of lipid peroxidation products, and which provoked reduced feed intake, decreased nutrient digestibility, impaired growth, pronounced oxidative stress, and even toxic effects. It has therefore been argued that the lipid-lowering action of dietary oxidized lipids was not due to specific compounds of dietary oxidized lipids but rather due to these confounding effects, notably a reduced intake of digestible energy from the diets. To avoid these confounding effects, in more recent studies only moderately oxidized lipids were used (Table 12.1) and a controlled feeding system, in which

TABLE 12.1
Fat Source and Heating Regime for Preparation of Oxidized Lipids Used in the Different Animal Studies

Fat Source	Heating Regime	Reference
Soybean oil	205°C, 6 h with wheat flour dough sheets in the oil, 4 consecutive days	Huang et al. (1988)
Soybean oil	130°C, 22 h	Eder and Kirchgessner (1998)
Lard:safflower oil (2:1, w/w)	150°C, 6 days	Eder (1999)
Soybean oil	205°C, 6 h with wheat flour dough sheets in the oil, 4 consecutive days	Chao et al. (2001)
Sunflower oil	98°C, 48 h	Ammouche et al. (2002)
Sunflower oil:lard (1:1, w/w)	50°C, 38 days	Eder et al. (2003a,b)
Sunflower oil:lard (1:1, w/w)	105°C, 81 h	Eder et al. (2003a,b)
Sunflower oil:lard (1:1, w/w)	190°C, 24 h	Eder et al. (2003a,b)
Sunflower oil	55°C, 42 days	Eder et al. (2003b)
Soybean oil	205°C, 6 h with wheat flour dough sheets in the oil, 4 consecutive days	Chao et al. (2004)
Sunflower oil:lard (1:1, w/w)	50°C, 38 days	Sülzle et al. (2004)
Sunflower oillLinseed oil (80:20, w/w)	50°C, 16 days	Brandsch and Eder (2004)
Sunflower oil	60°C, 25 days	Ringseis et al. (2007)
Sunflower oil	60°C, 25 days	Ringseis et al. (2007)
Sunflower oil	60°C, 25 days	Ringscis et al. (2007)
Heated hydrogenated palm fat:fresh sunflower oil (92:8, w/w)	170°C, 48 h with repeated frying of French fries for 6 min every 30 min	Kämmerer et al. (2011)
Soybean oil	205°C, 6 h with potato sticks in the oil, 4 consecutive days	Liu and Li (1998)
Sunflower oil	180°C, 24 h	Luci et al. (2007)
Sunflower oil:lard (1:1, w/w)	55°C, 49 days	Eder et al. (2004)

experimental animals of control and treatment groups received identical amounts of feed, was applied (Brandsch and Eder 2004; Eder 1999; Eder et al. 2003a,b, 2004; Eder and Kirchgessner 1998; Kämmerer et al. 2011a; Luci et al. 2007b; Ringseis et al. 2007a,b,c; Sülzle et al. 2004). In addition, the heating-induced losses of antioxidants and PUFA occurring during preparation of the oxidized lipids were replaced in these recent studies by supplementing the oxidized lipid diets with synthetic antioxidants (vitamin E) and PUFA-rich fats. This ensured that the results of these studies were not biased by a lower intake of antioxidants or PUFA in the treatment group receiving the oxidized lipids. Despite these precautions, the lipid-lowering actions of dietary oxidized lipids could still be observed in these studies indicating that the lipid-lowering effects of oxidized lipids were caused by specific compounds of oxidized lipids.

The mechanism behind the lipid-lowering phenomenon of dietary oxidized lipids became clear when hepatic PPARα was shown to be targeted by the administration of oxidized lipids prepared by heating soybean oil at 205°C with frying of potato sticks for 6 h/day on 4 consecutive days (Chao et al. 2001). Activation of hepatic PPARα provides a plausible explanation for this effect, because it causes upregulation of a comprehensive set of genes regulating most aspects of lipid catabolism (fatty acid uptake and oxidation) in the liver, thereby, leading to a catabolism of fatty acids and decreased TAG concentrations in liver and plasma. This first observation in rats (Chao et al. 2001) could be confirmed in subsequent studies in rats, mice, and even pigs (Chao et al. 2004; Kämmerer et al. 2011a; Luci et al. 2007b; Ringseis et al. 2007a,b,c; Sülzle et al. 2004). However, critical evaluation of these studies in a recent review (Ringseis and Eder 2011) revealed that the activation of PPARα by dietary oxidized lipids is stronger in the liver of rodents than in pigs, which provides an explanation for the stronger lipid-lowering effect of dietary oxidized lipids in rodents than in non-rodent species. One important reason for this species-dependence is the variation in the expression level of PPARα in tissues between these species. In rodents, PPARα is abundantly expressed in tissues, and activation of PPARα not only induces many genes involved in various metabolic pathways but also causes severe peroxisome proliferation in the liver (Peters et al. 2005). In contrast to rodents, PPARα activators do not induce peroxisome proliferation in the liver of guinea pigs, pigs, monkeys, and humans, because these species express PPARα in the liver at lower levels and the response of many genes to PPARα activation is much weaker than in rodents (Holden and Tugwood 1999). Thus, effects related to PPARα activation observed in rodents cannot be directly applied for pigs or humans. However, given that pigs express PPARα in the liver at a similar level to humans (Luci et al. 2007a), and the response to PPARα activators is quite similar between pigs and humans (Cheon et al. 2005), it is possible that oxidized lipids also cause a PPARα response in the liver of humans consuming oxidized lipids. Additionally, it has been proposed by Eder et al. (2003a) that oxidized lipids prepared by heating a sunflower oil:lard mixture at different temperatures (50–190°C) and time periods (38 d, 81 h, 24 h) inhibit activation of sterol regulatory element-binding protein (SREBP)-1c in the liver, which is a transcription factor acting as a master regulator of genes involved in fatty acid synthesis (lipogenic enzymes). This assumption is based on earlier findings that oxidized lipids prepared by heating sunflower oil at 130°C for 22 h reduce activity and gene expression, respectively, of lipogenic enzymes in rats (Eder and Kirchgessner 1998). Inhibition of SREBP-1c activation by dietary oxidized lipids is likely also a PPARα-dependent effect because it was recently shown that activation of PPARα reduces activation of SREBP-1 and fatty acid synthesis (König et al. 2009). Convincing data from Koch et al. (2007a) suggest that the cholesterol-lowering effect of dietary oxidized lipids is mediated by inhibiting activation of hepatic SREBP-2—another SREBP isoform preferentially activating genes involved in cellular cholesterol uptake and cholesterol synthesis—thereby reducing cellular cholesterol uptake and synthesis in the liver, which leads to a decreased secretion of cholesterol via lipoproteins into the blood. In the study by Koch et al. (2007a) the oxidized lipids were prepared by heating sunflower oil in a domestic fryer at 60°C for 25 days. Like inhibition of SREBP-1c activation, inhibition of SREBP-2 activation by dietary oxidized lipids involves a PPARα-dependent effect.

Koch et al. (2007a) found that the dietary oxidized lipids cause upregulation of the insulin-induced gene (Insig)-1 in the liver in a PPARα-dependent manner. Insigs are membrane proteins that reside in the endoplasmic reticulum and play a central role in the regulation of SREBP activation, because they prevent the translocation of inactive SREBPs from the endoplasmic reticulum to the Golgi, where proteolytic activation of SREBPs and subsequent release of transcriptionally active forms of SREBPs occur (Yabe et al. 2002). As a result, the synthesis of cholesterol declines in response to PPARα activation.

INHIBITION OF ALCOHOL-INDUCED FATTY LIVER DEVELOPMENT

Chronic alcohol abuse is a serious health-associated problem in developed countries as it favors the development of alcoholic liver disease (ALD). The onset of ALD is characterized by the development of fatty liver (steatosis), which is found in more than 90% of chronic heavy drinkers (Yip and Burt, 2006). While alcoholic fatty liver readily reverses upon alcohol abstinence, the ALD progresses with ongoing alcohol abuse to more severe pathologies like alcoholic steatohepatitis, fibrosis, cirrhosis, and finally alcoholic hepatocellular carcinoma (HCC) (Siegmund and Brenner, 2005). These severe pathologies are irreversible and the only treatment for such progressed damage is liver transplantation. Whereas alcoholic fatty liver has long been considered to be pathologically inert, more recent research indicates that alcoholic fatty liver favors ALD progression, and the development of fibrosis and cirrhosis (Harrison and Diehl 2002). Thus, prevention or treatment of alcoholic fatty liver is a reasonable approach for preventing the later stage ALD pathologies such as fibrosis, cirrhosis, and HCC. The development of alcoholic fatty liver has been linked to the dysregulation of two signaling pathways resulting in increased fat accumulation in the liver: blockade of the fatty acid catabolizing PPARα pathway and activation of the lipogenic SREBP pathway (Fischer et al. 2003; You et al. 2002). The key role of the impaired PPARα function in alcoholic fatty liver development has been demonstrated by the finding that pharmacological PPARα activators prevent alcoholic fatty liver development by restoring the disturbed PPARα function, and stimulating the transcription of PPARα target genes involved in fatty acid catabolism resulting in a higher rate of fatty acid β-oxidation and reduced TAG accumulation in the liver (Crabb et al. 2004; Fischer et al. 2003). Activation of the SREBP-1 pathway in the liver has been shown in a rodent model of alcohol fatty liver (You et al. 2002). Since SREBP-1 is a major transcriptional regulator of genes involved in fatty acid and TAG synthesis, its activation contributes to enhanced lipid synthesis and the development of fatty liver. The important role of SREBP-1 for alcoholic fatty liver development is shown by the fact that SREBP-1 knockout mice are protected against alcohol-induced fatty liver (You et al. 2002). In light of the fact that administration of dietary oxidized lipids or components of dietary oxidized lipids were shown to activate PPARα, inhibit SREBP activation and decrease TAG concentration in the liver (see section "Lowering of Plasma and Liver Lipids"), we have recently investigated in a rat model of ALD whether administration of oxidized lipids prepared by heating sunflower oil at 60°C for 25 days, like treatment with fibrates, are capable of preventing alcoholic fatty liver development (Ringseis et al. 2007b). Our study clearly showed that alcohol-induced TAG accumulation in the liver is markedly reduced in rats by simultaneous administration of dietary oxidized lipids relative to fresh fat indicating that dietary oxidized lipids are indeed capable of preventing alcoholic fatty liver development. As a plausible mechanism of action we found that administration of oxidized lipids but not fresh fat during ethanol feeding resulted in the induction of PPARα target genes involved in fatty acid oxidation even in the presence of ethanol. These findings suggested that the oxidized lipids-induced expression of PPARα target genes enhanced the capacity of the liver to oxidize fatty acids and, thus, counteracted TAG accumulation and the diminished PPARα function during ethanol feeding (Ringseis et al. 2007b). While activation of the SREBP pathway by alcohol feeding could be demonstrated in our study, which is in line with earlier reports (You et al. 2002), we found only a slight inhibition of the SREBP pathway in the liver of the rats by dietary oxidized lipids. This suggests that the protective effect of dietary oxidized lipids on alcoholic fatty liver development is mainly due to the activation of hepatic PPARα.

ELEVATION OF HEPATIC CARNITINE CONCENTRATION

Several studies from the 1970s and 1980s repeatedly demonstrated that energy deprivation and treatment with fibrates cause a strong increase in the concentration of carnitine in the liver of rats (Brass and Hoppel 1978; McGarry et al. 1975; Paul and Adibi 1979). Carnitine plays an important role in lipid and energy metabolism by acting as a shuttling molecule for the translocation of long-chain fatty acids from the cytosol into the mitochondrial matrix, where β-oxidation occurs. The mechanism underlying this phenomenon remained obscure until it was shown in 2006 for the first time that activation of hepatic PPARα, which is caused by both energy deprivation and fibrate treatment, stimulates the expression of genes involved in carnitine uptake and carnitine synthesis in liver cells (Luci et al. 2006). In subsequent experiments, convincing evidence for the PPARα-dependence of this effect could be provided by demonstrating that the energy deprivation- or fibrate-induced increase in hepatic carnitine concentration occurs only in wild-type mice but not in PPARα knockout mice (Koch et al. 2008; Van Vlies et al. 2007). Moreover, it was shown recently that the gene encoding the carnitine transporter novel organic cation transporter 2 (OCTN2/SLC22A5) contains a functional PPARα binding site in its regulatory region and is therefore directly activated by PPARα (Wen et al. 2010). Based on the knowledge that dietary oxidized lipids cause activation of hepatic PPARα (see section "Lowering of Plasma and Liver Lipids") and that PPARα activation influences carnitine homeostasis, Koch et al. (2007b) explored whether administration of oxidized lipids prepared by heating sunflower oil in a domestic fryer at 60°C for 25 days to rats increases liver carnitine concentrations and activates genes involved in carnitine homeostasis. The study from Koch et al. (2007b) revealed that administration of the oxidized lipids for 6 days results in elevated carnitine concentrations and transcript levels of the carnitine transporter OCTN2 in the liver of the rats, indicating that oxidized lipids indeed have the same effect on carnitine homeostasis as energy deprivation and treatment with fibrates. Thus, it is very likely that this effect of dietary oxidized lipids is mediated also by the activation of hepatic PPARα.

In addition, Koch et al. (2007b) observed that dietary oxidized lipids stimulate the expression of a further carnitine transporter, namely OCTN1, in the liver. This effect, however, has been suggested to contribute less to the oxidized lipids-induced elevation of hepatic carnitine content because OCTN1 has a lower carnitine transport activity than OCTN2 and is therefore considered physiologically less important (Tamai et al. 1998).

Besides carnitine uptake from blood, the hepatic carnitine content is influenced by hepatic carnitine biosynthesis, which is facilitated by a cascade of four distinct enzymatic reactions through which 6-N-trimethyllysine (TML), the substrate for carnitine biosynthesis, is converted stepwise into carnitine. Studies in rats demonstrated that both energy deprivation and fibrate treatment elevate the hepatic concentration of TML (Davis and Hoppel 1986; Luci et al. 2008; Ringseis et al. 2007d). Given that carnitine synthesis begins with the enzymatic conversion of TML, the availability of TML has been considered to be rate limiting for carnitine biosynthesis (Rebouche 2004). In fact, TML is subsequently converted into γ-butyrobetaine (BB), which itself is rapidly further converted into carnitine due to the large capacity of the liver to convert BB into carnitine (Rebouche 1983). It is, therefore, likely that carnitine synthesis is enhanced by PPARα activation through stimulating lysosomal and proteasomal degradation of proteins, which leads to the release of TML (LaBadie et al. 1976). The observation that both energy deprivation and fibrate treatment stimulate proteolysis (Paul and Adibi 1980) supports this assumption. Thus, one might speculate that the oxidized lipids in the study from Koch et al. (2007b) have stimulated carnitine synthesis in the liver by increasing the release of TML. In addition, it is possible that the increased expression of OCTN2 by dietary oxidized lipids caused a stimulation of hepatic carnitine synthesis through increasing the uptake of BB from plasma into the liver, because BB is a good substrate for OCTN2 (Tamai et al. 1998) and the liver has a high capacity to convert BB into carnitine (Vaz and Wanders 2002).

Besides the study in rats, one study in pigs reported that dietary oxidized lipids influence parameters of carnitine homeostasis (Varady et al. 2012b). Like in rats, the dietary oxidized lipids were shown to cause upregulation of OCTN2 in pig liver (Varady et al. 2012b), even though this effect was less pronounced than in rats. The effect of oxidized lipids on genes involved in carnitine synthesis in the liver was not reported in this study (Varady et al. 2012b). The less strong effect of dietary oxidized lipids on OCTN2 gene expression in pigs is probably due to the species difference with regard to hepatic PPARα expression (see section "Lowering of Plasma and Liver Lipids"). However, due to this species specificity, the pig is an excellent model for humans, because the hepatic expression level of PPARα is comparable to that in humans (Luci et al. 2007a). It is worth mentioning that in the pig study rapeseed oil was used for preparing the oxidized lipids because rapeseed oil is a commonly used fat for deep frying of foods in households and restaurants, at least in Mediterranean EU countries. In addition, the oxidation of the rapeseed oil was carried out by heating it in a domestic fryer at a temperature of 175°C for 72 h in order to reflect most accurately the practical situation of deep frying in human nutrition. Thus, the pig study allows the extrapolation of results from pigs to humans with only few limitations. Given this and the above-mentioned similarities between pigs and humans, it is not unlikely that a dietary regime rich in fried foods containing oxidized lipids induces hepatic OCTN2 expression and increases hepatic carnitine concentration also in humans.

INHIBITION OF ATHEROSCLEROTIC PLAQUE DEVELOPMENT

Like in hepatocytes, PPARα is also abundantly expressed in the cells of the vascular wall, such as endothelial cells (ECs), SMCs, and monocytes/macrophages. Notably, activation of PPARα in these cells causes antiatherogenic effects through modulating the expression of several genes implicated in atherosclerosis development, which results in reduced monocyte recruitment to ECs (Marx et al. 2004), decreased SMC proliferation and migration (Gizard et al. 2005), and enhanced cholesterol removal from macrophages (Chinetti et al. 2001). These direct atheroprotective effects in connection with the above-mentioned lipid-lowering effects of PPARα activators (see section "Lowering of Plasma and Liver Lipids") are mainly causative for the inhibitory effect of pharmacological PPARα activators on atherosclerosis (Ericsson et al. 1997; Hennuyer et al. 2005; Li et al. 2004; Rubins et al. 1999). Since dietary oxidized lipids have pronounced PPARα activating and lipid-lowering properties, the hypothesis that dietary oxidized lipids inhibit atherosclerosis in the low-density-lipoprotein receptor-deficient (LDLR$^{-/-}$) mouse model was recently tested (Kämmerer et al. 2011a). At first glance, the hypothesis from Kämmerer et al. (2011a) appears surprising given that dietary oxidized lipids are widely considered to have detrimental effects on human health (Khan-Merchant et al. 2002; Staprans et al. 1996; Steinberg 1997), and earlier reports even demonstrated that administration of dietary oxidized lipids to experimental animals has pro-atherogenic effects (Greco and Mingrone 1990; Kaunitz et al. 1965; Kritchevsky and Tepper 1967; Staprans et al. 1996). The latter, however, is likely explained by confounding effects resulting from the use of strongly abused oxidized lipids (as discussed in section "Lowering of Plasma and Liver Lipids"), which are not allowed for reuse as frying fat according to EU legislation and which induce excessive oxidative stress (Izaki et al. 1984; Liu and Huang 1995), a factor that is known to promote the development of atherosclerosis (Victor et al. 2009). In the study of Kämmerer et al. (2011a) a heated hydrogenated palm fat—a typical fat used for deep-frying in German restaurants—was used as oxidized lipids. For heating of the hydrogenated palm fat a realistic deep-frying regime in a domestic fryer was applied, during which a portion of 70 g French fries was deep-fried for 6 min every 30 min at a temperature of 170°C for 48 h. Analysis of the levels of lipid oxidation products in the oxidized lipids indicated that this heating regime caused a moderate degree of oxidation indicating suitability for reuse of the frying fat. In addition, the dietary fats (fresh fat and oxidized lipids) in the study from Kämmerer et al. (2011a) were equalized for their fatty acid composition by using fat mixtures, and vitamin E concentrations in the diets were adjusted. After 14 weeks, feeding the oxidized hydrogenated palm fat as part of a hyperlipidemic diet (20% fat, 0.15% cholesterol) clearly inhibited atherosclerosis as demonstrated by a markedly reduced lesion size and strongly decreased lipid and collagen contents in the aortic root. As one important mechanism contributing to this effect in LDLR$^{-/-}$ mice, the observed PPARα-mediated reduction in plasma cholesterol and TAG concentrations comes into question (Kämmerer et al. 2011a), because elevated blood lipid concentrations are known risk factors for the development of atherosclerosis. However, the observation from Kämmerer et al. (2011a) that PPARα expression was also markedly elevated in the aortic root lesions of LDLR$^{-/-}$ mice fed the oxidized lipids suggests that direct activation of PPARα in the vasculature has also contributed to this beneficial effect of dietary oxidized lipids. The elevated protein expression of PPARα in the atherosclerotic lesions of oxidized lipids-treated LDLR$^{-/-}$ mice is considered a beneficial effect because inhibition of atherosclerosis development is associated with an increased PPARα expression in the atherosclerotic plaque and the aorta, respectively (Toomey et al. 2006). PPARα activation in the vasculature has been demonstrated to inhibit atherosclerotic plaque formation due to repression of pro-inflammatory gene expression (Chinetti et al. 2001; Gizard et al. 2005; Marx et al. 2004) and inhibition of SMC proliferation and SMC migration into the arterial intima (Gizard et al. 2005). Immunohistochemical analysis of aortic root sections revealed that the pro-inflammatory protein VCAM-1, which mediates monocyte attachment to the luminal surface of the vessel wall and is necessary for subsequent infiltration of the subendothelial space by monocyte-derived macrophages, and the SMC-specific marker α-actin, a marker of SMC accumulation in the arterial wall, were less expressed in LDLR$^{-/-}$ mice fed the dietary oxidized lipids. In addition, the lipid and collagen contents in the aortic root of the LDLR$^{-/-}$ mice were also reduced by the dietary oxidized lipids. Given that SMC are the major collagen-producing cells in atherosclerotic lesions and collagens substantially contribute to lesion volume (Katsuda and Kaji 2003), it is not unlikely that the reduced SMC accumulation is causative for the decrease in collagen content and lesion size in LDLR$^{-/-}$ mice fed the oxidized lipids.

Although it is difficult to ascribe the PPARα activating property and the antiatherogenic effects of the dietary oxidized lipids in the study of Kämmerer et al. (2011a) to individual chemical compounds due to the complex nature of dietary oxidized lipids, hydroxy and hydroperoxy fatty acids as well as cyclic fatty acid monomers are the likely candidates (see section "Lipid Oxidation Products and PPAR Signaling"). In line with this, Litvinov et al. (2010) reported recently that administration of azelaic acid, an end product of linoleic acid peroxidation whose structure is shown in Figure 12.2, reduces atherosclerosis development in LDLR$^{-/-}$ mice. Moreover, an animal experiment from Garelnabi et al. (2008) revealed that feeding a diet supplemented with 13-HPODE reduces plasma TAG concentrations indicating that oxidized fatty acids are indeed the mediators of the lipid-lowering effects of dietary oxidized lipids. In a cell culture study it could be shown that 13-HODE reduces cholesterol content in murine RAW264.7 macrophages by stimulating apoA-I-dependent cholesterol efflux (Kämmerer et al. 2011b). This effect has to be considered beneficial because excessive accumulation of cholesterol by macrophages in the arterial wall promotes atherosclerosis. Using PPAR antagonists it could be further demonstrated that the 13-HODE-induced increase in cholesterol efflux from macrophages is mediated by a PPAR-dependent upregulation of LXRα and cholesterol transporters (ABCA1, ABCG1, SR-BI) (Kämmerer et al. 2011b), which operate on cholesterol export to extracellular acceptors such as apoA-I/HDL. Moreover, several cell culture studies revealed potent anti-inflammatory effects of oxidized n-3 fatty acids and oxidation products of n-6 fatty acids, like 4-HNE, mediated by inhibition of NF-κB in endothelial cells and monocytes, respectively, which were partially shown to be mediated via a PPARα-dependent mechanism (section "Lipid Oxidation Products and NF-κB Signaling"). In contrast, in another feeding experiment with LDLR$^{-/-}$ mice, administration of 13-HODE did not inhibit but even promoted atherosclerosis development (Khan-Merchant et al. 2002). The latter observation might be explained by the fact that the concentration of pure 13-HODE in the study of Khan-Merchant et al. (2002) was markedly higher than that in the oxidized lipids used in the study of Kämmerer et al. (2011a) and that at high concentration this hydroxy fatty acid may exert unfavorable effects, like induction of strong oxidative stress.

MODULATION OF STRESS AND INFLAMMATORY RESPONSES

NF-κB plays a central role in inflammatory settings through regulating the transcription of large set of genes involved in the inflammatory process (Jones et al. 2005). Its extraordinary role becomes apparent when considering that activation of NF-κB in certain cell types or tissues contributes to the development of inflammation-associated pathologies, like atherosclerosis, whereas inhibition of NF-κB is accompanied by a reduced development of such pathologies (Jones et al. 2005). Interestingly, PPARα and PPARγ activation inhibits NF-κB activation and thereby causes repression of transcription of inflammatory genes and attenuates the inflammatory process. This transrepression activity probably constitutes the mechanistic basis for the well-documented anti-inflammatory properties of PPAR agonists and explains that pharmacological PPAR ligands markedly reduce inflammation in experimental animal models of inflammation and different clinical settings with inflammatory involvement, like atherosclerosis (Chinetti et al. 2001; Gizard et al. 2005; Marx et al. 2004). Owing to the fact that dietary oxidized lipids have a strong PPARα activation activity *in vivo*, we have recently studied the effect of oxidized lipids representing a 92:8-mixture of heated (170°C, 48 h) hydrogenated palm fat and fresh sunflower oil on NF-κB signaling in a mouse model of atherosclerosis, the LDLR$^{-/-}$ mouse (Kämmerer et al. 2011a). Apart from the fact that the dietary oxidized lipids caused an inhibition of atherosclerosis development in LDLR$^{-/-}$ mice (see section "Inhibition of Atherosclerotic Plaque Development"), we observed that PPARα in the vasculature was activated and the protein level of the inflammatory NF-κB target gene VCAM-1, an adhesion molecule responsible for monocyte attachment to the vascular endothelium, was clearly reduced in mice of the oxidized lipids group (Kämmerer et al. 2011a). This indicates that NF-κB-induced inflammatory responses in the vasculature can be inhibited by feeding oxidized lipids via a mechanism involving PPARα-mediated repression of NF-κB. As already mentioned in section "Inhibition of Atherosclerotic Plaque Development", it is difficult to ascribe this effect of dietary oxidized lipids to specific components of dietary oxidized lipids. However, it is likely that oxidized fatty acids are responsible for this effect because they have been shown to be activators of PPARα and to inhibit stimulus-induced NF-κB activation and NF-κB target gene expression in different cell culture models (Chaudhary et al. 2004; Huang et al. 1997; Majkova et al. 2011; Mishra et al. 2004; Musiek et al. 2008; Marantos et al. 2008; Page et al. 1999).

In contrast, in a recent study with pigs (Ringseis et al. 2007e), which were fed two different diets containing either fresh fat or oxidized lipids prepared by 24 h heating at 200°C for a period of 4 wk, NF-κB activity and inflammatory gene expression in intestinal epithelial cells were not altered by dietary oxidized lipids, although the oxidized lipids caused a moderate activation of PPARγ signaling in the intestinal epithelial cells of the pigs. One explanation for the lack of effect of dietary oxidized lipids on NF-κB activity and NF-κB target gene expression in this study might be the comparatively slight PPARγ activation by the dietary oxidized lipids, which was presumably insufficient to mediate transrepression of NF-κB. However, the failure of dietary oxidized lipids to inhibit NF-κB may be also explained by the fact that the basal inflammatory state in the intestinal epithelium of the pigs was rather low, and a reduction of inflammatory indices in normal healthy animals, as used in this pig study, is expected to be only marginal. It is noteworthy that in a further study with mice (Varady et al. 2011) that were administered either a fresh fat (unheated sunflower oil) or oxidized lipids (sunflower oil heated in a domestic fryer at 190°C for 48 h) by gavage, activation of NF-κB in the intestinal mucosa was clearly demonstrated, despite strong activation of PPAR signaling in the intestinal mucosa. In this context it has to be considered that, unlike in the study from Ringseis et al. (2007e), the dietary oxidized lipids used in the mice study (Varady et al. 2011) were strongly oxidized as indicated by high concentrations of lipid peroxidation products and these peroxidation products are efficiently taken up by the intestinal epithelial cells (Penumetcha et al. 2000; Staprans et al. 1993), where they act as potent inducers for the generation of reactive oxygen species (Liu and Lee 1998). As reactive oxygen species are potent stimuli for the activation of NF-κB (Schoonbroodt and Piette 2000), it is therefore very likely that the ingestion of lipid peroxidation products and the resulting oxidative stress in the intestinal mucosa of the mice has led to the activation of NF-κB. Thus, the results from the mice study (Varady et al. 2011) suggest that activation of PPAR signaling by dietary oxidized lipids in the intestinal mucosa is obviously not capable of blocking the oxidative stress-induced activation of NF-κB. The failure of dietary oxidized lipids to cause induction of oxidative stress and to activate NF-κB in the intestinal mucosa in the pig study (Ringseis et al. 2007e) might be explained by the fact that the oxidized lipids were fed as part of a normal diet, whereas in the mice study (Varady et al. 2011) the dietary oxidized lipids were administered by gavage. In the latter case, the intestinal mucosa is probably more directly exposed to the ingested oxidized lipids than in the case that the oxidized lipids are virtually diluted by the other feed components. A further reason for the lack of response in the pig study might be that the amount of oxidized lipids administered to the animals when related to their body weights was clearly higher in the mice study (Varady et al. 2011) than in the pig study (Ringseis et al. 2007e).

The assumption that oxidative stress is the stimulus for NF-κB activation by dietary oxidized lipids is strengthened by the observation from the mice study (Varady et al. 2011) but also by a further pig study (Varady et al. 2012a) showing that dietary oxidized lipids lead to an activation of nuclear factor erythroid-derived 2-like 2 (Nrf2) signaling in the intestine and liver. Nrf2 is a transcription factor that is located in the cytosol when the cell is not exposed to any activating stimuli. However, in response to oxidative stress Nrf2 is dissociated from inhibitory proteins and translocates into the nucleus, where it binds to specific DNA sequences called antioxidant response element (ARE) in the regulatory region of target genes which have mainly functions in antioxidant defense and phase II detoxification (Dhakshinamoorthy et al. 2000; Kaspar et al. 2009; Niture et al. 2010). Thus, Nrf2 is regarded as the most important pathway in the cell to protect cells against oxidative stress. In both, the mice (Varady et al. 2011) and the pig study (Varady et al. 2012a), nuclear concentration of Nrf2 and expression of Nrf2 target involved in antioxidant defense (e.g., SOD1, GPX1, TXNR1, HO-1) and phase II detoxification (e.g., NQO1, MGST1, UGT1A1) were clearly elevated by the dietary oxidized lipids indicating that oxidative stress as a critical stimulus for Nrf2 activation was probably induced by dietary oxidized lipids. However, since it was reported that specific oxidized fatty acid acids are able to directly activate Nrf2 by initiating dissociation of Keap1 (Gao et al. 2007), the possibility that Nrf2 activation is directly induced by specific oxidation products present in the dietary oxidized lipids cannot be excluded. Induction of Nrf2 in the liver of pigs fed oxidized lipids can be interpreted as an adaptive response of the liver to cope with oxidative stress induced by administration of oxidized lipids, thereby, preventing reactive oxygen species-mediated damage. Nrf2 activation is generally regarded as a favorable effect since it results in the upregulation of a wide spectrum of antioxidant, cytoprotective and detoxifying genes, thus, protecting the cell against reactive oxygen species and toxic compounds.

CONCLUSIONS

Results from studies with experimental animals suggest that dietary oxidized lipids induce several biological effects that are mediated by specific lipid oxidation products contained in oxidized lipids, such as 9-HODE, 13-HODE, 13-HPODE, and CFAM, by interfering with intracellular signaling pathways, such as PPAR, NF-κB, and Nrf2 signaling. Despite the widely shared view that oxidized lipids are detrimental for health, at least some of these biological effects induced by oxidized lipids, like blood lipid-lowering effects and inhibition of alcohol-induced fatty liver development, have to be considered as beneficial for human health. Other effects such as activation of stress-signaling pathways are presumably also favorable, since it leads to an activation of antioxidant and cytoprotective systems, but the direct implication for human health cannot be assessed. Nevertheless, the beneficial effects of dietary oxidized lipids must not be interpreted in the sense that dietary oxidized lipids are generally a health-promoting dietary component without any risk for adverse effects. It rather suggests that dietary oxidized lipids are a mixture of chemically distinct substances, some of which exhibit a significant biological activity.

ABBREVIATIONS

ALD	alcoholic liver disease
BB	γ-butyrobetaine
CFAM	cyclic fatty acid monomers
DHA	docosahexaenoic acid
ECs	endothelial cells

EPA	eicosapentaenoic acid
HCC	hepatocellular carcinoma
HETE	hydroxy-eicosatetraenoic acid
HNE	hydroxynonenal
HODE	hydroxy-octadecadienoic acid
HPODE	hydroperoxy-octadecadienoic acid
Insig	insulin-induced gene
LDLR-/-	low density-lipoprotein receptor-deficient
NF-κB	nuclear factor-kappa B
Nrf2	nuclear factor erythroid-derived 2-like 2
OCTN	novel organic cation transporter
PPAR	peroxisome proliferator-activated receptor
PPREs	peroxisome proliferator response elements
RXR	retinoid X receptor
SMCs	smooth muscle cells
SREBP	sterol regulatory element-binding protein
TAG	triacylglycerol
TML	6-N-trimethyllysine

REFERENCES

Ammouche, A., F. Rouaki, A. Bitam, and M.M. Bellal. 2002. Effect of ingestion of thermally oxidized sunflower oil on the fatty acid composition and antioxidant enzymes of rat liver and brain in development. *Annals of Nutrition and Metabolism* 46: 268–275.

Brandsch, C., and K. Eder. 2004. Effects of peroxidation products in thermoxidised dietary oil in female rats during rearing, pregnancy and lactation on their reproductive performance and the antioxidative status of their offspring. *British Journal of Nutrition* 92: 267–275.

Brass, E.P., and C.L. Hoppel. 1978. Carnitine metabolism in the fasting rat. *Journal of Biological Chemistry* 253: 2688–93.

Bretillon, L., S.E. Alexson, F. Joffre, B. Pasquis, and J.L. Sébédio. 2003. Peroxisome proliferator-activated receptor α is not the exclusive mediator of the effects of dietary cyclic FA in mice. *Lipids* 38: 957–63.

Chao, P.M., C.Y. Chao, F.J. Lin, and C.J. Huang. 2001. Oxidized frying oil up-regulates hepatic acyl-CoA oxidase and cytochrome P450 4A1 genes in rats and activates PPARα. *Journal of Nutrition* 131: 3166–74.

Chao, P.M., S.C. Hsu, F.J. Lin, Y.J. Li, and C.J. Huang. 2004. The up-regulation of hepatic acyl-CoA oxidase and cytochrome P450 4A1 mRNA expression by dietary oxidized frying oil is comparable between male and female rats. *Lipids* 39: 233–238.

Chaudhary, A., A. Mishra, and S. Sethi. 2004. Oxidized omega-3 fatty acids inhibit pro-inflammatory responses in glomerular endothelial cells. *Nephron. Experimental Nephrology* 97: e136–45.

Cheon, Y., T.Y. Nara, M.R. Band, J.E. Beever, M.A. Wallig, and M.T. Nakamura. 2005. Induction of overlapping genes by fasting and a peroxisome proliferator in pigs: Evidence of functional PPARα in nonproliferating species. *American Journal of Physiology. Regulatory Integrative and Comparative Physiology* 288: R1525–35.

Chinetti, G., S. Lestavel, V. Bocher, A.T. Remaley, B. Neve, I.P. Torra, E. Teissier et al. 2001. PPAR-α and PPAR-γ activators induce cholesterol removal from human macrophage foam cells through stimulation of the ABCA1 pathway. *Nature Medicine* 7: 53–8.

Choe, E., and D.B. Min. 2007. Chemistry of deep-fat frying oils. *Journal of Food Science* 72: R77–86.

Coleman, J.D., K.S. Prabhu, J.T. Thompson, P.S. Reddy, J.M. Peters, B.R. Peterson, C.C. Reddy, and J.P. Vanden Heuvel. 2007. The oxidative stress mediator 4-hydroxynonenal is an intracellular agonist of the nuclear receptor peroxisome proliferator-activated receptor-β/δ (PPARβ/δ). *Free Radical Biology & Medicine* 42: 1155–64.

Corcos Benedetti, P., M. Di Felice, V. Gentili, B. Tagliamonte, and G. Tomassi. 1990. Influence of dietary thermally oxidized soybean oil on the oxidative status of rats of different ages. *Annals of Nutrition and Metabolism* 34: 221–31.

Crabb, D.W., A. Galli, M. Fischer, and M. You. 2004. Molecular mechanisms of alcoholic fatty liver: Role of peroxisome proliferator-activated receptor α. *Alcohol* 34: 35–8.

Davis, A.T., and C.L. Hoppel 1986. Effect of starvation on the disposition of free and peptide-linked trimethyllysine in the rat. *Journal of Nutrition* 116: 760–7.

Delerive, P., C. Furman, E. Teissier, J.C. Fruchart, P. Duriez, and B. Staels. 2000. Oxidized phospholipids activate PPARα in a phospholipase A2-dependent manner. *FEBS Letters* 471: 34–8.

Desvergne, B., and W. Wahli. 1999. Peroxisome proliferator-activated receptors: Nuclear control of metabolism. *Endocrine Reviews* 20: 649–88.

Dhakshinamoorthy, S., D.J. 2nd Long, and A.K. Jaiswal. 2000. Antioxidant regulation of genes encoding enzymes that detoxify xenobiotics and carcinogens. *Current Topics in Cellular Regulation* 36: 201–16.

Dwarakanath, R.S., S. Sahar, M.A. Reddy, D. Castanotto, J.J. Rossi, and R. Natarajan. 2004. Regulation of monocyte chemoattractant protein-1 by the oxidized lipid, 13-hydroperoxyoctadecadienoic acid, in vascular smooth muscle cells via nuclear factor-κB (NF-κB). *Journal of Molecular and Cellular Cardiology* 36: 585–95.

Eder, K. 1999. The effects of a dietary oxidized oil on lipid metabolism in rats. *Lipids* 34: 717–25.

Eder, K., A. Sülzle, P. Skufca, C. Brandsch, and F. Hirche. 2003a. Effects of dietary thermoxidized fats on expression and activities of hepatic lipogenic enzymes in rats. *Lipids* 38: 31–8.

Eder, K., and M. Kirchgessner. 1998. The effect of dietary vitamin E supply and a moderately oxidized oil on activities of hepatic lipogenic enzymes in rats. *Lipids* 33: 277–83.

Eder, K., U. Keller, and C. Brandsch. 2004. Effects of a dietary oxidized fat on cholesterol in plasma and lipoproteins and the susceptibility of low-density lipoproteins to lipid peroxidation in guinea pigs fed diets with different concentrations of vitamins E and C. *International Journal for Vitamin and Nutrition Research* 74: 11–20.

Eder, K., U. Keller, F. Hirche, and C. Brandsch. 2003b. Thermally oxidized dietary fats increase the susceptibility of rat LDL to lipid peroxidation but not their uptake by macrophages. *Journal of Nutrition* 133: 2830–7.

Ericsson, C.G., J. Nilsson, L. Grip, B. Svane, and A. Hamsten. 1997. Effect of bezafibrate treatment over five years on coronary plaques causing 20% to 50% diameter narrowing (the Bezafibrate Coronary Atherosclerosis Intervention Trial [BECAIT]). *American Journal of Cardiology* 80: 1125–9.

Fischer, M., M. You, M. Matsumoto, and D.W. Crabb. 2003. Peroxisome proliferators-activated receptor α (PPARα) agonist treatment reverses PPARα dysfunction and abnormalities in hepatic lipid metabolism in ethanol-fed mice. *Journal of Biological Chemistry* 278: 27997–8004.

Forman, B.M., J. Chen, and R.M. Evans. 1997. Hypolipidemic drugs, polyunsaturated fatty acids, and eicosanoids are ligands for peroxisome proliferator-activated receptors α and γ. *Proceedings of the National Academy of Sciences of the United States of America* 94: 4312–7.

Gao, L., J. Wang, K.R. Sekhar, H. Yin, N.F. Yared, S.N. Schneider, S. Sasi et al. 2007. Novel n-3 fatty acid oxidation products activate Nrf2 by destabilizing the association between Keap1 and Cullin3. *Journal of Biological Chemistry* 282: 2529–37.

Garelnabi, M., K. Selvarajan, D. Litvinov, N. Santanam, and S. Parthasarathy. 2008. Dietary oxidized linoleic acid lowers triglycerides via APOA5/APOCIII dependent mechanisms. *Atherosclerosis* 199: 304–9.

Gizard, F., C. Amant, O. Barbier, S. Bellosta, R. Robillard, F. Percevault, H. Sevestre et al. 2005. PPARα inhibits vascular smooth muscle cell proliferation underlying intimal hyperplasia by inducing the tumor suppressor p16INK4a. *Journal of Clinical Investigation* 115: 3228–38.

Göttlicher, M., E. Widmark, Q. Li, and J.A. Gustafsson. 1992. Fatty acids activate a chimera of the clofibric acid-activated receptor and the glucocorticoid receptor. *Proceedings of the National Acadamy of Sciences of the United States of America* 89: 4653–7.

Greco, A.V., and G. Mingrone. 1990. Serum and biliary lipid pattern in rabbits feeding a diet enriched with unsaturated fatty acids. *Experimental Pathology* 40: 19–33.

Guallar-Castillón, P., F. Rodríguez-Artalejo, N.S. Fornés, J.R. Banegas, P.A. Etxezarreta, E. Ardanaz, A. Barricarte et al. 2007. Intake of fried foods is associated with obesity in the cohort of Spanish adults from the European Prospective Investigation into Cancer and Nutrition. *American Journal of Clinical Nutrition* 86: 198–205.

Harrison, S.A., and A.M. Diehl. 2002. Fat and the liver—A molecular overview. *Seminars in Gastrointestinal Disease* 13: 3–16.

Hennuyer, N., A. Tailleux, G. Torpier, H. Mezdour, J.C. Fruchart, B. Staels, and C. Fiévet. 2005. PPARα, but not PPARγ, activators decrease macrophage-laden atherosclerotic lesions in a nondiabetic mouse model of mixed dyslipidemia. *Arteriosclerosis Thrombosis and Vascular Biology* 25: 1897–1902.

Holden, P.R., and J.D. Tugwood. 1999. Peroxisome proliferator-activated receptor α: Role in rodent liver cancer and species differences. *Journal of Molecular Endocrinology* 22: 1–8.

Huang, C.J., N.S. Cheung, and V.R. Lu. 1988. Effects of deteriorated frying oil and dietary protein levels on liver microsomal enzymes in rats. *Journal of the American Oil Chemists' Society* 65: 1796–1803.

Huang, Z.H., E.J. Bates, J.V. Ferrante, C.S. Hii, A. Poulos, B.S. Robinson, and A. Ferrante. 1997. Inhibition of stimulus-induced endothelial cell intercellular adhesion molecule-1, E-selectin, and vascular cellular adhesion molecule-1 expression by arachidonic acid and its hydroxy and hydroperoxy derivatives. *Circulation Research* 80: 149–58.

Itoh, T., L. Fairall, K. Amin, Y. Inaba, A. Szanto, B.L. Balint, L. Nagy, K. Yamamoto, and J.W. Schwabe. 2008. Structural basis for the activation of PPARγ by oxidized fatty acids. *Nature Structural & Molecular Biology* 15: 924–31.

Izaki, Y., S. Yoshikawa, and M. Uchiyama. 1984. Effect of ingestion of thermally oxidized frying oil on peroxidative criteria in rats. *Lipids* 19: 324–31.

Jones, W.K., M. Brown, M. Wilhide, S. He, and X. Ren. 2005. NF-κB in cardiovascular disease: Diverse and specific effects of a "general" transcription factor? *Cardiovascular Toxicology* 5: 183–202.

Kämmerer, I., R. Ringseis, and K. Eder. 2011a. Feeding a thermally oxidised fat inhibits atherosclerotic plaque formation in the aortic root of LDL receptor-deficient mice. *British Journal of Nutrition* 105: 190–9.

Kämmerer, I., R. Ringseis, R. Biemann, G. Wen, and K. Eder. 2011b. 13-Hydroxy linoleic acid increases expression of the cholesterol transporters ABCA1, ABCG1 and SR-BI and stimulates apoA-I-dependent cholesterol efflux in RAW264.7 macrophages. *Lipids in Health and Disease* 10: 222.

Kaspar, J.W., S.K. Niture, and A.K. Jaiswal. 2009. Nrf2:INrf2 (Keap1) signaling in oxidative stress. *Free Radicals in Biology and Medicine* 47: 1304–9.

Katsuda, S., and T. Kaji. 2003. Atherosclerosis and extracellular matrix. *Journal of Atherosclerosis and Thrombosis* 10: 267–74.

Kaunitz, H., R.E. Johnson, and L. Pegus. 1965. A long-term nutritional study with fresh and mildly oxidized vegetable and animal fats. *Journal of the American Oil Chemists' Society* 42: 770–4.

Khan-Merchant, N., M. Penumetcha, O. Meilhac, and S. Parthasarathy. 2002. Oxidized fatty acids promote atherosclerosis only in the presence of dietary cholesterol in low-density lipoprotein receptor knockout mice. *Journal of Nutrition* 132: 3256–62.

Koch, A., B. König, G.I. Stangl, and K. Eder. 2008. PPARα mediates transcriptional upregulation of novel organic cation transporters-2 and -3 and enzymes involved in hepatic carnitine synthesis. *Experimental Biology and Medicine (Maywood)* 233: 356–65.

Koch, A., B. König, J. Spielmann, A. Leitner, G.I. Stangl, and K. Eder. 2007a. Thermally oxidized oil increases the expression of insulin-induced genes and inhibits activation of sterol regulatory element-binding protein-2 in rat liver. *Journal of Nutrition* 137: 2018–23.

Koch, A., B. König, S. Luci, G.I. Stangl, and K. Eder. 2007b. Dietary oxidised fat up-regulates the expression of organic cation transporters in liver and small intestine and alters carnitine concentrations in liver, muscle and plasma of rats. *British Journal of Nutrition* 98: 882–9.

König, B., A. Koch, J. Spielmann, C. Hilgenfeld, F. Hirche, G.I. Stangl, and K. Eder. 2009. Activation of PPARα and PPARγ reduces triacylglycerol synthesis in rat hepatoma cells by reduction of nuclear SREBP-1. *European Journal of Pharmacology* 605: 23–30.

König, B., and K. Eder. 2006. Differential action of 13-HPODE on PPARα downstream genes in rat Fao and human HepG2 hepatoma cell lines. *Journal of Nutritional Biochemistry* 17: 410–8.

Krey, G., O. Braissant, F. L'Horset, E. Kalkhoven, M. Perroud, M.G. Parker, and W. Wahli. 1997. Fatty acids, eicosanoids, and hypolipidemic agents identified as ligands of peroxisome proliferator-activated receptors by coactivator-dependent receptor ligand assay. *Molecular Endocrinology* 11: 779–91.

Kritchevsky, D., and S.A. Tepper. 1967. Cholesterol vehicle in experimental atherosclerosis, part 9: Comparison of heated corn oil and heated olive oil. *Journal of Atherosclerosis Research* 7: 647–51.

LaBadie, J., W.A. Dunn, and N.N. Jr. Aronson. 1976. Hepatic synthesis of carnitine from protein-bound trimethyl-lysine. Lysosomal digestion of methyl-lysine-labelled asialofetuin. *Biochemical Journal* 160: 85–95.

Li, A.C., C.J. Binder, A. Gutierrez, K.K. Brown, C.R. Plotkin, J.W. Pattison, A.F. Valledor et al. 2004. Differential inhibition of macrophage foam-cell formation and atherosclerosis in mice by PPARα, β/δ, and γ. *Journal of Clinical Investigation* 114: 1564–76.

Limor, R., O. Sharon, E. Knoll, A. Many, G. Weisinger, and N. Stern. 2008. Lipoxygenase-derived metabolites are regulators of peroxisome proliferator-activated receptor gamma-2 expression in human vascular smooth muscle cells. *American Journal of Hypertension* 21: 219–23.

Litvinov, D., K. Selvarajan, M. Garelnabi, L. Brophy, and S. Parthasarathy. 2010. Anti-atherosclerotic actions of azelaic acid, an end product of linoleic acid peroxidation, in mice. *Atherosclerosis* 209: 449–54.

Liu, J.F., and C.J. Huang. 1995. Tissue α-tocopherol retention in male rats is compromised by feeding diets containing oxidized frying oil. *Journal of Nutrition* 125: 3071–80.

Liu, J.F., and Y.W. Lee. 1998. Vitamin C supplementation restores the impaired vitamin E status of guinea pigs fed oxidized frying oil. *Journal of Nutrition* 128: 116–22.

Luci, S., B. Giemsa, H. Kluge, and K. Eder. 2007a. Clofibrate causes an upregulation of PPAR-α target genes but does not alter expression of SREBP target genes in liver and adipose tissue of pigs. *American Journal of Physiology. Regulatory Integrative and Comparative Physiology* 293: R70–7.

Luci, S., B. König, B. Giemsa, S. Huber, G. Hause, H. Kluge, G.I. Stangl, and K. Eder. 2007b. Feeding of a deep-fried fat causes PPARα activation in the liver of pigs as a non-proliferating species. *British Journal of Nutrition* 97: 872–82.

Luci, S., F. Hirche, and K. Eder. 2008. Fasting and caloric restriction increases mRNA concentrations of novel organic cation transporter-2 and carnitine concentrations in rat tissues. *Annals of Nutrition and Metabolism* 52: 58–67.

Luci, S., S. Geissler, B. König, A. Koch, G.I. Stangl, F. Hirche, and K. Eder. 2006. PPARα agonists up-regulate organic cation transporters in rat liver cells. *Biochemical and Biophysical Research Communications* 350: 704–8.

Majkova, Z., J. Layne, M. Sunkara, A.J. Morris, M. Toborek, and B. Hennig. 2011. Omega-3 fatty acid oxidation products prevent vascular endothelial cell activation by coplanar polychlorinated biphenyls. *Toxicology and Applied Pharmacology* 251: 41–9.

Mandard, S., M. Müller, and S. Kersten. 2004. Peroxisome proliferator receptor α target genes. *Cellular and Molecular Life Sciences* 61: 393–416.

Marantos, C., V. Mukaro, J. Ferrante, C. Hii, and A. Ferrante. 2008. Inhibition of the lipopoly-saccharide-induced stimulation of the members of the MAPK family in human monocytes/macrophages by 4-hydroxynonenal, a product of oxidized omega-6 fatty acids. *American Journal of Pathology* 173: 1057–66.

Marantos, C., V. Mukaro, J. Ferrante, C. Hii, and A. Ferrante. 2008. Inhibition of the lipopoly-saccharide-induced stimulation of the members of the MAPK family in human monocytes/macrophages by 4-hydroxynonenal, a product of oxidized omega-6 fatty acids. *American Journal of Pathology* 173: 1057–66.

Martin, J.C., C. Caselli, S. Broquet, P. Juanéda, M. Nour, J.L. Sébédio, and A. Bernard. 1997. Effect of cyclic fatty acid monomers on fat absorption and transport depends on their positioning within the ingested triacylglycerols. *Journal of Lipid Research* 38: 1666–79.

Martin, J.C., F. Joffre, M.H. Siess, M.F. Vernevaut, P. Collenot, M. Genty, and J.L. Sébédio. 2000. Cyclic fatty acid monomers from heated oil modify the activities of lipid synthesizing and oxidizing enzymes in rat liver. *Journal of Nutrition* 130: 1524–30.

Marx, N., H. Duez, J.C. Fruchart, and B. Staels. 2004. Peroxisome proliferator-activated receptors and atherogenesis: Regulators of gene expression in vascular cells. *Circulation Research* 94: 1168–78.

McGarry, J.D., C. Robles-Valdes, and D.W. Foster. 1975. Role of carnitine in hepatic ketogenesis. *Proceedings of the National Academy of Sciences of the United States of America* 72: 4385–8.

Mishra, A., A. Chaudhary, and S. Sethi. 2004. Oxidized omega-3 fatty acids inhibit NF-κB activation via a PPARα-dependent pathway. *Arteriosclerosis Thrombosis and Vascular Biology* 24: 1621–7.

Moreira, R.G., M.E. Castell-Perez, and M.A. Barrufet. 1999. Oil absorption in fried foods. In: *Deep-Fat Frying: Fundamentals and Applications*. Chapman & Hall, Gaithersburg, MD, pp. 179–221.

Muga, S.J., P. Thuillier, A. Pavone, J.E. Rundhaug, W.E. Boeglin, M. Jisaka, A.R. Brash, and S.M. Fischer. 2000. 8S-lipoxygenase products activate peroxisome proliferator-activated receptor α and induce differentiation in murine keratinocytes. *Cell Growth & Differentiation* 11: 447–54.

Musiek, E.S., J.D. Brooks, M. Joo, E. Brunoldi, A. Porta, G. Zanoni, G. Vidari et al. 2008. Electrophilic cyclopentenone neuroprostanes are anti-inflammatory mediators formed from the peroxidation of the omega-3 polyunsaturated fatty acid docosahexaenoic acid. *Journal of Biological Chemistry* 283: 19927–35.

Nagy, L., P. Tontonoz, J.G. Alvarez, H. Chen, and R.M. Evans. 1998. Oxidized LDL regulates macrophage gene expression through ligand activation of PPARγ. *Cell* 93: 229–40.

Natarajan, R., M.A. Reddy, K.U. Malik, S. Fatima, and B.V. Khan. 2001. Signaling mechanisms of nuclear factor-kappab-mediated activation of inflammatory genes by

13-hydroperoxyoctadecadienoic acid in cultured vascular smooth muscle cells. *Arteriosclerosis, Thrombosis and Vascular Biology* 21: 1408–13.

Niture, S.K., J.W. Kaspar, J. Shen, and A.K. Jaiswal. 2010. Nrf2 signaling and cell survival. *Toxicology and Applied Pharmacology* 244: 37–42.

Nolen, G.A., J.C. Alexander, and N.R. Artman. 1967. Long-term rat feeding study with used frying fats. *Journal of Nutrition* 93: 337–48.

Page, S., C. Fischer, B. Baumgartner, M. Haas, U. Kreusel, G. Loidl, M. Hayn, H.W. Ziegler-Heitbrock, D. Neumeier, and K. Brand. 1999. 4-Hydroxynonenal prevents NF-κB activation and tumor necrosis factor expression by inhibiting IκB phosphorylation and subsequent proteolysis. *Journal of Biological Chemistry* 274: 11611–8.

Paul, H.S., and S.A. Adibi. 1979. Paradoxical effects of clofibrate on liver and muscle metabolism in rats. Induction of myotonia and alteration of fatty acid and glucose oxidation. *Journal of Clinical Investigation* 64: 405–12.

Paul, H.S., and S.A. Adibi. 1980. Leucine oxidation and protein turnover in clofibrate-induced muscle protein degradation in rats. *Journal of Clinical Investigation* 65: 1285–93.

Penumetcha, M., N. Khan, and S. Parthasarathy. 2000. Dietary oxidized fatty acids: An atherogenic risk? *Journal of Lipid Research* 41: 1473–80.

Peters, J.M., C. Cheung, and F.J. Gonzalez. 2005. Peroxisome proliferator-activated receptor-α and liver cancer: Where do we stand? *Journal of Molecular Medicine* 83: 774–85.

Poling, C.E., E. Eagle, E.E. Rice, A.M. Durand, and M. Fisher. 1970. Long-term responses of rats to heat-treated dietary fats. IV. Weight gains, food and energy efficiencies, longevity and histopathology. *Lipids* 5: 128–36.

Rebouche, C. J. 1983. Effect of dietary carnitine isomers and gamma-butyrobetaine on L-carnitine biosynthesis and metabolism in the rat. *Journal of Nutrition* 113: 1906–13.

Rebouche, C.J. 2004. Kinetics, pharmacokinetics, and regulation of L-carnitine and acetyl-L-carnitine metabolism. *Annals of the New York Academy of Sciences* 1033: 30–41.

Ringseis, R., A. Gutgesell, C. Dathe, C. Brandsch, and K. Eder. 2007a. Feeding oxidized fat during pregnancy up-regulates expression of PPARα-responsive genes in the liver of rat fetuses. *Lipids in Health and Disease* 6: 6.

Ringseis, R., A. Muschick, and K. Eder. 2007b. Dietary oxidized fat prevents ethanol-induced triacylglycerol accumulation and increases expression of PPARα target genes in rat liver. *Journal of Nutrition* 137: 77–83.

Ringseis, R., and K. Eder. 2011. Regulation of genes involved in lipid metabolism by dietary oxidized fat. *Molecular Nutrition and Food Research* 55: 109–21.

Ringseis, R., C. Dathe, A. Muschick, C. Brandsch, and K. Eder. 2007c. Oxidized fat reduces milk triacylglycerol concentrations by inhibiting gene expression of lipoprotein lipase and fatty acid transporters in the mammary gland of rats. *Journal of Nutrition* 137: 2056–2061.

Ringseis, R., G. Wen, and K. Eder. 2012. Regulation of genes involved in carnitine homeostasis by PPARα across different species (rat, mouse, pig, cattle, chicken, and human). *PPAR Research* 2012: 868317.

Ringseis, R., N. Piwek, and K. Eder. 2007e. Oxidized fat induces oxidative stress but has no effect on NF-kappaB-mediated proinflammatory gene transcription in porcine intestinal epithelial cells. *Inflammation Research* 56: 118–25.

Ringseis, R., S. Pösel, F. Hirche, and K. Eder. 2007d. Treatment with pharmacological peroxisome proliferator-activated receptor α agonist clofibrate causes upregulation of organic cation transporter 2 in liver and small intestine of rats. *Pharmacological Research* 56: 175–83.

Rubins, H.B., S.J. Robins, D. Collins, C.L. Fye, J.W. Anderson, M.B. Elam, F.H. Faas et al. 1999. Veterans Affairs High-Density Lipoprotein Cholesterol Intervention Trial Study Group. Gemfibrozil for the secondary prevention of coronary heart disease in men with low levels of high-density lipoprotein cholesterol. *New England Journal of Medicine* 341: 410–8.

Ruef, J., M. Moser, C. Bode, W. Kübler, and M.S. Runge. 2001. 4-Hydroxynonenal induces apoptosis, NF-κB-activation and formation of 8-isoprostane in vascular smooth muscle cells. *Basic Research in Cardiology* 96: 143–50.

Sayon-Orea, C., M. Bes-Rastrollo, F.J. Basterra-Gortari, J.J. Beunza, P. Guallar-Castillon, C. de la Fuente-Arrillaga, and M.A. Martinez-Gonzalez. 2013. Consumption of fried foods and weight gain in a Mediterranean cohort: The SUN project. *Nutrition, Metabolism and Cardiovascular Diseases* 23: 144–50.

Schoonbroodt, S., and J. Piette. 2000. Oxidative stress interference with the nuclear factor-κb activation pathways. *Biochemical Pharmacology* 60: 1075–83.

Sebedio, J.L., and A. Grandgirard. 1989. Cyclic fatty acids: Natural sources, formation during heat treatment, synthesis and biological properties. *Progress in Lipid Research* 28: 303–36.

Sethi, S., O. Ziouzenkova, H. Ni, D.D. Wagner, J. Plutzky, and T.N. Mayadas. 2002. Oxidized omega-3 fatty acids in fish oil inhibit leukocyte-endothelial interactions through activation of PPARα. *Blood* 100: 1340–6.

Siegmund, S.V., and D.A. Brenner DA. 2005. Molecular pathogenesis of alcohol-induced hepatic fibrosis. *Alcoholism, Clinical and Experimental Research* 29(11 Suppl): 102S-9S.

Staprans, I., J.H. Rapp, X.M. Pan, and K.R. Feingold. 1993. The effect of oxidized lipids in the diet on serum lipoprotein peroxides in control and diabetic rats. *Journal of Clinical Investigation* 92: 638–643.

Staprans, I., J.H. Rapp, X.M. Pan, D.A. Hardman, and K.R. Feingold. 1996. Oxidized lipids in the diet accelerate the development of fatty streaks in cholesterol-fed rabbits. *Arteriosclerosis Thrombosis and Vascular Biology* 16: 533–8.

Steinberg, D. 1997. Lewis A Conner Memorial Lecture. Oxidative modification of LDL and atherogenesis. *Circulation* 95: 1062–71.

Sülzle, A., F. Hirche, and K. Eder. 2004. Thermally oxidized dietary fat upregulates the expression of target genes of PPARα in rat liver. *Journal of Nutrition* 134: 1375–83.

Tamai, I., R. Ohashi, J. Nezu, H. Yabuuchi, A. Oku, M. Shimane, Y. Sai, and A. Tsuji. 1998. Molecular and functional identification of sodium ion-dependent, high affinity human carnitine transporter OCTN2. *Journal of Biological Chemistry* 273: 20378–82.

Toomey, S., B. Harhen, H.M. Roche, D. Fitzgerald, and O. Belton. 2006. Profound resolution of early atherosclerosis with conjugated linoleic acid. *Atherosclerosis* 187: 40–49.

Toschi, T.G., A. Cosat, and G. Lercker. 1997. Gas chromatographic study on high-temperature thermal degradation products of methyl linoleate hydroperoxides. *Journal of the American Oil Chemist's Society* 74: 387–91.

Van Vlies, N., S. Ferdinandusse, M. Turkenburg, R.J. Wanders, and F.M. Vaz. 2007. PPARα-activation results in enhanced carnitine biosynthesis and OCTN2-mediated hepatic carnitine accumulation. *Biochimica et Biophysica Acta* 1767: 1134–42.

Varady, J., K. Eder, and R. Ringseis. 2011. Dietary oxidized fat activates the oxidative stress-responsive transcription factors NF-κB and Nrf2 in intestinal mucosa of mice. *European Journal of Nutrition* 50: 601–9.

Varady, J., D.K. Gessner, E. Most, K. Eder, and R. Ringseis. 2012a. Dietary moderately oxidized oil activates the Nrf2 signaling pathway in the liver of pigs. *Lipids Health Disease* 11: 31.

Varady, J., R. Ringseis, and K. Eder. 2012b. Dietary moderately oxidized oil induces expression of fibroblast growth factor 21 in the liver of pigs. *Lipids Health Disease* 11: 34.

Vaz, F.M., and R.J. Wanders. 2002. Carnitine biosynthesis in mammals. *Biochemical Journal* 361: 417–29.

Victor, V.M., N. Apostolova, R. Herance, A. Hernandez-Mijares, and M. Rocha. 2009. Oxidative stress and mitochondrial dysfunction in atherosclerosis: Mitochondria-targeted antioxidants as potential therapy. *Current Medicinal Chemistry* 16: 4654–67.

Waku, T., T. Shiraki, T. Oyama, Y. Fujimoto, K. Maebara, N. Kamiya, H. Jingami, and K. Morikawa. 2009. Structural insight into PPARγ activation through covalent modification with endogenous fatty acids. *Journal of Molecular Biology* 385: 188–99.

Wen, G., R. Ringseis, and K. Eder. 2010. Mouse OCTN2 is directly regulated by peroxisome proliferator-activated receptor α (PPARα) via a PPRE located in the first intron. *Biochemical Pharmacology* 79: 768–76.

Wilson, R., K. Lyall, L. Smyth, C.E. Fernie, and R.A. Riemersma. 2002. Dietary hydroxy fatty acids are absorbed in humans: Implications for the measurement of "oxidative stress" in vivo. *Free Radical Biology & Medicine* 32: 162–8.

Yabe, D., M.S. Brown, and J.L. Goldstein. 2002. Insig-2, a second endoplasmic reticulum protein that binds SCAP and blocks export of sterol regulatory element-binding proteins. *Proceedings of the National Academy of Sciences of the United States of America* 99: 12753–8.

Yip, W.W., and A.D. Burt. 2006. Alcoholic liver disease. *Seminars in Diagnostic Pathology* 23: 149–60.

Yokoi, H., H. Mizukami, A. Nagatsu, H. Tanabe, and M. Inoue. 2010. Hydroxy monounsaturated fatty acids as agonists for peroxisome proliferator-activated receptors. *Biological & Pharmaceutical Bulletin* 33: 854–61.

Yoshida, H., and G. Kajimoto. 1989. Effect of dietary vitamin E on the toxicity of autoxidized oil to rats. *Annals of Nutrition and Metabolism* 33: 153–61.

You, M., M. Fischer, M.A. Deeg, and D.W. Grabb. 2002. Ethanol induces fatty acid synthesis pathways by activation of sterol regulatory element-binding protein (SREBP). *Journal of Biological Chemistry* 277: 29342–7.

13 Oxidized and Nitrated Lipid Interactions with the Keap1–Nrf2 Pathway

Emilia Kansanen and Anna-Liisa Levonen

CONTENTS

INTRODUCTION

Many endogenous electrophilic signaling mediators are products of lipid peroxidation. Even though the accumulation of electrophiles can cause major tissue damage and cellular dysfunction in oxidative stress-related diseases such as neurodegeneration, diabetes, and atherosclerosis, moderate concentrations of lipid oxidation products act as signaling messengers activating cytoprotective signaling. In this chapter, we briefly introduce the Keap1–Nrf2 pathway, the key cytoprotective response upon electrophilic stress, and summarize how electrophilic products of lipid peroxidation can trigger pathway activation. We also discuss the interplay of the Keap1–Nrf2 pathway with other stress signaling pathways, and the therapeutic potential of fatty acid-derived electrophiles.

KEAP1–NRF2 PATHWAY

The Keap1–Nrf2 pathway regulates the expression of multiple cytoprotective genes in response to oxidative or electrophilic stress. Keap1 is a redox-sensitive protein that in the presence of an oxidative or electrophilic stimulus allows the activation of transcription factor Nrf2. Once activated, Nrf2 is translocated to the nucleus, binds to the *cis*-acting antioxidant response element (ARE) also called the electrophile response element (EpRE) of the regulatory regions of target genes, and drives their expression.

TRANSCRIPTION FACTOR NRF2

Nuclear factor-E2-related factor 2 (Nrf2) is a bZip transcription factor and a member of the Cap "n" Collar family of regulatory proteins. Nrf2 mediates the cellular response to electrophiles and oxidants via binding to an enhancer element in the promoter regions of cytoprotective genes (Itoh et al., 1997). Nrf2 protein consists of six Neh (Nrf2–ECH homology) domains (Neh1–Neh6) (Figure 13.1b). Neh1 contains CNC basic region for DNA binding and a leucine zipper structure for dimerization with small Maf proteins (Itoh et al., 1997). Neh3 (Nioi et al., 2005), Neh4 (Itoh et al., 1997), and Neh5 (Zhang et al., 2007) are transactivation domains. Neh2 is the domain that negatively regulates the transcriptional activity of Nrf2 via binding to the inhibitor protein Keap1 (Itoh et al., 1999). Nrf2 heterodimerizes with small Maf proteins and the Nrf2–Maf heterodimer binds to ARE, also known as the electrophile response element (EpRE), in the regulatory regions of cytoprotective genes. ARE was initially characterized to have a consensus core sequence TGACnnnGC (Rushmore et al., 1991), which was further characterized and extended to TMAnnRTGAYnnnGCRwwww (Wasserman and Fahl, 1997), although the existence of a universally applicable consensus sequence has been questioned (Nioi et al., 2005). Nrf2 preferably binds ARE with small Maf proteins, but other transcription factors such as Nrf1, Nrf3, BTB, and CNC homology (Bach) 1 and 2 have also been implicated in the regulation of ARE-dependent gene expression (Dhakshinamoorthy et al., 2005; Levy et al., 2009; Venugopal and Jaiswal, 1996, 1998; Yang et al., 2009). Small Maf proteins can also form homodimers to repress ARE-dependent transcription (Kannan et al., 2012).

TARGET GENES OF NRF2

Target genes of Nrf2 contain an ARE sequence in their regulatory region (Rushmore et al., 1991) and have been identified by gene-expression profiling of Nrf2 knockout ($Nrf2^{-/-}$) mice (Kwak et al., 2003; Lee et al., 2003). These genes include detoxification enzymes such as NAD(P)H quinone oxidoreductase 1 (NQO1) and some glutathione

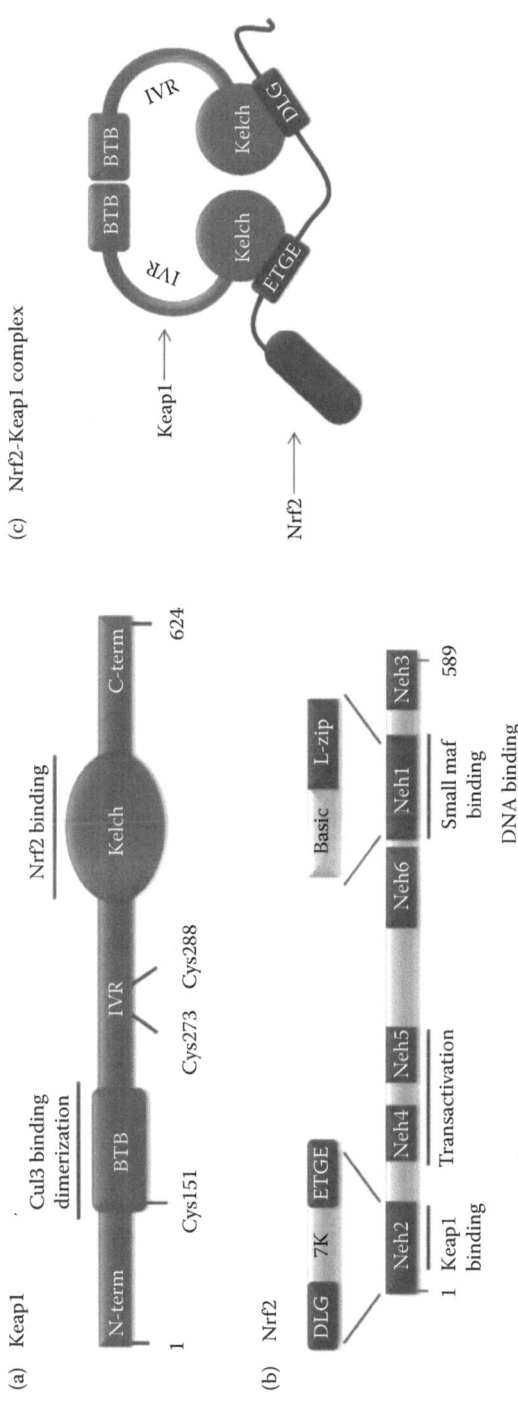

FIGURE 13.1 Structures of Keap1 and Nrf2. (a) Keap1 consists of 624 amino acid residues and has five domains. (b) Nrf2 consists of 589 amino acids and has six evolutionarily highly conserved domains, Neh1-6. Neh2 domain contains ETGE and DLG motifs, which are required for the interaction with Keap1, and a hydrophilic region of lysine residues (7 K), which is indispensable for the Keap1-dependent polyubiquitination and degradation of Nrf2. (c) Nrf2 interacts with two molecules of Keap1 through its Neh2 ETGE and DLG motifs. Both ETGE and DLG bind to similar sites on the bottom surface of the Keap1 Kelch motif.

S-transferases (GSTs), antioxidant proteins, including heme oxygenase-1 (HMOX1) and thioredoxin-1, as well as enzymes involved in GSH synthesis, such as glutamate–cysteine ligase (GCL). Furthermore, Nrf2 regulates proteasomal and chaperone proteins, indicating that Nrf2 has an important role also in the reparation and removal of damaged proteins.

Keap1 as a Sensor for Electrophiles

ARE activating agents were originally screened based on their ability to activate the ARE reporter construct and increase NQO1 activity (Friling et al., 1990, 1992; Prestera et al., 1993). A common feature for these chemical compounds is that they can react with sulfhydryls such as cysteine thiols (Dinkova-Kostova et al., 2002). The discovery of the cysteine-rich Keap1 in 1999 (Itoh et al., 1999) and the finding that ARE activating chemicals bind Keap1 (Dinkova-Kostova et al., 2002) confirmed the role of Keap1 as the sensor of the Nrf2 pathway. Keap1 protein consists of five domains, the N-terminal domain, Broad complex, Tramtrack, and Bric-à-Brac (BTB) domain, the intervening region (IVR), the Kelch domain, also known as double glycine repeats, and C-terminal domain (Figure 13.1a). Keap1 is a homodimer that dimerizes through the BTB domain and forms a structure that resembles a "cherry-bob" (Ogura et al., 2010) (Figure 13.1c). In the cell, Keap1 is tethered to the cytoskeletal actin through the Kelch domain, which is also responsible for the Nrf2 binding of Keap1 (Itoh et al., 1999).

Keap1 responds to various stimuli including electrophiles, hydroperoxides, and heavy metals leading to nuclear accumulation of Nrf2 and the subsequent activation of cytoprotective enzymes. Several research groups have identified the specific pattern of Keap1 cysteines modified in response to different ARE-activating chemicals. The most frequently reported cysteine targets, identified by mass spectrometry (MS) methods, include C151, C257, C273, C288, C297, and C613 (Dinkova-Kostova et al., 2002; Eggler et al., 2005; Hong et al., 2005a, b; Hu et al., 2011; Kansanen et al., 2011; Kobayashi et al., 2009; Luo et al., 2007). In addition to MS experiments, affinity purification using biotinylated electrophilic compounds and the avidin/streptavidin pull-down approach have been used to investigate the adduction of Keap1 by electrophiles (Hosoya et al., 2005; Levonen et al., 2004; Oh et al., 2008). These experiments have shown that cysteines in the IVR are critical for binding of electrophiles to Keap1, because the deletion of IVR but not the BTB domain or the Kelch domain abolishes binding (Hosoya et al., 2005). Interestingly, there are differences in the patterns of Keap1 cysteines modified in response to different electrophiles suggesting that there is a specific "cysteine code" for each electrophile or group of electrophiles (Kobayashi et al., 2009).

In addition to proteomic analyses of Keap1 cysteine modifications, the functional importance of individual Keap1 cysteines has been extensively investigated. The current data suggest that three cysteine residues C151, C273, and C288 are functionally the most important in Keap1. The IVR cysteines C273 and C288 are important for the ability of Keap1 to repress Nrf2 under basal conditions. The mutation of these cysteines results in the inability of Keap1 to inhibit ARE activation by Nrf2 both under basal and electrophile-induced conditions, assessed by reporter assays (Kansanen et al., 2011; Kobayashi et al., 2006; Levonen et al., 2004; Wakabayashi et al., 2004; Zhang and Hannink, 2003). In addition, the lack of C273 in Keap1 diminished the levels of both cytoplasmic and nuclear Nrf2 protein as well as the ubiquitination of Nrf2 (Zhang et al., 2004). The ability of C273 and C288 to repress Nrf2 was further confirmed *in vivo* with a transgenic complementation rescue model. This study revealed that a single mutation of C273 or C288 could inhibit the ability of Keap1 to repress Nrf2 *in vivo* under unstressed conditions resulting in the accumulation of Nrf2 protein and elevated levels of Nrf2 target genes (Yamamoto et al., 2008), confirming the *in vitro* findings. In addition to IVR cysteines C273 and C288, Keap1 BTB cysteine C151 has also an important, but very different role in the Keap1–Nrf2–ARE system. While C273 and C288 are critical for Keap1 to inhibit Nrf2 under basal conditions, C151 is a critical residue for a subset of Nrf2 activators (Kobayashi et al., 2009; McMahon et al., 2010; Yamamoto et al., 2008). However, lipid-derived electrophilic species nitro-oleic acid (OA-NO$_2$) (Kansanen et al., 2011), prostaglandin A$_2$ (PGA$_2$), 15d-PGJ$_2$ (Kobayashi et al., 2009), and the heavy metal compound arsenite (Wang et al., 2008) can activate Nrf2 independent of Keap1 C151.

PROPOSED MECHANISMS FOR THE KEAP1–NRF2 PATHWAY ACTIVATION

It is evident that the modification of Keap1 cysteines leads to the accumulation of Nrf2 in the nucleus and expression of Nrf2 target genes. The mechanism for Nrf2 nuclear enrichment is not fully understood but several models have been proposed. According to the first model for Nrf2 activation, it was thought that under basal conditions, Nrf2 was sequestered in the cytoplasm by the actin-bound Keap1. During electrophilic stress, the modification of cysteine residues in Keap1 would release Nrf2 allowing it to translocate to the nucleus, bind to ARE, and drive the expression of target genes (Dinkova-Kostova et al., 2002). However, subsequently it was found that inducers were unable to dissociate Keap1–Nrf2 binding (Eggler et al., 2005), and Nrf2 is rapidly ubiquitinated and degraded under basal conditions (Zhang and Hannink, 2003) requiring novel models to be developed.

Dissociation of Keap1 and Cul3

In basal conditions, Keap1 functions as an adaptor protein in the Cul3 (Cullin3)-based E3 ligase complex, resulting in rapid ubiquitination and subsequent degradation of Nrf2 (Cullinan et al., 2004; Kobayashi et al., 2004; Zhang et al., 2004) (Figure 13.2a). According to the Keap1–Cul3 dissociation model, the binding of Keap1 and Cul3 is disrupted in response to electrophiles, leading to the escape of Nrf2 from the ubiquitination system (Figure 13.2b). This has been confirmed with various Nrf2 inducers including sulforaphane (SFN) (Kansanen et al., 2011; Wang et al., 2008; Zhang et al., 2004), *N*-iodoacetyl-*N*-biotinylhexylenediamine (IAB) (Rachakonda et al., 2008), and (tert-butylhydroquinone) tBHQ (Wang et al., 2008). Modifications of C151 appear to decrease Cul3 interaction, leading to inhibition of Keap1-dependent ubiquitination of Nrf2 (Zhang et al., 2004). It has been postulated that in addition to electrophile adduction to C151, a bulky modification, such as cysteine to tryptophan substitution, at this site would cause conformational changes that alter Cul3 binding allowing Nrf2 to escape proteosomal degradation

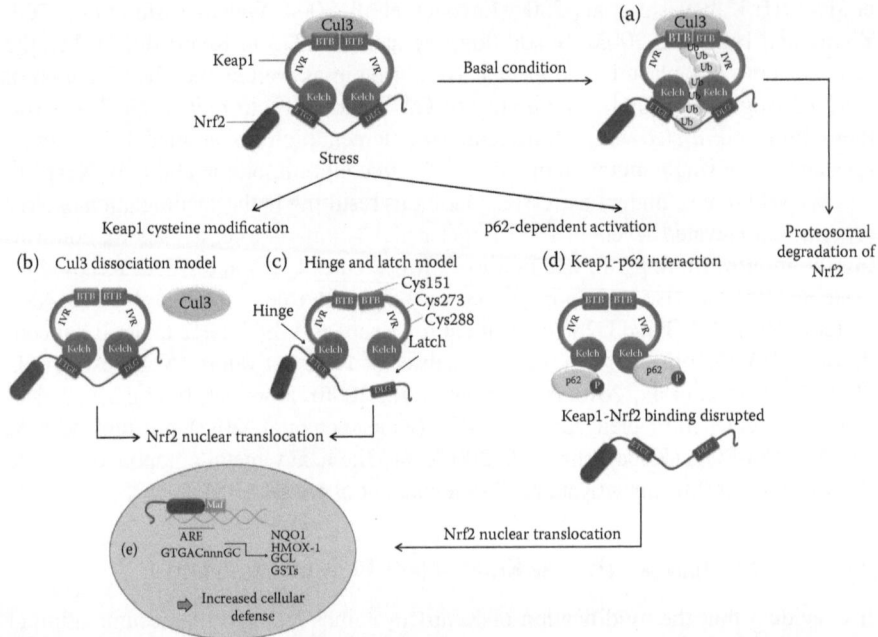

FIGURE 13.2 Keap1–Nrf2 pathway activation. (a) In basal conditions, two Keap1 molecules bind to Nrf2 and Nrf2 is polyubiquitylated by the Cul3-based E3 ligase complex. The polyubiquitylation results in rapid Nrf2 degradation by the proteasome. Under stress conditions, inducers modify the Keap1 cysteines leading to the inhibition of Nrf2 ubiquitylation via dissociation of the inhibitory complex. (b) In the Keap1–Cul3 dissociation model, the binding of Keap1 and Cul3 is disrupted in response to electrophiles, leading to the escape of Nrf2 from the ubiquitination system. (c) According to the hinge and latch model, modification of specific Keap1 cysteine residues leads to conformational changes in Keap1 resulting in the detachment of the Nrf2 DLG motif from Keap1. Ubiquitination of Nrf2 is disrupted but the binding with the ETGE motif remains. (d) In p62-dependent activation, p62 binds to the Kelch domain of Keap1 leading to attenuated Nrf2 ubiquitination, which leads to Nrf2 nuclear accumulation. (e) In all models, the newly synthesized Nrf2 proteins translocate into the nucleus, bind to ARE and drive the expression of Nrf2 target genes such as NQO1, HMOX1, GCL, and GSTs.

(Eggler et al., 2009). When Nrf2 escapes ubiquitination, it can translocate to the nucleus and drive target gene expression (Figure 13.2e). However, it has also been shown that Nrf2 can be activated without the need for dissociation of Keap1 and Cul3 in response to Nrf2 activators including OA-NO$_2$, 15d-PGJ$_2$ (Kansanen et al., 2011), and arsenite (Wang et al., 2008).

HINGE AND LATCH MODEL

The Neh2 domain of Nrf2 has two different motifs that bind Keap1, the ETGE and the DLG motifs, and this results in a Keap1–Nrf2 complex with 1:2 stoichiometry with two binding sites in Nrf2 (Tong et al., 2006a) (Figure 13.1c). The two-site binding is beneficial for ubiquitination of Nrf2, because the lysine residues are aligned between the Keap1 binding ETGE and DLG motifs (Figure 13.1a, 7Ks). According to the hinge and latch model, during electrophilic stress, modification of specific Keap1 cysteines leads to conformational changes in Keap1, resulting in the detachment of the weaker binding DLG motif from Keap1 (Latch). In this setting, ubiquitination of Nrf2 is disturbed but the binding with the ETGE motif remains (Hinge) (Tong et al., 2006a, 2006b, 2007) (Figure 13.2c). Similar to the Keap1–Cul3 dissociation model, Nrf2 escapes the degradation and is able to translocate to the nucleus (Figure 13.2e).

Other Models for Keap1–Nrf2 Pathway Activation

In addition to Keap1 cysteine modifications, also other models for activation of Keap1–Nrf2 pathway have been suggested. For example, multiple protein kinases are implicated in the regulation of Nrf2 activity. Protein kinase C (PKC) can phosphorylate Nrf2 serine 40, which results in the release of Nrf2 from Keap1 leading to increased ARE-dependent gene expression (Bloom and Jaiswal, 2003; Huang et al., 2000).

Moreover, recent reports have detected novel proteins that compete with the binding of Nrf2 to Keap1. SQSTM1/p62 is shown to activate the Keap1–Nrf2 pathway via an interaction with Keap1 through a motif within the C-terminal region resembling the ETGE motif in Nrf2 (Jain et al., 2010; Komatsu et al., 2010; Taguchi et al., 2012). Phosphorylation of p62 at S351 in response to As(III) results in increased interaction of p62 and Keap1 leading to nuclear accumulation of Nrf2 and subsequent Nrf2 target gene expression (Ichimura et al., 2013) (Figure 13.2d).

In addition, Keap1 nucleocytoplasmic shuttling has been suggested as an alternative model for Nrf2 activation (Niture and Jaiswal, 2009; Sun et al., 2007). According to this model, Keap1 enters the nucleus and removes Nrf2 under both basal and induced conditions. The inducers inhibit the nuclear entry of Keap1 allowing the transcription of Nrf2-dependent genes. However, Keap1 is predominantly located in the cytoplasm and only 5% of Keap1 is present in the nucleus under both basal and induced conditions (Watai et al., 2007). This suggests that while nucleocytoplasmic shuttling may occur, it might not have a significant impact on the electrophile-induced activation of Nrf2. In addition, it has been proposed that in response to inducers, instead of Nrf2, Keap1 becomes the target for ubiquitination (Hong et al., 2005a; Zhang et al., 2005).

KEAP1–NRF2 PATHWAY ACTIVATING OXIDIZED AND NITRATED LIPIDS

SHORT-CHAIN ALDEHYDES

In the lipid peroxidation reactions, multiple short-chain aldehydes such as acrolein, malondialdehyde, 4-hydroxy-2-nonenal (4-HNE), and 4-oxo-2-nonenal (4-ONE) are formed (West and Marnett, 2006). The signaling functions of 4-HNE are the most extensively studied among these. 4-HNE can induce Nrf2 nuclear accumulation and the expression of Nrf2-dependent antioxidant enzymes in several cell lines such as endothelial cells (Ishikado et al., 2010), mouse macrophages and smooth muscle cells (Ishii et al., 2004), rat cardiomyocyes (Zhang et al., 2010), and in human pigment epithelial cells (Chen et al., 2009). In addition to Keap1 cysteine modification (Levonen et al., 2004; McMahon et al., 2010), the mechanism for Keap1–Nrf2 pathway activation in response to 4-HNE may include phosphatidylinositol 3 kinase (PI3K) pathway (Chen et al., 2009). In addition to Keap1–Nrf2 pathway, 4-HNE may also induce cytoprotection via the heat-shock response (HSR) (Jacobs and Marnett, 2007).

CYCLOPENTENONE PROSTAGLANDINS

Enzymatic lipid peroxidation of arachidonic acid by cyclo-oxygenase leads to the production of several prostaglandins. Prostaglandin D_2 (PGD2) is a major COX product in many tissues and a precursor for the J series of cyclopentenone prostaglandins (cyPGs) such as 15-deoxy-Δ-^{12,14}-prostaglandin J_2 (15d-PGJ$_2$). It is a ligand for peroxisome proliferator-activated receptor γ (PPAR-γ), a nuclear receptor that regulates adipocyte differentiation and metabolic homeostasis. In addition, 15d-PGJ$_2$ mediates cytoprotection via activation of Nrf2-dependent detoxifying enzymes (Kansanen et al., 2009a). As an electrophile, 15d-PGJ$_2$ covalently binds to signaling proteins mediating anti-inflammatory actions in the cell. With regards to the Keap1–Nrf2 pathway, the modification of Keap1 cysteine residues 273 and 288 by 15d-PGJ$_2$ leads to pathway activation (Kobayashi et al., 2009; Levonen et al., 2004). The contribution of the pathway in resolving inflammation has been demonstrated using a carrageenan-induced pleurisy model and antibody-based methods. Itoh et al. showed COX2-dependent accumulation of 15d-PGJ$_2$ in peritoneal macrophages (Itoh et al., 2004). In this model, both COX2 inhibition by a specific inhibitor, NS-398, and Nrf2 deficiency resulted in the attenuated induction of the Nrf2 target gene peroxiredoxin I as well as delayed resolution of inflammation. Similar findings were also reported using a carrageenan-induced acute lung injury model (Itoh et al., 2004). It has also been proposed that PGD$_2$ derivatives mediate atheroprotective actions of laminar shear stress in endothelial cells (Hosoya et al., 2005) and that endogenous prostaglandin D2 and its metabolites protect against cardiac ischemia–reperfusion injury via Nrf2 through combined effects of PGD$_2$ on PGF2α receptor (FP) as well as FP-independent effects of 15d-PGJ$_2$ (Katsumata et al., 2014).

ELECTROPHILIC OXO-DERIVATIVES

Recently, a new class of electrophilic oxo-derivatives (EFOX) was identified from activated macrophages. EFOXs are produced from omega-3 fatty acids in reactions mediated by the COX-2 enzyme. They are electrophiles that can inhibit the production of cytokines and induce the Keap1–Nrf2 pathway assessed by nuclear accumulation of Nrf2 and increased expression of Nrf2 target genes HMOX1, GCLM, and NQO1 (Groeger and Freeman, 2010).

OXIDIZED PHOSPHOLIPIDS

Unsaturated fatty acids esterified to phospholipids are also targets for lipid peroxidation. Oxidative fragmentation of phospholipids such as 1-palmitoyl-2-arachidonoyl-*sn*-glycero-3-phosphocholine (PAPC) generates several oxidation products, which mediate both inflammatory and anti-inflammatory actions (Bochkov et al., 2010). The oxidation products of PAPC (oxPAPC) have a complete phospholipid structure with a glycerol backbone-linked polar head groups and two fatty acid residues of which sn2-positioned fatty acid chain is oxidized. OxPAPC has been found in minimally oxidized LDL, in atherosclerotic lesions from cholesterol-fed rabbits (Watson et al., 1997), and there are natural antibodies to these oxidized phospholipids found in the sera of hypercholes-terolemic apoE$^{-/-}$ mice (Horkko et al., 1996) as well as LDLR$^{-/-}$ mice, that may have atheroprotective functions (Tsimikas et al., 2001).

OxPAPC can modulate multiple intracellular signaling cascades. In addition to inflammatory properties such as induction of monocyte binding to the endothelial cell layer (Leitinger et al., 1999; Watson et al., 1997) and the expression of several adhesion molecules and cytokines (Furnkranz et al., 2005; Subbanagounder et al., 2002), oxidized phospholipids can act as anti-inflammatory mediators. For example, oxPAPC blocks the acute inflammatory reactions induced by LPS (Bochkov et al., 2002; Leitinger et al., 1999) by inhibiting LPS to act through its receptors, toll-like receptors 2 and 4 (Walton et al., 2003).

In addition, oxPAPC activates the Keap1–Nrf2-signaling pathway in human primary endothelial cells and in murine arteries (Jyrkkanen et al., 2008). In this study, the electrophilic oxidation products of oxPAPC, 1-palmitoyl-2-5,6-epoxy isopros-tane E2-*sn*-glycero-3-phosphocholine (PEIPC) and hydroperoxide phospholipid (PAPCOOH), were identified as the Nrf2 activating species in oxPAPC, and the electrophilic *sn*-2 side chain was found to be critical for the activation of Nrf2 target genes. In addition to Nrf2-regulated expression of antioxidant enzymes, oxPAPC also increases the amount of glutathione (GSH), a major cellular antioxidant, as well as the expression of GCL, a gene that encodes the rate-limiting enzyme in the synthesis of GSH (Gargalovic et al., 2006a; Jyrkkanen et al., 2008; Moellering et al., 2002).

The mechanism for activation of Nrf2-dependent antioxidant expression by oxPAPC is presently unknown. However, as the electrophilic property of oxPAPC is critical for Nrf2 activation (Jyrkkanen et al., 2008), it is likely that a direct Keap1 modification would be the mechanism for Nrf2 activation by electrophilic oxidation products of PAPC. This hypothesis is supported by studies where protein targets of oxidized 1-palmitoyl-2-arachidonoyl-*sn*-glycero-phosphatidylethanolamine (PAPE) were investigated (Gugiu et al., 2008; Springstead et al., 2012). These studies highlight that oxPAPE can form tight interactions with intracellular proteins (Gugiu et al., 2008) and that these interactions can be blocked by *N*-ethylmaleimide and gene expression changes attenuated by *N*-acetylcysteine (Springstead et al., 2012), strengthening the notion that reactions with cysteine residues in proteins is important for signaling functions. Importantly, in these studies, labeled oxPAPE in which biotin was coupled to the primary amine of PAPE was found to have intracellular targets such as H-ras, indicating that phospholipase cleavage of bioactive lipid in *sn*-2 position is not necessary for biological activity (Gugiu et al., 2008; Springstead et al., 2012).

NITRATED FATTY-ACIDS

In addition to being targets for direct lipid peroxidation, unsaturated fatty acids can react with $\cdot$NO and $\cdot$NO-derived species, such as ONOO$^-$, yielding a variety of oxidized and nitrated derivates such as nitrated oleic (OA-NO$_2$), linoleic (LNO$_2$), linolenic, arachidonic, and eicosapentaenoic acids (Lima et al., 2003; Rubbo et al., 1994). Nitro-fatty acids OA-NO$_2$ and LNO$_2$ have been detected *in vivo* at low nanomolar concentrations from plasma and tissue, and the concentrations are significantly elevated in inflammatory conditions such as ischemic preconditioning (IPC) and myocardial I/R (Nadtochiy et al., 2009; Rudolph et al., 2010; Tsikas et al., 2009). Nitro-fatty acids can modulate specific signaling cascades by electrophilic adduction of biological targets. These compounds can alter the functions of regulatory proteins and enzymes via post-translational modification of the susceptible nucleophilic amino acids (Baker et al., 2007; Batthyany et al., 2006). The direct molecular targets of nitro-fatty acids include Keap1 (Kansanen et al., 2011), PPAR-γ (Schopfer et al., 2010), p65 subunit of NF-κB (Cui et al., 2006), glyceraldehyde 3-phosphate dehydrogenase (GADPH) (Cui et al., 2006), GSH (Baker et al., 2007), xanthine oxidoreductase (XO) (Kelley et al., 2008), and matrix metalloproteinases (MMP) proMMP-7 and proMMP-9 (Bonacci et al., 2011). Of nitro-fatty acids, conjugated linoleic acid (CLA) has been identified as the primary endogenous substrate for fatty acid nitration both *in vitro* as well as *in vivo* (Bonacci et al., 2012), and pharmacologic administration of stable nitro-fatty acid species such as nitro-oleic acid may be useful in metabolic and inflammatory disorders (Delmastro-Greenwood et al., 2014).

The anti-inflammatory properties of nitro-fatty acids are mediated via a direct modification of Keap1 in the Nrf2 signaling pathway (Kansanen et al., 2011) and the p65 subunit of NF-κB (Cui et al., 2006). In addition to the activation of Nrf2 signaling in vascular endothelial cells (Kansanen et al., 2009b), heme oxygenase-1 (HMOX1), a target gene of Nrf2, is postulated to mediate various protective effects of OA-NO$_2$ in the vasculature. HMOX1 is upregulated by nitro-fatty acids in multiple cell types *in vitro* including human aortic endothelial cells (HAECs) (Wright et al., 2006), vascular smooth muscle cells (Khoo et al., 2010), rat aortic smooth muscle cells (Cole et al., 2009), and mouse macrophages (Cui et al., 2006). In addition, in a mouse model where endoluminal injury to the femoral artery was induced by an angioplasty wire, the administration of OA-NO$_2$ significantly inhibited neointimal hyperplasia as measured by the intima/media ratio and intimal area. The inhibition of neointimal formation was partly reversed when the injury was induced in HMOX1 knockout mice. These results indicate that HMOX1 is involved in reduced neointimal thickening (Cole et al., 2009). Even though the role of Nrf2 was not studied in these cases, the involvement of HMOX1 suggests that Nrf2 may mediate the protective effects of OA-NO$_2$ also *in vivo*.

ACTIVATION MECHANISM OF THE KEAP1–NRF2 PATHWAY BY OXIDIZED AND NITRATED LIPIDS

The data available to date suggest that oxidized and nitrated share the same activation mechanism for Keap1–Nrf2 pathway involving Keap1 cysteine modification. Keap1 has multiple sensing mechanisms for Nrf2 activation having a different "cysteine code" for each electrophile or group of electrophiles (Kobayashi et al., 2009; McMahon et al., 2010), While some classical Nrf2 inducers such as SFN are recognized through Keap1 C151, oxidized and nitrated lipids 4-HNE, 15d-PGJ$_2$, and

OA-NO$_2$ share the same activation mechanisms for Keap1 modification belonging to the C273/288-preferring group of inducers. These lipid derivates can activate the Keap1–Nrf2 system independent of Keap1 C151 (Kansanen et al., 2011; Kobayashi et al., 2009; McMahon et al., 2010). The fact that oxidized and nitrated lipids seem to activate Nrf2 pathway independent of Keap1-C151, Cul3 dissociation most likely does not mediate the Nrf2 nuclear translocation in response to 4-HNE, 15d-PGJ$_2$, and OA-NO$_2$ (Figure 13.3).

The specific pattern of Keap1 cysteine modification by oxPAPC has not been investigated to date, but it is known that an electrophilic *sn*-2 side chain in oxidation products of PAPC is critical for Keap1–Nrf2 pathway activation (Jyrkkanen et al., 2008). Furthermore, the importance of electrophilic properties was further studied

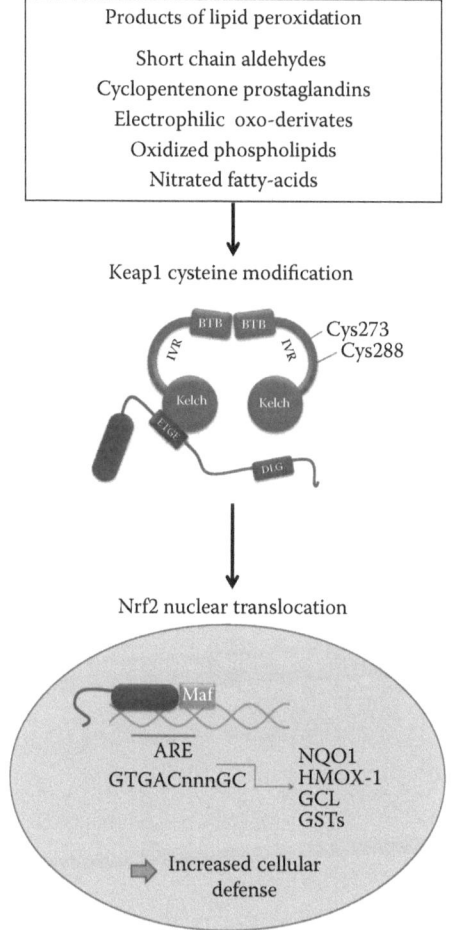

FIGURE 13.3 Mechanism for Keap1–Nrf2 pathway activation in response to lipid-derived electrophilic species. Lipid-derived electrophilic species induce Nrf2-dependent gene expression via modification of Keap1 C283 and C288.

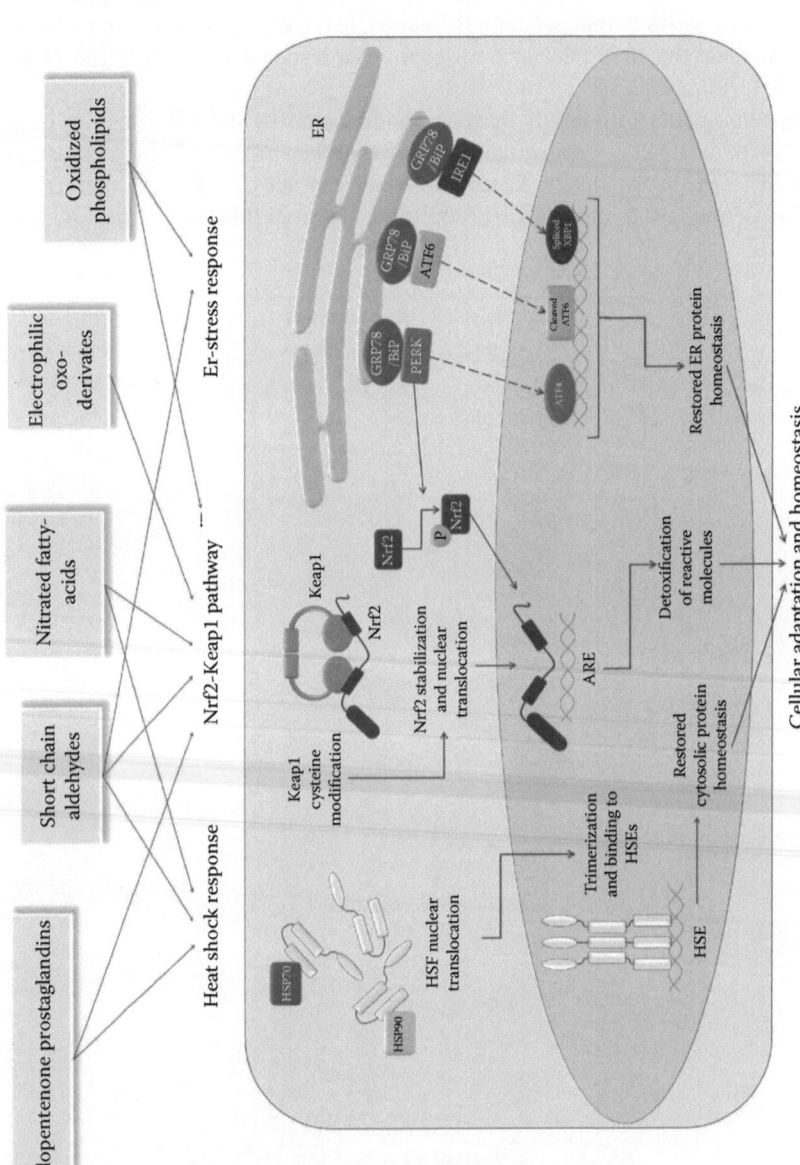

FIGURE 13.4

by Springstead et al., who concluded that electrophilic oxidation product of PAPC, PEIPC, accounts for almost all the binding of oxPAPC to cysteine (Springstead et al., 2012). Moreover, the authors showed that 15d-PGJ$_2$, PEIPC, and oxPAPC induce similar binding and gene-expression pattern in human endothelial cells suggesting that oxPAPC may also belong to the C273/288-preferring group of inducers of the Keap1–Nrf2 pathway.

However, also other Keap1–Nrf2 pathway activation mechanisms in response to oxPAPC have been suggested. Double-stranded RNA-activated protein kinase-like ER kinase (PERK) has been shown to phosphorylate Nrf2 resulting in nuclear accumulation of Nrf2 (Cullinan et al., 2003). PERK-dependent phosphorylation may partially explain the Keap1–Nrf2 pathway activation by oxPAPC. Even though PERK-dependent phosphorylation of Nrf2 has not been shown, treatment with oxPAPC leads to the activation of ER stress and subsequent activation of the unfolded protein response (Gargalovic et al., 2006b). Unfolded protein response leads to PERK-dependent phosphorylation and nuclear translocation of Nrf2 (Figure 13.4).

INTERPLAY WITH OTHER STRESS SIGNALING PATHWAYS

In addition to the Keap1–Nrf2-regulated stress response, cells have other integrated stress response pathways triggered by endogenous stimuli or environmental stresses. HSR primarily responds to protein misfolding in the cytoplasm. HSR is regulated at the transcriptional level by heat-shock factors such as Heat Shock Factor 1 (HSF1). It is suggested that chaperones HSP70 and HSP90 have the potential to inhibit HSF1 and they keep HSF1 in an inactive form. Upon activation, HSF1 is released from the complexes and it undergoes multistep processing involving post-translational modifications, nuclear enrichment, trimerization, and binding to heat-shock elements, resulting in transcription of a large family of heat-shock genes. Many of the proteins encoded by heat-shock genes function as chaperones, proteases, or other proteins essential for protection of the cell against proteotoxic stress (Kansanen et al., 2012). Nitrated and oxidized lipids such as OA-NO$_2$ (Kansanen et al., 2009b), 4-HNE (Jacobs and Marnett, 2007), and cyPGs (Vila et al., 2008) can induce the HSR (HSR; Figure 13.4). It is postulated that HSP70 may serve as a redox-sensor for HSR in response to thiol-reactive molecules (Wang et al., 2012). Intriguingly, many of the novel potent activators of the HSR identified by high-throughput

---

FIGURE 13.4   Stress-responsive pathways. In HSR, HSP70 and HSP90 are postulated to hold HSF1 monomers inactive in the cytoplasm. Upon activation, the monomers locate to nucleus, bind to heat-shock element (HSE) and induce the expression of heat-shock proteins that function to restore the protein homeostasis. Antioxidant response is initiated by cysteine modification of Keap1, which leads to stabilization and nuclear translocation of Nrf2. In the nucleus, Nrf2 binds to the ARE and drives the expression of antioxidant and detoxifying enzymes. ER-stress response consists of three pathways regulated by transcription factors ATF4, ATF6, and XBP1. Activation of these pathways involves a modification of the ER residing protein, GRP78/BiP. Target genes of ER-stress response function to restore the protein homeostasis in the ER. Together these three responses allow the cell to survive under conditions of stress.

chemical screening are electrophiles that can simultaneously activate the Keap1–Nrf2 pathway, indicating that these two stress response pathways converge at the level of the initial signal (Calamini et al., 2012; Santagata et al., 2012). Furthermore, genome-wide interrogation of HSF1 binding sites by chromatin immunoprecipitation sequencing (ChIP-seq) identified ARE sequences to be enriched in HSF1 binding genomic areas, indicating crosstalk of the two pathways at the epigenetic level as well (Mendillo et al., 2012).

ER is the site of secreted and transmembrane protein synthesis and modification. These functions are highly regulated and therefore sensitive to disruption. Conditions that disturb the ER homeostasis results in the activation of an adaptive signaling pathway called the ER-stress response. It is thought that the activation of the three arms of ER-stress response, regulated by transcription factors ATF4, ATF6, and XBP1, involves the modification of GRP79/BiP protein residing in the ER (Kansanen et al., 2012). 4-HNE and oxPAPC can trigger ER-stress response (Gargalovic et al., 2006b; West and Marnett, 2006), but there is no mechanism of action determined. However, it is known that both 4-HNE and oxidized PAPE can bind GRP78/BiP protein (Gugiu et al., 2008; Vila et al., 2008), suggesting that it might serve as the sensor for electrophiles (Figure 13.4).

THERAPEUTIC POTENTIAL OF LIPID-DERIVED ELECTROPHILIC SPECIES

The notion that endogenous cytoprotective genes could be induced via the activation of stress signaling pathways has attracted a lot of attention in drug development, as this approach is lucrative in many disease processes in which oxidative stress and inflammation play a role in the pathogenesis. In terms of clinical development, isothiocyanate sulforaphane (SFN), often administered in the form of broccoli sprout extract, has drawn the most attention, with a total of 26 clinical trials currently listed in the NIH clinical trial registry (clinicaltrials.gov), indications ranging from cancer chemoprevention to chronic obstructive pulmonary disease. The first Keap1–Nrf2 pathway activating electrophilic drug that has been approved by FDA EMA for clinical use is dimethyl fumarate. Its clinical indication is relapsing multiple sclerosis, and it has been shown to reduce the number of MS relapses and lesions detected by MRI and decrease the rate of disability progression in two phase III clinical trials (Fox et al., 2012; Gold et al., 2012).

With respect to fatty acid-derived electrophiles, cyPGs and their synthetic derivatives have been used in preclinical studies. In carrageenin-induced pleurisy model in rats, 15d-PGJ$_2$ has been shown to attenuate inflammation (Gilroy et al., 1999), and effect that has been proposed to be mediated via the Keap1–Nrf2 pathway (Itoh et al., 2004). Prostaglandin A2 has been used as an antitumor agent in mouse tumor models (reviewed 97). Interestingly, liposome encapsulated PGA$_2$ has been shown to have atheroprotective properties in LDL receptor-deficient mice (Homem de Bittencourt et al., 2007), but whether this is due to its Nrf2-mediated anti-inflammatory properties or other mechanisms is not known. Electrophilic neurite outgrowth-promoting prostaglandin (NEPP) compounds have been developed based on the chemical structures of cyclopentenone prostaglandins (Satoh et al., 2001), and one of these, NEPP11, was found to be protective against ischemia–reperfusion injury in mice via Nrf2 activation (Satoh et al., 2006). However, neither cyPGs nor their synthetic analogs have entered the clinical phase.

There is a wealth of preclinical evidence of the protective effects of nitroalkenes, OA-NO$_2$ in particular, in various animal models of metabolic and inflammatory disorders (reviewed in Francisco J. Schopfer, Cipollina, and Freeman 2011). These are now being developed for the treatment of diseases associated with kidney injury, inflammation and metabolic disorders, the initial clinical target being contrast imaging dye-induced nephropathy.

CONCLUSIONS

Herein, we summarized the current understanding of activation mechanisms and biological consequences of the Keap1–Nrf2 pathway to electrophilic lipids, and the interplay of the Nrf2 response with other stress signaling pathways. It is evident that none of the electrophilic lipid species characterized to date solely activate the Keap1–Nrf2 pathway, but have other protein targets as well. The pleiotropic nature of these species is important to bear in mind in drug development, as each species is likely to have a unique ability to activate different signaling pathways and cause toxicity. Advances in genome-wide methods to interrogate gene regulation as well as proteomics and mass spectrometry will be instrumental in assessing the biological effects and therapeutic potential of these species and may also lead to identification of novel biologically active electrophilic lipids with signaling functions.

ABBREVIATIONS

15d-PGJ$_2$	15-deoxy-Δ-^{12,14}-prostaglandin J$_2$
4-HNE	4-hydroxy-2-nonenal
4-ONE	4-oxo-2-nonenal
ARE	antioxidant response element
BTB	Broad complex, Tramtrack and Bric-à-Brac
ChIP-seq	chromatin immunoprecipitation sequencing
Cul3	Cullin3
cyPGs	cyclopentenone prostaglandins
EFOX	electrophilic oxo-derivatives
EpRE	electrophile response element
GADPH	glyceraldehyde 3-phosphate dehydrogenase
GCL	glutamate-cysteine ligase
GSH	glutathione
GSTs	glutathione S-transferases antioxidant proteins including
HMOX1	heme oxygenase-1
HSF1	heat-shock factor 1
IAB	N-iodoacetyl-N-biotinylhexylenediamine
IPC	ischemic preconditioning
IVR	the intervening region

MMP	matrix metalloproteinases
MS	mass spectrometry
Neh	Nrf2–ECH homology
NEPP	neurite outgrowth-promoting prostaglandin
NQO1	NAD(P)H quinone oxidoreductase 1
Nrf2	nuclear factor-E2-related factor 2
Nrf2$^{-/-}$	Nrf2 knockout
OA-NO2	nitro-oleic acid
PAPC	1-palmitoyl-2-arachidonoyl-*sn*-glycero-3-phosphocholine
PAPCOOH	hydroperoxide phospholipid
PAPE	1-palmitoyl-2-arachidonoyl-*sn*-glycero-phosphatidylethanolamine
PEIPC	1-palmitoyl-2-5,6-epoxy isoprostane E2-*sn*-glycero-3-phosphocholine
PGA$_2$	prostaglandin A$_2$
PGD$_2$	prostaglandin D$_2$
PI3K	phosphatidylinositol 3 kinase
PKC	protein kinase C
PPAR-γ	peroxisome proliferator-activated receptor γ
SFN	isothiocyanate sulforaphane
tBHQ	tert-butylhydroquinone

REFERENCES

Baker, L.M., Baker, P.R., Golin-Bisello, F., Schopfer, F.J., Fink, M., Woodcock, S.R., Branchaud, B.P., Radi, R., and Freeman, B.A. 2007. Nitro-fatty acid reaction with glutathione and cysteine. Kinetic analysis of thiol alkylation by a Michael addition reaction. *J. Biol. Chem.* 282, 31085–31093.

Batthyany, C., Schopfer, F.J., Baker, P.R., Duran, R., Baker, L.M., Huang, Y., Cervenansky, C., Branchaud, B.P., and Freeman, B.A. 2006. Reversible post-translational modification of proteins by nitrated fatty acids in vivo. *J. Biol. Chem.* 281, 20450–20463.

Bloom, D.A., and Jaiswal, A.K. 2003. Phosphorylation of Nrf2 at Ser40 by protein kinase C in response to antioxidants leads to the release of Nrf2 from INrf2, but is not required for Nrf2 stabilization/accumulation in the nucleus and transcriptional activation of antioxidant response element. *J. Biol. Chem.* 278, 44675–44682.

Bochkov, V.N., Kadl, A., Huber, J., Gruber, F., Binder, B.R., and Leitinger, N. 2002. Protective role of phospholipid oxidation products in endotoxin-induced tissue damage. *Nature* 419, 77–81.

Bochkov, V.N., Oskolkova, O.V., Birukov, K.G., Levonen, A.L., Binder, C.J., and Stockl, J. 2010. Generation and biological activities of oxidized phospholipids. *Antioxid. Redox. Signal.* 12, 1009–1059.

Bonacci, G., Baker, P.R.S., Salvatore, S.R., Shores, D., Khoo, N.K.H., Koenitzer, J.R., Vitturi, D.A. et al. 2012. Conjugated linoleic acid is a preferential substrate for fatty acid nitration. *J. Biol. Chem.* 287, 44071–44082.

Bonacci, G., Schopfer, F.J., Batthyany, C.I., Rudolph, T.K., Rudolph, V., Khoo, N.K., Kelley, E.E., and Freeman, B.A. 2011. Electrophilic fatty acids regulate matrix metalloproteinase activity and expression. *J. Biol. Chem.* 286, 16074–16081.

Calamini, B., Silva, M.C., Madoux, F., Hutt, D.M., Khanna, S., Chalfant, M.A., Saldanha, S.A. et al. 2012. Small-molecule proteostasis regulators for protein conformational diseases. *Nat. Chem. Biol.* 8, 185–196.

Chen, J., Wang, L., Chen, Y., Sternberg, P., and Cai, J. 2009. Phosphatidylinositol 3 kinase pathway and 4-hydroxy-2-nonenal-induced oxidative injury in the RPE. *Invest. Ophthalmol. Visual Sci.* 50, 936–942.

Cole, M.P., Rudolph, T.K., Khoo, N.K., Motanya, U.N., Golin-Bisello, F., Wertz, J.W., Schopfer, F.J. et al. 2009. Nitro-fatty acid inhibition of neointima formation after endoluminal vessel injury. *Circ. Res.* 105, 965–972.

Cui, T., Schopfer, F.J., Zhang, J., Chen, K., Ichikawa, T., Baker, P.R., Batthyany, C. et al. 2006. Nitrated fatty acids: Endogenous anti-inflammatory signaling mediators. *J. Biol. Chem.* 281, 35686–35698.

Cullinan, S.B., Gordan, J.D., Jin, J., Harper, J.W., and Diehl, J.A. 2004. The Keap1-BTB protein is an adaptor that bridges Nrf2 to a Cul3-based E3 ligase: Oxidative stress sensing by a Cul3-Keap1 ligase. *Mol. Cell Biol.* 24, 8477–8486.

Cullinan, S.B., Zhang, D., Hannink, M., Arvisais, E., Kaufman, R.J., and Diehl, J.A. 2003. Nrf2 is a direct PERK substrate and effector of PERK-dependent cell survival. *Mol. Cell Biol.* 23, 7198–7209.

Delmastro-Greenwood, M., Freeman, B.A., and Gelhaus Wendell, S. 2014. Redox-dependent anti-inflammatory signaling actions of unsaturated fatty acids. *Ann. Rev. Physiol.* 76, 79–105.

Dhakshinamoorthy, S., Jain, A.K., Bloom, D.A., and Jaiswal, A.K. 2005. Bach1 competes with Nrf2 leading to negative regulation of the antioxidant response element (ARE)-mediated NAD(P)H:quinone oxidoreductase 1 gene expression and induction in response to antioxidants. *J. Biol. Chem.* 280, 16891–16900.

Dinkova-Kostova, A.T., Holtzclaw, W.D., Cole, R.N., Itoh, K., Wakabayashi, N., Katoh, Y., Yamamoto, M., and Talalay, P. 2002. Direct evidence that sulfhydryl groups of Keap1 are the sensors regulating induction of phase 2 enzymes that protect against carcinogens and oxidants. *Proc. Natl. Acad. Sci. USA* 99, 11908–11913.

Eggler, A.L., Liu, G., Pezzuto, J.M., van Breemen, R.B., and Mesecar, A.D. 2005. Modifying specific cysteines of the electrophile-sensing human Keap1 protein is insufficient to disrupt binding to the Nrf2 domain Neh2. *Proc. Natl. Acad. Sci. USA* 102, 10070–10075.

Eggler, A.L., Small, E., Hannink, M., and Mesecar, A.D. 2009. Cul3-mediated Nrf2 ubiquitination and antioxidant response element (ARE) activation are dependent on the partial molar volume at position 151 of Keap1. *Biochem. J.* 422, 171–180.

Fox, R.J., Miller, D.H., Phillips, J.T., Hutchinson, M., Havrdova, E., Kita, M., Yang, M. et al. 2012. Placebo-controlled phase 3 study of oral BG-12 or glatiramer in multiple sclerosis. *New Engl. J. Med.* 367, 1087–1097.

Friling, R.S., Bensimon, A., Tichauer, Y., and Daniel, V. 1990. Xenobiotic-inducible expression of murine glutathione S-transferase Ya subunit gene is controlled by an electrophile-responsive element. *Proc. Natl. Acad. Sci. USA* 87, 6258–6262.

Friling, R.S., Bergelson, S., and Daniel, V. 1992. Two adjacent AP-1-like binding sites form the electrophile-responsive element of the murine glutathione S-transferase Ya subunit gene. *Proc. Natl. Acad. Sci. USA* 89, 668–672.

Furnkranz, A., Schober, A., Bochkov, V.N., Bashtrykov, P., Kronke, G., Kadl, A., Binder, B.R., Weber, C., and Leitinger, N. 2005. Oxidized phospholipids trigger atherogenic inflammation in murine arteries. *Arterioscler. Thromb. Vasc. Biol.* 25, 633–638.

Gargalovic, P.S., Imura, M., Zhang, B., Gharavi, N.M., Clark, M.J., Pagnon, J., Yang, W.P. et al. 2006a. Identification of inflammatory gene modules based on variations of human endothelial cell responses to oxidized lipids. *Proc. Natl. Acad. Sci. USA* 103, 12741–12746.

Gargalovic, P.S., Gharavi, N.M., Clark, M.J., Pagnon, J., Yang, W.P., He, A., Truong, A. et al. 2006b. The unfolded protein response is an important regulator of inflammatory genes in endothelial cells. *Arterioscler. Thromb. Vasc. Biol.* 26, 2490–2496.

Gilroy, D.W., Colville-Nash, P.R., Willis, D., Chivers, J., Paul-Clark, M.J., and Willoughby, D.A. 1999. Inducible cyclooxygenase may have anti-inflammatory properties. *Nat. Med.* 5, 698–701.

Gold, R., Kappos, L., Arnold, D.L., Bar-Or, A., Giovannoni, G., Selmaj, K., Tornatore, C. et al. 2012. Placebo-controlled phase 3 study of oral BG-12 for relapsing multiple sclerosis. *New Engl. J. Med.* 367, 1098–1107.

Groeger, A.L., and Freeman, B.A. 2010. Signaling actions of electrophiles: Anti-inflammatory therapeutic candidates. *Mol. Interv.* 10, 39–50.

Gugiu, B.G., Mouillesseaux, K., Duong, V., Herzog, T., Hekimian, A., Koroniak, L., Vondriska, T.M., and Watson, A.D. 2008. Protein targets of oxidized phospholipids in endothelial cells. *J. Lipid Res.* 49, 510–520.

Homem de Bittencourt, P.I., Lagranha, D.J., Maslinkiewicz, A., Senna, S.M., Tavares, A.M.V., Baldissera, L.P., Janner, D.R. et al. 2007. LipoCardium: Endothelium-directed cyclopentenone prostaglandin-based liposome formulation that completely reverses atherosclerotic lesions. *Atherosclerosis* 193, 245–258.

Hong, F., Freeman, M.L., and Liebler, D.C. 2005b. Identification of sensor cysteines in human Keap1 modified by the cancer chemopreventive agent sulforaphane. *Chem. Res. Toxicol.* 18, 1917–1926.

Hong, F., Sekhar, K.R., Freeman, M.L., and Liebler, D.C. 2005a. Specific patterns of electrophile adduction trigger Keap1 ubiquitination and Nrf2 activation. *J. Biol. Chem.* 280, 31768–31775.

Horkko, S., Miller, E., Dudl, E., Reaven, P., Curtiss, L.K., Zvaifler, N.J., Terkeltaub, R. et al. 1996. Antiphospholipid antibodies are directed against epitopes of oxidized phospholipids. Recognition of cardiolipin by monoclonal antibodies to epitopes of oxidized low density lipoprotein. *J. Clin. Invest.* 98, 815–825.

Hosoya, T., Maruyama, A., Kang, M.I., Kawatani, Y., Shibata, T., Uchida, K., Warabi, E., Noguchi, N., Itoh, K., and Yamamoto, M. 2005. Differential responses of the Nrf2-Keap1 system to laminar and oscillatory shear stresses in endothelial cells. *J. Biol. Chem.* 280, 27244–27250.

Hu, C., Eggler, A.L., Mesecar, A.D., and van Breemen, R.B. 2011. Modification of keap1 cysteine residues by sulforaphane. *Chem. Res. Toxicol.* 24, 515–521.

Huang, H.C., Nguyen, T., and Pickett, C.B. 2000. Regulation of the antioxidant response element by protein kinase C-mediated phosphorylation of NF-E2-related factor 2. *Proc. Natl. Acad. Sci. USA* 97, 12475–12480.

Ichimura, Y., Waguri, S., Sou, Y.-S., Kageyama, S., Hasegawa, J., Ishimura, R., Saito, T. et al. 2013. Phosphorylation of p62 activates the Keap1-Nrf2 pathway during selective autophagy. *Mol. Cell* 51, 618–631.

Ishii, T., Itoh, K., Ruiz, E., Leake, D.S., Unoki, H., Yamamoto, M., and Mann, G.E. 2004. Role of Nrf2 in the regulation of CD36 and stress protein expression in murine macrophages: Activation by oxidatively modified LDL and 4-hydroxynonenal. *Circ. Res.* 94, 609–616.

Ishikado, A., Nishio, Y., Morino, K., Ugi, S., Kondo, H., Makino, T., Kashiwagi, A., and Maegawa, H. 2010. Low concentration of 4-hydroxy hexenal increases heme oxygenase-1 expression through activation of Nrf2 and antioxidative activity in vascular endothelial cells. *Biochem. Biophys. Res. Commun.* 402, 99–104.

Itoh, K., Chiba, T., Takahashi, S., Ishii, T., Igarashi, K., Katoh, Y., Oyake, T. et al. 1997. An Nrf2/small Maf heterodimer mediates the induction of phase II detoxifying enzyme genes through antioxidant response elements. *Biochem. Biophys. Res. Commun.* 236, 313–322.

Itoh, K., Mochizuki, M., Ishii, Y., Ishii, T., Shibata, T., Kawamoto, Y., Kelly, V., Sekizawa, K., Uchida, K., and Yamamoto, M. 2004. Transcription factor Nrf2 regulates inflammation by mediating the effect of 15-deoxy-Delta(12,14)-prostaglandin j(2). *Mol. Cell Biol.* 24, 36–45.

Itoh, K., Wakabayashi, N., Katoh, Y., Ishii, T., Igarashi, K., Engel, J.D., and Yamamoto, M. 1999. Keap1 represses nuclear activation of antioxidant responsive elements by Nrf2 through binding to the amino-terminal Neh2 domain. *Genes Dev.* 13, 76–86.

Jacobs, A.T., and Marnett, L.J. 2007. Heat shock factor 1 attenuates 4-Hydroxynonenal-mediated apoptosis: Critical role for heat shock protein 70 induction and stabilization of Bcl-XL. *J. Biol. Chem.* 282, 33412–33420.

Jain, A., Lamark, T., Sjottem, E., Larsen, K.B., Awuh, J.A., Overvatn, A., McMahon, M., Hayes, J.D., and Johansen, T. 2010. p62/SQSTM1 is a target gene for transcription factor NRF2 and creates a positive feedback loop by inducing antioxidant response element-driven gene transcription. *J. Biol. Chem.* 285, 22576–22591.

Jyrkkanen, H.K., Kansanen, E., Inkala, M., Kivela, A.M., Hurttila, H., Heinonen, S.E., Goldsteins, G. et al. 2008. Nrf2 regulates antioxidant gene expression evoked by oxidized phospholipids in endothelial cells and murine arteries in vivo. *Circ. Res.* 103, e1–e9.

Kannan, M.B., Solovieva, V., and Blank, V. 2012. The small MAF transcription factors MAFF, MAFG and MAFK: Current knowledge and perspectives. *Biochim. Biophys. Acta* 1823, 1841–1846.

Kansanen, E., Bonacci, G., Schopfer, F.J., Kuosmanen, S.M., Tong, K.I., Leinonen, H., Woodcock, S.R. et al. 2011. Electrophilic nitro-fatty acids activate NRF2 by a KEAP1 cysteine 151-independent mechanism. *J. Biol. Chem.* 286, 14019–14027.

Kansanen, E., Jyrkkänen, H.-K., and Levonen, A.-L. 2012. Activation of stress signaling pathways by electrophilic oxidized and nitrated lipids. *Free Radic. Biol. Med.* 52, 973–982.

Kansanen, E., Jyrkkanen, H.K., Volger, O.L., Leinonen, H., Kivela, A.M., Hakkinen, S.K., Woodcock, S.R. et al. 2009b. Nrf2-dependent and -independent responses to nitro-fatty acids in human endothelial cells: Identification of heat shock response as the major pathway activated by nitro-oleic acid. *J. Biol. Chem.* 284, 33233–33241.

Kansanen, E., Kivela, A.M., and Levonen, A.L. 2009a. Regulation of Nrf2-dependent gene expression by 15-deoxy-Delta12,14-prostaglandin J2. *Free Radic. Biol. Med.* 47, 1310–1317.

Katsumata, Y., Shinmura, K., Sugiura, Y., Tohyama, S., Matsuhashi, T., Ito, H., Yan, X. et al. 2014. Endogenous prostaglandin D2 and its metabolites protect the heart against ischemia–reperfusion injury by activating Nrf2. *Hypertension* 63(1), 80–7.

Kelley, E.E., Batthyany, C.I., Hundley, N.J., Woodcock, S.R., Bonacci, G., Del Rio, J.M., Schopfer, F.J., Lancaster Jr., J.R., Freeman, B.A., and Tarpey, M.M. 2008. Nitro-oleic acid, a novel and irreversible inhibitor of xanthine oxidoreductase. *J. Biol. Chem.* 283, 36176–36184.

Khoo, N.K., Rudolph, V., Cole, M.P., Golin-Bisello, F., Schopfer, F.J., Woodcock, S.R., Batthyany, C. and Freeman, B.A. 2010. Activation of vascular endothelial nitric oxide synthase and heme oxygenase-1 expression by electrophilic nitro-fatty acids. *Free Radic. Biol. Med.* 48, 230–239.

Kobayashi, A., Kang, M.I., Okawa, H., Ohtsuji, M., Zenke, Y., Chiba, T., Igarashi, K., and Yamamoto, M. 2004. Oxidative stress sensor Keap1 functions as an adaptor for Cul3-based E3 ligase to regulate proteasomal degradation of Nrf2. *Mol. Cell Biol.* 24, 7130–7139.

Kobayashi, A., Kang, M.I., Watai, Y., Tong, K.I., Shibata, T., Uchida, K., and Yamamoto, M. 2006. Oxidative and electrophilic stresses activate Nrf2 through inhibition of ubiquitination activity of Keap1. *Mol. Cell Biol.* 26, 221–229.

Kobayashi, M., Li, L., Iwamoto, N., Nakajima-Takagi, Y., Kaneko, H., Nakayama, Y., Eguchi, M., Wada, Y., Kumagai, Y., and Yamamoto, M. 2009. The antioxidant defense system Keap1-Nrf2 comprises a multiple sensing mechanism for responding to a wide range of chemical compounds. *Mol. Cell Biol.* 29, 493–502.

Komatsu, M., Kurokawa, H., Waguri, S., Taguchi, K., Kobayashi, A., Ichimura, Y., Sou, Y.S. et al. 2010. The selective autophagy substrate p62 activates the stress responsive transcription factor Nrf2 through inactivation of Keap1. *Nat. Cell Biol.* 12, 213–223.

Kwak, M.K., Wakabayashi, N., Itoh, K., Motohashi, H., Yamamoto, M., and Kensler, T.W. 2003. Modulation of gene expression by cancer chemopreventive dithiolethiones through the Keap1-Nrf2 pathway. Identification of novel gene clusters for cell survival. *J. Biol. Chem.* 278, 8135–8145.

Lee, J.M., Calkins, M.J., Chan, K., Kan, Y.W., and Johnson, J.A. 2003. Identification of the NF-E2-related factor-2-dependent genes conferring protection against oxidative stress in primary cortical astrocytes using oligonucleotide microarray analysis. *J. Biol. Chem.* 278, 12029–12038.

Leitinger, N., Tyner, T.R., Oslund, L., Rizza, C., Subbanagounder, G., Lee, H., Shih, P.T. et al. 1999. Structurally similar oxidized phospholipids differentially regulate endothelial binding of monocytes and neutrophils. *Proc. Natl. Acad. Sci. USA* 96, 12010–12015.

Levonen, A.L., Landar, A., Ramachandran, A., Ceaser, E.K., Dickinson, D.A., Zanoni, G., Morrow, J.D. et al. 2004. Cellular mechanisms of redox cell signalling: Role of cysteine modification in controlling antioxidant defences in response to electrophilic lipid oxidation products. *Biochem. J.* 378, 373–382.

Levy, S., Jaiswal, A.K., and Forman, H.J. 2009. The role of c-Jun phosphorylation in EpRE activation of phase II genes. *Free Radic. Biol. Med.* 47, 1172–1179.

Lima, E.S., Di, M.P., and Abdalla, D.S. 2003. Cholesteryl nitrolinoleate, a nitrated lipid present in human blood plasma and lipoproteins. *J. Lipid Res.* 44, 1660–1666.

Luo, Y., Eggler, A.L., Liu, D., Liu, G., Mesecar, A.D., and van Breemen, R.B. 2007. Sites of alkylation of human Keap1 by natural chemoprevention agents. *J. Am. Soc. Mass Spectrom.* 18, 2226–2232.

McMahon, M., Lamont, D.J., Beattie, K.A., and Hayes, J.D. 2010. Keap1 perceives stress via three sensors for the endogenous signaling molecules nitric oxide, zinc, and alkenals. *Proc. Natl. Acad. Sci. USA* 107, 18838–18843.

Mendillo, M.L., Santagata, S., Koeva, M., Bell, G.W., Hu, R., Tamimi, R.M., Fraenkel, E., Ince, T.A., Whitesell, L., and Lindquist, S. 2012. HSF1 drives a transcriptional program distinct from heat shock to support highly malignant human cancers. *Cell* 150, 549–562.

Moellering, D.R., Levonen, A.L., Go, Y.M., Patel, R.P., Dickinson, D.A., Forman, H.J., and rley-Usmar, V.M. 2002. Induction of glutathione synthesis by oxidized low-density lipoprotein and 1-palmitoyl-2-arachidonyl phosphatidylcholine: Protection against quinone-mediated oxidative stress. *Biochem. J.* 362, 51–59.

Nadtochiy, S.M., Baker, P.R., Freeman, B.A., and Brookes, P.S. 2009. Mitochondrial nitroalkene formation and mild uncoupling in ischaemic preconditioning: Implications for cardioprotection. *Cardiovasc. Res.* 82, 333–340.

Nioi, P., Nguyen, T., Sherratt, P.J., and Pickett, C.B. 2005. The carboxy-terminal Neh3 domain of Nrf2 is required for transcriptional activation. *Mol. Cell Biol.* 25, 10895–10906.

Niture, S.K., and Jaiswal, A.K. 2009. Prothymosin-alpha mediates nuclear import of the INrf2/Cul3 Rbx1 complex to degrade nuclear Nrf2. *J. Biol. Chem.* 284, 13856–13868.

Ogura, T., Tong, K.I., Mio, K., Maruyama, Y., Kurokawa, H., Sato, C., and Yamamoto, M. 2010. Keap1 is a forked-stem dimer structure with two large spheres enclosing the intervening, double glycine repeat, and C-terminal domains. *Proc. Natl. Acad. Sci. USA* 107, 2842–2847.

Oh, J.Y., Giles, N., Landar, A., and rley-Usmar, V. 2008. Accumulation of 15-deoxy-delta(12,14)-prostaglandin J2 adduct formation with Keap1 over time: Effects on potency for intracellular antioxidant defence induction. *Biochem. J.* 411, 297–306.

Prestera, T., Holtzclaw, W.D., Zhang, Y., and Talalay, P. 1993. Chemical and molecular regulation of enzymes that detoxify carcinogens. *Proc. Natl. Acad. Sci. USA* 90, 2965–2969.

Rachakonda, G., Xiong, Y., Sekhar, K.R., Stamer, S.L., Liebler, D.C., and Freeman, M.L. 2008. Covalent modification at Cys151 dissociates the electrophile sensor Keap1 from the ubiquitin ligase CUL3. *Chem. Res. Toxicol.* 21, 705–710.

Rubbo, H., Radi, R., Trujillo, M., Telleri, R., Kalyanaraman, B., Barnes, S., Kirk, M., and Freeman, B.A. 1994. Nitric oxide regulation of superoxide and peroxynitrite-dependent lipid peroxidation. Formation of novel nitrogen-containing oxidized lipid derivatives. *J. Biol. Chem.* 269, 26066–26075.

Rudolph, V., Rudolph, T.K., Schopfer, F.J., Bonacci, G., Woodcock, S.R., Cole, M.P., Baker, P.R., Ramani, R., and Freeman, B.A. 2010. Endogenous generation and protective effects of nitro-fatty acids in a murine model of focal cardiac ischaemia and reperfusion. *Cardiovasc. Res.* 85, 155–166.

Rushmore, T.H., Morton, M.R., and Pickett, C.B. 1991. The antioxidant responsive element. Activation by oxidative stress and identification of the DNA consensus sequence required for functional activity. *J. Biol. Chem.* 266, 11632–11639.

Santagata, S., Xu, Y.-M., Wijeratne, E.M.K., Kontnik, R., Rooney, C., Perley, C.C., Kwon, H. et al. 2012. Using the heat-shock response to discover anticancer compounds that target protein homeostasis. *ACS Chem. Biol.* 7, 340–349.

Satoh, T., Furuta, K., Tomokiyo, K., Namura, S., Nakatsuka, D., Sugie, Y., Ishikawa, Y., Hatanaka, H., Suzuki, M., and Watanabe, Y. 2001. Neurotrophic actions of novel compounds designed from cyclopentenone prostaglandins. *J. Neurochem.* 77, 50–62.

Satoh, T., Okamoto, S.I., Cui, J., Watanabe, Y., Furuta, K., Suzuki, M., Tohyama, K., and Lipton, S.A. 2006. Activation of the Keap1/Nrf2 pathway for neuroprotection by electrophilic [correction of electrophillic] phase II inducers. *Proc. Natl. Acad. Sci. USA* 103, 768–773.

Schopfer, F.J., Cipollina, C., and Freeman, B.A. 2011. Formation and signaling actions of electrophilic lipids. *Chem. Rev.* 111, 5997–6021.

Schopfer, F.J., Cole, M.P., Groeger, A.L., Chen, C.S., Khoo, N.K., Woodcock, S.R., Golin-Bisello, F. et al. 2010. Covalent peroxisome proliferator-activated receptor gamma adduction by nitro-fatty acids: Selective ligand activity and anti-diabetic signaling actions. *J. Biol. Chem.* 285, 12321–12333.

Springstead, J.R., Gugiu, B.G., Lee, S., Cha, S., Watson, A.D., and Berliner, J.A. 2012. Evidence for the importance of OxPAPC interaction with cysteines in regulating endothelial cell function. *J. Lipid Res.* 53, 1304–1315.

Subbanagounder, G., Wong, J.W., Lee, H., Faull, K.F., Miller, E., Witztum, J.L., and Berliner, J.A. 2002. Epoxyisoprostane and epoxycyclopentenone phospholipids regulate monocyte chemotactic protein-1 and interleukin-8 synthesis. Formation of these oxidized phospholipids in response to interleukin-1beta. *J. Biol. Chem.* 277, 7271–7281.

Sun, Z., Zhang, S., Chan, J.Y., and Zhang, D.D. 2007. Keap1 controls postinduction repression of the Nrf2-mediated antioxidant response by escorting nuclear export of Nrf2. *Mol. Cell Biol.* 27, 6334–6349.

Taguchi, K., Fujikawa, N., Komatsu, M., Ishii, T., Unno, M., Akaike, T., Motohashi, H., and Yamamoto, M. 2012. Keap1 degradation by autophagy for the maintenance of redox homeostasis. *Proc. Natl. Acad. Sci. USA* 109, 13561–13566.

Tong, K.I., Katoh, Y., Kusunoki, H., Itoh, K., Tanaka, T., and Yamamoto, M. 2006b. Keap1 recruits Neh2 through binding to ETGE and DLG motifs: Characterization of the two-site molecular recognition model. *Mol. Cell Biol.* 26, 2887–2900.

Tong, K.I., Kobayashi, A., Katsuoka, F., and Yamamoto, M. 2006a. Two-site substrate recognition model for the Keap1–Nrf2 system: A hinge and latch mechanism. *Biol. Chem.* 387, 1311–1320.

Tong, K.I., Padmanabhan, B., Kobayashi, A., Shang, C., Hirotsu, Y., Yokoyama, S., and Yamamoto, M. 2007. Different electrostatic potentials define ETGE and DLG motifs as hinge and latch in oxidative stress response. *Mol. Cell Biol.* 27, 7511–7521.

Tsikas, D., Zoerner, A.A., Mitschke, A., and Gutzki, F.M. 2009. Nitro-fatty acids occur in human plasma in the picomolar range: A targeted nitro-lipidomics GC-MS/MS study. *Lipids* 44, 855–865.

Tsimikas, S., Palinski, W., and Witztum, J.L. 2001. Circulating autoantibodies to oxidized LDL correlate with arterial accumulation and depletion of oxidized LDL in LDL receptor-deficient mice. *Arterioscler. Thromb. Vasc. Biol.* 21, 95–100.

Venugopal, R., and Jaiswal, A.K. 1996. Nrf1 and Nrf2 positively and c-Fos and Fra1 negatively regulate the human antioxidant response element-mediated expression of NAD(P) H:quinone oxidoreductase1 gene. *Proc. Natl. Acad. Sci. USA* 93, 14960–14965.

Venugopal, R., and Jaiswal, A.K. 1998. Nrf2 and Nrf1 in association with Jun proteins regulate antioxidant response element-mediated expression and coordinated induction of genes encoding detoxifying enzymes. *Oncogene* 17, 3145–3156.

Vila, A., Tallman, K.A., Jacobs, A.T., Liebler, D.C., Porter, N.A., and Marnett, L.J. 2008. Identification of protein targets of 4-hydroxynonenal using click chemistry for ex vivo biotinylation of azido and alkynyl derivatives. *Chem. Res. Toxicol.* 21, 432–444.

Wakabayashi, N., Dinkova-Kostova, A.T., Holtzclaw, W.D., Kang, M.I., Kobayashi, A., Yamamoto, M., Kensler, T.W., and Talalay, P. 2004. Protection against electrophile and oxidant stress by induction of the phase 2 response: Fate of cysteines of the Keap1 sensor modified by inducers. *Proc. Natl. Acad. Sci. USA* 101, 2040–2045.

Walton, K.A., Cole, A.L., Yeh, M., Subbanagounder, G., Krutzik, S.R., Modlin, R.L., Lucas, R.M. et al. 2003. Specific phospholipid oxidation products inhibit ligand activation of toll-like receptors 4 and 2. *Arterioscler. Thromb. Vasc. Biol.* 23, 1197–1203.

Wang, X.J., Sun, Z., Chen, W., Li, Y., Villeneuve, N.F., and Zhang, D.D. 2008. Activation of Nrf2 by arsenite and monomethylarsonous acid is independent of Keap1-C151: Enhanced Keap1–Cul3 interaction. *Toxicol. Appl. Pharmacol.* 230, 383–389.

Wang, Y., Gibney, P.A., West, J.D., and Morano, K.A. 2012. The yeast Hsp70 Ssa1 is a sensor for activation of the heat shock response by thiol-reactive compounds. *Mol. Biol. Cell* 23, 3290–3298.

Wasserman, W.W., and Fahl, W.E. 1997. Functional antioxidant responsive elements. *Proc. Natl. Acad. Sci. USA* 94, 5361–5366.

Watai, Y., Kobayashi, A., Nagase, H., Mizukami, M., McEvoy, J., Singer, J.D., Itoh, K., and Yamamoto, M. 2007. Subcellular localization and cytoplasmic complex status of endogenous Keap1. *Genes Cells* 12, 1163–1178.

Watson, A.D., Leitinger, N., Navab, M., Faull, K.F., Horkko, S., Witztum, J.L., Palinski, W. et al. 1997. Structural identification by mass spectrometry of oxidized phospholipids in minimally oxidized low density lipoprotein that induce monocyte/endothelial interactions and evidence for their presence in vivo. *J. Biol. Chem.* 272, 13597–13607.

West, J.D., and Marnett, L.J. 2006. Endogenous reactive intermediates as modulators of cell signaling and cell death. *Chem. Res. Toxicol.* 19, 173–194.

Wright, M.M., Schopfer, F.J., Baker, P.R., Vidyasagar, V., Powell, P., Chumley, P., Iles, K.E., Freeman, B.A., and Agarwal, A. 2006. Fatty acid transduction of nitric oxide signaling: Nitrolinoleic acid potently activates endothelial heme oxygenase 1 expression. *Proc. Natl. Acad. Sci. USA* 103, 4299–4304.

Yamamoto, T., Suzuki, T., Kobayashi, A., Wakabayashi, J., Maher, J., Motohashi, H., and Yamamoto, M. 2008. Physiological significance of reactive cysteine residues of Keap1 in determining Nrf2 activity. *Mol. Cell Biol.* 28, 2758–2770.

Yang, H., Ramani, K., Xia, M., Ko, K.S., Li, T.W.H., Oh, P., Li, J., and Lu, S.C. 2009. Dysregulation of glutathione synthesis during cholestasis in mice: Molecular mechanisms and therapeutic implications. *Hepatology (Baltimore, MD)* 49, 1982–1991.

Zhang, D.D., and Hannink, M. 2003. Distinct cysteine residues in Keap1 are required for Keap1-dependent ubiquitination of Nrf2 and for stabilization of Nrf2 by chemopreventive agents and oxidative stress. *Mol. Cell Biol.* 23, 8137–8151.

Zhang, D.D., Lo, S.C., Cross, J.V., Templeton, D.J., and Hannink, M. 2004. Keap1 is a redox-regulated substrate adaptor protein for a Cul3-dependent ubiquitin ligase complex. *Mol. Cell Biol.* 24, 10941–10953.

Zhang, D.D., Lo, S.C., Sun, Z., Habib, G.M., Lieberman, M.W., and Hannink, M. 2005. Ubiquitination of Keap1, a BTB-Kelch substrate adaptor protein for Cul3, targets Keap1 for degradation by a proteasome-independent pathway. *J. Biol. Chem.* 280, 30091–30099.

Zhang, J., Hosoya, T., Maruyama, A., Nishikawa, K., Maher, J.M., Ohta, T., Motohashi, H. et al. 2007. Nrf2 Neh5 domain is differentially utilized in the transactivation of cytoprotective genes. *Biochem. J.* 404, 459–466.

Zhang, Y., Sano, M., Shinmura, K., Tamaki, K., Katsumata, Y., Matsuhashi, T., Morizane, S. et al. 2010. 4-Hydroxy-2-nonenal protects against cardiac ischemia-reperfusion injury via the Nrf2-dependent pathway. *J. Mol. Cell. Cardiol.* 49, 576–586.

Section III

Oxidized Lipids in
Pathology and Disease

14 Role of Oxidized Phospholipids in Cardiovascular Disease

Gregor Leibundgut, Adam Taleb,
and Sotirios Tsimikas

CONTENTS

INTRODUCTION

Atherosclerosis is a chronic inflammatory disease where both innate and adaptive immune responses are intimately involved in its pathogenesis. Oxidation-specific epitopes (OSE) represent danger-associated molecular patterns (DAMPs) that promote inflammation and cell death. Oxidized phospholipids (OxPL) are well-studied OSE generated during oxidation of low-density lipoprotein (OxLDL) and of cells undergoing apoptosis. OxPL are highly immunogenic, pro-inflammatory and are present in atherosclerotic lesions of animals and humans, particularly in pathologically defined vulnerable and disrupted plaques. OxPL are important contributors to early and late events in atherogenesis by activating pro-inflammatory genes, leading to inflammatory cascades in the vessel wall. The interaction of lipid accumulation, generation of OxPL, activation of inflammatory processes, recruitment of inflammatory cells, endothelial dysfunction, platelet activation, and thrombosis ultimately leads to plaque progression and clinical events. Innate and adaptive immune mechanisms play a central role throughout these events, resulting in atherosclerotic lesions having many features of a chronic inflammatory disease. Over the last decade, it has been demonstrated by a variety of studies that OxPL are carried by lipoprotein (a)

[Lp(a)], and that this component of Lp(a) may be responsible for many of its clinical sequelae. Elevated plasma levels of OxPL on apolipoprotein B-100 (OxPL/apoB), which primarily reflect the biological activity of Lp(a), are positively associated with endothelial dysfunction, angiographically defined coronary artery disease, progression of carotid and femoral atherosclerosis, predict new cardiovascular events in healthy subjects and reclassify one third of patients to higher or lower clinical risk categories. OxPL are also present on plasminogen, which, unexpectedly, are associated with enhanced fibrinolysis and potentially reduced atherothrombosis risk. The translation of OxPL research from cell culture and animal studies to the clinical frontier suggests that OxPL are viable clinical biomarkers, and may be targets in molecular imaging applications and for therapeutic approaches.

OxPL AND INNATE AND ADAPTIVE IMMUNITY

It is well accepted that innate and adaptive immune mechanisms modulate atherosclerosis [1,2] and immune responses to OSEs, such as OxPLs, play a central role in the development of atherosclerosis. [3] Following lipid accumulation in the vessel wall, atherosclerosis develops when the net balance between subsets of immune cells and their products, such as proinflammatory and anti-inflammatory subsets of macrophages, B-1 cells and B-2 cells and their respective IgM and IgG antibodies, T effector cells and Tregs and co-stimulatory and co-inhibitory molecules tilts toward a pro-inflammatory phenotype [4].

OSEs are a major class of atherosclerosis relevant antigens and define oxidative modifications of the breakdown of polyunsaturated fatty acids that can modify LDL particles, apoptotic cells, and other proteins in the vessel wall. OSEs in both the lipid phase and covalently bound to apolipoprotein B-100 (apoB), are present in animal and human atherosclerotic lesions and are a major cause of inflammation in atherosclerosis. OSEs share molecular or immunological identity with pathogen-associated molecular patterns (PAMPs) on microbial pathogens, as well as DAMPs on apoptotic cells and OxLDL. OSEs are recognized by a common set of innate pattern recognition receptors (PRRs) which have been evolutionary selected to protect against pro-inflammatory properties of OSEs and microbial pathogens (Figure 14.1) [3].

The innate immune system is responsible for the first line of defense against PAMPs and DAMPs where it generates inflammatory responses. Innate immune function is comprised within PRRs present as SRs on dendritic cells and monocyte/macrophages, toll-like receptors, natural antibodies (NAbs) present in the germ line that are preserved and accentuated by natural selection, and by effector proteins such as C-reactive protein (CRP) and complement factor H (CFH). OSEs accumulate in the vessel wall in response to a variety of risk factors, particularly hypercholesterolemia, and become potent pro-inflammatory, disease-specific antigens. OSEs are a major target of innate IgM NAbs in both mice and humans, and 15–30% of all such natural antibodies bind OSEs. [5] Interestingly, a high prevalence of these NAbs are IgM antibodies directed to malondialdehyde–protein adducts, suggesting an important role in homeostasis sustained by an evolutionary pressure of organisms having high levels of such antibodies. Innate IgMs, such as the natural antibody E06, which binds to the PC moiety of OxPL present on OxLDL or apoptotic cells, inhibits the

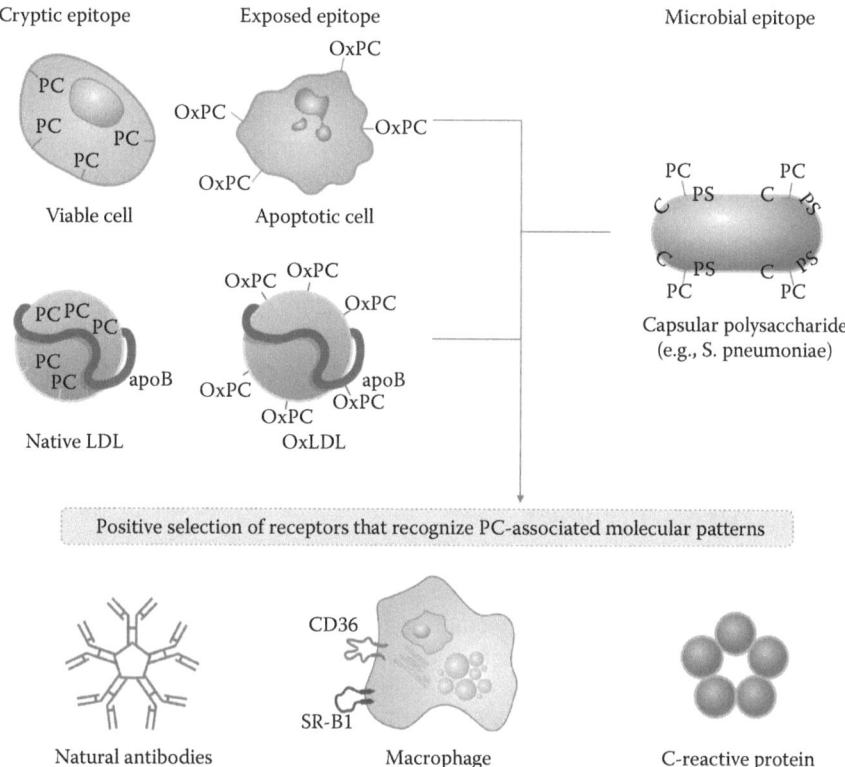

FIGURE 14.1 Pattern recognition of danger-associated molecular patterns (DAMPs) and microbial pathogen-associated molecular patterns (PAMPs). Using the example of the phosphocholine (PC) epitope, this diagram illustrates the hypothesis of the emergence and positive selection of multiple pattern recognition receptors (PRRs) that recognize common epitopes, shared by modified self and microbial pathogens. Oxidation of plasma membrane phospholipids in apoptotic cells alters the conformation of the PC head group, yielding an exposed epitope, accessible to recognition by macrophage scavenger receptors, natural antibodies (Nabs), and C-reactive protein (CRP). These PRRs were selected to clear apoptotic cells from developing or regenerating tissues. Recognition by the same receptors of the PC epitope of capsular polysaccharide in Gram-positive bacteria (e.g., *S. pneumoniae*), which is not part of a phospholipid, strengthened positive selection of these PRRs and probably helped to select additional strong proinflammatory components to PRR-dependent responses. Oxidized lipoproteins, prevalent in humans as a result of dyslipidemia and impact of environmental factors and in experimental animals, carry oxidized phospholipids (OxPLs) with the PC epitope exposed in an analogous manner to that of apoptotic cells, which leads to recognition by PRRs and initiation of innate immune responses. The balance between proinflammatory responses of cellular PRRs and atheroprotective roles of NAbs plays an important role in the development of atherosclerosis. (Reprinted with permission from Miller YI et al. *Circulation Research* 2011;108:235–248.)

binding of OxLDL to macrophage SRs, and are atheroprotective. In support of this concept, epidemiological studies suggest high titers of IgM autoantibodies to OSE are associated with reduced risk of cardiovascular disease (CVD) [6,7].

Activation of PRRs results in release of pro-inflammatory cytokines that leads to removal of antigens but also further activates the immune system. Recognition of DAMPs by the innate immune system also activates the adaptive immune system, comprised of subsets of T cells and B2 cells that mediate the recruitment of lymphocytes to provide more specific responses to such antigens. Although these are initially protective mechanisms, if chronic antigenic stimulation occurs or regulation of such responses tips in favor of pro-inflammatory responses, a state of chronic inflammation will ensue. If this is superimposed on uncontrolled clinical risk factors of atherogenesis, then a maladaptive inflammatory response will occur leading to development and progression of atherosclerotic lesions and clinical sequelae.

EFFECTS OF OxPL ON CELLS ASSOCIATED WITH ATHEROSCLEROSIS

ENDOTHELIAL CELLS

In the initiation of early atherosclerotic lesions, OxPLs such as 1-palmitoyl-2-(5-oxovaleroyl)-sn-glycero-3-phosphorylcholine (POVPC) and 1-palmitoyl-2-glutaroyl-sn-glycero-3-phosphorylcholine (PGPC) present in mmLDL, stimulate endothelial cells (ECs) to secrete chemoattractants MCP-1 and IL-8 [8]. Macrophage migration then ensues through upregulation of connecting segment (CS-1) fibronectin by POVPC and vascular cell adhesion molecule 1 (VCAM-1) and E-selectin expression by PGPC [9] 1-palmitoyl-2-epoxyisoprostaneE2-sn-glycero-3-phosphorylcholine (PEIPC) binds to the prostaglandin E2 receptor (EP2) [10] on ECs and induces additional chemokines IL-6, CCL3, CCL4, and VEGF [11,12]. OxPLs also stimulate ECs to generate reactive oxygen species (ROS) through binding to VEGFR2 and recruitment of Rac1 to the NOx4 complex, leading to enhanced oxidative stress and atherosclerosis progression.

MACROPHAGES

Apart from lipid accumulation and foam cell transformation, macrophages play a major role in the development of atherosclerotic lesions [13], and a variety of distinct receptors are involved in immunogenic and proinflammatory processes [14]. Comprehensive reviews are available on all-known macrophage receptors [15,16], corresponding immune recognition [17], and the impact of macrophage receptor activation on atherosclerosis [18,19]. Other important functions of macrophages include clearance of apoptotic cells [20]. Here, we focus on mechanisms including macrophage receptors that recognize OxPLs [21]. Among the eight known classes of scavenger receptors (SRs), SR-A1 and CD36 play a dominant role in the uptake of OxLDL, although their relative contributions in vivo remain uncertain [22]. Intracellular trafficking of SRs is mediated through the PPAR family of nuclear receptors (Figure 14.2). Of the class A scavenger receptors that are largely expressed

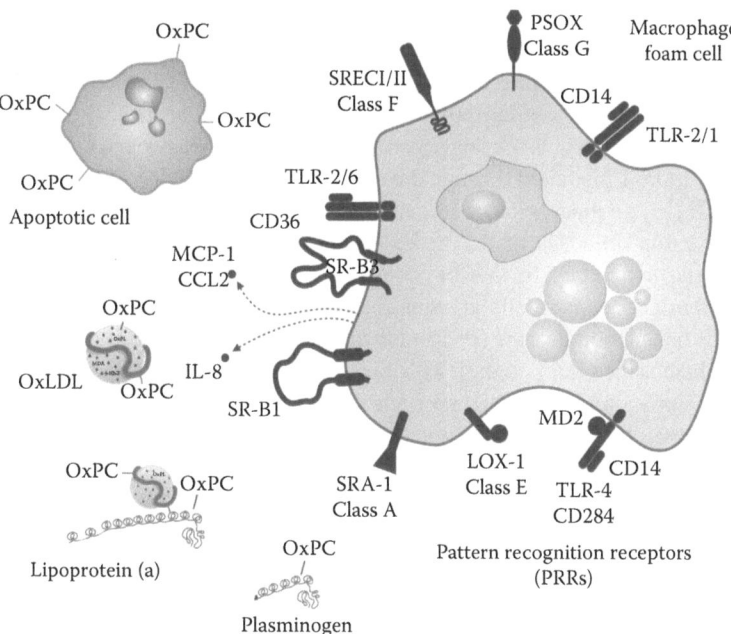

FIGURE 14.2 Sources of OxPL and receptors on macrophages. OxPC on the surface of apoptotic cells and from the lipid- and protein-phase of oxidized low-density lipoprotein (OxLDL) are recognized by macrophages via membrane-bound pattern recognition receptors (PRRs).

on macrophages, SR-A1 has the ability to bind OxPL, is upregulated by OxLDL[23], macrophage colony-stimulating factor (M-CSF) [24], and downregulated by tumor necrosis factor α (TNF-α) [25], N-acetylcysteine[26], interferon-γ (IFN-γ) [27], and transforming growth factor-β1 (TGF-β1) [28]. SR-A1 recognizes OxPL covalently bound to lysine residues of apoB [29] of OxLDL and on apolipoprotein (a) [apo(a)] [30–32].

Class B scavenger receptors on macrophages that recognize OxPL include SR-B1 (CLA-1) [33] and CD36 (SR-B3) [34]. PPARα, PPARγ [35], and testosterone induce SR-B1 expression but OxLDL [36], TNF-α, interleukin-1 (IL-1), and lipopolysaccharides (LPS) reduce SR-B1 expression. CD36 acts as a major internalization receptor for OxLDL [37], is upregulated by OxLDL [23,38], interleukin-4 (IL-4), and M-CSF, whereas LPS, IFN-γ [39], TGF-β1, and TGF-β2 decrease CD36 cell surface expression [40]. SR-B1, is postulated to be atheroprotective and CD36 to be pro-atherogenic and it was shown to upregulate other SRs and promote foam cell formation [41]. CD68 (Macrosialin) of class D SRs on macrophages binds OxLDL [42] and is upregulated by OxLDL [23] M-CSF, GM-CSF, and PMA and downregulated by IFN-γ. Unlike CD36, CD68 does not play a significant role in the internalization of OxLDL.

The lectin-like oxidized LDL receptor-1 (LOX-1) is an endothelial-specific SR of class E that mediates cholesterol uptake [43] and foam cell formation [44]. It can also be detected on macrophages and its expression is induced by OxLDL [45], TNF-α, endothelin-1 [46], and angiotensin II [47]. LOX-1 binds to a variety of ligands including OxLDL [48] and apoptotic cells [49]. SNPs within the LOX-1 gene (*OLR1*) have been linked to cardiovascular [50] and peripheral artery disease [51] in large patient populations. Recent evidence suggests the regulation of SR-A1 and CD36 in macrophages by LOX-1 mediated through the nuclear receptor PPAR-γ [52]. The class F SR SREC-I/II [53] on macrophages is increased in response to LPS [54] and downregulated under the influence of IL-1α, IL-1β, and TNF-α [55]. SREC-I binds OxLDL with similar specificity to that of SR-A1, however, does not internalize modified LDL, and might act as a cellular signaling conductor [56]. SR-PSOX is a class G SR that binds phosphatidylserine (PS) and oxidized lipoproteins and is identical with the soluble chemokine CXCL16 [57]. This SR is increased by stimulation of TNF-α and IFN-γ [58], binds and internalizes PS and OxLDL. Elevated levels of SR-PSOX have been demonstrated in atherosclerotic lesions [59] and may be used as a biomarker for acute coronary syndrome (ACS) [60,61].

Oxidized phosphatidylserine (OxPS) and phosphatidylcholine (OxPC) derivatives serve as ligands for macrophage uptake of apoptotic cells [5,62] OxLDL, OxPL, and lipoprotein (a) [Lp(a)], the preferential carrier of OxPLs in human plasma, [31], all trigger TLR2/6-dependent apoptosis in macrophages [63,64]. Interestingly, CD36 may serve as a function analogous to CD14 in the activation of the CD36-TLR2/6 complex [65] with apoptotic cells accumulating in CD36 null mice. [66].

Recently, a water-soluble POVPC–peptide adduct has been designed [67], which provides a stable substrate for future studies and may be used to target imaging agents or drugs to macrophages and enables macrophage-specific diagnostic and therapeutic possibilities.

PATHOPHYSIOLOGICAL ROLE OF OXPL ON LIPOPROTEIN (A)

Lp(a) was first described by Kare Berg in 1963 [68] and later found to be associated with CAD[69]. Lp(a) comprises of apo(a) covalently bound to apoB via a disulfide bond on kringle IV type 9 (KIV-9) to a site near the low-density lipoprotein (LDL) receptor binding site of apoB [70]. Apo(a) is encoded by the *LPA* gene present on the long arm of chromosome 6 and is highly homologous to the plasminogen (*PLG*) gene[71]. Apo(a) has evolved from duplication of the plasminogen gene and is only found in humans and apes, and in an unrelated version comprising kringle III in the European hedgehog [72–74]. Apo(a) expanded and diversified into 10 different KIV types (types 1–10) with multiple and variable copies of KIV-2, K-V but with an inactive protease domain[75]. Lp(a) plasma levels are largely (>90%) genetically determined with small isoforms being associated with higher plasma concentrations. Epidemiological, genetic association, and Mendelian randomization studies strongly suggest that Lp(a) is a causal genetic risk factor for CVD [76–81].

A physiological role for Lp(a) remains to be identified. Lp(a) may be atherogenic through its LDL moiety, but also through pro-inflammatory and pro-atherogenic effects of apo(a). Interactions of Lp(a) with endothelial cells and atherosclerotic lesions are mediated through potent lysine binding sites on the kringle structures of its apolipoprotein, particularly KIV-10 [82]. Apo(a) binds to a variety of cell surfaces, matrix proteins, and mediates cholesterol transport that play a role in tissue healing, innate immunity, and atherothrombosis [83]. Apo(a) is also pathologically associated with increased endothelial cell permeability, expression of adhesion molecules, promotion of smooth muscle cell (SMC) proliferation, enhancement of monocyte recruitment and macrophage foam cell formation, upregulation of release of pro-inflammatory cytokines, and antifibrinolytic effects *in vitro* [84]. Finally, Lp(a), as opposed to other lipoproteins, has been identified as a preferential carrier of pro-inflammatory and proatherogenic OxPLs [31,82] that, in clinical studies, reflect a key biological factor in the pathogenesis of Lp(a) atherogenicity and prediction of clinical events[85]. Lp(a) contains atherogenic OxPL both in its LDL-like moiety (lipid phase) as well as covalently bound to apo(a) [30,31,82]. Determinants of binding of OxPL on apo(a) and Lp(a) include an intact lysine binding site (LBS) on kringle IV-10 which is unique to human apo(a) and plasminogen, but not present in Lp(a) of other species, all of which have modifications of the lysine binding pocket of KIV-10 [82]. OxPL on Lp(a) have been shown to induce IL-8 expression [30] and, importantly, both OxPL and Lp(a) mediate macrophage apoptosis, a key component of plaque vulnerability [86], in endoplasmic reticulum-stressed macrophages by signaling through the CD36/Toll-like receptor 2 pathway [64]. Additionally, MCP-1 binds to Lp(a) in OxLDL in human plasma, and retains its ability to recruit monocytes, consistent with enhancing trafficking in the vessel wall [87]. Finally, apo(a) and OxPL epitopes are significantly enriched in pathologically defined human vulnerable plaques, consistent with a clinically relevant pathological role in the expression cardiovascular disease and events [88].

Lp(a) is being increasingly recognized as a causal cardiovascular risk factor. The European Atherosclerosis Society Consensus Panel recommended the screening of patients at intermediate or high risk of cardiovascular disease or CAD for Lp(a) levels, with the desirable level being <50 mg/dL [80]. The National Lipid Association also provided a consensus document on screening moderate-to-high risk patients for Lp(a) levels, in support to the European Atherosclerosis Society. Recently, a specific antisense oligonucleotide directed to KIV-II repeats was described that lowers apo(a) mRNA and apo(a) plasma levels in apo(a) transgenic mice [89–91]. Lp(a) has become a target of therapy for reducing cardiovascular risk and several approaches are in development or in phase I studies, including antisense oligonucleotides [90,89]. In a fist-in-man Phase I study, Viney et al. showed that subcutaneous administration of 50–300 mg of ISIS-APO(a)Rx was generally well tolerated with an acceptable safety profile with dose-dependent reductions in Lp(a) levels [92]. Mean percent change from baseline values in the 100, 200, and 300 mg multiple-dose groups were 40%, 59%, and 78%, respectively, 1 week after the last dose. In addition, reductions in Lp(a) correlated with reductions in levels of OxPL/apoB. Phase 2 trials in patients with high Lp(a) levels are underway. This robust Lp(a) lowering will allow testing of clinical hypotheses that lowering of Lp(a) levels will reduce risk of CVD as well as calcific aortic stenosis, based on recent GWAS data showing LPA gene and specifically SNP rs10455872 was the only genetic predictor of aortic valve calcification and aortic stenosis[93,88].

CLINICAL EVIDENCE FOR THE ROLE OF OxPL
IN CARDIOVASCULAR DISEASE

We have developed an ELISA to measure OxPL on all lipoproteins, with the most validated assay being measuring OxPL on apolipoprotein B-100 (OxPL/apoB). The OxPL/apoB assay has been validated as a predictor of cardiovascular disease and cardiovascular events in a variety of clinical studies (Figure 14.3) [94]. OxPL/apoB levels have been reported in 32 clinical studies, totaling 12,099 patients and 17,586 assays, in a variety of settings with CVD and are summarized in Table 14.1. A few of the key studies are described below.

OxPL/apoB were found to increase following ACS and in iatrogenic "ACS equivalents," like uncomplicated percutaneous coronary interventions (PCI) where there is disruption of the atherosclerotic plaque with the balloon/stent catheter. In a prospective ACS study [97], it was demonstrated that OxPL/apoB levels increase immediately after an acute myocardial infarction (AMI) and return to baseline over the next 7 months. Similarly, uncomplicated PCI results in an acute increase (over 1–2 h) in OxPL/apoB levels immediately following PCI, supporting the hypothesis of OxPL release from the atherosclerotic plaques upon rupture. Post procedure, these OxPL are approximately equally present on non-Lp(a)-ApoB and on Lp(a), but by 6 h primarily on Lp(a), consistent with a strong affinity and transfer of OxPL by Lp(a), as suggested by *in vitro* transfer studies [31]. This has been further demonstrated in studies of patients undergoing percutaneous interventions in coronary saphenous vein graft, carotid peripheral and renal arteries, showing direct evidence of the presence of OxPL by LC-MS/MS and immunohistochemistry in material captured in distal protection devices [123].

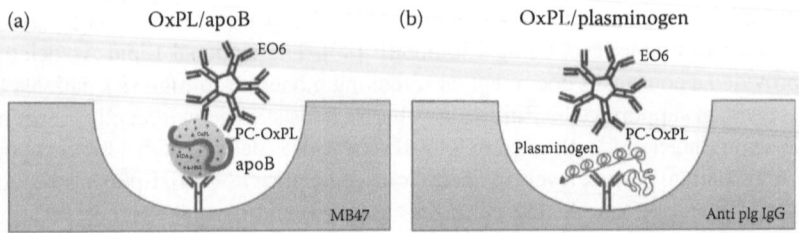

FIGURE 14.3 Enzyme-linked immunosorbent assays for the detection of OxPL on apoB and plasminogen. (a) Microtiter well plates are coated with the murine monoclonal antibody MB47, which captures all apoB containing particles (LDL, Lp(a), and VLDL). Human plasma is added, so apoB particles are bound to MB47. OxPL on apoB-100 are then detected with biotinylated murine monoclonal antibody EO6. The assay is designed in such way that each well captures equal amounts of apoB-100 particles irrespective of the plasma apoB levels. In human plasma, only 10–15% of EO6 immunoreactivity is on OxLDL, 85–90% is on Lp(a). (b) Most of the OxPL content present in human plasma is on plasminogen. To measure OxPL on plasminogen (OxPL/plasminogen) levels, microtiter well plates are incubated with a mouse monoclonal anti-human plasminogen IgG antibody, human plasma added, and OxPL content determined with biotinylated EO6 as the detection antibody. This assay normalizes all wells to the same amount of plasminogen and is therefore independent of plasma plasminogen levels.

TABLE 14.1

Clinical Studies Examining the Role of OxPL/apoB and OxPL/plasminogen in Cardiovascular Disease

Study	Year	Study Name	Patient Population	Patients/Samples	Outcome on OxPL/apoB and Clinical Events
Wu et al. [95]	1999		Borderline hypertension	146/146	Increased in borderline hypertension and may reflect early vascular changes
Penny et al. [96]	2001	UCSD Regression Study	Hypercholesterolemic patients undergoing quantitative angiography before/after lipid-lowering therapy	29/54	Related significantly to the severity of endothelial dysfunction and was the single most powerful independent risk factor
Tsimikas et al. [97]	2003	ACS	Acute coronary syndromes AMI, unstable angina, stable CAD, healthy controls	66/272	Increase after AMI and unstable angina
Tsimikas et al. [98]	2004	Toronto PCI	Patients with stable angina pectoris undergoing PCI	141/1269	Increase immediately after PCI and return to baseline after 6 h
Segev et al. [99]	2005	Toronto PCI			No relationship to restenosis.
Tsimikas et al. [100]	2004	MIRACL	Impact of Atorvastatin in ACS	2341/4682	Increase 9.6% with atorvastatin 80 mg/day.
Fraley et al. [101]	2009	MIRACL			Baseline levels varied according to specific CVD risk factors and were largely independent of inflammatory biomarkers
Silaste et al. [102]	2004		Healthy young women	37/74	Increase 19–27% with low-fat, high vegetable diet
Tsimikas et al. [103]	2005	MAYO	Coronary angiography	504/504	Strong and graded association with presence and extent of CAD
Tsimikas et al. [104]	2014	MAYO IL1	Coronary angiography and events		Angiographically defined CAD and future events mediated by OxPL/apoB and/or Lp(a) are greatest in patients with highly inflammatory IL1(+) genotypes

(Continued)

TABLE 14.1 (*Continued*)
Clinical Studies Examining the Role of OxPL/apoB and OxPL/plasminogen in Cardiovascular Disease

Study	Year	Study Name	Patient Population	Patients/ Samples	Outcome on OxPL/apoB and Clinical Events
Tsimikas et al. [105]	2006	BRUNECK	Random sample of 40–79-year-old males and females) in population	765/1436	Predict presence and progression of carotid and femoral atherosclerosis
Kiechl et al. [106]	2007	BRUNECK			Predict 10-year CVD event rates independently of traditional risk factors, hsCRP and FRS
Tsimikas et al. [7]	2012	BRUNECK			Predict 15-year CVD and stroke risk
Rodenburg et al. [107]	2006		Children with familial hypercholesterolemia and unaffected siblings	256/512	Increase 29% with Step II AHA diet Increase 49% with Pravastatin 40 mg/day
Bossola et al. [108]	2007		End-stage renal failure patients undergoing chronic hemodialysis	52/104	Reduced in end-stage renal failure patients following hemodialysis
Bossola et al. [109]	2011		End-stage renal failure patients undergoing chronic hemodialysis	52/52	In chronic hemodialysis patients, OxPL/apoB, Lp(a), and OxLDL biomarkers are not associated with CVD
Ky et al. [110]	2008	PROXI	Hypercholesterolemic patients were randomized to different type and dose of statin	120/240	Increased 26% with Pravastatin 40 mg/day and 20% with Atorvastatin 80 mg/day
Choi et al. [111]	2008	REVERSAL	Patients with CAD who underwent coronary IVUS and assigned to statin therapy	214/428	Increased 48% with Atorvastatin 80 mg/day and 39% with Pravastatin 40 mg/day. No relationship to IVUS parameters
Tsimikas et al. [112]	2009	Dallas Heart Study	Multiethnic, probability-based sample of the Dallas County population	3481/3481	Vary according to race/ethnicity, are independent of cardiovascular risk factors and are inversely associated with apo(a) isoform size
Tsimikas et al. [113]	2010	EPIC-Norfolk Study	45–79-years-old healthy males and females prospectively followed for 6 years	2160/2160	The highest tertiles are associated with higher risk of CAD events

Reference	Year	Study	Population	N	Findings
Budoff et al. [114]	2009	Aged garlic study	Asymptomatic patients with CAD treated with aged garlic extract plus supplement followed with coronary artery calcium scan (CAC)	60/120	Increase OxPL/apoB with aged garlic extract predicts lack of CAC progression
Ahmadi et al. [115]	2010	Aged garlic study			Increase OxPL/apoB with aged garlic extract correlates with improvement in vascular function
Arai et al. [116]	2010	I4399M SNP	Carriers and non-carriers of I4399M single-nucleotide LPA polymorphism	174/174	Elevated in carriers than in noncarriers, while patients with small apolipoprotein(a) isoforms had the highest OxPL/apoB levels
Faghihnia et al. [117]	2010	CHORI	Healthy subjects consuming a high-fat low-carbohydrate (HFLC) diet and low-fat high-carbohydrate (LFHC) diet	63/126	OxPL/apoB and OxPL/apo(a) are increased by a LFHC diet
Leibundgut et al. [32]	2012		Plasminogen from commercially available preparations, plasma from chimpanzees, gorillas, bonobos, cynomolgus monkeys, wild-type, $apoE-/-$, $LDLR-/-$, and Lp(a)-transgenic mice, healthy humans, and patients with FH, stable CVD, and AMI	58/107	Plasminogen contains covalently bound OxPL that influence fibrinolysis. OxPL/plasminogen represent a second major plasma pool of OxPL and may have pathophysiological implications in AMI and atherothrombosis
Arai et al. [118]	2012		Pre- and post-apheresis blood samples from 18 patients with FH and with low, intermediate, or high Lp(a) levels	18/36	Apheresis significantly reduced levels of OxPL and Lp-PLA$_2$ on apoB and Lp(a) (50–75%), particularly in patients with intermediate and high Lp(a) levels
Fefer et al. [119]	2012		Subjects with CTO, with SCD, and following PCI	41/246	Successful PCI of CTOs results in a slower increase in OxPL/apoB and Lp(a) compared to non-CTO vessels.
Holleboom et al. [120]	2012	LCAT gene mutation	Carriers of 2 mutant LCAT alleles, heterozygotes, and family controls	130/130	OxPL/apoB-containing lipoproteins were increased in heterozygotes (17%; $p < 0.001$) but not in carriers of two defective LCAT alleles
Yoshida et al. [121]	2013	VISION	Patients with hypercholesterolemia randomized to 12-week of pitavastatin or atorvastatin	42/84	Within-group changes from baseline to 12-week revealed significant increases in OxPL/apoB and reductions in small-dense LDL in both groups

(Continued)

TABLE 14.1 (*Continued*)
Clinical Studies Examining the Role of OxPL/apoB and OxPL/plasminogen in Cardiovascular Disease

Study	Year	Study Name	Patient Population	Patients/ Samples	Outcome on OxPL/apoB and Clinical Events
Bertoia et al. [122]	2013	Health Professionals Follow-up Study/ Nurses' Health Study	The study population included 2 parallel nested case–control studies of 143 men within the Health Professionals Follow-up Study (1994–2008) and 144 women within the Nurses' Health Study (1990–2010) with incident confirmed cases of clinically significant PAD, matched 1:3 to control subjects	1096/1096	OxPL/apoB were positively associated with risk of PAD in men and women
Leibundgut et al. [82]	2013		Human, bonobo, chimpanzee, gorilla, rhesus and cynomolgous monkey plasma	29/29	OxPL as detected by E06 immunoreactivity are present on human Lp(a) but not in gorilla, rhesus, and cynomolgous monkey plasma
Ravandi et al. [123]	2014		Patients undergoing coronary vein graft, renal artery, carotid and peripheral interventions with distal protection devices	24/24	OxPLs and oxidized cholesteryl esters are strongly present in plaque debris trapped by distal protection devices
TOTAL: 32				12,099/17,586	

Abbreviations: ACS: acute coronary syndrome; AHA: American Heart Association; CAD: coronary artery disease; CHORI: Children's Hospital Oakland Research Institute; CVD: cardiovascular disease; EPIC: European Prospective Investigation of Cancer; FRS: Framingham Risk Score; hsCRP: high-sensitivity C-reactive protein; IVUS: intravascular ultrasound; MI: myocardial infarction; MIRACL: myocardial ischemia reduction with aggressive cholesterol lowering; OxPL: oxidized phospholipid; PCI: percutaneous coronary intervention; PROXI: Pravastatin and Atorvastatin on markers of oxidative stress in hypercholesterolemic humans; REVERSAL: reversal of atherosclerosis with aggressive lipid lowering; SNP: single-nucleotide polymorphism; UCSD: University of California at San Diego; Lp(a): lipoprotein (a); OxLDL: oxidized low-density lipoprotein; FH: familial hypercholesterolemia; AMI: acute myocardial infarction; Lp-PLA2: lipoprotein-associated phospholipase A2; CTO: chronic total coronary occlusion; SCD: sudden cardiac death; LCAT: lecithin–cholesterol acyltransferase; PAD: peripheral artery disease.

OxPL/apoB levels are associated with the presence and extent of CAD [103]. For example, in 504 patients undergoing clinically indicated coronary angiography, elevated levels of OxPL/apoB strongly correlated with the extent of angiographically documented CAD, which was defined as >50% stenosis and measured as one-, two-, or three-vessel disease. In patients <60 years old, OxPL/apoB remained an independent predictor of CAD, even with Lp(a) in the model, suggesting that the atherogenic effect of OxPL in younger populations goes beyond its binding affinity to Lp(a), probably through other proinflammatory pathways.

Elevated OxPL/apoB levels are also associated with peripheral artery disease (PAD). For example, the Bruneck study, a prospective population-based study of 40–79-year-old men and women, revealed an association with the presence, extent and development of carotid and femoral atherosclerosis [105], which was strongest in patients with small apo(a) isoforms and highest Lp(a) concentration. In a nested case–control study within the Health Professionals Follow-up Study and the Nurses' Health Study strengthened the evidence supporting OxPL/apoB positive association with PAD [97]. The risk of developing PAD was almost double in the highest tertile group compared to the lowest tertile, and each 1-standard deviation increase in OxPL/apoB was followed by a 37% increase in PAD. This study also showed a similar association for Lp(a).

OxPL/apoB and Lp(a) have clinical prognostic utility in predicting future death, AMI, stroke and transient ischemic attack and revascularization (Figure 14.4).

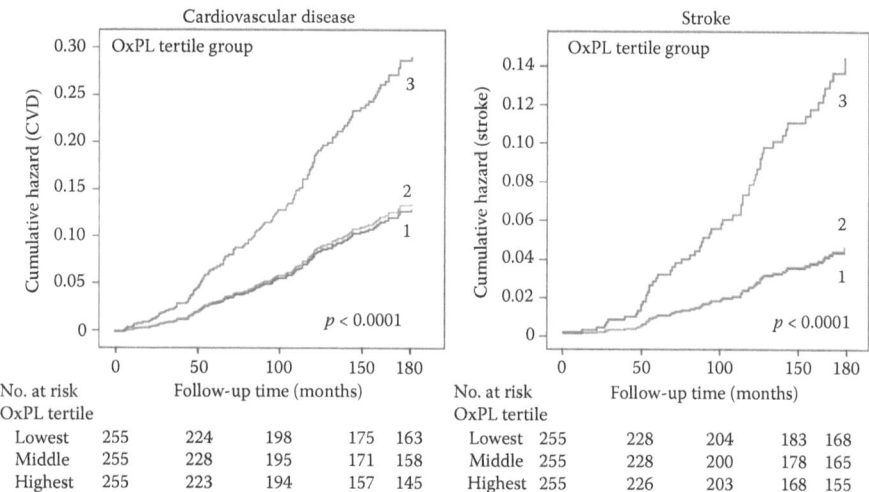

FIGURE 14.4 Clinical risk prediction of OxPL/apoB. Cumulative hazard curves for CVD incidence and stroke incidence by OxPL/apoB tertile groups, 1995–2010. The median OxPL/apoB level (relative light units [RLUs]) for the lowest tertile was 2908 (range 1584–3631); for the middle tertile, it was 4862 (range 3632–8124); and for the highest tertile, it was 18,830 (range 8125–79,541). There were 138 cases of incident cardiovascular disease (CVD) and 60 cases of incident stroke. Y-axis shown in light blue indicates a range from 0 to 0.15. (Reprinted with permission from Tsimikas S et al. *Journal of the American College of Cardiology* 2012;60:2218–2229.)

Subjects from the general community in the highest tertile of OxPL/apoB at base-line had a significantly higher risk of cardiovascular events than those in the lowest tertile, with hazard ratio being 2.4 (95%-CI: 1.3–4.3) over a 15-year pro-spective follow-up period [7]. These findings were confirmed in the EPIC-Norfolk study, which was a prospective, case-controlled study consisting of 45–79-year-old healthy men and women followed up for 6 years [113]. Both studies demonstrated the enhanced predictive value over traditional risk factors and the Framingham Risk Score. Measuring OxPL/apoB levels in each tertile of FRS allowed further risk stratification and added predictive value, with the most important being in the medium risk tertile, potentially providing more accurate assessment of treatment options. In addition, measuring OxPL/apoB levels reclassified approximately one third of patients initially in the intermediate category into either lower or higher risk categories, which would allow clinicians to fine tune clinical decision making [79].

In view of the potential role of OxPL with pro-inflammatory pathways, we recently evaluated the genetic variations in the IL-1 region that are known to be associated with increased levels of inflammatory mediators and their influence on CAD risk mediated by OxPL and Lp(a). IL-1 genotypes, OxPL/apoB and Lp(a) lev-els were measured in 499 patients undergoing coronary angiography. The compos-ite genotype termed IL-1(+) was defined by three single-nucleotide polymorphisms (SNPs) in the IL-1 gene cluster associated with higher levels of pro-inflammatory cytokines. All other IL-1 genotypes were termed IL-1(–). Among IL-1(+) patients, the highest quartile of OxPL/apoB was significantly associated with a higher risk of CAD compared to the lowest quartile (OR 2.84, $p = 0.001$). This effect was accentuated in patients ≤60 years old (OR 7.03, $p < 0.001$). In IL-1(–) patients, OxPL/apoB levels showed no association with CAD. The interaction was signifi-cant for OxPL/apoB (OR 1.99, $p = 0.004$) and Lp(a) (OR 1.96, $p < 0.001$) in IL-1(+) versus IL-1(–) groups for patients ≤60 years old but not for patients >60 years old. In IL-1(+) patients ≤60 years old, after adjusting for established risk factors, high sensitivity C-reactive protein and Lp(a), OxPL/apoB remained an indepen-dent predictor of CAD. IL-1(+) patients above the median OxPL/apoB presented to the cardiac catheterization laboratory a mean of 3.9 years earlier ($p = 0.002$) and had worse 4-year event-free survival (death, MI, stroke, and revascularization) compared to other groups ($p = 0.006$) [104]. This study suggests that IL-1 geno-type status can stratify population risk for CAD and cardiovascular events medi-ated by OxPL. These data suggest a clinically relevant biological link between pro-inflammatory IL-1 genotypes, oxidation of phospholipids, Lp(a), and genetic predisposition to CAD and cardiovascular events (Figure 14.5). These findings have important implications in the CANTOS trial studying the effect of a mono-clonal antibody to interleukin 1β and other studies investigating anti-inflammatory therapies, and suggest that subgroups of patients with IL-1(+) genotypes and ele-vated OxPL/apoB and Lp(a) will be both at higher risk and most likely to benefit, whereas lower risk patients may not benefit [124].

Several therapeutic interventions are shown to modify OxPL/apoB and Lp(a) levels, not always in the assumed and predictable direction. The initial hypothesis was that OxPL/apoB and Lp(a) levels would increase during hypercholesterolemia

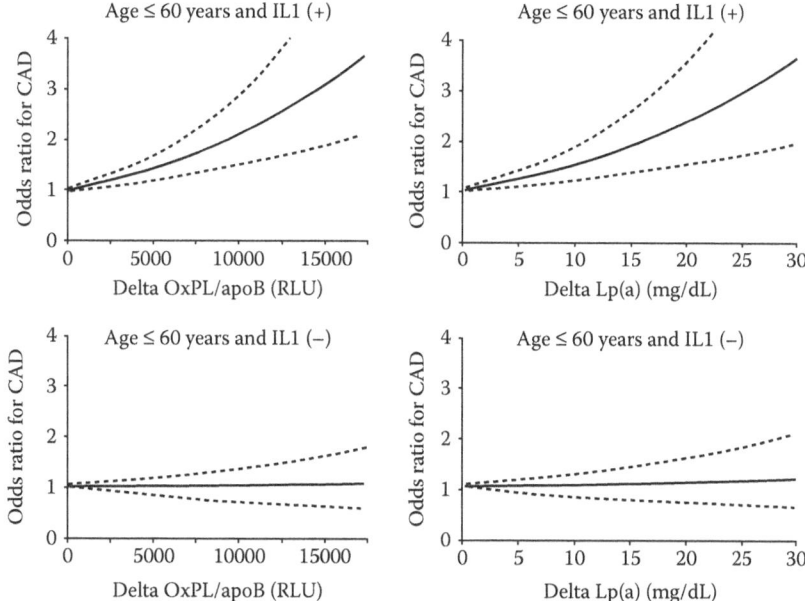

FIGURE 14.5 IL1, OxPL, Lp(a), and CAD. Odds ratios (OR) (solid line) and 95% confidence intervals (dashed lines) for CAD were calculated in a logistic regression model. In this model, risk associated with an incremental increase of each risk factor ranging from 0 (i.e., an odds ratio of 1) to the value equal to the difference between the 75th and 25th percentiles of the risk factors. The analysis was performed on patients ≤60 years of age stratified as IL-1(+) or IL-1(–). (Reprinted with permission from Tsimikas S et al. *Journal of the American College of Cardiology* 2014;63:1724–1734.)

and atherosclerosis progression and would decrease during atherosclerosis regression. Surprisingly, the results were the opposite in both animals [125] and humans [107,110,111]. In animals on regression diets [125], mainly by switching from a high-fat to a low-fat/cholesterol diet, young women on low-fat diets [102], Step II AHA diets [106] and different statins [75,100, 101, 107,110,126,127], an increase in OxPL/apoB and Lp(a) has been noted. Although the mechanisms of this are not understood yet, insights are obtained from immunostaining of atherosclerotic lesions in New Zealand white rabbits (which do not have Lp(a)) and cynomolgus monkey (which have Lp(a) but no OxPL immunoreactivity on Lp(a)) [125], models of atherosclerosis during dietary-induced regression, showing that concomitant with an increase in OxPL/apoB levels in the plasma, there is loss of OxPL in atherosclerotic plaques, consistent with a flux of OxPL from the vessel wall to the circulation [125]. Similar findings were seen in human studies with statin use [110,111] aged garlic supplement [114] and low-fat diet [117]. In the Myocardial Ischemia Reduction with Aggressive Cholesterol Lowering (MIRACL) trial, both OxPL/apoB and Lp(a) levels increased at 16 weeks after initiation of atorvastatin, while the rate of recurrent clinical events decreased [100] Further studies are needed to understand the mechanisms behind this and whether the increase in OxPL/apoB is a predictor of events, or whether it reflects a nonclinically significant change in OxPL levels following what appears to be beneficial therapies.

OxPL ON PLASMINOGEN

Plasminogen plays a key role in the fibrinolytic system and has also been implicated in other important pathophysiological properties, including tissue remodeling, angiogenesis, embryogenesis, tumor metastasis, infections, wound healing, and leukocyte migration [128]. Activation of plasminogen to plasmin that mediates fibrinolysis is regulated by tissue plasminogen activator (t-PA) and plasminogen activator inhibitor 1 (PAI-1). Plasminogen deficiency has been shown to accelerate the development of atherosclerotic lesions in apoE–/– mice [129]. The accumulation of fibrin supports cell adhesion, migration, and proliferation of inflammatory cells and the net effect appears to be independent from its fibrinolytic function. Other mechanisms may contribute to the acceleration of atherosclerosis in plasminogen-deficient mice [130].

Oxidative stress plays an important role in regulating the expression of adhesion molecules. TNF-α induces the production of reactive oxygen species by ECs and smooth muscle cells (SMCs) and the expression of adhesion molecules such as ICAM-1 and VCAM-1 [131]. A recent study demonstrated that human plasminogen directly inhibits the TNF-α-induced expression of ICAM-1 and VCAM-1 [132]. Interestingly, the effect of plasminogen was more potent than that of kringle I–III [133] and angiostatin [134], which is comprised of kringle I–IV. The data suggest that plasminogen might serve as an antioxidant and anti-inflammatory agent.

Plasminogen was recently found to be a key activator of OxLDL-driven macrophage gene and CD36 expression via leukotriene B4 and promote foam cell formation [135]. Plasminogen exerts divergent effects in atherothrombosis. There is evidence for early antiatherogenic effects [129] (e.g., extracellular matrix degradation, activation of MMPs, enhanced fibrinolysis [32]) that may be dominated by proatherogenic activity [136] (e.g., recruitment of inflammatory cells [137], neointima formation [138], or enhanced lipid uptake [135]) at later stages of lesion development [139]. It is to note that degradation of extracellular matrix again contributes to plaque instability at a very late stage of disease progression.

We and others have discovered that plasminogen binds OxPLs covalently [32,140]. Plasminogen and apolipoprotein (a) [apo(a)] are highly homologous. Due to structural homology to apo(a) plasminogen has been found to also bind OxPL covalently and represents a second major plasma pool of OxPL. Our group has demonstrated that Lp(a) and plasminogen are the two main protein carriers of OxPL in plasma and that the pools of OxPL on Lp(a) and plasminogen are separate and distinct. Whereas OxPL/apoB correlate with and vary widely according to Lp(a) levels, OxPL/plasminogen are present in a narrow range and correlate only with plasminogen levels and not with Lp(a) levels. Importantly, the OxPL on plasminogen facilitate fibrinolysis *in vitro*, which would be a putative protective effect on atherothrombosis. For example, enzymatic removal of OxPL from plasminogen resulted in a longer lysis time for fibrin clots *in-vitro* indicating that OxPL is a key functional parameter for fibrinolysis and may also be involved in the pathogenesis or destabilization of atherosclerotic lesions (Figure 14.6). In clinical studies, OxPL/

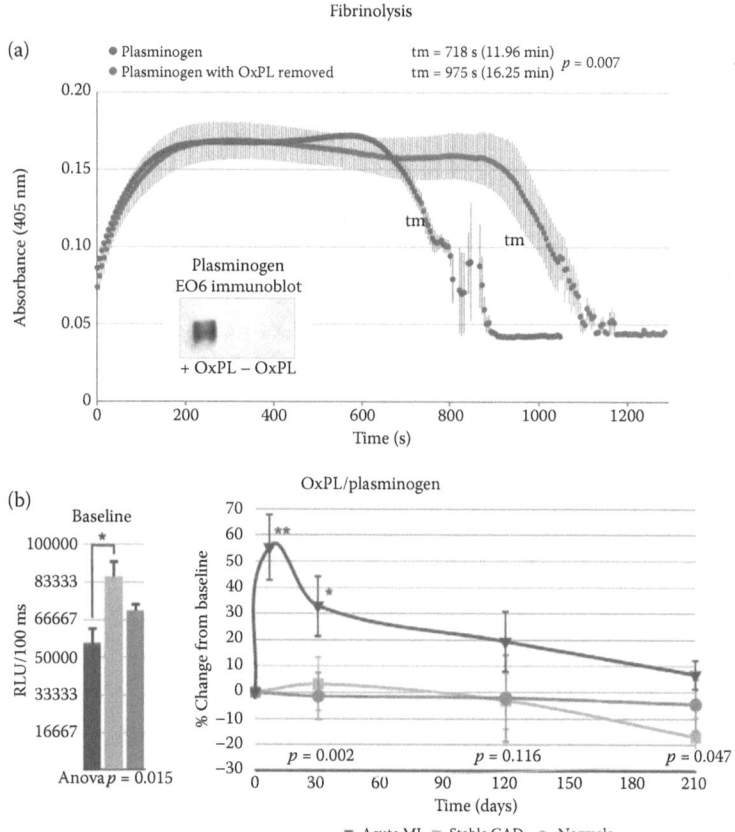

FIGURE 14.6 OxPL on plasminogen affect fibrinolysis and are elevated after acute MI. (a) *In vitro* fibrin clot lysis of native plasminogen containing OxPL and plasminogen with OxPL removed enzymatically. *In vitro* clot lysis assay assessing the ability of plasminogen to degrade fibrin clots. Native plasminogen containing OxPL and plasminogen with OxPL enzymatically removed (inset) with phospholipase A_2 were used. In this system, thrombin-induced clot formation occurs within the first 2 min and is marked by an initial rapid increase in turbidity, as measured by absorbance at 405 nm. Subsequent clot lysis is indicated by a rapid return of the turbidity signal to baseline levels. The parameter tm (transition midpoint) is taken as the standard measure of lysis time and is defined as the time point on the lysis curve that is halfway between the minimum and maximum excursions. The curves represent the mean ± SEM of three separate experiments with measurement of absorbance at 405 nm every 5 s. (b) Change in OxPL/plasminogen in healthy subjects, patients with stable CAD and AMI. Shown are changes in OxPL/plasminogen over a 7-month period in patients after (AMI) ($n = 8$), in patients with stable CAD ($n = 17$), and in healthy subjects ($n = 18$). The p values at the bottom of the figures represent the discharge (average of 4 days for the AMI group) and 30-, 120-, and 210-day differences between groups at each time point. *$p < 0.05$ and **$p < 0.01$ represent Bonferroni posttest for changes within groups over time. (Reprinted with permission from Leibundgut G et al. *Journal of the American College of Cardiology* 2012;59:1426–1437.)

plasminogen levels increased acutely over the first month in patients following AMI, implying that OxPL/plasminogen may be a novel biomarker of ACS [32,141]. As opposed to OxPLs on Lp(a), which predict increased cardiovascular risk, the predictive risk of OxPL/plasminogen remains to be proven in future studies.

CONCLUSION

OxPL are immunogenic, pro-inflammatory, and pro-atherogenic. *In vitro* studies have provided significant insights into how OxPL may mediate atherothrombosis. The serendipitous finding that OxPL are primarily present on Lp(a) and plasminogen in plasma has opened novel avenues in trying to understand the pathophysiology of Lp(a) in mediating clinical cardiovascular disease. OxPL on Lp(a) are clearly pro-atherogenic and can be used as robust biomarkers providing unique clinical information. With the strong data that Lp(a) may be a causal risk factor for cardiovascular disease, its relationship to OxPL, makes both potential targets of therapy. The clinical implications of OxPL on plasminogen have not been evaluated yet, but in preliminary findings suggest enhancement of fibrinolysis. Future clinical trials in this arena are waited to test these interesting hypotheses in humans.

ABBREVIATIONS

ACS	acute coronary syndrome
AMI	acute myocardial infarction
apo(a)	apolipoprotein (a)
apoB	apolipoprotein B-100
CFH	complement factor H
CS-1	connecting segment
CVD	cardiovascular disease
DAMPs	danger-associated molecular patterns
ECs	endothelial cells
EP2	prostaglandin E2 receptor
IFN-γ	interferon-γ
IL-1	interleukin-1
IL-4	interleukin-4
LDL	low density lipoprotein
Lp(a)	lipoprotein (a)
M-CSF	macrophage colony-stimulating factor
mmLDL	minimally modified LDL
Nabs	natural antibodies
OSE	oxidation-specific epitopes
OxLDL	oxidized low-density lipoprotein
OxPC	oxidized phosphatidylcholine
OxPL	oxidized phospholipids
OxPL/apoB	OxPL on apolipoprotein B-100
OxPS	oxidized phosphatidylserine

PAD	peripheral artery disease
PAI-1	plasminogen activator inhibitor 1
PAMPs	pathogen associated molecular patterns
PCI	percutaneous coronary interventions
PEIPC	1-palmitoyl-2-epoxyisoprostaneE2-sn-glycero-3-phosphorylcholine
PGPC	1-palmitoyl-2-glutaroyl-*sn*-glycero-3-phosphorylcholine
POVPC	1-palmitoyl-2-(5-oxovaleroyl)-*sn*-glycero-3-phosphorylcholine
PRRs	pattern recognition receptors
PS	phosphatidylserine
ROS	reactive oxygen species
SMC	smooth muscle cell
SNPs	single nucleotide polymorphisms
SRs	scavenger receptors
TGF-β1	transforming growth factor-β1
tLOX-1	lectin-like oxidized LDL receptor-1
TNF-α	tumor necrosis factor α
t-PA	tissue plasminogen activator
VCAM-1	vascular cell adhesion molecule 1

REFERENCES

1. Libby P, Ridker PM, Hansson GK. Progress and challenges in translating the biology of atherosclerosis. *Nature* 2011;473:317–325.
2. Hansson GK, Hermansson A. The immune system in atherosclerosis. *Nature Immunology* 2011;12:204–212.
3. Miller YI, Choi S-H, Wiesner P et al. Oxidation-specific epitopes are danger-associated molecular patterns recognized by pattern recognition receptors of innate immunity. *Circulation Research* 2011;108:235–248.
4. Lichtman AH, Binder CJ, Tsimikas S, Witztum JL. Adaptive immunity in atherogenesis: New insights and therapeutic approaches. *The Journal of Clinical Investigation* 2013;123:27–36.
5. Chou M-Y, Fogelstrand L, Hartvigsen K et al. Oxidation-specific epitopes are dominant targets of innate natural antibodies in mice and humans. *The Journal of Clinical Investigation* 2009;119:1335–1349.
6. Ravandi A, Boekholdt SM, Mallat Z et al. Relationship of IgG and IgM autoantibodies and immune complexes to oxidized LDL with markers of oxidation and inflammation and cardiovascular events: Results from the EPIC-Norfolk Study. *The Journal of Lipid Research* 2011;52:1829–1836.
7. Tsimikas S, Willeit P, Willeit J et al. Oxidation-specific biomarkers, prospective 15-year cardiovascular and stroke outcomes, and net reclassification of cardiovascular events. *Journal of the American College of Cardiology* 2012;60:2218–2229.
8. Watson AD, Leitinger N, Navab M et al. Structural identification by mass spectrometry of oxidized phospholipids in minimally oxidized low density lipoprotein that induce monocyte/endothelial interactions and evidence for their presence in vivo. *The Journal of Biological Chemistry* 1997;272:13, 597–13, 607.
9. Leitinger N, Tyner TR, Oslund L et al. Structurally similar oxidized phospholipids differentially regulate endothelial binding of monocytes and neutrophils. *Proceedings of the National Academy of Sciences of the United States of America* 1999;96:12, 010–12, 015.

10. Li R, Mouillesseaux KP, Montoya D et al. Identification of prostaglandin E2 receptor subtype 2 as a receptor activated by OxPAPC. *Circulation Research* 2006;98:642–650.
11. Bochkov VN. Inflammatory profile of oxidized phospholipids. Thrombosis and haemostasis 2007;97:348–354.
12. Afonyushkin T, Oskolkova OV, Philippova M et al. Oxidized phospholipids regulate expression of ATF4 and VEGF in endothelial cells via NRF2-dependent mechanism: Novel point of convergence between electrophilic and unfolded protein stress pathways. *Arteriosclerosis, Thrombosis, and Vascular Biology* 2010;30:1007–1013.
13. Moore KJ, Tabas I. Macrophages in the pathogenesis of atherosclerosis. *Cell* 2011;145:341–355.
14. Steinberg D, Witztum JL. Oxidized low-density lipoprotein and atherosclerosis. *Arteriosclerosis, Thrombosis, and Vascular Biology* 2010;30:2311–2316.
15. Plüddemann A, Neyen C, Gordon S. Macrophage scavenger receptors and host-derived ligands. *Methods* 2007;43:207–217.
16. Murphy JE, Tedbury PR, Homer-Vanniasinkam S, Walker JH, Ponnambalam S. Biochemistry and cell biology of mammalian scavenger receptors. *Atherosclerosis* 2005;182:1–15.
17. Taylor PR, Martinez-Pomares L, Stacey M, Lin H-H, Brown GD, Gordon S. Macrophage receptors and immune recognition. *Annual Review of Immunology* 2005;23:901–944.
18. Yan Z-q, Hansson GK. Innate immunity, macrophage activation, and atherosclerosis. *Immunological Reviews* 2007;219:187–203.
19. Hartvigsen K, Chou M-Y, Hansen LF et al. The role of innate immunity in atherogenesis. *The Journal of Lipid Research* 2009;50 Suppl:S388–93.
20. Thorp E, Subramanian M, Tabas I. The role of macrophages and dendritic cells in the clearance of apoptotic cells in advanced atherosclerosis. *European Journal of Immunology* 2011;41:2515–2518.
21. Berliner JA, Leitinger N, Tsimikas S. The role of oxidized phospholipids in atherosclerosis. *The Journal of Lipid Research* 2009;50 Suppl:S207–12.
22. Witztum JL. You are right too! *The Journal of Clinical Investigation* 2005;115:2072–2075.
23. Yoshida H, Quehenberger O, Kondratenko N, Green S, Steinberg D. Minimally oxidized low-density lipoprotein increases expression of scavenger receptor A, CD36, and macrosialin in resident mouse peritoneal macrophages. *Arteriosclerosis, Thrombosis, and Vascular Biology* 1998;18:794–802.
24. de Villiers WJ, Fraser IP, Hughes DA, Doyle AG, Gordon S. Macrophage-colony-stimulating factor selectively enhances macrophage scavenger receptor expression and function. *The Journal of Experimental Medicine* 1994;180:705–709.
25. Hsu HY, Nicholson AC, Hajjar DP. Inhibition of macrophage scavenger receptor activity by tumor necrosis factor-alpha is transcriptionally and post-transcriptionally regulated. *The Journal of Biological Chemistry* 1996;271:7767–7773.
26. Svensson L, Norén K, Wiklund O, Lindmark H, Ohlsson B, Hultén LM. Inhibitory effects of N-acetylcysteine on scavenger receptor class A expression in human macrophages. *Journal of Internal Medicine* 2002;251:437–446.
27. Geng YJ, Hansson GK. Interferon-gamma inhibits scavenger receptor expression and foam cell formation in human monocyte-derived macrophages. *The Journal of Clinical Investigation* 1992;89:1322–1330.
28. Bottalico LA, Wager RE, Agellon LB, Assoian RK, Tabas I. Transforming growth factor-beta 1 inhibits scavenger receptor activity in THP-1 human macrophages. *The Journal of Biological Chemistry* 1991;266:22, 866–22, 871.
29. Steinbrecher UP, Lougheed M, Kwan WC, Dirks M. Recognition of oxidized low density lipoprotein by the scavenger receptor of macrophages results from derivatization of apolipoprotein B by products of fatty acid peroxidation. *The Journal of Biological Chemistry* 1989;264:15, 216–15, 223.

30. Edelstein C, Pfaffinger D, Hinman J et al. Lysine-phosphatidylcholine adducts in kringle V impart unique immunological and potential pro-inflammatory properties to human apolipoprotein(a). *The Journal of Biological Chemistry* 2003;278:52, 841–52, 847.

31. Bergmark C, Dewan A, Orsoni A et al. A novel function of lipoprotein [a] as a preferential carrier of oxidized phospholipids in human plasma. *The Journal of Lipid Research* 2008;49:2230–2239.

32. Leibundgut G, Arai K, Orsoni A et al. Oxidized phospholipids are present on plasminogen, affect fibrinolysis, and increase following acute myocardial infarction. *Journal of the American College of Cardiology* 2012;59:1426–1437.

33. Mineo C, Shaul PW. Functions of scavenger receptor class B, type I in atherosclerosis. *Current Opinion in Lipidology* 2012;23:487–493.

34. Podrez EA, Poliakov E, Shen Z et al. Identification of a novel family of oxidized phospholipids that serve as ligands for the macrophage scavenger receptor CD36. *The Journal of Biological Chemistry* 2002;277:38, 503–38, 516.

35. Nicholson AC. Expression of CD36 in macrophages and atherosclerosis: The role of lipid regulation of PPARgamma signaling. *Trends in Cardiovascular Medicine* 2004;14:8–12.

36. Han J, Nicholson AC, Zhou X, Feng J, Gotto AM, Hajjar DP. Oxidized low density lipoprotein decreases macrophage expression of scavenger receptor B-I. *The Journal of Biological Chemistry* 2001;276:16, 567–16, 572.

37. Kunjathoor VV, Febbraio M, Podrez EA et al. Scavenger receptors class A-I/II and CD36 are the principal receptors responsible for the uptake of modified low density lipoprotein leading to lipid loading in macrophages. *The Journal of Biological Chemistry* 2002;277:49, 982–49, 988.

38. Han J, Hajjar DP, Febbraio M, Nicholson AC. Native and modified low density lipoproteins increase the functional expression of the macrophage class B scavenger receptor, CD36. *The Journal of Biological Chemistry* 1997;272:21, 654–21, 659.

39. Nakagawa T, Nozaki S, Nishida M et al. Oxidized LDL increases and interferon-gamma decreases expression of CD36 in human monocyte-derived macrophages. *Arteriosclerosis, Thrombosis, and Vascular Biology* 1998;18:1350–1357.

40. Han J, Hajjar DP, Tauras JM, Feng J, Gotto AM, Nicholson AC. Transforming growth factor-beta1 (TGF-beta1) and TGF-beta2 decrease expression of CD36, the type B scavenger receptor, through mitogen-activated protein kinase phosphorylation of peroxisome proliferator-activated receptor-gamma. *The Journal of Biological Chemistry* 2000;275:1241–1246.

41. Rahaman SO, Lennon DJ, Febbraio M, Podrez EA, Hazen SL, Silverstein RL. A CD36-dependent signaling cascade is necessary for macrophage foam cell formation. *Cell Metabolism* 2006;4:211–221.

42. Ramprasad MP, Terpstra V, Kondratenko N, Quehenberger O, Steinberg D. Cell surface expression of mouse macrosialin and human CD68 and their role as macrophage receptors for oxidized low density lipoprotein. *Proceedings of the National Academy of Sciences of the United States of America* 1996;93:14, 833–14, 838.

43. Kumano-Kuramochi M, Xie Q, Kajiwara S, Komba S, Minowa T, Machida S. Lectin-like oxidized LDL receptor-1 is palmitoylated and internalizes ligands via caveolae/raft-dependent endocytosis. *Biochemical and Biophysical Research Communications* 2013;434:594–599.

44. Xu S, Ogura S, Chen J, Little PJ, Moss J, Liu P. LOX-1 in atherosclerosis: Biological functions and pharmacological modifiers. *Cellular and Molecular Life Sciences: CMLS* 2013;70:2859–2872.

45. Li D, Mehta JL. Upregulation of endothelial receptor for oxidized LDL (LOX-1) by oxidized LDL and implications in apoptosis of human coronary artery endothelial cells: Evidence from use of antisense LOX-1 mRNA and chemical inhibitors. *Arteriosclerosis, Thrombosis, and Vascular Biology* 2000;20:1116–1122.

46. Morawietz H, Duerrschmidt N, Niemann B, Galle J, Sawamura T, Holtz J. Induction of the oxLDL receptor LOX-1 by endothelin-1 in human endothelial cells. *Biochemical and Biophysical Research Communications* 2001;284:961–965.

47. Morawietz H, Rueckschloss U, Niemann B et al. Angiotensin II induces LOX-1, the human endothelial receptor for oxidized low-density lipoprotein. *Circulation* 1999;100:899–902.

48. Moriwaki H, Kume N, Sawamura T et al. Ligand specificity of LOX-1, a novel endothelial receptor for oxidized low density lipoprotein. *Arteriosclerosis, Thrombosis, and Vascular Biology* 1998;18:1541–1547.

49. Oka K, Sawamura T, Kikuta K et al. Lectin-like oxidized low-density lipoprotein receptor 1 mediates phagocytosis of aged/apoptotic cells in endothelial cells. *Proceedings of the National Academy of Sciences of the United States of America* 1998;95:9535–9540.

50. Knowles JW, Assimes TL, Boerwinkle E et al. Failure to replicate an association of SNPs in the oxidized LDL receptor gene (OLR1) with CAD. BMC medical genetics 2008;9:23.

51. Fukui M, Tanaka M, Senmaru T et al. LOX-1 is a novel marker for peripheral artery disease in patients with type 2 diabetes. *Metabolism: Clinical and Experimental* 2013;62:935–938.

52. Dai Y, Su W, Ding Z et al. Regulation of MSR-1 and CD36 in macrophages by LOX-1 mediated through PPAR-γ. *Biochemical and Biophysical Research Communications* 2013;431:496–500.

53. Adachi H, Tsujimoto M, Arai H, Inoue K. Expression cloning of a novel scavenger receptor from human endothelial cells. *The Journal of Biological Chemistry* 1997;272:31, 217–31, 220.

54. Tamura Y, Osuga J-i, Adachi H et al. Scavenger receptor expressed by endothelial cells I (SREC-I) mediates the uptake of acetylated low density lipoproteins by macrophages stimulated with lipopolysaccharide. *The Journal of Biological Chemistry* 2004;279:30, 938–30, 944.

55. Adachi H, Tsujimoto M. Characterization of the human gene encoding the scavenger receptor expressed by endothelial cell and its regulation by a novel transcription factor, endothelial zinc finger protein-2. *The Journal of Biological Chemistry* 2002;277:24, 014–24, 021.

56. Ishii J, Adachi H, Aoki J et al. SREC-II, a new member of the scavenger receptor type F family, trans-interacts with SREC-I through its extracellular domain. *The Journal of Biological Chemistry* 2002;277:39, 696–39, 702.

57. Shimaoka T, Kume N, Minami M et al. Molecular cloning of a novel scavenger receptor for oxidized low density lipoprotein, SR-PSOX, on macrophages. *The Journal of Biological Chemistry* 2000;275:40, 663–40, 666.

58. Abel S, Hundhausen C, Mentlein R et al. The transmembrane CXC-chemokine ligand 16 is induced by IFN-gamma and TNF-alpha and shed by the activity of the disintegrin-like metalloproteinase ADAM10. *Journal of Immunology* 2004;172:6362–6372.

59. Minami M, Kume N, Shimaoka T et al. Expression of SR-PSOX, a novel cell-surface scavenger receptor for phosphatidylserine and oxidized LDL in human atherosclerotic lesions. *Arteriosclerosis, Thrombosis, and Vascular Biology* 2001;21:1796–1800.

60. Mitsuoka H, Toyohara M, Kume N et al. Circulating soluble SR-PSOX/CXCL16 as a biomarker for acute coronary syndrome -comparison with high-sensitivity C-reactive protein. *Journal of Atherosclerosis and Thrombosis* 2009;16:586–593.

61. Jansson AM, Aukrust P, Ueland T et al. Soluble CXCL16 predicts long-term mortality in acute coronary syndromes. *Circulation* 2009;119:3181–3188.

62. Chou M-Y, Hartvigsen K, Hansen LF et al. Oxidation-specific epitopes are important targets of innate immunity. *Journal of Internal Medicine* 2008;263:479–488.

63. Stewart CR, Stuart LM, Wilkinson K et al. CD36 ligands promote sterile inflammation through assembly of a Toll-like receptor 4 and 6 heterodimer. *Nature Immunology* 2010;11:155–161.
64. Seimon TA, Nadolski MJ, Liao X et al. Atherogenic lipids and lipoproteins trigger CD36-TLR2-dependent apoptosis in macrophages undergoing endoplasmic reticulum stress. *Cell Metabolism* 2010;12:467–482.
65. Hoebe K, Georgel P, Rutschmann S et al. CD36 is a sensor of diacylglycerides. *Nature* 2005;433:523–527.
66. Greenberg ME, Sun M, Zhang R, Febbraio M, Silverstein R, Hazen SL. Oxidized phosphatidylserine-CD36 interactions play an essential role in macrophage-dependent phagocytosis of apoptotic cells. *The Journal of Experimental Medicine* 2006;203:2613–2625.
67. Turner WW, Hartvigsen K, Boullier A, Montano EN, Witztum JL, VanNieuwenhze MS. Design and synthesis of a stable oxidized phospholipid mimic with specific binding recognition for macrophage scavenger receptors. *Journal of Medicinal Chemistry* 2012;55:8178–8182.
68. Berg K. A new serum type system in man—The Lp system. *Acta Pathologica et Microbiologica Scandinavica* 1963;59:369–382.
69. Berg K, Dahlen G, Frick MH. Lp(a) lipoprotein and pre-beta1-lipoprotein in patients with coronary heart disease. *Clinical Genetics* 1974;6:230–235.
70. Dubé JB, Boffa MB, Hegele RA, Koschinsky ML. Lipoprotein(a): More interesting than ever after 50 years. *Current Opinion in Lipidology* 2012;23:133–140.
71. McLean JW, Tomlinson JE, Kuang WJ et al. cDNA sequence of human apolipoprotein(a) is homologous to plasminogen. *Nature* 1987;330:132–137.
72. Makino K, Abe A, Maeda S, Noma A, Kawade M, Takenaka O. Lipoprotein(a) in non-human primates. Presence and characteristics of Lp(a) immunoreactive materials using anti-human Lp(a) serum. *Atherosclerosis* 1989;78:81–85.
73. Laplaud PM, Beaubatie L, Rall SC, Luc G, Saboureau M. Lipoprotein[a] is the major apoB-containing lipoprotein in the plasma of a hibernator, the hedgehog (*Erinaceus europaeus*). *The Journal of Lipid Research* 1988;29:1157–1170.
74. Lawn RM, Schwartz K, Patthy L. Convergent evolution of apolipoprotein(a) in primates and hedgehog. *Proceedings of the National Academy of Sciences of the United States of America* 1997;94:11, 992–11, 997.
75. Tsimikas S. High-dose statins prior to percutaneous coronary intervention: A paradigm shift to influence clinical outcomes in the cardiac catheterization laboratory. *Journal of the American College of Cardiology* 2009;54:2164–2166.
76. Collaboration ERF, Erqou S, Kaptoge S et al. Lipoprotein(a) concentration and the risk of coronary heart disease, stroke, and nonvascular mortality. *The Journal of the American Medical Association* 2009;302:412–423.
77. Kamstrup PR, Tybjaerg-Hansen A, Steffensen R, Nordestgaard BG. Genetically elevated lipoprotein(a) and increased risk of myocardial infarction. *The Journal of the American Medical Association* 2009;301:2331–2339.
78. Clarke R, Peden JF, Hopewell JC et al. Genetic variants associated with Lp(a) lipoprotein level and coronary disease. *The New England Journal of Medicine* 2009;361:2518–2528.
79. Tsimikas S, Hall JL. Lipoprotein(a) as a potential causal genetic risk factor of cardiovascular disease: A rationale for increased efforts to understand its pathophysiology and develop targeted therapies. *Journal of the American College of Cardiology* 2012;60:716–721.
80. Nordestgaard BG, Chapman MJ, Ray K et al. Lipoprotein(a) as a cardiovascular risk factor: Current status. *European Heart Journal* 2010;31:2844–2853.
81. Kronenberg F, Utermann G. Lipoprotein(a): Resurrected by genetics. *Journal of Internal Medicine* 2013;273:6–30.

82. Leibundgut G, Scipione C, Yin H et al. Determinants of binding of oxidized phospholipids on apolipoprotein (a) and lipoprotein (a). *The Journal of Lipid Research* 2013;54:2815–2830.

83. Koschinsky ML, Marcovina SM. Structure–function relationships in apolipoprotein(a): Insights into lipoprotein(a) assembly and pathogenicity. *Current Opinion in Lipidology* 2004;15:167–174.

84. Spence JD, Koschinsky ML. Mechanisms of lipoprotein(a) pathogenicity: Prothrombotic, proatherosclerotic, or both? *Arteriosclerosis, Thrombosis, and Vascular Biology* 2012;32:1550–1551.

85. Tsimikas S, Witztum JL. The role of oxidized phospholipids in mediating lipoprotein(a) atherogenicity. *Current Opinion in Lipidology* 2008;19:369–377.

86. Ravandi A, Harkewicz R, Leibundgut G. Identification of oxidized phospholipids and cholesteryl esters in embolic protection devices post percutaneous coronary, carotid and peripheral interventions in humans (abstr). *Arteriosclerosis, Thrombosis, and Vascular Biology*, Scientific Sessions. April 28–30, 2011; Chicago, IL.

87. Wiesner P, Tafelmeier M, Chittka D et al. MCP-1 binds to oxidized LDL and is carried by lipoprotein(a) in human plasma. *The Journal of Lipid Research* 2013;54:1877–1883.

88. van Dijk RA, Kolodgie F, Ravandi A et al. Differential expression of oxidation-specific epitopes and apolipoprotein(a) in progressing and ruptured human coronary and carotid atherosclerotic lesions. *The Journal of Lipid Research* 2012;53:2773–2790.

89. Merki E, Graham M, Taleb A et al. Antisense oligonucleotide lowers plasma levels of apolipoprotein (a) and lipoprotein (a) in transgenic mice. *Journal of the American College of Cardiology* 2011;57:1611–1621.

90. Kolski B, Tsimikas S. Emerging therapeutic agents to lower lipoprotein (a) levels. *Current Opinion in Lipidology* 2012;23:560–568.

91. Davidson MH, Ballantyne CM, Jacobson TA et al. Clinical utility of inflammatory markers and advanced lipoprotein testing: Advice from an expert panel of lipid specialists. *Journal of Clinical Lipidology* 2011;5:338–367.

92. Viney N, Graham M, Crooke R, Huges S, Singleton W. Evaluation of Isis tapo(a) Rx, an antisense inhibitor to apolipoprotein(a), in healthy volunteers. *Circulation* 2013;128:A14196.

93. Thanassoulis G, Campbell CY, Owens DS et al. Genetic associations with valvular calcification and aortic stenosis. *The New England Journal of Medicine* 2013;368:503–512.

94. Taleb A, Witztum JL, Tsimikas S. Oxidized phospholipids on apoB-100-containing lipoproteins: A biomarker predicting cardiovascular disease and cardiovascular events. *Biomarkers in Medicine* 2011;5:673–694.

95. Wu R, de Faire U, Lemne C, Witztum JL, Frostegård J. Autoantibodies to OxLDL are decreased in individuals with borderline hypertension. *Hypertension* 1999;33:53–59.

96. Penny WF, Ben-Yehuda O, Kuroe K et al. Improvement of coronary artery endothelial dysfunction with lipid-lowering therapy: Heterogeneity of segmental response and correlation with plasma-oxidized low density lipoprotein. *Journal of American College of Cardiology* 2001;37:766–774.

97. Tsimikas S, Bergmark C, Beyer RW et al. Temporal increases in plasma markers of oxidized low-density lipoprotein strongly reflect the presence of acute coronary syndromes. *Journal of the American College of Cardiology* 2003;41:360–370.

98. Tsimikas S, Lau HK, Han K-R et al. Percutaneous coronary intervention results in acute increases in oxidized phospholipids and lipoprotein(a): Short-term and long-term immunologic responses to oxidized low-density lipoprotein. *Circulation* 2004;109:3164–3170.

99. Segev A, Strauss BH, Witztum JL, Lau HK, Tsimikas S. Relationship of a comprehensive panel of plasma oxidized low-density lipoprotein markers to angiographic

restenosis in patients undergoing percutaneous coronary intervention for stable angina. *American Heart Journal* 2005;150:1007–1014.

100. Tsimikas S, Witztum JL, Miller ER et al. High-dose atorvastatin reduces total plasma levels of oxidized phospholipids and immune complexes present on apolipoprotein B-100 in patients with acute coronary syndromes in the MIRACL trial. *Circulation* 2004;110:1406–1412.

101. Fraley AE, Schwartz GG, Olsson AG et al. Relationship of oxidized phospholipids and biomarkers of oxidized low-density lipoprotein with cardiovascular risk factors, inflammatory biomarkers, and effect of statin therapy in patients with acute coronary syndromes: Results from the MIRACL (Myocardial Ischemia Reduction with Aggressive Cholesterol Lowering) trial. *Journal of the American College of Cardiology* 2009;53:2186–2196.

102. Silaste M-L, Rantala M, Alfthan G et al. Changes in dietary fat intake alter plasma levels of oxidized low-density lipoprotein and lipoprotein(a). *Arteriosclerosis, Thrombosis, and Vascular Biology* 2004;24:498–503.

103. Tsimikas S, Brilakis ES, Miller ER et al. Oxidized phospholipids, Lp(a) lipoprotein, and coronary artery disease. *The New England Journal of Medicine* 2005;353:46–57.

104. Tsimikas S, Duff GW, Berger PB et al. Pro-inflammatory interleukin-1 genotypes potentiate the risk of coronary artery disease and cardiovascular events mediated by oxidized phospholipids and lipoprotein (a). *Journal of the American College of Cardiology* 2014;63:1724–1734.

105. Tsimikas S, Kiechl S, Willeit J et al. Oxidized phospholipids predict the presence and progression of carotid and femoral atherosclerosis and symptomatic cardiovascular disease: Five-year prospective results from the Bruneck study. *Journal of the American College of Cardiology* 2006;47:2219–2228.

106. Kiechl S, Willeit J, Mayr M et al. Oxidized phospholipids, lipoprotein(a), lipoprotein-associated phospholipase A2 activity, and 10-year cardiovascular outcomes: Prospective results from the Bruneck study. *Arteriosclerosis, Thrombosis, and Vascular Biology* 2007;27:1788–1795.

107. Rodenburg J, Vissers MN, Wiegman A et al. Oxidized low-density lipoprotein in children with familial hypercholesterolemia and unaffected siblings: Effect of pravastatin. *Journal of the American College of Cardiology* 2006;47:1803–1810.

108. Bossola M, Tazza L, Merki E et al. Oxidized low-density lipoprotein biomarkers in patients with end-stage renal failure: Acute effects of hemodialysis. *Blood Purification* 2007;25:457–465.

109. Bossola M, Tazza L, Luciani G, Tortorelli A, Tsimikas S. OxPL/apoB, lipoprotein(a) and OxLDL biomarkers and cardiovascular disease in chronic hemodialysis patients. *Journal of Nephrology* 2011;24:581–588.

110. Ky B, Burke A, Tsimikas S et al. The influence of pravastatin and atorvastatin on markers of oxidative stress in hypercholesterolemic humans. *Journal of the American College of Cardiology* 2008;51:1653–1662.

111. Choi S-H, Chae A, Miller E et al. Relationship between biomarkers of oxidized low-density lipoprotein, statin therapy, quantitative coronary angiography, and atheroma: Volume observations from the REVERSAL (Reversal of Atherosclerosis with Aggressive Lipid Lowering) study. *Journal of the American College of Cardiology* 2008;52:24–32.

112. Tsimikas S, Clopton P, Brilakis ES et al. Relationship of oxidized phospholipids on apolipoprotein B-100 particles to race/ethnicity, apolipoprotein(a) isoform size, and cardiovascular risk factors: Results from the Dallas Heart Study. *Circulation* 2009;119:1711–1719.

113. Tsimikas S, Mallat Z, Talmud PJ et al. Oxidation-specific biomarkers, lipoprotein(a), and risk of fatal and nonfatal coronary events. *Journal of the American College of Cardiology* 2010;56:946–955.

114. Budoff MJ, Ahmadi N, Gul KM et al. Aged garlic extract supplemented with B vitamins, folic acid and L-arginine retards the progression of subclinical atherosclerosis: A randomized clinical trial. *Preventive Medicine* 2009;49:101–107.

115. Ahmadi N, Tsimikas S, Hajsadeghi F et al. Relation of oxidative biomarkers, vascular dysfunction, and progression of coronary artery calcium. *The American Journal of Cardiology* 2010;105:459–466.

116. Arai K, Luke MM, Koschinsky ML et al. The I4399M variant of apolipoprotein(a) is associated with increased oxidized phospholipids on apolipoprotein B-100 particles. *Atherosclerosis* 2010;209:498–503.

117. Faghihnia N, Tsimikas S, Miller ER, Witztum JL, Krauss RM. Changes in lipoprotein(a), oxidized phospholipids, and LDL subclasses with a low-fat high-carbohydrate diet. *The Journal of Lipid Research* 2010;51:3324–3330.

118. Arai K, Orsoni A, Mallat Z et al. Acute impact of apheresis on oxidized phospholipids in patients with familial hypercholesterolemia. *The Journal of Lipid Research* 2012;53:1670–1678.

119. Fefer P, Tsimikas S, Segev A et al. The role of oxidized phospholipids, lipoprotein (a) and biomarkers of oxidized lipoproteins in chronically occluded coronary arteries in sudden cardiac death and following successful percutaneous revascularization. *Cardiovascular Revascularization Medicine: Including Molecular Interventions* 2012;13:11–19.

120. Holleboom AG, Daniil G, Fu X et al. Lipid oxidation in carriers of lecithin:cholesterol acyltransferase gene mutations. *Arteriosclerosis, Thrombosis, and Vascular Biology* 2012;32:3066–3075.

121. Yoshida H, Shoda T, Yanai H et al. Effects of pitavastatin and atorvastatin on lipoprotein oxidation biomarkers in patients with dyslipidemia. *Atherosclerosis* 2013;226:161–164.

122. Bertoia ML, Pai JK, Lee J-H et al. Oxidation-specific biomarkers and risk of peripheral artery disease. *Journal of the American College of Cardiology* 2013;61:2169–2179.

123. Ravandi A, Leibundgut G, Hung M-Y et al. Release and capture of bioactive oxidized phospholipids and oxidized cholesteryl esters during percutaneous coronary and peripheral arterial interventions in humans. *Journal of the American College of Cardiology* 2014;63:1961–1971.

124. Ridker PM, Thuren T, Zalewski A, Libby P. Interleukin-1β inhibition and the prevention of recurrent cardiovascular events: Rationale and design of the Canakinumab Anti-inflammatory Thrombosis Outcomes Study (CANTOS). *American Heart Journal* 2011;162:597–605.

125. Tsimikas S, Aikawa M, Miller FJ et al. Increased plasma oxidized phospholipid:apolipoprotein B-100 ratio with concomitant depletion of oxidized phospholipids from atherosclerotic lesions after dietary lipid-lowering: A potential biomarker of early atherosclerosis regression. *Arteriosclerosis, Thrombosis, and Vascular Biology* 2007;27:175–181.

126. Choi SH, Chae A, Chen CH, Merki E, Shaw PX, Tsimikas S. Emerging approaches for imaging vulnerable plaques in patients. *Current Opinion in Biotechnology* 2007;18:73–82.

127. Choi SH, Chae A, Miller E et al. Relationship between biomarkers of oxidized low-density lipoprotein, statin therapy, quantitative coronary angiography, and atheroma: Volume observations from the REVERSAL (Reversal of Atherosclerosis with Aggressive Lipid Lowering) study. *Journal of the American College of Cardiology* 2008;52:24–32.

128. Castellino FJ, Ploplis VA. Structure and function of the plasminogen/plasmin system. *Thrombosis and Haemostasis* 2005;93:647–654.

129. Xiao Q, Danton MJ, Witte DP et al. Plasminogen deficiency accelerates vessel wall disease in mice predisposed to atherosclerosis. *Proceedings of the National Academy of Sciences of the United States of America* 1997;94:10, 335–10, 340.

130. Plow EF, Hoover-Plow J. The functions of plasminogen in cardiovascular disease. *Trends in Cardiovascular Medicine* 2004;14:180–186.
131. Marui N, Offermann MK, Swerlick R et al. Vascular cell adhesion molecule-1 (VCAM-1) gene transcription and expression are regulated through an antioxidant-sensitive mechanism in human vascular endothelial cells. *The Journal of Clinical Investigation* 1993;92:1866–1874.
132. Chang PC, Wu HL, Lin HC, Wang KC, Shi GY. Human plasminogen kringle 1–5 reduces atherosclerosis and neointima formation in mice by suppressing the inflammatory signaling pathway. *Journal of Thrombosis and Haemostasis: JTH* 2010;8:194–201.
133. Raskopf E, Gerceker S, Vogt A, Standop J, Sauerbruch T, Schmitz V. Plasminogen fragment K1–3 inhibits expression of adhesion molecules and experimental HCC recurrence in the liver. *International Journal of Colorectal Disease* 2009;24:837–844.
134. Moulton KS, Vakili K, Zurakowski D et al. Inhibition of plaque neovascularization reduces macrophage accumulation and progression of advanced atherosclerosis. *Proceedings of the National Academy of Sciences of the United States of America* 2003;100:4736–4741.
135. Das R, Ganapathy S, Mahabeleshwar GH et al. Macrophage gene expression and foam cell formation are regulated by plasminogen. *Circulation* 2013;127:1209–1218-e1–16.
136. Kremen M, Krishnan R, Emery I et al. Plasminogen mediates the atherogenic effects of macrophage-expressed urokinase and accelerates atherosclerosis in apoE-knockout mice. *Proceedings of the National Academy of Sciences of the United States of America* 2008;105:17, 109–17, 114.
137. Ploplis VA, French EL, Carmeliet P, Collen D, Plow EF. Plasminogen deficiency differentially affects recruitment of inflammatory cell populations in mice. *Blood* 1998;91:2005–2009.
138. Carmeliet P, Moons L, Ploplis V, Plow E, Collen D. Impaired arterial neointima formation in mice with disruption of the plasminogen gene. *The Journal of Clinical Investigation* 1997;99:200–208.
139. Carmeliet P, Collen D. Role of the plasminogen/plasmin system in thrombosis, hemostasis, restenosis and atherosclerosis evaluation in transgenic animals. *Trends in Cardiovascular Medicine* 1995;5:117–122.
140. Edelstein C, Pfaffinger D, Yang M, Hill JS, Scanu AM. Naturally occurring human plasminogen, like genetically related apolipoprotein(a), contains oxidized phosphatidylcholine adducts. *Biochimica et Biophysica Acta* 2010;1801:738–745.
141. Leibundgut G, Strauss BH, Tsimikas S. Effect of uncomplicated PCI on plasminogen and oxidized phospholipids present on plasminogen in patients with stable angina. *Journal of the American College of Cardiology*, 2013:S0735–1097.

15 Cholesterol Oxidation Products in the Initiation, Progression, and Fate of Atherosclerotic Lesions

*Giuseppe Poli, Simona Gargiulo, Paola Gamba,
Barbara Sottero, Fiorella Biasi,
and Gabriella Leonarduzzi*

CONTENTS

INTRODUCTION

Cholesterol oxidation products are undoubtedly among the lipid oxidation products that most likely contribute to the development of atherosclerotic lesions in human large- and medium-sized arteries. They are known collectively as oxysterols, and their importance stems not only from the quantitative impact of cholesterol and its esters on the lipid-metabolism-dependent effects on atherosclerosis, but also from the various significant biochemical properties exhibited by this class of compounds.

Oxysterols are a family of 27-carbon molecules that originate from the oxidation of cholesterol by both enzymatic and nonenzymatic mechanisms. Compared to cholesterol, they contain an additional hydroxyl, epoxide, or ketone group in the sterol nucleus, and/or a hydroxyl group in the side chain (for molecular structure and principal biochemical properties, see Leonarduzzi et al. 2002, 2005a; Schroepfer 2000). Figure 15.1 lists the oxysterols of enzymatic and nonenzymatic origin that are of interest in human pathophysiology. For many years, oxysterols were considered chiefly for the physiological roles that they play, in particular in bile acid synthesis, steroid hormone biosynthesis, sterol transport, and gene regulation. More recently, reports have begun to focus on their potential contribution to the pathogenesis and progression of human diseases in which hypercholesterolemia is a primary risk factor and that are associated with inflammation, specifically atherosclerosis and Alzheimer's disease. The recent recognition of cholesterol oxides, like 24-, 25-, and 27-hydroxycholesterol (24-OH, 25-OH, and 27-OH), as very good ligands of

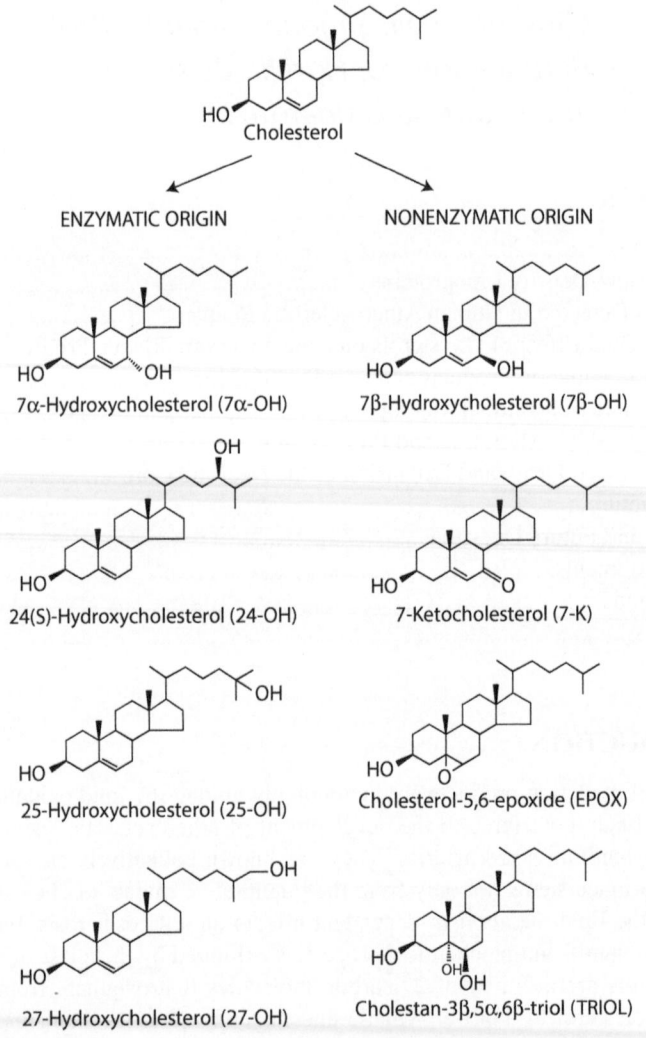

FIGURE 15.1   The principal oxysterols of human pathophysiological interest.

a variety of physiologically important nuclear receptors has meant that emphasis has again shifted to the beneficial aspects of oxysterols' biochemical effects. Some significant examples will be discussed. Most likely, the tale of oxysterols has not yet been told in full. New chapters will emerge, relating to physiology and pathology, the relative importance of the two depending on the scenario as a whole, the types of cells and tissues involved, oxysterol concentrations, the co-presence of other agents/factors, and last but not least, fashions in scientific trends.

As far as the implication of oxysterols in the pathogenesis of atherosclerosis is concerned, we here analyze, as critically as possible, the most recent relevant literature, keeping in mind that, in human pathophysiology, antithetic events and processes continually co-occur. The appropriate tuning of these processes can lead to disease stabilization or progression: in the case of atherosclerosis, pro-inflammatory *versus* anti-inflammatory reactions, or pro-apoptotic *versus* antiapoptotic stimuli. In the recent past, the emphasis is either only on the physiological or on the pathological effects of oxysterols, this chapter attempts to provide a comprehensive approach to the overall pathophysiological significance of cholesterol oxidation products in relation to atherosclerosis.

In this connection, the type and concentration of oxysterols that are actually detectable in oxidized low-density lipoproteins (oxLDL) and, above all, those detectable in atherosclerotic plaques, must first be considered.

OXIDIZED LOW-DENSITY LIPOPROTEINS

A wealth of scientific observations has accumulated over the years, pointing to oxidation of both the lipid and the protein component of low-density lipoprotein (LDL) plasma lipoproteins as the key event in triggering and sustaining the development of atherosclerosis. Of the various plasma lipoproteins, LDL is the lipoprotein that tends to accumulate within the arterial sub-intimal space, mainly after having been oxidized or otherwise modified (Levitan et al. 2010).

Looking only at the most recent papers to analyze oxysterols possibly stemming from LDL oxidation, five compounds of this class emerge as being produced in human LDL following *in vitro* $CuSO_4$-mediated oxidation. These are 7β-hydroxycholesterol (7β-OH), 7-ketocholesterol (7-K), 7α-hydroxycholesterol (7α-OH), cholesterol-5β,6β-epoxide (β-EPOX), and cholesterol-5α,6α-epoxide (α-EPOX) (Levitan et al. 2010; Shentu et al. 2012). In particular, 7β-OH and 7-K were already significantly increased in minimally modified LDL (mmLDL) compared to unoxidized LDL. From the quantitative standpoint, when LDL was incubated for 8 h in the presence of $CuSO_4$, almost one quarter of the total cholesterol was oxidized to oxysterols, while after 24 h of LDL *in vitro* oxidation, the total oxysterols measured accounted for one third of the total cholesterol (Shentu et al. 2012). It is significant in this connection that only trace amounts of side-chain oxidized cholesterol products, namely 25-OH, 27-OH, and 24-OH, were detectable in plasma oxLDL (Brown and Jessup 1999), whereas 27-OH is prevalent in atheromatous plaques (Levitan et al. 2010). Of note, all the oxysterols recognized to be present *in vivo* are mainly produced by the intracellular sterol hydroxylase systems and not *via* the nonenzymatic pathways (Patel et al. 1996). This observation questions the reliability of Cu^{2+}-oxLDL model; indeed, even if metal ions may accumulate in the advanced atherosclerotic plaque and metal-containing proteins (e.g., ceruloplasmin) may physiologically induce *in vivo* LDL oxidation, experimental confirmation of that is still lacking (Yoshida and Kisugi 2010). Conversely, diverse oxysterols might be produced by cholesterol auto-oxidation leading to circulating mmLDL with low affinity for macrophage scavenger receptors but still significant affinity for the native LDL ones (Sevanian et al. 1997; Yoshida and Kisugi 2010).

Another important finding is that mmLDL contains mainly phospholipid oxidation products, while the oxysterol content increases proportionally with the oxidation rate (Shentu et al. 2012). Taken together, these very recent experimental reports and reviews suggest a primary role for oxysterols in the progression of atherosclerosis, rather than in its initiation. However, the marked biochemical changes that oxysterols have been found to bring about in endothelial cells (ECs) suggest that the possibility that cholesterol oxides make some contribution to endothelial dysfunction in atherosclerosis should not be discarded *a priori*.

OXYSTEROLS DETECTED IN HUMAN ATHEROSCLEROTIC PLAQUES

Before analyzing the contribution that these products might make to each step of the atherosclerotic process, it appears essential to examine the qualitative and quantitative data available concerning the presence of cholesterol oxidation products within the plaque. The fact that 27-OH, of enzymatic origin, is by far the most abundant oxysterol in advanced plaques was reported first by Björkhem and colleagues (Björkhem et al. 1994), and then by Brown and Jessup, who also showed that 80% of plaque oxysterols were in the esterified form, and that 7-K was second in terms of abundance, at least in carotid and femoral lesions (Brown and Jessup 1999). Two years later, a careful measurement of the oxysterol content of normal human arteries, fatty streaks, and fibrotic plaques further proved that whereas normal arteries contain trace amounts of cholesterol oxides, these compounds have already accumulated significantly in the initial atherosclerotic lesions; this study also showed that, alongside 27-OH, all the other main oxysterols recovered in oxLDL, namely (in decreasing order) 7-K, 7α-OH, 7β-OH, β-EPOX, and α-EPOX, were consistently present both in fatty streaks and in advanced lesions of carotid arteries (Garcia-Cruset et al. 2001). Vaya and colleagues confirmed that 27-OH was by far the most abundant oxysterol in human carotid and coronary plaques, also reporting a similar mean concentration (0.25 ± 0.1 μg/mg wet weight tissue) (Vaya et al. 2001). Quantitative analyses reported by the same group pointed to 7-K as the second most abundant oxysterol, both in coronary and in carotid plaques (0.08 ± 0.04 μg/mg wet weight tissue), the unoxidized cholesterol mean values being about 50–52 μg/mg wet weight tissue. Further quantitative analyses of oxysterol concentrations in human atherosclerotic plaques were subsequently reported (Helmschrodt et al. 2013; Leonarduzzi et al. 2007).

The available data on the oxysterol content of plaques are lacking in consistency, which likely depends on the wide variability of cholesterol deposition in different lesions in a single individual, on the stage of development of the plaques examined, and possible also on the different sensitivities of the analytical methods used. However, despite quantitative discrepancies, the same trend emerges from these studies, with 27-OH followed by 7-K as the most abundant oxysterol in atherosclerotic lesions followed in turn by 7α- and 7β-OH. Besides producing further consistent evidence, a very recent report on the accumulation of oxysterols in stable ($n = 13$) *versus* unstable ($n = 6$) human carotid plaques showed no significant difference in 27-OH and 7-K accumulation in the two types of lesion (Vaya 2013).

IS THERE A DUAL EFFECT OF OXYSTEROLS ON ATHEROSCLEROTIC PLAQUE PROGRESSION?

Until very recently, research into the involvement of cholesterol oxidation products in the pathogenesis of atherosclerosis was mainly driven by the marked pro-inflammatory and pro-apoptotic effects that some members of this class of compounds exhibit (Poli et al. 2009; Vejux and Lizard 2009), whereas no attention was paid to possible anti-inflammatory or antiapoptotic signals generated by oxysterols. Nevertheless, an increased body of literature confirms the capability of oxysterols to interact with different nuclear receptors, which are known to play a main role in driving the anti-inflammatory and antiapoptotic cellular responses, thus suggesting a possible involvement of oxysterols even in these pathways.

Among those receptors, liver X receptors (LXRs) recognize certain oxysterols as primary natural ligands (Viennois et al. 2011); these receptors were initially discovered as orphan receptors, whereas they are now recognized to be transcription factors with a critical role in regulating lipid synthesis and transport in a number of tissues and organs. The oxysterols that have shown high affinity for LXRs are 20-hydroxycholesterol (20-OH), 22-hydroxycholesterol (22-OH), 24-OH, 25-OH, and 27-OH (Bensinger and Tontonoz 2008; Fu et al. 2001), which are all compounds generated by the enzymatic oxidation of cholesterol on the side chain.

Very importantly, a marked antiatherosclerotic effect of LXR upregulation or of genetic manipulation, has been consistently reported in a number of mice models of atherosclerosis (Calkin and Tontonoz 2010; Honzumi et al. 2011; Töröcsik et al. 2009). Figure 15.2 schematically shows the main anti-inflammatory effects exerted by LXR agonists on vascular ECs, smooth muscle cells (SMCs), and macrophages (Calkin and Tontonoz 2010). Briefly, LXR agonists favor reverse cholesterol transport and apoptotic cell elimination within the vascular wall, counteract vasoconstriction stimuli, and downregulate the major pro-inflammatory cytokines and chemokines.

Taking this evidence together, it thus appears likely that oxysterols exercise a dual effect on the multistep (and multifactorial) process of atherogenesis: oxysterols of enzymatic origin tend to exert anti-inflammatory effects, while oxysterols of nonenzymatic origin tend to induce inflammation (Vaya 2013). However, there are exceptions to these trends. For instance, α-EPOX, a cholesterol oxide of nonenzymatic origin, has also been recognized to be a good LXR ligand (Berrodin et al. 2010), and 27-OH has been shown to enhance inflammation and oxidative stress in cells of the macrophage lineage, as well as destabilizing plaques by upregulating matrix metalloprotease-9 (MMP-9) (Gargiulo et al. 2014). Most likely, the pro-inflammatory effects of 27-OH might be the result of activation of different signaling pathways, involving the transcription factor nuclear factor-κB (NF-κB) rather than LXR, as well as other oxysterols do (Leonarduzzi et al. 2005b, 2010).

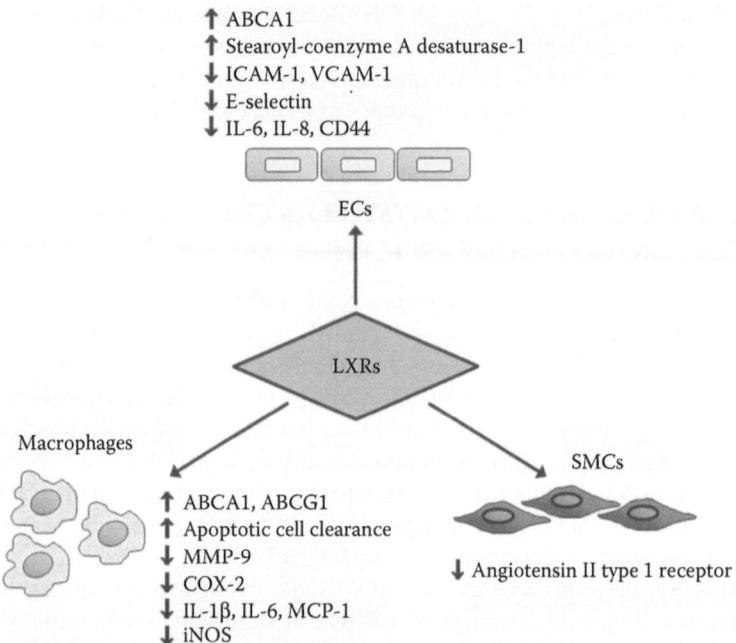

FIGURE 15.2 Anti-inflammatory effects exerted by liver X receptor (LXR) agonists on vascular endothelial cells (ECs), smooth muscle cells (SMCs) and macrophages. ABC, ATP-binding cassette transporter; ICAM-1, intercellular adhesion molecule-1; VCAM-1, vascular cell adhesion molecule-1; IL, interleukin; MMP-9, matrix metalloprotease-9; COX-2, cyclo-oxygenase-2; MCP-1, monocyte chemotactic protein-1; iNOS, inducible nitric oxide synthase.

On the basis of current knowledge of oxysterol-mediated molecular signaling and effects, it might be preliminarily concluded that this very interesting class of cholesterol oxides may trigger both pro-inflammatory and anti-inflammatory stimuli; in the case of the development of atherosclerotic lesions, these two actions occur because oxysterols can modulate more than one signaling pathway, and more than one transcription factor. The overall result most likely depends on a variety of conditions, including the cellular and molecular environment, the absolute and relative concentrations of different oxysterols of pathophysiological interest, and the co-presence of other modulators of inflammation-related signaling.

Bearing in mind this state of affairs, the potential contribution of oxysterols to each of the main steps in the atherosclerotic process will now be examined, focusing chiefly on the relevant literature published in the last 3–4 years, and referring to certain reviews for older data (Poli et al. 2009; Töröcsik et al. 2009; Vejux and Lizard 2009).

CHOLESTEROL OXIDATION PRODUCTS AND ECs DYSFUNCTION

Interposed between the blood and tissues, ECs act to ensure body homeostasis; however, their position makes them highly vulnerable. Several conditions, including blood hypertension, hyperglycemia, hyperlipidemia, and smoking, may induce EC dysfunction by modulating their constitutive gene expression, increasing the permeability of the endothelium and leading to surface expression of cell adhesion molecules, chemokines, and so on. Lipid oxidation products other than oxysterols, such as oxidized phospholipids and reactive aldehydes, are probably more involved than cholesterol oxides in this very early step of atherogenesis, as a very recent review stresses (Usatyuk and Natarajan 2012). Still, cholesterol oxides might contribute to the overall modulation of EC function and dysfunction by generating either pro- and anti-inflammatory stimuli.

In oxLDL, due to their marked hydrophobicity nonesterified oxysterols are preferentially located on the particle's surface; this is a privileged localization for interacting with the vascular milieu, and thus with ECs, not only in the plasma but also in the sub-intimal space. In the sub-intimal space, mmLDL, having transmigrated through the endothelium by either active or passive mechanisms, may accumulate and be further oxidized. This increases the amount of reactive oxidized lipids present, and might affect the phenotype and function of various vascular cells, naturally including ECs (Figure 15.3). This hypothesis appears consistent with the general statement that accumulation of oxLDL at the sub-endothelial level leads to plaque formation.

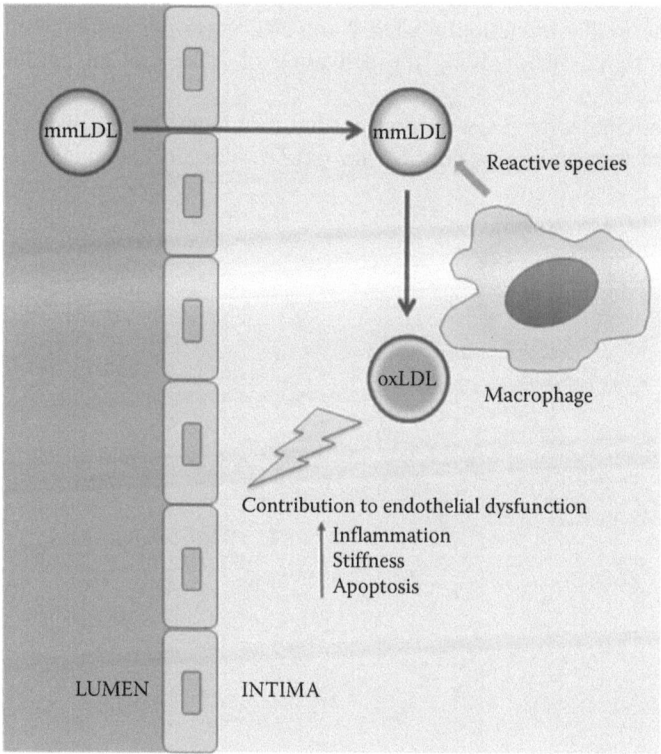

FIGURE 15.3 Endothelial changes due to minimally modified/oxidized low-density lipoproteins (mm/oxLDL) are more likely induced by micelles that accumulate in the sub-intimal space of arterial vessels.

Experimental findings that have accumulated over recent years, and that support a significant pro-inflammatory action of oxysterols on the vascular endothelium, point to this class of compounds as being able to stimulate all the steps in leukocyte migration from the blood stream to the sub-intimal space; they apparently achieve this through the overexpression of selectins (rolling), intercellular adhesion molecules (adhesion), and chemokines (transmigration) (Poli et al. 2009) (Figure 15.4).

A recent study aimed to investigate the possible pro-inflammatory effect of cholestan-3β,5α,6β-triol (TRIOL) on human umbilical vein endothelial cells (HUVECs); the findings showed net upregulation of the expression and synthesis of cyclooxygenase-2 (COX-2), and of protein levels of endothelial nitric oxide synthase (eNOS), two enzymes recognized to play major roles in inflammation (Liao et al. 2009). The only concern about this report is that TRIOL is only detectable, and inconsistently so, in human plasma (Brown and Jessup 1999) and has in fact never yet been detected in atheromas (Garcia-Cruset et al. 2001; Leonarduzzi et al. 2007; Vaya et al. 2001). One can speculate that in *in vivo* conditions the extracellular TRIOL may act by binding to plasma membrane receptors, thus making it unnecessary inside cells; alternatively, TRIOL might be intracellularly modified into active metabolites, an event that would also explain its lack in plaque samples. Nevertheless, both hypotheses have not yet been proven. Despite that, the demonstration that induction of COX-2 and eNOS was dependent upon a TRIOL-dependent signaling involving the activation of NF-κB is an interesting result (Liao et al. 2009).

Using bovine aortic ECs, oxysterols, and to a lesser extent oxidized phospholipids, were found to be responsible for oxLDL-induced endothelial stiffness and

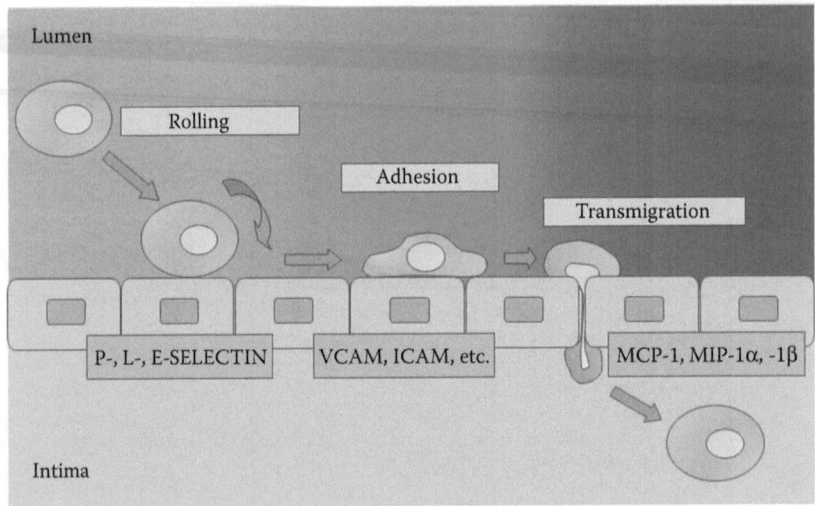

FIGURE 15.4 Oxysterols may upregulate all steps in the transmigration of leukocytes through the endothelial layer lining the arterial vessels. VCAM-1, vascular cell adhesion molecule-1; ICAM-1, intercellular adhesion molecule-1; MCP-1, monocyte chemotactic protein-1; MIP, monocyte inflammatory protein.

lack of elasticity, a condition responsible for the endothelial layer's increased sensitivity to shear stress (Byfield et al. 2006). In particular 7-K and 7α-OH, but also 27-OH, were effective in inducing endothelial stiffness, while the effects of 7β-OH, 25-OH, and α- and β-EPOX were minimal or null (Shentu et al. 2012). Interestingly, at least with regard to oxysterol-mediated endothelial stiffness, no functional distinction between nonenzymatically and enzymatically derived oxysterols, nor between sterol-ring oxidized oxysterols and side-chain oxidized oxysterols, could be made.

Oxysterols might also contribute to endothelial dysfunction through a pro-apoptotic effect. A great bulk of experimental data supports the ability of a number of oxysterols of pathophysiological interest to stimulate programmed death pathways in vascular cells (Vejux and Lizard 2009). More recently, in a study on HUVECs, it was shown that the marked stimulation of the mitochondrial pathway of cell death by 7-K, and also by 7β-OH, involves upregulation of superoxide generation, release of the von Willebrand factor (VWF), and release of the lysosomal protein cathepsin D. VWF certainly plays a crucial role in platelet adhesion to the endothelium, and cathepsin D is known to be involved in protein degradation and cell death (Li et al. 2011).

In connection with the possible involvement of LXR agonists in maintaining the homeostasis of the endothelium, one of the two isoforms of LXR namely LXRβ, has been shown to co-localize with estrogen α receptors on the lipid rafts of ECs, and to interact with the latter receptors in preserving cell integrity (Ishikawa et al. 2013). This study did not employ oxysterols to stimulate LXR activity, but demonstrated indirectly that LXR activation might also have a nonnuclear function. In addition, oxysterols that are recognized as LXR ligands may signal to the nucleus of target cells through molecular pathways that do not involve LXRs as transcription factors. In this connection, a recent study has shown that the hyper-regulation of interleukin (IL)-8 and intercellular adhesion molecule-1, as induced in HUVECs by 22-OH and by 24,25-epoxycholesterol (25-EPOX), was not decreased even slightly when the cells were transfected with small interfering RNAs (siRNAs) targeting LXRα and/ or LXRβ (Morello et al. 2009). Also in this case, it can be argued that the two oxysterols tested, while being good LXR ligands, have not been found in oxLDL nor in atheromas. However, the lack of effect of LXR knock-down on the pro-inflammatory action displayed by the cholesterol oxides employed, provides further evidence of the existence of different signaling pathways that this class of compounds might, in principle, be able to trigger and sustain.

Finally, an important question is related to the degree of expression of LXRs in the different areas of the arterial tree. Is this expression uniform throughout, or does it differ in different areas? In relation to this very important point, a recent study, carried out on mouse aorta, showed that LXRs and their target genes are more expressed in straight sections of the arterial wall, where the blood flow is steady and laminar, while they are markedly downregulated in areas of turbulent flow, such as the atherosclerotic-prone aortic arch (Zhu et al. 2008).

With regard to the involvement of cholesterol oxides in the initiation of atherosclerotic lesions, the current state of research indicates that they can indeed "modulate" endothelial dysfunction and that their pro-inflammatory effects appear to be prominent in atherosclerotic-prone areas of the arterial tree.

OXYSTEROLS AND SUB-INTIMAL MACROPHAGES

It is quite clear that macrophages play a key role in the pathogenesis of atheroma, that is, in atherosclerotic lesions (Ley et al. 2011). Attracted by chemotactic factors, and favored by the expression of adhesion molecules on ECs, monocytic cells move from the blood stream into the sub-intimal space, where they are induced by various stimuli to differentiate into macrophages and, subsequently into foam cells. SMCs also contribute to the formation of foam cells. The crucial event that differentiates monocytic cells and SMCs into cells that can take up and accumulate oxLDL is the expression of scavenger receptor of class B CD36 on the cell surface (Figure 15.5).

Of the oxidized lipids present in LDL, oxysterols were the first to be shown to induce CD36 overexpression in cells of the macrophage lineage (Leonarduzzi et al. 2008). Among the various components of the oxysterol mixture that can overexpress CD36, at the concentrations present in the mixture, 7α-OH and α-EPOX were those with the strongest effect. It is possible that other scavenger receptors contribute to this process; however, the knock-down of CD36 in oxysterol-treated macrophages, by means of pre-incubation with anti-CD36-specific antibodies, prevented the formation of foam cells (Leonarduzzi et al. 2008).

The molecular mechanism underlying the oxysterol-induced stimulation of CD36 expression and synthesis in macrophages has been shown to be based on activation of the protein kinase Cδ (PKCδ) and of the extracellular signal-regulated protein kinase (ERK) signaling pathway; this activation leads to the upregulation of peroxisome proliferator-activated receptor γ (PPARγ). This was confirmed by the effective prevention of oxysterol-induced CD36 upregulation in experiments employing selective chemical inhibitors or antagonists (Leonarduzzi et al. 2008).

Besides the generation of foam cells, oxLDL-driven activation of cells of the macrophage lineage within the arterial sub intima, and specifically driven by oxysterols, leads to the synthesis of a great variety of bioactive molecules. Because of the universally recognized role of inflammation in the development and possible complication of atherosclerotic plaques, for many years research has focused on the production of

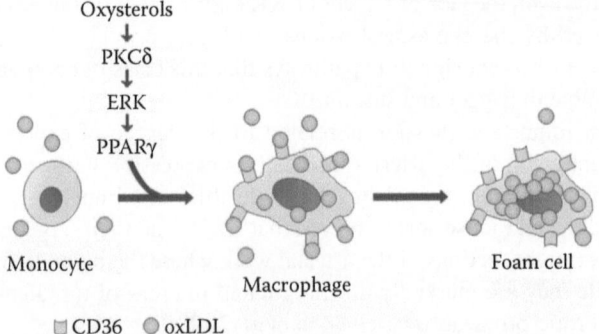

FIGURE 15.5 Oxysterols induce foam cell formation through upregulation of CD36 receptor. PKCδ, protein kinase Cδ; ERK, extracellular signaling-regulated kinase; PPARγ, peroxisome proliferator-activated receptor γ; oxLDL, oxidized LDL.

pro-inflammatory molecules. However, we now know that the overall inflammatory process taking place within the diseased arterial wall is the steady-state result of two opposing forces, that is, pro-inflammatory reactions and anti-inflammatory reactions.

Importantly, the prevalence of either inflammatory or anti-inflammatory stimuli generated by activated macrophages depends on the cell phenotype prevailing at any point in time. To make the complex story simple, macrophages can either express a pro-inflammatory phenotype (macrophage type I or M1) or a net anti-inflammatory, pro-angiogenetic and pro-fibrogenic phenotype (macrophage type II or M2) (Leitinger and Schulman 2013; Mantovani et al. 2009).

A careful analysis of the functional levels of inflammation-related genes in macrophages of normal intima, and of those in atherosclerotic plaque, showed that in normal intima a macrophage phenotype prevails that is characterized by the activation of genes mediating reverse cholesterol transport and contrasting macrophage attraction and accumulation in the sub-intimal space. Conversely, in the altered intima of diseased arteries, cholesterol efflux is somewhat depressed and several pro-inflammatory genes tend to be overexpressed (Waldo et al. 2008).

In atherosclerotic lesions M1 and M2 macrophage phenotypes represent two extreme opposing conditions. Depending on the concomitant presence of different environmental signals, a number of intermediate phenotypic conditions come about. A reliable representation of the complex question of macrophage modulation, with the eventual activation of pro- or antiatherogenic programs, is that of a "macrophage balance" between pro-inflammatory and anti-inflammatory cytokine production (Mantovani et al. 2009). This concept is schematically illustrated in Figure 15.6.

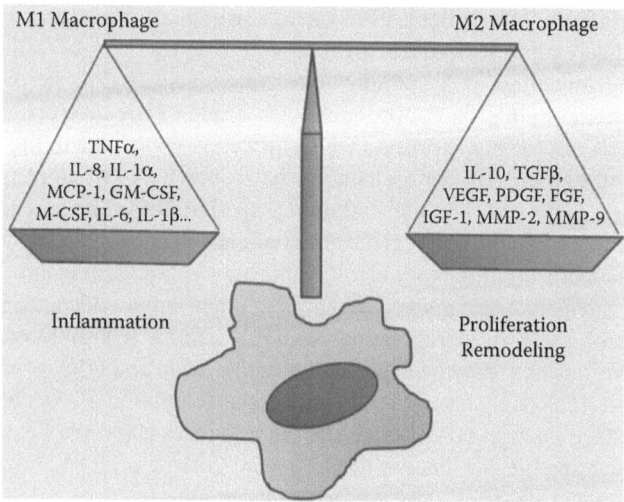

FIGURE 15.6 The macrophage balance between the two extreme M1 and M2 phenotypes. TNFα, tumor necrosis factor α; IL, interleukin; MCP-1, monocyte chemotactic protein-1; GM-CSF, granulocyte–macrophage colony-stimulating factor; M-CSF, macrophage colony-stimulating factor; TGFβ, transforming growth factor β; VEGF, vascular endothelial growth factor; PDGF, platelet-derived growth factor; FGF, fibroblast growth factor; IGF-1, insulin-like growth factor-1; MMP, matrix metalloprotease.

Induction of a M1 macrophage phenotype is mediated by interferon γ (IFNγ), high mobility group box protein 1 (HMGB1), and tumor necrosis factor α (TNFα), as well as by all ligands of Toll-like Receptor 2 (TLR2) such as glycolipids and lipo-proteins (bacteria) and zymosan (fungi), and all ligands of TLR4 such as lipopoly-saccharides from Gram-negative bacteria, fibrinogen, or hyaluronic acid fragments. Moreover, among environmental pro-inflammatory stimuli within the sub-intimal space, oxLDL undoubtedly plays a very important role. The oxysterols, a class of compounds that may trigger and sustain inflammation, at least by enhancing the expression of the redox-sensitive transcription factor NF-κB, are quantitatively important in oxLDL (Leonarduzzi et al. 2005b). Administered to colonic epithe-lial cells (CaCo2 cell line), another oxysterol of pathophysiological interest, namely 7β-OH, strongly enhanced the expression of key inflammatory and chemotactic cytokines, including IL-1α, IL-6, IL-8, IL-23, monocyte chemotactic protein-1 (MCP-1), and transforming growth factor β1 (TGFβ1); in all cases this action was stronger than that of unoxidized cholesterol (Mascia et al. 2010). Cytokines includ-ing IL-6 and IL-8 and TNFα were recently found to be upregulated by a mixture of oxysterols, or by individual cholesterol oxides, in THP-1 (Yehuda et al. 2011) and U937 promonocytic cells (Gargiulo et al. 2014). Of note, all those cytokines except TGFβ1 recognize NF-κB as the main transcription factor.

On the other side of the "macrophage balance," cytokines such as IL-4, IL-10, and IL-13 have a net anti-inflammatory effect, as also do LXR ligands. As certain oxysterols, but not all of them, are good LXR ligands, the idea that all cholesterol oxides activate anti-inflammatory programs risks becoming a firm but erroneous belief. In order to avoid this bias, the recent review by Olkkonen on the potential pro- and anti-inflammatory effects of oxysterols in cells of the macrophage lineage is recommended reading (Olkkonen 2012).

To date, the "dynamic" concept may be accepted, whereby oxysterols of nonen-zymatic origin are mainly on the pro-atherogenic side of the macrophage balance, while oxysterols of enzymatic origin may significantly contribute to the antiathero-genic side, with one *proviso*: several studies have shown 27-OH to trigger, at least in macrophages, both pro- and anti-inflammatory signaling. The mechanisms that may modulate or switch the atherogenic effect of a given cholesterol oxide remain to be clarified. The pro-inflammatory effect displayed by 27-OH in macrophagic cells might be dependent upon the marked overproduction of superoxide and H_2O_2 caused by the oxysterol itself, by upregulating NADPH oxidase (Gargiulo et al. unpub-lished data). Moreover, it may be expected that superoxide and H_2O_2 overproduction induced by 27-OH strongly favors the nonenzymatic generation of oxysterols, at least in the macrophage lineage; this is a pathway that produces cholesterol oxides that are capable of activating NF-kB, but not LXRs.

Alongside oxidative stress, another factor that might affect the balance between pro-inflammatory and anti-inflammatory effects of plaque macrophages is the quan-titative and/or qualitative modulation of the cytoplasmic pool of oxysterol binding proteins; these belong to a family of carriers involved in sterol uptake and transport. Impairment of this protein pool might favor the intracellular accumulation of oxys-terols, thus interfering not only in lipid transport but also in pro- or anti-inflamma-tory programs, or possibly even in cell fate decisions (Olkkonen 2012).

Another interesting aspect of the interaction of oxysterols and M1 and M2 macrophages has very recently emerged: it has been shown that 7-K, a poor ligand of LXRs, influences and modulates the macrophage phenotype, specifically by enhancing the pro-inflammatory effects of M1 macrophages, and switching M2 macrophages to an intermediate phenotype, characterized by increased production of pro-inflammatory cytokines and chemokines. Human peripheral blood monocytes were induced to express either the M1 or the M2 phenotype, by a 6-day incubation with granulocyte–monocyte colony stimulating factor (GM-CSF), or monocyte colony stimulating factor (M-CSF), respectively. Following incubation with 7-K, 30 µM final concentration, M1 macrophages doubled their synthesis of IL-6, IL-10, TNFα, RANTES, and CC chemokine ligand 4 (CCL4), and significantly augmented production of IL-8 and MCP-1; M2 macrophages treated with the oxysterol markedly induced the synthesis of all mentioned pro-inflammatory cytokines, whose production in untreated M2 cells was confirmed to be very low or only moderate. 7-K treatment of M2 macrophages exerted a weak effect on the already-elevated IL-10 production, but conversely strongly upregulated the synthesis and activity of MMP-9 (Buttari et al. 2013).

INDUCTION OF SMCs MIGRATION AND PROLIFERATION

Only some of the SMCs actually involved in plaque formation derive from the artery's tonaca media. A significant fraction is generated directly within the subintimal space by stimulated proliferation of the SMC sub-population resident there (Gordon et al. 1990).

Migratory and proliferative activities of vascular SMCs are modulated by many different molecules, including IL-1 among the promoters, and TGFβ1 among the inhibitors, that is, two cytokines whose expression and synthesis appear to be strongly stimulated in oxysterol-treated monocytes and macrophages (Lizard et al. 1997; Leonarduzzi et al. 2001).

In the authors' laboratory, when human artery SMCs in primary culture were challenged with a similar oxysterol mixture, cell incorporation of tritiated thymidine increased significantly, likely *via* the downregulation of cyclin-dependent kinase inhibitor p21 and p27 proteins, whereas cell viability remained unaffected (Leonarduzzi et al. unpublished results).

A mitogenic effect has been reported for 7-K and α-EPOX on rat aortic SMCs in primary culture, at pathophysiologically compatible concentrations. Both cell proliferation and cell migration were strongly enhanced by the two oxysterols, likely through upregulation of epidermal growth factor (EGF) receptors and the ERK and phosphoinositide 3-kinase (PI3K)/Akt pathways, *via* a signal transduction step known to be involved in SMC activation by a variety of stimuli (Liao et al. 2010). Another interesting finding resulting from the same experimental work, and related to the observed enhancement of cell migration, was the ability of both 7-K and α-EPOX to overexpress MMP-2 and MMP-9, and to stimulate their enzymatic activity (Liao et al. 2010). Thus, the increased presence and activity of gelatinases 2 and 9 in the sub-intimal space is likely due not only to macrophages but also to SMCs.

INDUCTION OF CELL DEATH AND EXPANSION OF THE NECROTIC CORE

A large and growing body of literature indicates that certain oxysterols, under given experimental conditions, exhibit strong pro-apoptotic properties against all types of vascular cells implicated in atheroma formation and development (Berthier et al. 2004; Pedruzzi et al. 2004; Poli et al. 2009; Vejux and Lizard 2009).

Apoptosis of cells of the macrophage lineage is undoubtedly a primary mechanism of expansion of the necrotic core, because monocytes, macrophages and foam cells are located centrally in the atherosclerotic lesion. Moreover, the increased clearance of SMCs that occurs during progression of the atherosclerotic process, has two consequences: on the one hand it contributes to enlargement of the necrotic area inside the atheroma, while on the other, it is a major cause of plaque instability, because the network of myofibroblasts and connective matrix that surrounds the atheroma is thereby demolished. Very recent reports confirm that increased cytoplasmic Ca^{2+} concentration is a very early event in oxysterol-induced programmed cell death of both macrophages (Mackrill 2011) and SMCs (Appukuttan et al. 2013). Exposure of human promonocytic cells (U937 cell line) to either 7β-OH or β-EPOX upregulates intracellular levels of Ca^{2+}, and triggers apoptosis (Lordan et al. 2009); this pathway had previously been found to be upregulated in the same type of cells following incubation with a biologically relevant mixture of oxysterols (Leonarduzzi et al. 2007). Again, apoptosis induced in vascular SMCs isolated from rat aorta, then challenged with 7-K or 25-OH, was confirmed to be initiated by overproduction of reactive oxygen species (ROS), and cytochrome c release from mitochondria (Appukuttan et al. 2013).

PLAQUE RUPTURE

A sustained inflammatory process together with extensive cell remodeling and loss are key mechanisms of plaque expansion and progression. As the atherosclerotic lesion develops, the risk of instability increases, the latter condition predisposing to plaque rupture and atherothrombosis. The upregulation of MMP-2 and MMP-9 expression and activity, in rat aortic SMCs challenged with either 7-K or α-EPOX, was mentioned above (Liao et al. 2010). Again, with regard to the macrophage lineage, 7-K has been found to upregulate MMP-9 in human blood monocytes differentiated into M2 macrophages by pre-incubation with M-CSF (Buttari et al. 2013).

The oxysterol-induced upregulation of MMP-9 levels in macrophagic cells has been comprehensively investigated in a human promonocytic cell line (U937); the study simultaneously examined expression and levels of the natural tissue inhibitors of the protease, as well as the signal transduction underlying protease enhancement (Gargiulo et al. 2011). Cell aliquots were incubated for up to 24–48 h in the presence of an oxysterol mixture of composition similar to that detectable in advanced human carotid plaques, a treatment already demonstrated to differentiate the promonocytic cells into macrophages (Leonarduzzi et al. 2008). While challenge with the oxysterol mixture markedly stimulated expression, synthesis and enzymatic activity of MMP-9 in the macrophages, the same treatment modified neither expression nor synthesis of tissue inhibitors of metalloprotease-1 (TIMP-1) or -2 (TIMP-2) (Gargiulo et al. 2011).

This marked change in the balance between MMP-9 and TIMPs production in macrophages might significantly contribute to atherosclerotic plaque destabilization and rupture. The MMP-9/TIMP-1 ratio has been found to be consistently increased in carotid plaques from patients who underwent endarterectomy, but was significantly higher in disrupted than in nondisrupted plaques (Higashikata et al. 2006; Sapienza et al. 2005).

To return to the above study in which U937 cells were challenged with an oxysterol mixture of pathophysiological relevance, analysis of the molecular mechanisms triggered by the mixture leading to MMP-9 upregulation showed that a concatenated series of events occurred, namely superoxide and H_2O_2 overproduction due to NADPH-oxidase activation and mitochondria dysfunction, upregulation of mitogen-activated protein kinase (MAPK) signaling, and upregulation of activator protein-1 (AP-1)- and NF-κB-DNA binding.

Notably, at the concentrations present in the mixture, 27-OH and 7α-OH, were the main molecules responsible for the observed MMP-9 overexpression (Gargiulo et al. 2011). The same two cytokines have very recently been shown to strongly enhance the synthesis of TNFα, in the THP-1 human macrophagic cell line; this oxysterol-mediated induction of a pro-inflammatory microenvironment was interpreted as an important contribution to plaque instability (Kim et al. 2013).

SUMMARY AND FUTURE QUESTIONS

Undoubtedly oxysterols, cholesterol oxidation products consistently detected in both early and stable and unstable advanced atherosclerotic lesions, play a key role in determining the fate of atherosclerosis. As we have discussed in this chapter, oxysterols, by far more bioreactive than unoxidized cholesterol, may trigger both pro-inflammatory and anti-inflammatory stimuli through more than a single signaling pathway and transcription factor modulation. As extensively reported in literature, in the arterial tract affected by the atherosclerotic process, oxysterols predominantly play a pro-inflammatory role, most likely in all principal steps of the multiphasic and multifactorial process of arterial degeneration. They contribute to endothelial dysfunction, and they are key players in foam cell formation and in the expansion of the plaque necrotic core; moreover, they might have a primary role in plaque destabilization through a marked upregulation of MMP-9 expression, synthesis, and activity. In this connection, focusing on the emerging role recognized by the macrophage M1 and M2 phenotypes in the development of atherosclerosis, it must be underlined how some of the oxysterols detected in fibrotic plaques might significantly contribute to monocyte differentiation and macrophagic phenotype modulation with a consequent upregulation of a variety of inflammatory-related genes.

Conversely, at least in relation to atherosclerosis, the anti-inflammatory properties of these compounds have not yet been proven although they cannot be ruled out, as suggested by a growing body of evidence, which links oxysterols to important anti-inflammatory mediators, such as the nuclear receptors LXRs. However, it is important to note that the expression of LXRs is downregulated in areas of the arterial tree more prone to develop atheromas. As a consequence, in these arterial tracts a dominant inflammatory process is induced by one or more cardiovascular risk factors through the involvement of pro-inflammatory transcription peptides, such as NF-κB, instead of LXRs, which might be either activated by oxysterols.

Moreover, as described in this chapter, it has been fully established that oxysterols also exert a pro-apoptotic stimuli against the various types of vascular cells, thus likely contributing to the expansion of the necrotic core of the lesion and to its instability. Conversely, a possible antiapoptotic action of oxysterols may be only speculated since it has not been experimentally supported so far.

On this basis, a versatile even if contradictory role of oxysterols may emerge, which poses the question of how they actually affect the atherosclerotic lesion development and which factors might modulate their behavior. In this connection, how much does it depend on the nature of oxysterols themselves (e.g., chemical structure, enzymatic or nonenzymatic origin) and how much on extrinsic factors (e.g., stage of disease progression, cell phenotype, presence of other bioactive compounds)?

In conclusion, to better understand the involvement of oxysterols in the pathogenesis of atherosclerosis a deeper knowledge of their potentially dual behavior is still necessary and deserves further consideration and experimental supporting.

ACKNOWLEDGMENTS

The authors wish to thank the University of Turin and the CRT Foundation, Turin, for supporting the research here quoted.

ABBREVIATIONS

α-EPOX	cholesterol-5α,6α-epoxide
β-EPOX	cholesterol-5β,6β-epoxide
7α-OH	7α-hydroxycholesterol
7β-OH	7β-hydroxycholesterol
7-K	7-ketocholesterol
20-OH	20-hydroxycholesterol
22-OH	22-hydroxycholesterol
24-OH	24-hydroxycholesterol
25-EPOX	24,25-epoxycholesterol
25-OH	25-hydroxycholesterol
27-OH	27-hydroxycholesterol
ABC	ATP-binding cassette transporter
AP-1	activator protein-1
CCL	CC chemokine ligand
COX	cyclooxygenase
EC	endothelial cell
EGF	epidermal growth factor
eNOS	endothelial nitric oxide synthase
ERK	extracellular signal-regulated protein kinase

FGF	fibroblast growth factor
GM-CSF	granulocyte–monocyte colony stimulating factor
ICAM	intercellular adhesion molecule
IFN	interferon
IGF	insulin-like growth factor
IL	interleukin
iNOS	inducible nitric oxide synthase
HMGB	high mobility group box protein
HUVEC	human umbilical vein endothelial cell
LDL	low-density lipoprotein
LXR	liver X receptor
M-CSF	monocyte colony stimulating factor
M1 and M2	type I and type II macrophage
MAPK	mitogen-activated protein kinase
MCP	monocyte chemotactic protein
MIP	monocyte inflammatory protein
mmLDL	minimally modified LDL
MMP	matrix metalloprotease
NF-κB	nuclear factor-κB
oxLDL	oxidized LDL
PDGF	platelet-derived growth factor
PI3K	phosphoinositide 3-kinase
PKC	protein kinase C
PPAR	peroxisome proliferator-activated receptor
ROS	reactive oxygen species
siRNA	small interfering RNA
SMC	smooth muscle cell
TGF	transforming growth factor
TIMP	tissue inhibitor of metalloprotease
TLR	Toll-like receptor
TNF	tumor necrosis factor
TRIOL	cholesterol-3β,5α,6β-triol
VCAM	vascular cell adhesion molecule
VEGF	vascular endothelial growth factor
VWF	Von Willebrand factor

REFERENCES

Appukuttan, A., S. A. Kasseckert, S. Kumar et al. 2013. Oxysterol-induced apoptosis of smooth muscle cells is under the control of a soluble adenylyl cyclase. *Cardiovasc Res* 99 (4):734–42.

Bensinger, S. J., and P. Tontonoz. 2008. Integration of metabolism and inflammation by lipid-activated nuclear receptors. *Nature* 454 (7203):470–7.

Berrodin, T. J., Q. Shen, E. M. Quinet, M. R. Yudt, L. P. Freedman, and S. Nagpal. 2010. Identification of 5α, 6α-epoxycholesterol as a novel modulator of liver X receptor activity. *Mol Pharmacol* 78 (6):1046–58.

Berthier, A., S. Lemaire-Ewing, C. Prunet et al. 2004. Involvement of a calcium-dependent dephosphorylation of BAD associated with the localization of Trpc-1 within lipid rafts in 7-ketocholesterol-induced THP-1 cell apoptosis. *Cell Death Differ* 11 (8):897–905.

Björkhem, I., O. Andersson, U. Diczfalusy et al. 1994. Atherosclerosis and sterol 27-hydroxy-lase: Evidence for a role of this enzyme in elimination of cholesterol from human macrophages. *Proc Natl Acad Sci USA* 91 (18):8592–6.

Brown, A. J., and W. Jessup. 1999. Oxysterols and atherosclerosis. *Atherosclerosis* 142 (1):1–28.

Buttari, B., L. Segoni, E. Profumo et al. 2013. 7-Oxo-cholesterol potentiates pro-inflammatory signaling in human M1 and M2 macrophages. *Biochem Pharmacol* 86 (1):130–7.

Byfield, F. J., S. Tikku, G. H. Rothblat, K. J. Gooch, and I. Levitan. 2006. OxLDL increases endothelial stiffness, force generation, and network formation. *J Lipid Res* 47 (4):715–23.

Calkin, A. C., and P. Tontonoz. 2010. Liver X receptor signaling pathways and atherosclerosis. *Arterioscler Thromb Vasc Biol* 30 (8):1513–8.

Fu, X., J. G. Menke, Y. Chen et al. 2001. 27-Hydroxycholesterol is an endogenous ligand for liver X receptor in cholesterol-loaded cells. *J Biol Chem* 276 (42):38378–87.

Garcia-Cruset, S., K. L. Carpenter, F. Guardiola, B. K. Stein, and M. J. Mitchinson. 2001. Oxysterol profiles of normal human arteries, fatty streaks and advanced lesions. *Free Radic Res* 35 (1):31–41.

Gargiulo, S., P. Gamba, G. Testa, F. Biasi, G. Poli, and G. Leonarduzzi. 2015. Relation between TLR4/NF-kB signaling pathway activation by 27-hydroxycholesterol and 4-hydroxynonenal, and atherosclerotic plaque instability. Accepted for publication to *Aging Cell*.

Gargiulo, S., B. Sottero, P. Gamba, E. Chiarpotto, G. Poli, and G. Leonarduzzi. 2011. Plaque oxysterols induce unbalanced up-regulation of matrix metalloproteinase-9 in macrophagic cells through redox-sensitive signaling pathways: Implications regarding the vulnerability of atherosclerotic lesions. *Free Radic Biol Med* 51 (4):844–55.

Gordon, D., M. Reidy, E. Benditt, and S. Schwartz. 1990. Cell proliferation in human coronary arteries. *Proc Natl Acad Sci USA* 87 (12):4600–4.

Helmschrodt, C., S. Becker, J. Schröter et al. 2013. Fast LC–MS/MS analysis of free oxysterols derived from reactive oxygen species in human plasma and carotid plaque. *Clin Chim Acta* 425c:3–8.

Higashikata, T., M. Yamagishi, T. Higashi et al. 2006. Altered expression balance of matrix metalloproteinases and their inhibitors in human carotid plaque disruption: Results of quantitative tissue analysis using real-time RT-PCR method. *Atherosclerosis* 185 (1):165–72.

Honzumi, S., A. Shima, A. Hiroshima, T. Koieyama, and N. Terasaka. 2011. Synthetic LXR agonist inhibits the development of atherosclerosis in New Zealand White rabbits. *Biochim Biophys Acta* 1811 (12):1136–45.

Ishikawa, T., I. S. Yuhanna, J. Umetani et al. 2013. LXRβ/estrogen receptor-α signaling in lipid rafts preserves endothelial integrity. *J Clin Invest* 123 (8):3488–97.

Kim, S. M., H. Jang, and Y. Son. 2013. 27-Hydroxycholesterol induces production of tumor necrosis factor-alpha from macrophages. *Biochem Biophys Res Commun* 430 (2):454–9.

Leitinger, N., and I. G. Schulman. 2013. Phenotypic polarization of macrophages in atherosclerosis. *Arterioscler Thromb Vasc Biol* 33 (6):1120–6.

Leonarduzzi, G., E. Chiarpotto, F. Biasi, and G. Poli. 2005a. 4-Hydroxynonenal and cholesterol oxidation products in atherosclerosis. *Mol Nutr Food Res* 49 (11):1044–9.

Leonarduzzi, G., P. Gamba, S. Gargiulo et al. 2008. Oxidation as a crucial reaction for cholesterol to induce tissue degeneration: CD36 overexpression in human promonocytic cells treated with a biologically relevant oxysterol mixture. *Aging Cell* 7 (3):375–82.

Leonarduzzi, G., P. Gamba, B. Sottero et al. 2005b. Oxysterol-induced up-regulation of MCP-1 expression and synthesis in macrophage cells. *Free Radic Biol Med* 39 (9):1152–61.

Leonarduzzi, G., S. Gargiulo, and P. Gamba et al. 2010. Molecular signaling operated by a diet-compatible mixture of oxysterols in up-regulating CD36 receptor in CD68 positive cells. *Mol Nutr Food Res* 54 (Suppl 1):S31–41.

Leonarduzzi, G., G. Poli, B. Sottero, and F. Biasi. 2007. Activation of the mitochondrial pathway of apoptosis by oxysterols. *Front Biosci* 12:791–9.

Leonarduzzi, G., A. Sevanian, B. Sottero et al. 2001. Up-regulation of the fibrogenic cytokine TGF-beta1 by oxysterols: A mechanistic link between cholesterol and atherosclerosis. *FASEB J* 15 (9):1619–21.

Leonarduzzi, G., B. Sottero, and G. Poli. 2002. Oxidized products of cholesterol: Dietary and metabolic origin, and proatherosclerotic effects. *J Nutr Biochem* 13 (12):700–10.

Levitan, I., S. Volkov, and P. V. Subbaiah. 2010. Oxidized LDL: Diversity, patterns of recognition, and pathophysiology. *Antioxid Redox Signal* 13 (1):39–75.

Ley, K., Y. I. Miller, and C. C. Hedrick. 2011. Monocyte and macrophage dynamics during atherogenesis. *Arterioscler Thromb Vasc Biol* 31 (7):1506–16.

Li, W., M. Ghosh, S. Eftekhari, and X. M. Yuan. 2011. Lipid accumulation and lysosomal pathways contribute to dysfunction and apoptosis of human endothelial cells caused by 7-oxysterols. *Biochem Biophys Res Commun* 409 (4):711–6.

Liao, P. L, Y. W. Cheng, C. H. Li, Y. L. Lo, and J. J. Kang. 2009. Cholesterol-3-beta, 5-alpha, 6-beta-triol induced PI(3)K-Akt-eNOS-dependent cyclooxygenase-2 expression in endothelial cells. *Toxicol Lett* 190 (2):172–8.

Liao, P. L., Y. W. Cheng, C. H. Li, Y.T. Wang, and J. J. Kang. 2010. 7-Ketocholesterol and cholesterol-5alpha, 6alpha-epoxide induce smooth muscle cell migration and proliferation through the epidermal growth factor receptor/phosphoinositide 3-kinase/Akt signaling pathways. *Toxicol Lett* 197 (2):88–96.

Lizard, G., S. Lemaire, S. Monier, S. Gueldry, D. Néel, and P. Gambert. 1997. Induction of apoptosis and of interleukin-1beta secretion by 7beta-hydroxycholesterol and 7-ketocholesterol: Partial inhibition by Bcl-2 overexpression. *FEBS Lett* 419 (2–3):276–80.

Lordan, S., N. M. O'Brien, and J. J. Mackrill. 2009. The role of calcium in apoptosis induced by 7beta-hydroxycholesterol and cholesterol-5beta,6beta-epoxide. *J Biochem Mol Toxicol* 23 (5):324–32.

Mackrill, J. J. 2011. Oxysterols and calcium signal transduction. *Chem Phys Lipids* 164 (6):488–95.

Mantovani, A., C. Garlanda, and M. Locati. 2009. Macrophage diversity and polarization in atherosclerosis: A question of balance. *Arterioscler Thromb Vasc Biol* 29 (10):1419–23.

Mascia, C., M. Maina, E. Chiarpotto, G. Leonarduzzi, G. Poli, and F. Biasi. 2010. Proinflammatory effect of cholesterol and its oxidation products on CaCo-2 human enterocyte-like cells: Effective protection by epigallocatechin-3-gallate. *Free Radic Biol Med* 49 (12):2049–57.

Morello, F., E. Saglio, A. Noghero et al. 2009. LXR-activating oxysterols induce the expression of inflammatory markers in endothelial cells through LXR-independent mechanisms. *Atherosclerosis* 207 (1):38–44.

Olkkonen, V. M. 2012. Macrophage oxysterols and their binding proteins: Roles in atherosclerosis. *Curr Opin Lipidol* 23 (5):462–70.

Patel, R. P., U. Diczfalusy, S. Dzeletovic, M. T. Wilson, and V. M. Darley-Usmar. 1996. Formation of oxysterols during oxidation of low density lipoprotein by peroxynitrite, myoglobin, and copper. *J Lipid Res* 37 (11):2361–71.

Pedruzzi, E., C. Guichard, V. Ollivier et al. 2004. NAD(P)H oxidase Nox-4 mediates 7-ketocholesterol-induced endoplasmic reticulum stress and apoptosis in human aortic smooth muscle cells. *Mol Cell Biol* 24 (24):10703–17.

Poli, G., B. Sottero, S. Gargiulo, and G. Leonarduzzi. 2009. Cholesterol oxidation products in the vascular remodeling due to atherosclerosis. *Mol Aspects Med* 30 (3):180–9.

Sapienza, P., L. di Marzo, V. Borrelli et al. 2005. Metalloproteinases and their inhibitors are markers of plaque instability. *Surgery* 137 (3):355–63.

Shentu, T. P., D. K. Singh, M-J. Oh et al. 2012. The role of oxysterols in control of endothelial stiffness. *J Lipid Res* 53 (7): 1348–58.

Schroepfer, G. J. Jr. 2000. Oxysterols: Modulators of cholesterol metabolism and other processes. *Physiol Rev* 80 (1):361–554.

Sevanian, A., G. Bittolo-Bon, G. Cazzolato et al. 1997. LDL- is a lipid hydroperoxide-enriched circulating lipoprotein. *J Lipid Res* 38 (3):419–28.

Töröcsik, D., A. Szanto, and L. Nagy. 2009. Oxysterol signaling links cholesterol metabolism and inflammation via the liver X receptor in macrophages. *Mol Aspects Med* 30 (3):134–52.

Usatyuk, P. V., and V. Natarajan. 2012. Hydroxyalkenals and oxidized phospholipids modulation of endothelial cytoskeleton, focal adhesion and adherens junction proteins in regulating endothelial barrier function. *Microvasc Res* 83 (1):45–55.

Vaya, J. 2013. The association between biomarkers in the blood and carotid plaque composition-focusing on oxidized lipids, oxysterols and plaque status. *Biochem Pharmacol* 86 (1):15–8.

Vaya, J., M. Aviram, S. Mahmood et al. 2001. Selective distribution of oxysterols in atherosclerotic lesions and human plasma lipoproteins. *Free Radic Res* 34 (5):485–97.

Vejux, A., and G. Lizard. 2009. Cytotoxic effects of oxysterols associated with human diseases: Induction of cell death (apoptosis and/or oncosis), oxidative and inflammatory activities, and phospholipidosis. *Mol Aspects Med* 30 (3):153–70.

Viennois, E., A. J. Pommier, K. Mouzat et al. 2011. Targeting liver X receptors in human health: Deadlock or promising trail? *Expert Opin Ther Targets* 15 (2):219–32.

Waldo, S. W., Y. Li, C. Buono et al. 2008. Heterogeneity of human macrophages in culture and in atherosclerotic plaques. *Am J Pathol* 172 (4):1112–26.

Yehuda, H., A. Szuchman-Sapir, S. Khatib, R. Musa, and S. Tamir. 2011. Human atherosclerotic plaque lipid extract promotes expression of proinflammatory factors in human monocytes and macrophage-like cells. *Atherosclerosis* 218 (2):339–43.

Yoshida, H., and R. Kisugi. 2010. Mechanisms of LDL oxidation. *Clin Chim Acta* 411 (23–24):1875–82.

Zhu, M., Y. Fu, Y. Hou et al. 2008. Laminar shear stress regulates liver X receptor in vascular endothelial cells. *Arterioscler Thromb Vasc Biol* 28 (3):527–33.

16 Lipid Peroxidation and Age-Related Neurodegenerative Disorders

Tanea T. Reed, Zachariah P. Sellers,
and D. Allan Butterfield

CONTENTS

INTRODUCTION

The brain controls the body's functions through cell signaling and other regulatory pathways. Lipids are hydrophobic biomolecules composed of long hydrocarbon chains. Classes of lipids include but are not limited to phospholipids, triacylglycerols, steroids, and eicosanoids. The brain is abundant in polyunsaturated fatty acids, making lipids vulnerable to oxidation as the brain consumes more than 30% of inspired oxygen. The decomposition of ω-6 and omega ω-3 fatty acids generates several lipid peroxidation products including 4-hydroxy-2-nonenal and 4-hydroxy-2-hexenal, respectively. Mitochondria, the "powerhouse" for ATP production in the cell, are also vulnerable to lipid peroxidation. The process of lipid peroxidation is demonstrated in several neurodegenerative disorders including Alzheimer's disease (AD), Parkinson's disease (PD), amyotrophic sclerosis, and Huntington's disease (HD). This chapter will describe the role of lipid peroxidation in neurodegenerative disease mechanisms as well as the role of mitochondrial involvement in brain disorders.

REDOX PROTEOMICS, LIPID PEROXIDATION, AND NEURODEGENERATIVE DISORDERS

The proteome, a term coined by Wilkins in 1995, can be described as the complete set of proteins expressed in an organism or cell. The study of the proteome is also known as proteomics. Through proteomics, a protein is described in terms of its name, isoelectric point, mass, and amino acid sequence. This technique is helpful in determining the protein identification, modification(s), and function. Protein spots are excised and digested with trypsin, which cleaves proteins at the C-terminal side of Arg and Lys residues. These protein fragments are then analyzed using mass spectrometry in order to yield the protein's identification as well as potential locations of posttranslational modifications. As proteomics is a very time consuming and expensive effort, typically the peptide sequence data generated by mass spectrometry is not confirmed. This can lead to a lower confidence for posttranslational modification. If the peptide sequence data is confirmed, this results in a higher validity for posttranslational modification. In all studies described in this chapter, peptide sequence data was not confirmed. Proteomics is one of the fastest growing areas of biomedical science and is widely being used to determine potential biomarkers in neurodegenerative diseases (Abdi et al. 2006; Herrmann and Obeid 2011; Trojanowski and Hampel 2011). These biomarkers may be used for future disease diagnosis and in the early stages of these disorders, thereby improving patient care overall.

LIPID PEROXIDATION

Lipid peroxidation is a process involving free radical chain reactions that continue until terminated. Oxidation of phospholipids, triacylglycerols, steroids, and eicosanoids can result in protein dysfunction via several mechanisms involving lipid bilayer rearrangement to a "lipid whisker model," in which oxidized fatty acids ("whiskers") are projected into the aqueous layer (Greenberg et al. 2008). This new orientation causes the oxidized fatty acids to interact with cell surface receptors such as those involved in inflammation including PGE2 and CD36 (Deigner and Hermetter 2008; Catala 2012). The process of lipid peroxidation can occur in six steps (Figure 16.1). A carbon-centered radical is formed through free radical attack of an allylic hydrogen linked to a C-centered free radical (step 1). Paramagnetic oxygen rapidly reacts with this radical to generate peroxyl radicals (step 2). Lipid hydroperoxides are formed though the peroxyl radicals continuously reacting with additional allylic H atoms (step 3). These lipid hydroperoxides can decompose to produce multiple reactive

(1) $LH + X\bullet \rightarrow L\bullet + XH$

(2) $L\bullet + O_2 \rightarrow LOO\bullet$

(3) $LOO\bullet + LH \rightarrow LOOH + L\bullet$

(4) $LOO\bullet + LOO\bullet \rightarrow LOOL$ (nonradical) $+ O_2$

(5) Vitamin $E + LOO\bullet \rightarrow$ Vitamin $E\bullet + LOOH$

(6) $GSH +$ Vitamin $E\bullet \rightarrow GSH\bullet +$ Vitamin E

FIGURE 16.1   An example of lipid peroxidation.

products including several α, β-unsaturated aldehydes, including 4-hydroxy-2-non-enal (HNE), 4-hydroxy-2-hexenal (4-HHE), acrolein, and malondialdehyde (Figure 16.2). Via Michael addition, these highly reactive aldehydes, covalently bind with the amino acids cysteine, lysine, and histidine to form stable adducts (Figure 16.3). These protein modifications cause protein dysfunction and reduced enzymatic activity in several proteins including serine–threonine kinase LKB1 (Calamaras et al. 2012), mitochondrial aconitase (Liu et al. 2013), peptidyl-propyl *cis/trans* isomerase (Aluise et al. 2013), sirtuin 3 (Fritz et al. 2011), aldehyde dehydrogenase (Picklo et al. 2011), tubulin (Stewart et al. 2007), and actin (Aldini et al. 2005). However, for certain biomolecules such as cyclopentenone, prostaglandins, and peroxisome

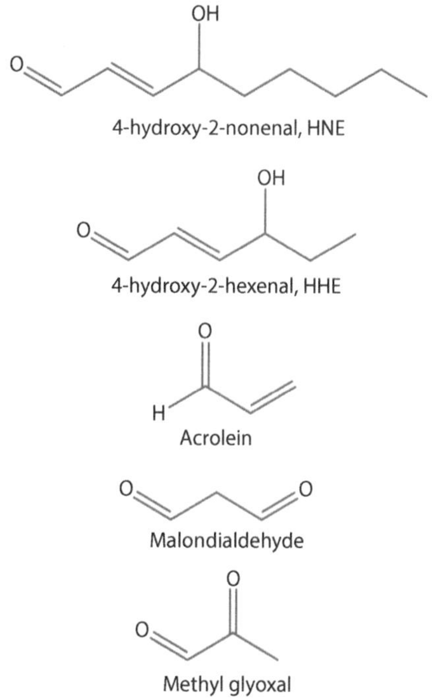

4-hydroxy-2-nonenal, HNE

4-hydroxy-2-hexenal, HHE

Acrolein

Malondialdehyde

Methyl glyoxal

FIGURE 16.2   Structures for 4-hydroxy-2-nonenal (HNE), 4-hydroxyl-2-hexenal (HHE), acrolein, and malondialdehyde, and methyl glyoxal.

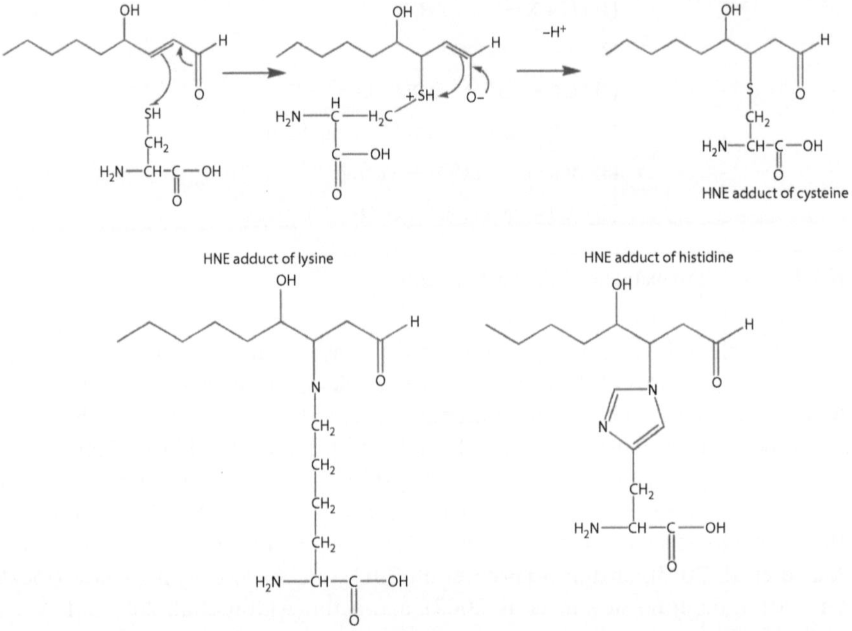

FIGURE 16.3 Covalent modification for amino acid residues by Michael addition.

proliferator-activated receptor (PPAR) γ, this lipid peroxidation may trigger various cell signaling pathways (Almeida et al. 2009; Domingues et al. 2013). Lipid peroxidation can be terminated by two radicals reacting to form a nonradical and oxygen (Figure 16.1, step 4). Vitamin E (α-tocopherol), a "chain breaking" antioxidant, can abrogate propagation steps of lipid peroxidation (step 5). Liu showed that a mixture of tocopherols (γ and δ) had a stronger inhibitory effect on lipid peroxidation than α-tocopherol alone (Liu et al. 2002). When the hydrogen is abstracted in step one, an alpha tocopherol radical forms that can be reverted back to vitamin E by ascorbic acid (vitamin C) or glutathione, both potent antioxidants (step 6).

Although malondialdehyde and HNE represent the major products of lipid peroxidation (Esterbauer et al. 1991), other aldehyde products including acrolein, isoprostanes, neuroprostanes, and glyoxals (Calingasan et al. 1999; Montine et al. 1999; Onyango 2012) are formed. Malondialdehyde (MDA) and HNE cause cell toxicity. The antioxidant tripeptide, glutathione (GSH), has been shown to detoxify HNE in cells (Subramaniam et al. 1997; Joshi et al. 2007). Increased levels of HNE cause membrane damage, disruption of Ca^{2+} homeostasis, and cell death (Esterbauer et al. 1991), while elevated levels of this alkenal cause glutamate transport impairment thereby leading to potential excitotoxicity (Lauderback et al. 2001; Lovell et al. 2012). Both increased acrolein and malondialdehyde can result in loss of phospholipid asymmetry (Castegna et al. 2004), glucose and glutamate uptake impairment (Lovell et al. 2000), and nucleotide excision repair inhibition (Feng et al. 2006).

Acrolein, an electrophilic molecule, has been shown to inhibit glutamate and glucose uptake in primary neuronal cultures thereby demonstrating its effect on cellular transport (Lovell et al. 2000). Isoprostanes, another class of lipid peroxidation by-products, were first introduced by Roberts and Morrow in 1990 (Morrow et al. 1990; Roberts and Morrow 1994). These molecules are produced by the nonenzymatic free-radical-mediated oxidation of arachidonic acid *in vivo*. As the enzyme cyclo-oxygenase is not required for isoprostane synthesis, there are a large number of iso-prostane molecules and their metabolites generated (Roberts and Morrow 1995). Isoprostanes play a central role in smooth muscle contraction, pain sensation, platelet function, and cell proliferation similar to that of prostaglandins (Roberts and Morrow 1997; Morrow and Roberts 1997). Structurally, isoprostanes differ from prostaglan-dins by a *cis* modification on the cyclopentyl ring (Figure 16.4). Neuroprostanes, derived from the nonenzymatic oxidation of *cis*-4,7,10,13,16,19-docosahexanoic acid (DHA), were first identified by Roberts in 1998 (Roberts et al. 1998). Since DHA contains more double bonds than arachidonic acid, neuroprostanes have a higher degree of unsaturation than isoprostanes, thereby increasing their cellular reactivity. Neuroprostanes possess anti-inflammatory properties, and also inhibit proteosome activity (Davies et al. 2002; Musiek et al. 2008). Like other lipid peroxidation prod-ucts, isoprostanes and neuroprostanes can be used as markers of oxidative stress and are being investigated as prospective biomarkers for disease (Greco et al. 2000; Roberts and Morrow 2000).

However, both iso- and neuroprostanes can undergo further reactions to form highly reactive iso- and neuroketals, respectively (Roberts and Morrow 2002).

Glyoxals, a class of organic molecules containing a dialdehyde, have antiviral activity (De Bock et al. 1957). They have been shown to be lipid peroxidation products from lipid and glucose metabolism. Increased levels of glyoxals, specifi-cally methyl glyoxal, have been associated in rheumatoid arthritis (Mukhopadhyay et al. 2007) and diabetes (Kilhovd et al. 2003) via depletion of antioxidant

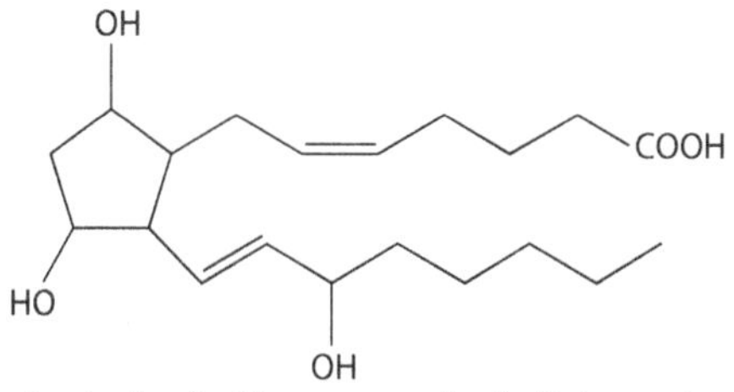

Prostaglandin F2$_\alpha$, an example of a F$_2$ isoprostane

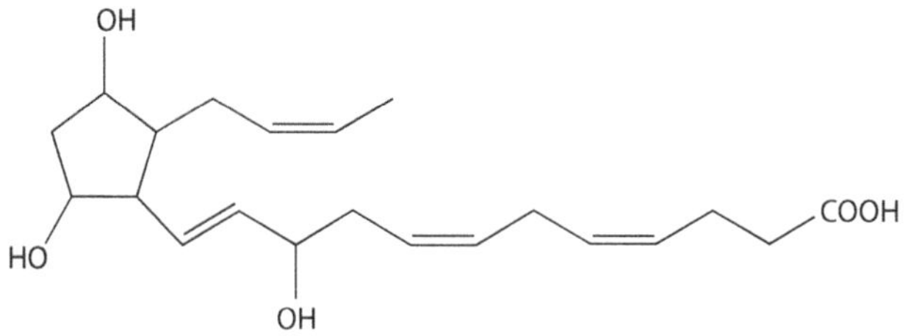

Chemical structure for a series 10 – F$_4$ neuroprostane

FIGURE 16.4   Structure of an isoprostane and a neuroprostane.

defense proteins, sirtuin 1 and advance glycation end product (AGE) receptor 1
(Cai et al. 2012).

IMPORTANCE OF LIPID PEROXIDATION AND ITS EFFECTS
IN NEURODEGENERATIVE DISEASES

Lipids are readily available throughout the body; therefore, lipid peroxidation is a
common phenomenon. Currently, there are no absolute diagnostic biomarkers for
the early diagnosis of any neurodegenerative disease. As lipid peroxidation can have
detrimental effects on proteins, DNA, RNA, and mitochondria, the study of this
oxidative modification and its role on neurodegenerative diseases is paramount. This
chapter will highlight the role of lipid peroxidation in neurodegenerative disease
mechanisms as well as the role of mitochondrial involvement in brain disorders.

ALZHEIMER'S DISEASE

AD is characterized by neurodegeneration that progresses through several stages,
leading to cognitive decline, physical debilitation, dementia, and ultimately death.
Pathologically, AD is evidenced by several disease hallmarks including senile
plaques (SP), neurofibrillary tangles (NFT), neutropil threads (NT), and loss of
synapses. Conclusive diagnosis of the presence of AD is determined post-mortem
specifically by the presence and distribution of NTs and NFTs, where advancement
of the disease correlates with an increased prevalence of these two pathological hall
marks according to the Braak staging method (Braak and Braak 1991).

The progression of AD from preclinical AD (PCAD) to amnestic mild cognitive
impairment (MCI), early AD (EAD), and finally late-stage AD (LAD) is indexed in
part by Braak staging from stage I to stage VI and by several markers of oxidative
stress that are elevated throughout each of these disease stages (Sultana et al. 2013).
In fact, oxidative stress occurs prior to the onset of dementia. Of the various markers
of oxidative stress that exist in AD, products of lipid peroxidation in particular have
been identified at each stage of the disease (Nunomura et al. 2001; Bradley et al.
2012; Sultana et al. 2013). These facts, coupled with the intricate relationship between
pathological hallmarks and oxidative stress, serve to highlight the significance of lipid
peroxidation in the progression and conceivably in the development of this disease.

Lipid Peroxidation during AD Progression

The preclinical stage of AD exists with significant NFT and NT manifestation
resulting in a Braak stage of III or IV (Jicha et al. 2012). However, despite AD neuropathology, these patients do not present any signs of cognitive impairment or loss
of hippocampal neurons (West et al. 2004). It is at this stage that brain cells may be
defending themselves against the effects of these pathological hallmarks, as expression proteomics techniques have identified upregulation of proteins associated with
cellular defense mechanisms (Aluise et al. 2010).

Low sample availability has limited research efforts into the presence of oxidative
stress in the PCAD brain thus far. However, extractable and protein-bound forms

of the lipid peroxidation products, HNE and acrolein, have been characterized in the PCAD brain, as have other markers of oxidative stress from this disease stage and from its transition into MCI (Aluise et al. 2010; Bradley et al. 2010). The ω-3 unsaturated fatty acid peroxidation product, 4-hydroxy-2-hexenal (HHE) was also recently described as being elevated both in extractable and protein-bound forms in hippocampus from subjects with PCAD (Bradley et al. 2012), but in inferior parietal lobule, elevated oxidative stress was not observed (Aluise et al. 2010).

AD progresses from PCAD to MCI as cognitive ability begins to decline without inhibiting activities of daily living (Petersen 2004). This period in the progression of AD includes a subtype diagnosis, termed amnestic MCI, which is distinguished by greater memory impairment (Dubois et al. 2007). While, like PCAD, Braak scores also often lie between stages III or IV in MCI, there is also significant loss of hippocampal synapses and significantly lower levels of the dendritic spine plasticity marker drebrin (DRB). Symptoms of memory impairment are not necessarily indicative of further disease progression as MCI patients can revert back to a normal state (Larrieu et al. 2002; Scheff et al. 2006, 2007; Markesbery 2010; Counts et al. 2012). The prevalence of oxidative damage is increased during MCI compared to PCAD, which has been confirmed by a direct comparison between the inferior parietal lobes of amnestic MCI brains and the PCAD brains using redox proteomics analysis of protein carbonyls (Aluise et al. 2011). Similarly, lipid peroxidation also is more established in brain in MCI, with F_2-isoprostanes (F_2-IsoP), F_4-neuroprostanes (F_4-NP), isoprostane 8,12,-iso-iPF$_{2\alpha}$-VI, protein-bound 4-hydroxy-2-nonenal (HNE), acrolein, malondialdehyde (MDA), and thiobarbituric acid-reactive substances (TBARS) being identified (Pratico et al. 2002; Keller et al. 2005; Markesbery et al. 2005; Butterfield et al. 2006; Williams et al. 2006). Specifically, the proteins beta-actin, pyruvate kinase, ATP synthase, initiation factor alpha, and elongation factor Tu are HNE-modified in amnestic MCI, and ATP synthase has significantly lowered activity during this stage of the disease (Reed et al. 2008).

For those patients whose clinical symptoms of cognitive decline progress beyond MCI, the accompanying neuropathological marks increase by way of NFT formation in the frontal lobe, temporal lobe, subiculum, and amygdala, leading to Braak scores of IV–V (Markesbery et al. 2006). This stage of AD progression is referred to as EAD. Loss of hippocampal synapses, loss of striatum radiatum volume, and absence of the dendritic spine plasticity marker postsynaptic DRB is also more profound in brain of subjects with EAD brain than in subjects with MCI (Scheff et al. 2006, 2007; Counts et al. 2012). Owing to the neuropathological similarities between MCI and EAD, it has even been suggested that amnestic MCI is equivalent to EAD (Morris et al. 2001; Markesbery et al. 2006).

Like MCI brain, EAD brain also has significant malondialdehyde, TBARS, acrolein, and HNE incidence (Keller et al. 2005; Williams et al. 2006). Significant levels of protein-bound HNE has also been characterized in EAD brain, with the proteins manganese superoxide dismutase (Mn-SOD), alpha enolase, F1 ATPase (alpha subunit), dihydropyriminidase-related protein 2 (DRP-2), triose phosphate isomerase (TPI), and malate dehydrogenase (MDH) carrying HNE covalent modifications (Perluigi et al. 2009). It is important to note that activities of Mn-SOD and α-enolase are decreased and the activity of MDH is increased in EAD brain (Perluigi et al. 2009).

Mn-SOD is an important antioxidant enzyme for the cell due to its role in detoxifying the free radical species superoxide (O_2^-), so HNE modification of this protein makes the cell more vulnerable to free radical attack. Alpha enolase facilitates the penultimate step of glycolysis by catalyzing the conversion of 2-phosphoglycerate into phosphoenolpyruvate. With HNE modification of alpha enolase, the cell is at risk of inadequate ATP stores due to inhibited production of pyruvate for fueling the citric acid cycle. Similarly, HNE modification of ATPase can lead to inhibited ATP formation due to the direct role of this enzyme in ATP synthesis. Triose phosphate isomerase catalyzes the reversible conversion of dihydroxyacetone phosphate to glyceraldehyde-3-phosphate in glycolysis and MDH catalyzes the oxidation of malate to oxaloacetate, so HNE modification of these proteins can also lead to lower ATP production.

The final stage of AD, known as LAD, carries Braak scores of IV–VI and a substantial presence of oxidative stress markers (Castegna et al. 2002a,b, 2003; Reed et al. 2008; Butterfield et al. 2013; Sultana et al. 2013). Clinically, this results in severe disability due to dementia and a great deal of required care. Similar to EAD, this dementia can be attributed at least in part due to the significant synapse loss and enlarged synapses found in LAD (Scheff et al. 1990).

An elevated significant burden of lipid peroxidation is carried by LAD patients. Specifically, the lipid peroxidation products HNE, malondialdehyde, acrolein, isoprostane $8,12,-iso-iPF_{2\alpha}$-VI, F_2-IsoP, and F_4-NP (Yao et al. 2003; Markesbery et al. 2005; Lauderback et al. 2001). In addition, several proteins bear modification by HNE in LAD, including glutathione-S-transferase (GST), multidrug-resistant protein 1 (MRP1), ATP synthase (ATPase), glutamine synthetase (GS), peroxiredoxin 6 (PRX6), aldolase, Mn-SOD, alpha enolase, aconitase, alpha tubulin, and DRP-2 (Sultana and Butterfield 2004; Perluigi et al. 2009). A summary of HNE-modified proteins throughout the progression of AD is depicted in Figure 16.5.

Glutathione S-transferase catalyzes the conjugation of glutathione to toxic species such as HNE, so HNE modification of GST could inhibit HNE detoxification. Similarly, MRP1 exports glutathione conjugates such as HNE-bound glutathione out of the cell, so modification of MRP1 could also hinder HNE detoxification. ATP synthase, Complex V in the electron transport chain, is responsible for the bulk of ATP production in oxidative phosphorylation. Oxidative modification of this enzyme greatly impairs ATP production, which can be used to maintain pumps and ion homoeostasis. Glutamine synthetase catalyzes the conversion of synaptic glutamate into glutamine to prevent excitotoxicity; therefore, HNE modification of GS could lead to excitotoxic conditions that can contribute to cell death. Peroxiredoxins detoxify H_2O_2 and play a role in cell differentiation and apoptosis. These antioxidant proteins contain a reactive cysteine residue which can become oxidized to sulfenic acid. Peroxiredoxin VI, a 1-Cys peroxiredoxin, acts as a secondary messenger for cytokines and growth factors. It can form a complex with glutathione S-transferase, which can alter enzymatic activity in both proteins (Ralat et al. 2006). HNE modification of PRX6 could prevent adequate detoxification of peroxide species, severely limiting cellular antioxidant capabilities. Aldolase, a glycolytic enzyme, catalyzes the cleavage of fructose-1,6-bisphosphate to dihydroxyacetone phosphate and glyceraldehyde-3-phosphate, while aconitase catalyzes the conversion of citrate into isocitrate during the citric acid cycle. HNE modification of these metabolic proteins

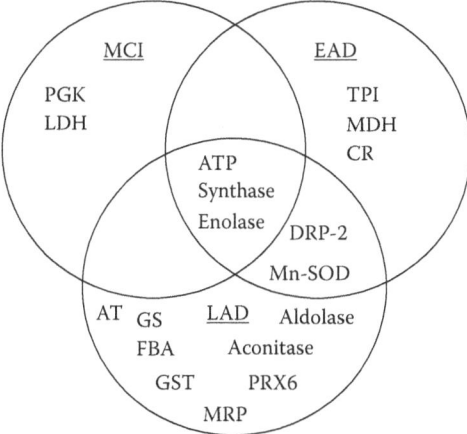

FIGURE 16.5 The lipid peroxidation product 4-hydroxynonenal modifies several proteins throughout the progression of AD (Sultana and Butterfield 2004; Reed et al. 2008, 2009; Perluigi et al. 2009). (PGK = phosphoglycerate kinase, LDH = lactate dehydrogenase, TPI = triose phosphate isomerase, MDH = malate dehydrogenase, CR = carbonyl reductase, Mn-SOD = manganese superoxide dismutase, FBA = fructose bisphosphate aldolase, GST = glutathione-*S*-transferase, PRX6 = peroxiredoxin 6, MRP = multidrug resistant protein 6, GS = glutamine synthetase, AT = alpha tubulin).

could limit ATP production at a time when cellular defense is weakened. Alpha tubulin and DRP-2 are involved in structural functions and intercellular communication. HNE modification of these proteins could severely impede maintenance of the structural integrity of the cell and its communication with the surrounding neurons (Perluigi et al. 2009). Consequently, modification of key proteins by lipid peroxidation product, HNE plays key roles in the pathogenesis and progression of AD.

Parkinson's Disease

PD presents clinically with a variety of mobility difficulties, including resting tremor, muscle rigidity, and bradykinesia and in late stages of the disease cognitive deficits often are observed. Although depression is observed in almost half of PD patients and development of dementia occurs in about 80% of patients, clinical diagnosis of idiopathic PD depends on four different factors: asymmetric resting tremor, asymmetric rigidity, asymmetric bradykinesia, and response to anti-PD drugs (Cedarbaum and McDowell 1987; Ward and Gibb 1990; Calne et al. 1992; Steece-Collier et al. 2002). While clinical diagnosis is separated into three categories based on the existence of those factors: clinically possible, clinically probably, and clinically definite. A diagnosis of clinically possible is given to patients showing only one of the three listed motor symptoms, while clinically probable patients show two of the motor symptoms (Bhidayasiri and Reichmann 2013). Clinically definite patients show two of the motor symptoms, but must also have shown response to drugs that are used to treat PD, such as levodopa, illustrated in Figure 16.6 (Ward and Gibb 1990; Solla et al. 2010). Although definitive diagnosis can only be determined post-mortem, clinical

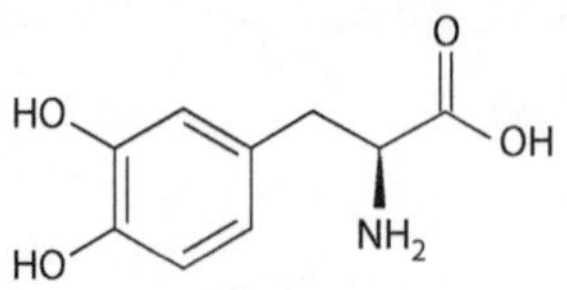

FIGURE 16.6 Levodopa, a drug used to treat PD.

diagnoses such as these are good predictors of the existence of idiopathic PD pathology (Hughes et al. 2002).

Neuropathologically, PD is hallmarked by Lewy bodies, small protein aggregates containing fibrils of the protein α-synuclein, which are found inside neurons and are believed to play a significant role in the neurodegeneration that occurs during PD. The substantia nigra is the primary locus of this neurodegeneration, with the caudal nigra losing about half of its neurons by the time symptoms appear (Fearnley and Lees 1991). Loss has also been seen before forebrain degeneration in early PD (Ziegler et al. 2013). Because this region is heavily involved in dopamine production, as shown in Figure 16.7, it plays a pivotal role in motor coordination. Therefore, the loss of dopaminergic neurons can account for the depression, dementia, and physical symptoms present in PD patients. Lewy bodies have also been found in other regions of the PD brain including the amygdale, vagal nucleus, cortex, locus ceruleus, and the peripheral autonomic nervous system. The decay of these regions may contribute to the nonmotor symptoms of PD such as the various dysfunctions of the autonomic nervous system and the psychiatric, sensory, and sleep maladies (Wakabayashi and Takahashi 1997; Braak et al. 2003; Samii et al. 2004).

While the exact role(s) of Lewy bodies in the development and progression of PD remain unclear, the Lewy body proteins, α-synuclein and ubiquitin, are associated with the progression of oxidative stress that is evidenced by several biomarkers in the PD brain. Indeed, changes in antioxidant enzyme expression and reduction of available reduced glutathione (GSH) indicate a significant role of oxidative stress

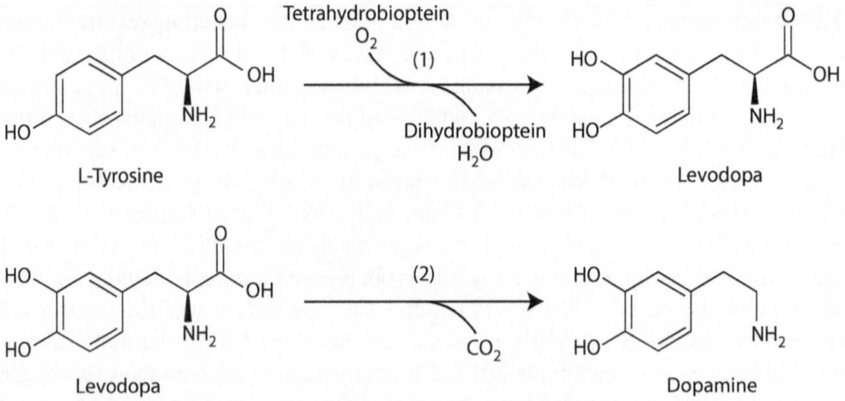

FIGURE 16.7 Tyrosine hydroxylase (1) and dopa decarboxylase (2) catalyze the synthesis of dopamine from the amino acid L-tyrosine.

in the PD brain. Both α-synuclein and ubiquitin, which are implicated in protein aggregation mechanisms, are known to be affected by oxidative stress (Jungmann et al. 1993; Hashimoto et al. 1999; Zhang et al. 1999; McNaught and Jenner 2001; McNaught et al. 2002; Hoglinger et al. 2003). Due in part to the limitations in studying the PD brain, such as the large number of variables influencing patient testing and the requirements of post-mortem analysis, elucidating these kinds of relationships that influence PD pathologies is not a simple task. However, the development of animal models has become very helpful in this regard, since various toxins and genetic predispositions are known to spur PD pathologies and symptoms.

Animal Models

The use of animal models offers researchers excellent opportunities for studying aspects of PD without the need for lengthy patient involvement or patient mortality. These aspects can include physical symptoms such as the motor dysfunctions that are necessary for clinical diagnoses of idiopathic PD, the nonmotor symptoms such as emotional and memory-related defects, as well as the neurochemical and neurobiological aspects of PD including blood flow, glucose consumption, and oxidative stress markers. Animal models also enable researchers to refine drug development techniques before commencing human trials, offering insight into the details of drug treatments on individual components of the disease, and these models are especially important for testing more invasive treatment methods such as stem cell therapies and deep brain stimulation (Pardal and Lopez-Barneo 2012; Sutton et al. 2013). Using animals to model PD generally depends on the use of two different approaches to inducing PD pathology—toxins and genetics (Rana et al. 2013).

Upon the discovery of a small group of heroin-abusing patients who had developed PD-like symptoms following the use of a new "synthetic heroin" sold in California in the early 1980s, it was revealed that this substance was a mixture of the compounds 1-methyl-4-phenyl-4-propionoxy-piperidine (MPPP) and 1-methyl-4-phenyl-1,2,5,6-tetrahydropyridine (MPTP), and researchers proposed that MPTP induced PD pathology (Langston et al. 1983). MPPP and MPTP are illustrated in Figure 16.8. Since this discovery, MPTP has served as a useful compound for inducing PD pathology in

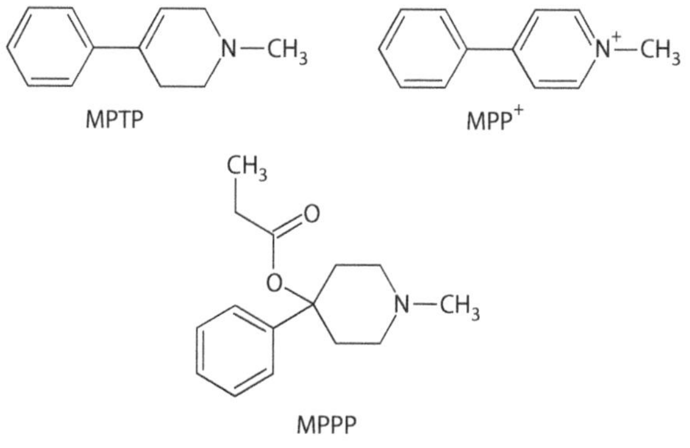

FIGURE 16.8   MPTP and MPPP were discovered in a mixture that induced PD-like symptoms. MPP^+ is a toxic product of MPTP metabolism.

animal models for study. Due to its ability to cross the blood–brain barrier, MPTP is taken up by astrocytes and metabolized by monoamine oxidase B into 1-methyl-4-phenylpyridinium (MPP+), which is thought to displace dopamine from vesicles for metabolism into toxic compounds such as 3,4-dihydroxyphenylacetaldehyde (DOPAL) (Jackson-Lewis et al. 2012; Rana et al. 2013). Although MPP+ acutely increases oxidative stress and dysfunction of mitochondrial complex I, a slow progression of the motor and nonmotor symptoms of PD allows for the use of longitudinal studies to assess a relatively realistic progression of PD and is generally favored by PD researchers over other toxins such as 6-hydroxydopamine (6-OHDA) (Jackson-Lewis et al. 2012; Rana et al. 2013).

The neurotoxin 6-OHDA, illustrated in Figure 16.9, is also capable of inducing Parkinson-like symptoms, and its use in animal models of PD predates the discovery of MPTP's neurotoxic properties (Breese et al. 2005). Like MPTP, 6-OHDA also is able to model development of PD symptoms and is thus useful for studying the progression of symptoms following the death of neurons containing tyrosine hydroxylase (Jackson-Lewis et al. 2012; Rana et al. 2013). However, 6-OHDA cannot cross the blood–brain barrier, so direct injection is required and this creates a quick development of symptoms when injection occurs in the substantia nigra pars compacta, and a slower development of symptoms when injection occurs in the striatum (Jackson-Lewis et al. 2012; McDowell and Chesselet 2012; Rana et al. 2013). Using these injection methods, 6-OHDA has proven to be a useful toxin for models of PD by producing a reduction in glucose metabolism in the striatum, modified cognition, and symptoms of depression and special memory loss (Ferro et al. 2005; Tadaiesky et al. 2008; McDowell and Chesselet 2012).

A number of genes have been associated with the development of PD, and progress in transgenic animal models has enabled the study the independent relationships between many of these genes and PD symptoms (Jackson-Lewis et al. 2012). Genetic mutations found in familial PD account for approximately 10% of the PD patient population (Jackson-Lewis et al. 2012). While these mutations may not influence PD development in the majority of patients, the identification of genes associated with PD has provided avenues of research into the roles of certain proteins that may have otherwise remained hidden. This is beneficial to advancements in novel methods of gene therapy, and animal models produced by this method have a more mild form of PD that progresses in a more realistic fashion (McDowell and Chesselet 2012). This area of study is fairly new in comparison to toxin-induced PD development, but it offers a unique perspective on events that occur in the PD brain and recent developments in gene knockout research have given rise to triple gene knockout mouse

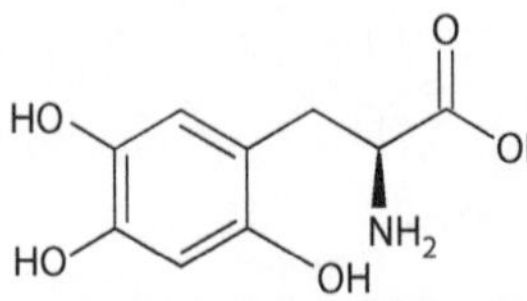

FIGURE 16.9 The compound 6-OHDA is a neurotoxin that is capable of inducing PD symptoms.

models of PD and triple transgenic mouse models of AD (Halagappa et al. 2007; Greten-Harrison et al. 2010). However, genetic models of PD do not faithfully mimic the pathology of PD, that is, there is no loss of dopaminergic neurons. Although animal models are a useful tool in studying PD, there are limitations in using such models. Multiple models can exhibit symptoms similar to PD; however it is unclear if the underlying cause in these models is the same.

Lipid Peroxidation in PD

Oxidative stress is a key aspect of PD pathology, with several oxidative stress biomarkers being elevated and important antioxidant enzymes having altered expression of the PD brain and animal models of PD (Surendran and Rajasankar 2010). Due in part to the high concentration of polyunsaturated fatty acids in the brain, lipid peroxidation products make up a significant portion of the elevated oxidative stress biomarkers discovered thus far (Surendran and Rajasankar 2010). These lipid peroxidation products include 4-hydroxy-2-nonenal (HNE)-modified proteins and malondialdehyde, which are both elevated in the substantia nigra of the PD brain (Dexter et al. 1989; Yoritaka et al. 1996). In addition, the role of HNE in the progression of PD appears to be more involved than simply a by-product of other neurodegenerative mechanisms. Elevated HNE levels are also known to circulate in the cerebrospinal fluid and plasma of PD patients, and it is hypothesized that these levels relate to the rise in HNE-glutathione conjugates and decrease in free glutathione that circulates in brainstems of MPTP-treated mice (Selley 1998). Indeed, while the function of oxidative stress in the pathogenesis and progression of PD remains unclear, research indicates that lipid peroxidation plays a significant role in the neurodegeneration that occurs in PD and much of this neurodegeneration could be facilitated by the lipid peroxidation product, HNE (Di Domenico et al. 2012).

AMYOTROPHIC LATERAL SCLEROSIS

Amyotrophic lateral sclerosis (ALS), also known as Lou Gehrig's disease, is a progressive neurodegenerative disorder in which all voluntary muscle movement is lost within 1–5 years. Discovered by Jean Martin-Charcot in 1869, this disorder rapidly progresses as death occurs within 2–5 years after the first symptoms occur. Motor neurons are greatly affected in ALS resulting in muscle weakness, atrophy, and spasticity (Cleveland and Rothstein 2001). ALS has two forms: inherited (familial) and sporadic, of which about 90% are sporadic. Of the 10% familial cases, 20% are attributed to mutations in the antioxidant enzyme, Cu/Zn superoxide dismutase which leads to toxicity, mitochondrial dysfunction, and protein aggregation (Nirmalananthan and Greensmith 2005). Other genetic mutations in the sigma 1 receptor, senataxin, and alsin and have been identified in juvenile ALS (Bowling et al. 1993; Strong et al. 2005; Al-Saif et al. 2011). These proteins are involved in cell signaling, membrane transport, DNA repair, and RNA processing (Topp et al. 2005; Panzeri et al. 2006; Suraweera et al. 2009; Hirano et al. 2011). Patients with ALS significantly have high levels of glutamate in their serum which implies that in the CNS there may be an excitotoxic effect on neurons, thereby eventually causing cell death demonstrated by the rapid progression of the disease. DNA damage has also been displayed in ALS by elevated levels of 8-hydroxy-2′-deoxyguanosine (8-OHdG) in a regional brain study in the G93A-SOD1 transgenic mouse (Aguirre et al. 2005). This transgenic mouse model, in which an Ala for Gly substitution occurs at position 93, has been thoroughly investigated in ALS research. This mouse overexpresses mutant human SOD1 and exhibits the age-dependent motor neuronal characteristics associated with amyotrophic lateral sclerosis making it an excellent model to study this neurodegenerative disorder.

Lipid Peroxidation and Redox Proteomics in ALS

The G93A-SOD1 mouse model is the most well-established animal model for amyotrophic lateral sclerosis. Levels of HNE are elevated in this mouse model (Perluigi et al. 2005). Intraperitoneal treatment with melatonin did not lower HNE levels, but opposingly raised oxidative stress levels (Dardiotis et al. 2013); however, treatment with the *Pulsatilla koreana* root extract, SK-PC-B70M, showed a decrease in levels of malondialdehyde and 4-hydroxy-2-nonenal in the ventral horn of G93A-SOD1 mice spinal cord (Seo et al. 2011). Through proteomics, there were three HNE modified proteins identified in this disorder: alpha enolase, dihydropyrimidinase-related protein 2 (DRP-2), and heat-shock protein 70 (Perluigi et al. 2005).

As noted above, alpha enolase catalyzes the dehydration of 2-phosphoglycerate to phosphoenolpyruvate in glycolysis, an ATP generating process in the cytoplasm. The brain encompasses a minute amount of mass in the body but accounts for 20% of glucose metabolism and over 30% of inspired oxygen. Since glycolysis uses glucose as its initial substrate, glucose metabolism is an important factor in energy production. Glucose metabolism is required for proper brain function and nominal interruption of glucose metabolism causes brain dysfunction and memory loss (Meier-Ruge et al. 1994; Heller and Macdonald 1996; Hoyer 1996). There are three isoforms of enolase, which require a functional dimer of two of these isoforms. PET studies revealed reduced cerebral glucose utilization in ALS brain (Dalakas et al. 1987), which could be attributed to cognitive decline and frontal-lobe dysfunction observed in ALS patients (Abrahams et al. 1996). HNE modification of this protein supports the hypothesis that altered energy metabolism is prominent in ALS and a common thread in neurodegenerative disease. Decreased levels of cellular ATP at nerve terminals may lead to loss of synapses and synaptic function. Both consequences can affect action potential propagation which may contribute to reduced muscle function, a symptom of this debilitating disorder. The reduction in ATP levels can also negatively affect the function of ion-motive ATPases used in maintenance of cell potential and in transport systems such as glucose (Guo et al. 2000) and neuronal glutamate, both of which are altered in ALS (Kanai and Hediger 2003). Glyoxal levels are increased in ALS spinal cord, which inactivates the glutamate transporter, GLT-1 as well (Kawaguchi et al. 2005). However, recently riluzole, the only FDA-approved ALS treatment, has been recently shown to enhance glucose transport by phosphorylating AMP activate protein kinase and its downstream target, AS-160 (Daniel et al. 2013).

DRP-2 is a member of the dihydropyrimidinase-related protein family involved in axonal outgrowth and pathfinding through transmission and modulation of extracellular signals (Hamajima et al. 1996; Kato et al. 1998). DRP-2, also known as collapsin response mediator protein 2 (CRMP-2), can induce growth cone collapse by rho-kinase phosphorylation (Goshima et al. 1995; Arimura et al. 2000). It can also bind to tubulin heterodimers and bundle microtubules as carriers to promote microtubule assembly and dynamics (Gu and Ihara 2000; Fukata et al. 2002). This protein has also been shown to play a role in voltage-gated calcium channel regulation (Wang et al. 2010). Similarly, CRMP4a, which plays a pivotal role in axonal degeneration and cell death, was upregulated in the G93A-SOD1 mouse model (Duplan et al. 2010). ALS is a motor neuron disease and by definition affects the motor neurons, which are composed of actin microfilaments and microtubules. The HNE modification of DRP-2 provides a rationale for the poor muscle control, a symptom of this progressive neurodegenerative disorder. Altered sulfur metabolism, leading to modulation of CRMP-2 has been reported in the G93A mouse model of ALS (Hensley et al. 2006).

Heat-shock proteins act as molecular chaperones to aid in the protein refolding or to guide damaged proteins to the proteosome. They are involved in combating stress by protecting proteins from denaturation (Calabrese et al. 2003). They also vary in size and are named according to their molecular weight. In addition, heat-shock proteins have been found to be differentially expressed in spinal cord and skeletal muscle in the G93A-SOD transgenic mouse model (Wei et al. 2013). HspB8 has been found to play a role in the removal of misfolded proteins (Crippa et al. 2010), while other heat-shock proteins including Hsp27, Hsp 70, and Hsp1 have all been identified in the progression of ALS (Brown 2007; Sharp et al. 2008; Stetler et al. 2009; Brettschneider et al. 2010). Heat-shock protein 71 has been found to be nitrated in the ALS mouse model demonstrating that impairment of these chaperone proteins may exacerbate protein misfolding and proteosomal overload in neurodegenerative diseases, further illustrating the importance of functioning heat-shock proteins in the cell (Casoni et al. 2005).

HUNTINGTON'S DISEASE

HD is a progressive, autosomal, dominant, inherited neurodegenerative disorder in which multiple CAG (Gln) repeats occur in the huntingtin protein. Normal individuals typically have 6–31 glutamine repeats; whereas HD individuals have 36–82 glutamine repeats. The r6/2 transgenic mouse model of HD expresses exon1 of the human *htt* gene, which contains about 150 CAG repeats. These mice develop symptoms typical of those represented in HD including poor motor coordination and cognitive decline (Li et al. 2005). Although the function of the huntingtin protein is unclear, it has been shown to interact in vesicle transport, cell signaling, a possible cytoskeletal integrity (DiFiglia et al. 1995; DiProspero et al. 2004; Truant et al. 2006; Colin et al. 2008; Li et al. 2008). HD was first fully described by George Huntington in 1872 upon evaluation of a Long Island family exhibiting chorea, a disorder in which involuntary jerky limb movements, increased eye blinking, and fidgeting are exhibited. Although progressive, this neurological disease takes approximately 10 years to fully manifest into bradykinesia and rigidity. The juvenile form has a much lower incidence, but more rapid and virulent course than the inherited form. Muscle movement and cognition decline over time. Neuronal degeneration throughout the basal ganglia and striatum is apparent, but brainstem, cortex, and thalamus eventually become involved as HD progresses.

Lipid Peroxidation in HD

A significant increase of HNE adducts is observed in the striatum and found to be colocalized in huntingtin inclusions (Lee et al. 2010). Malondialdehyde is elevated in HD brain (Browne et al. 1999). DNA damage is also present in this disorder, as levels of 8-OHdG are increased in HD patients (Chen et al. 2007). Increased oxidative stress and mitochondrial dysfunction are observed in blood elements, consistent with the notion of altered energy metabolism in this neurodegenerative disorder (Chen et al. 2007; Sorolla et al. 2008). Consistent with this suggestion, defects in energy metabolism have been observed in presymptomatic and symptomatic HD individuals using positron emission topography (Turner and Schapira 2010). Altered energy metabolism may be a contributing factor to the physical characteristics of this disorder.

Redox Proteomics of HNE-Bound Proteins in HD

Redox and expression proteomic experiments have been performed in the transgenic r6/2 mouse model and human HD brain. Several proteins were identified as being differentially expressed in *S. nigra*, the brain region involved in motor coordination, eye blinking, learning, and reward processing (Chen et al. 2012). These proteins include, but are not limited to: actin, creatine kinase BB, glucose-6-phosphate dehydrogenase, NADH dehydrogenase, sirtuin 2, and 14-3-3 proteins ε, η, and θ (Table 16.1).

Actin plays a central role in maintaining cellular integrity, morphology, and the structure of the plasma membrane. Actin microfilaments play a role in the neuronal membrane cytoskeleton by maintaining the distribution of membrane proteins, and segregating axonal and dendritic proteins (Battaini et al. 1999). Moreover, actin is involved in the elongation of the growth cone and loss of function of actin could play a role in the loss of synapse and neuronal communication documented in AD (Masliah et al. 1994). Lipid peroxidation of actin could lead to loss of membrane cytoskeletal structure, decreased membrane fluidity, and trafficking of synaptic proteins and mitochondria.

TABLE 16.1
Proteins Identified to be Differentially Expressed in HD Brain

Protein	Function
Actin	Cytoskeletal integrity
Creatine kinase BB	Energy metabolism
Glucose-6-phosphate dehydrogenase	Energy metabolism
NADH dehydrogenase	Energy metabolism
Sirtuin 2	Sterol biosynthesis
14-3-3- ε	Signal transduction
14-3-3- η	Signal transduction
14-3-3- θ	Signal transduction

Creatine kinase BB, the brain-specific isoform of creatine kinase (CK), catalyzes the phosphorylation of creatine to phosphocreatine using ATP as a substrate. Creatine kinase, a reversible enzyme, can also generate ATP by way of phosphocreatine hydrolysis. ATP is in constant demand due to its use in signal transduction and ion gradient maintenance. CK has a dosage-dependent reduced enzymatic activity when chemically reacted with physiological concentrations of HNE (Eliuk et al. 2007). This lipid peroxidation yields reduced energy production, poor signal transduction, and ionic imbalance as a consequence, all of which are detrimental to neurons.

Glucose-6-phosphate dehydrogenase (G-6-PDH), the first enzyme in the pentose phosphate pathway, leads to production of the enzymatic co-factor, NADPH. NADPH is also used in cholesterol, fatty acid, and steroid synthesis. It supplies reducing equivalents of glutathione reductase to maintain the levels of the antioxidant, glutathione, in red blood cells. Lipid peroxidation of red blood cells was significantly elevated in G-6-PDH(−/−) subjects compared to G-6-PDH(+/−) subjects (Clemens et al. 1985; Gurbuz et al. 2004) resulting in decreased levels of glutathione. Lipid peroxidation of G-6-PDH can cause reduced antioxidant defense, thereby causing increased oxidative stress.

NADH dehydrogenase, Complex I of the electron transport chain, is used to transfer electrons from NADH to coenzyme CoQ_{10}. Lipid peroxidation of this specific enzyme can lead to disruption of ion homeostasis and maintenance of ion gradients, a common occurrence in mitochondrial dysfunction via reduced ATP production. Sirtuins are multifunctional enzymes that play a role in energy metabolism, stress response, and genomic stability. Sirtuins, classified in five different groups based on their functions and enzymatic activities, are being evaluated as therapeutic agents in the treatment of neurodegenerative disorders such as AD, HD, and PD (Herskovits and Guarente 2013). Sirtuins have been found to delay progression of HD *in vivo* and *in vitro*. Sirtuin 2, a cytoplasmic tubulin deacetylating protein located in the cytoplasm, possesses deacetylase activity. Oxidative modification of sirtuin 2 may result in increased protein aggregation, a hallmark of HD (Sorolla et al. 2011).

Scaffolding proteins interact with multiple proteins in order to connect proteins to various components in order to perpetuate signaling cascades. The 14-3-3 family of scaffolding proteins interact with multiple proteins used in signaling pathways including, among others, the Bcl-2 proteins, Bad and Bax, both involved in promotion of apoptosis, as well as cell cycle via cyclin-dependent kinases. These proteins frequently bind to phosphoserine and phosphotyrosine residues and there is evidence of 14-3-3 proteins binding to tau and stimulating kinases thereby playing a pivotal role in neurodegenerative diseases. The 14-3-3 family of proteins has multiple isoforms based on their elution profile via HPLC. Specifically, 14-3-3 proteins ε, η, and θ are differentially expressed in the *S. nigra* of HD patients. The mutation of these specific proteins shows a possible correlation with the neurological disorder Miller–Dieker syndrome, which will be discussed below. 14-3-3-ε, localized in the molecular layer of the cerebellum, has been shown to interact with protein kinase A, calmodulin, and protein kinase C (Luk et al. 1999; Oriente et al. 2005; Gu et al. 2006). Zuo discovered that this protein plays a role in the mitogen-associated protein kinase signaling cascade in which DNA repair, cell cycle regulation, and protein synthesis were affected (Zuo et al. 2010).

Miller–Dieker syndrome is a rare genetic disorder in which facial abnormalities, growth retardation, and brain structural malformations, leading to cognitive decline, are exhibited. Patients with Miller–Dieker syndrome have a continuous deletion of chromosome 17p13.3, which codes for tyrosine-3-monooxygenase/tryptophan 5-monooxygenase activation protein, ε polypeptide, a protein used to make the scaffolding protein, 14-3-3-ε. A deletion in both 14-3-3-η and platelet-activating factor acetylhydrolase 1b can be the causative genetic causative factor in Miller–Dieker syndrome, thereby perpetuating a decline in cognitive function (Capra et al. 2012).

HIV is a retrovirus in which the virus uses the body's cellular machinery to transform RNA to DNA. The number of helper T cells is dramatically reduced thereby lowering the body's immune response. GP-120 is essential to the virus binding to the cell surface receptor for entry into the cell. 14-3-3-θ is specifically implicated in the G_2 checkpoint for cell cycle arrest for the HIV-1 VPR (viral protein R) accessory protein, which is associated with cell death since VPR is necessary for viral infection of helper T cells (Bolton et al. 2008). 14-3-3-θ can bind to several cyclin-dependent kinases that regulate the cell cycle leading to overall neurodegeneration in this immune disorder. Lipid peroxidation of the 14-3-3 family of proteins may cause reduced interaction with protein kinases, which can result in reduced signal transduction by which protein synthesis, cell cycle regulation, and DNA repair may be affected.

Involuntary muscle control is a staple of HD signs and symptoms that may involve the oxidative modification of several energy-related proteins, namely aconitase, α- and γ-enolase, creatine kinase BB, and glyceraldehyde-3-phosphate dehydrogenase (GAPDH) as indicated in Table 16.2 (Sorolla et al. 2008). There is an upregulation of glycolytic proteins in the first two weeks of development in r6/2 mice, which coincides with the peak in protein alteration also demonstrated during this time (Zabel et al. 2009). Glial fibrillary acidic protein (GFAP), heat-shock protein 90 (HSP90), and voltage-dependent anion channel 1 (VDAC1) were also identified as being oxidatively modified by HNE in human HD brain and the r6/2 mouse model (Perluigi et al. 2005). Peroxiredoxins and glutathione peroxidase are upregulated in this neurodegenerative disease, likely acting as a compensatory repair mechanism. Levels of lipid peroxidation are increased and antioxidant enzyme activity is reduced in these systems which can lead to further oxidative stress and physical symptoms of this disorder.

TABLE 16.2
Oxidatively Modified Proteins in HD Brain

Protein	Function
Aconitase	Energy metabolism
Alpha enolase	Energy metabolism
Gamma enolase	Energy metabolism
Creatine kinase BB	Energy metabolism
Glyceraldehyde-3-phosphate dehydrogenase	Energy metabolism
Glial fibrillary acidic protein	Structural integrity
Voltage-dependent anion channel 1	Calcium ion homeostasis
Heat-shock protein 90	Stress response

Aconitase (ACO) is a mitochondrial protein that acts as the second enzyme in the TCA cycle contributing to overall energy production. Aconitase activity is significantly reduced in the striatum and cortex of HD individuals thereby yielding in overall decrease of ATP production (Sorolla et al. 2008). Mitochondrial dysfunction, which is observed in HD, can occur via the oxidative modification of aconitase (Petrozzi et al. 2007). Decreased ATP production due to lipid peroxidation can lead to altered energy metabolism and reduced gluconeogenesis, which have been associated with HD pathogenesis (Josefsen et al. 2010).

As described above, enolase catalyzes the dehydration of 2-phosphoglycerate to phosphoenolpyruvate and is a frequent target of oxidative modification in several neurodegenerative diseases including: AD (Castegna et al. 2002), ALS (Ekegren et al. 2006), HD (Sorolla et al. 2008), and PD (De Iuliis et al. 2005; Poon et al. 2005). Although the main function for enolase is energy metabolism, this enzyme also plays a role in the MAPK/ERK to promote cell survival and neurite outgrowth (Sousa et al. 2005; Butterfield and Lange 2009; Hafner et al. 2012). Oxidative modification of enolase has a significant effect not only on energy production, but also on cell survival, both of which are imperative for healthy cellular function.

Glyceraldehyde-3-phosphate dehydrogenase (GAPDH), the first enzyme in the ATP generating stage of glycolysis, phosphorylates the substrate, glyceraldehyde-3-phosphate to 1,3-bisphosphoglycerate. Catalytically, a cysteine residue in the active site forms a thioester intermediate after addition of NAD^+. This reaction is essential as it is the only glycolytic reaction that produces NADH, an important biochemical cofactor. This intermediate also reacts with an adjacent histidine residue in the active site, and His is highly susceptible to HNE adduction, which leads to protein dysfunction and overall lowered energy generation (Ishii et al. 2003; Perluigi et al. 2005). This enzyme has the propensity to be S-nitrosylated and can contribute to inhibition of gluconeogenesis (Mohr et al. 1996). Although a primary role of GAPDH involves energy metabolism, and equally important, is its role as a NO sensor (Hara et al. 2006) and in apoptosis (Lee et al. 2012), among other functions (Butterfield et al. 2010).

Glial fibrillary acidic protein (GFAP), a marker for astrocytic activation, is the major intermediate filament protein found in astrocytes. Astrocytes are glial cells that provide support to neurons in the brain and regulate the chemical and extracellular environment. Upon activation, astrocytes repair cellular damage, mount an inflammatory response, and activate microglia. The main function of GFAP is maintaining cytoskeletal integrity and synaptic plasticity (Middeldorp and Hol 2011). GFAP has a lowered expression in the hypothalamus of transgenic HD rat model compared to wild type (Cong et al. 2012). As the hypothalamus is involved in temperature regulation, eating, sleep, and emotional behavior, modification of GFAP by HNE is consistent with the disturbance of these functions in HD patients (Weydt et al. 2006; Arnulf et al. 2008; Videnovic et al. 2009; Ille et al. 2011). Lipid peroxidation can impair muscle movement by reducing synaptic plasticity and cytoskeletal integrity, which are severely impaired in HD.

HSP90 is a member of the heat-shock protein family. This proteosomal-related protein is upregulated in 12 week, fully symptomatic, r6/2 transgenic mice (Zabel et al. 2009). Lipid peroxidation is elevated as a function of age. This upregulation likely is a compensatory regulatory mechanism to repair oxidative damage displayed at this stage of the disease.

The mitochondrial permeability transition pore (MPTP) plays an essential role in cell death by regulating Ca^{2+} levels in mitochondria and by facilitating transport of ATP and other moieties. The MPTP is composed of several proteins including voltage-dependent anion channel 1, adenine nucleotide translocase, and cyclophilin D. Excess calcium triggers opening of the pore, which is consistent in the neuropathology of various central nervous system injuries and traumatic brain/spinal cord injury (Sullivan et al. 2005). Opening of the pore results in cytochrome c release, an apoptotic signal (Barrientos et al. 2011). VDAC1, a component of the MPTP, is essential in metabolite import and export into the mitochondria. This protein is involved in Ca^{2+} homeostasis, as well as the release of cytochrome c, caspases, and other critical apoptotic-related proteins (Lan et al. 2010; De Stefani et al. 2012). VDAC1 controls superoxide anion leakage from the mitochondria, which is imperative in regulating the production of reactive oxygen species (Han et al. 2003). Lipid peroxidation contributes to the increased oxidative stress and neurodegeneration observed in HD brain.

MITOCHONDRIAL INVOLVEMENT IN BRAIN DISORDERS

Mitochondria are the powerhouse of ATP production in the cell, and through oxidative phosphorylation, are involved in approximately 90% of the ATP generated. Respiratory bursts can cause mitochondrial leakage and result in elevated levels of oxidative stress parameters, as evidenced in the microglial lesions of AD (Lassmann 2011), PD (Qian and Flood 2008), and other inflammatory CNS disorders (Qian et al. 2010). All complexes in the electron transport chain can undergo oxidation modification in these neurodegenerative disorders, causing mitochondrial chain impairment thereby reducing levels of ATP (Jenner 1993; Schapira 1999; Moran et al. 2012; Lunnon et al. 2012). Not only are these complexes impaired, but several key mitochondrial enzymes including alpha ketoglutarate dehydrogenase (Gibson et al. 1998, 2012), dihydrolipoyl S- succinyltransferase (DLST) (Dumont et al. 2009), pyruvate dehydrogenase, and isocitrate dehydrogenase (Bubber et al. 2005) have reduced activity or deficiency in AD. The aforementioned enzymes, with the exception of DLST are directly involved in the TCA cycle. DLST uses the TCA intermediate succinyl-CoA as a substrate and liberates the CoA that can be used to make acetyl CoA, thus perpetuating the TCA cycle and increasing ATP production indirectly. Impairment of these mitochondrial enzymes results in reduced ATP production causing altered glutamate and glucose transport as well as ion gradient maintenance. Mitochondrial dysfunctional is exhibited in ALS by calcium and iron dysregulation (Jeong et al. 2009; Tradewell et al. 2011) and defects in autophagy, leading to motor neuron death that can be attributed to the motor deficient condition demonstrated in this neurological disorder (Ferrucci et al. 2011). Specifically, Complex I is deficient in lymphocytes in ALS patients (Ghiasi et al. 2012), consistent with the altered energy metabolism observed in neurodegenerative disorders. Mitochondrial dysfunction is an early event and precedes other lysosomal and proteosomal abnormalities *in vitro* in the SH-SY5Y neuroblastoma cell line, a model used to study PD *in vitro* (Yong-Kee et al. 2012). Likewise, mitochondrial impairment is common in several genetic models of PD including those with mutations in the genes that code for α-synuclein, parkin, DJ-1, PTEN-induced kinase 1 (PINK1), and leucine-rich repeat kinase 2 (LRRK2), all common in familial PD patients (Trancikova et al. 2012). Mitochondrial DNA damage has been observed *in vitro* in HD models (Siddiqui et al. 2012). Recently, a plausible mechanism linking impairment of the mitochondrial disulfide relay system, which is involved in the transport of cysteine-rich proteins into the mitochondria, changes in mitochondrial morphology and reduced activities of Complex I, IV, and V, resulting in progressive neurodegeneration in HD was formulated (Napoli et al. 2013).

CONCLUSIONS

Lipid peroxidation induces free radical chain reactions (Figure 16.1). As the brain is abundant in both oxygen and PUFA-containing lipids, this organ is especially vulnerable to lipid peroxidation. This phenomenon can affect phospholipids as well as disrupting the lipid bilayer. Lipid peroxidation by-products, such as HHE and HNE, can covalently attach to amino acid residues, thereby leaving proteins dysfunctional. As the "powerhouse" of the cell, mitochondria are also susceptible to lipid peroxidation. This mitochondrial dysfunction can cause an overall reduction in ATP production that can affect ion homeostasis, pump maintenance, and transport systems. Lipid peroxidation in neurodegenerative diseases such as AD, PD, ALS, and HD, can result in loss of protein function and structure leading to memory loss, and muscle rigidity. Figure 16.10 shows that elevated levels of 4-hydroxy-2-nonenal are observed

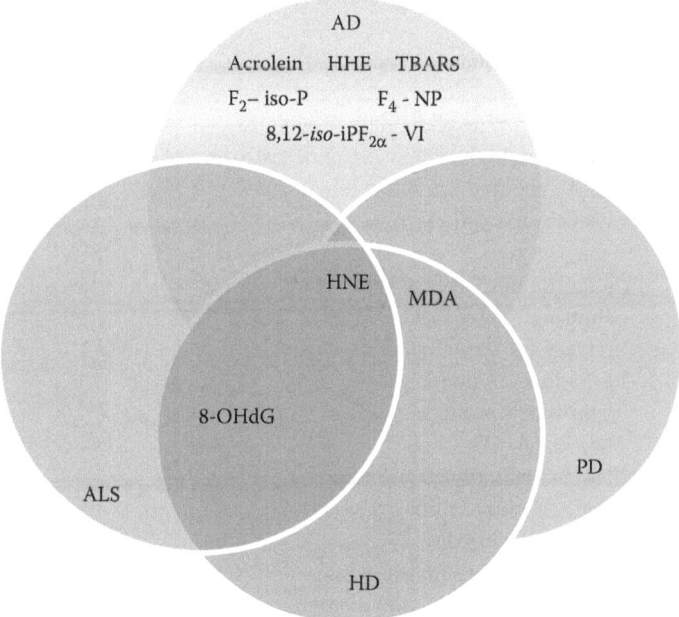

FIGURE 16.10 Venn diagram to illustrate the importance of lipid peroxidation as a causative factor in the pathogenesis of AD, PD, ALS, and HD.

in all aforementioned disorders, while at least two markers for lipid peroxidation are shared in at least two of these neurodegenerative disorders. Lipid peroxidation has also been shown to exacerbate disease progression via increasing oxidative damage and neurodegeneration. The identification of biomarkers for disease diagnosis via proteomics and use of pharmacological interventions may lead to the plausible alleviation of lipid peroxidation and treatment of these debilitating disorders.

ABBREVIATIONS

4-HHE	4-hydroxy-2-hexenal
6-OHDA	6-hydroxydopamine
8-OHdG	8-hydroxy-2'-deoxyguanosine
ACO	aconitase
AD	Alzheimer's disease
AGE	advance glycation end product
ALS	Amyotrophic lateral sclerosis
ATPase	ATP synthase
CK	creatine kinase
CRMP-2	collapsin response mediator protein 2
DHA	cis-4,7,10,13,16,19-docosahexanoic acid
DLST	dihydrolipoyl S- succinyltransferase
DOPAL	3,4-dihydroxyphenylacetaldehyde
DRB	drebrin
DRP-2	dihydropyriminidase-related protein 2
EAD	early AD
F2-IsoP	F2-isoprostanes
F4-NP	F4-neuroprostanes
G-6-PDH	glucose-6-phosphate dehydrogenase
GAPDH	glyceraldehyde-3-phosphate dehydrogenase
GFAP	glial fibrillary acidic protein
GS	glutamine synthetase
GSH	glutathione
GST	glutathione-S-transferase
HD	Huntington's disease
HIV	human immunodeficiency virus
HNE	4-hydroxy-2-nonenal
HSP90	heat shock protein 90
LAD	late stage AD
LRKK2	leucine-rich repeat kinase 2
MCI	mild cognitive impairment
MDA	malondialdehyde
MDH	malate dehydrogenase
Mn-SOD	manganese superoxide dismutase
MPP+	1-methyl-4-phenylpyridinium
MPPP	1-methyl-4-phenyl-4-propionoxy-piperidine
MPTP	1-methyl-4-phenyl-1,2,5,6-tetrahydropyridine

MPTP   mitochondrial permeability transition pore
MRP1   multidrug resistant protein 1
NFT   neurofibrillary tangles
NT   neutropil threads
$O_2^{\cdot-}$ superoxide
PCAD   preclinical AD
PD   Parkinson's disease
PINK1   PTEN-induced kinase 1
PPAR   peroxisome proliferator-activated receptor
PRX6   peroxiredoxin 6
SP   senile plaques
TBARS   thiobarbituric acid-reactive substances
TPI   triose phosphate isomerase
VDAC1   voltage-dependent anion channel 1

REFERENCES

Abdi, F., J. F. Quinn, J. Jankovic et al. Detection of biomarkers with a multiplex quantitative proteomic platform in cerebrospinal fluid of patients with neurodegenerative disorders. *J Alzheimers Dis* 9(3), 2006:293–348.

Abrahams, S., L. H. Goldstein, J. J. Kew et al. Frontal lobe dysfunction in amyotrophic lateral sclerosis. A PET study. *Brain* 119(Pt 6) (1), 1996:2105–20.

Aguirre, N., M. F. Beal, W. R. Matson et al. Increased oxidative damage to DNA in an animal model of amyotrophic lateral sclerosis. *Free Radic Res* 39(4), 2005:383–8.

Al-Saif, A., F. Al-Mohanna, and S. Bohlega. A mutation in sigma-1 receptor causes juvenile amyotrophic lateral sclerosis. *Ann Neurol* 70(6), 2011:913–9.

Aldini, G., I. Dalle-Donne, G. Vistoli et al. Covalent modification of actin by 4-hydroxy-*trans*-2-nonenal (HNE): LC-ESI-MS/MS evidence for Cys374 Michael adduction. *J Mass Spectrom* 40(7), 2005:946–54.

Almeida, M., E. Ambrogini, L. Han et al. Increased lipid oxidation causes oxidative stress, increased peroxisome proliferator-activated receptor-gamma expression, and diminished pro-osteogenic Wnt signaling in the skeleton. *J Biol Chem* 284(40), 2009:27438–48.

Aluise, C. D., R. A. Robinson, T. L. Beckett et al. Preclinical Alzheimer disease: Brain oxidative stress, Abeta peptide and proteomics. *Neurobiol Dis* 39(2), 2010:221–8.

Aluise, C. D., R. A. Robinson, J. Cai et al. Redox proteomics analysis of brains from subjects with amnestic mild cognitive impairment compared to brains from subjects with preclinical Alzheimer's disease: Insights into memory loss in MCI. *J Alzheimers Dis* 23(2), 2011:257–69.

Aluise, C. D., K. Rose, M. Boiani et al. Peptidyl-prolyl cis/trans-isomerase A1 (Pin1) is a target for modification by lipid electrophiles. *Chem Res Toxicol* 26(2), 2013:270–9.

Arimura, N., N. Inagaki, K. Chihara et al. Phosphorylation of collapsin response mediator protein-2 by Rho-kinase. Evidence for two separate signaling pathways for growth cone collapse. *J Biol Chem* 275(31), 2000:23973–80.

Arnulf, I., J. Nielsen, E. Lohmann et al. Rapid eye movement sleep disturbances in Huntington disease. *Arch Neurol* 65(4), 2008:482–8.

Barrientos, S. A., N. W. Martinez, S. Yoo et al. Axonal degeneration is mediated by the mitochondrial permeability transition pore. *J Neurosci* 31(3), 2011:966–978.

Battaini, F., A. Pascale, L. Lucchi et al. Protein kinase C anchoring deficit in postmortem brains of Alzheimer's disease patients. *Exp Neurol* 159(2), 1999:559–64.

Bhidayasiri, R., and H. Reichmann. Different diagnostic criteria for Parkinson disease: What are the pitfalls?. *J Neural Transm* 120(4), 2013:619–25.

Bolton, D. L., R. A. Barnitz, K. Sakai et al. 14-3-3 theta binding to cell cycle regulatory factors is enhanced by HIV-1 Vpr. *Biol Direct* 3(3), 2008:17.

Bowling, A. C., J. B. Schulz, R. H. Brown, Jr. et al. Superoxide dismutase activity, oxidative damage, and mitochondrial energy metabolism in familial and sporadic amyotrophic lateral sclerosis. *J Neurochem* 61(6), 1993:2322–5.

Braak, H., and E. Braak. Neuropathological stageing of Alzheimer-related changes. *Acta Neuropathol* 82(4), 1991:239–59.

Braak, H., K. Del Tredici, U. Rub et al. Staging of brain pathology related to sporadic Parkinson's disease. *Neurobiol Aging* 24(2), 2003:197–211.

Bradley, M. A., W. R. Markesbery, and M. A. Lovell. Increased levels of 4-hydroxynonenal and acrolein in the brain in preclinical Alzheimer disease. *Free Radic Biol Med* 48(12), 2010:1570–6.

Bradley, M. A., S. Xiong-Fister, W. R. Markesbery et al. Elevated 4-hydroxyhexenal in Alzheimer's disease (AD) progression. *Neurobiol Aging* 33(6), 2012:1034–44.

Breese, G. R., D. J. Knapp, H. E. Criswell et al. The neonate-6-hydroxydopamine-lesioned rat: A model for clinical neuroscience and neurobiological principles. *Brain Res Brain Res Rev* 48(1), 2005:57–73.

Brettschneider, J., V. Lehmensiek, H. Mogel et al. Proteome analysis reveals candidate markers of disease progression in amyotrophic lateral sclerosis (ALS). *Neurosci Lett* 468(1), 2010:23–7.

Brown, I. R. Heat shock proteins and protection of the nervous system. *Ann N Y Acad Sci* 1113, 2007:147–58.

Browne, S. E., R. J. Ferrante, and M. F. Beal. Oxidative stress in Huntington's disease. *Brain Pathol* 9(1), 1999:147–63.

Bubber, P., V. Haroutunian, G. Fisch et al. Mitochondrial abnormalities in Alzheimer brain: Mechanistic implications. *Ann Neurol* 57(5), 2005:695–703.

Butterfield, D. A., S. S. Hardas, and M. L. Lange. Oxidatively modified glyceraldehyde-3-phosphate dehydrogenase (GAPDH) and Alzheimer's disease: Many pathways to neurodegeneration. *J Alzheimer's Dis* 20(2), 2010:369–93.

Butterfield, D. A., and M. L. Lange. Multifunctional roles of enolase in Alzheimer's disease brain: Beyond altered glucose metabolism. *J Neurochem* 111(4), 2009:915–33.

Butterfield, D. A., T. Reed, M. Perluigi et al. Elevated protein-bound levels of the lipid peroxidation product, 4-hydroxy-2-nonenal, in brain from persons with mild cognitive impairment. *Neurosci Lett* 397(3), 2006:170–3.

Butterfield, D. A., A. M. Swomley, and R. Sultana. Amyloid beta-peptide (1-42)-induced oxidative stress in Alzheimer disease: Importance in disease pathogenesis and progression. *Antioxid Redox Signal* 19(8), 2013:823–35.

Cai, W., M. Ramdas, L. Zhu et al. Oral advanced glycation endproducts (AGEs) promote insulin resistance and diabetes by depleting the antioxidant defenses AGE receptor-1 and sirtuin 1. *Proc Natl Acad Sci USA* 109(39), 2012:15888–93.

Calabrese, V., G. Scapagnini, C. Colombrita et al. Redox regulation of heat shock protein expression in aging and neurodegenerative disorders associated with oxidative stress: A nutritional approach. *Amino Acids* 25(3–4), 2003:437–44.

Calamaras, T. D., C. Lee, F. Lan et al. Post-translational modification of serine/threonine kinase LKB1 via adduction of the reactive lipid species 4-hydroxy-*trans*-2-nonenal (HNE) at lysine residue 97 directly inhibits kinase activity. *J Biol Chem* 287(50), 2012:42400–6.

Calingasan, N. Y., K. Uchida, and G. E. Gibson. Protein-bound acrolein: A novel marker of oxidative stress in Alzheimer's disease. *J Neurochem* 72(2), 1999:751–6.

Calne, D. B., B. J. Snow, and C. Lee. Criteria for diagnosing Parkinson's disease. *Ann Neurol* 32(Suppl 2), 1992:S125–7.

Capra, V., M. Mirabelli-Badenier, M. Stagnaro et al. Identification of a rare 17p13.3 duplication including the BHLHA9 and YWHAE genes in a family with developmental delay and behavioural problems. *BMC Med Genet* 13(13), 2012:93.

Casoni, F., M. Basso, T. Massignan et al. Protein nitration in a mouse model of familial amyotrophic lateral sclerosis: Possible multifunctional role in the pathogenesis. *J Biol Chem* 280(16), 2005:16295–304.

Castegna, A., M. Aksenov, M. Aksenova et al. Proteomic identification of oxidatively modified proteins in Alzheimer's disease brain. Part I: Creatine kinase BB, glutamine synthase, and ubiquitin carboxy-terminal hydrolase L-1. *Free Radic Biol Med* 33(4), 2002:562–71.

Castegna, A., M. Aksenov, V. Thongboonkerd et al. Proteomic identification of oxidatively modified proteins in Alzheimer's disease brain. Part II: Dihydropyrimidinase-related protein 2, alpha-enolase and heat shock cognate 71. *J Neurochem* 82(6), 2002:1524–32.

Castegna, A., C. M. Lauderback, H. Mohmmad-Abdul et al. Modulation of phospholipid asymmetry in synaptosomal membranes by the lipid peroxidation products, 4-hydroxynonenal and acrolein: Implications for Alzheimer's disease. *Brain Res* 1004(1–2), 2004:193–7.

Castegna, A., V. Thongboonkerd, J. B. Klein et al. Proteomic identification of nitrated proteins in Alzheimer's disease brain. *J Neurochem* 85(6), 2003:1394–401.

Catala, A. Lipid peroxidation modifies the picture of membranes from the Fluid Mosaic Model to the Lipid Whisker Model. *Biochimie* 94(1), 2012:101–9.

Cedarbaum, J. M., and F. H. McDowell. Sixteen-year follow-up of 100 patients begun on levodopa in 1968: Emerging problems. *Adv Neurol* 45, 1987:469–72.

Chen, C. M., Y. R. Wu, M. L. Cheng et al. Increased oxidative damage and mitochondrial abnormalities in the peripheral blood of Huntington's disease patients. *Biochem Biophys Res Commun* 359(2), 2007:335–40.

Chen, S., F. F. Lu, P. Seeman et al. Quantitative proteomic analysis of human substantia nigra in Alzheimer's disease, Huntington's disease and multiple sclerosis. *Neurochem Res* 37(12), 2012:2805–13.

Clemens, M. R., H. Einsele, and H. D. Waller. The fatty acid composition of red cells deficient in glucose-6-phosphate dehydrogenase and their susceptibility to lipid peroxidation. *Klin Wochenschr* 63(13), 1985:578–82.

Cleveland, D. W., and J. D. Rothstein. From Charcot to Lou Gehrig: Deciphering selective motor neuron death in ALS. *Nat Rev Neurosci* 2(11), 2001:806–19.

Colin, E., D. Zala, G. Liot et al. Huntingtin phosphorylation acts as a molecular switch for anterograde/retrograde transport in neurons. *EMBO J* 27(15), 2008:2124–34.

Cong, W. N., H. Cai, R. Wang et al. Altered hypothalamic protein expression in a rat model of Huntington's disease. *PLoS One* 7(10), 2012:e47240.

Counts, S. E., B. He, M. Nadeem et al. Hippocampal drebrin loss in mild cognitive impairment. *Neurodegener Dis* 10(1–4), 2012:216–9.

Crippa, V., S. Carra, P. Rusmini et al. A role of small heat shock protein B8 (HspB8) in the autophagic removal of misfolded proteins responsible for neurodegenerative diseases. *Autophagy* 6(7), 2010:958–60.

Dalakas, M. C., J. Hatazawa, R. A. Brooks et al. Lowered cerebral glucose utilization in amyotrophic lateral sclerosis. *Ann Neurol* 22(5), 1987:580–6.

Daniel, B., O. Green, O. Viskind et al. Riluzole increases the rate of glucose transport in L6 myotubes and NSC-34 motor neuron-like cells via AMPK pathway activation. *Amyotroph Lateral Scler Frontotemporal Degener* 14(5–6), 2013:434–43.

Dardiotis, E., E. Panayiotou, M. L. Feldman et al. Intraperitoneal melatonin is not neuroprotective in the G93ASOD1 transgenic mouse model of familial ALS and may exacerbate neurodegeneration. *Neurosci Lett* 548(2), 2013:170–5.

Davies, S. S., V. Amarnath, K. S. Montine et al. Effects of reactive gamma-ketoaldehydes formed by the isoprostane pathway (isoketals) and cyclooxygenase pathway (levuglandins) on proteasome function. *FASEB J* 16(7), 2002:715–7.

De Bock, C. A., J. Brug, and J. N. Walop. Antiviral activity of glyoxals. *Nature* 179(4562), 1957:706–7.

De Iuliis, A., J. Grigoletto, A. Recchia et al. A proteomic approach in the study of an animal model of Parkinson's disease. *Clin Chim Acta* 357(2), 2005:202–9.

De Stefani, D., A. Bononi, A. Romagnoli et al. VDAC1 selectively transfers apoptotic Ca^{2+} signals to mitochondria. *Cell Death Differ* 19(2), 2012:267–73.

Deigner, H. P., and A. Hermetter. Oxidized phospholipids: Emerging lipid mediators in pathophysiology. *Curr Opin Lipidol* 19(3), 2008:289–94.

Dexter, D. T., C. J. Carter, F. R. Wells et al. Basal lipid peroxidation in substantia nigra is increased in Parkinson's disease. *J Neurochem* 52(2), 1989:381–9.

Di Domenico, F., R. Sultana, A. Ferree et al. Redox proteomics analyses of the influence of co-expression of wild-type or mutated LRRK2 and Tau on C. elegans protein expression and oxidative modification: Relevance to Parkinson disease. *Antioxid Redox Signal* 17(11), 2012:1490–506.

DiFiglia, M., E. Sapp, K. Chase et al. Huntingtin is a cytoplasmic protein associated with vesicles in human and rat brain neurons. *Neuron* 14(5), 1995:1075–81.

DiProspero, N. A., E. Y. Chen, V. Charles et al. Early changes in Huntington's disease patient brains involve alterations in cytoskeletal and synaptic elements. *J Neurocytol* 33(5), 2004:517–33.

Domingues, R. M., P. Domingues, T. Melo et al. Lipoxidation adducts with peptides and proteins: Deleterious modifications or signaling mechanisms?. *J Proteomics* 92(1), 2013:110–31.

Dubois, B., H. H. Feldman, C. Jacova et al. Research criteria for the diagnosis of Alzheimer's disease: Revising the NINCDS-ADRDA criteria. *Lancet Neurol* 6(8), 2007:734–46.

Dumont, M., D. J. Ho, N. Y. Calingasan et al. Mitochondrial dihydrolipoyl succinyltransferase deficiency accelerates amyloid pathology and memory deficit in a transgenic mouse model of amyloid deposition. *Free Radic Biol Med* 47(7), 2009:1019–27.

Duplan, L., N. Bernard, W. Casseron et al. Collapsin response mediator protein 4a (CRMP4a) is upregulated in motoneurons of mutant SOD1 mice and can trigger motoneuron axonal degeneration and cell death. *J Neurosci* 30(2), 2010:785–96.

Ekegren, T., J. Hanrieder, S. M. Aquilonius et al. Focused proteomics in post-mortem human spinal cord. *J Proteome Res* 5(9), 2006:2364–71.

Eliuk, S. M., M. B. Renfrow, E. M. Shonsey et al. active site modifications of the brain isoform of creatine kinase by 4-hydroxy-2-nonenal correlate with reduced enzyme activity: Mapping of modified sites by Fourier transform-ion cyclotron resonance mass spectrometry. *Chem Res Toxicol* 20(9), 2007:1260–8.

Esterbauer, H., R. J. Schaur, and H. Zollner. Chemistry and biochemistry of 4-hydroxynonenal, malonaldehyde and related aldehydes. *Free Radic Biol Med* 11(1), 1991:81–128.

Fearnley, J. M., and A. J. Lees. Ageing and Parkinson's disease: Substantia nigra regional selectivity. *Brain* 114(Pt 5) (5), 1991:2283–301.

Feng, Z., W. Hu, L. J. Marnett et al. Malondialdehyde, a major endogenous lipid peroxidation product, sensitizes human cells to UV- and BPDE-induced killing and mutagenesis through inhibition of nucleotide excision repair. *Mutat Res* 601(1–2), 2006:125–36.

Ferro, M. M., M. I. Bellissimo, J. A. Anselmo-Franci et al. Comparison of bilaterally 6-OHDA- and MPTP-lesioned rats as models of the early phase of Parkinson's disease: Histological, neurochemical, motor and memory alterations. *J Neurosci Methods* 148(1), 2005:78–87.

Ferrucci, M., F. Fulceri, L. Toti et al. Protein clearing pathways in ALS. *Arch Ital Biol* 149(1), 2011:121–49.

Fritz, K. S., J. J. Galligan, R. L. Smathers et al. 4-Hydroxynonenal inhibits SIRT3 via thiol-specific modification. *Chem Res Toxicol* 24(5), 2011:651–62.

Fukata, Y., T. J. Itoh, T. Kimura et al. CRMP-2 binds to tubulin heterodimers to promote microtubule assembly. *Nat Cell Biol* 4(8), 2002:583–91.

Ghiasi, P., S. Hosseinkhani, A. Noori et al. Mitochondrial complex I deficiency and ATP/ADP ratio in lymphocytes of amyotrophic lateral sclerosis patients. *Neurol Res* 34(3), 2012:297–303.

Gibson, G. E., H. L. Chen, H. Xu et al. Deficits in the mitochondrial enzyme alpha-ketoglutarate dehydrogenase lead to Alzheimer's disease-like calcium dysregulation. *Neurobiol Aging* 33(6), 2012:1121 e13–24.

Gibson, G. E., K. F. Sheu, and J. P. Blass. Abnormalities of mitochondrial enzymes in Alzheimer disease. *J Neural Transm* 105(8–9), 1998:855–70.

Goshima, Y., F. Nakamura, P. Strittmatter et al. Collapsin-induced growth cone collapse mediated by an intracellular protein related to UNC-33. *Nature* 376(6540), 1995: 509–14.

Greco, A., L. Minghetti, and G. Levi. Isoprostanes, novel markers of oxidative injury, help understanding the pathogenesis of neurodegenerative diseases. *Neurochem Res* 25(9–10), 2000:1357–64.

Greenberg, M. E., X. M. Li, B. G. Gugiu et al. The lipid whisker model of the structure of oxidized cell membranes. *J Biol Chem* 283(4), 2008:2385–96.

Greten-Harrison, B., M. Polydoro, M. Morimoto-Tomita et al. alphabetagamma-Synuclein triple knockout mice reveal age-dependent neuronal dysfunction. *Proc Natl Acad Sci USA* 107(45), 2010:19573–8.

Gu, Y., and Y. Ihara. Evidence that collapsin response mediator protein-2 is involved in the dynamics of microtubules. *J Biol Chem* 275(24), 2000:17917–20.

Gu, Y. M., Y. H. Jin, J. K. Choi et al. Protein kinase A phosphorylates and regulates dimerization of 14-3-3 epsilon. *FEBS Lett* 580(1), 2006:305–10.

Guo, Z., M. S. Kindy, I. Kruman et al. ALS-linked Cu/Zn-SOD mutation impairs cerebral synaptic glucose and glutamate transport and exacerbates ischemic brain injury. *J Cereb Blood Flow Metab* 20(3), 2000:463–8.

Gurbuz, N., O. Yalcin, T. A. Aksu et al. The relationship between the enzyme activity, lipid peroxidation and red blood cells deformability in hemizygous and heterozygous glucose-6-phosphate dehydrogenase deficient individuals. *Clin Hemorheol Microcirc* 31(3), 2004:235–42.

Hafner, A., N. Obermajer, and J. Kos. gamma-Enolase C-terminal peptide promotes cell survival and neurite outgrowth by activation of the PI3K/Akt and MAPK/ERK signalling pathways. *Biochem J* 443(2), 2012:439–50.

Halagappa, V. K., Z. Guo, M. Pearson et al. Intermittent fasting and caloric restriction ameliorate age-related behavioral deficits in the triple-transgenic mouse model of Alzheimer's disease. *Neurobiol Dis* 26(1), 2007:212–20.

Hamajima, N., K. Matsuda, S. Sakata et al. A novel gene family defined by human dihydropyrimidinase and three related proteins with differential tissue distribution. *Gene* 180(1–2), 1996:157–63.

Han, D., F. Antunes, R. Canali et al. Voltage-dependent anion channels control the release of the superoxide anion from mitochondria to cytosol. *J Biol Chem* 278(8), 2003:5557–63.

Hara, M. R., M. B. Cascio, and A. Sawa. GAPDH as a sensor of NO stress. *Biochim Biophys Acta* 1762(5), 2006:502–9.

Hashimoto, M., L. J. Hsu, Y. Xia et al. Oxidative stress induces amyloid-like aggregate formation of NACP/alpha-synuclein in vitro. *Neuroreport* 10(4), 1999:717–21.

Heller, S. R., and I. A. Macdonald. The measurement of cognitive function during acute hypoglycaemia: Experimental limitations and their effect on the study of hypoglycaemia unawareness. *Diabet Med* 13(7), 1996:607–15.

Hensley, K., M. Mhatre, S. Mou et al. On the relation of oxidative stress to neuroinflammation: Lessons learned from the G93A-SOD1 mouse model of amyotrophic lateral sclerosis. *Antioxid Redox Signal* 8(11–12), 2006:2075–87.

Herrmann, W., and R. Obeid. Biomarkers of neurodegenerative diseases. *Clin Chem Lab Med* 49(3), 2011:343–4.

Herskovits, A. Z., and L. Guarente. Sirtuin deacetylases in neurodegenerative diseases of aging. *Cell Res* 23(6), 2013:746–58.

Hirano, M., C. M. Quinzii, H. Mitsumoto et al. Senataxin mutations and amyotrophic lateral sclerosis. *Amyotroph Lateral Scler* 12(3), 2011:223–7.

Hoglinger, G. U., G. Carrard, P. P. Michel et al. Dysfunction of mitochondrial complex I and the proteasome: Interactions between two biochemical deficits in a cellular model of Parkinson's disease. *J Neurochem* 86(5), 2003:1297–307.

Hoyer, S. Oxidative metabolism deficiencies in brains of patients with Alzheimer's disease. *Acta Neurol Scand Suppl* 165(1), 1996:18–24.

Hughes, A. J., S. E. Daniel, Y. Ben-Shlomo et al. The accuracy of diagnosis of parkinsonian syndromes in a specialist movement disorder service. *Brain* 125(Pt 4), 2002:861–70.

Ille, R., A. Schafer, W. Scharmuller et al. Emotion recognition and experience in Huntington disease: A voxel-based morphometry study. *J Psychiatry Neurosci* 36(6), 2011:383–90.

Ishii, T., E. Tatsuda, S. Kumazawa et al. Molecular basis of enzyme inactivation by an endogenous electrophile 4-hydroxy-2-nonenal: Identification of modification sites in glyceraldehyde-3-phosphate dehydrogenase. *Biochemistry* 42(12), 2003:3474–80.

Jackson-Lewis, V., J. Blesa, and S. Przedborski. Animal models of Parkinson's disease. *Parkinsonism Relat Disord* 18(Suppl 1) (1), 2012:S183–5.

Jenner, P. Altered mitochondrial function, iron metabolism and glutathione levels in Parkinson's disease. *Acta Neurol Scand Suppl* 146(1), 1993:6–13.

Jeong, S. Y., K. I. Rathore, K. Schulz et al. Dysregulation of iron homeostasis in the CNS contributes to disease progression in a mouse model of amyotrophic lateral sclerosis. *J Neurosci* 29(3), 2009:610–9.

Jicha, G. A., E. L. Abner, F. A. Schmitt et al. Preclinical AD Workgroup staging: Pathological correlates and potential challenges. *Neurobiol Aging* 33(3), 2012:622 e1–622 e16.

Josefsen, K., S. M. Nielsen, A. Campos et al. Reduced gluconeogenesis and lactate clearance in Huntington's disease. *Neurobiol Dis* 40(3), 2010:656–62.

Joshi, G., S. Hardas, R. Sultana et al. Glutathione elevation by gamma-glutamyl cysteine ethyl ester as a potential therapeutic strategy for preventing oxidative stress in brain mediated by *in vivo* administration of adriamycin: Implication for chemobrain. *J Neurosci Res* 85(3), 2007:497–503.

Jungmann, J., H. A. Reins, C. Schobert et al. Resistance to cadmium mediated by ubiquitin-dependent proteolysis. *Nature* 361(6410), 1993:369–71.

Kanai, Y., and M. A. Hediger. The glutamate and neutral amino acid transporter family: Physiological and pharmacological implications. *Eur J Pharmacol* 479(1–3), 2003:237–47.

Kato, Y., N. Hamajima, H. Inagaki et al. Post-meiotic expression of the mouse dihydropyrimidinase-related protein 3 (DRP-3) gene during spermiogenesis. *Mol Reprod Dev* 51(1), 1998:105–11.

Kawaguchi, M., N. Shibata, S. Horiuchi et al. Glyoxal inactivates glutamate transporter-1 in cultured rat astrocytes. *Neuropathology* 25(1), 2005:27–36.

Keller, J. N., F. A. Schmitt, S. W. Scheff et al. Evidence of increased oxidative damage in subjects with mild cognitive impairment. *Neurology* 64(7), 2005:1152–6.

Kilhovd, B. K., I. Giardino, P. A. Torjesen et al. Increased serum levels of the specific AGE-compound methylglyoxal-derived hydroimidazolone in patients with type 2 diabetes. *Metabolism* 52(2), 2003:163–7.

Lan, C. H., J. Q. Sheng, D. C. Fang et al. Involvement of VDAC1 and Bcl-2 family of proteins in VacA-induced cytochrome c release and apoptosis of gastric epithelial carcinoma cells. *J Dig Dis* 11(1), 2010:43–9.

Langston, J. W., P. Ballard, J. W. Tetrud et al. Chronic Parkinsonism in humans due to a product of meperidine-analog synthesis. *Science* 219(4587), 1983:979–80.

Larrieu, S., L. Letenneur, J. M. Orgogozo et al. Incidence and outcome of mild cognitive impairment in a population-based prospective cohort. *Neurology* 59(10), 2002:1594–9.

Lassmann, H. Mechanisms of neurodegeneration shared between multiple sclerosis and Alzheimer's disease. *J Neural Transm* 118(5), 2011:747–52.

Lauderback, C. M., J. M. Hackett, F. F. Huang et al. The glial glutamate transporter, GLT-1, is oxidatively modified by 4-hydroxy-2-nonenal in the Alzheimer's disease brain: The role of Abeta1-42. *J Neurochem* 78(2), 2001:413–6.

Lee, J., B. Kosaras, S. J. Del Signore et al. Modulation of lipid peroxidation and mitochondrial function improves neuropathology in Huntington's disease mice. *Acta Neuropathol* 121(4), 2010:487–98.

Lee, S. Y., J. H. Kim, H. Jung et al. Glyceraldehyde-3-phosphate, a glycolytic intermediate, prevents cells from apoptosis by lowering S-nitrosylation of glyceraldehyde-3-phosphate dehydrogenase. *J Microbiol Biotechnol* 22(4), 2012:571–3.

Li, J. Y., N. Popovic, and P. Brundin. The use of the R6 transgenic mouse models of Huntington's disease in attempts to develop novel therapeutic strategies. *NeuroRx* 2(3), 2005:447–64.

Li, X., E. Sapp, A. Valencia et al. A function of huntingtin in guanine nucleotide exchange on Rab11. *Neuroreport* 19(16), 2008:1643–7.

Liu, M., R. Wallin, A. Wallmon et al. Mixed tocopherols have a stronger inhibitory effect on lipid peroxidation than alpha-tocopherol alone. *J Cardiovasc Pharmacol* 39(5), 2002:714–21.

Liu, Q., D. C. Simpson, and S. Gronert. Carbonylation of mitochondrial aconitase with 4-hydroxy-2-(E)-nonenal: Localization and relative reactivity of addition sites. *Biochim Biophys Acta* 1834(6), 2013:1144–54.

Lovell, M. A., M. A. Bradley, and S. X. Fister. 4-Hydroxyhexenal (HHE) impairs glutamate transport in astrocyte cultures. *J Alzheimers Dis* 32(1), 2012:139–46.

Lovell, M. A., C. Xie, and W. R. Markesbery. Acrolein, a product of lipid peroxidation, inhibits glucose and glutamate uptake in primary neuronal cultures. *Free Radic Biol Med* 29(8), 2000:714–20.

Luk, S. C., S. M. Ngai, S. K. Tsui et al. In vivo and *in vitro* association of 14-3-3 epsilon isoform with calmodulin: Implication for signal transduction and cell proliferation. *J Cell Biochem* 73(1), 1999:31–5.

Lunnon, K., Z. Ibrahim, P. Proitsi et al. Mitochondrial dysfunction and immune activation are detectable in early Alzheimer's disease blood. *J Alzheimers Dis* 30(3), 2012:685–710.

Markesbery, W. R. Neuropathologic alterations in mild cognitive impairment: A review. *J Alzheimers Dis* 19(1), 2010:221–8.

Markesbery, W. R., R. J. Kryscio, M. A. Lovell et al. Lipid peroxidation is an early event in the brain in amnestic mild cognitive impairment. *Ann Neurol* 58(5), 2005:730–5.

Markesbery, W. R., F. A. Schmitt, R. J. Kryscio et al. Neuropathologic substrate of mild cognitive impairment. *Arch Neurol* 63(1), 2006:38–46.

Masliah, E., M. Mallory, L. Hansen et al. Synaptic and neuritic alterations during the progression of Alzheimer's disease. *Neurosci Lett* 174(1), 1994:67–72.

McDowell, K., and M. F. Chesselet. Animal models of the non-motor features of Parkinson's disease. *Neurobiol Dis* 46(3), 2012:597–606.

McNaught, K. S., L. M. Bjorklund, R. Belizaire et al. Proteasome inhibition causes nigral degeneration with inclusion bodies in rats. *Neuroreport* 13(11), 2002:1437–41.

McNaught, K. S., and P. Jenner. Proteasomal function is impaired in substantia nigra in Parkinson's disease. *Neurosci Lett* 297(3), 2001:191–4.

Meier-Ruge, W., C. Bertoni-Freddari, and P. Iwangoff. Changes in brain glucose metabolism as a key to the pathogenesis of Alzheimer's disease. *Gerontology* 40(5), 1994:246–52.

Middeldorp, J., and E. M. Hol. GFAP in health and disease. *Prog Neurobiol* 93(3), 2011:421–43.

Mohr, S., J. S. Stamler, and B. Brune. Posttranslational modification of glyceraldehyde-3-phosphate dehydrogenase by S-nitrosylation and subsequent NADH attachment. *J Biol Chem* 271(8), 1996:4209–14.

Montine, T. J., M. F. Beal, M. E. Cudkowicz et al. Increased CSF F2-isoprostane concentration in probable AD. *Neurology* 52(3), 1999:562–5.

Moran, M., D. Moreno-Lastres, L. Marin-Buera et al. Mitochondrial respiratory chain dysfunction: Implications in neurodegeneration. *Free Radic Biol Med* 53(3), 2012:595–609.

Morris, J. C., M. Storandt, J. P. Miller et al. Mild cognitive impairment represents early-stage Alzheimer disease. *Arch Neurol* 58(3), 2001:397–405.

Morrow, J. D., K. E. Hill, R. F. Burk et al. A series of prostaglandin F2-like compounds are produced *in vivo* in humans by a non-cyclooxygenase, free radical-catalyzed mechanism. *Proc Natl Acad Sci USA* 87(23), 1990:9383–7.

Morrow, J. D., and L. J. Roberts. The isoprostanes: Unique bioactive products of lipid peroxidation. *Prog Lipid Res* 36(1), 1997:1–21.

Mukhopadhyay, S., S. Sen, B. Majhi et al. Methyl glyoxal elevation is associated with oxidative stress in rheumatoid arthritis. *Free Radic Res* 41(5), 2007:507–14.

Musiek, E. S., J. D. Brooks, M. Joo et al. Electrophilic cyclopentenone neuroprostanes are anti-inflammatory mediators formed from the peroxidation of the omega-3 polyunsaturated fatty acid docosahexaenoic acid. *J Biol Chem* 283(29), 2008:19927–35.

Napoli, E., S. Wong, C. Hung et al. Defective mitochondrial disulfide relay system, altered mitochondrial morphology and function in Huntington's disease. *Hum Mol Genet* 22(5), 2013:989–1004.

Nirmalananthan, N., and L. Greensmith. Amyotrophic lateral sclerosis: Recent advances and future therapies. *Curr Opin Neurol* 18(6), 2005:712–9.

Nunomura, A., G. Perry, G. Aliev et al. Oxidative damage is the earliest event in Alzheimer disease. *J Neuropathol Exp Neurol* 60(8), 2001:759–67.

Onyango, A. N. Small reactive carbonyl compounds as tissue lipid oxidation products; and the mechanisms of their formation thereby. *Chem Phys Lipids* 165(7), 2012:777–86.

Oriente, F., F. Andreozzi, C. Romano et al. Protein kinase C-alpha regulates insulin action and degradation by interacting with insulin receptor substrate-1 and 14-3-3 epsilon. *J Biol Chem* 280(49), 2005:40642–9.

Panzeri, C., C. De Palma, A. Martinuzzi et al. The first ALS2 missense mutation associated with JPLS reveals new aspects of alsin biological function. *Brain* 129(Pt 7), 2006:1710–9.

Pardal, R., and J. Lopez-Barneo. Neural stem cells and transplantation studies in Parkinson's disease. *Adv Exp Med Biol* 741(3), 2012:206–16.

Perluigi, M., H. F. Poon, W. Maragos et al. Proteomic analysis of protein expression and oxidative modification in r6/2 transgenic mice: A model of Huntington disease. *Mol Cell Proteomics* 4(12), 2005:1849–61.

Perluigi, M., H.F. Poon, K. Hensley et al. Proteomic analysis of 4-hydroxy-2-nonenal-modified proteins in G93A-SOD1 transgenic mice—A model of familial amyotrophic lateral sclerosis. *Free Radic Biol Med* 38(7), 2005:960–8.

Perluigi, M., R. Sultana, G. Cenini et al. Redox proteomics identification of 4-hydroxynonenal-modified brain proteins in Alzheimer's disease: Role of lipid peroxidation in Alzheimer's disease pathogenesis. *Proteomics Clin Appl* 3(6), 2009:682–693.

Petersen, R. C. Mild cognitive impairment as a diagnostic entity. *J Intern Med* 256(3), 2004:183–94.

Petrozzi, L., G. Ricci, N. J. Giglioli et al. Mitochondria and neurodegeneration. *Biosci Rep* 27(1–3), 2007:87–104.

Picklo, M. J., A. Azenkeng, and M. R. Hoffmann. Trans-4-oxo-2-nonenal potently alters mito-chondrial function. *Free Radic Biol Med* 50(2), 2011:400–7.

Poon, H. F., M. Frasier, N. Shreve et al. Mitochondrial associated metabolic proteins are selectively oxidized in A30P alpha-synuclein transgenic mice—A model of familial Parkinson's disease. *Neurobiol Dis* 18(3), 2005:492–8.

Pratico, D., C. M. Clark, F. Liun et al. Increase of brain oxidative stress in mild cognitive impairment: A possible predictor of Alzheimer disease. *Arch Neurol* 59(6), 2002:972–6.

Qian, L., and P. M. Flood. Microglial cells and Parkinson's disease. *Immunol Res* 41(3), 2008:155–64.

Qian, L., P. M. Flood, and J. S. Hong. Neuroinflammation is a key player in Parkinson's dis-ease and a prime target for therapy. *J Neural Transm* 117(8), 2010:971–9.

Ralat, L. A., Y. Manevich, A. B. Fisher et al. Direct evidence for the formation of a complex between 1-cysteine peroxiredoxin and glutathione S-transferase pi with activity changes in both enzymes. *Biochemistry* 45(2), 2006:360–72.

Rana, A. Q., M. S. Masroor, and A. S. Khan. A review of methods used to study cognitive deficits in Parkinson's disease. *Neurol Res* 35(1), 2013:1–6.

Reed, T., M. Perluigi, R. Sultana et al. Redox proteomic identification of 4-hydroxy-2-non-enal-modified brain proteins in amnestic mild cognitive impairment: Insight into the role of lipid peroxidation in the progression and pathogenesis of Alzheimer's disease. *Neurobiol Dis* 30(1), 2008:107–20.

Reed, T. T., W. M. Pierce, W. R. Markesbery et al. Proteomic identification of HNE-bound proteins in early Alzheimer disease: Insights into the role of lipid peroxidation in the progression of AD. *Brain Res* 1274(1), 2009:66–76.

Roberts, L. J., II, T. J. Montine, W. R. Markesbery et al. Formation of isoprostane-like com-pounds (neuroprostanes) *in vivo* from docosahexaenoic acid. *J Biol Chem* 273(22), 1998:13605–12.

Roberts, L. J., II, and J. D. Morrow. Isoprostanes. Novel markers of endogenous lipid peroxi-dation and potential mediators of oxidant injury. *Ann N Y Acad Sci* 744(5), 1994:237–42.

Roberts, L. J., II, and J. D. Morrow. The isoprostanes: Novel markers of lipid peroxidation and potential mediators of oxidant injury. *Adv Prostaglandin Thromboxane Leukot Res* 23(2), 1995:219–24.

Roberts, L. J., II, and J. D. Morrow. The generation and actions of isoprostanes. *Biochim Biophys Acta* 1345(2), 1997:121–35.

Roberts, L. J., II, and J. D. Morrow. Products of the isoprostane pathway: Unique bioactive compounds and markers of lipid peroxidation. *Cell Mol Life Sci* 59(5), 2002:808–20.

Roberts, L. J., and J. D. Morrow. Measurement of F(2)-isoprostanes as an index of oxidative stress in vivo. *Free Radic Biol Med* 28(4), 2000:505–13.

Samii, A., J. G. Nutt, and B. R. Ransom. Parkinson's disease. *Lancet* 363(9423), 2004:1783–93.

Schapira, A. H. Mitochondrial involvement in Parkinson's disease, Huntington's disease, hereditary spastic paraplegia and Friedreich's ataxia. *Biochim Biophys Acta* 1410(2), 1999:159–70.

Scheff, S. W., S. T. DeKosky, and D. A. Price. Quantitative assessment of cortical synaptic density in Alzheimer's disease. *Neurobiol Aging* 11(1), 1990:29–37.

Scheff, S. W., D. A. Price, F. A. Schmitt et al. Synaptic alterations in CA1 in mild Alzheimer disease and mild cognitive impairment. *Neurology* 68(18), 2007:1501–8.

Scheff, S. W., D. A. Price, F. A. Schmitt et al. Hippocampal synaptic loss in early Alzheimer's disease and mild cognitive impairment. *Neurobiol Aging* 27(10), 2006:1372–84.

Selley, M. L. (E)-4-hydroxy-2-nonenal may be involved in the pathogenesis of Parkinson's disease. *Free Radic Biol Med* 25(2), 1998:169–74.

Seo, J. S., I. S. Baek, Y. H. Leem et al. SK-PC-B70M alleviates neurologic symptoms in G93A-SOD1 amyotrophic lateral sclerosis mice. *Brain Res* 1368, 2011:299–307.

Sharp, P. S., M. T. Akbar, S. Bouri et al. Protective effects of heat shock protein 27 in a model of ALS occur in the early stages of disease progression. *Neurobiol Dis* 30(1), 2008:42–55.

Siddiqui, A., S. Rivera-Sanchez, R. Castro Mdel et al. Mitochondrial DNA damage is associated with reduced mitochondrial bioenergetics in Huntington's disease. *Free Radic Biol Med* 53(7), 2012:1478–88.

Solla, P., A. Cannas, F. Marrosu et al. Therapeutic interventions and adjustments in the management of Parkinson disease: Role of combined carbidopa/levodopa/entacapone (Stalevo). *Neuropsychiatr Dis Treat* 6, 2010:483–90.

Sorolla, M. A., C. Nierga, M. J. Rodriguez-Colman et al. Sir2 is induced by oxidative stress in a yeast model of Huntington disease and its activation reduces protein aggregation. *Arch Biochem Biophys* 510(1), 2011:27–34.

Sorolla, M. A., G. Reverter-Branchat, J. Tamarit et al. Proteomic and oxidative stress analysis in human brain samples of Huntington disease. *Free Radic Biol Med* 45(5), 2008: 667–78.

Sousa, L. P., B. M. Silva, B. S. Brasil et al. Plasminogen/plasmin regulates alpha-enolase expression through the MEK/ERK pathway. *Biochem Biophys Res Commun* 337(4), 2005:1065–71.

Steece-Collier, K., E. Maries, and J. H. Kordower. Etiology of Parkinson's disease: Genetics and environment revisited. *Proc Natl Acad Sci USA* 99(22), 2002:13972–4.

Stetler, R. A., Y. Gao, A. P. Signore et al. HSP27: Mechanisms of cellular protection against neuronal injury. *Curr Mol Med* 9(7), 2009:863–72.

Stewart, B. J., J. A. Doorn, and D. R. Petersen. Residue-specific adduction of tubulin by 4-hydroxynonenal and 4-oxononenal causes cross-linking and inhibits polymerization. *Chem Res Toxicol* 20(8), 2007:1111–9.

Strong, M. J., S. Kesavapany, and H. C. Pant. The pathobiology of amyotrophic lateral sclerosis. A proteinopathy?. *J Neuropathol Exp Neurol* 64(8), 2005:649–64.

Subramaniam, R., F. Roediger, B. Jordan et al. The lipid peroxidation product, 4-hydroxy-2-*trans*-nonenal, alters the conformation of cortical synaptosomal membrane proteins. *J Neurochem* 69(3), 1997:1161–9.

Sullivan, P. G., A. G. Rabchevsky, P. C. Waldmeier et al. Mitochondrial permeability transition in CNS trauma: Cause or effect of neuronal cell death?. *J Neurosci Res* 79(1–2), 2005:231–9.

Sultana, R., and D. A. Butterfield. Oxidatively modified GST and MRP1 in Alzheimer's disease brain: Implications for accumulation of reactive lipid peroxidation products. *Neurochem Res* 29(12), 2004:2215–20.

Sultana, R., M. Perluigi, and D. Allan Butterfield. Lipid peroxidation triggers neurodegeneration: A redox proteomics view into the Alzheimer disease brain. *Free Radic Biol Med* 62(2), 2013:157–69.

Suraweera, A., Y. Lim, R. Woods et al. Functional role for senataxin, defective in ataxia oculomotor apraxia type 2, in transcriptional regulation. *Hum Mol Genet* 18(18), 2009:3384–96.

Surendran, S., and S. Rajasankar. Parkinson's disease: Oxidative stress and therapeutic approaches. *Neurol Sci* 31(5), 2010:531–40.

Sutton, A. C., W. Yu, M. E. Calos et al. Deep brain stimulation of the substantia nigra pars reticulata improves forelimb akinesia in the hemiparkinsonian rat. *J Neurophysiol* 109(2), 2013:363–74.

Tadaiesky, M. T., P. A. Dombrowski, C. P. Figueiredo et al. Emotional, cognitive and neurochemical alterations in a premotor stage model of Parkinson's disease. *Neuroscience* 156(4), 2008:830–40.

Topp, J. D., D. S. Carney, and B. F. Horazdovsky. Biochemical characterization of Alsin, a Rab5 and Rac1 guanine nucleotide exchange factor. *Methods Enzymol* 403(2), 2005:261–76.

Tradewell, M. L., L. A. Cooper, S. Minotti et al. Calcium dysregulation, mitochondrial pathology and protein aggregation in a culture model of amyotrophic lateral sclerosis: Mechanistic relationship and differential sensitivity to intervention. *Neurobiol Dis* 42(3), 2011:265–75.

Trancikova, A., E. Tsika, and D. J. Moore. Mitochondrial dysfunction in genetic animal models of Parkinson's disease. *Antioxid Redox Signal* 16(9), 2012:896–919.

Trojanowski, J. Q., and H. Hampel. Neurodegenerative disease biomarkers: Guideposts for disease prevention through early diagnosis and intervention. *Prog Neurobiol* 95(4), 2011:491–5.

Truant, R., R. Atwal, and A. Burtnik. Hypothesis: Huntingtin may function in membrane association and vesicular trafficking. *Biochem Cell Biol* 84(6), 2006:912–7.

Turner, C., and A. H. Schapira. Mitochondrial matters of the brain: The role in Huntington's disease. *J Bioenerg Biomembr* 42(3), 2010:193–8.

Videnovic, A., S. Leurgans, W. Fan et al. Daytime somnolence and nocturnal sleep disturbances in Huntington disease. *Parkinsonism Relat Disord* 15(6), 2009:471–4.

Wakabayashi, K., and H. Takahashi. Neuropathology of autonomic nervous system in Parkinson's disease. *Eur Neurol* 38(Suppl 2), 1997:2–7.

Wang, Y., J. M. Brittain, S. M. Wilson et al. Emerging roles of collapsin response mediator proteins (CRMPs) as regulators of voltage-gated calcium channels and synaptic transmission. *Commun Integr Biol* 3(2), 2010:172–5.

Ward, C. D., and W. R. Gibb. Research diagnostic criteria for Parkinson's disease. *Adv Neurol* 53(4), 1990:245–9.

Wei, R., A. Bhattacharya, R. T. Hamilton et al. Differential effects of mutant SOD1 on protein structure of skeletal muscle and spinal cord of familial amyotrophic lateral sclerosis: Role of chaperone network. *Biochem Biophys Res Commun* 438(1), 2013:218–23.

West, M. J., C. H. Kawas, W. F. Stewart et al. Hippocampal neurons in pre-clinical Alzheimer's disease. *Neurobiol Aging* 25(9), 2004:1205–12.

Weydt, P., V. V. Pineda, A. E. Torrence et al. Thermoregulatory and metabolic defects in Huntington's disease transgenic mice implicate PGC-1alpha in Huntington's disease neurodegeneration. *Cell Metab* 4(5), 2006:349–62.

Williams, T. I., B. C. Lynn, W. R. Markesbery et al. Increased levels of 4-hydroxynonenal and acrolein, neurotoxic markers of lipid peroxidation, in the brain in mild cognitive impairment and early Alzheimer's disease. *Neurobiol Aging* 27(8), 2006:1094–9.

Yao, Y., V. Zhukareva, S. Sung et al. Enhanced brain levels of 8,12-iso-iPF2alpha-VI differentiate AD from frontotemporal dementia. *Neurology* 61(4), 2003:475–8.

Yong-Kee, C. J., E. Sidorova, A. Hanif et al. Mitochondrial dysfunction precedes other subcellular abnormalities in an *in vitro* model linked with cell death in Parkinson's disease. *Neurotox Res* 21(2), 2012:185–94.

Yoritaka, A., N. Hattori, K. Uchida et al. Immunohistochemical detection of 4-hydroxynonenal protein adducts in Parkinson disease. *Proc Natl Acad Sci U S A* 93(7), 1996:2696–701.

Zabel, C., L. Mao, B. Woodman et al. A large number of protein expression changes occur early in life and precede phenotype onset in a mouse model for huntington disease. *Mol Cell Proteomics* 8(4), 2009:720–34.

Zhang, J., G. Perry, M. A. Smith et al. Parkinson's disease is associated with oxidative damage to cytoplasmic DNA and RNA in substantia nigra neurons. *Am J Pathol* 154(5), 1999:1423–9.

Ziegler, D. A., J. S. Wonderlick, P. Ashourian et al. Substantia nigra volume loss before basal forebrain degeneration in early Parkinson disease. *JAMA Neurol* 70(2), 2013:241–7.

Zuo, S., Y. Xue, S. Tang et al. 14-3-3 epsilon dynamically interacts with key components of mitogen-activated protein kinase signal module for selective modulation of the TNF-alpha-induced time course-dependent NF-kappaB activity. *J Proteome Res* 9(7), 2010:3465–78.

17 Association of Oxidative Stress and Lipids with Risk Factors of Metabolic Syndrome

Sampath Parthasarathy, Chandrakala
Aluganti Narasimhulu, Irene Fernandez Ruiz,
Kathryn Young, Vivek Krishna Pulakazhi Venu,
Xueting Jiang, and Aladdin Riad

CONTENTS

INTRODUCTION

The term "Metabolic Syndrome" (MS) refers to a group or cluster of risk factors, which together raise the risk of heart disease and related diseases, such as diabetes and stroke [1–3]. Recent advances seem to indicate that the definition may extend to other diseases, including colon cancer and other noncardiovascular diseases [4,5]. Thus, there is an increased awareness of the risk factors that may increase our propensity to acquire many different disease pathologies. The risk factors (Figure 17.1)

FIGURE 17.1   Risk factors.

that are typically thought about in the context of the MS are large waistline or waist circumference (WC), elevated plasma triglyceride (TG) levels, dysfunctional or low levels of high-density lipoprotein (HDL), high blood pressure (BP), higher plasma glucose (GL) levels, dysfunctional endothelium (DE), and inflammatory macrophages [6–12]. Family history of cardiovascular diseases, gender, race, ethnicity, dietary and nutritional factors, and other factors, including the use of certain drugs also may contribute to the profile of MS [13,14]. It is suggested that at least three metabolic risk factors be present to be diagnosed with MS.

Each of these factors, in turn, could be influenced by a plethora of additional risk factors, such as smoking, low physical activity, hormonal imbalance, fat composition of diet and stress, and so on. Most of us have or are likely to develop, some of these risk factors as we age. Thus, a more vigorous approach to defining and identifying MS is needed. However, in lieu of the increasing incidence of cardio-metabolic diseases around the globe and the increasing number of young subjects who develop these risk factors, it is important to consider potential means of reducing the profile that contributes to MS. These diverse symptoms, on the surface, are unlikely to have a common etiology; however, they may have common features which may influence the course of the pathology.

OXIDATIVE STRESS AND INFLAMMATION

There have been considerable speculations on the role of oxidants in affecting the risk factors associated with MS. Many, if not all, the components are suggested to be associated with excess oxygen radical production [15]. In this chapter, the connection or the involvement of oxidative stress in MS will be discussed. Similarly, many of these symptoms are also associated with enhanced presence of inflammatory components. Oxidative stress is a loose term and is often difficult to define. Many reviews have defined it as a result of tilting the balance between pro-oxidant and antioxidant defense systems in the body toward more pro-oxidant conditions [16–18].

Inflammation is an equally vague term, often indicated by the increased presence of certain chemokines, cytokines, and their receptors as well as the generation of prostanoids and pro-inflammatory lipids. The activation of phospholipases and the generation of inflammatory prostaglandins, other oxidized lipids, and lipid markers are also indicative of a pro-inflammatory state. However, the mere increase in these compounds alone cannot be taken as an indication of abnormal pathology, because such increases could be transient or could merely represent an activation of the body's defense mechanism. For example, simple infection, skin abrasions, exposure to toxins or even exposure to smog could trigger transient inflammation and the elevation of these markers. Furthermore, while most, if not all, cell types are capable of generating oxidants and inflammatory lipids, generally blood cells (neutrophils, monocytes, lymphocytes, and other minor cell types) and their tissue counterparts (e.g., tissue macrophages) are suggested to be key players in inflammation. Liver, endothelium and other vascular cells, and many other cell types could contribute either directly or indirectly to the recruitment and retention of inflammatory cells by way of generation of chemotactic factors. Traditionally, oxidative stress, oxidized lipids and inflammation are all considered deleterious, although there is considerable evidence in the literature to suggest that often oxidative stress itself promotes the induction of genes responsible for antioxidant defense, some oxidized lipids have anti-inflammatory and protective effects, and that many inflammatory cytokines [19,20] might be involved in beneficial processes in the body, such as wound healing and angiogenesis [21,22].

In addition, there are many "anti-inflammatory" cytokines that are also produced by the same cell types that produce pro-inflammatory mediators. In recent years, specific "phenotypes" of macrophages namely M1 and M2 have been recognized as pro- and anti-inflammatory, respectively [23–26]. Typically, pro-inflammatory cytokines are suggested to be produced by M1 and anti-inflammatory cytokines by M2 types of macrophages [27]. However, such concepts have not been extended to oxidative and antioxidative stress. There is a huge void of information with respect to when an oxidative stress could be beneficial and when it becomes pathological. This chapter also will delineate some of the associations between oxidative stress and inflammation.

As MS is on the rise around the globe, one would wonder whether oxidative stress and inflammatory stress follows a similar pattern. At this point, it is premature to predict whether antioxidant supplementation would alter the course of disease progression, as there are too many unknowns. The disturbances in the balance between pro- and antioxidant status in the body could be at the subcellular, cellular, extracellular, and at the blood plasma level. It could be localized at a tissue level or could denote an overall body's response to inadequate coping with an excess of oxidants. While most literature on the topic addresses the production of oxidants by the body due to metabolic perturbations, the ability of foreign, for example, environmental particles, as well as dietary products to promote oxidative stress as well as to tilt the balance in either a pro- or antioxidant direction has also been realized [28–31]. More importantly, many nutraceuticals, prescription drugs, exposure to UV light, environmental particles, socio-psychological factors, and many other factors could tilt the balance.

HIGH BLOOD GLUCOSE AND OXIDATIVE STRESS

Oxidative stress has long been implicated in diabetes, particularly in type 2 diabetes [32]. However, whether reactive oxygen species (ROS) play a causative role in diabetic complications or they are just by-products of increased glucose concentrations are yet to be ascertained. High blood or extracellular or intracellular levels of glucose could generate oxidative stress in many ways. Spontaneous and enzymatic glucose oxidation has been known to generate oxygen radicals and hydrogen peroxide (H_2O_2) [33]. While bacteria may have the enzyme glucose oxidase that would generate H_2O_2, whether other oxidases in the mammalian system would act on glucose has not been determined. Glucose is not a very reactive sugar; however, a high concentration of glucose modifies lysine and other free amino groups resulting in the formation of advanced glycation end products (AGE) [34–36]. Such AGE modification of many proteins has been noted, particularly that of low-density lipoprotein (LDL) [37]. Similar modification of amino group containing lipids AGE-phosphatidyl ethanolamine (AGE-PtdEtn) has also been noted [38].

The AGE-modified proteins/lipids could induce oxidative stress by directly interacting with target cells and inducing oxidases, such as NADPH oxidase [39]. Conversely, as most lysine modifications are immunogenic [40], such antigen–antibody complexes could also trigger oxidative stress [41]. It has been noted that exposure of insulin-secreting cells, HIT-T15 to high glucose concentrations affects the expression of insulin gene transcriptional repressor and the enhancer binding protein β [42]. Generation of ROS was observed in the presence of the reducing sugar resulting in the decreased transcription of the insulin gene. Antioxidants, such as N-acetyl-cysteine (NAC) reversed the effects [42]. Treatment of Zucker diabetic fatty (ZDF) rats with either aminoguanidine or NAC restored insulin mRNA level to the control levels, as well as insulin content and insulin secretion [43]. Such treatment also restored normal blood glucose levels. *In vitro* treatment of isolated islets with NAC and aminoguanidine also diminished the accumulation of AGE, restored the aminoguanidin [44].

Excess free fatty acids (FFA) cause β cell death, a process in which the activation of cytokines and nuclear factor NFκB might be involved [45]. It is currently well established that the activation of these might involve the induction of oxidative stress [46–51]. Many chemokines and cytokine genes have a NFκB response element and are known to be induced by oxidative stress. Peroxidized fatty acids and various forms of oxidized low-density lipoprotein (Ox-LDL) have been shown to induce the synthesis of monocyte chemotactic protein 1 (MCP-1: a chemokine which recruits inflammatory monocytes) as well as many cytokines [52–56]. However, many of these actions are also mimicked by AGE-modified proteins [57], which may suggest that modified or damaged proteins also could contribute to these effects, in addition to oxidants. For example, the protein component of oxidized LDL also has been shown to induce the synthesis of IL-1α by arterial macrophages [58]. An important factor in the link between oxidized lipids and metabolic syndrome may be the scavenger receptor CD36, which is known to take up long-chain fatty acids, oxidized phospholipids, and Ox-LDL, and may lead to signaling *via* PPARs. Several studies have implicated CD36 as a factor leading to insulin resistance in macrophages in pro-inflammatory situations such as hyperlipidemia or glucose bolus [59].

HYPERTRIGLYCERIDEMIA

One of the lipid abnormalities that has been associated with T2DM is an increased plasma TG levels [60,61]. Most fasting plasma TG is carried in the form of very low-density lipoprotein (VLDL) [62,63]. Increased secretion of VLDL by the liver, and decreased clearance of VLDL and intestinally derived chylomicrons result in prolonged plasma retention of TG-rich particles in the plasma that lead to the accumulation of TG-rich lipoproteins. Studies by Chisolm and coworkers [64] as well by Henriksen et al. [65] over 4 decades showed that diabetic serum was toxic to cells and such toxicity could be inhibited by antioxidants. They also showed most of the toxicity resided in the TG-rich lipoproteins.

Whether oxidants affect the lipase action or promote the synthesis and release of TG-rich lipoproteins has not been determined. It is also possible that TG containing peroxidized fatty acid components are resistant to lipolysis. Earlier studies have demonstrated that antioxidants enhanced the (hepatic) lipase actions [66]; however, these studies have not been corroborated or confirmed by more robust studies. In contrast, phospholipases seem to act more robustly on peroxidized or oxidatively tailored phospholipids [67–70] although the enzyme itself could be inactivated by oxidants [70]. Recently it was reported that the protein disulfide isomerase, a redox-sensitive enzyme, could contribute to the endoplasmic reticulum-associated degradation of apoB through its chaperone activity [71].

VLDL is metabolized *via* intermediate density lipoprotein (IDL) to LDL in the plasma by the action of lipoprotein lipase [72]. Hepatic lipase may also contribute to the formation of TG-rich LDL particles [73]. The increased production/decreased clearance of VLDL also may result in the increased production of precursors of small dense LDL (sd-LDL) particles [57,74,75]. Such sd-LDL particles have been shown to contribute to vascular diseases, as LDL receptors have a reduced affinity for sd-LDL, and sd-LDL are more vulnerable to oxidative modification [76–78]. Numerous clinical studies have shown increased CVD in subjects with high prevalence of sd-LDL [79]. In animal models and in human diabetic conditions, sd-LDL levels are elevated [80,81] and such accumulation appears to be accompanied by decreased paraoxonase 1 (PON1) activity. PONs are antioxidant enzymes that are known to "detoxify" H_2O_2 and lipid peroxides [82,83]. The presence of PON1 in lipoproteins also protects the lipoproteins against oxidation [84]. Interestingly, PON1 also hydrolyzes the thioester bond of homocysteine, a known risk factor for atherosclerosis [85].

HYPERTENSION

Oxidative stress has been suggested to contribute to the etiology of hypertension [86–90]. Subjects with increased blood pressure have been noted to have decreased plasma antioxidants [91–93]. Epidemiological and clinical trial data also suggest that antioxidant-rich diets appear to reduce blood pressure and may reduce cardiovascular risk [94]. A direct involvement of oxidants in the hypertensive response of the vasculature as well as the consequence of oxidative inactivation of the potent vasodilator, nitric oxide (NO) has been implicated [95]. Oxidative stress has been implicated both as a causative as well as a consequence of hypertension.

NADPH oxidase, induced by vasodepressor molecules such as angiotensin II [96–98], abnormal mitochondria oxidant generation [99,100], xanthine oxidase [101], endothelium-derived nitric oxide synthase [102,103], cyclo-oxygenase and lipo-oxygenase [104–106] and other enzymes have been suggested to contribute toward oxygen radical and lipid peroxide formation during hyperglycemia as well as in other conditions in which blood pressure is increased. Superoxide anion, lipid peroxide, specific 12-lipoxygenase products, either alone or via products formed with nitric oxide (e.g $ONOO^-$: peroxynitrite) have been assumed to play a role in hypertension [107–109].

Nitric oxide synthesized by NO synthase (NOS), is a potent vasodilator, and helps to maintain blood pressure within a normal range [110]. Increased levels of ROS can reduce NO concentrations, resulting in an increase in blood pressure. Whether ROS or high glucose could directly affect NO synthase gene expression/activity has not been studied in detail.

The interaction of angiotensin with the endothelium and the response of NADPH oxidase have been noted as one of the most critical pathways in the production of oxygen radicals [111]. Consequently, agents that inhibit the renin–angiotensin system, such as angiotensin-converting enzyme (ACE) inhibitors and angiotensin II type 1 (AT1) receptor blockers; have considerable benefit in hypertension and heart failure [112–114]. Studies from several laboratories have shown that many AT1 receptor blockers (e.g., losartan and irbesartan) may have actions against oxidants and may reduce inflammation in humans [115–118]. Similarly, many ACE inhibitors, such as Captopril, have been recognized to have antioxidant activities [119–122].

Many antioxidants have been studied for their ability to lower blood pressure. Alpha lipoic acid has been reported to lower blood pressure in animals and in humans [99,123]. However, whether it is due to a direct antioxidant effect or the result of improvement in mitochondrial functions cannot yet be ascertained. Almost all studies with alpha lipoic acid are performed with mixed isomers. It remains to be seen whether a natural isomer would have additional beneficial effects.

It would appear that AngII and AGE-modified protein (as well as high glucose) mediated vascular damage may be accompanied by oxidative stress. Lowering blood pressure seems to lower levels of oxidative stress in the body [124], but a direct demonstration that lowering levels of oxidative stress may lower high blood pressure is lacking. Using antioxidant-rich sesame oil, Sankar and coworkers demonstrated that sesame oil lowered pulmonary hypertension and reduced the dose of medication [125,126]. The antiatherosclerotic actions of sesame oil and of sesamol and its derivatives also have been described [127,128] although it remains to be determined whether these actions are exclusively due to antioxidant effects.

Recently there has been a widespread interest in thiol antioxidants, glutathione, and related enzymes, hydrogen sulfide, sesame oil components, L-arginine (L-Arg) and other nitric oxide modulators [129]. While the former compounds may have direct antioxidant actions, L-Arg is both a direct precursor of nitric oxide as well as an antioxidant [110]. These compounds may assist in lowering high blood pressure.

Physical activity, or exercise, has long been known to play a role in lowering the risk of hypertension as well as many of the other risk factors [130]. However, as pointed out in many studies, exercise is known to lower plasma antioxidant levels and increase oxidative stress [131–133]. A model in which sustained exercise-induced oxidative stress would induce antioxidant defense and afford long-term antioxidant defense has been suggested [27,134–136].

LOW AND DYSFUNCTIONAL HDL

HDL particles are quite heterogeneous, encompassing a range of sizes and densities; they also contain multiple surface apolipoproteins (Apo A1, AII, AIV, AV, CII, CIII, D, E, F, H, J, L, M) varying concentrations of cholesterol and phospholipids, different types of lipid transfer proteins (CETP, PLTP) as well as various antioxidant or pro-oxidant enzymes (PON1, LCAT, GPx, PAF-AH), now recognized as significant factors in the balance of pro- and antiatherogenic metabolism [137,138]. Several properties of functional and dysfunctional HDL have been compared in Table 17.1.

Diverse clinical circumstances associated with oxidation, inflammation, protein carbamylation, and enzymatic degradation or advanced glycation, can alter the functionality of HDL, converting normal, cardioprotective HDL into dysfunctional HDL

TABLE 17.1
Properties of Functional and Dysfunctional HDL

Functional HDL	Dysfunctional HDL
• Considered as "good cholesterol"	• Considered as "bad cholesterol"
• Anti-inflammatory	• Pro-inflammatory
• Anti-oxidant	• Pro-oxidant
• Anti-thrombotic	• Pro-thrombotic
• Anti-apoptotic	• Pro-apoptotic
• Anti-atherogenic	• Pro-atherogenic
• Efficient RTC	• Impaired RTC
• Protective and healing activities on ECs	• No such effect
• Vasodilatory	• No such effect
• Detoxification of oxidized lipids	• Not observed
• Inhibits monocyte chemotaxis	• Enhances monocyte chemotaxis
• Inhibits LDL oxidation	• _____

[139–143]. The atherogenic capacity of HDL can be limited *via* enrichment of the lipoprotein with the enzymes group IIA secretory phospholipase A2 and myeloperoxidase, and other factors such as triglycerides, ceruloplasmin, serum amyloid A, and haptoglobin–hemoglobin complex [144,145].

The antioxidative activity of HDL is demonstrated through multiple mechanisms [146]. Removal and inactivation of lipid hydroperoxides (LOOH), which accumulate during LDL oxidation, may represent the central mechanism accounting for the antioxidative properties of HDL. Indeed, HDL-associated enzymes, including PON1 [147], PAF-AH [148], and lecithin: cholesterol acyltransferase [149], may play a key role in the inactivation of LOOH by HDL. In addition, ApoA-I can prevent the formation of oxidized lipids in LDL by removing seeding molecules of LOOH from LDL [150]. The impairment of HDL antioxidative capacity in MS subjects was solely observed at later stages of LDL oxidation, suggesting that the mechanisms involved in the inactivation of oxidized lipids (i.e., LOOH removal and hydrolysis) have become dysfunctional [151].

WAIST CIRCUMFERENCE/OBESITY

Waist circumference (WC) is a major risk factor for all metabolic disorders. WC is a parameter which provides an estimate of body girth at the level of the abdomen, a proxy for visceral adipose deposition. There are different means by which adiposity can be quantified. Overweight and obesity are traditionally characterized by body mass index (BMI), but the ratio of waist circumference-to-hip circumference (WHR) is now considered to be the best measure of central obesity [152]. The criteria for assessing WHR are given in Table 17.2. Abdominal obesity, characterized by the accumulation of excess visceral fat, is strongly correlated with an increased risk for several metabolic diseases. In general, obesity is known to be correlated to an increase in arterial pressure, cardiac output, metabolic rate, heart rate, glomerular filtration rate, renal sympathetic activity, sodium reabsorption, renal tubular reabsorption, and plasma renin activity, all of which are thought to contribute to essential hypertension, another component of MS [153].

Evidence suggests that oxidative stress may be the mechanistic link between abdominal obesity and related metabolic complications. Generally, mitochondria are a major site of intracellular ROS production, due to electron leakage along the respiratory chain. However, ROS may also arise from plasma membrane systems, endoplasmic

TABLE 17.2
Types of Measurement of Obesity

BMI	Body mass index	Degree of overweight/obesity measures body fat and muscle mass
WHR	Waist hip ratio	Abdominal fat in relation to body size
SAD	Sagittal abdominal diameter	Visceral fat
WC	Waist circumference	Fat in and around the abdomen
WHtR	Waist-to-Height ratio	Waist circumference at the umbilical level at standing height

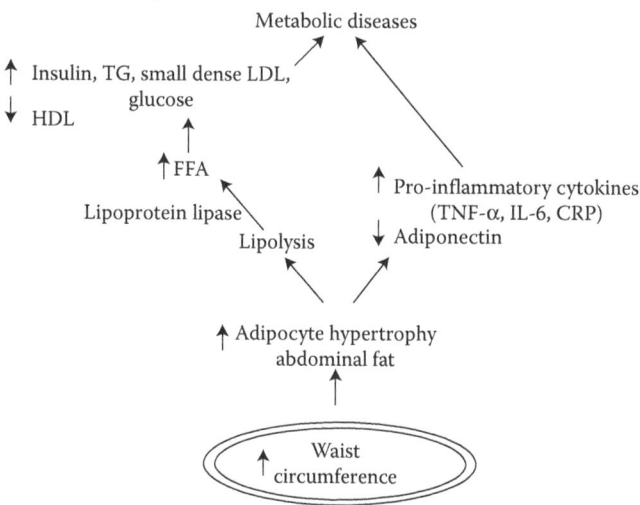

FIGURE 17.2 Abdominal fat as an integral risk factor for metabolic disorders.

reticulum, lysosomes, peroxisomes, and cytosolic enzymes. Animal, clinical, and epi-demiological studies have shown that obesity is associated with an altered redox state and increased metabolic risk [154–159]. Oxidative stress can be both a consequence and a cause of obesity (Figure 17.2). It has been identified that chronic hypernutrition, high fat/high carbohydrate (HFHC) meals, and trans-fatty acids stimulate intracel-lular pathways leading to oxidative stress through multiple biochemical mechanisms [28,160,161]. ROS and specific metabolic products could affect a plethora of devel-opments associated with fat accumulation, inflammation, and eventually, abdominal obesity. ROS and these metabolic products likely influence nuclear receptors, such as PPARs and LXR, transcription factors (e.g., NFκB), protein kinases, agonists and antagonists of the aforementioned factors, as well as suppress antioxidant enzymes such as paraoxonases, and promote synthesis of various pro-inflammatory cytokines.

PHYSICAL INACTIVITY

Physical inactivity is an important underlying cause of the actual manifestation of the disease [162]. Because physical inactivity is a highly modifiable risk factor, much emphasis has been placed upon fitness as a clinical intervention [163]. An increase in physical activity not only results in improved overall cardiovascular health, but also correlates to a reduction of other MS risk factors [164]. AHA guidelines recom-mend a minimum of 30 min of moderate intensity physical activity several times a week [165,166]. Increased physical activity and fitness have been attributed to increased HDL cholesterol, decreased LDL, and overall improved lipid profile [167]. Conversely, decreased physical activity results in an atherogenic lipid profile coupled with increased risk for type 2 diabetes mellitus [168], factors perpetuating the risk for development of MS [162]. Individuals who rated themselves as being at least moderately active during leisure time have a significantly reduced chance of having three or more risk factors for metabolic diseases, compared to those who rated themselves as inactive [168]. Physical activity has been shown to offset the progression of cardiovascular disease in animal models, even in models, genetically predisposed to developing cardiovascular complications [169,170]). Lifestyle changes, including exercise, serve an important preventative role in reducing the onset of disease and serve as a key primary intervention for individuals affected by MS.

Exercise and oxidative stress has been a topic of great interest for several decades. Vigorous physical activity increases oxygen consumption and increases oxygen and nutrient availability to cells and tissues. Paradoxically, exercise also induces oxidative stress and depletes antioxidants. Such negative effects would appear to negate its positive value. This paradox has been addressed in several studies [171–173] and it would appear that while beginning exercise could promote oxidative stress, sustained and long-term exercise appears to result in the induction of antioxidant enzymes. For example, Meilhac et al. reported that male mice subjected to exercise showed increased expression of arterial catalase and nitric oxide synthase [134] together with a great reduction in atherosclerosis. Alarmingly, supplementation of vitamin E diminished the induction of antioxidant enzymes and prevented the exercise-induced reduction in lesions [134]. Interestingly, there appears to be a gender difference in the response to exercise-induced oxidative stress.

DIET

In addition to these major risk factors for MS, several less obvious risk factors exist. As mentioned, many drugs increase blood glucose levels (e.g., some antidepressants and psychotic drugs). Recently, monosodium glutamate (MSG), an amino acid, which enhances the flavor of food, has attracted attention as a potential inducer of many symptoms associated with MS [174]. Epidemiological and animal studies seem to support the contention [175,176], and even a mouse model for metabolic syndrome has resulted from these studies [177]. Glutamate has numerous functions in the body. It is one of the component amino acids of glutathione (GSH), an important antioxidant. However, high levels of glutamate may promote oxidative stress and deplete GSH and other antioxidant levels [178]. Additional authors suggest that MSG might be a critical factor in the initiation of obesity and nonalcoholic steatohepatitis, and restricting caloric intake might modulate the progression of disease to fibrosis and inflammation [179]. It is currently not known whether the effects of glutamate in promoting MS are solely dependent on its ability to induce oxidative stress.

SMOKING

Smoking is considered a risk factor for several diseases associated with oxidative stress [180–183]. Cigarette smoke contains NO and NO_2, as well as several lipid peroxy radicals and free radicals [184]. All these components may serve as substrates for oxidation and generate $ONOO^-$, likely furthering the cascade of oxidative damage [184,185]. Smokers also have lower amounts of plasma antioxidants, such as ascorbate and vitamin E [186,187] and the consumption of antioxidants attenuate the production of inflammatory mediators [188]. Evidence also suggests that smoking is directly and indirectly involved in the oxidation of LDL and thus, the progression of atherosclerosis [189–193].

MS IN RELATION TO COLON CANCER AND ALZHEIMER'S DISEASE

With metabolic syndrome, an increased risk of colon cancer exists [194,195]. The mechanism linking these two conditions is thought to relate to either insulin resistance or elevated C-peptide [196]. Additionally, high levels of glucose in the body, common in metabolic syndrome patients, create oxidative stress in the form of ROS, which subsequently has the potential to damage DNA and disrupt additional physiological processes [197]. This damage to DNA is capable of inducing carcinogenesis [198].

Alzheimer's disease (AD) is a neurodegenerative disease common among the elderly. It has been speculated that free radicals generated during oxidative stress are pathologically important in AD. Evidence also suggests that amyloid beta, tau, and APOE play a critical role in inducing oxidative stress in AD [199]. Elevated oxidative stress can cause damage of mitochondrial membranes and a dysfunctional electron transport chain [200]. The mechanisms by which metabolic syndrome correlates with AD are still unknown. However, biochemical alterations observed in MS-like induction of chronic inflammation and oxidative stress, induction of insulin and leptin resistance, impairment of endothelial cell function, hyperglycemia-related increase in AGEs, and microvascular injury may signify a pathological connection between metabolic syndrome and neurological disorders.

TAKE-HOME MESSAGE

Although many studies suggest a role for oxidative stress in risk factors associated with metabolic syndrome, more studies are needed to define its role and the causes, as well as the possible contributions of oxidized lipids. While antioxidant therapy might be seen as a remedy, studies from exercise may suggest otherwise. More mechanistic insights are warranted, and strategies to induce the body's own antioxidant defense need to be developed. This chapter covers only the major risk factors of metabolic diseases associated with oxidative stress. However, genetic deficiencies may affect antioxidant enzyme systems, enhancing the susceptibility to oxidative stress. Gender differences in oxidative stress have been observed due to the phenolic hydroxyl group of estradiol. Oxidative stress is also correlated to the aging process, creating dysfunctional organ physiology and function.

ABBREVIATIONS

ACE	Angiotensin converting enzyme
ACS	Acyl Co-A synthetase
AD	Alzheimer's disease
AGE	Advanced glycation endproduct
Ang II	Angiotensin II
APO	Apolipoprotein
AT1	Angiotensin type1 receptor

BP	Blood pressure
CETP	Cholesteryl ester transfer protein
CRP	C-reactive protein
DE	Dysfunctional endothelium
FATP	Fatty acid transfer protein
FFA	Free fatty acid
GL	Glucose
GLP	Glucagon-like peptide
GPx	Glutathione peroxidase
HDL	High-density lipoprotein
HFHC	High-fat high carbohydrate
IL	Interleukin
LCAT	Lecithin cholesterol acyl transferase
LOOH	Lipid peroxide
LPL	Lipoprotein lipase
LXR	Liver x receptor
MS	Metabolic syndrome
NAC	*N*-acetyl cysteine
NO	Nitric oxide
NOS	Nitric oxide synthase
Ox-LDL	Oxidized low-density lipoprotein
PAF-AH	Platelet-activating factor acetyl hydrolase
PLTP	Phospholipid transfer protein
PtdEtn	Phosphatidyl ethanolamine
PPAR	Peroxisome proliferator-activated receptor
PON1	Paraoxonase1
ROS	Reactive oxygen species
sd-LDL	Small dense low-density lipoprotein
T2DM	Type 2 diabetes mellitus
TG	Triglycerides
TNF	Tumor necrotic factor
TRL	Triglyceride-rich lipoprotein
VLDL	Very-low-density lipoprotein
WC	Waist circumference
WHR	Waist-to-hip ratio
ZDF	Zucker diabetic fatty

REFERENCES

1. *Diabetes Atlas*, second edition, International Diabetes Federation, Belgium, 2003.
2. UKPDS Group. UK Prospective Diabetes Study 17: A nine-year update of a randomized, controlled trial on the effect of improved metabolic control on complications in non-insulin-dependent diabetes mellitus. *Ann. Intern. Med.* 1996;124:136 45.
3. Hu, G., Qiao, Q., Tuomilehto, J. et al. Plasma insulin and cardiovascular mortality in non-diabetic European men and women: A meta-analysis of data from eleven prospective studies. The DECODE Insulin Study Group. *Diabetologia* 2004;47:1245r56.

4. Zimmet, P., Alberti, K.G.M.M., Shaw, J. Global and societal implications of the diabetes epidemic. *Nature* 2001;414:782–7.

5. Carey, V.J., Walters, E.E., Colditz, G.A. et al. Body fat distribution and risk of non-insulin-dependent diabetes in women: The Nurses' Health Study. *Am. J. Epidemiol.* 1997;145:614–19.

6. Lemieux, I., Pascot, A., Couillard, C. et al. Hypertriglyceridemic waist: A marker of the atherogenic metabolic triad (hyperinsulinemia; hyperapolipoprotein B; small, dense LDL) in men? *Circulation* 2000;102:179–84.

7. Berneis, K.K., and Krauss, R.M. Metabolic origins and clinical significance of LDL heterogeneity. *J. Lipid Res.* 2000;43:1363–79.

8. National Cholesterol Education Program Expert Panel on Detection, E., and Treatment of High Blood Cholesterol in, A. Third report of the National Cholesterol Education Program (NCEP) expert panel on detection, evaluation, and treatment of high blood cholesterol in adults (Adult Treatment Panel III) final report. *Circulation* 2002;106:3143–421.

9. Chobanian, A.V., Bakris, G.L., Black, H.R. et al. Blood Institute Joint National Committee on Prevention, D. E., Treatment of High Blood, P., and National High Blood Pressure Education Program Coordinating, C. The seventh report of the joint National Committee on Prevention, Detection, Evaluation, and Treatment of High Blood Pressure: The JNC 7 report. *JAMA* 2003;289:2560–72.

10. Aronson, D., and Rayfield, E.J. How hyperglycemia promotes atherosclerosis: Molecular mechanisms. *Cardiovasc. Diabetol.* 2002;1:1.

11. Schachinger, V., Britten, M.B., and Zeiher, A.M. Prognostic impact of coronary vasodilator dysfunction on adverse long-term outcome of coronary heart disease. *Circulation* 2000;101:1899–906.

12. Pearson, T.A., Mensah, G.A., Alexander, R.W. et al. Markers of inflammation and cardiovascular disease: Application to clinical and public health practice: A statement for healthcare professionals from the Centers for Disease Control and Prevention and the American Heart Association. *Circulation* 2003;107:499–511.

13. Brunzell, J.D., Davidson, M., Furberg, C.D. et al. Lipoprotein management in patients with CMR: Consensus conference report from the American Diabetes Association and the American College of Cardiology Foundation. *J. Am. Coll. Cardiol.* 2008;51:1512–24.

14. Sowers, J.R., Whaley-Connell, A., Hayden, M.R. The role of overweight and obesity in the cardiorenal syndrome. *Cardiorenal Med.* 2011;1:5–12.

15. Chandrakala Aluganti Narasimhulu, Xueting Jiang, Zhaohui Yang, Krithika Selvarajan and Sampath Parthasarathy. Is there a connection between inflammation and oxidative stress. *Chronic Inflammation, Molecular Pathophysiology, Nutritional and Therapeutic Interventions*, CRC Press, Boca Raton, 2012: 139–52.

16. Sampath, P., Nalini, S., Sumathi, R. and Olivier, M. Oxidants and antioxidants in atherogenesis: An appraisal. *J. Lipid Res.*, 1999;40:2143–57.

17. Sampath, P., Achuthan, R., Mahdi, O.G., and Nalini, S. Oxidized low-density lipoprotein. *Methods Mol Biol.* 2010;610:403–17.

18. Nemat, K., Yadollah, S. and Mahdi, M. Chronic inflammation and oxidative stress as a major cause of age-related diseases and cancer. *Recent Patents on Inflammation and Allergy Drug Discov.* 2009;3:73–80.

19. Li, Y., Ma, Q. G., Zhao, L. H., Wei, H., Duan, G. X., Zhang, J. Y., and Ji, C. Effects of lipoic acid on immune function, the antioxidant defense system, and inflammation-related genes expression of broiler chickens fed aflatoxin contaminated diets. [Research Support, Non-U.S. Gov't]. *Int. J. Mol. Sci.* 2014;15(4):5649–62.

20. Zingg, J.M., Hasan, S.T., and Meydani, M. Molecular mechanisms of hypolipidemic effects of curcumin. *Biofactors* 2013;39(1):101–21.

21. Auf dem Keller, U., Kumin, A., Braun, S., and Werner, S. Reactive oxygen species and their detoxification in healing skin wounds. [Research Support, Non-U.S. Gov't Review]. *J. Invest. Dermatol. Symp Proc.* 2006;11(1):106–11.

22. Mitjavila, M.T., and Moreno, J.J. The effects of polyphenols on oxidative stress and the arachidonic acid cascade. Implications for the prevention/treatment of high prevalence diseases. *Biochem Pharmacol.* 2012;84(9):1113–22.

23. Hansson, G.K., Robertson, A.K., Soderberg-Naucler, C. Inflammation and atherosclerosis. *Ann. Rev. Pathol.* 2006;1:297–329.

24. Takahashi, K., Takeya, M., Sakashita, N. Multifunctional roles of macrophages in the de-velopment and progression of atherosclerosis in humans and experimental animals. *Med. Electron Microsc.: Off. J. Clin. Electron Microsc. Soc. Japan* 2002;35:179–203.

25. Shalhoub, J., Falck-Hansen, M., Davies, A., Monaco, C. Innate immunity and monocyte-macrophage activation in atherosclerosis. *J. Inflamm* 2011;8:1–17.

26. Fairweather, D., Cihakova, D. Alternatively activated macrophages in infection and auto-immunity. *J. Autoimmunity* 2009;33:222–30.

27. Cho, D.I., Kim, M.R., Jeong, H.Y., Jeong, H.C., Jeong, M.H., Yoon, S.H., Kim, Y.S, Ahn, Y. Mesenchymal stem cells reciprocally regulate the M1/M2 balance in mouse bone marrow-derived macrophages. *Exp. Mol. Med.* 2014;46:e70.

28. Penumetcha, M., Khan, N., Parthasarathy, S. Dietary oxidized fatty acids: An atherogenic risk? *J. Lipid Res.* 2000;41(9):1473–80.

29. Sies, H. Stahl, W. and Sevanian, A. Nutritional, dietary and postprandial oxidative stress. *J. Nutr.* 2005;135(5):969–72.

30. Khan-Merchant, N.1, Penumetcha, M., Meilhac, O., Parthasarathy, S. Oxidized fatty acids promote atherosclerosis only in the presence of dietary cholesterol in low-density lipoprotein receptor knockout mice. *J. Nutr.* 2002;132(11):3256–62.

31. Bayani Uttara, Ajay V. Singh, Paolo Zamboni, and Mahajan, R.T. Oxidative stress and neurodegenerative diseases: A review of upstream and downstream antioxidant therapeutic options. *Curr Neuropharmacol.* 2009;7(1):65–74.

32. Rosen, P., Nawroth, P.P., King, G., Moller, W., Tritschler, H.J., Packer, L. The role of oxidative stress in the onset and progression of diabetes and its complications: A summary of a Congress Series sponsored by UNESCO-MCBN, the American Diabetes Association, and the German Diabetes Society. *Diabetes Metab. Res. Rev.* 2001;17:189–891.

33. Hunt J.V., Smith C.C.T., Wolff S.P. Autoxidative glycosylation and possible involvement of peroxides and free radicals in LDL modification by glucose. *Diabetes* 1990;39:1420–4.

34. Croze, M.L., and Soulage, C.O. Potential role and therapeutic interests of myo-inositol in metabolic diseases. *Biochimie.* 2013;95(10):1811–27.

35. Hofmann, B., Yakobus, Y., Indrasari, M., Nass, N., Santos, A.N., Kraus, F.B., Simm, A. RAGE influences the development of aortic valve stenosis in mice on a high fat diet. *Exp. Gerontol.* 2014, doi: 10.1016/j.exger.2014.05.001. [Epub ahead of print].

36. Kanazawa, I., and Sugimoto, T. [The relationship between bone and glucose/lipid metabolism]. [Review]. *Clin. Calcium* 2013;23(2):181–8.

37. Hunt, K.J., Baker, N., Cleary, P., Backlund, J.Y., Lyons, T., Jenkins, A., Lopes-Virella, M.F. Oxidized LDL and AGE-LDL in circulating immune complexes strongly predict progression of carotid artery IMT in type 1 diabetes. *Atherosclerosis* Dec 2013;231(2):315–22.

38. Vay, D., Vidali, M., Allochis, G., Cusaro, C., Rolla, R., Mottaran, E., Albano, E. Antibodies against advanced glycation end product Nepsilon-(carboxymethyl) lysine in healthy controls and diabetic patients. [Research Support, Non-U.S. Gov't]. *Diabetologia* 2000;43(11):1385–8.

39. Parthasarathy, S., Quinn, M.T., and Steinberg, D. Is oxidized low density lipoprotein involved in the recruitment and retention of monocyte/macrophages in the artery wall during the initiation of atherosclerosis? *Basic Life Sci.* 1988;49:375–80.

40. Boyle, J.J., Harrington, H.A., Piper, E., Elderfield, K., Stark, J., Landis, R.C., and Haskard, D.O. Coronary intraplaque hemorrhage evokes a novel atheroprotective macrophage phenotype. [Research Support, Non-U.S. Gov't]. *Am. J. Pathol.* 2009;174(3):1097–108. doi: 10.2353/ajpath.2009.080431

41. Lopes-Virella, M.F., Baker, N.L., Hunt, K.J., Lachin, J., Nathan, D., and Virella, G. Oxidized LDL immune complexes and coronary artery calcification in type 1 diabetes. *Atherosclerosis* 2011;214(2):462–67.

42. Matsuoka, T., Kajimoto, Y., Watada, H. et al. Glycationdependent, reactive oxygen species-mediated suppression of the insulin gene promoter activity in HIT cells. *J Clin Invest* 1997;99(1):144144.

43. Tanaka, Y., Gleason, C.E., Tran, P.O. et al. Prevention of glucose toxicity in HIT-T15 cells and Zucker diabetic fatty rats by antioxidants. *Proc. Natl. Acad. Sci. USA* 1999;96(19):10857–62.

44. Kaneto, H., Fujii, J., Myint, T. et al. Reducing sugars trigger oxidative modification and apoptosis in pancreatic beta-cells by provoking oxidative stress through the glycation reaction. *Biochem J* 1996;320(Pt 3):855–63.

45. Kharroubi, I., Ladriere, L., Cardozo, A.K. et al. Free fatty acids and cytokines induce pancreatic beta-cell apoptosis by different mechanisms: Role of nuclear factor-kappaB and endoplasmic reticulum stress. *Endocrinology* 2004;145(11):5087–96.

46. Fridlyand, L.E., Philipson, L.H. Does the glucose dependent insulin secretion mechanism itself cause oxidative stress in pancreatic beta-cells? *Diabetes* 2004;53(8):1942–8.

47. Ceriello, A., Motz, E. Is oxidative stress the pathogenic mechanism underlying insulin resistance, diabetes, and cardiovascular disease? The common soil hypothesis revisited. *Arterioscler Thromb Vasc Biol* 2004;24(5):816–23.

48. Evans, J.L., Goldfine, I.D., Maddux, B.A. et al. Are oxidative stress-activated signaling pathways mediators of insulin resistance and beta-cell dysfunction? *Diabetes* 2003;52(1):1.

49. Evans, J.L., Goldfinc, I.D., Maddux, B.A. et al. Oxidative stress and stress-activated signaling pathways: A unifying hypothesis of type 2 diabetes. *Endocr Rev* 2002;23(5):599–622.

50. Moore, P.C., Ugas, M.A., Hagman, D.K. et al. Evidence against the involvement of oxidative stress in fatty acid inhibition of insulin secretion. *Diabetes* 2004;53(10):2610–6.

51. Steinberg, D., Witztum, J.L. Lipoproteins and atherogenesis; current concepts. *J. Am. Med. Assoc.* 1990;264:3047–52.

52. Ross, R. The pathogenesis of atherosclerosis; a perspective for the 1990. *Nature* 1993;362:801–9.

53. Goldstein, J.L., Ho Y.K., Basu S.K., Brown M.S. 1979. Binding site on macrophages that mediates uptake and degradation of acetylated low density lipoprotein, producing massive cholesterol deposition. *Proc. Natl. Acad. Sci. USA*, 76:333–7.

54. Brown, M.S., Basu, S.K., Falck, J.R., Ho, Y.K., Goldstein, J.L. 1980 The scavenger cell pathway for lipoprotein degradation: Specificity of the binding site that mediates the uptake of negatively-charged LDL by macrophages. *J. Supramol Struct.* 13:67–81.

55. Swirski, F.K., Pittet, M.J., Kircher, M.F., Aikawa, E., Jaffer, F.A., Libby, P., and Weissleder, R. 2006. Monocyte accumulation in mouse atherogenesis is progressive and proportional to extent of disease. *Proc. Natl. Acad. Sci. U. S. A* 103, 10340–5.

56. Lipton, B., Parthasarathy, S., Ord, V., Clinton, S., Libby, P., Rosenfeld M. Components of the protein fraction of oxidized low density lipoprotein stimulates interleukin-1α production by rabbit arterial macrophage-derived foam cells. *J. Lipid Res.* 1995;36:2232–42.

57. Shi Du Yan, S., Schmidt, A.M., Anderson, G.M., Zhang, J., Brett, J., Zou, Y.S., Pinsky, D. and Stern, D. Sn enhanced cellular oxidant stress by the interaction of advanced glycation end products with their receptors binding proteins. *J. Biol. Chem.* 1994;269(13):9889–97.
58. Jacobson, T.A., Miller, M., Schaefer, E.J. Hypertriglyceridemia and cardiovascular risk reduction. *Clin. Ther.* 2007;29(5):763–77.
59. Kennedy, D.J., and Kashyap, S.R. Pathogenic role of scavenger receptor CD36 in the metabolic syndrome and diabetes. *Metabolic Syndrome and Related Disorders* 2011;9(4):239–49.
60. Ginsberg, H.N., Huang, L.S. The insulin resistance syndrome: Impact on lipoprotein metabolism and atherothrombosis. *J. Cardiovasc Risk.* 2000;7(5):325–31.
61. Ebara, T., Conde, K., Kako, Y., Liu, Y., Xu, Y., Ramakrishnan, R., Goldberg, I.J., Shachter, N.S. Delayed catabolism of apoB-48 lipoproteins due to decreased heparan sulfate proteoglycan production in diabetic mice. *J. Clin. Invest.* 2000;105(12):1807–18.
62. Memon, R.A., Fuller, J., Moser, A.H., Smith, P.J., Grunfeld, C., Feingold, K.R. Regulation of putative fatty acid transporters and Acyl-CoA synthetase in liver and adipose tissue in ob/ob mice. *Diabetes.* 1999;48(1):121–7.
63. Chisolm, G.M., Irwin, K.C., and Penn, M.S. Lipoprotein oxidation and lipoprotein-induced cell injury in diabetes. *Diabetes* 1992;41(2):61–6.
64. Henriksen, T., Evensen, S. A., and Carlander, B. Injury to cultured endothelial cells induced by low density lipoproteins: Protection by high density lipoproteins. *Scandinavian J. Clin. Laboratory Invest.* 1979;39:369–75.
65. Pritchard, K.A., Jr., Patel, S.T., Karpen, C.W., Newman, H.A., and Panganamala, R.V. Triglyceride-lowering effect of dietary vitamin E in streptozocin-induced diabetic rats. Increased lipoprotein lipase activity in livers of diabetic rats fed high dietary vitamin E. *Diabetes* 1986;35:278–81.
66. Sevanian, A., Stein, R.A., Mead, J.F. Metabolism of epoxidized phosphatidylcholine by phospholipase A2 and epoxide hydrolase. *Lipids* 1981;16(11):781–9.
67. Kim, E.H., Sevanian, A. Hematin- and peroxide-catalyzed peroxidation of phospholipid liposomes. *Arch. Biochem. Biophys.* 1991;288(2):324–30.
68. UP Steinbrecher, S., Parthasarathy, D.S., Leake, J.L., Witztum, D. Steinberg modification of low density lipoprotein by endothelial cells involves lipid peroxidation and degradation of low density lipoprotein phospholipids. *Proc. Natl Academy Sci.* 1984; 81(12);3883–7.
69. S. Parthasarathy, Barnett, J., Fong, L.G. High-density lipoprotein inhibits the oxidative modification of low-density lipoprotein. *Biochim. Biophys. Acta,* 1990;1044:275–83.
70. Sarah G., Guo, L., Fisher, E.A., and Brodsky, J.L. Protein disulfide isomerases contribute differentially to the endoplasmic reticulum-associated degradation of apolipoprotein B and other substrates *Mol. Biol. Cell.* 2012;23(4):520–532.
71. Austin, M.A., King, M.C., Vranizan, K.M. et al. Atherogenic lipoprotein phenotype. A proposed genetic marker for coronary heart disease risk. *Circulation* 1990;82(2):495–506.
72. Havel, R.J. The formation of LDL: Mechanisms and regulation. *J. Lipid Res.* 1984;25:1570–6.
73. Murdoch, S.J., Breckenridge, W.C. Influence of lipoprotein lipase and hepatic lipase on the transformation of VLDL and HDL during lipolysis of VLDL. *Atherosclerosis.* 1995;118(2):193–212.
74. Austin, M.A., Breslow, J.L., Hennekens, C.H. et al. Low density lipoprotein subclass patterns and risk of myocardial infarction. *JAMA* 1988;260(13):1917–21.
75. Campos, H., Genest J.J. Jr, Blijlevens E. et al. Low density lipoprotein particle size and coronary artery disease. *Arterioscler Thromb* 1992;12(2):187–95.
76. Ginsberg, H.N., Huang, L.S. The insulin resistance syndrome: Impact on lipoprotein metabolism and atherothrombosis. *J. Cardiovasc Risk.* 2000;7(5):325–31.

77. Rizzo, M., Rini, G.B., Berneis, K. The clinical relevance of LDL size and subclasses modulation in patients with type-2 diabetes. *Exp. Clin. Endocrinol. Diabetes.* 2007;115(8):477–82.

78. Mehta, P.K., Griendling, K.K. Angiotensin II cell signaling: Physiological and pathological effects in the cardiovascular system. *Am. J. Physiol. Cell Physiol.* 2007;292:C82–97.

79. Hirayama, S., and Miida, T. Small dense LDL: An emerging risk factor for cardiovascular disease. *Clin. Chim. Acta*, 2012;414:215–24.

80. Al-Mashhadi, R.H., Sorensen, C.B., Kragh, P.M. et al. Familial hypercholesterolemia and atherosclerosis in cloned minipigs created by DNA transposition of a human PCSK9 gain-of-function mutant. *Sci. Transl. Med.*, 2013;5(166):166ra1.

81. Dreon, D.M., and Krauss, R.M. Diet-gene interactions in human lipoprotein metabolism. *J. Am. Coll. Nutr.*, 1997;16(4):313–24.

82. Bionaz, M., Trevisi, E., Calamari, L., Librandi, F., Ferrari, A., and Bertoni, G. Plasma paraoxonase, health, inflammatory conditions, and liver function in transition dairy cows. *J. Dairy Sci.*, 2007;90(4):1740–50.

83. Bortolasci, C.C., Vargas, H.O., Souza-Nogueira, A. et al. Lowered plasma paraoxonase (PON)1 activity is a trait marker of major depression and PON1 Q192R gene polymorphism-smoking interactions differentially predict the odds of major depression and bipolar disorder. *J. Affective Disord.* 2014;159:23–30.

84. Aviram, M., and Rosenblat, M. Paraoxonases 1, 2, and 3, oxidative stress, and macrophage foam cell formation during atherosclerosis development. *Free Radic. Biol. Med.* 2004;37(9):1304–16.

85. Borowczyk, K., Shih, D.M., and Jakubowski, H. Metabolism and neurotoxicity of homocysteine thiolactone in mice: Evidence for a protective role of paraoxonase 1. *J. Alzheimers Dis.* 2012;30(2):225–31.

86. Ushio-Fukai M., Alexander R.W. Reactive oxygen species as mediators of angiogenesis signaling: Role of NAD(P)H oxidase. *Mol. Cell. Biochem.* 2004;264:85–97.

87. Ushio-Fukai, M. Redox signaling in angiogenesis: Role of NADPH oxidase. *Cardiovasc. Res.* 2006;71:226–35.

88. Touyz, R.M. Reactive oxygen species as mediators of calcium signaling by angiotensin II: Implications in vascular physiology and pathophysiology. *Antioxid. Redox. Signal.* 2005;7:1302–14.

89. Beckman, J.S., Beckman, T.W., Chen, J., Marshall, P.A., Freeman, B.A. Apparent hydroxyl radical production by peroxynitrite: Implications for endothelial injury from nitric oxide and superoxide. *Proc. Natl. Acad. Sci. USA.* 1990;87:1620–4.

90. Mikhail, M.S., Anyaegbunam, A., Garfinkel, D., Palan, P.R., Basu, J., and Romney, S.L. Preeclampsia and antioxidant nutrients: Decreased plasma levels of reduced ascorbic acid, α-tocopherol, and beta-carotene in women with preeclampsia. *Am. J. Obstet. Gynecol.* 1994;171(1):150–7.

91. Russo, C., Olivieri, O., Girelli, D., Faccini, G., Zenari, M.L., Lombardi, S., and Corrocher, R. Antioxidant status and lipid peroxidation in patients with essential hypertension. *J. Hypertens.* 1998;16(9):1267–71.

92. Edwards, R.L., Lyon, T., Litwin, S.E., Rabovsky, A., Symons, J.D., and Jalili, T. Quercetin reduces blood pressure in hypertensive subjects. *J. Nutr.* 2007;137(11):2405–11.

93. Rodriguez-Iturbe, B., Zhan, C-D., Quiroz, Y., Sindhu, R.K., and Vaziri, N.D. Antioxidant-rich diet relieves hypertension and reduces renal immune infiltration in spontaneously hypertensive rats. *Hypertension* 2003;41:341–6.

94. Forstermann, U. Oxidative stress in vascular disease: Causes, defense mechanisms, and potential therapies. *Nat. Clin. Pract. Cardiovasc. Med.* 2008;5:338–49.

95. Griendling, K.K., Minieri, C.A., Ollerenshaw, J.D., and Alexander, R.W. Angiotensin II stimulates NADH and NADPH oxidase activity in cultured vascular smooth muscle cells. *Circ. Res.* 1994;74:1141–8.

96. Zafari, A.M., Ushio-Fukai, M., Akers, M., Yin, Q., Shah, A., Harrison, D.G., Taylor, W.R., and Griendling, K.K. Role of NADH/NADPH oxidase-derived H2O2 in angiotensin II-induced vascular hypertrophy. *Hypertension* 1998;32:488–95.

97. Zhang, H., Schmeisser, A., Garlichs, C.D., Plotze, K., Damme, U., Mugge, A., and Daniel, W.G. Angiotensin II-induced superoxide anion generation in human vascular endothelial cells: Role of membrane-bound NADH-/NADPH-oxidases. *Cardio. Res.* 1999;44(1):215–22.

98. K Dikalova, A.E., Bikineyeva, A.T., Budzyn, K., Nazarewicz, R.R., McCann, L., Lewis, W., Harrison, D.G., and Dikalov, S.I. Therapeutic targeting of mitochondrial superoxide in hypertension. *Circ. Res.* 2010;107:106–16.

99. Midaoui, A.E.L., Elimadi, A., Wu, L., Haddad, P.S., and de Champlain, J. Lipoic acid prevents hypertension, hyperglycemia, and the increase in heart mitochondrial superoxide production. *Am. J. Hypertens.* 2003;16(3):173–9.

100. Suzuki, H., DeLano, F.A., Parks, D.A., Jamshidi, N., Granger, D.N., Ishii, H., Suematsu, M., Zweifach, B.W., and Schmid-Schonbein, G.W. Xanthine oxidase activity associated with arterial blood pressure in spontaneously hypertensive rats. *PNAS* 1998;95(8):4754–9.

101. Landmesser, U., Dikalov, S., Price, S.R., McCann, L, Fukai, T., Holland, S.M., Mitch, W.E., and Harrison, D.G. Oxidation of tetrahydrobiopterin leads to uncoupling of endothelial cell nitric oxide synthase in hypertension. *J. Clin. Invest.* 2003;111(8):1201–9.

102. Forstermann, U. and Munzel, T. Endothelial nitric oxide synthase in vascular disease: From marvel to menace. *Circulation* 2006;113:1708–14.

103. Nozawa, K., Tuck, M.L., Golub, M., Eggena, P., Nadler, J.L., and Stern, N. Inhibition of lipoxygenase pathway reduces blood pressure in renovascular hypertensive rats. *AJP-Heart* 1990;259(6):H1774–80.

104. Hao, C-M. and Breyer, M.D. Hypertension and cyclooxygenase-2 inhibitors: Target: The renal medulla. *Hypertension* 2004;44:396–7.

105. Aw, T-J., Haas, S.J., Liew, D., and Krum, H. Meta-analysis of cyclooxygenase-2 inhibitors and their effects on blood pressure. *JAMA Int. Med.* 2005;165(5):490–6.

106. Q Cabassi, A., Dumont, E.C., Girouard, H., Bouchard, J-F., Le Jossec, M., Lamontagne, D., Besner J-G., and de Champlain J. 2001. Effects of chronic N-acetylcysteine treatment on the actions of peroxynitrite on aortic vascular reactivity in hypertensive rats. *J. Hypertens.* 19(7):1233–44.

107. Wu, R., Millette, E., Wu, L., and de Champlain, J. Enhanced superoxide anion formation in vascular tissues from spontaneously hypertensive and desoxycorticosterone acetate-salt hypertensive rats. *J. Hypertens.* 2001;19(4):741–8.

108. Russo, C., Olivieri, O., Girelli, D., Faccini, G., Zenari, M.L., Lombardi, S., and Corrocher, R. Antioxidant status and lipid peroxidation in patients with essential hypertension. *J. Hypertens.* 1998;16(9):1267–71.

109. Gradman, A.H., Arcuri, K.E., Goldberg, A.I. et al. A randomized placebo-controlled, double-blind study of various doses of losartan potassium compared with enalapril maleate in patients with essential hypertension. *Hypertension* 1995;25:1345t2550.

110. Mahesh, S. Joshi, T. Bruce Ferguson, Fruzsina K. Johnson, Robert A. Johnson, Sampath Parthasarathy, and Jack R. Lancaster. Receptor-mediated activation of nitric oxide synthesis by arginine in endothelial cells. *PNAS* 2007;104:9982–7.

111. Rajagopalan, S., Kurz, S., Munzel, T., Harrison, D.G. Angiotensin II-mediated hypertension in the rat increases vascular superoxide production via membrane NADH/NADPH oxidase activation. *J. Clin. Invest.* 1996;97:1916n23.

112. McIntyre, M., Caffe, S.E., Michalak, R.A., Reid, J.L. Losartan, an orally active AT1 receptor antagonist: A review of its efficacy and safety in essential hypertension. *Pharmacol Ther* 1997;74:181–94.

113. Pitt, B., Segal, R., Martinez, F.A. et al. Randomized trial of losartan versus captopril in patients over 65 with heart failure. *Lancet* 1997;349:747–52.
114. Cheetham, C., Collis, J., O'Driscoll G. et al. Losartan, an AT1 receptor antagonist, improves endothelial function in non-insulin-dependent diabetes. *J. Am. Coll. Card.* 2000;36:1461 6.
115. Vanhoutte, P.M., Boulanger, C.M., Mombouli, J.V. Endothelium-derived relaxing factors and converting-enzyme inhibition. *Am. J. Cardiol.* 1995;76:3E–12E.
116. Hornig, B., Kohler, C., Drexler, H. Role of bradykinin in mediating vascular effects of angiotensin converting enzyme inhibitors in humans. *Circulation* 1997;95:482b6.
117. Berkenboom, G. Bradykinin and the therapeutic actions of angiotensin converting enzyme inhibitors. *Am. J. Cardiol.* 1998;82:11Si13S.
118. Pfeffer, J.M., Pfeffer, M.A., Braunwald, E. Influence of chronic captopril therapy on the infarcted left ventricle of the rat. *Circ. Res.* 1985;57:84–95.
119. Pfeffer, M.A., Pfeffer, J.M., Steinberg, C., Jinn P. Survival after an experimental myocardial infarction: Beneficial effects of long-term therapy with captopril. *Circulation* 1985;72:406–12.
120. Jugdutt, B.I., Schwarz-Michorowski, B.L., Khan M.I. Effects of long-term captopril therapy on left ventricular remodelling and function during healing in canine myocardial infarction. *J. Am. Coll. Cardiol.* 1992;19:71321.
121. Mancini, G.B.J., Henry, G.C., Macaya, C.B. et al. Angiotensin-converting enzyme inhibition with quinapril improves endothelial vasomotor dysfunction in patients with coronary artery disease. The TREND (Trial on Reversing Endothelial Dysfunction) Study. *Circulation* 1996;94:258–65.
122. Keaton, A.K., White, C.R., and Berecek, K.H. Captopril treatment and its withdrawal prevents impairment of endothelium-dependent responses in the spontaneously hypertensive rat. *Clin. Exp. Hypertens.* 1998;20:847–66.
123. Vasdev, S., Ford, C.A., Parai, S., Longerich, L., Gadag, V. Dietary alpha-lipoic acid supplementation lowers blood pressure in spontaneously hypertensive rats. *J. Hypertension* 2000;18(5):567–73.
124. Grossman, E. Does increased oxidative stress cause hypertension? *Diabetes Care* 2008;31:S185–9.
125. Sankar, D., Sambandam, G., Rao, M.R., and Pugalendi, K.V. Modulation of blood pressure, lipid profiles and redox status in hypertensive patients taking different edible oils. *Clinica Chimica Acta* 2005;355:97–104.
126. Sankar, D., Rao, M.R., Sambandam, G., and Pugalendi, K.V. A pilot study of open label sesame oil in hypertensive diabetics. *J. Med. Food* 2006;9(3):408–12.
127. Bhaskaran, S., Santanam, N., Penumetcha, M., Parthasarathy, S. Inhibition of atherosclerosis in low-density lipoprotein receptor-negative mice by sesame oil. *J. Med. Food* 2006;9(4):487–90.
128. Ying, Z., Kherada, N., Kampfrath, T. et al. A modified sesamol derivative inhibits progression of atherosclerosis. Arterioscler. Thromb. *Vasc. Biol.* 2011;31(3):536–42.
129. Wu, G., Morris, S.M. Arginine metabolism: Nitric oxide and beyond. *Biochem J*, 1998;336:1–7.
130. Wallace, J.P. Exercise in hypertension. *Sports Med*, 2003;33(8):585–98.
131. Wetzstein, C.J., Shern-Brewer, R.A., Santanam, N., Green, N.R., White-Welkley, J.E., Parthasarathy, S. Does acute exercise affect the susceptibility of low density lipoprotein to oxidation? *Free Radic. Biol. Med.* 1998;24:679i82.
132. Shern-Brewer, R., Santanam, N., Wetzstein, C., White-Welkley, J., Parthasarathy, S. Exercise and cardiovascular disease: A new perspective. Arterioscler. *Thromb. Vasc. Biol.* 1998;18:1181 87.
133. Parthasarathy, S., Khan-Merchant, N., Penumetcha, M., Santanam, N. Oxidative stress in cardiovascular disease. *J. Nucl. Cardiol.* 2001;8:379 89.

134. Meilhac, O., Ramachandran, S., Chiang, K., Santanam, N, and Parthasarathy, S. Role of arterial wall antioxidant defense in beneficial effects of exercise on atherosclerosis in mice. *ATVB* 2001;21:1681–8.

135. Robin Shern-Brewer, Nalini Santanam, Carla Wetzstein, Jill White-Welkley, Sampath Parthasarathy Arteriosclerosis. *Thrombosis, and Vascular Biol.* 1998;18:1181–7.

136. Chen Wei, Meera Penumetcha, Mahdi Garelnabi, Ya-Guang Liu, Nalini Santanam, and Sampath Parthasarathy, Could exercise influence cardiovascular diseases via the generation of natural PPAR ligands? *Front. Biosci.* 2007;12:8.

137. Eren, N. Yilmaz, O. Aydin, High density lipoprotein and it's dysfunction. *Open Biochem. J.* 2012;6:78–93.

138. Eugene A. Podrez Anti-oxidant properties of high-density lipoprotein and atherosclerosis. *Clin. Exp. Pharmacol. Physiol.* 2010;37(7):71925.

139. Yilmaz, N. Relationship between paraoxonase and homocysteine: Crossroads of oxidative diseases. *Arch. Med. Sci.* 2012;8:138–53.

140. Corina Serban, Danina Muntean, Dimitri P Mikhailids, Peter P Toth and Maciej Banach. Dysfunctional HDL: The journey from savior to slayer. *Clin. Lipidol.* 2014;9(1):49–59.

141. Miriam Lee-Rueckert and Petri T. Kovanen. Extracellular modifications of HDL *in vivo* and the emerging concept of proteolytic inactivation of preb-HDL. *Current Opinion in Lipidology* 2011;22:394–402.

142. Mohamad N., Srinivasa T.R., Brian J.V.L., Anantharamaiah G.M., and Alan M.F. The role of dysfunctional HDL in atherosclerosis. *J. Lipid Res.*, 2009; 50:S145–9.

143. deGoma, E.M., deGoma, R.L. and Rader, D. J. Beyond high-density lipoprotein cholesterol levels. Evaluating highdensity lipoprotein function as influenced by novel therapeutic approaches. *J. Am. College Cardiol.*, 200851(23), 2199–211.

144. Asleh, R., Blum, S., Kalet-Litman, S. et al. Correction of HDL dysfunction in individuals with diabetes and the haptoglobin 2–2 genotype. *Diabetes*, 2008, 57, 2794–2800. *Arch. Med. Sci.*, 2012, 0, 138 53

145. Rao, H.B. G, V.S., Kakkar, V.V. Friend turns foe: Transformation of anti inflammatory HDL to proinflammatory HDL during acute-phase response. *Cholesterol* 2011;274629.

146. Van Lenten, B.J., Navab, M., Shih, D., Fogelman, A.M., Lusis, A.J. The role of high-density lipoproteins in oxidation and inflammation. *Trends. Cardiovasc Med.* 2001;11:155–61.

147. Durrington, P.N., Mackness, B., Mackness, M.I. Paraoxonase and atherosclerosis. *Arterioscler Thromb Vasc Biol* 2001;21:473–80.

148. Tsimihodimos, V., Karabina, S.A., Tambaki, A.P., Bairaktari, E., Goudevenos, J.A., Chapman, M.J., Elisaf, M., Tselepis, A.D. Atorvastatin preferentially reduces LDL-associated platelet-activating factor acetylhydrolase activity in dyslipidemias of type IIA and type IIB. *Arterioscler Thromb Vasc Biol* 2002;22:306–11.

149. Goyal, J., Wang, K., Liu, M., Subbaiah, P.V. Novel function of lecithin cholesterol acyltransferase. *J. Biol. Chem.* 1997;272:16231–9.

150. Navab, M., Hama, S.Y., Anantharamaiah, G.M., Hassan, K., Hough, G.P., Watson, A.D., Reddy, S.T., Sevanian, A., Fonarow, G.C., Fogelman, A.M. Normal high density lipoprotein inhibits three steps in the formation of mildly oxidized low density lipoprotein: Steps 2 and 3. *J. Lipid. Res.* 2000;41:1495–508.

151. Hansel, B., Giral, P., Nobecourt, E., Chantepie, S., Bruckert, E., Chapman, M.J., Kontush, A. Metabolic syndrome is associated with elevated oxidative stress and dysfunctional dense high-density lipoprotein particles displaying impaired antioxidative activity. *J. Clin. Endocrinol. Metab.* 2004;89(10):4963–71.

152. World Health Organization. Obesity: Preventing and managing the global epidemic. *Report of a WHO Consultation on Obesity.* Geneva: World Health Organization, 1998.

153. Hall, J.E. The kidney, hypertension, and obesity. *Hypertension* 2003;41(3 Pt 2):625–33.

154. Warolin, J., Coenen, K.R., Kantor, J.L., Whitaker, L.E., Wang, L., Acra, S.A., Roberts, L.J., 2nd; Buchowski, M.S. The relationship of oxidative stress, adiposity and metabolic risk factors in healthy Black and White American youth. *Pediatr. Obes.* 2013;9(1):43–52.

155. Tran, B., Oliver, S., Rosa, J., Galassetti, P. Aspects of inflammation and oxidative stress in pediatric obesity and type 1 diabetes: An overview of ten years of studies. *Exp. Diabetes Res.* 2012; 2012:683680.

156. Krzystek-Korpacka, M., Patryn, E., Boehm, D., Berdowska, I., Zielinski, B., Noczynska, A. Advanced oxidation protein products (AOPPs) in juvenile overweight and obesity prior to and following weight reduction. *Clin. Biochem.* 2008;41:943–9.

157. Codoñer-Franch, P.; Tavárez-Alonso, S.; Murria-Estal, R.; Tortajada-Girbés, M.; Simó-Jordá, R.; Alonso-Iglesias, E. Elevated advanced oxidation protein products (AOPPs) indicate metabolic risk in severely obese children. *Nutr. Metab. Cardiovasc. Dis.* 2012, 22:237–243.

158. Hermsdorff, H.H., Barbosa, K.B., Volp, A.C., Puchau, B., Bressan, J., Zulet, M.A., Mart A. Advanced oxidation protein products (AOPPs) in juvenile overweight and obesity prior to and following weigntioxidant capacity, and central adiposity indicators. *Eur. J. Prev. Cardiol.* 2012;41(12):943–49.

159. Karaouzene, N., Merzouk, H., Aribi, M., Merzouk, S.A., Berrouiguet, A.Y., Tessier, C., Narce, M. Effects of the association of aging and obesity on lipids, lipoproteins and oxidative stress biomarkers: A comparison of older with young men. *Nutr. Metab. Cardiovasc. Dis.* 2011;21:792–9.

160. Dandona, P., Ghanim, H., Chaudhuri, A., Dhindsa, S., Kim, S.S. Macronutrient intake induces oxidative and inflammatory stress: Potential relevance to atherosclerosis and insulin resistance. *Exp. Mol. Med.* 2010;42:245–53.

161. Serra, D., Mera, P., Malandrino, M.I., Mir, J.F., Herrero, L. Mitochondrial fatty acid oxidation in obesity. *Antioxid. Redox Signal.* 2012;1797(6-7):1195–1202. doi:10.1089/ars.2012.4875

162. Grundy, S.M., Cleeman, J.I., Daniels, S.R. et al. Diagnosis and management of the MS: An American Heart Association/National Heart, Lung, and Blood Institute Scientific Statement. *Circulation* 2005;112, 2735–52.

163. Grundy, S.M., Hansen, B., Smith, S.C., Jr. et al. Clinical management of MS: Report of the American Heart Association/National Heart, Lung, and Blood Institute/American Diabetes Association conference on scientific issues related to management. *Arteriosclerosis, Thrombosis, and Vascular Biology* 2004;24:e19–24.

164. Thompson, P.D., Buchner, D., Pina, I.L. et al. American Heart Association Council on Clinical Cardiology Subcommittee on Exercise, R., Prevention, American Heart Association Council on Nutrition, P. A., and Metabolism Subcommittee on Physical, A. Exercise and physical activity in the prevention and treatment of atherosclerotic cardiovascular disease: A statement from the Council on Clinical Cardiology (Subcommittee on Exercise, Rehabilitation, and Prevention) and the Council on Nutrition, Physical Activity, and Metabolism (Subcommittee on Physical Activity), Circulation 2003;107:3109–16.

165. Franklin, B.A., Kahn, J.K., Gordon, N.F., and Bonow, R.O. A cardioprotective polypill? Independent and additive benefits of lifestyle modification. *Am. J. Cardiol.* 2004;94:162–6.

166. Sternfeld, B., Sidney, S., Jacobs, D.R., Jr., Sadler, M.C., Haskell, W.L., and Schreiner, P.J. Seven-year changes in physical fitness, physical activity, and lipid profile in the CARDIA study. Coronary Artery Risk Development in Young Adults. *Annals of Epidemiology* 1999;9:25–33.

167. Jones, D.A., Ainsworth, B.E., Croft, J.B., Macera, C.A., Lloyd, E.E., and Yusuf, H.R. Moderate leisure-time physical activity: Who is meeting the public health recommendations? A national cross-sectional study. *Arch. Family Med.* 1998;7:285–9.

168. Gustat, J., Srinivasan, S.R., Elkasabany, A., and Berenson, G.S. Relation of self-rated measures of physical activity to multiple risk factors of insulin resistance syndrome in young adults: The Bogalusa Heart Study. *J. Cli. Epidemiol.* 2002;55:997–1006.

169. Matsumoto, Y., Adams, V., Jacob, S., Mangner, N., Schuler, G., and Linke, A. 2010 Regular exercise training prevents aortic valve disease in low-density lipoprotein-receptor-deficient mice. *Circulation* 2010;121:759–67.

170. Pellegrin, M., Alonso, F., Aubert, J. F. et al. Swimming prevents vulnerable atherosclerotic plaque development in hypertensive 2-kidney, 1-clip mice by modulating angiotensin II type 1 receptor expression independently from hemodynamic changes. *Hypertension* 2009;53, 782–9.

171. Leaf, D.A., Kleinman, M.T., Hamilton, M., and Deitrick, R.W. The exercise-induced oxidative stress paradox: The effects of physical exercise training. *Am. J. Med. Sci.* 1999;317(5):295–300.

172. Leeuwenburgh, C. and Heinecke, J.W. Oxidative stress and antioxidants in exercise. *Curr. Med. Chem.* 2001;8(7):829–38.

173. Radak, Z., Chung, H.Y., Koltai, E., Taylor, A.W., and Goto, S. Exercise, oxidative stress and hormesis. *Ageing Res. Rev.* 2008;7(1):34–42.

174. Diniz, Y.S., Faine, L.A., Galhardi, C.M., Rodrigues, H.G., Ebaid, G.X., Burneiko, R.C., Cicogna, A.C., and Novelli, E.L.B. Monosodium glutamate in standard and high-fiber diets: Metabolic syndrome and oxidative stress in rats. *Nutrition* 2005;21(6):749–55.

175. Yamazaki, Y., Usui, I., Kanatani, Y. et al. Treatment with SRT1720, a SIRT1 activator, ameliorates fatty liver with reduced expression of lipogenic enzymes in MSG mice. *Am. J. Physiol.-Endo. Metab.* 2009;297:E1179–86.

176. Insawang, T., Selmi, C., Chaon, U. et al. Monosodium glutamate (MSG) intake is associated with the prevalence of metabolic syndrome in a rural Thai population. *Nutr. Metab.* 2012;9:50–5.

177. Nagata, M., Suzuki, W., Iizuka, S., Tabuchi, M., Maruyama, H., Takeda, S., Aburada, M., and Miyamoto, K. 2006. Type 2 diabetes mellitus in obese mouse model induced by monosodium glutamate. *Exp. Anim.* 2006;55(2):109–15.

178. Parthasarathy, S. How safe is monosodium glutamate? Exploring the link to obesity, metabolic disorders, and inflammatory disease. *J. Med. Food* 2014;17(5):1–1.

179. Fujimoto, M., Tsuneyama, K., Nakanishi, Y. et al. A dietary restriction influences the progression but not the initiation of MSG-induced nonalcoholic steatohepatitis. *J. Med. Food* 2014;17(3):374–83.

180. Pryor, W.A. Cigarette smoke radicals and the role of free radicals in chemical carcinogenicity. *Environ Health Perspect* 1997;105(suppl 4):87582.

181. Chalmers, A. Smoking and oxidative stress. *Am. J. Clin. Nutr.* 1999;69:572.

182. Cross, C.E., Van der Vliet, A., Eiserich, J.P. Cigarette smokers and oxidant stress: A continuing mystery. *Am. J. Clin. Nutr.* 1998;67:184N85.

183. Burke, A., Fitzgerald, G.A. Oxidative stress and smoking-induced vascular injury. *Prog. Cardiovasc. Dis.* 2003;46:79v90.

184. Eiserich, J.P., van der Vliet, A., Handelman, G.J., Halliwell, B., Cross C.E. Dietary antioxidants and cigarette smoke-induced biomolecular damage: A complex interaction. *Am. J. Clin. Nutr.* 1995;62(suppl):1490S–500S.

185. Padmaja, 5, Huie, R.E. The reaction of nitric oxide with organic peroxyl radicals. *Biochem Biophys Res. Commun.* 1993:195:539–44.

186. Kallner, A.B., Hartmann, D., Hornig, D.H. On the requirements of ascorbic acid in man: Steady-state turnover and body pools in smokers. *Am. J. Clin. Nutr.* 1981;34:1347u55.

187. Munro, L.H., Burton, G., Kelly, F.J. Plasma RRR—tocopherol concentrations are lower in smokers than in non-smokers after ingestion of a similar oral load of this antioxidant vitamin. *Clin Sci* 1997;92:87c93.

188. Heitzer, T., Just, H., Munzel, T. Antioxidant vitamin C improves endothelial dysfunction in chronic smokers. *Circulation* 1996;94:6nd.

189. Yokode, M., Nagano, Y., Arai, H., Ueyama, K., Ueda, Y., Kita, T. Cigarette smoke and lipoprotein modification. A possible interpretation for development of atherosclerosis. *Ann. N Y Acad Sci.* 1995;748:294–300.

190. Yamaguchi Y., Matsuno S., Kagota S., Haginaka J., Kunitomo M. Oxidants in cigarette smoke extract modify low-density lipoprotein in the plasma and facilitate atherogenesis in the aorta of Watanabe heritable hyperlipidemic rabbits. *Atherosclerosis* 2001;156:109i17.

191. Bloomer, R.J. Decreased blood antioxidant capacity and increased lipid peroxidation in young cigarette smokers compared to nonsmokers: Impact of dietary intake. *Nutr. J.* 2007;6:39.

192. Frei B., Forte T.M., Ames B.N., Cross C.E. Gas phase oxidants of cigarette smoke induce lipid peroxidation and changes in lipoprotein properties in human blood plasma. *Protective Effects of Ascorbic Acid. Biochem. J.* 1991;277:133 8.

193. Nalini S., Sanchezb, R., Hendlerb, S., Sampath P. Aqueous extracts of cigarette smoke promote the oxidation of low density lipoprotein by peroxidases. *FEBS Lett.* 1997;414(3):549–51.

194. Ahmed, R.L., Schmitz, K.H., Anderson, K.E., Rosamond, W.D., Folsom, A.R. The metabolic syndrome and risk of incident colorectal cancer. *Cancer* 2006;107:28–36.

195. Bowers, K., Albanes D., Limburg P. et al. A prospective study of anthropometric and clinical measurements associated with insulin resistance syndrome and colorectal cancer in male smokers. *Am. J. Epidemiol* 2006;164:652–4.

196. Giovannucci, E. 2007. Metabolic syndrome, hyperinsulinemia, and colon cancer: A review. *Am. J. Clin. Nutr.* 86(3):836S–42S.

197. Cowey, S. and Hardy, R.W. The metabolic syndrome: A high-risk state for cancer? *Am. J. Pathol.* 2006;168(5):1505–22.

198. Lane, M.A., Black, A., Handy, A., Tilmont, E.M., Ingram, D.K., Roth, G.S. Caloric restriction in primates. *Ann. NY Acad. Sci.* 2001;928:287–95.

199. Farooqui, Akhlaq A. Metabolic syndrome an important risk factor for stroke. *Alzheimer Disease, and Depression* 2013, XX, 412.

200. Diana L.-C., Karla C., Danira T.-R. Diana Franco-Bocanegra, and Victoria Campos—Peña Oxidative Stress and Metabolic Syndrome: Cause or Consequence of Alzheimer's Disease? *Oxidative Med. Cell. Longevity* 2014;2014:1–11.

18 Oxidized Lipid Products and Carcinogenesis

Françoise Guéraud

CONTENTS

INTRODUCTION

Lipid peroxidation is a general process that occurs when unsaturated (i.e., prone to oxidation) lipids are in the presence of oxidant compounds. This can happen within living organisms, in food, or during food processing, giving rise to similar products, wherever the process takes place. So the compounds produced are often qualified as "endogenous lipid oxidation products," because they are a consequence of endogenous oxidative stress, but it is important to bear in mind that they can also be considered as "contaminants" or "food neoformed products" because they can enter the body through the digestive tract. This consideration is especially important for digestive diseases and cancer, because the digestive tract is the primary target organ for these compounds. For this reason, there is, in this chapter, a special focus on lipid peroxidation products and colorectal cancer.

Cancer is a multistep process involving an initial mutation in a critical gene. The mutation of proto-oncogenes can lead to their activation and the subsequent uncontrolled proliferation of cells. The mutation of tumor suppressor genes can lead to the survival of cells bearing DNA damage. Genes involved in senescence control or in DNA repair are also critical. This initial event needs additional mutations on critical genes to drive the cell/tissue into a cancerous state. Promoting agents will increase the clonal expansion of the cells, for instance by increasing the growth selective advantage of transformed cells, without any effect on genetic material. This chapter describes the genotoxic, mutagenic, and promoting effect of lipid peroxidation-derived aldehydes.

STRUCTURE AND REACTIVITY OF LIPID PEROXIDATION PRODUCTS

Lipid peroxidation of polyunsaturated fatty acids yields a group of reactive unsaturated aldehydic compounds (enals) such as 2-alkenals (acrolein [Acr], crotonaldehyde [Cro], 2-hexenal), 4-hydroxy-2-alkenals (4-hydroxy-2-nonenal [HNE], 4-hydroxy-2-hexenal [HHE]), and ketoaldehydes (malondialdehyde [MDA], glyoxal, 4-oxo-2-nonenal [ONE]) (Uchida 2003) (Figure 18.1).

The special reactivity of all these compounds is due to their two reaction centers: The aldehydic function on C1 and the conjugated system of the C2–C3 double bond and the CO carbonyl group, making the C3 prone to nucleophilic attack. In 4-hydroxy-2-alkenals, the electron-withdrawing 4-hydroxy group renders the

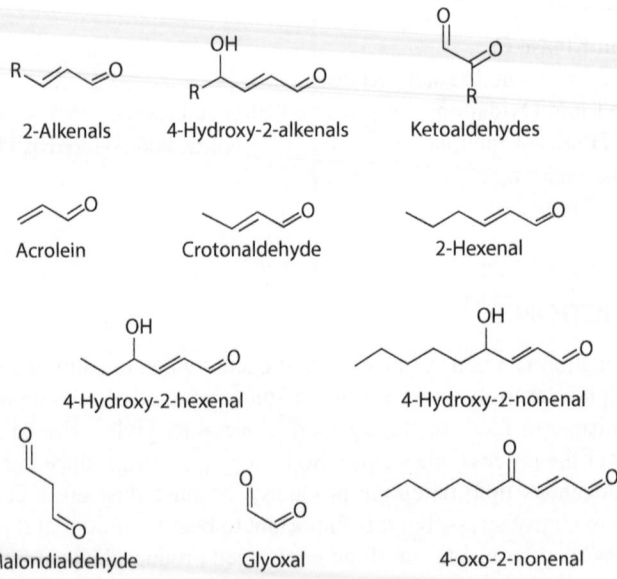

FIGURE 18.1 Chemical structure of reactive lipid oxidation enals.

electrophilicity of the C3 even more important. For these reasons, all these compounds are able to make covalent binding with biomolecules such as proteins, DNA, and phospholipids with different reactivity, also depending on the alkyl chain length. For reviews, see Esterbauer et al. (1991) and Uchida (2003).

Among these compounds, MDA and HNE are the most abundant species and HNE is one of the most toxic. They are the most studied for their chemical and biological activities.

LIPID PEROXIDATION ORIGINATING ENALS MAKE ADDUCTS WITH DNA: THEY ARE GENOTOXIC AND MUTAGENIC AGENTS

EXOCYCLIC ADDUCTS

α,β Unsaturated bifunctional aldehydes (enals) can make exocyclic adducts with DNA. Exocyclic adducts are DNA adducts with an additional 5- or 6-membered ring, depending on the adduct type (Figure 18.2).

Two types of exocyclic adducts were reported when nucleosides or isolated DNA were exposed to aldehydic lipid oxidation products (Chung et al. 1996). Propano-type adducts, with a 6-membered additional ring, come from a direct adduction of the reactive product to deoxyguanosine by a Michael addition mechanism followed by ring closure, leading to the formation of HNE-dGuo, Acr-dGuo, Cro-dGuo, all existing as different isomers, in the case of HNE, acrolein, and crotonaldehyde, respectively. The reactivity of these compounds toward deoxyguanosine depends on the length of the alkyl chain, with the simplest enal (acrolein) being the most reactive. Etheno-type adducts, with a 5-membered additional ring, come from an indirect adduction of the reacting compound to cytosine, adenine, and guanine after its activation through the formation of an epoxide intermediate that is more reactive to DNA than the parent compound, followed by adduction to the nucleobase. In the case of substituted adducts, the side chain of these exocyclic adducts is specific to the reacting

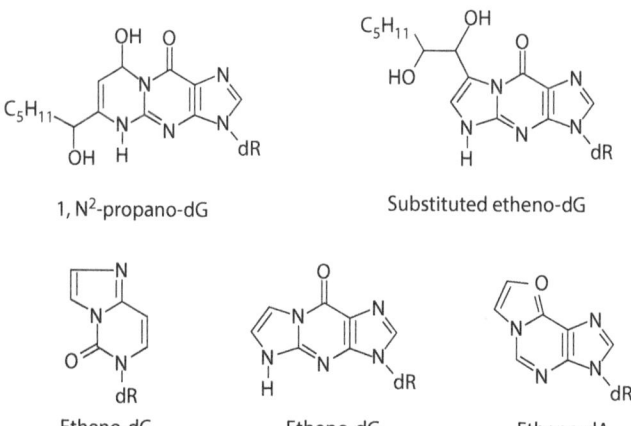

1, N²-propano-dG   Substituted etheno-dG

Etheno-dC   Etheno-dG   Etheno-dA

FIGURE 18.2 Chemical structure of substituted and unsubstituted exocyclic adducts.

electrophilic compound that is covalently linked to the DNA base, but in the case of unsubstituted adducts, the side chain is lost, so it is difficult to identify the structure of the reacting compound. However, Etheno-dG was reported to be a potential biomarker for DNA damage induced by HNE and related "alkenals" (Sodum and Chung 1988). MDA can also give rise to cyclic adducts, particularly M_1G (Marnett 1999).

It is important to note that the modifications generated by those lipid oxidation products contribute nearly to the same extent to DNA damage than the direct oxidized bases (Winczura et al. 2012). These lipid peroxidation aldehydes–DNA adducts have been reported *in vivo* in rodent and human DNA, in a wide variety of organs and tissue. For most of them, they can be found at a basal state (Marnett 1999; Nair et al. 1999, 2007), but their concentration is increased in the case of oxidative stress due, for instance, to inflammatory processes (Nair et al. 2007), but also in the case of PUFA-rich diet (Fang et al. 2007). For etheno-adducts, most of the studies report the presence of unsubstituted adducts, making the identification of the reactant enal impossible. However, a substituted etheno-adduct specific to the lipid oxidation product 4-oxo-nonenal has been found in greater amounts in the small intestine of mice prone to intestinal cancer (Min mice) and overexpressing the enzyme COX-2 involved in inflammatory processes than in the small intestine of control mice (Williams et al. 2006).

GENOTOXICITY AND MUTAGENICITY

All of these adducts are highly mutagenic. Unsubstituted etheno-adducts induce transversions and transitions while substituted etheno and propano-adducts strongly inhibit DNA synthesis, due to their bigger structure (Winczura et al. 2012). Those adducts are repaired by two different pathways, namely base excision repair (BER) and nucleotide excision repair (NER). BER is specific for the repair of small base lesions including oxidation and alkylation of small molecules that give only little distortion in the DNA helix, like unsubstituted etheno-adducts, while NER is dedicated to the repair of bulky adducts such as substituted etheno- or propano-adducts (Chung et al. 2003; Winczura et al. 2012). Some authors showed that HNE DNA adducts were more mutagenic and toxic in NER-deficient cells than in NER-proficient cells (Feng et al. 2003). Bad repair leads to heritable mutations. Interestingly, HNE preferentially makes adducts in p53 mutational hotspots, which are DNA sequences that have high susceptibility to mutations (Hu et al. 2002), and these adducts were shown to be propano-adducts (Chung et al. 2003; Feng et al. 2003). Mutation of this tumor suppressor gene at this site is a common feature in numerous human cancers, particularly in hepatocellular carcinoma.

In primary cells (namely hepatocytes), which may reflect the *in vivo* situation more adequately than prokaryotic systems or cancerous cell lines, Eclk et al. reported that HNE treatment induced increased levels of sister chromatid exchanges (SCE) at low concentration treatments (0.1 μM), and of micronuclei and chromosomal aberrations at higher dosages (1 and 10 μM) (Eckl 2003; Eckl et al. 1993). Sister chromatid exchange, micronuclei and chromosomal aberrations are classical tests used to measure chromosomal damage and genotoxicity. Similar results were obtained in cloned porcine cerebral endothelial cells representing the blood–brain barrier (Karlhuber et al. 1997). Those treatments are representative of HNE concentrations *in vivo*, as pathophysiological concentrations of HNE were reported to be between 1 and 25 µM (Forman et al. 2008).

It is generally accepted that oxidative stress/lipid peroxidation conditions existing during inflammatory processes often linked to cancer development (Reuter et al. 2010), are accompanied by an increase of endogenous DNA lesions. However, exogenous lipid oxidation products, which can be found in the diet (Gasc et al. 2007; Munasinghe et al. 2003; Surh and Kwon 2003, 2005) or formed during digestion could also participate in this DNA damage burden in the intestinal tract, especially in relation to colorectal cancer, the digestive tract being the primary target of these compounds.

FOCUS ON LIPID OXIDATION PRODUCTS AND COLORECTAL CANCER: DNA ADDUCTS

Colorectal cancer is most commonly found in Western countries. Nutritional factors seem to play an important role in its promotion and prevention. Among factors incriminated in its development, red and processed meats are, together with alcohol consumption, the only food factors that were qualified by the World Cancer Research Fund (WCRF) as "convincing" for increasing the risk of colorectal cancer (WCRF/ AICR 2007). Red meat is characterized by a high concentration of heme iron. Heme iron was reported to induce DNA damage in primary colon cells and in the HT29 cell line (Glei et al. 2006) or in the CaCo2 cell line (Ishikawa et al. 2010). These authors attributed the effect of heme iron to the production of hydroxyl radical and hydrogen peroxide. However, a possible subsequent production of lipid oxidation enals and their involvement in DNA damage cannot be ruled out as specific lipid peroxidation DNA etheno-adducts were shown to be increased in SW480 cells treated with hemoglobin together with linoleic acid hydroperoxide, a precursor of HNE (Angeli et al. 2011). Moreover, those lipid oxidation enals are much more diffusible than hydroxyl radical or hydrogen peroxide and could be regarded as toxic messengers of heme iron in the nucleus. These results show that DNA damage to colon cells could also originate from luminal lipid peroxidation and breakdown enals. HNE has been found in foodstuffs containing heme-iron (red and processed meat) and in food fried in thermally oxidized oils (Gasc et al. 2007; Seppanen and Csallany 2002, 2004). HNE was present in higher amounts in fecal waters of rats fed on a red-meat-rich diet as compared to controls (Pierre et al. 2007).

LIPID PEROXIDATION ORIGINATING ENALS MAKE ADDUCTS WITH PROTEINS, AND PLAY A ROLE IN CELLULAR SIGNALING PATHWAYS, AT PATHOLOGICAL AND AT PHYSIOLOGICAL CONCENTRATIONS: A ROLE IN CANCER PROMOTION?

REACTIVITY WITH PROTEINS

Proteins and peptides are the main group of biomolecules targeted by secondary lipid oxidation products. Protein modification is the major way for lipid oxidation

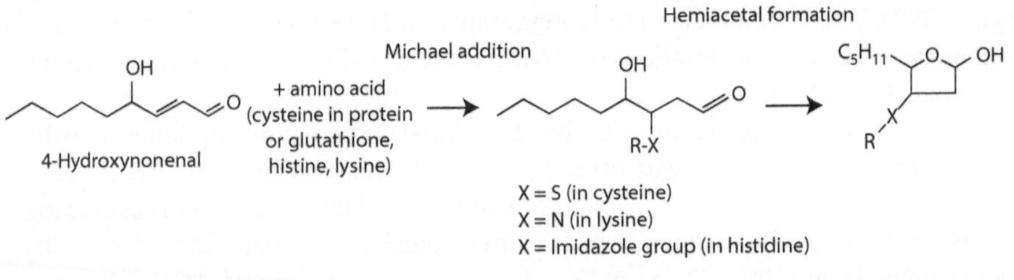

FIGURE 18.3 Michael addition of amino acid residues on HNE and subsequent hemiacetal formation.

products to play a biological role. Two types of reaction can occur for the formation of protein/peptide adducts.

The first one is a nucleophilic attack on the electropositive C3 by Michael addition (Figure 18.3). This can be achieved by cysteine in its thiolate form (in glutathione or proteins), by the imidazole group of histidine and the ε-amino group of lysine, with a further stabilization as cyclic hemiacetal in the case of 4-hydroxy-2-alkenals. Cysteine residues are by far the most reactive, followed by histidine and lysine residues. Adduct formation of HNE with protein cysteines was reported to be reversible in vitro, in the presence of physiological concentrations of glutathione, possibly providing a protective mechanism (Carbone et al. 2005), or a role in cellular signaling (Petersen and Doorn 2004).

The second one is a Schiff base formation with lysine residues (Figure 18.4), giving a very stable pyrrole adduct. Interestingly, protein cross-linking can occur, involving two protein-based lysine ε-amino groups, through Michael addition on C3 and Schiff base formation on C1 (Uchida 2003), giving rise to the formation of fluorophores. The polarity and the pH of the micro-environment, together with the accessibility of the amino residues in the protein are also important for the formation of protein adducts (Poli et al. 2008b).

OCCURRENCE OF LIPID OXIDATION PRODUCTS–PROTEIN ADDUCTS IN CANCER

There is a considerable amount of literature describing the detection of HNE–protein adducts. Those adducts are present under physiological conditions and their level is increased in animal models of oxidative stress but also under pathological conditions associated with oxidative stress, such as inflammatory diseases (Poli et al. 2008a;

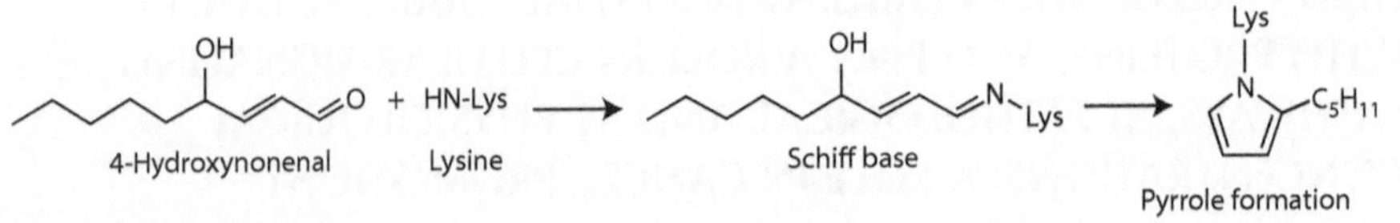

FIGURE 18.4 Schiff base formation of HNE with the ε-amino group of lysine and subsequent pyrrole formation.

Uchida 2003). However, the presence of those adducts as a cause of the disease or only as a consequence of the accompanying oxidative stress is still a matter of debate. In an animal model of hepatic oxidative injury, the formation of HNE adducts with specific proteins appears very early, several hours before clinical and histopathological signs, suggesting a possible mechanistic role of HNE (Petersen and Doorn 2004). Lipid oxidation protein adducts are often used as biomarkers of lipid peroxidation/oxidative stress. They can be measured by immunological or mass spectrometry methods. For a review, see Spickett (2013).

It is established that cancer cell growth is associated with an increased H_2O_2 production. However, concerning lipid oxidation products–protein adducts in cancer cells, literature shows opposite results, depending on cell type and cancer stage (Barrera 2012). In hepatoma cells, the level of free HNE and TBARS is decreased when compared to less transformed cells (Hammer et al. 1997; Rossi and Cecchini 1983). Some authors demonstrated that this decrease could be explained by an increased activity of enzymes metabolizing those lipid oxidation products, leading to increased excretion (Canuto et al. 1993; Tjalkens et al. 1999). The same decrease has been observed in kidney tumors (Oberley et al. 1999). On the contrary, some studies report more protein adducted to lipid oxidation products in cancer cells (Skrzydlewska et al. 2001), with a link to the grade of malignancy in astrocytomas and ependymomas (Juric-Sekhar et al. 2009). Some authors reported an increase in HNE–protein adducts in breast cancer with the degree of malignancy, but the contrary with 8-hydroxydeoxyguanosine, a marker of oxidative DNA damage (Karihtala et al. 2011). These discrepancies could be related to differences in lipid oxidation metabolizing capacities, differences in the polyunsaturated fatty acid composition of membranes, and the presence or absence of accompanying inflammatory processes (Barrera 2012).

PROTEIN ADDUCTS AND FUNCTIONS

Most of the time, modification of proteins by lipid peroxidation enals impairs their function. A wide array of proteins is reported to be modified by HNE (Poli et al. 2008b). However, the function of some proteins appears to be activated by this compound: for instance, caspase 3, PDGFRβ, EGFR, and some PKCβ isozymes (Poli et al. 2008b). The altered proteins are removed by the proteasomal pathway responsible for most intracellular protein degradation. Interestingly, proteasome was reported to be a target for HNE modification, with a subsequent decrease in proteolysis and further cellular degeneration (Grune and Davies 2003).

LIPID OXIDATION PRODUCTS, CELLULAR PATHWAYS AND CANCER

The presence of critical cysteines, histidines, or lysines in catalytic sites or binding site or regulatory domains of proteins involved in cellular pathways make them prone to adduction to reactive lipid oxidation products. As a consequence, those compounds cannot be regarded as only mere degradation products of lipid peroxidation but as real secondary messengers of oxidative stress, playing a role in numerous cellular signaling pathways. In this respect, HNE has been and is still extensively studied. For reviews, see Uchida (2003), Forman et al. (2008), Poli et al. (2008b), and Forman (2010).

Nrf2/Keap1-ARE Pathway

It is of interest to note that most of lipid oxidation product metabolizing enzymes, such as glutathione S-transferases (GSTs) that catalyze their conjugation to glutathione (GSH), aldo–keto reductases (AKRs) and aldehyde dehydrogenases (ALDHs) that respectively reduce and oxidize the aldehyde function are under the dependence of the nuclear factor E2-related factor 2 (Nrf2)/antioxidant responsive element (ARE) pathway. This pathway is strongly involved in lipid oxidation product biotransformation and excretion, but it is also linked to cancer development and therapy.

Nrf2 is a transcription factor from the cap'n'collar family of leucine-zipper proteins (b-ZIP), that plays an important role in the response to cellular stress by initiating the transcription of numerous genes involved in antioxidant defenses and detoxication (Li and Kong 2009; Moi et al. 1994). Under basal conditions, Nrf2 is sequestered in the cytosol by its inhibitory binding partner Kelch-like ECH-associated protein 1 (Keap1) (Itoh et al. 1999): Keap1 dimers bind Nrf2 and recruit a Cullin-3-based E3 ubiquitin ligase (Cul3), which provokes Nrf2 ubiquitination and its further proteasomal degradation (Kobayashi et al. 2004). Under cellular stress conditions, this association is disrupted, Nrf2 is released in the cytosol, stabilized and can undergo nuclear translocation. The disruption of the Nrf2/Keap1/Cul3 complex can be achieved by the modification of critical cysteine residues in Keap1, which act as cytosolic redox sensors, and which in turn induce a conformational modification of Keap1 and the subsequent Nrf2 release (Dinkova-Kostova et al. 2002) and phosphorylation of Nrf2 by protein kinase C (PKC) (Huang et al. 2002) and Akt (Niture et al. 2009) in a sequential fashion. Once in the nucleus, Nrf2 combines with other proteins such as small Maf proteins (Katsuoka et al. 2005) or Jun (Niture et al. 2009), and binds the ARE, which is present in the promoter region of numerous antioxidant enzyme genes, to induce their transcription (Itoh et al. 1997). Those other proteins may play a role in the nuclear stabilization of Nrf2 (Li et al. 2008). Among the target genes of the Nrf2/Keap1 pathway: NAD(P)H quinone oxidoreductase 1 (NQO1) that catalyzes the reduction of highly reactive quinones; the two subunits of the glutamate–cysteine ligase involved in glutathione synthesis; heme-oxygenase-1 (HO-1), which catalyzes the transformation of heme into the antioxidants biliverdin and bilirubin, carbon monoxide and iron; some phase II conjugating enzymes such as glutathione S-transferases (GSTs), which catalyze the conjugation of GSH to a variety of electrophilic substrates, UDP-glucuronosyltransferases (UGTs), which catalyze the conjugation of glucuronic acid to a variety of compounds; some membrane transporters such as multidrug resistance-associated proteins (MRPs) that are involved in the efflux of various compounds (Wu et al. 2012).

As redox sensing is one of the ways to activate the Nrf2/Keap1 pathway, reactive oxygen species or H_2O_2, but also electrophilic compounds such as tert-butylhydroquinone or oltipraz and numerous natural compounds present in fruits or vegetables such as sulforaphane (Kensler et al. 2013), curcumin, resveratrol, pterostilbene (Ramkumar et al. 2013) are known to induce the transcription of Nrf2 target genes through modification of Keap1-sensitive cysteines and subsequent Nrf2 nuclear translocation (Dinkova-Kostova and Talalay 2008; Hayes et al. 2010). Those electrophilic compounds were shown to have beneficial properties in numerous inflammatory diseases and even in cancer in experimental models.

Nrf2 is considered as the major switch turning on cellular defenses against oxidative stress. For this reason, it has been classically considered to be a tumor suppressor compound due to its cytoprotective properties. As a result, Nrf2-deficient mice are much more sensitive to inflammatory diseases and cancer development than their wild-type counterparts (Chan and Kan 1999; Ramos-Gomez et al. 2001). Oral administration of dextran sodium sulfate, an pro-inflammatory agent, induced more preneoplastic lesions in the colon of Nrf2-deficient mice, in which colon carcinogenesis has been initiated by azoxy-methane, than in wild-type controls (Khor et al. 2008; Osburn et al. 2007). Nrf2 loss has also been associated with increased metastasis formation (Satoh et al. 2010).

However, the cytoprotective effect of the Nrf2/Keap1 pathway can have adverse effects, in the case of cancer treatment, because it participates in chemoresistance (Sporn and Liby 2012), through the enhanced expression of enzymes such as GSTs or HO-1 and efflux transporters that are able to inactivate or export drugs used for cancer chemotherapy. Moreover, under certain pathological conditions, that is, in cancer cells, expression of the signal cannot return to a basal state and overactivation of the system can occur. This overactivation has been reported to be involved in tumorigenesis. For these reasons, Nrf2 can also be defined as a proto-oncogene (Shelton and Jaiswal 2013) whereas its inhibitor Keap1 is regarded as a tumor suppressor gene (Ganan-Gomez et al. 2013)

Overactivation of the system leads to the overexpression of cytoprotective genes and gives a selective advantage to tumor cells, particularly when those cells are in contact with oxygen or pollutants such as lung tumor cells, squamous carcinoma cells (Kim et al. 2010). Many of the genes that are classically reported to be protective may also have some other effects related to tumorigenesis: For instance, HO-1 has been reported to promote angiogenesis in pancreatic cancer and to inhibit apoptosis in chronic myelogenous leukemia (Mayerhofer et al. 2004; Sunamura et al. 2003).

Beside the role in inducing antioxidant target genes, Nrf2 by itself can promote cell growth and proliferation and decrease apoptosis through the regulation of other transcription factors. For instance, Nrf2 modulates proliferation through the epidermal growth factor receptor EGFR-MEK1/2-ERK axis (Yamadori et al. 2012). Nrf2 was reported to induce Notch1, which is involved in cell proliferation (Wakabayashi et al. 2010). Nrf2 can also play a role in metabolic reprogramming (Mitsuishi et al. 2012). Reprogramming energy metabolism is for cancer cells a way to switch their energy from maintenance to anabolism, to be able to proliferate. However, to activate this pathway, Nrf2 needs to accumulate in levels higher than those necessary for the transcription of antioxidant genes and requires being in a tissue with a proliferative potential, two features that coexist in cancer cells (Ganan-Gomez et al. 2013; Mitsuishi et al. 2012). Another way for cancer cells to survive and proliferate is to escape their own apoptotic pathways. For instance, the Nrf2/Keap1 pathway is connected with the Bcl-2 antiapoptotic protein through an interaction with Keap1, leading to apoptosis inhibition when Keap1 is altered (Niture and Jaiswal 2011). In the same way, the tumor suppressor gene p53, which increases apoptosis in case of DNA damage and which regulates negatively the transcriptional activation of Nrf2 target genes, is often mutated in cancer cells, leading to exacerbation of Nrf2 effects together with a decrease in apoptosis (Ganan-Gomez et al. 2013). Moreover, Nrf2 can activate the inhibitor of p53 (You et al. 2011).

This overexpression of Nrf2 can be explained by some mutations in the Keap1 (Padmanabhan et al. 2006; Shibata et al. 2008a) or Nrf2 (Shibata et al. 2008b) genes, as it has been reported in several types of cancers (Hayes and McMahon 2009).

HNE, and other alkenals, due to their electrophilic properties are activators of the Nrf2/Keap1 pathway (Poli et al. 2008b). HNE was reported to induce the transcription of glutamate cysteine ligase through activation of the Nrf2/Keap1–ARE pathway (Zhang et al. 2007) and GST (Tjalkens et al. 1998). More recently, Huang and coworkers reported the rapid nuclear translocation of Nrf2 upon HNE treatment in HeLa cells and the enhanced transcription of GSTA4, HO-1, and aldo-ketone reductase (AKR) 1C1 (Huang et al. 2012). By this way, those compounds could play a promoting role in cancer development.

Nf-κB Pathway

NF-κB (nuclear factor kappa-light-chain-enhancer of activated B cells) is a family of transcription factors involved in cellular response to inflammation and oxidative stress and in immune response to infections. As Nrf2, NF-κB is a rapid acting transcription factor that is present in the cell in an inactive state, sequestered in the cytosol by its inhibitor IκB. Following stimulation, the inhibitor is phosphorylated by IKK (IκB kinase) and subsequently degraded by the proteasome. NF-κB is then rapidly translocated into the nucleus, where it binds to DNA to induce the transcription of a variety of genes that bear a specific NF-κB binding sequence and that control inflammatory and immune response, cell survival, or proliferation. As a master regulator of cell proliferation, the deregulation of this pathway has been involved in the development of cancer (Fan et al. 2013; Li et al. 2013). In cancer cells, NF-κB is often constitutively active. This can be due to mutation on gene encoding NF-κB itself or its regulators. On the other hand, NF-κB could also act as a cell sensitizer to apoptosis through the Fas pathway. As Nrf2, NF-κB has a complex relationship with cancer.

HNE was reported to prevent NF-κB activation by directly inhibiting IκB phosphorylation (Ji et al. 2001; Page et al. 1999), while other authors reported an activation of the pathway with a low HNE dosage (1 μM) (Ruef et al. 2001). In fact, the effect of HNE on the NF-κB pathway is dependent on the cell type and HNE dosage (Poli et al. 2008b). It is noteworthy that NF-κB also has critical cysteine sensors that are important for its DNA binding (Nishi et al. 2002) and that an effect of electrophilic compounds on these cysteines cannot be ruled out.

Protein Kinase C

Protein kinase Cs (PKCs) are enzymes involved in the intracellular transduction of signals taking part in differentiation, proliferation, and apoptosis. The effect of HNE on PKC activity depends on its concentration and the PKC isoforms (Forman et al. 2008). PKCs possess several critical cysteines in the regulatory domain and at the catalytic site that may be targeted by HNE (Gopalakrishna and Jaken 2000). PKC is an indirect way for HNE to activate the Nrf2–ARE pathway (Huang et al. 2002; Numazawa et al. 2003) and the nuclear binding of AP-1 (Nitti et al. 2002), another transcription factor involved in inflammatory response and under the influence of oxidative stress.

Mitogen-Activated Protein Kinases

The mitogen-activated protein kinase (MAPK) family is a group of serine-threonine kinases that include extracellular regulated kinases (ERKs), c-Jun N-terminal kinases (JNKs) and p38 MAPKs, which are also involved in the intracellular transduction of signals involved in differentiation, proliferation, and apoptosis (Chang and Karin 2001; Poli et al. 2008b). These kinases are particularly sensitive to modulation by lipid oxidation products such as HNE (Forman et al. 2008). JNKs are directly activated by HNE by adduction on critical histidine residues (Parola et al. 1998). JNK signaling is involved in apoptosis. Other kinases such as p38 MAPK are involved in the HNE-induced expression of COX-2, an enzyme involved in inflammatory processes (Kumagai et al. 2004). Interestingly, HNE was shown to induce proliferation in mouse aortic smooth muscle cells when cells were originating from young animals, but was shown to induce cytotoxicity when cells were originating from old ones, through ERK activation (Lee et al. 2006).

FOCUS ON LIPID OXIDATION PRODUCTS AND COLORECTAL CANCER: PROMOTING EFFECT OF HNE: AN INTERPLAY BETWEEN APC MUTATION AND NRF2/ARE PATHWAY

Heme iron in red meat can promote carcinogenesis because it increases cell proliferation in the colonic mucosa, through lipoperoxidation and/or cytotoxicity of fecal water (Sesink et al. 1999). This promotion would be related to cytotoxicity of a solubilized factor in the colon lumen, that is neither heme itself, nor its metabolites and nor even the oxygen radicals that would be formed due to the catalytic activity of heme iron (Sesink et al. 1999). Thiobarbituric acid reacting substances (TBARS), reflecting the presence of lipid oxidation-derived aldehydes, are dramatically increased in the feces of heme-fed rats and of red-meat-fed rats (Pierre et al. 2003, 2004, 2008; Sesink et al. 1999). Moreover, in the urine of heme-fed rats and of red-meat-fed rats, the concentration of the major metabolite of HNE, the mercapturic acid of 1, 4 dihydroxynonene (DHN-MA), increases dramatically (Pierre et al. 2006). HNE, and related aldehydes originating from lipid oxidation, could be the cytotoxic factor invoked by Sesink et al. (1999) to explain the relationship between colorectal cancer promotion and heme iron. *In vivo*, HNE has been found in fecal water of beef-fed rats (Pierre et al. 2007). The author's group showed that HNE was more cytotoxic to normal mouse epithelial colon cells than to premalignant epithelial cells that were mutated on the *Apc* gene, a frequent and early event in human colorectal cancer. Likewise, fecal water of meat-fed rats had the same differential effect on the two cell lines. Interestingly, HNE was present in the fecal water of those rats (Pierre et al. 2008). We thus proposed that HNE, or other aldehydes produced by dietary lipid oxidation, could "select" *Apc* mutated cells, when present in the fecal water of rats, as it is the case when rats are fed on diets containing heme-iron as hemin or red meat. These secondary products of dietary lipid peroxidation could be the missing link between red meat and increased CRC risk evidenced in epidemiological studies. The molecular basis of this differential cytotoxicity exerted by lipid oxidation products can be explained at least in part by an increased metabolizing capacity of *Apc* mutated cells toward HNE (Baradat et al. 2011) that could be due to a constitutively activated Nrf2 pathway. This result could be related to the decrease in HNE–protein adducts in colon biopsies compared to the surrounding normal tissue observed by Zanetti et al. (2003) and to the increased metabolizing capacities described by Canuto et al. (1993) during rat liver carcinogenesis.

CONCLUSION

In vivo, the adduction of lipid oxidation products, particularly HNE, to DNA and proteins occurs under basal conditions and is increased in case of oxidative stress and inflammation that often accompany cancer development. However, it is difficult to determine if these adducts really play a role in carcinogenesis or if they are only a consequence of oxidative stress. So, the effect of HNE and related lipid oxidation on cancer development is a matter of debate that differs with the model. *In vitro*, HNE appears to have rather an antiproliferative effect and an apoptotic effect on cancer cells, but these effects can be opposite in normal cells (Barrera 2012). Moreover, HNE effects are highly dependent upon cell type and treatment time and concentration, probably due to metabolic capacity differences. Cancer cells and normal cells may be very different in this respect, probably due to the activation of major cellular cytoprotective pathway such as Nfr2. Cancer cells may not be the best model to study the involvement of lipid oxidation products on early phases of cancer development: It is important to study the effect of lipid oxidation products in normal or preneoplastic cells that represent the physiological targets of those compounds when considering early cancer development, while cancer cells are the pertinent targets when considering cancer therapy. It is also important to bear in mind that the digestive tract may represent a special tissue with respect to lipid oxidation products, due to their presence in food (Seppanen and Csallany 2004) together with variable concentrations of natural antioxidants. For this specific aspect, *in vivo* studies will surely provide interesting insights.

REFERENCES

Angeli, J. P., C. C. Garcia, F. Sena, F. P. Freitas, S. Miyamoto et al. 2011. Lipid hydroperoxide-induced and hemoglobin-enhanced oxidative damage to colon cancer cells. *Free Radic Biol Med* 51: 503–515.

Baradat, M., I. Jouanin, S. Dalleau, S. Tache, M. Gieules et al. 2011. 4-Hydroxy-2(E)-nonenal metabolism differs in Apc(+/+) cells and in Apc(Min/+) cells: It may explain colon cancer promotion by heme iron. *Chem Res Toxicol* 24: 1984–1993.

Barrera, G. 2012. Oxidative stress and lipid peroxidation products in cancer progression and therapy. *ISRN Oncol* 2012: 137289.

Canuto, R. A., G. Muzio, M. Maggiora, M. E. Biocca, and M. U. Dianzani, 1993. Glutathione-S-transferase, alcohol dehydrogenase and aldehyde reductase activities during diethyl-nitrosamine-carcinogenesis in rat liver. *Cancer Lett* 68: 177–183.

Carbone, D. L., J. A. Doorn, Z. Kiebler, and D. R. Petersen, 2005. Cysteine modification by lipid peroxidation products inhibits protein disulfide isomerase. *Chem Res Toxicol* 18: 1324–1331.

Chan, K., and Y. W. Kan, 1999. Nrf2 is essential for protection against acute pulmonary injury in mice. *Proc Natl Acad Sci USA* 96: 12731–12736.

Chang, L., and M. Karin, 2001. Mammalian MAP kinase signalling cascades. *Nature* 410: 37–40.

Chung, F. L., H. J. Chen, and R. G. Nath, 1996. Lipid peroxidation as a potential endogenous source for the formation of exocyclic DNA adducts. *Carcinogenesis* 17: 2105–2111.

Chung, F. L., J. Pan, S. Choudhury, R. Roy, W. Hu et al. 2003. Formation of *trans*-4-hydroxy-2-nonenal- and other enal-derived cyclic DNA adducts from omega-3 and omega-6 polyunsaturated fatty acids and their roles in DNA repair and human p53 gene mutation. *Mutat Res* 531: 25–36.

Dinkova-Kostova, A. T., W. D. Holtzclaw, R. N. Cole, K. Itoh, N. Wakabayashi et al. 2002. Direct evidence that sulfhydryl groups of Keap1 are the sensors regulating induction of phase 2 enzymes that protect against carcinogens and oxidants. *Proc Natl Acad Sci USA* 99: 11908–11913.

Dinkova-Kostova, A. T., and P. Talalay, 2008. Direct and indirect antioxidant properties of inducers of cytoprotective proteins. *Mol Nutr Food Res* 52 (Suppl 1): S128–138.

Eckl, P. M., 2003. Genotoxicity of HNE. *Mol Aspects Med* 24: 161–165.

Eckl, P. M., A. Ortner, and H. Esterbauer, 1993. Genotoxic properties of 4-hydroxyalkenals and analogous aldehydes. *Mutat Res* 290: 183–192.

Esterbauer, H., R. J. Schaur, and H. Zollner, 1991. Chemistry and biochemistry of 4-hydroxynonenal, malonaldehyde and related aldehydes. *Free Radic Biol Med* 11: 81–128.

Fan, Y., R. Mao, and J. Yang, 2013. NF-kappaB and STAT3 signaling pathways collaboratively link inflammation to cancer. *Protein Cell* 4: 176–185.

Fang, Q., J. Nair, X. Sun, D. Hadjiolov, and H. Bartsch, 2007. Etheno-DNA adduct formation in rats gavaged with linoleic acid, oleic acid and coconut oil is organ- and gender specific. *Mutat Res* 624: 71–79.

Feng, Z., W. Hu, S. Amin, and M. S. Tang, 2003. Mutational spectrum and genotoxicity of the major lipid peroxidation product, *trans*-4-hydroxy-2-nonenal, induced DNA adducts in nucleotide excision repair-proficient and -deficient human cells. *Biochemistry* 42: 7848–7854.

Forman, H. J., 2010. Reactive oxygen species and alpha, beta-unsaturated aldehydes as second messengers in signal transduction. *Ann N Y Acad Sci* 1203: 35–44.

Forman, H. J., J. M. Fukuto, T. Miller, H. Zhang, A. Rinna et al. 2008. The chemistry of cell signaling by reactive oxygen and nitrogen species and 4-hydroxynonenal. *Arch Biochem Biophys* 477: 183–195.

Ganan-Gomez, I., Y. Wei, H. Yang, M. C. Boyano-Adanez, and G. Garcia-Manero, 2013. Oncogenic functions of the transcription factor Nrf2. *Free Radic Biol Med* 65: 750–64.

Gasc, N., S. Tache, E. Rathahao, J. Bertrand-Michel, V. Roques et al. 2007. 4-hydroxynonenal in foodstuffs: Heme concentration, fatty acid composition and freeze-drying are determining factors. *Redox Rep* 12: 40–44.

Glei, M., S. Klenow, J. Sauer, U. Wegewitz, K. Richter et al. 2006. Hemoglobin and hemin induce DNA damage in human colon tumor cells HT29 clone 19A and in primary human colonocytes. *Mutat Res* 594: 162–171.

Gopalakrishna, R., and S. Jaken, 2000. Protein kinase C signaling and oxidative stress. *Free Radic Biol Med* 28: 1349–1361.

Grune, T., and K. J. Davies, 2003. The proteasomal system and HNE-modified proteins. *Mol Aspects Med* 24: 195–204.

Hammer, A., M. Ferro, H. M. Tillian, F. Tatzber, H. Zollner et al. 1997. Effect of oxidative stress by iron on 4-hydroxynonenal formation and proliferative activity in hepatomas of different degrees of differentiation. *Free Radic Biol Med* 23: 26–33.

Hayes, J. D., and M. McMahon, 2009. NRF2 and KEAP1 mutations: Permanent activation of an adaptive response in cancer. *Trends Biochem Sci* 34: 176–188.

Hayes, J. D., M. McMahon, S. Chowdhry, and A. T. Dinkova-Kostova, 2010. Cancer chemoprevention mechanisms mediated through the Keap1-Nrf2 pathway. *Antioxid Redox Signal* 13: 1713–1748.

Hu, W., Z. Feng, J. Eveleigh, G. Iyer, J. Pan et al. 2002. The major lipid peroxidation product, *trans*-4-hydroxy-2-nonenal, preferentially forms DNA adducts at codon 249 of human p53 gene, a unique mutational hotspot in hepatocellular carcinoma. *Carcinogenesis* 23: 1781–1789.

Huang, H. C., T. Nguyen, and C. B. Pickett, 2002. Phosphorylation of Nrf2 at Ser-40 by protein kinase C regulates antioxidant response element-mediated transcription. *J Biol Chem* 277: 42769–42774.

Huang, Y., W. Li, and A. N. Kong, 2012. Anti-oxidative stress regulator NF-E2-related factor 2 mediates the adaptive induction of antioxidant and detoxifying enzymes by lipid peroxidation metabolite 4-hydroxynonenal. *Cell Biosci* 2: 40.

Ishikawa, S., S. Tamaki, M. Ohata, K. Arihara, and M. Itoh, 2010. Heme induces DNA damage and hyperproliferation of colonic epithelial cells via hydrogen peroxide produced by heme oxygenase: A possible mechanism of heme-induced colon cancer. *Mol Nutr Food Res* 54: 1182–1191.

Itoh, K., T. Chiba, S. Takahashi, T. Ishii, K. Igarashi et al. 1997. An Nrf2/small Maf heterodimer mediates the induction of phase II detoxifying enzyme genes through antioxidant response elements. *Biochem Biophys Res Commun* 236: 313–322.

Itoh, K., N. Wakabayashi, Y. Katoh, T. Ishii, K. Igarashi et al. 1999. Keap1 represses nuclear activation of antioxidant responsive elements by Nrf2 through binding to the amino-terminal Neh2 domain. *Genes Dev* 13: 76–86.

Ji, C., K. R. Kozak, and L. J. Marnett, 2001. IkappaB kinase, a molecular target for inhibition by 4-hydroxy-2-nonenal. *J Biol Chem* 276: 18223–18228.

Juric-Sekhar, G., K. Zarkovic, G. Waeg, A. Cipak, and N. Zarkovic, 2009. Distribution of 4-hydroxynonenal-protein conjugates as a marker of lipid peroxidation and parameter of malignancy in astrocytic and ependymal tumors of the brain. *Tumori* 95: 762–768.

Karihtala, P., S. Kauppila, U. Puistola, and A. Jukkola-Vuorinen, 2011. Divergent behaviour of oxidative stress markers 8-hydroxydeoxyguanosine (8-OHdG) and 4-hydroxy-2-nonenal (HNE) in breast carcinogenesis. *Histopathology* 58: 854–862.

Karlhuber, G. M., H. C. Bauer, and P. M. Eckl, 1997. Cytotoxic and genotoxic effects of 4-hydroxynonenal in cerebral endothelial cells. *Mutat Res* 381: 209–216.

Katsuoka, F., H. Motohashi, T. Ishii, H. Aburatani, J. D. Engel et al. 2005. Genetic evidence that small Maf proteins are essential for the activation of antioxidant response element-dependent genes. *Mol Cell Biol* 25: 8044–8051.

Kensler, T. W., P. A. Egner, A. S. Agyeman, K. Visvanathan, J. D. Groopman et al. 2013. Keap1-nrf2 signaling: A target for cancer prevention by sulforaphane. *Top Curr Chem* 329: 163–177.

Khor, T. O., M. T. Huang, A. Prawan, Y. Liu, X. Hao et al. 2008. Increased susceptibility of Nrf2 knockout mice to colitis-associated colorectal cancer. *Cancer Prev Res (Phila)* 1: 187–191.

Kim, Y. R., J. E. Oh, M. S. Kim, M. R. Kang, S. W. Park et al. 2010. Oncogenic NRF2 mutations in squamous cell carcinomas of oesophagus and skin. *J Pathol* 220: 446–451.

Kobayashi, A., M. I. Kang, H. Okawa, M. Ohtsuji, Y. Zenke et al. 2004. Oxidative stress sensor Keap1 functions as an adaptor for Cul3-based E3 ligase to regulate proteasomal degradation of Nrf2. *Mol Cell Biol* 24: 7130–7139.

Kumagai, T., N. Matsukawa, Y. Kaneko, Y. Kusumi, M. Mitsumata et al. 2004. A lipid peroxidation-derived inflammatory mediator: Identification of 4-hydroxy-2-nonenal as a potential inducer of cyclooxygenase-2 in macrophages. *J Biol Chem* 279: 48389–48396.

Lee, T. J., J. T. Lee, S. K. Moon, C. H. Kim, J. W. Park et al. 2006. Age-related differential growth rate and response to 4-hydroxynonenal in mouse aortic smooth muscle cells. *Int J Mol Med* 17: 29–35.

Li, W., and A. N. Kong, 2009. Molecular mechanisms of Nrf2-mediated antioxidant response. *Mol Carcinog* 48: 91–104.

Li, W., S. Yu, T. Liu, J. H. Kim, V. Blank et al. 2008. Heterodimerization with small Maf proteins enhances nuclear retention of Nrf2 via masking the NESzip motif. *Biochim Biophys Acta* 1783: 1847–1856.

Li, X., A. B. Abdel-Mageed, D. Mondal, and E. Kandil, 2013. The nuclear factor kappa-B signaling pathway as a therapeutic target against thyroid cancers. *Thyroid* 23: 209–218.

Marnett, L. J., 1999. Lipid peroxidation-DNA damage by malondialdehyde. *Mutat Res* 424: 83–95.

Mayerhofer, M., S. Florian, M. T. Krauth, K. J. Aichberger, M. Bilban et al. 2004. Identification of heme oxygenase-1 as a novel BCR/ABL-dependent survival factor in chronic myeloid leukemia. *Cancer Res* 64: 3148–3154.

Mitsuishi, Y., K. Taguchi, Y. Kawatani, T. Shibata, T. Nukiwa et al. 2012. Nrf2 redirects glucose and glutamine into anabolic pathways in metabolic reprogramming. *Cancer Cell* 22: 66–79.

Moi, P., K. Chan, I. Asunis, A. Cao, and Y. W. Kan, 1994. Isolation of NF-E2-related factor 2 (Nrf2), a NF-E2-like basic leucine zipper transcriptional activator that binds to the tandem NF-E2/AP1 repeat of the beta-globin locus control region. *Proc Natl Acad Sci USA* 91: 9926–9930.

Munasinghe, D. M., K. Ichimaru, T. Matsui, K. Sugamoto, and T. Sakai, 2003. Lipid peroxidation-derived cytotoxic aldehyde, 4-hydroxy-2-nonenal in smoked pork. *Meat Sci* 63: 377–380.

Nair, J., A. Barbin, I. Velic and H. Bartsch, 1999. Etheno DNA-base adducts from endogenous reactive species. *Mutat Res* 424: 59–69.

Nair, U., H. Bartsch, and J. Nair, 2007. Lipid peroxidation-induced DNA damage in cancer-prone inflammatory diseases: A review of published adduct types and levels in humans. *Free Radic Biol Med* 43: 1109–1120.

Nishi, T., N. Shimizu, M. Hiramoto, I. Sato, Y. Yamaguchi et al. 2002. Spatial redox regulation of a critical cysteine residue of NF-kappa B in vivo. *J Biol Chem* 277: 44548–44556.

Nitti, M., C. Domenicotti, C. d'Abramo, S. Assereto, D. Cottalasso et al. 2002. Activation of PKC-beta isoforms mediates HNE-induced MCP-1 release by macrophages. *Biochem Biophys Res Commun* 294: 547–552.

Niture, S. K., A. K. Jain, and A. K. Jaiswal, 2009. Antioxidant-induced modification of INrf2 cysteine 151 and PKC-delta-mediated phosphorylation of Nrf2 serine 40 are both required for stabilization and nuclear translocation of Nrf2 and increased drug resistance. *J Cell Sci* 122: 4452–4464.

Niture, S. K., and A. K. Jaiswal, 2011. INrf2 (Keap1) targets Bcl-2 degradation and controls cellular apoptosis. *Cell Death Differ* 18: 439–451.

Numazawa, S., M. Ishikawa, A. Yoshida, S. Tanaka, and T. Yoshida, 2003. Atypical protein kinase C mediates activation of NF-E2-related factor 2 in response to oxidative stress. *Am J Physiol Cell Physiol* 285: C334–342.

Oberley, T. D., S. Toyokuni, and L. I. Szweda, 1999. Localization of hydroxynonenal protein adducts in normal human kidney and selected human kidney cancers. *Free Radic Biol Med* 27: 695–703.

Osburn, W. O., B. Karim, P. M. Dolan, G. Liu, M. Yamamoto et al. 2007. Increased colonic inflammatory injury and formation of aberrant crypt foci in Nrf2-deficient mice upon dextran sulfate treatment. *Int J Cancer* 121: 1883.–1891.

Padmanabhan, B., K. I. Tong, T. Ohta, Y. Nakamura, M. Scharlock et al. 2006. Structural basis for defects of Keap1 activity provoked by its point mutations in lung cancer. *Mol Cell* 21: 689–700.

Page, S., C. Fischer, B. Baumgartner, M. Haas, U. Kreusel et al. 1999. 4-Hydroxynonenal prevents NF-kappaB activation and tumor necrosis factor expression by inhibiting IkappaB phosphorylation and subsequent proteolysis. *J Biol Chem* 274: 11611–11618.

Parola, M., G. Robino, F. Marra, M. Pinzani, G. Bellomo et al. 1998. HNE interacts directly with JNK isoforms in human hepatic stellate cells. *J Clin Invest* 102: 1942–1950.

Petersen, D. R., and J. A. Doorn, 2004. Reactions of 4-hydroxynonenal with proteins and cellular targets. *Free Radic Biol Med* 37: 937–945.

Pierre, F., A. Freeman, S. Tache, R. Van der Meer, and D. E. Corpet, 2004. Beef meat and blood sausage promote the formation of azoxymethane-induced mucin-depleted foci and aberrant crypt foci in rat colons. *J Nutr* 134: 2711–2716.

Pierre, F., G. Peiro, S. Tache, A. J. Cross, S. A. Bingham et al. 2006. New marker of colon cancer risk associated with heme intake: 1, 4-Dihydroxynonane mercapturic acid. *Cancer Epidemiol Biomarkers Prev* 15: 2274–2279.

Pierre, F., R. Santarelli, S. Tache, F. Gueraud, and D. E. Corpet, 2008. Beef meat promotion of dimethylhydrazine-induced colorectal carcinogenesis biomarkers is suppressed by dietary calcium. *Br J Nutr* 99: 1000–1006.

Pierre, F., S. Tache, F. Gueraud, A. L. Rerole, M. L. Jourdan et al. 2007. Apc mutation induces resistance of colonic cells to lipoperoxide-triggered apoptosis induced by faecal water from haem-fed rats. *Carcinogenesis* 28: 321–327.

Pierre, F., S. Tache, C. R. Petit, R. Van der Meer, and D. E. Corpet, 2003. Meat and cancer: Haemoglobin and haemin in a low-calcium diet promote colorectal carcinogenesis at the aberrant crypt stage in rats. *Carcinogenesis* 24: 1683–1690.

Poli, G., F. Biasi, and G. Leonarduzzi, 2008a. 4-Hydroxynonenal-protein adducts: A reliable biomarker of lipid oxidation in liver diseases. *Mol Aspects Med* 29: 67–71.

Poli, G., R. J. Schaur, W. G. Siems, and G. Leonarduzzi, 2008b. 4-Hydroxynonenal: A membrane lipid oxidation product of medicinal interest. *Med Res Rev* 28: 569–631.

Ramkumar, K. M., T. V. Sekar, K. Foygel, B. Elango, and R. Paulmurugan, 2013. Reporter protein complementation imaging assay to screen and study nrf2 activators in cells and living animals. *Anal Chem* 85: 7542–7549.

Ramos-Gomez, M., M. K. Kwak, P. M. Dolan, K. Itoh, M. Yamamoto et al. 2001. Sensitivity to carcinogenesis is increased and chemoprotective efficacy of enzyme inducers is lost in nrf2 transcription factor-deficient mice. *Proc Natl Acad Sci USA* 98: 3410–3415.

Reuter, S., S. C. Gupta, M. M. Chaturvedi, and B. B. Aggarwal, 2010. Oxidative stress, inflammation, and cancer: How are they linked? *Free Radic Biol Med* 49: 1603–1616.

Rossi, M. A., and G. Cecchini, 1983. Lipid peroxidation in hepatomas of different degrees of deviation. *Cell Biochem Funct* 1: 49–54.

Ruef, J., M. Moser, C. Bode, W. Kubler, and M. S. Runge, 2001. 4-hydroxynonenal induces apoptosis, NF-kappaB-activation and formation of 8-isoprostane in vascular smooth muscle cells. *Basic Res Cardiol* 96: 143–150.

Satoh, H., T. Moriguchi, K. Taguchi, J. Takai, J. M. Maher et al. 2010. Nrf2-deficiency creates a responsive microenvironment for metastasis to the lung. *Carcinogenesis* 31: 1833–1843.

Seppanen, C. M., and A. S. Csallany, 2002. Formation of 4-hydroxynonenal, a toxic aldehyde, in soybean oil at frying temperature. *J Am Oil Chem Soc* 79: 1033–1038.

Seppanen, C. M., and A. S. Csallany, 2004. Incorporation of the toxic aldehyde 4-hydroxy-2-*trans*-nonenal into food fried in thermally oxidized soybean oil. *J Am Oil Chem Soc* 81: 1137–1141.

Sesink, A. L., D. S. Termont, J. H. Kleibeuker, and R. Van der Meer, 1999. Red meat and colon cancer: The cytotoxic and hyperproliferative effects of dietary heme. *Cancer Res* 59: 5704–5709.

Shelton, P., and A. K. Jaiswal, 2013. The transcription factor NF-E2-related factor 2 (Nrf2): A protooncogene? *FASEB J* 27: 414–423.

Shibata, T., A. Kokubu, M. Gotoh, H. Ojima, T. Ohta et al. 2008a. Genetic alteration of Keap1 confers constitutive Nrf2 activation and resistance to chemotherapy in gallbladder cancer. *Gastroenterology* 135: 1358–1368, 1368 e1351–1354.

Shibata, T., T. Ohta, K. I. Tong, A. Kokubu, R. Odogawa et al. 2008b. Cancer related mutations in NRF2 impair its recognition by Keap1-Cul3 E3 ligase and promote malignancy. *Proc Natl Acad Sci USA* 105: 13568–13573.

Skrzydlewska, E., A. Stankiewicz, M. Sulkowska, S. Sulkowski, and I. Kasacka, 2001. Antioxidant status and lipid peroxidation in colorectal cancer. *J Toxicol Environ Health A* 64: 213–222.

Sodum, R. S., and F. L. Chung, 1988. 1, N2-ethenodeoxyguanosine as a potential marker for DNA adduct formation by *trans*-4-hydroxy-2-nonenal. *Cancer Res* 48: 320–323.

Spickett, C. M., 2013. The lipid peroxidation product 4-hydroxy-2-nonenal: Advances in chemistry and analysis. *Redox Biol* 1: 145–152.

Sporn, M. B., and K. T. Liby, 2012. NRF2 and cancer: The good, the bad and the importance of context. *Nat Rev Cancer* 12: 564–571.

Sunamura, M., D. G. Duda, M. H. Ghattas, L. Lozonschi, F. Motoi et al. 2003. Heme oxygenase-1 accelerates tumor angiogenesis of human pancreatic cancer. *Angiogenesis* 6: 15–24.

Surh, J., and H. Kwon, 2003. Simultaneous determination of 4-hydroxy-2-alkenals, lipid peroxidation toxic products. *Food Additives Contam* 20: 325–330.

Surh, J., and H. Kwon, 2005. Estimation of daily exposure to 4-hydroxy-2-alkenals in Korean foods containing n-3 and n-6 polyunsaturated fatty acids. *Food Additives Contam* 22: 701–708.

Tjalkens, R. B., L. W. Cook, and D. R. Petersen, 1999. Formation and export of the glutathione conjugate of 4-hydroxy-2, 3-E-nonenal (4-HNE) in hepatoma cells. *Arch Biochem Biophys* 361: 113–119.

Tjalkens, R. B., S. W. Luckey, D. J. Kroll, and D. R. Petersen, 1998. Alpha,beta-unsaturated aldehydes increase glutathione S-transferase mRNA and protein: Correlation with activation of the antioxidant response element. *Arch Biochem Biophys* 359: 42–50.

Uchida, K., 2003. 4-Hydroxy-2-nonenal: A product and mediator of oxidative stress. *Prog Lipid Res* 42: 318–343.

Wakabayashi, N., S. Shin, S. L. Slocum, E. S. Agoston, J. Wakabayashi et al. 2010. Regulation of notch1 signaling by nrf2: Implications for tissue regeneration. *Sci Signal* 3: ra52.

World Cancer Research Fund/American Institute for Cancer Research, 2007 *Food, Nutrition, Physical Activity, and the Prevention of Cancer: A Global Perspective.* Washington DC: AICR.

Williams, M. V., S. H. Lee, M. Pollack, and I. A. Blair, 2006. Endogenous lipid hydroperoxide-mediated DNA-adduct formation in min mice. *J Biol Chem* 281: 10127–10133.

Winczura, A., D. Zdzalik, and B. Tudek, 2012. Damage of DNA and proteins by major lipid peroxidation products in genome stability. *Free Radic Res* 46: 442–459.

Wu, K. C., J. Y. Cui, and C. D. Klaassen, 2012. Effect of graded Nrf2 activation on phase-I and -II drug metabolizing enzymes and transporters in mouse liver. *PLoS One* 7: e39006.

Yamadori, T., Y. Ishii, S. Homma, Y. Morishima, K. Kurishima et al. 2012. Molecular mechanisms for the regulation of Nrf2-mediated cell proliferation in non-small-cell lung cancers. *Oncogene* 31: 4768–4777.

You, A., C. W. Nam, N. Wakabayashi, M. Yamamoto, T. W. Kensler et al. 2011 Transcription factor Nrf2 maintains the basal expression of Mdm2: An implication of the regulation of p53 signaling by Nrf2. *Arch Biochem Biophys* 507: 356–364.

Zanetti, D., G. Poli, B. Vizio, B. Zingaro, E. Chiarpotto et al. 2003. 4-hydroxynonenal and transforming growth factor-beta1 expression in colon cancer. *Mol Aspects Med* 24: 273–280.

Zhang, H., N. Court, and H. J. Forman, 2007. Submicromolar concentrations of 4-hydroxynonenal induce glutamate cysteine ligase expression in HBE1 cells. *Redox Rep* 12: 101–106.

Index

A

AA, see Arachidonic acid (ARA)
ACE, see Angiotensin-converting enzyme (ACE)
Acetylated-COX-2 (ac-COX-2), 48
Aconitase (ACO), 347
Acrolein (Acr), 122, 332–333, 388
ACS, see Acute coronary syndrome (ACS)
Actin, 344
Activator protein-1 (AP-1), 323
Acute coronary syndrome (ACS), 286
Acute myocardial infarction (AMI), 288
AD, see Alzheimer's disease (AD)
Adenyl ayclase, 52
Adipose PLA_2 (AdPLA_2), 46
Advanced glycation end products (AGE), 8, 366
    RAGEs, 155
    receptors for, 155
AEA, see Endocannabinoid anandamide (AEA)
AGE, see Advanced glycation end products (AGE)
AGE-phosphatidyl ethanolamine (AGE-PtdEtn), 366
Age-related macular degeneration (AMD), 203, 223
AIF, see Apoptosis-inducing factor (AIF)
AKRs, see Aldo–keto reductases (AKRs)
ALA, see α-linoleic acid (ALA)
Alcohol-induced fatty liver development
        inhibition, 239–240
Alcoholic liver disease (ALD), 239
ALD, see Alcoholic liver disease (ALD)
Aldehyde dehydrogenases (ALDHs), 394
Aldehydes, 121–122
ALDHs, see Aldehyde dehydrogenases (ALDHs)
Aldo–keto reductases (AKRs), 51, 394, 396
Alkenals, 121, 122, 388
    adducts, 123–125
    carbon chain cleavage to, 35–36
    protein lipoxidation by, 128
Alkoxyl radical products, 31–33
1-O-alkyl-2-acetyl-glycerophosphatidylcholine,
        see Platelet-activating factor (PAF)
Alkyl–acyl precursor species, 199
α-chlorofatty aldehyde, 88; see also Halogenated
        lipids
    biological targets and metabolites, 89–91
    metabolism, 88
α-EPOX, see Cholesterol-5α,6α-epoxide
        (α-EPOX)
α-halofatty aldehyde liberation
    biological production in leukocytes, 85–87

chemical identity, 84–85
in vivo inflammatory conditions, 85–87
mechanisms for action of halogenated lipids,
        87–89
7α-hydroxycholesterol (7α-OH), 164, 311
α-iodofatty aldehyde, 84
α-linoleic acid (ALA), 46
α-tocopherol, 31
ALS, see Amyotrophic lateral sclerosis (ALS)
ALX/FPR2 receptors, 185
Alzheimer's disease (AD), 330, 334, 373
    lipid peroxidation, 334–337
    MS relation, 373
    by neurodegeneration, 334
Amadori rearrangement, 8
AMD, see Age-related macular degeneration
        (AMD)
AMI, see Acute myocardial infarction (AMI)
AMP-activated protein kinase (AMPK), 108
Amyotrophic lateral sclerosis (ALS), 341
    lipid peroxidation, 342, 343
    redox proteomics, 342, 343
Angiotensin-converting enzyme (ACE), 368
Angiotensin II type 1 (AT1), 368
Animal models, 339
    aspects of PD, 339
    MPPP and MPTP, 339
    neurotoxin 6-OHDA, 340
Anti-inflammatory cytokines, 365
Antigen-presenting cells (APCs), 153
Antioxidant response element (ARE), 107, 246,
        256, 394
Antioxidants, 366
AP-1, see Activator protein-1 (AP-1)
APCs, see Antigen-presenting cells (APCs)
Apolipoprotein B-100 (apoB), 282
Apoptosis-inducing factor (AIF), 159
Arachidonate-containing phospholipids, full
        chain oxidized products of, 11
Arachidonic acid (ARA), 4, 25, 46, 108, 174, 175
    COXs mediating hydroxylation, 5
    enzymatic conversion, 177
    LX and ATL, 178
    signaling, 110
ARE, see Antioxidant response element (ARE)
Aspirin, 48
Aspirin-triggered lipoxins (ATL), 177
    aspirin involvement, 179
    biosynthesis, 178
    transcellular formation, 178

405

413

CONTINUOUS CREATION

A Biological Concept of the Nature of Matter

By

WILFRED BRANFIELD

Volume 3

Routledge
Taylor & Francis Group

LONDON AND NEW YORK

First published in 1950

This edition first published in 2009 by
Routledge
2 Park Square, Milton Park, Abingdon, Oxon, OX14 4RN

Simultaneously published in the USA and Canada
by Routledge
270 Madison Avenue, New York, NY 10016

*Routledge is an imprint of the Taylor & Francis Group, an informa
business*

© 1950 Routledge & Kegan Paul

British Library Cataloguing in Publication Data
A catalogue record for this book is available from the British Library

Library of Congress Cataloging in Publication Data
A catalog record for this book has been requested

ISBN 10: 0-415-42029-6 (Set)
ISBN 10: 0-415-47438-8 (Volume 3)

ISBN 13: 978-0-415-42029-7 (Set)
ISBN 13: 978-0-415-47438-2 (Volume 3)

Publisher's Note
The Publisher has gone to great lengths to ensure the quality of this
reprint but points out that some imperfections in the original copies
may be apparent.

Disclaimer
The Publishers have made every effort to trace copyright holders and
would welcome correspondence from those they have been unable to
contact.

CONTINUOUS CREATION

A Biological Concept of the Nature of Matter

BY

WILFRED BRANFIELD

INTRODUCTION BY
PROFESSOR FREDERIC WOOD JONES,
D.Sc., F.R.S., F.R.C.S.

LONDON

ROUTLEDGE & KEGAN PAUL LIMITED

BROADWAY HOUSE; 68–74 CARTER LANE, E.C.

First published 1950

DEDICATION

To all those who fostered Adult Education movements in the past, and to the memory of Jonathan Barber, for forty years President and Leader of the Men's Adult School, Button Lane, Sheffield.

Printed in Great Britain by Butler & Tanner Ltd., Frome and London

CONTENTS

ILLUSTRATIONS

INTRODUCTION

SHOULD a man, being curious concerning the relation of life to the Universe, consult the writings of those philosophers who have given thought to the matter, he will find that he is permitted much latitude in his views. Should the dictum of Jeans, that life is so trivial an accident in the Universe that its habitat may be compared to " a millionth part of a grain of sand out of all the sea sand in the world ", seem to him no better than a ridiculous guess, he may turn to Samuel Butler. Butler made a guess of a very different kind, for he declared that it was justifiable to regard " every atom in the Universe as living ". Between these two dicta there is freedom of choice enough for any man. The author of *Continuous Creation* makes no excursions into the Universe. He remains with his feet planted firmly on Earth, and writes simply of the homely constituents of the only part of the Universe with which any of us may claim much familiarity. But though he writes of familiar things his thesis concerning their origins, their transformation and their ultimate destinies is anything but familiar.

We all know, though perhaps the knowledge has never greatly impressed us, that the whole great mass of the vast rolling Chalk Downs is the direct product of living creatures. We are accustomed to regard the hidden wealth of the coal deposits all over the world as having been amassed by the agency of life. But though all educated people know these things with taught familiarity, so rare a product of education is curiosity that we have all been content to let the matter rest there. To Branfield belongs the credit of possessing that hardihood for carrying the question to its logical conclusion. If all the chalk and all the coal and the limestones and a host of other geological formations are

ix

manifestly the products of an endless series of generations of living things, what then of the rest of the constituents of Earth ?   Are all of them, lifeless, inert and changed though they may be, the products of life ?   And in the end the question must be squarely faced : What is the relation of life to matter ?   Was the inanimate matter of which Earth is composed conjured somehow out of space by some cosmic miracle, and, when it had become sufficiently complex in its structure, did it give birth to something that represented life at its lowest ebb ?   Or has life itself been the pioneer in the business, and is the inert matter of Earth the ultimate product of living forces ?   It may be said that Branfield's thesis that life is the precursor of the matter of Earth is merely another academic excursion which, though perhaps of interest to scientific men, has no message for the rank and file of humanity.   But let there be no doubt about it, his main ideas entail far more than mere academic argument.   If he is justified in his claims we are done with the stultifying belief that we live on an Earth, created somehow in cataclysm and destined to run down to a cold and lifeless disintegration. We are done with the belief in the ridiculously inconsiderable sphere of life in cosmic affairs. We are liberated from the thraldom of cold, material, academic chemistry and physics to escape into the sunlight and the magic of radiant energy.   Instead of origin in cosmic upheaval, followed by inevitable degradation and ultimate dissolution, we see continuous creation by which " the material of the sun riding like fairies on sunbeams " is still building our Earth " by the labours of the creatures that dwell thereon ".

F. Wood Jones.

PREFACE

THIS work is an adventure in metabiological thought. An attempt is made to find the solution of the problem of the Nature of Life, and of the relations existing between matter and light, and the part that these seemingly unlike yet inter-connected elements play in the life and development of every creature. A biological concept for the whole of Nature is put forward as the solution of the problem.

The work is, in part, unorthodox from the mechanistic point of view, but it should not be unwelcome because of that. Until quite recently, nuclear fission was unorthodox, and atomic physics had only theoretic interest. In this world of accelerated evolution, we know not what next we shall have to accept as factual. The work is written by a layman interested in science, and is written for laymen in non-technical language as far as is possible, in the hope that some of them may become more interested in biology and the constitution of living things and their evolution, in geology and the history of the earth and in the creative aspect of the World of Nature.

The writer acknowledges his debt with gratitude to all the authors quoted in the text.

W. B.

DERBY.

CHAPTER I

INTRODUCTION

I am ; how little more I know !
Whence came I ? Whither do I go ?
A centred self, which feels and is ;
A cry between the silences !
JOHN G. WHITTIER. *Questions of Life.*

HOW very little do any of us know about ourselves
after a century of intensive study and research in
biology and other branches of science ? We do
not even know the nature of life. We see life arriving,
we know not whence ; and soon departing, we know not
whither. Life may be a real and basal form of existence,
not being created *de novo* but being called out, as it were,
from an inexhaustible reservoir, and animating matter
for a time and then leaving it, returning whence it came.
There is much speculation as to *how* life is associated with
matter ; it may be indirect, with the cohesive force of
the ether as the link that binds them in one seeming
whole. Life appears (to us) to need a material vehicle
in and through which to express itself, and to display
its activities. This material vehicle, the intricate,
presumably chemical substance called protoplasm, is in
a state of continual flux, a stream of " becoming " rather
than something that *is*.

Each individual life begins as a single cell or egg. All
the diverse forms of life that are known to-day have
evolved from some primitive ancestral type, developing
and adapting to changing conditions, and, step by step,
down the ages of the geological past, enregistering
successive structural adaptations in the hereditary
factors. The fertilised egg or ovum is the starting point
of the germ or embryo, which is endowed with certain
specific powers in virtue of which the germ-cell proceeds
to divide and again divide, and grow in the process,
gradually building up a body resembling that of the

13

parent. In the development of the egg the embryo recapitulates the history of its kind ; it is one of the charms of embryology to read the life history of the race to which the organism belongs, as it repeats in the germ-plasm, step by step, the successive gains of racial evolution, as in a moving picture.

All living creatures seek food morsels, engulf and digest them. Almost nothing is known of the method by which the individual converts the ingested moiety into the substance of its own body. Parts of the body are continually being destroyed, or worn and wasted, and replaced by new protoplasm which is continually being built up and assimilated into the living substance of the body. This power to build up a basic organic pattern from the food, or from the energy derived there-from, is the same whether it be food, light or any other " experience ". The technical name for this complete transformation, and co-ordination of structure and function, is metabolism.

Unfortunately there is no common agreement among the experts as to " how " life maintains its material vehicle. Many of the mechanistic school are convinced that there is nothing in biology but what is physical mechanism under somewhat complex circumstances. For a century, farmers have been induced to treat crop plants as test-tube recipients of chemical drugs, and so they have overlooked the fact that life and growth are physiological—not chemical—phenomena. " The chemical analysis of soils for the determination of the nutritional elements available to plants has engaged chemists since their earliest study of the soil, but the system ' plant-soil ' is almost an unexplored field of study." (1)

There is what amounts to a conspiracy of silence about one of the earliest experiments in plant physiology. About the year 1600, a Flemish chemist and physiologist, Jean Baptiste van Helmont, planted a willow in an earthenware pot containing 200 lbs. of soil. The pot was carefully protected from dust, and regularly watered.

After five years the weight of the willow, and of the soil was again determined. The tree had increased by 164 lbs.; the soil had lost only 2 ozs.

That Helmont concluded that water alone had caused the increase in weight of the tree, and believed it possible that water may be the only ultimate constituent of all matter, does not vitiate the experiment. His conclusions were no more in error than the modern explanation that the increase in weight is due to carbon dioxide alone or with the aid of soil mineral matter. The man Helmont is greater than his critics—verbose or silent. He did discover carbon dioxide, and gave it, quite appropriately, the name " gas sylvestre ", because he obtained it by burning charcoal. The recent versions of the reactions taking place in the green leaf are as far removed from the whole truth as Helmont's, and with very much less justification.

The crux of the experiment was that the tree did not increase in weight at the expense of the soil by imbibing soluble mineral matter. The experiment, like so many others, was too much like a test-tube trial in a laboratory. Such tests really defeat the purpose intended, and are incapable of yielding such results as an ecological approach would produce. Trees grown in isolation and plants in pots are not free agents in their reactions to their environment. The dead leaves are swept up, or blown away, and lost to fertility (their natural purpose), and the production of a carpet of humus. It is surprising that only 2 ozs. of humus was lost, under the imposed conditions. In nature, trees flourish by their association in a forest. There, the soil is shaded ; the micro-organisms are neither frozen nor scorched nor yet washed out, but are protected by leaf-fall. There, smoothly, silently, at relatively low temperatures, the living tree carries on its remarkable synthesis, absorbing and fixing sunlight and, by its help or agency, in some subtle way, builds up from such simple substances as water and carbon dioxide, carbohydrates and large molecules of " biologically determined " patterns, and incorporates them into its tissues, whereby the tree grows and increases in stature. For too long have students of plant nutrition been influenced by the idea that fertility is determined by the chemistry of the soil.

All life of the kind we know can only exist under suitable conditions of light and temperature ; trees and men are able to live because the earth receives the right amount of radiation from the sun. The scientific pundits have formulated the physical principle known as the second law of thermo-dynamics which avers that the world is running down, like a clock, gradually losing heat in the process, during which the temperature will fall so low as to make life impossible. (How the world-clock was wound up, " in the beginning ", and then left to run down, has not been elucidated.)

Just what is this " radiation " on which all life depends? Radiation is entirely an electronic phenomenon ! An electron drops in its tiny orbit in blazing solar hydrogen, and $8\frac{1}{2}$ minutes later an electron in a complex molecule in a green leaf leaps up just as high as the solar electron had leaped down. The energy transferred from the sun traversed space as electric and magnetic vibrations. The energy is obvious, for the heat when the radiation is arrested by the body is familiar. Light rays simulate some of the properties of a projectile. Light must be, in some way, related to, and be inherent in, matter. It will assist our understanding if we consider some recent developments in atomic physics.

The sudden use of the atomic bomb has quickened everybody's interest in the atom and in its content of energy. Until recently, matter has been regarded as dead inert stuff, and present in the world chaotically, before the creation. Suddenly, we are made to realise that there are tremendous stores of energy in the stuff, awaiting ways and means of release, and methods of control for the benefit of mankind. It is a great disappointment that the first utilisation of atomic energy should be of a destructive type, but let us make no mistake, there will be bigger and much more important discoveries yet to come.

Atomic energy is present in all matter, but the substance from which it has been released in the bomb is the radio-active element, uranium. The discovery of the fission of uranium, and its application in the atomic bomb, is no isolated event, but follows a series of discoveries which, since the beginning of the century, has been the basis of the modern science of physics. In the nineteenth century it was regarded as a cardinal point that the atoms of any element were discrete, indestructible particles, which could in no way be changed or converted into those of another.

Then in 1896, the French physicist, H. Becquerel, found that crystals of the double sulphate of uranium emitted radiation of unknown type, which affected photographic plates. Two years later, Pierre and Marie Curie isolated radium from uranium ore, which showed, to a much greater degree, this same property of emitting radiation, and it was clear that the phenomenon of " radio-activity ", as it was called, was altogether different from those associated with normal chemical reactions between atoms. Part of this radiation, the so-called " *alpha*-rays ", consists of helium atoms carrying a positive charge of electricity, and these have been found to be of the greatest value for further exploration of the structure of atoms. It was research on the penetration of matter by " *alpha*-rays " which led Rutherford, at Manchester University in 1911, to the fundamental discovery that the whole mass of each atom is concentrated in a minute central nucleus which carried a positive electric charge. Round this nucleus revolved elementary negative electric charges—the " electrons " in numbers sufficient to neutralise the positive charge of the nucleus. The physical and chemical properties of any atom are determined by the electrons revolving round the nucleus. The atoms of nearly all the elements are stable and it is only in radio-active elements that spontaneous disintegration of the nucleus takes place. In

B

1919, at Manchester, Rutherford broke up the nuclei of atoms of nitrogen gas by collision with the charged *alpha*-particles from radium, and the nuclei of two other atomic elements—oxygen and hydrogen—were formed. This development was pursued in the following years by Rutherford and Chadwick, who found that many other light elements could be transmuted in a similar way. In each case a proton (hydrogen nucleus) was ejected, and generally the process of transmutation was accompanied by the release of a considerable amount of energy. It thus appeared that the proton was a common constituent of atomic nuclei and one of the fundamental particles of which matter is built up. The proton is the unit of weight in the sphere of atomic physics, its weight is taken to be 1. The proton is used again to provide a unit of electric charge. The electric charge of a proton is positive, and its amount is defined as one unit. The first and simplest atom consists of one proton with one " attendant " electron bearing its negative charge ; this is called the hydrogen atom.

Electrons are minute electrified particles moving around the nucleus in all directions, completely enveloping it. They can be disturbed by heating the matter (metal, for instance) to incandescence (thermionic emission), also when certain metals are exposed to light, negatively charged electrons are emitted (photo-electric effect). Electrons are always of the same kind, and are now regarded as another essential material, along with protons, of which atoms of matter are made. In the photo-electric cell, the number of electrons liberated depends on the intensity of the illumination, that is, the amount of radiant energy per unit area. Light is measured in Planck's " quanta " or " photons " as they are called. Whenever a ray exchanges energy with matter, a photon is absorbed. When an electron is discharged, there is increased heat-motion in the vicinity (measured in atomic distances, of course), which causes intense oscillations of atoms.

The discovery of quanta seems to require of scientific

thought one of the greatest changes in orientation that it has ever had to make. (13) " There is considerable justification for regarding photons as being of the same nature as electrons and protons, but without their electric charges. The fundamental distinction that electrons and protons carry electric charges, whereas photons do not, is enough in itself to account for many of the differences of their properties. Both carry energy and mass. For a charged particle to carry a finite amount of energy, it must move more slowly than light, whereas for a photon to carry a finite amount of energy, it must move at precisely the speed of light." Sir James Jeans (23) stated this before the neutron had been discovered, and there is little doubt he would have included it in his statement. " The facts admit of a far wider interpretation," Jeans continued, " the whole of the available observational evidence indicates that this relation is true throughout all nature."

Then, in 1932, while working in the Cavendish laboratory, Cambridge, Chadwick discovered the neutron. It, like the photon, has no electric charge, and has a weight of 1. The element helium has two electrons enveloping a nucleus containing two protons and two neutrons. It is found, in all elements except hydrogen, that nuclei must contain at least as many neutrons as protons. The heavier atoms may contain helium nuclei (*alpha*-particles) as well as protons and neutrons.

The discovery of the neutron was of great practical value in that its lack of electric charge make it a perfect projectile for carrying out nuclear transformations. Chadwick caused nuclear changes in lithium, carbon and nitrogen.

The quantities of matter that undergo transmutation by artificial bombardment are very minute indeed. If in the atomic pile a gram of chalk is subjected to bombardment for a period of a week, about one part of the calcium in 1000 million will have been transformed to its radio-active isotope. The hope underlying all atomic research is that a source of usable energy may be revealed. If we could convert one pound of hydrogen into the next element, helium, and control the process, enormous energy could be released. " If atom-building takes place at all, anywhere in Nature, in the universe, in inter-stellar space, there is one such atom-building act that is more fundamental than all others, and that also must take place more frequently than all others, namely, the formation of helium out of hydrogen, because we have abundant evidence that all the elements are actually built-up out of hydrogen and helium, and that helium is built up of four atoms of hydrogen, so that hydrogen-to-helium transformation should take place much more frequently than any other." (21). The energy of this transformation is computed by Prof. Millikan to be of the order of 25 million volts. It is deserving of very serious thought, and research.

Up to the present no process of atom-building has been found in progress in the realm of inanimate matter. Yet we have grown accustomed to think of the modern age as rich in power : it has seemed so, only by a prodigal use of coal. Coal is a legacy from the bygone past. It represents the accumulation of solar energy over almost incredible periods of time. All the technological developments that our civilisation has yet contrived, are to augment our natural supplies of energy out of capital, under the delusion that it was supporting itself out of income. The beds of coal suggest that ancient forests absorbed more energy than was needed for their physiological activities, and stored it in their woody tissues. May not this be a pointer directing us to a force that opposes the second law of thermo-dynamics ? If it were found that living organisms could build up matter, we should be compelled to form a new conception concerning the nature of life and of matter, and the relation of life to the structure of the atom and its electrical components.

CHAPTER II

LIFE AND MATTER

The present matter of the universe cannot have existed for ever . . . [this] leads us to contemplate a definite event, or a series of events, or a continuous process of creation of matter sometime not infinitely remote. In some way matter which had not previously existed, came, or was brought, into being.
SIR JAMES JEANS, *The Universe Around Us*, p. 327.

MATTER and life are utterly unlike each other in our thoughts ; our concept of life is as vague as our concept of matter is hard and rigid. Yet in our experience they seem inexplicably linked together. The phenomena of growth and reproduction seem to hold the very secret of life and to bring us close to the secret of matter. (2)   It seems as if a power had its seat in living things that could, by some inherent faculty, not only maintain and extend itself, but also subdue surrounding forces.   We perceive that in each variety of organism, there is a character and identity of its own, and it builds up matter to correspond with or to represent that identity. (2)   Further, this definite character is handed down from one generation to another.

It is through some system of evolution that creation works.   It is not tenable to set up matter as a separate and independent reality, already in the world when creative evolution began to work.   A revered physicist, the late Sir Oliver Lodge, advanced the hypothesis that " all life is dependent upon ether tremors ; matter is responsive to them, and that they may have been instrumental in bringing matter into existence ". (2)   " We live ", said Sir Oliver, " in a universe of which we know very little ; we eke out our knowledge by precarious reasoning."   Matter has been shown to be an electrical phenomenon ; light also is an electrical phenomenon ; both light and matter are electro-magnetic ; matter is discrete, radiation is emitted and absorbed in packets, or quanta.   Radiation is a means of transmitting energy from place to place.   We live in such a stream of energy, coming from the sun and cosmic space ; it is essential to the display of every form of life.   Without this radiant energy, life could not enter into relation with matter. (2)   It is only detected by its effects when it has arrived ; it then affects the retina of our eyes, our photographic plates, and the surface of our skins, setting up chemical and other changes with which we are (more or less) familiar.   It operates on the green parts of plants, and thereby renders possible the whole realm of vegetation.   A plant or tree is an incarnation of solar energy ; and the complete understanding of vegetative processes is impossible without taking this space-energy into account.   Just as a charged body was to Faraday the termination of an electric field, so a vegetable organism is the termination of a luminiferous field. (16)

Let us first see the way light affects the retina of our eyes, before going further into photo-synthetic phenomena.

Light, as an optical phenomenon, is the most intense experience of man.   The subtle and mysterious emanation or radiation issuing from the gigantic orb of the sun, and entering the eye, creates for man a world of colour and of form, and lays the foundation of his sensuous existence.   The sunlight enters the eyeball and in a moment the light is mysteriously translated into sight which is the greatest part of the *conscious* being of man.*   Primitive man recognised that the light revealed the world to him, and warmed his body, and ripened his crops, and he worshipped the sun as the source of all beneficence.   His philosophy was simple : " God said, Let there be Light, and there was Light."   Man's evolution is far more extraordinary than the first chapter of Genesis used to lead people to suppose ; the story of

---

* The cerebral centre and the eye evolved co-ordinately as parts of the whole organism.

man is far more wonderful than the wonders of physical science ; the translation of light rays by his eyes into a vision of colour, with near objects in relief, and with the distant " back-ground " all in focus, is far more wonderful than the prism's separation of the component colours of the ray of light that falls on one face of the prism.

The eyes not only focus the rays of light, but form a picture of the outer world that is projected on to a sensitive plate at the back—the retina ; in this it might be compared to a photographic camera. Yet the eye is an organ that has evolved ; it has come down to us over countless centuries, from dim beginnings in the lowest organisms that first evinced the faculty of sight, continually modified by small increments in accordance with the changing needs of an innumerable succession of creatures, some of them living in water, some in air, others on land ; so that it now possesses a great range of inherited faculties, and is, in its complex structure, a record of its own extraordinary history. It must be remembered that sight was, in its origin, a differentiation of touch. The light, falling on the generally sensitive surface of a primitive organism, provoked a tactile irritation, which, in the course of evolution, becomes a specialised sense at some small area of the surface as what we know as sight or vision. Thus has life evolved —after ages of struggle—this wonderful organ the eye, so that the living organism could see where it was going, and so avoid the dangers that threatened its existence ; it is a good example of the working of life's incessant aspiration to higher organisation, to wider, deeper, intenser consciousness and clearer understanding. The force in living creatures has achieved wonders in this striving for greater consciousness of their surroundings and of themselves.

Returning to the vegetative processes under the influence of sunlight on the green leaves, we find again, wonderful and mysterious changes or processes of a synthetic type. The way in which the solar energy reaches the earth is of interest. It does not travel as heat, it comes through space which is cold. Heat is generated when the ether waves are arrested ; they cause evaporation where they encounter the wide expanse of the oceans ; of that portion that is arrested by the earth, a large part is absorbed by vegetation, the leaf surfaces of which are able by photo-synthesis to carry out reactions of a nature that enables individual plants to accrete into their own structure otherwise seemingly alien material, over which the plant exercises specific control, building it up into definite localised forms of a specific type according to the plant's own peculiarity. We have very little knowledge of the substance (if it be a substance) that fills space, but we know it is the subject of luminiferous vibrations ; we know that it transmits energy by waves which are electric and magnetic. We should be prepared to find that the elusive ether of space has other functions to fulfil than those of which we know. To recognise it as a tremendous vehicle of energy is only a beginning. This " space-energy " must be taken into account in any attempt to understand vegetative processes. A vegetable organism, as the termination of a luminiferous field of energy, is, in an essential sense, a charged body, and potentially different in consequence. The radiant energy, by interaction with the vital energy possessed by living plants, amplifies the physiological functions and the organism synthesises material that is incorporated in the plants' plasma substance. One of such processes is the synthesis of carbohydrates in the leaves of plants. Much has been written about the process of building up sugar in the leaf tissues, but actually very little is beyond controversy. So far as is known, photo-synthesis does not take place indis-criminately in the leaf cells, but centres round or upon certain bodies lying in the cytoplasm. The bodies are known as chloroplasts, and it is on these bodies that the green chlorophyll is fixed. It has been found that photo-synthesis reaches a maximum rate at about the same light intensity, and is not dependent on the chlorophyll content of the leaves.* The reaction may be expressed thus :

$$\text{Light Energy} + 6CO_2 + 6H_2O = C_6H_{12}O_6 + 6O_2$$

Light Energy + 6 Molecules of Carbon Dioxide + 6 Mols. of water = 1 Mol. of Sugar + 6 Mols. of Oxygen

The formula only expresses the beginning and the end of a series of complicated reactions, in which water plays an important role, and it is not pretended to illustrate the process of carbon assimilation. It is part only of many photo-physiological processes or functions that take place in the active cells, by which they and the whole plant grow and develop. Lying in the plasma of the plant cells are a number of other products of the physiological activities of the organism, such as proteid granules, oil drops, colouring matter (dyes), odorous ethereal substances (scents and flavourings) alkaloids (potent drugs) and many others, all products of photosynthesis, and each representing only part of a chain of complex processes. Besides all these there are varying small amounts of mineral matter—potash, soda, lime, magnesia, iron, phosphorus, sulphur, silica, etc.

It has been demonstrated very ably by the Russian scientist Krascheninnikoff that only a portion of the carbon dioxide absorbed by the plant is present as carbohydrate and that the gain in dry weight during a period of photo-synthesis is greater than can be accounted for by the CO_2 absorbed or the carbohydrate manufactured. (12) In bamboo and the tobacco plant the carbohydrate represents only 75 per cent of the dry weight increase.

The photo-synthetic process is endothermic, and the requisite energy is supplied by the respiratory system of

* Besides chlorophyll, many plants possess the red pigment *Carotin*, and the yellow *Xanthophyll*. These substances play some part in photo-synthetic activity and absorption of radiant energy, varying at different stages of growth in the leaves of plants. The presence of these substances is more evident in the autumn when the green chlorophyll ceases to function.

the plant. All the cells respire. So long as they are alive, hydrogen-transfer reactions which involve the release of energy proceed in them. In some of these actions oxidation processes break down carbohydrate or fat to CO_2 and water. Little is actually known of the process of oxidation of sugar or protein at ordinary temperatures of plant environment, but we must conclude that respiration is inextricably connected with the process we call life, and is involved in the synthesis of more complex substances, by the absorption of ions, for the building up and maintenance of protoplasm in the living condition. As respiration is slowed or quickened, so are the general bio-chemical processes (i.e. the metabolism) of the plant.

The connection between transpiration and the mineral constituents of plants is even less apparent. Formerly it was assumed to be a simple process of diffusion through cell membranes. The wide differences in mineral content of different kinds of plants growing in the same soil environment is disconcerting, if it is accepted that the minerals are imbibed with the dilute soil solution by the roots and conveyed by the stem capillaries up to the leaves, where the excess water is transpired from the leaves. It is now found that transpiration has nothing to do with the absorption of mineral salts. Tobacco seedlings have been grown in the open where transpiration was normal, and a second series under calico to reduce transpiration. At the end of a given time the mineral content of the two sets was determined. The ash of the shade series was 11·2 per cent of the dry weight, compared with 9·2 per cent for those grown in the open. The latter plants had transpired 30 per cent more water than the shade series. (4)

The very small amounts of mineral matter in soil solutions consist of the partially broken-down products of decay of the previous season's plant life, still in the natural colloid form, and they play little part beyond retaining moisture. The addition of minerals in soluble form to the soil often proves toxic to the plants it is intended to benefit. As long ago as 1856–73, Lawes and Gilbert at Rothamsted found that magnesia salts had an ill effect on grasses, and the magnesia content of the ash of the grasses was much lower than that of the control plants grown on another plot without the magnesia salts. It is evident that the mineral content can only be correlated with the growth of the plants.

The chlorophyll pigment has the following formula for its constitution :

$$Chlorophyll\ a = C_{55}H_{72}O_5N_4Mg$$
$$,, \qquad b = C_{55}H_{70}O_6N_4Mg$$

Gortner, in his *Bio-Chemistry* (1929), gives the ash content of chlorophyll to be 4·5 per cent of pure magnesium oxide ; it must be borne in mind that results obtained by destructive methods can offer at best only an approximation of the composition of the living chlorophyll.

In plant culture-waters entirely devoid of magnesium salts plants will grow, the chloroplasts will be rich in the green magnesium-containing chlorophyll, and the photosynthetic processes will continue throughout the daylight hours. Whence is the magnesium obtained ? Chlorophyll itself—a first requisite for the plant's well-being—is also a product of the physiological processes. Photosynthesis is apparently " enzymatic " in character, one reaction being an energiser of still further reactions. That factor above all other factors must be present—the plant must be alive and growing. In two leaves of a plant, one of which is old, the other young, though both are equally radiated, only in the latter will there be a transformation of radiant energy into bio-chemical energy. (12) In the young and active leaf, the chlorophyll is not a kind of test tube in which carbohydrates are built up from carbon dioxide and water. The chlorophyll substance is used up in the process and new chlorophyll is formed. Chlorophyll itself is transient, ever-changing and reforming, and this is the peculiar characteristic of all living matter. Though chlorophyll has the formula $C_{55}H_{72}O_5N_4Mg$ in the morning and in the afternoon, it is not composed of the self-same atoms of carbon, hydrogen and the other elements all the while. Every atom constituting the chlorophyll molecules becomes, in succession, part of other molecular structures, and is engaged in other processes—an endless chain of renewal and replacement phenomena that is described by the word metabolism. The chemistry of growth and development involving long series of most complicated reactions, each one being set in motion by its predecessor and, in its turn, starting the next, is something outside the experience of the chemist. In the development and maintenance of a living organism, co-ordination is paramount ; the development of each part can be shown to be dependent on that of other parts. Yet all these processes are co-ordinated in such a manner that the organism develops as a whole ; and the total result is greater than the sum of the numerous processes. (3)

Only by breaking the " eventful " chain, by cutting down and drying the plant, for instance, will the process stop and the atomic structures become fixed in that state they happen to be in at the time the chain is broken. The " forces that fashion " can no longer act on the substance, life having gone out of it, and it falls to the category of inanimate matter. Then, not only mineral matter, as magnesium, silicon, potash, etc., remain stable as elements and permanent in quantity, but compound structures, like sugar, whose sweetness can only be realised by the tongue, and cannot be explained by a thousand words ; and alkaloids, with valuable curative properties, alike remain unchanged and so reliable in serving man's needs that we remember them as chemical substances and forget the " vital " processes that fabricated them.* When sugar is packeted and

* The appropriate time to " break the chain " of bio-chemical events depends upon what is required. If poppies are grown for the opium they yield, the whole operation of collecting must be completed in the few days during which the capsules are capable of yielding the drug ; this is while

labelled with some well-known manufacturer's name and trade-mark, it appears as much a manufacturer's product as, say, ironmongery, and as far removed from cultivated fields and the biography of plants. In the process of sugar extraction and refining, chemical agents are used in decolourising and bleaching, heat concentrates the extracted juice to the crystallising stage, carbon dioxide gas is blown through the liquid, but none of the operations disturb the sugar molecules or alter their structure. It is said to be inert.

On the other hand, it is the transience of matter in the living state that makes it so difficult to speak of its chemical constituents and their proportions, for the methods by which the results are obtained involve the destruction of cell substance, and at death, protoplasm, as such, ceases to exist. Also, there is evidence for the view that in many processes involved in metabolism, it is the electrical properties of atoms that are the important qualities, and not their chemical properties, as such. What have appeared to us, hitherto, to be chemical reactions involved in physiological processes, and which have been grouped together and named " biochemistry ", are not chemical in the commonly accepted sense at all, but something beyond chemistry, in the realm of the atomic and sub-atomic transforming phenomena, and are beyond the limits of chemical laws. In the sub-atomic region to which the new physics

the unripe capsules yield readily to pressure of the fingers, and the green colour has gone only a shade lighter. The milky juice or latex exudes from incisions made in the poppy-heads during the afternoon and evening of one day, and is scraped off the heads next morning on to leaves. These are pressed together to form a mass of the latex, which is then air-dried.

Chemically, opium is a gum-resin containing many alkaloids, with meconic and other acids, and the ordinary constituents of a plant juice. Its composition varies greatly : alkaloids-morphine 4-15 per cent, narcotine 2·5-9·0 per cent, with thebaine, codeine, narceine, and papaverine, of each 0·5 to 4·0 per cent, and cryptopine, landanine, pseudomorphine, codamine, meconine, protopine, lantho-pine, papaveramine, and others, all in small amount. They exist free or in combination with meconic, lactic, sulphuric, and phosphoric acids. There is also about 8 per cent of saccharine matter, and about 35 per cent of gum, resin, fat, albumen, etc., and various inorganic bases. In the air-dried latex the alkaloids are stable compounds. Poppy seed yields 35-41 per cent of a sweet oil containing no narcotic principle (vide *Chambers's Encyclopædia*).

introduces us, all concepts familiar to us in chemistry are useless and new ones have to be created.

The so-called photo-chemical phenomenon involved in carbohydrate synthesis is primarily photo-electric. To get an insight into it we must regard the action of light falling on the leaves of plants in a manner analogous to the action of light on the photo-electric cell used in television. Under normal conditions the atoms constituting the light-sensitive metal of the cell vibrate at their own peculiar frequency; but when electro-magnetic rays of light impinge on the surface of the metal the atoms vibrate much faster, and some electrons lose their hold upon the nuclear positive charge, are wrested from the atoms, and are ejected at high velocity by the electric component of the light energy, thereby providing a means (in the photo-electric cell) for changing light rays into electric impulses. This is known as the photo-electric effect.

In the interaction of radiation and matter as depicted in the operation of the photo-electric cell we recognise intuitively a something that lies at the foundation of world structure, and not merely a fortuitous sub-atomic disarrangement. The phenomenon is not fully understood, but it can be directed, and used, as in television, fluorescence phenomena, photography, thermo-couples and photo-chemistry.

The chloroplasts in the bio-plasm of the leaves of plants are more sensitive to light than the metals used in the photo-electric cell. The living cell is the seat of intricate inter-atomic reactions, displacing electrons and facilitating nuclear coalescence. Carbon dioxide is readily absorbed by the moist surfaces of the leaf cells, and diffuses toward the chloroplasts in the ionised state, that is, carrying an electric charge—a characteristic of solutions in water. To this the water itself contributes: each hydrogen atom gives an electron to the oxygen atom, so that the neighbourhood of the oxygen atom will be under the influence of its acquired negative charges, and the neighbourhood of the hydrogen atoms will be under that of their positive charges. In this ionised state the chemical bonds are loosened, and the whole of the plasma-substance is in a greater degree of " excited " instability.

Hydrogen ions play a predominating part in vital processes, including respiration. The far-reaching influence of this ion in such processes is associated with its unique constitution, consisting of one positively charged atomic nucleus, unaccompanied by any electron. The hydrogen ion, being the smallest particle, has the smallest atomic radius of any material structure, and through the absence of any negative electrons is able to approach other negatively charged particles more closely than any other positive ion. The hydrogen ion consequently possesses a high degree of absorbability in unstable atomic structures, when enhanced by hydrogenase-enzyme activation.

All radiant energy is due to atomic disturbances. The simplest interpretation of solar radiation is that light quanta, or photons as they are named, originate out of the transformation of hydrogen nuclei into helium, by a chain of reactions in which four protons disappear and one *alpha*-particle and two positive electrons are created, for each helium atom formed. The helium atom weighs less than the four protons, and the difference is emitted as radiant energy, or light quanta. The state of instability of atoms in the sun is due to the extreme degrees of heat and pressure. The state of instability present in protoplasm is due to the fact that it is living substance always in a process of change, of " becoming " something other than it is at every instant. Thus radiation is absorbed, each photon of energy tipping the delicate balance in the direction that meets the co-ordinated needs of the organism. The process of absorption of radiation by living protoplasm, transforming energy to matter, is a " reconstitution " of the material of the sun in the living substance.

Instead of the fantastic cosmological upheaval one or two thousand million years ago, by which the earth-fragment was expelled from the sun, we have here a picture of the material of the sun riding like fairies on sunbeams throughout that long vista of time ; and still riding, and building our home, the earth, by the labours of the creatures that dwell thereon.   *This is the biological concept of the origin of matter, and for the whole of nature.*

Within the dynamic ionised bioplasm incident light rays maintain a continuous process of creation of matter —a synthesis additional to and beyond that of sugar formation—to meet the plant's peculiar need.   The dominating factor in determining needs, is, of course, the species to which the plant belongs.   A study of several species of *Lemna* proved each to possess a definite chemical composition, within very narrow limits.   This in itself points to the quantitive chemical composition as being a specific characteristic, controlled by enzymes.

Of the plant's mineral requirements magnesium meets its first need for the production of chlorophyll.   It is built up of hydrogen nuclei and neutrons by a process the details of which are as yet unknown, but possibly by accretion to carbon nuclei.   It has already been noted that more carbon dioxide is absorbed by the leaves of a plant than is accounted for by the carbohydrate produced.

That carbon itself can be fabricated was proved by Mme. Curie Joliot in France when she bombarded boron with *alpha*-particles, a radio-active isotope of nitrogen was formed, which changed to the stable isotope of carbon.   That carbon—the organic atom—can be produced artificially is of far-reaching importance. *

But there is, according to Prof. W. F. Libby and his

* If a beryllium target is bombarded with *alpha*-particles, some of the particles are absorbed by beryllium nuclei, and form unstable compound nuclei ; these immediately split up into carbon nuclei by throwing off a neutron from each :

Beryllium	Helium Nucleus	Carbon (unstable)	Neutron
$_4Be^9$	$_2He^4$	$_6C^{12}$	$_1N^1$

If boron is bombarded with *alpha*-particles, radio-active nitrogen is formed, which is again transformed within fifteen minutes into carbon.

colleagues at the University of Chicago, more radio-active carbon in the bodies of human beings, animals and plants than physicists are likely to make in the laboratory. We carry in our bodies the end-result of those atomic disturbances—solar, stellar and interstellar—that solar and cosmic radiation signify. Prof. Libby was impressed by the fact that the neutrons released by cosmic radiation were themselves very efficient agents for nuclear transmutation. One probable change, in his opinion, derived from calculation based on the number of neutrons per unit area at sea-level, was the transmutation of nitrogen (mass 14) into radio-carbon (of nearly the same mass), according to the equation :

$$N^{14} + \text{neutron} \rightarrow C^{14} + \text{proton. (30)}$$

There are five isotopes of carbon having atomic weights of 10 to 14. Two of them, C^{12} and C^{13}, exist quite stable in nature, the proportion of C^{13} in natural carbon being about 1 in 100. The other isotopes are radio-active.

Cosmic radiation is at least a tenth of the total radiant energy arrested by the surface of the earth.

That the mineral magnesium of the chlorophyll in the cells of the growing plant is in the nascent, unstable and changeful state is shown by analyses of cereal plants at different stages of growth. The content per cent, of magnesium (expressed as oxide, MgO), in the mineral residue is approximately the same between narrow limits in both young and mature plants, whereas the silica content rises persistently throughout the life of the plant. Further, the minerals are built up in the different parts of the plant discriminatively on some directive principle. Wheat plants are good examples of this phenomena. The analyses given on page 34 were taken at intervals during the growing period.

The rate of increase of the silica slows down as the plant reaches maturity, when chlorophyll formation slows down toward the ripe stage. It is characteristic of the wheat plant in the ripe stage to have arranged its silica constituent so that there is 67 per cent in the straw,

PARTS PER 100 OF THE MINERAL ASH (7)

		MgO	SiO$_2$	K$_2$O	
Winter-wheat, Autumn sown	June 13th	2·3	49·3	31·9	Whole plant
	July 5th	2·7	60·5	22·4	
	Aug. 1st	2·3	67·8	16·5	
Do. the year following	May 2nd	1·7	42·6	36·2	Do.
	June 15th	2·0	62·5	21·0	
	July 29th	2·7	71·3	13·0	
Winter-wheat, another variety	May 2nd	1·2	41·8	33·1	Do.
	June 15th	2·2	58·5	23·0	
	July 29th	2·9	70·9	13·4	
Do. another variety	May 6th	1·8	28·7		Do.
	June 14th	2·1	42·1		
	Aug. 1st	2·1	69·1		

82 per cent in the husk, but none in the grain. The magnesia in the growing plant ranges from 1·8 per cent to 2·7 per cent, but in the ripe grain which is free of silica, there is 12·0 per cent of magnesia. It will be observed that while the active chlorophyll constituent remains at an approximately constant level, the silica increases in amount in proportion to its growth and development.*

It has been found, quite recently, by spectroscopic analysis, that in the growing point of wheat plants (and of barley and oats), there is a relatively substantial amount of that rare, radio-active and light-sensitive

* Neutrons fired on to either Al, Si and Ph form a radio-active isotope of aluminium $_{13}$Al^{28} in accord with the following schemes :

$$_{13}\text{Al}^{27} + _0\text{N}^1 \longrightarrow _{13}\text{Al}^{28}$$
$$_{14}\text{Si}^{28} + _0\text{N}^1 \longrightarrow _{13}\text{Al}^{28} + _2\text{H}^1$$
$$_{15}\text{Ph}^{31} + _0\text{N}^1 \longrightarrow _{13}\text{Al}^{28} + _2\text{He}^4$$

In every case the same final product is obtained, as the radio-active $_{13}$Al^{28} has a short life and breaks down to form silicon $_{14}$Si^{28} by emission of an electron.

The same *final* result is obtained if magnesium is fired on by *alpha*-particles :

$$_{12}\text{Mg}^{25} + _2\text{He}^4 \longrightarrow _{13}\text{Al}^{28} + _1\text{H}^1$$

Such transmutations seem to indicate that the structure of the atom of silicon is more stable than other elements that make up the mineral framework of many species of plant life.

mineral rubidium. It is found to be in relatively greater amount than potassium in the growing point, compared with the ratio in lower parts of the plant. (5) Further, in every straw, sodium increases in quantity from the growing point downwards progressively, while the potash decreases.* It is a strange coincidence that the metals commonly used for their light-sensitive properties should be found in the active growing point of cereal plants, and that each of them decrease as the plant ages.

The focus of all development is the growing point. Anyone who carefully dissects a suitable bud, peeling off the successively smaller leaves, may finally see with the naked eye or with a simple lens a pearly cone of semitransparent tissue at the tip of the stem. This is the growing point itself, which possesses theoretically unlimited formative powers. The cell-substance of the growing point is in a state of metastasis in the highest degree, and its powers of fabrication of new material are limited only by its specific needs for growth and well-being. There is an orderly process of synthesis and growth, at different rates as it changes its form. We can have no idea how it is controlled until we understand the mutual reactions and interactions by which the cytoplasm is synthesised. It is probable that in the growing point the unstable isotopes of potassium and rubidium are built up by the peculiar method of " condensing " radiation possessed by plants ; and, being radio-active, the atoms of potassium and rubidium slowly break down to elements more stable, by such means augmenting the organism's available energy for its life-processes. When we consider this aspect of photo-synthesis carefully we see that in the question of the nature of life, the validity

* In Bucharest, V. G. Bossie found that silica which amounts to nearly one half of the total ash of the wheat plant continued to increase more or less regularly until maturity.

Bossie analysed stems and roots of wheat plants : from the 26th day to the 36th day the aerial organs of a hundred wheat plants lost 0·187 gram of potash, whereas the roots only accumulated 0·001 gram in the same time. (*Lab. de Chimie Anal.* No. 4, 1934.)

of our physical and chemical conception of the Universe is involved.

The wheat plant so disposes of the products of its activities that nearly two-thirds of the mature crop is garnered in the form of straw, a little more than one-third as grain, while the husk or chaff amounts to about 7 per cent of the whole. The average mineral constituents of the straw are :

K_2O	Na_2O	CaO	MgO	Fe_2O_3	P_2O_5	SO_3	SiO_2
13·6	1·4	5·8	2·0	0·6	4·8	2·4	67·5%

The ash amounts to 5 per cent to 6 per cent of the dry straw. The mineral matter in the grain is :

K_2O	Na_2O	CaO	MgO	Fe_2O_3	P_2O_5
33·4	3·2	4·3	12·0	2·2	44·0%

The ash is less than 2 per cent of the dry grain. The mineral content of the husk is :

K_2O	NaO_2	CaO	MgO	Fe_2O_3	P_2O_5	SiO_2
9·1	1·8	1·9	1·3	0·4	4·3	81·2%

The ash is 10·7 per cent of the dry chaff. (7)

What co-ordination in the development of each part is here displayed. There is something of discrimination and the gift of choice in the way the mineral atoms are disposed by the plant in its stem, husk and seed to meet its own peculiar need.

Sometimes red clover is sown on the land with winter-wheat and they grow side by side, sharing the same rain and sunshine, with the same soil environment ; yet the clover weaves a different atomic pattern. The mineral elements found in red clover at the full flowering stage being :

K_2O	Na_2O	CaO	MgO	F_2O_3	P_2O_5	SO_3	SiO_2	Cl.
32·1	2·0	35·2	10·9	1·1	9·9	3·0	2·4	3·9 (7)

In comparing this analysis with those for the wheat plant it will be observed that there is a very wide difference in the silica content ; the clover is a complete organism with one-thirtieth part of what is required to support the wheat stem. On the other hand, the clover needs six times the amount of calcium that is found in the wheat plant. No more may it be said that the mineral constituents are obtained by imbibition of soil water by the roots ; or at least, not while it can be stated— " Practically nothing fundamental is known about the relation of the plant to the soil." (1)

The statement quoted was made under the sub-heading : " The Available Plant Food of Soils " after admitting that : " It is a rather sad fact that all that can be said under ' recent advances ' on this matter of soil analysis for mineral plant foods is that it is *hoped* to establish arbitrary laboratory methods the results of which can be accepted." (1) More recently it has been stated : " A soil solution is not a simple saturated system of minerals and water, but instead is an ever-changing system, *biologically controlled* by the activities of soil micro-organisms and of the higher plants growing in the soil." *

Between the plant and the innumerable other tenants of the soil—mostly microscopic—there is an intimate and complex reciprocal relationship that expresses itself in various ways, and promises biological results of the deepest interest, and which we must observe and not hinder. The uppermost layer of soil in field and forest consists of an intimate mixture of mineral grains— mostly inert and insoluble—with animal remains and excrementa, and casts of earthworms, insect chitinous casts, and the remains of previous vegetation in the form of leaves, fragments of twigs, fruits, seeds, etc., in varying stages of decomposition by the action of millions of bacteria, and of fungal mycelia in symbiotic associations with the plants.

The analyses given are of cultivated plants and were, for that reason, easier to obtain. Such plants are grown to be harvested and do not appear to have geological value. The ash of the ripe straw is only 5 per cent to 6 per cent, and if the straw was left to decay on the field,

* Hoagland, Prof. D. R. " Inorganic Nutrition of Plants " (Lectures at Harvard Univ., 1944, p. 7).

it would add practically nothing to the soil. " If plants are allowed to grow and cast their seeds, then to wither and die, and their seeds to spring up, in successive generations, these will in time become changed into the earth's substance and you will behold its increase ", as Leonardo da Vinci affirmed over four hundred years ago. So an example of plant life that does possess geological value ought to be included. Good examples possessing this qualification, are the microscopic aquatic plants known as Diatoms. These plants are found in all fresh and marine waters of the globe that are capable of sustaining life at all. Each plant, in addition to its vegetable cell structure, possesses an outer shield or frustule of nearly pure silica, instead of the usual cellulose that forms the cell-walls of other plants.

> Things are *organic*, we confess,
> If they contain some " quanta "
> Of carbon atoms—more or less
> But not if made of silica.

Yet the frustules of diatoms are as distinctly organic as other plant tissues that consist of cellulose. The living diatoms are clothed with a membrane of muco-siliceous matter of a pectin nature. Outside this siliceous membrane there is a thin transparent pellicle, in vital connection with the living substance within, while the internal substance consists of normal plant protoplasm with nucleus and chloroplasts. Some species of diatoms grow in colonies in masses of mucus, while others grow in long gelatinous filaments resembling *Confervæ*. Constituting the pasture of the " meadows of the ocean " they are directly or indirectly the food of all the larger forms of marine life. The silica, besides being in combination with gelatinous organic matter, is also associated with traces of alumina and iron, yet 65 per cent of the weight of the plant is silica. Carbo-hydrates amount to 22 per cent, protein 10 per cent, with oil-drops. There are thousands of different kinds and almost as many different shapes of shields. Some are round like a pill-box with lid and bottom pierced with ultra-microscopic holes spaced in geometric patterns ; others are triangular, or square, or boat-shaped, and all are marvellously ornamented, and yet within the dimensions of a point. The plants reproduce by division, the daughter cells quickly secrete for themselves new shields by their physiological processes operating with and directing photo-synthesis. Although so exceedingly minute, by their rapid multiplication and accumulation they fill up lakes and harbours, form islands in estuaries, and in other ways contribute to the geological history of the earth. Diatomaceous earth is a name given to clean deposits of the skeletal remains of these plants. Microscopic examination of clays, muds, and brick-earths reveal their organic nature, being packed with diatom shields and other plant remains, lying in a matrix of colloidal silica, the accumulation of mucus and gelatinous filaments, with which some kinds of diatoms surround themselves. Plate II (*a*) is a microlithograph of the Tripoli Deposit of Richmond, Virginia, executed by Prof. Ehrenberg. Tripoli has long been known under the name of Infusiorial Earth or Mountain Meal, and is used for polishing metals.

One of the most striking and characteristic diatomaceous deposits exists in North America. The River Columbia, in its course at Place-du-Camp, runs between two precipices 700 to 800 feet high composed of porcelain clay 500 feet thick, covered over by a layer of compact basalt. Dr. Bailey has shown that this apparently argillaceous deposit is entirely composed of fresh-water Infusoria. (48) Its perfect freedom from sand shows that it is not a drift but has been formed on the spot. With such an example before us, suggesting the creation of silica by vegetative processes, a biological concept for Geology becomes a rational proposition. Life, when properly understood, will be found to be a fundamental thing in the universe, being the manipulative agent that ties those knots in the ether, concentrating radiant energy into nuclei of the kind that is designated as matter.

Just as it is astonishing to find that the diatoms with

cell walls sheathed in silica exist in water which contains no trace of silicic acid, and develop normally under conditions which render the absorption of silica impossible : so the equally astonishing fact has to be acknowledged that the water-lily plant, if dried and burned, leaves an ash one-third of the weight of which consists of common salt, yet neither the water in which it grew, nor the mud into which it plunged its roots contained a trace of that compound. That sodium chloride would be found in seaweeds and sea-grasses, and such plants as thrive on salt-marshes, is not surprising, as a commonly held idea is that plants extract such mineral salts as are found in the tissue and plasma substance from the surrounding medium—water or soil. When we learn that such fresh-water plants as bulrushes, water-lilies and duckweed, and such marsh and bog plants as lady-ferns, sphagnum moss and rushes, and land plants like may-flowers, rye-grass, nettles, black oats and mangolds, all contain chlorine * with sodium and other bases, surprise gives place to wonder ; and doubt, as to whether we really understand the conditions of growth even of seaweeds. Among the sea-grasses and algæ, all of which grow in the same salt-water medium, the chlorine, sodium and potassium contents vary tremendously in the different species, and in no instance are mineral salts present in the same ratios in the plant as in surrounding water. Curious as it may seem, plants that grow on soil soaked with brine are in danger of suffering from drought. Such a soil is said to be physiologically dry, offering almost desert conditions to the plants dwelling on it.

Plants that grow on saline soil are known as halophytes, and they are nearly all succulent plants with thick fleshy leaves. Halophytes are a special form of

---

* Chlorine per cent of ash : Bulrush 22 ; Water-lily 23 ; Sphagnum Peat Moss 12 ; Fern (*Aspidium Felix*) 12·6 ; Nettle 10 ; Mangold 18 ; Black Oats—straw 9·5 ; Clover (red) 3·9.

Sulphur per cent (as trioxide SO_3) of Ash : Bulrush 3·3 ; Water-lily 1·6 ; Sphagnum Peat Moss 4·3 ; Fern (*Aspidium Felix*) 6·5 ; Nettle 6·1 ; Mangold 4·5 ; Black Oats—straw 3·7 ; Clover (red) 3·0.

xerophytes—that is, they are adapted to a precarious water-supply, in a similar manner to desert plants, such as cacti. Many halophytes grow equally on a non-saline soil and still retain their saliferous character; but the soil and climate must be dry, which denotes their xerophytic character.

As chlorine (and sulphur) are common constituents of plant and animal tissues there is no need to postulate "secret mineral springs" as a source of chloride (or sulphur) salts; all that is necessary is the endless succession of life-forms. It is clear that it is the soluble mineral remains of such organic life that is carried by the rivers to the sea. Erasmus Darwin in his *The Temple of Nature* adds a note to the effect that "as the salt of the sea has been gradually accumulating, being washed down into it from the recrements of animal and vegetable bodies; the sea must originally have been as fresh as river water".

*   *   *   *   *

Enquiry into the effects of photo-synthesis among the lower classes of the animal world is complicated by symbiotic phenomena. Symbiosis is a partnership or co-operation between an animal and a plant living together to their mutual well-being; it is the very opposite to parasitism. Embedded in the sarcode—the gelatinous material of the soft body of the animal—lie a number of minute plants known as *Zoochlorellæ* and *Zoo-Xanthellæ*. By digestion and respiration the animal provides waste nitrogenous matter and carbon dioxide for the plants, and in return the plants give up oxygen and some proportion of the carbohydrate synthesised to the animal in a process of mutual exchange. By this means the photo-synthesis of plants is carried forward into the realm of animal life. This phenomena of symbiosis is an important factor in evolution, there is much evidence for the hypothesis that organic evolution is co-evolutionary in character—the world of living things moves forward as a whole. These lower classes of animals produce in their life cycle a greater proportion of mineral matter relatively to their body-weight than do the higher types of animals. While the skeleton of a mammal is internal and provides a flexible support for the body, the skeleton of the mollusca and invertebrate animals is external, surrounding or covering and protecting their soft bodies. The shells of molluscs weigh about as much as their bodies. Being widely distributed, they are of great geological interest.

Every new development in the animal or vegetable kingdom changes the environment of creatures belonging to the other kingdom, to the advantage of those species that respond to the changed conditions and adapt themselves. In a manner still obscure, animal, vegetable and cosmic evolution proceeds along parallel paths, as if each were a co-ordinating part of the whole process of evolution.

Of the lowly class of Rhizopoda, let us examine two only of the creatures included in this group—*Foraminifera*, which produced the vast chalk deposits; and *Radiolaria*, so abundant in some siliceous rocks and chert deposits. The body of each consists of a single cell which carries out all the activities of a living animal. No difference can be detected in the jelly-like substance of their bodies, either by the microscope or chemical analysis; both live in the same seas, in a seemingly similar environment, feeding on such food as the water of the seas provides. Yet the Foraminifera build houses of many chambers for themselves with carbonate of lime, and the Radiolaria produce lattice-work protective helmets or floats of silica, with every possible variation of ornament. Both these creatures have this in common, they are coloured with the green and yellow pigments of chlorophyll and xanthophyll that react symbiotically with their physiological processes; but one expresses its individuality in silica, and the other in carbonate of lime, and in this respect show a similarity in their reactions to their environment as do the wheat plant and clover to theirs.

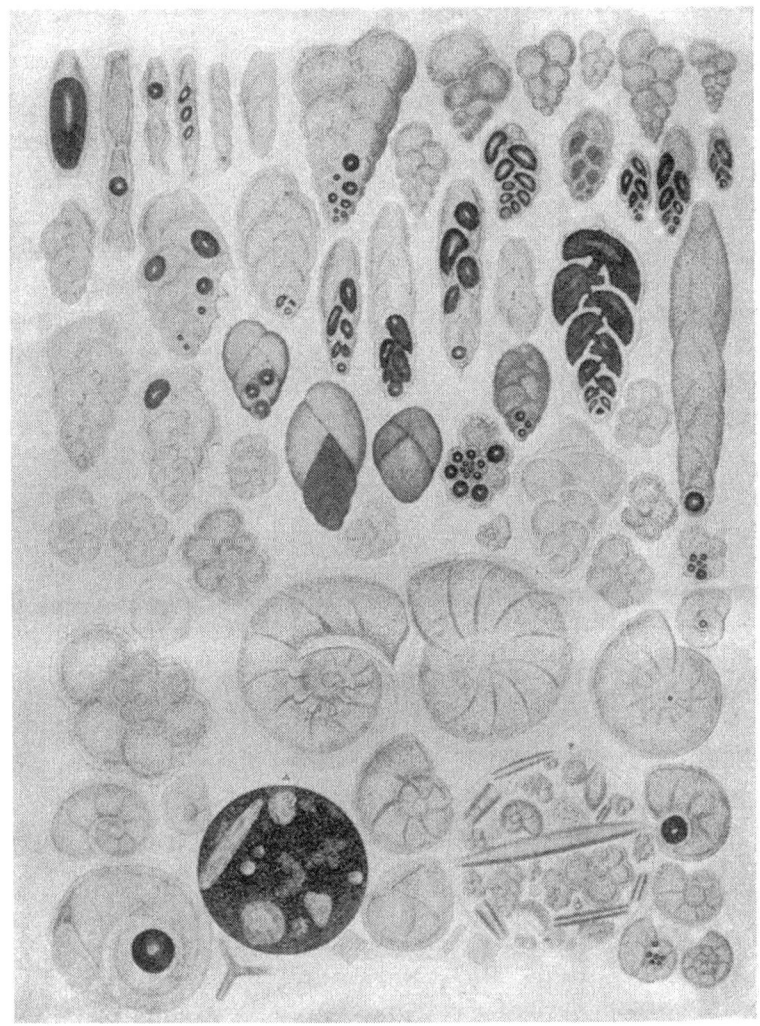

[By Ehrenberg

PLATE I.

Gravesend Chalk. A and B mass views magnified about 100. Separate
figures magnified about 320

[face p. 42

After death, the skeletal coverings sink to the floor of the ocean, where they form a dominant component of the mud or ooze. In the turning of the wheel of Time the ooze may become elevated above the waters and solidify to a hard rock. The cliffs of Dover were formed in this manner, inch by inch, through endless years, by the incessant death of Foraminifera. (See Plate I, micro-lithograph of chalk.)

If we avoid emphasising the organic origin of the siliceous rock for a moment, it is none the less plain to geologists that the world before geological history was populated almost solely by algal plants and sponges that secreted silica in their tissues ; so predominantly siliceous was the plant and animal life that have left traces of their existence, that it can be claimed that the dawn of geological history was siliceous in character.

And then an evolutionary change in the life-forms occurred that was so widespread over the globe that evidence of the change is found so far apart as Arctic America and Australia, Asia and the Argentine ; and everywhere attended by the same characteristic feature of being distinctly calcareous.

Limestone consists almost entirely of shells of molluscs, whole and broken, crowded together and cemented into a solid mass. Molluscs make their shell by the activity of a gland, named the " mantle ". This organ secretes the calcareous colloid, depositing it upon the edge of the shell to enlarge it. The colloid soon loses part of its moisture and hardens. The shell is thickened by the secretion of nacre—a lustrous and iridescent " mother-of-pearl "—on the inner surface of the shell. The nacre is intimately compounded with a substance of a horny nature.

The molluscs are supposed to abstract the carbonate of lime from the sea-water. This can be no longer accepted since : (a) The earliest molluscs had horny shells, " because there was no lime ". Now, the more calcareous debris is deposited on the ocean floor, and stored in limestone masses, coral reefs, chalk cliffs and downs, the more carbonate of lime remains in the ever-deepening and widening seas. There are 48 million square miles of ocean floor covered with a deep deposit of nearly pure calcareous ooze. (b) It is contrary to Nature's economy for animals to take in, for physiological activities, waste products of those processes : carbon dioxide (CO_2) is formed in the process of respiration. The mollusc can provide 44 per cent of the shell by its breathing exercises. (c) The carbon dioxide is dissolved in, or incorporated with, the blood pigment of the animal ; there it must react with water (H_2O) to form the carbonic acid (H_2CO_3) ; this is brought about by an enzyme. This acid—known as the anion of the calcium carbonate—seems to be of equal importance with the calcium (called the cation). To illustrate by an analogy—it is known that interference with normal anion production in the common fowl, due to pathological conditions, even though lime is present, leads to the laying of " shell-less " eggs.

The " ability " of the molluscs that built up the Carboniferous Limestone masses, to fabricate " mother-of-pearl " emulsion, had been acquired because such a skeletal support was an improvement on the chitinous skeletal covering in which their less skilled ancestors housed themselves. It is strange how tardy biologists and evolutionists are to recognise the development of this quality in molluscan life, and yet credit these creatures with marvellous " selective " power for carbonate of lime from among the group of minerals dissolved in sea-water. Selectivity, by those forces beloved by physicists—osmosis, surface-tension, permeability of cell-membrane and what not, leave so much unexplained. They should appreciate any attempt to analyse that residuum.

If the same agencies are still at work, as have been in progress during the long succession of geological events, we should expect to be able to trace recent growth of new geological formations comparable to those referred to above. Agassiz gives many illustrations of this growth, as for example : " The pteropods and heteropods are by far the most common pelagic forms of the mollusca, the dead tests of the former being literally heaped up in beds on the bottom of the sea in deep water. The dredge often came up completely choked with pteropod shells, showing what an important part they play in building up the deep-sea deposits. In former geological periods, when there were gastropods allied to pteropods of gigantic size, their effect in forming bottom accumulations must have been still greater." (8) Additional evidence of this phenomenon, gathered by Murray, Semper and Agassiz, shows that great submarine plateaux have gradually been built up in the Gulf of Mexico and the Caribbean by the decay of animal life upon banks which lie directly in the path of the Great Atlantic equatorial currents. Such currents carry along their course the pelagic life that serves as food for the animals, and they therefore congregate in the path of these currents also. Hence all the elements required for the accumulation of submerged land are present. When such banks rise to a certain level they become the foundation upon which reefs are formed. The coral polyp is another representative of this new kind of organism that invaded the algal world, which secretes a calcareous skeleton and prefers a colonial existence. The colony is a living sheet of organised animal matter having many mouths. The gelatinous tissue bearing the mouth and sensitive tentacles rises above the stony mass, and disappears on alarm into recesses in the coral. Fabre states that coral polyps have built one-fifth of the globe. In the Pacific and Indian Oceans there is still great activity, forming numerous coral islands. The Great Barrier Reef of Australia consists of a whole series of submarine mountain ranges, twelve hundred miles long, almost without interruption, and up to ninety miles across its widest part.

Corals do not build these great reefs unaided. The coral polyps grow over one another, but they leave unnumbered holes and cracks. There are two primarily biological forces working together, each helping to shape the reefs. Besides the animal builders, there are lime-secreting plants—*Nullipores*, and about a dozen species of *Lithothamnioneæ*—that play a great part in reef construction, by cementing coral, shells, calcareous sand and tiny skeletons into an impenetrable mass. The animal builders " house " the plants in their endoderm cells, the plants co-operating and reciprocating by supplying oxygen and carbohydrates to their hosts, the coral polyps. (20) Photo-synthesis gives an extraordinary efficiency to the reef-building organisms, causing an incredible rapidity of formation of limestone. The coral-building plants, besides secreting carbonate of lime in their tissues, also secrete magnesium carbonate up to about 18 per cent. This is in sharp contrast with the pure carbonate of lime deposited by the animal builders. As the amount of magnesium salts in sea-water is three times that of lime salts, there is an unexplainable preference for lime on the part of the polyps, on any theory of absorption. There could be no dissolved lime salts in the sea when corals were first evolved and began their task of island building, yet they grew in profusion and soon became predominant over a large part of the globe. Considering the solvent action of water, the wonder is not that there are so few fossils to tell the story of evolution and the birth of matter, but that there are so many that have withstood erosion through countless ages. The water dissolves projecting parts and reprecipitates crystals of carbonate of lime in the interstices of the rock, thus producing a more or less crystalline and compact texture to the coral-rock face.

Corals have only about 2 per cent of animal matter which when fresh gives a strong reaction for phosphoric acid.

The coral polyps continue to work upwards until they reach low-water mark, when their further upward progress is checked. Corallines, or limestone algæ, grow in abundance upon the dead fields of corals which have reached the surface, and add to their mass. Alex. Agassiz describes the Florida Reefs—these consist of numerous low islands separated by narrow channels from the mainland or united by mud flats that are bare at low-water. Mangrove trees grow luxuriantly, forming a belt of jungle. Around the dense vegetation additional sand and mud soon collect and gradually build up extensive islands covered with a thick tangle of mangrove and other plants. The sand consists of broken coral, and fragments of shells derived from the remains of the hosts of invertebrates that once lived on the active coral reef, and forms a coral breccia or the different grades of coral oölite. (8)

* * * * *

The part that radiation plays in the growth and development of the higher or vertebrate animal organism, including the human, is easier to measure. That there is a process of photo-physiology akin to photosynthesis is demonstrable. The building up of bone in the gelatinous meshwork of the skeletal frame is one of the most beautiful and mysterious processes, and can be watched by means of X-ray photography. The cure of rickets in infants and the young of domesticated animals by exposing their limbs to sunlight is part of the routine of medical practice. The radiation of sick protoplasm by ultra-violet rays has become a specialised branch of clinical treatment. The radiant energy interacts with the molecular energy of the bioplasm, and through the physiological processes builds up the anatomical form, the physiological processes seem to be directed by some invisible guide in the path that it follows. For a long time it was thought that rickets was due to malnutrition or wrong feeding, but it has been ably proved that rickets is not due to errors in diet, but is a " disease of darkness ". Dr. A. F. Hess of New York has shown that it is essentially a winter disease. He demonstrated that a few minutes daily exposure to ultra-violet radiation will double the amount of phosphorus in a baby's blood in a fortnight, and healthy new bone begins to be laid down, as can be seen by daily observance of its little legs and wrists, by the aid of X-rays. (28). The evidence is exact, abundant, indisputable. The skin is not merely an integument, it is an organ producing something of the nature of hormones under the influence of light.

The new knowledge, gained in recent years, on nutrition problems, particularly of the " vitamins ", possesses significance in this, that the previously accepted theory of values of different kinds of food, based as it was on purely chemical research, is found to be erroneous. The effect of irradiated milk on rickets plainly points to this conclusion. These " accessory food factors " as Hopkins (6) termed them, possess some vital principle that promotes the assimilation of food—the " anabolic fuel " of the metabolic furnace of the organism. Any wholesome food may be consumed and utilised by the organism in the presence of these " accessory food factors ", and the energy derived from the process of assimilation is used for its own peculiar requirements, in its protoplasmic activities, and in expressing its own characteristic in creative effort, in growth, and in reproduction of its kind. The structure of an organism does not depend on the kind of food taken in, but on some directive principle. (11) For example, consider reindeer feeding on the lichen *Cladonia rangifernia*, or reindeer-moss ; this lichen is rich in lichen starch, but is very low in minerals of all kinds, and especially low in the bone-forming minerals lime and phosphates. Its ash analysis is as follows : Ash 1·1 per cent of dry matter,

K_2O	Na_2O	CaO	MgO	Fe_2O_3	P_2O_5	SO_3	SiO_2	Al_2O_3
9·5	1·1	10·9	1·0	0·2	2·8	1·5	70·3	1·9%

The analysis is conspicuous for its extremely low phosphate content, and there is something mysterious in such feeding if rickets is not a common disease among reindeer. Yet rickets is practically unknown to these creatures. On the other hand, rickets is very common among children that are fed almost exclusively on white bread. The mineral analysis of wheat is as follows : Ash 1·95 per cent of dry matter,

K_2O	Na_2O	CaO	MgO	Fe_2O_3	P_2O_5
33·4	3·2	4·3	12·0	2·2	44·0%

From this it is plain that the value of food based on chemical analysis has been overrated, and the phenomenon of feeding not understood.

In the later Victorian years anæmia was very prevalent among girls and young women in England. They wore high-necked dresses, tightly laced corsets and very long skirts ; their leisure was spent in indoor pursuits—crochet, knitting, embroidery and dancing. The novels of the period sang the praises of their delicate white and almost transparent skins. Their languor gave cause for alarm, and medical aid was sought ; medicine containing iron was commonly prescribed. Still the anæmic condition persisted. Then someone more daring invented the low-necked blouse, open V-shape at front. This became very popular ; tight-lacing was soon discarded, the skirts made shorter, and outdoor pursuits such as tennis and walking began to be favoured.

No one saw the anæmia go, it crept away like a thief in the night. The next generation of young women added inches to their stature. It was not iron that the anæmic women needed ; they were oxygen-starved, robbed of the cheapest of all food factors—fresh air—by their indoor habits and tight-lacing.

Much has been claimed for the " mineral basis of life " and the need of minerals, including iodine, in the diet and in medicine. It had been used medically for a long time in the treatment of goitre. The discovery in 1895 that iodine is present in the thyroid gland stimulated the " mineral basis " school of enquiry. The writer admits that he was influenced by it, for some years. Needing medical aid, he was ordered cod-liver oil ; and to impress its value, the doctor said that cod-liver oil owed all its curative properties to the small amount of iodine which it contained. The writer's

D

health was soon restored.   But investigation and experiment convinced him that too much was claimed for the trace of iodine.   The oil contained vitamins unspoiled by heat, and was, therefore, a perfect and nourishing food.   To insist that only the iodine mattered is to make the part greater than the whole.   The thyroid-adrenal glandular system has had too many vital metabolic functions to carry through, in the eons of time that it has played its part in the series of transformations from fish to man, to be dependent solely on a moiety of iodine.   If chronic poverty prevails, or pathological conditions are present ; if the tin-opener and the frying-pan are the only utensils in the kitchen, and only white bread " fortified " by added chalk, in the pantry, then doses of iodine are almost a necessity ; but good food, fresh milk and vegetables, fruit, brown bread and butter, good water and fresh air are the gateway to health and well-being.

It begins to appear that it is possible to be overfed by having access to an abundance of food, yet undernourished by the lack of balance in the diet, lack of pure air, and the absence of the accessory food factors that promote the transfer of energy.*   The old idea that a man or animal *is* what he eats—that is, his life and channel of activity depend on something inherent to the food consumed—is deposed by the idea that food is utilised on some directive principle by the organism.   A dog devours bones to such an extent that it ought to

* Remarkable evidence on the deterioration of the native population of South Africa owing to defective nutrition has just been given to the Native Labour Committee by Dr. E. H. Cluver, deputy chief health officer of the Union.   We take the following extracts from the report in the *Cape Times* of February 4 :

" Dr. Cluver asserted that at least three-quarters of the total population were not in a position to obtain a sufficiency of protective foods, such as milk, meat, fruit and vegetables, to maintain themselves in reasonable health.

" Dr. Cluver said it had amazed him how the South African natives built up their bodies on a diet which had an *almost complete lack of lime*. He was not condemning South Africa's staple product.   The mealie was a very good food, but it was essential to have it supplemented by a minimum amount of protective foods." (Vide *Manchester Guardian*, Wkly Ed., Feb. 25, 1938.)

build up a skeletal frame like an elephant's if the kind of food were the deciding factor ; and the herbivorous elephant, grazing and browsing on the leaves and bark of trees, should possess rabbit-like qualities and structure, were it not contrary to its nature to be small and weak ; the elephant, living as befits its peculiar identity, secures the energy its food provides, and this, in conjunction with the physiological processes as a whole, promotes the synthesis of such constituents as are requisite to the formation of its massive framework, and the building of muscle of such great strength. The same bloodstream bathes and nourishes both bone and muscle, but only the bone is ossified. Bone must not be regarded as a dense dead structure, the essential nature of the process of bone formation is much more than the deposition of lime and phosphate material, which thereby converts gristle into mineral matter. Bone is a living structure in which blood vessels are developed in the cartilaginous substance. The building of the bones is a prolonged and elaborate process, and is carried out by free amœboid cells, akin to the white blood corpuscles. These little cells are carriers and layers of the bone mineral matter in the cartilaginous meshes of the prefiguring matrix. Seen on a motion film photographic record, it looks like team-work, chalky spicules of bone-in-the-making shoot across the screen, as if labourers were raising scaffolding. The scene suggests purposive behaviour by individual cells, and still more colonies of cells. The spicules are laid down at the right place, and at the right time, to bring about the formation of elements of the right size and shape, with reference to what they will ultimately become. The first signs of ossification appear in the osteoblastic layer investing the middle region of the cartilage, where the bone is formed of fine interwoven fibres cemented together. Eventually, the original cartilage round which ossification takes place is eroded, and the marrow cavity is formed.

Sometimes, the bones do not harden as they should, but under the influence of sunlight, or ultra-violet rays, an anti-rackitic substance is formed in the skin, and its absorption into the blood cures the rickets.   Irradiation has an effect on the calcium and phosphorus metabolism, and is believed to increase the iron and iodine in the blood ; it causes pigmentation in the skin ; and increases the number of white corpuscles in the blood.   In some way not yet understood, the metabolism of carbohydrates is intimately connected with the growth and development of bone.   It is found that diseases of bone —rickets, osteomalacia, etc.—are associated with disturbances in sugar assimilation.   Sugar passes through an intermediate stage of combination with phosphoric acid, which the cell breaks down, and in those cells concerned in the formation of bone (named osteoblasts by physiologists) the phosphate structure is fixed in the cartilaginous meshwork in a new combination.   The metabolism of phosphorus and its combinations with fats, sugar and proteins, and the measured effects of precise quanta of radiation on human subjects, offers a wide and interesting field for further investigation, and would lead to more exact knowledge of the synthesis of this element.

The *Bio-Chemical Journal*, 1924, gave information of Robinson's discovery of an enzyme in calcifying bone which transmutes phosphorus in combination with sugar into phosphorus in combination with calcium. [" An enzyme " :—it should be explained that the word enzyme means " in yeast ".   " Enzyme ", " zymase ", " diastase ", are names for substances produced by living organisms for the purpose of starting or assisting specific chemical reactions in the organism.]

Yeasts are minute living plants, composed of a bag or sac of the same composition as the substance of wood-cellulose ; the semi-fluid substance within the sac in a vegeto-animal protein.   All yeasts have the ability to change sugar into alcohol and carbon dioxide.   In the process, there is also formed some glycerine, succinic acid and acetaldehyde, which demonstrates that the fermentation and decomposition of sugar implies co-ordinated enzyme action, and reveals the essential complexity of the fermentation process. The metabolic activity of the yeast cell is the operating factor in the process, and includes reproduction of the yeast plants.

An average sample of yeast contains 1×10^{-7} per cent of uranium.*

Phosphates play an important part in the fermentation of sugar by yeast. The yeasts synthesise and hydrolyse hexose-phosphate; all the sugar passes through this intermediate stage in the process of formation of alcohol and carbon dioxide. The conception that sugars must pass through an intermediate combination with phosphoric acid before they are broken down by the cell is not restricted to fermentation by yeast. It would indicate a narrow outlook if bio-chemists did not realise that, on an evolutionary basis alone, so fundamental a process as the breaking down of hexose-phosphate must, in its essential steps, be the same in plants and animals. In plants, there is a bio-chemical relationship between the vitamin C and phosphatases. Muscle tissue possesses the same power to synthesise or to hydrolyse hexose-phosphate as is possessed by yeast.

It was from his researches on fermentation by living yeast cells that Pasteur conceived the idea of the microbial cause of epidemic disease in plants, animals and man.

*   *   *   *   *

Are we then ready to admit honestly that we do not know how the cell-contents of plants, or of animals including man are controlled? The most important single factor affecting plants is the soil water that bathes the plants' roots. Water is as indispensable as light and warmth; but the cause of the upward movement of water in the trunk or stem is unknown. The most divergent views are held, not one of which has proved capable of satisfactory demonstration. The relative importance of capillarity, the lifting power of evaporation, or of osmotic pressure, it is impossible to assess,

* *Annual Review of Bio-Chemistry*, Vol. XV.

with our present incomplete knowledge of the facts. The effect of transpiration upon the rate of movement would seem to bespeak great importance for this factor, though it is difficult to prove this. (31)

Of animal products, how little is known of the process involved in our daily milk supply. Our knowledge of the cytology of milk secretion is very meagre indeed. In milk synthesis it seems clear that the substances from which milk is made must be brought to the mammary gland by the circulating blood. Yet three constituents of milk are not present, as such, in the bloodstream; these are milk-fat, casein and lactose. The two last named are produced nowhere in nature except in the cells of the mammary gland.

Cows tied up in their byres during the winter months, unable to obtain appetising fresh green grass, need a stimulus if the milk yield is to be maintained. For this reason they receive, in addition to their hay ration, a quantity of sliced mangolds smeared with treacle. They seem to enjoy it too. Treacle is an uncrystallisable residue from the evaporation of the sweet juice of the sugar beet to obtain the sugar.

Do we know much more about our own body-fluid, the liquor sanguinis, with its red and white corpuscles? The colouring matter is pronounced to be a combination of iron and protoplasm known as hæmoglobin. The red corpuscles are the gas-carriers of the blood, conveying the oxygen we breathe in from the air to all parts of the body, via the lungs. A minute and scarcely measurable increase in the hydrogen ion concentration in the blood excites the respiratory centre to intense activity. Similar minute alterations in the concentration of water or sugar, or of sodium chloride, have a corresponding influence on the secretory action of the kidneys. The respiratory centre is so extremely sensitive to any increase or diminution of the partial pressure of carbon dioxide in the blood that a variation of 1·5 mm. of mercury will double the rate of breathing if + or up; and make the breathing painful and difficult if the variation is downward. The stimulus to which the respiratory centre responds is the difference in hydrogen ion concentration, the regulation of which is mainly the function of the kidneys.*

Reference has already been made to that pathological condition of bloodlessness known as anæmia, the treatment for which by the medical faculty has always been iron ; what may be the effect of a change of air ? There are not many accumulated assets resulting from attempts to climb Mt. Everest, and the relating of an advantage gained in such a pursuit may fittingly be narrated here.

Sir Francis Younghusband had observed : " As the different expeditions proceeded man found to his satisfaction that his organism did adapt itself to some extent to the higher regions. As he exercised his organism to the extreme limit, so did his capacities to some extent increase." But men must be given time to adjust themselves to the changed conditions at each new altitude. The 1921 expedition had shown that, when breathing became difficult, a resting period to become " acclimatised " enabled the body to adapt itself to the changed conditions by the production of extra blood corpuscles ; after a few days the climbers found themselves able to put up with conditions which they had previously found unsupportable. At five miles high the blood count had actually doubled ! † To establish the principle of the body's ability to double the number of blood corpuscles at high altitudes is an observation of the highest scientific value, and a tremendous addition to our meagre store of facts.

At such high altitudes climbers have a craving for sugar to " stoke-up " the blood. For this reason such sweet things as " bulls' eyes ", those striped lumps of candy that helped to sweeten our school-days, are included in the food-packs of the climbers.

* Haldane, J. S. " Mechanism, Life, and Personality ", d/d at Guy's Hosp., 1913.
† Snaith, Stanley. *At Grips with Everest*, 1937.

A man, at his birth, enters this world weighing about seven pounds. At maturity of growth and manhood he will weigh as much as twenty times seven pounds. During his life of activity he will often absorb a few units of energy from a hasty meal, and by sheer power of his will expend as much as ten times the energy absorbed, defying and reversing the second law of thermo-dynamics (a law of the mechanistic philosophy) in accomplishing things that are desirable to him.

In passing out of this life he will leave monuments to his creativeness in houses or bridges built, in roads and railways made, in tunnels bored through mountains, or in ships fabricated in steel and riveted, to sail the wide oceans ; or he will leave works of the creative arts ; all of which benefit the next generation—his works live on after he has ceased to be. Just as the bees gather more honey than they can need, and know not whether they will eat the honey they harvest ; so man sows, knowing not if he will reap, and " rejoices in his own work, for creativeness is his portion ".

In departing, he will leave in that twenty times seven pounds of lifeless flesh, enough iron to make a few rivets, enough lime in the skeletal bones to make a little mortar or to sweeten some sour soil, enough phosphates and nitrogen compounds, potash and sulphur to meet the needs of many millions of those micro-organisms that serve in the kitchen of plant communities fabricating nutriment for the roots while the plants distil their photo-synthetic ambrosia in the laboratories of the leaves, to feed the next generation of men. There seems nothing with which we can compare this strange world of organisms with its cycle and web of life, that has its foundations laid on carbohydrates, the outstanding product of photo-synthesis.

To summarise : By feeding, breathing, assimilating, the living organism receives energy. The technical term for all the combined processes of its activity is metabolism, which word means change, or exchange. That the essential thing exchanged is material is no longer tenable. Any atom of phosphorus (or other element) is as good as any other of its kind, providing it is part of a compound food particle, grown on fertile soil, and suitable " fuel " for the " metabolic " fire. With the energy derived, the organism absorbs light and heat rays, and by a photo-synthetic process builds the product into its cell-substance.

Examples have been given of a progressive series of analyses of wheat plants during growth; and of the different make-up in the tissues of clover grown on the same site as the wheat plants. It has been shown that the characteristic minerals found in an animal organism do not depend on the kind of food consumed, but on a sufficiency of sunlight and vitamins (which are themselves products of radiant energy, one such product—carotene—is the precursor of vitamin A). The fact that certain accessory food factors acting as vitamins in the animal organism also have important functions in plants, indicates a similarity in the metabolism of plant and animal cells.

Those interested in chemistry have pictured for us the analyst in his laboratory holding up a test-tube to the light, and seeing in it Nature's deepest secret. The sketch here given, in barest outline, of the living cell, is not that of a test-tube, but of a laboratory and synthesist in one, manipulating those cosmic forces, electrical and radiant, that are the part and counterpart of matter.

Whether in the analyst's test-tube or in the cells of the living organism, given the conditions necessary for the process, there will be an end-product separated out, bearing witness to what has occurred. The writer prefers, therefore, to call the principal witness for the biological origin of matter—the accumulated products of past life as revealed by Geological Sequence,

CHAPTER III

GEOLOGICAL SEQUENCE

I look at the geological record as a history of the world imperfectly kept, and written in a changing dialect, of this history we possess the last volume alone, relating only to two or three countries. Of this volume, only here and there a short chapter has been preserved, and of each page, only here and there a few lines. . . .

CHARLES DARWIN.
Origin of Species, Chap. 10.

A CENTURY and a half ago the grandfather of Charles Darwin having observed the gradual evolution of the young animal or plant from its egg or seed on the one hand, and that the greater part of the earth had been formed out of organic recrements, as the immense beds of limestone, chalk and marble from the shells of fish, and the extensive provinces of clay, ironstone and coal which had been produced by organic life; he was led to imagine that the world itself had likewise its infancy, and its gradual progress to maturity. (17) He declared that " the world itself might have been generated rather than created; that is, it might have been produced from very small beginnings, increasing by the activity of its inherent principles, rather than by a sudden evolution of the whole of the Almighty fiat ". (17) Unfortunately Erasmus Darwin could not satisfy his contemporaries with his physico-philosophical theories. There is nothing in Charles Lyell's *Principles of Geology* to indicate that Dr. Darwin had influenced geologists in laying down the principles on which this important branch of scientific enquiry was founded. One of the commonly accepted principles was that the substance of the earth had remained constant in quantity (except for slight increments of meteoric dust) throughout geological time; the first solidified crust that formed as the plutonic stage passed—the so-called primary rocks—had been broken down by destructive agents to form the secondary rocks ; and the secondary rocks, again, by breaking down, provided the material to form more recent strata. Lyell himself affirmed that " the extent and thickness of our sedimentary formations are the result and the measure of the denudation which the earth's crust has elsewhere undergone ". (18) Charles Darwin repeated and endorsed it in his *Origin of Species*, and it would be correct to say that all geologists have accepted it without question since. Yet some secondary formations are not chemically the same as the primary rocks. Those sedimentary formations that have been produced by erosion and denudation " elsewhere " are generally a commixture of multifarious rock substances possessing no marked chemical characters. On the other hand, the deposit laid down on the floor of the great Carboniferous Sea bears not only strongly marked chemical peculiarities that no earlier rocks possessed, but in such a high degree of purity that is astounding under any world conditions of deposition, and which defies explanation by any theory of breaking down and reforming of older rock masses. The sea in which the Carboniferous limestone was laid down covered almost the whole of the globe. It reached from the far north to Brazil and Australia. It covered the United States from east to west as well as the continent of Asia. In some areas the thickness of the limestone deposit exceeds 6,000 feet. The question as to the position of those other continents, by the denudation of which those vast masses were produced, has remained an unsolved problem to this day. This vast and widespread deposit of limestone presents a further problem in that it consists almost wholly of the aggregated remains of shells of molluscs, with corals and crinoids, crowded and crushed together, with the abundant tests of foraminifera, the plates and spines of sea-urchins, and the lace-like fronds of " sea-mats " and other polyzoa, instead of grains of rock torn from earlier-formed rock masses, and worn by the vicissitudes of time. The problem is still further complicated

by the fact that the primary rocks are distinctly
siliceous : the Carboniferous limestone marks the end
of a long siliceous era, and the opening of a calcareous
one ; it marks not only a physical and chemical change
on the floors of the seas at a particular period of the
earth's history, it reveals a great biological development
also. The active life in the seas at this period was
different. The Cambrian, the Silurian and the Lower
Devonian Systems form the first botanical epoch of
Professor Schimper, to which he gave the title the
" Period of Thalassophytes ", because seaweeds were
almost solely recognised in these formations. Lower
Cambrian fossils are of organisms that had only a
chitinous protective covering ; middle Cambrian fossils
possessed chitinous tests only slightly strengthened by
carbonate of lime. Prof. Daly has suggested that this
is " because the necessary supply of lime salts was not
available for them ". (41) If it was not then available,
from whence did the Carboniferous organisms obtain
it ? The Carboniferous was a period of corals, crinoids
and molluscs, the discarded skeletons, fossil stems, shells
and tests of which laid down the limestone masses.
" The tenants perish but their cells remain," wrote
Erasmus Darwin in *The Temple of Nature*. Are we not
justified in our deduction that the physical, the chemical
and the organic changes manifested in the limestone
deposits are simply several aspects of the phenomena of
Evolution ? " Fossiliferous strata contain, entombed
within them, the floras and faunas of bygone ages. We
ascend the stream of time, as in our study of the relations
of super-position we descend deeper and deeper through
the different strata, in which lies revealed before us a
past world of animal and vegetable life. . . . In some
cases these organised structures have been preserved
perfect in the minutest details of tissues, integument, and
articulated parts, whilst, in others, the animal has left
nothing but the remains of its undigested food in
coprolites—the fossilised excrement. In the lias of
Lyme Regis, the ink bag of the sepia, or cuttlefish, has

been so wonderfully preserved, that the material, which myriads of years ago might have served the animal to conceal itself from its enemies, still yields the colour with which its image may be drawn." (15) Excrements of the *ichthyosauri* have been found in such abundance in England that, to use Buckland's expression, " they lay like potatoes scattered in the ground ".

In short, the application of botanical and zoological evidence to determine the relative age of rocks, indicates one of the most glorious epochs of modern geognosy. Palæontological investigations have imparted a vivifying breath of grace and diversity to the science of the structure of the earth. (15)

Superimposed on the Carboniferous limestone, but separated therefrom by sandstone—the Millstone Grit—lie the coal measures. Coal is composed of compressed and " mineralised " vegetation. In Britain each layer of coal is usually underlain by a bed of clay, or of shale. There can be little doubt that each bed of clay is an old soil, while the coal lying upon it represents the matted growth of vegetation which that soil supported. Hence the association of a layer of clay and a coal seam furnishes distinct evidence of a terrestrial surface. Perhaps the nearest analogy in the present world to the vegetation of the Carboniferous period is supplied by the mangrove swamps of tropical coasts. There is no controversy among physicists and philosophers, geologists and mineralogists, plutonists and neptunists, catastrophists and uniformitarians, or the nebula school of Laplace about the origin of the coal measures. All of them agree that the coal beds are the result of solar energy reacting with the physiologic processes of primeval vegetation. They all agree that in this one case, the solar energy was not " thrown off" from the sun in one vulcanic convulsion, but arrived silently, like fairies riding on sunbeams through some millions of years, co-operating in creative synthesis with the plant tissues in co-ordinated growth activities. They may even agree that the deposition of matted vegetation was not due to an isolated interposition of supernatural forces, but the necessary result of the operation of natural law. The biological concept for the origin of matter demands that the coal beds and the limestone masses are both the result of common causes.

Scarcely any palæontological discovery is more striking than the fact that the forms of life change almost simultaneously throughout the world. Thus the European chalk formation can be recognised in many different regions, as in North and South America, South Africa and in India. The fact of the forms of life changing simultaneously in this large sense at distant parts of the world suggest that the phenomenon is dependent on general laws which govern the whole animal kingdom, and is not owing to mere changes of ocean currents or of climate. The greatest single factor tending to induce change has been the continual increase in the density of the respiratory medium. The long period known as archæan and early Cambrian was predominantly botanical, pouring oxygen into the earth's early attenuated atmosphere, an end-product of the physiological processes of the vegetation. Ever since, and down to the present day, there has always been a preponderance of the vegetable kingdom over the animal world.

The earliest known fishes, known as *Ganoid*, because of the enamelled scales and plates of bone in which they were encased, are found in deposits of the Silurian seas, and corals began laying down those limestone masses that form so large a part of the formations that are superimposed on the primary siliceous rocks.

Amphibia appear in the Carboniferous, and Reptiles in the Permian formations ; the first Mammals appear in the Upper Trias. These are historical facts of the first rank, briefly and roughly outlined. The inhabitants of the world that belong to the main line of evolutionary succession are, at each successive period, higher in the scale of organisation, and their structure has become more specialised.

By the death of successive generations of organisms that have peopled the earth and seas, vast accumulations of skeletal debris form successive platforms, and it is the nature of the organisms predominating in any area and on any platform that determines the chemical nature of the mineral deposit at that site. Further, there can be traced in the successive platforms evidence of the evolution of the creatures from primitive to more specialised types. From this fact alone palæontology seems irrational without a biological concept of geology, and Dr. Darwin's conception of the world's growth from small beginnings clearly demonstrates the correlation of biology, palæontology and geology.

Species belonging to different genera and classes have not, necessarily, changed at the same rate, or in the same degree. In the older Tertiary beds a few shells may be found of still living creatures in the midst of a multitude of forms that are extinct. The Silurian *Lingula* differs very little from the species of this genus living at the present time; whereas most of the other Silurian molluscs have changed greatly. Foraminifera have not progressed in organisation since the Laurentian epoch, but this does not present a valid argument against the idea of evolutionary progress. (37) There are some types of organisms, most microscopic, that remain, for countless ages, the " handmaids " of Creation, and the fittest to meet some need in Nature's social economy, that are organised for simple conditions of life, and what could be better adapted for this end than these lowly Protozoa ? Under very simple conditions of life a high organisation would be of no service—possibly would be of actual disservice as being of a more delicate nature, and more liable to injury. But to suppose that most of the many now-existing low forms have not advanced since their first dawn of life would be extremely rash. The Foraminifera had progressed wonderfully far for single-celled organisms, during a period before the Laurentian epoch, possibly as remote as the Silurian is now. (37)

The microscope has proved to be as necessary to the study of geology as it is to biology and medicine. Those organisms that we term microscopic occupy the largest space in geological deposits in consequence of their rapid propagation. Some idea of the minuteness of some of these " specks of life " may be obtained from a statement that the polishing-stone deposit of Bilin, 14 feet thick, consists entirely of the siliceous shells of *Gaillonellæ* ; a cubic inch of the stone contains forty-one thousand millions ! (27)

*   *   *   *   *

Coming to more recent formations, what is revealed ? In the district of London when excavations are made for drainage, building tunnels for tube-railways, or other purposes, there are sometimes found many feet below the level of the present streets mosaic pavements and founda- tions, together with earthen vessels, bronze implements, coins and other relics of Roman time. Now if we knew nothing from actual authentic history, of the existence of such people as the Romans, or of their former presence in England, these discoveries deep beneath the surface of modern London would prove that long before the present streets were made, the site was occupied by a civilised race that employed bronze and iron for the useful pur- poses of life, and had a metal coinage, and showed some artistic skill in its pottery and ornaments. But down beneath the alluvium wherein the Roman remains are embedded, lie gravels and sands from which rudely- fashioned implements of flint have been obtained. From this we further learn that, before the civilised metal-using people appeared, an earlier race had been there, which employed weapons and instruments of roughly chipped flint. (14)

In 1925, during excavating work in Leadenhall Street, E.C., for the new Lloyd's building, workmen found a skull of an " early human " being in the blue marly clay, 42 feet below the ground surface.

Further down still, all evidence of even the users of

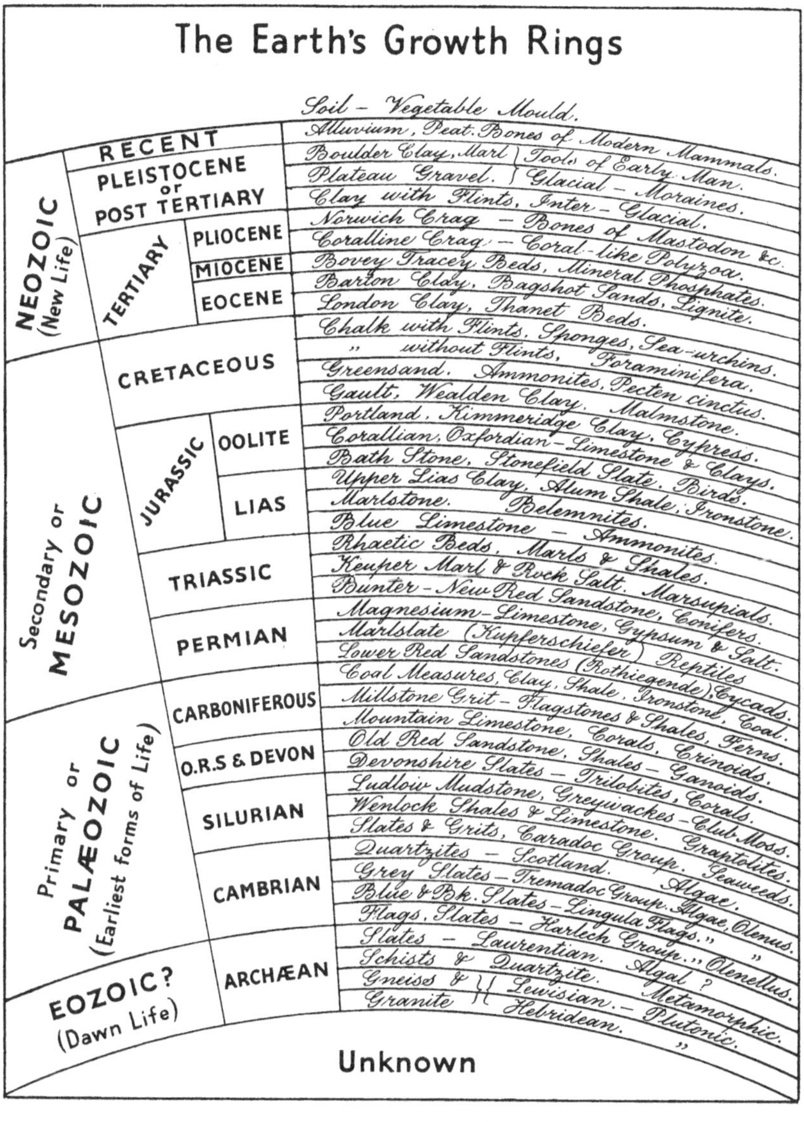

The Earth's Growth Rings

			Soil – Vegetable Mould.
NEOZOIC (New Life)	RECENT		Alluvium, Peat. Bones of Modern Mammals.
	PLEISTOCENE or POST TERTIARY		Boulder Clay, Marl. } Tools of Early Man.
			Plateau Gravel. } Glacial – Moraines.
			Clay with Flints, Inter – Glacial.
	TERTIARY	PLIOCENE	Norwich Crag – Bones of Mastodon &c.
			Coralline Crag – Coral-like Polyzoa.
		MIOCENE	Bovey Tracey Beds, Mineral Phosphates.
		EOCENE	Barton Clay, Bagshot Sands, Lignite.
			London Clay, Thanet Beds.
Secondary or MESOZOIC	CRETACEOUS		Chalk with Flints, Sponges, Sea-urchins.
			„ without Flints, Foraminifera.
			Greensand. Ammonites, Pecten cinctus.
			Gault, Wealden Clay. Malmstone.
	JURASSIC	OOLITE	Portland, Kimmeridge Clay, Cypress.
			Corallian, Oxfordian – Limestone & Clays.
			Bath Stone, Stonefield Slate, Birds.
		LIAS	Upper Lias Clay, Alum Shale, Ironstone.
			Marlstone. Belemnites.
			Blue Limestone – Ammonites.
	TRIASSIC		Rhaetic Beds, Marls & Shales.
			Keuper Marl & Rock Salt. Marsupials.
			Bunter – New Red Sandstone, Conifers.
	PERMIAN		Magnesium – Limestone, Gypsum & Salt.
			Marlslate (Kupferschiefer) Reptiles.
			Lower Red Sandstones (Rothiegende) Cycads.
Primary or PALÆOZOIC (Earliest forms of Life)	CARBONIFEROUS		Coal Measures, Clay, Shale, Ironstone, Coal.
			Millstone Grit – Flagstones & Shales, Ferns.
			Mountain Limestone, Corals, Crinoids.
	O.R.S & DEVON		Old Red Sandstone. Shales – Canoids.
			Devonshire Slates – Trilobites, Corals.
	SILURIAN		Ludlow Mudstone, Greywackes – Club Moss.
			Wenlock Shales & Limestone. Graptolites.
			Slates & Grits, Caradoc Group. Seaweeds.
	CAMBRIAN		Quartzites – Scotland. Algae.
			Grey Slates – Tremadoc Group. Algae. Olenus
			Blue & Bk. Slates – Lingula Flags. „ Olenellus.
			Flags, Slates – Harlech Group. „ „
EOZOIC? (Dawn Life)	ARCHÆAN		Slates – Laurentian. Algal?
			Schists & Quartzite. Metamorphic.
			Gneiss & } Lewisian – „
			Granite } Hebridean. Plutonic. „

Unknown

FIG. I.

A Biological Concept for the Origin of the Crust of the Earth

flints disappear, while yet lower still by some hundreds of feet, after passing through brick-earths and marls full of infusorial remains, through the London clay, and the Woolwich beds, the floor of the Cretaceous Sea is reached, in which the flints themselves are found embedded, lying in some semblance of order as if they had grown there.

Geologists find everywhere a sequence after this manner, one formation succeeding another, but with organic remains slightly different, yet related, the one to the other. Wherever a particular formation is found there are the same characteristic features revealed, so that particular horizons can be distinguished from other horizons in any part of the world. Further than this, stratigraphical succession is always in the same order—Silurian is always below the Carboniferous Limestone, Cretaceous always above the Jurassic, and so on. When it is a question of marine deposits the geologist in Australia and his colleagues in Russia and America all know whether it is " early-life ", " secondary life ", or " newer-life " deposits that he is examining; and such terms as " Carboniferous Limestone ", " Jurassic ", " Cretaceous ", are known in all parts of the world, and have become naturalised wherever geology is studied. (10)

The diagram of "growth rings" of the earth, Fig. 1, gives the main- and the sub-divisions of the geological record as commonly accepted, along with the typical formations belonging to each sub-division.

It will be observed that many of the formations belonging to the *Neozoic* or New-Life series, down to and including the Cretaceous, are to be found in the London district. The diagram is not drawn with any attempt at scale, either of thickness of any formation or of a sub-division, or of the time taken to produce any or all. The highest formation in the Carboniferous sub-division, shown by a line and marked Coal Measures, represents hundreds of seams of coal, each indicating a former platform of terrestrial vegetation and separated from the

E

succeeding seam by bands of clay, shale, sandstone, iron-stone, etc. The earth's crust consists of many thousands of layers which are comparable to the successive skins of a giant onion, except that the earth's layers differ in lithological characters and the dominant organic forms resting therein. Hugh Miller regarded these buried platforms as ranged like the sheets of a work in the course of printing, that, after being stamped by the pressman, are then placed over one another in a pile.

Fresh-water limestone is laid down in lakes by the same slow process of alternations of life and death among the creatures inhabiting the water. Scrope (9) provides a very good example : " Deep depressions in the surface of the granitic plateau in Auvergne, Central France, have been occupied from early Tertiary time by fresh-water lakes. These have since been drained, leaving proofs of their former existence in strata of clay, marl, limestone and sand frequently containing the remains of fresh-water molluscs. There are 600 to 700 feet of chalky marl ; this marl is thinly foliated, a character which arises from the innumerable thin shells or carapace valves of that minute animal known as *Cypris*, which is known to moult its integuments periodically, differing in this respect from the conchiferous molluscs. On other points flattened steps of charæ, or myriads of small paludinæ or other fresh-water shells, may be observed by the microscope to occasion this foliation, which is carried to such a degree that twenty or thirty laminæ may often be counted in the thickness of an inch.

" Interstratified with the marls are beds of fresh-water limestone. The most remarkable form is that known as ' Indusial ', from the cases or indusiæ of caddisworms. It is well known that certain varieties of the Phryganeæ, or caddis flies, are in the habit, when in their caterpillar stage, of clothing their bodies with a cylindrical case composed entirely of minute river shells of some single species—helices, mytili, planorbes or other—united by glutinous filaments, and disposed in some sort of order. These habitations are quitted when the insects' metamorphosis is completed, and on the banks of rivers or marshes frequented by them heaps of such empty cases may be observed. If we suppose them in this state to be cemented together by a solution of part of the shells, they will assume the appearance of this remarkable rock which we find in Auvergne, composing repeated strata of considerable bulk, alternating through several hundred feet with the more ordinary marls.

" More than a hundred minute paludina shells may be counted round a single tube, and ten or twelve tubes may be packed together irregularly in a cubic inch of the rock. When it is added that repeated strata of this kind 8 or 10 feet thick appear to have covered many square miles, some idea may be formed of the countless myriads of minute animals which must have formerly lived and died on the bottom or shores of this extensive lake !

" There are ossiferous deposits in this lake belonging to the same period as the indusial limestone, and include Carnivora, Insectivora, Rodentia and Ungulata, among which are rhinoceros, crocodiles, tortoises, fish, etc." (9)

Charæ, whose flattened stems are referred to as present in the marl, is the green alga, Stonewort (*Charæ fœtida*). It makes a large contribution to marl deposits, over 30 per cent of its dry weight is mineral matter, and of this 95 per cent is lime. Among other aquatic plants that add their quota to marl formation are pondweed (*Elodia canadensis*) with 17·5 per cent mineral ash, more than one-third of which is lime, and the discoid " duck-weed " that covers the surface of ponds—*Lemna trisulea* and *L. minor*, with 12·8 per cent of its dry weight as mineral residue, over one-fifth of this being lime. Both of these weeds contribute to the redness of the marl by their iron content, 10 per cent of their mineral ash being oxide of iron. On page 69 analyses of these and other plants are given.

The Lakes Winnipeg, Winnipegosis and others in Manitoba are remnants of an immense glacial lake that American geologists have named Lake Agassiz. It extended unto that part of the United States which forms the basin of the Red River. This ancient lake, covering over 100,000 square miles, became filled up largely by the remains of its own vegetation. It now forms a fertile valley over fifty miles wide.

The presence of iron in green vegetation is so universal that, when released by decomposition of the tissues, it stains the soil brown or red-brown according to the degree of its oxidation. " Iron ores are so common, and close enough to the surface to be mined easily, that an iron industry may be developed in any region." * West of Lake Superior, in Northern Minnesota, lies the Mesobi deposits, said to be the richest iron deposit in the world. The ore lies just below the surface of humus consisting of decayed pine needles. Wherever forests have stood for long periods of time, there, below the forest soil, will be found a layer of ironstone.

The " mineral phosphates " of Norwich, known as Norwich Crag, are bone-beds, the remains of the mastodon, elephant, rhinoceros, hippopotamus, etc., of the Tertiary Period. Other deposits are known in Belgium and Germany ; in Canada and South America. All the deposits are as distinctly organic as the coal measures or the peat-beds. The Tertiary period may very well be termed the Phosphate Age, for, with the passing of the huge mammals, the proportion of phosphates in the make-up of their successors grows less in the hard skeletal frame : the frame being smaller ; on the other hand, the brain is larger and its content of phosphorus higher.

There are large reserves of peat laid down in this Recent Period. There are 100 million acres in the U.S.A. Peat bogs cover one-fifth of Ireland—4,500 square miles. Water accumulates on the impervious boulder clay which was deposited on the limestone of the Central Plain in the Ice Age, forming marshes in which mosses and bog plants grow till they form a solid mass of matted roots. The stems of forest trees and the bones of extinct animals are found embedded in peat bogs.

* *Ency. Brit.*, 14th Ed.

ANALYSIS OF PLANTS OF MOOR, BOG, HEATH, MARSH, POND, SAND-DUNES, AND SALT MARSH

Parts in 100 of Ash

	% Ash	K_2O	Na_2O	CaO	MgO	Fe_2O_3	P_2O_5	SO_3	SiO_2	Cl	Al_2O_3	MnO
Heathers (11)	—	12·9	6·2	21·5	9·3	4·1	6·7	4·1	29·7	2·4	2·1	3·9
Cotton Grass *Eriph. Vagin.*		30·1	2·5	11·0	4·6	4·0	6·4	2·3	33·8	1·7		
Mosses (7) *Sphag. Hypnum*, etc.	2·6	13·5	8·4	11·5	4·9	11·8	4·5	5·0	28·8	5·0	3·4	
Bracken *Pteris Aquil.*	6·8	19·4	4·8	12·5	2·3	3·9	5·1	1·8	43·6	6·2	—	trace
Ferns (8) *Aspidium Osmunda*	18·7	35·6	4·0	12·3	6·9	1·6	8·2	3·5	20·3	7·9	—	trace
Horsetails *Equis arvense*	10·1	19·2	0·5	17·2	2·8	0·7	2·8	10·2	41·7	6·3		
Water-lily *Nymphaea alba*	12·8	18·5	25·9	24·3	3·4	0·3	3·3	1·6	0·6	23·1		
Duckweed *Lemna tres.*	17·5	18·3	4·1	21·9	6·6	9·6	11·3	7·9	16·0	5·5		
Pondweed *Elodia canadensis*	31·3	18·1	8·6	37·2	4·4	10·2	8·9	5·3	9·3	3·1		
Stonewort *Chara fœtida*	22·1	0·8	0·4	95·3	1·0	0·1	0·5	0·4	1·2	0·2		
Marram Grass *Ps. arenaria*	15·5	18·7	40·5	3·7	6·1	0·9	2·3	4·7	4·0	36·5	1·03	I
Aster *Tripolium*	22·7	6·4	47·9	5·0	1·7	1·3	2·9	4·3	0·8	42·1	0·3	0·67
Grass wrack *Zostera Marina*	14·9	5·2	11·4	30·1	1·6	1·9	5·4	3·1	25·4	15·1		
Fucus		13·0	22·9	13·6	8·1	0·8	3·1	21·5	2·1	17·9		

Analysis from *Aschen Analysen*, Emil Wolff. (7)

Peat is a veritable treasure-house to the archæologist, yielding stone and bronze tools, ancient pottery, foundations of dwellings, etc.

Turning now to soil, the present and last of the series of formations, and searching for a starting-point where we may be certain it is a new and not an already made one, we may choose a region lying between Java and Sumatra, in the Dutch East Indies ; there is a small volcanic island, Krakatua, in lat. 6° S., that was erupted in 1883. The forest-clad island was converted into a lifeless desert of laval ashes varying from 90 to 200 feet in thickness. Not a vestige of soil could be traced on the island.

Very soon, blue-green algæ covered the barren waste and began to prepare the way for lichens and such-like plants of simple type ; then mosses appeared. Three years later a few ferns had established themselves, to be followed soon by more ferns and plants of larger growth. In fifty years a new soil has been developed and the island reclothed with forest.

Lichens are usually the first real plants to cover bare stony surfaces, hiding the nakedness of the rock with their shaggy tufts, grey or greenish yellow, or orange-red in colour. Whether at Krakatua with its tropical heat and 60 inches rainfall per annum, or in the frigid zones, they play a principal part in Nature's economy. The " reindeer moss ", *Cladonia rangifernia*, is a lichen that grows in the polar regions and the tundra of Asia. It is a leathery and somewhat cartilaginous substance. Another lichen, a species of *Gyrophora*, is the " tripe de roche " of arctic voyagers, and the food of Canadian hunters. Capt. John Franklin in his *Journeys to the Polar Seas*, 1819–22, refers to it as a glutinous substance, which, with fish, made a tolerable supper ; " it was not of the most choice kind, yet good enough for hungry men ".

Mosses follow lichens, and by their more vigorous growth and mineral residue mingling in their bulky carbonaceous matter, carry the soil formation a stage further, their alumina residue making for plasticity, and higher moisture-retaining properties. Ferns follow if the environment is moist and shady, or sphagnum moss with reeds and sedges if swamp conditions prevail. In a relatively short time a soil is produced that will support seed-bearing plants and trees. Here is depicted the course of development, very scantily outlined, that has taken place in many parts of Northern Europe, particularly in Denmark, for centuries, gradually raising the surface above water-level, and it is still developing.

In the making of a soil there has been no need to call in those destructive agents, weathering of rocks, the action of frost, etc. Actually the part that disintegration plays in a living universe is negligible, it is the negation of growth, and is a very small factor in the sum of forces that produce the dynamic surface layer of the earth. It is acknowledged as a fact that the soil acts as a protection for the rock beneath.

A soil, when complete, is much more than plant residues and vegetable mould. It is a vast unexplored realm of living things, the home of a host of microscopic organisms and their skeletons. In life they co-operate with the larger vegetation in some mysterious way, and in death make even a larger mineral contribution because of their numbers, and more rapid increase, than the larger life-forms. Ehrenberg (27) analysed the earth from the roots of hundreds of different plants of many kinds, and found that all were profuse with life-forms. In many samples nearly the whole mass consisted of the remains of microscopic organisms.

Just as lakes are filled up, and harbours choked, so on land the continual additions of organic remains bury cities and obliterate the evidence of older civilisations. In recent years much excavation work has been carried out by archæologists, at Maiden Castle, Dorset, and St. Albans, in England; and in Palestine, Egypt and other places chiefly connected with Bible history. In Gaza, four palaces were unearthed, " superimposed, the one on the other ". In Antioch, Jericho, and in Samaria, there have been found, deeply buried, ancient buildings, with paintings and other works of art, pottery and implements. American archæologists excavating the site of Troy have discovered nine successive settlements. The first and most ancient settlement of Trojans is of immense antiquity. In the process of excavation, " islands " of undisturbed soil were left at intervals, which contained the complete stratification from the ninth and latest " city " to the first. These are as interesting to geologists as they are to archæologists ; evidence is presented that the sixth settlement— identified as the Troy of Homer—was destroyed by earthquake (about 1350 B.C.).

In Athens, excavators found a carved boundary stone of marble on which was inscribed " Boundary of the Agora ". This stone, marking the limit of the ancient market-place, was still standing *in situ* where it had been originally placed by its makers, at the meeting of two roads, one of which led directly up to the Acropolis. Above the level at which the stone stood were some 15 feet of earth, deposited by the accumulations of history. A section of the deposit that still stands behind the stone gives us an epitome of the history of Athens. The cross-section of antiquity thus revealed is very similar to geological cross-sections, in that all are authentic records of history.

Charles Darwin attributed this burying phenomena to the work of earthworms, which by bringing up their casts undermined the walls, causing them to sink. Notwithstanding 300 pages of argument in his *Vegetable Mould and Earthworms*, 1881, his hypothesis is unconvincing. Following are quotations from two of his illustrations.

Writing of the discovery of the old Roman City of Uriconium at Wroxeter in Shropshire, Darwin states : " The walls with such deep foundations (14 feet) cannot have subsided, . . . hence it is very difficult to account for their being now completely covered with earth." Nor was he any more assured in the fields within the old walls : " The mould was thickest on and close to the summit of the field. . . . The field slopes down at an angle of rather more than 2°, and I should have expected that the mould, from being washed down during heavy rain, would have been thicker in the lower than in the upper part, but this was not the case in two out of the three trenches here dug." On the summit the mould was 40 inches thick, on the slope near the top it was 26–28 inches, while at the lower part of the field the mould was only 15–17 inches in thickness.

The other illustration is a case of reclaimed land. " A piece of waste, swampy land was enclosed, drained, ploughed, harrowed and thickly covered in the year 1822 with burnt marl and cinders. It was sowed with grass seeds, and now supports a tolerably good but coarse pasture. Holes were dug in this field in 1837, or 15 years after its reclamation. The turf was half an inch thick, beneath which there was a layer of vegetable mould $2\frac{1}{2}$ inches in thickness, full of fragments of burnt marl, conspicuous from their red colour, and of cinders. . . . Beneath this layer, and at a depth of $4\frac{1}{2}$ inches from the surface, the original black, peaty, sandy soil with a few pebbles were encountered. . . . I am surprised that a greater quantity had not been brought up, for in the closely underlying black peaty soil there were many worms."

From the figures given it will be seen that the average annual increase of thickness was about $\frac{1}{5}$ inch. This and many other examples that Darwin related gives the general average increase in mould (which Darwin attributed to the work of earthworms) at 1·9 inches to 2·2 inches in ten years.

If the increase in mould is attributed to all the organisms living and dead, on and under the surface, as is the claim made in this thesis, Darwin's figures give us useful data for comparison with other data respecting denudation. Giekie estimated that the basin of the Mississippi suffered denudation by the river carrying silt to the sea equal to $\frac{1}{6000}$th part of a foot annually. It will at once be seen that the annual increments of vegetable mould are one hundred times greater than the decrements caused by the Mississippi.

The Ganges brings down much more solid matter than the American river. Ehrenberg found by microscopic analysis of the mud that one-quarter consisted of fresh-water organisms, besides such a large amount of organic lime that it may well be that half the whole mass of mud may be formed by microscopic organisms. In the suspended solids he found 129 different kinds of organic life ; one half of the forms were of vegetative character, the remainder being mobile. (27)

When a river reaches the sea much of its mud-load may be quickly deposited, forming mud-banks or deltas at the mouth or estuary ; this is partly because the current is checked and partly because the salt water causes coagulation of the fine mud particles. The River Thames is a good example. Millions of years ago the London clay was carried down by the river at full spate, due to the tropical rains of the early Tertiary period. Sheppey Island was laid down as a mud-bank. Its low cliffs are rich in fruits and seeds and other remains of terrestrial vegetation. (40) These are of sub-tropical affinities and include *Amygdalus*, *Gingko*, *Magnolia*, *Liquidamber*, etc. (38) Over two hundred species have been recorded—excluding diatoms—yet parts of Sheppey Island consist mainly of the triangular form of diatom, *Triceratium sarcophagus*, shown enormously enlarged (see Plate II (c)).

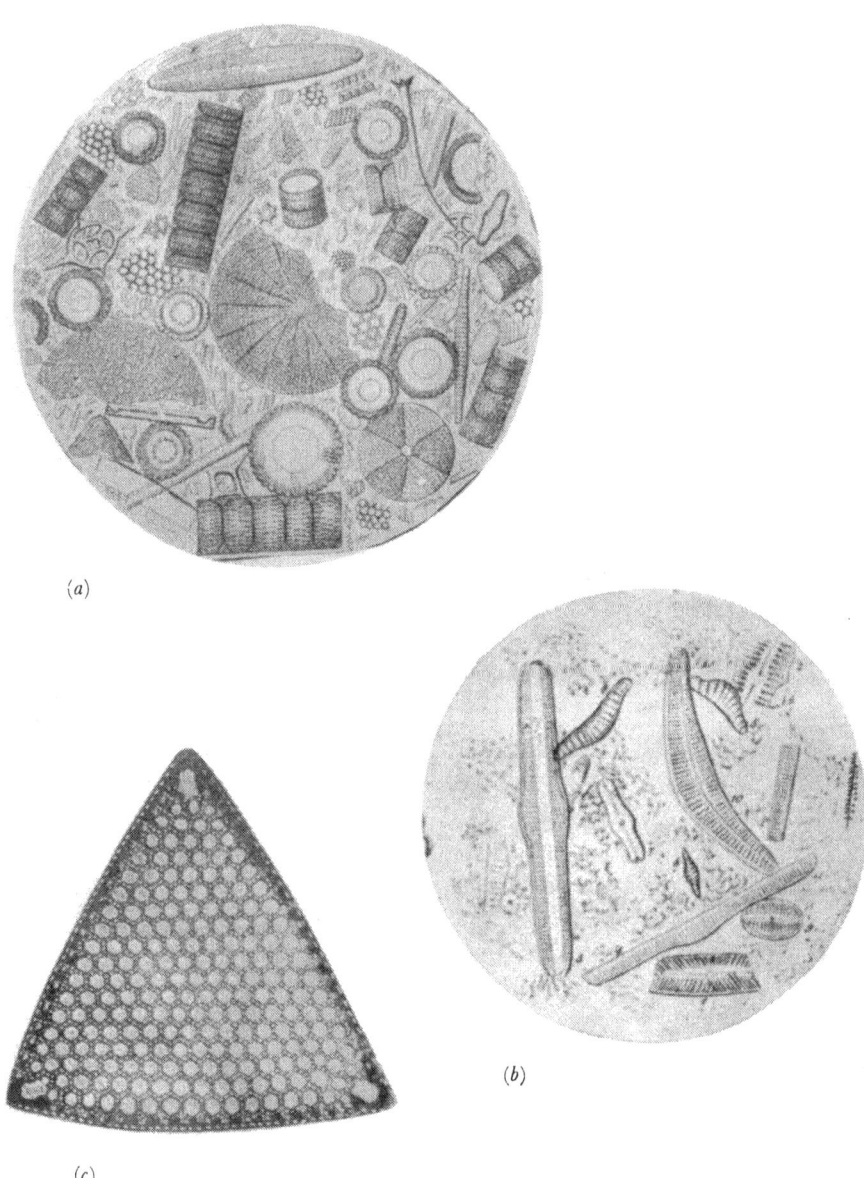

(a)

(b)

(c)

PLATE II.

(a) Tripoli Deposit. Richmond, Virginia. (Diatom Fossils) x300.

(b) Represents a sample taken from the upper layer of the Pumice Conglomerate of the " Hochsimmer ". The mass view is rich in larger infusoria, with round-celled fine pumice dust. *Eunotia gibba* and a *Cocconema* are prominent.

(c) *Triceratium Sarcophagus.* (Diatom.)

face p. 74

CHAPTER IV

GRANITOID ROCKS

Granite, in petrology, is the group name for a family of plutonic or deep-seated acidic igneous rocks characterized essentially by the presence of the minerals quartz, felspar and some ferro-magnesian mineral. Granites are wholly crystalline, and possess an irregular granular (granitoid) texture—whence the name from Latin *granum*, a grain.

Ency. Brit., 14th Ed.

WHILE the theory advanced in this thesis may be accepted for marl or limestone deposits, and for coral reef formation, granite and granitoid rocks seem to be a different proposition and in a separate category. We have to account for the so-called igneous rocks, that is, rock masses that were, somehow, formed by the agency of fire. These rocks are generally considered to be part of the primeval crust of the planet, which solidified from fusion. But " we are profoundly ignorant as to the conditions under which they arose ". (14)

" Granite to the geologist is classic ground : from its widespread limits, and its beautiful and compact texture, few rocks have been more anciently recognised. Granite has given rise, perhaps, to more discussion concerning its origin than any other formation. We generally see it constituting the fundamental rock, and, however formed, we know it is the deepest layer in the crust of this globe to which man has penetrated. The limit of man's knowledge in any subject possesses a high interest, which is perhaps increased by its close neighbourhood to the realms of imagination."

The above was entered in Charles Darwin's diary (36) on December 30th, 1834, after ascending one of some high hills on an island in the Chonos Archipelago, composed of abrupt masses of granite. There is no mention of igneous origin, but " however formed ", the

75

sight of them started thoughts on the limits of man's knowledge. In the 1850's Hugh Miller suggested a reform of geological nomenclature and classification : " Recent discoveries have unsettled almost every one of the characters and tests of the age of rocks. Neither hardness, crystalline structure or the absence of organic remains—hitherto described as the grand features of the primitive rocks—are now to be trusted." (26) The most abundant of the Archæan rocks is gneiss, passing on the one hand into granite, and on the other into micaceous and argillaceous schists. There is no difference as regards composition between gneiss and granite ; gneiss may be called a foliated granite. " Argillaceous schist also contains all the constituents of granite, potash not excepted." (15) The schists are interstratified with bands of hornblende, garnetiferous rocks, limestone, dolomite, serpentine, quartzite, graphite, hæmatite, etc. ; but the beds are thin and inconstant. The schists in some areas have passed through a process of granitisation. It is now demonstrable that fossiliferous formations, some of them older than the Cambrian strata, others belonging to the Oölitic period, and some even of Tertiary date, as in the Swiss Alps, have been converted into gneiss. The transformation has been effected by the influence of subterranean heat acting under great pressure, and aided by thermal water or steam and other gases permeating the porous rocks and giving rise to various chemical decompositions and new combinations, the whole of which action has been termed *plutonic* as expressing in one word all the modifying causes brought into play at great depths. (18)

" Based simply on analysis, evidence might be afforded of the igneous origin of the amphibolite rock " of Elephant Island, according to the " Quest " report of the *Shackleton-Rowett* Expedition to the South Atlantic, 1921–2 ; " but this view could not be sustained ", the report continued. " It is precisely the impure calcareous rocks which, losing their carbon dioxide content on metamorphism, give rise to mineral assemblages whose bulk analysis may correspond closely to that of an igneous rock."

No unquestionable relic of organic existence has been found in Archæan rocks, though the manner in which the limestone and graphite occur bears a slight resemblance to the limestone and coal in the fossil-bearing strata. There are traces of vegetable tissues, probably fucoidal, in the graphitic shales of New Brunswick. Occasionally, too, phosphatic nodules have been found which, under the microscope, are seen to have a fibrous texture.

On the continent of Europe, Archæan rocks have their greatest extension in Scandinavia, where they evidently belong to the same ancient land as that of which the Hebrides and Scottish Highlands are portions. In Canada they are estimated to occupy an area of more than two millions of square miles, stretching from the Arctic regions southwards to the great lakes. In Australia and New Zealand they cover large tracts of country. It thus appears that all over the world the oldest known rocks present a remarkable uniformity of character. (14)

Newer, perhaps, than Archæan gneiss, but older than the lowest Cambrian strata, are many stratified formations which are grouped under the common term pre-Cambrian. Included in the strata are beds of limestone and dolomite, often highly crystalline, and sometimes containing a considerable amount of graphite, and some of the shales are highly carbonaceous. The general character of these alternating carbonaceous, calcareous, and siliceous masses reminds the observer of rocks which have undoubtedly been formed by the agency of organic life. Moreover, there occur extensive deposits of iron-carbonate associated with chert, that again recall the results of the co-operation of plant and animal life. The large amount of carbon in some of the shales points likewise in the same direction. (14a)

In North America, where pre-Cambrian formations are developed extensively, they are estimated to attain a thickness of twelve miles, and have been regarded chronologically as quite equal to the whole of the rest of the geological record. (14a)

Immediately we leave the zone of the pre-Cambrian, the richness of the fauna distinguishes the younger rocks from the older. Trilobites swarmed in the early seas of lower Cambrian age, and in some cases sponges cover almost the whole surface of the beds. (19) The fossils of the Cambrian rocks obviously had a long series of ancestors ; it must therefore be regarded as certain that life had existed for a long series of ages before the Cambrian fauna appeared, in order that such well-advanced grades of organisation should then have been reached.

Species belonging to most of the stems or main divisions of the animal kingdom, excepting the vertebrates, suddenly appear in the lowest fossil-bearing rocks. "It cannot be doubted that all the Cambrian and Silurian trilobites are descended from some one crustacean, which must have lived long before the Cambrian age, and which probably differed greatly from any known animal. . . ." "It is indisputable that before the lowest Cambrian stratum was deposited, long periods elapsed, as long as, or probably far longer than the whole interval from the Cambrian age to the present day ; and that during these vast periods the world swarmed with living creatures." (37) In fact, any theory of evolution requires the existence of life so far back in the dim pre-geological past that life must have preceded the time at which the earth is supposed to have been cooling down from a very high temperature. Evolution and geology contravert the igneous hypothesis, and strongly support the organic origin of the earth. (19) The so-called igneous rocks are so closely related chemically to the Cambrian and later formations that their origin must be similar. They consist of the same kinds of minerals but these are differently arranged ; circumstances only are different. (19) Let any Cambrian, Silurian or later formation be subject to great heat, the dissolving power of super-heated water, bringing about recombinations and crystallisation, and all or any of those structural changes that are expressed in the word metamorphism, then they are indistinguishable from the fundamental rocks. The fundamental rocks contain a high percentage of silica, largely in the form of quartz, or in combination with magnesia, iron, potash, and in some that are metamorphosed argillaceous deposits alumina and some lime. The high silica-magnesia complex is a definite pointer to an origin that is mainly vegetable. What little limestone there is embedded in these fundamental masses is usually magnesian.

In the quartz of granite and metamorphosed rocks there are innumerable cavities in which the water is imprisoned that has exuded from the silica gel mass—the slimy, glairy siliceous sarcode—as it sets to a cheese-like mass, then horny, and finally hardens into stone. In a section of the rock, the cavities often resemble strings of bubbles, each minute, and sometimes only partly filled with liquid. In some cases, the liquid is water, usually containing saline matter ; in others, the saline solution is concentrated and crystals have formed ; in yet other cases, the liquid is a hydrocarbon, like the so-called mineral oil which is present in many deep-seated rocks ; in still further instances, the liquid is carbonic acid, which has been liquefied by great pressure.

Silica gels were studied by Graham over eighty years ago ; he observed that they always exuded liquid, while the gel stiffened and contracted. He named the exudation or sweating process *syneresis*. This part of Graham's work remained unnoticed until Ostwald, in 1915, again drew attention to it. The water content of a freshly made artificial gel is of the order of 90 per cent. The gel is comparatively rigid, and on standing exudes liquid and contracts until its volume is about 10 per cent of its original value. On further shrinkage the gel cracks and breaks up spontaneously. This syneresis or weeping takes place even under water. After the gel is set it cannot be brought back into the sol state—the process of setting being irreversible. The hard gel is still colloidal, but if heated to 300° C. crystallisation occurs. That is, on heating the gel becomes quartz-like.

Here we have an artificially prepared gel behaving in the same way as the natural gel from the plant tissues of the algæ. Both set hard as stone even under water. Quite a relatively small raising of temperature causes crystallisation. It is convincing evidence that the quartz-like rocks have an organic origin. The earth has not lacked internal heat to bring about the crystallisation that has been to many geologists the proof of their " igneous " source.

If two solutions are allowed to diffuse through a gel a system of periodically arranged precipitate layers occurs. When cut with a sharp knife, the gel shows a concentric banding, very similar to that of an agate. There can be no doubt that agates are formed in the " mineral realm ", by a similar " decomposition-coagulation " process. Sometimes agates are found that have been broken into pieces—brecciated, geologists call it ; and have been " mended " with the liquid colloidal silica out of which it was itself congealed.

Flint nodules consist of silica coloured with iron and other minerals. Often there are, entangled in the coagulated sarcode, spicules of the sponges that had produced the viscid sarcode by their activity when living.

Siliceous sponges are often found in great numbers in the Globigerina ooze off Santa Cruz. The whole mass of the mud is so thoroughly impregnated with spicules and with sponge sarcode as to be sticky and viscid. Alexander Agassiz writes in *Three Cruises of the Blake*— " The dredge must have plunged headlong into the ubiquitous sponge beds—the glairy mass like white of egg, with a multitude of spicules distributed like hair in mortar, throughout the mud. This amorphous substance which gives to the mud its viscidity is due to the presence of a mass of decomposed sarcode." It is evident that off Santa Cruz Nature is laying down at the present time the materials for a new chalk formation, in which flints may also grow if the conditions become again favourable.

Under conditions favourable to the process, the siliceous mucus loses its constitutional water and coagulates to a cheese-like mass which slowly hardens and crystallises to form the hard quartz. One of the most interesting entries in Edward Wilson's notebook of his journey with Scott to the South Pole, is that in which he recorded the finding of large quartzite pebbles at the base of some cliffs of light yellow sandstone, below the snow-tops of Mount Buckley. The pebbles were water-worn ; " in them were to be seen long stalks of vegetable origin with cellular markings in cast, and black crystalline coal fragments in the pits. Most of the bigger leaves are distinctly beech-like in shape, but smaller ". Such leaves and stalks, of the Tertiary Period, must have become embedded in the siliceous mucus, and the latter by syneresis had slowly hardened to quartzite. Crystalline quartz, having lost all marks of its organic origin, is classified by geologists as primeval rock.

Brecciated limestones and gravel conglomerate are frequently found cemented together by quartz as if it had flowed as a thin paste and filled the interstices of the mass and, exerting a styptic power as it sets hard, bound the fragments together. Even tender things like leaves that would break like burnt paper at the slightest touch have made nests for themselves, as if the quartz had been soft as wool ; every vein of the leaves being traced in the quartz. (46)

These enclosures in quartz may be compared to what has always seemed unsurpassed among Nature's wonders ; many examples of enclosures in amber have been found, of leaves, blossoms, insects, small crustaceans and fragments of vegetable matter that have been identified as belonging to yew, juniper, oak, poplar, beech, etc., as well as the amber-pine, *Pinites succifer.*

This amber-tree abounded in resin to a degree far surpassing that of any of the coniferous trees of our own time.

There were forests of these amber-pines in the south-western parts of what is now the bed of the Baltic Sea.

It seems strange that so hard a mineral as quartz can possibly be an end-product of the life cycle of such filamentous organisms as algæ, or the mucus of glass sponges, but we must not forget that the diamond is harder still, yet Sir Isaac Newton described it as " an unctuous substance coagulated ". The diamond is a form of carbon that burns in oxygen to carbonic acid, leaving but little ash, and this consists of silica with a trace of oxide of iron. The ash possesses a cellular structure which again indicates its vegetable origin— a gum probably.*

Darwin (36) described " streams of stones " lying in valleys in the Falkland Islands, consisting of great angular fragments of white quartz rock. The blocks are not water-worn, nor thrown together in irregular piles, but are spread out into level sheets. The width of these sheets varies from a few hundred feet to a mile, in crossing which it was necessary to jump from one pointed stone to another. " Never did any scene, like these ' streams of stones ', so forcibly convey to my mind the idea of a convulsion, yet the progress of knowledge will probably some day give a simple explanation of this phenomenon."

Can it possibly be that a mass of colloidal silica accumulated in some inland basin had flowed out like a stream of water glass and set hard, contracted, and broken up into these angular fragments?

Quartz always contains microscopic crystallites very much like sponge spicules, as well as the strings of minute bubbles or cavities, containing liquid.

The making of synthetic quartz from silica gels by

* The problem of the precise organic origin of the diamond cannot be said to have been completely solved.

heating in an autoclave with potassium carbonate at 350° C. is now carried on, on a large scale.

*   *   *   *   *

Iron in the oxidised state is a constituent of granitoid rocks ; in fact, it is present in all rocks, finely diffused throughout the mass. It would be a matter for surprise if it were not so, be it considered from any of the many viewpoints on world-making. On the planetesimal or meteorite theory of growth of the earth, and the iron-core theory,* there should be a greater preponderance of iron in the earth's crust than is actually found there. On either theory the oldest rocks should be richer in iron than those more recently deposited. On the meteorite theory, the primary rocks should contain iron, in some proportions, dependent on the iron content of meteorites. The most abundant element in meteorites is iron, but silicon and oxygen are present in small proportions only ; whereas in the earth's crust oxygen forms one-half of the rocks' mass and silicon another quarter. In the earth's crust the presence of iron is readily observed by its power to colour the mass yellow, red-brown, or brown of an even texture or degree dependent on the amount of iron present and the state of its oxidation. This even dissemination of iron in the oxidised state is what we should expect if the rocks are the remains of past life, since it is present in all green vegetation. Iron deposits are frequently associated with carbonaceous matter, as carbonate of iron and " blackband "—a clay iron ore containing 20 per cent to 25 per cent of coal. There are many deposits of bog-iron ore that appear in every case to belong to peat formations. Prof. Ehrenberg has shown that these bog ores consist of innumerable articulated threads of a yellow ochre colour, composed of silica, argillaceous matter and peroxide of iron.

---

* The iron-core theory has been abandoned. According to Chamberlin's " earth-knot " theory, the new-born earth had a diameter of about 5,510 miles. By the accretion of *planetesimal* dust, the earth grew to a diameter of 8,100 miles, " at the end of its growing period ". (Sir J. A. Thomson, *The Gospel of Evolution*, Chap. II.)

These threads are the cases of a microscopic body named *Gaillonella ferruginea*, associated with the siliceous plates of other fresh-water algæ. (22) Layers of this iron ore occurring in peat bogs in Scotland are dug for steel-making. Ferruginous mud is dredged periodically from some of the lakes of Sweden and used for the production of Swedish iron of high quality. After a sufficient interval, dredging operations are renewed, and a new deposit of ore is found.

" How strange ", remarked Hugh Miller (26), " if the steel axe of the woodman should have once formed part of an ancient forest ! if after being a component part in the stems and twigs of a thick forest of arboraceous plants—then as an iron carbonate accumulating at the bottom of a morass of the coal measures, to be later found as a brown ore underlying a seam of coal, which is dug out and smelted by the aid of the coal, and fashioned into axes for the woodman in the forests of to-day." It seems easy to conceive how, as generation after generation of the primitive vegetation withered and slowly disintegrated, the minute mineral particles that they had contained would be carried downwards through the lighter stratum of soil by percolating waters, till, reaching the impermeable platform of tenacious clays beneath, they would gradually accumulate and at length bind its upper layer into a ferruginous stony crust. Clay iron stone seems to have owed its accumulation to such a process as here depicted. Ironmongery and sugar are closely related.

West of Lake Superior is the most fabulously rich iron ore deposit in the world. Seventy-five miles north-east of Duluth, in Northern Minnesota, lie the Mesabi ore deposits. The ore lies in loose deposits just below the surface of humus consisting of decayed pine needles.

Alumina, the oxide of aluminium, is a constituent of granitic rocks. The chief of the known sources of this element is that class of simple plants, the ancient club-mosses—known as the Lycopodia. In Archæan times there must have been many primitive plants, ancestral to club-mosses, that built up this element in their tissues and plasma substance. Club-mosses yield 25 per cent up of alumina. Algæ, lichens and some peat-mosses also yield this element.

The chief source of aluminium metal is bauxite, a hydrated oxide of aluminium. A notable feature of most of the deposits of this mineral is their close association with old land surfaces. Those deposits in British and Dutch Guiana, of the Gold Coast and in India, lie on almost level surfaces, which have long remained undisturbed under tropical conditions. In the U.S.S.R., the metal is extracted from lignite ashes.

Potash, also found in these rocks, is present in all vegetation. Seaweed is a rich source of this mineral, though growing in a medium containing little potash and much sodium salts. Yet the potash content usually exceeds that of sodium. In some species, it greatly exceeds the sodium content, and amounts to one-third of the total mineral residue.

The alga *Nitella*, a very simple plant, lives in pond water containing only five parts of sodium per million, and no potassium that can be detected ; inside the cell of *Nitella* there may be as much as 200 parts of sodium and 2,000 parts of potassium per million.

In human and animal muscle cells are potassium but no sodium, while the lymph which bathes them contains more sodium than potassium.

Combined with each of the elements named as constituents of the primary rocks is oxygen, which represents about one-half of the whole. In combination with silicon it is the hard quartz ; with aluminium and silicon it is the hard granite, and the plastic clay ; with iron it gives the earth its rich brown tone and bricks their warm red colour ; with magnesium in the chlorophyll it *is* the green fields and hedges ; with hydrogen it is the sea, the river, and the refreshing rain. As part of the atmosphere it is the " breath of life ". Yet it is supposed that the young earth had no atmosphere.

The simple plant life, by absorbing carbonic acid,* and emitting oxygen, an excretory product—through age after age, has gradually created our atmosphere, containing free oxygen. " It seems at first surprising ", to Sir James Jeans (34), " that oxygen figures so largely in the earth's atmosphere, in view of its readiness to enter into chemical combinations with other substances. We know, however, that vegetation is continually discharging oxygen into the atmosphere, and it has often been suggested that the oxygen of the earth's atmosphere may be mainly or entirely of vegetable origin." Herbert Spencer (35) first suggested this nearly a century ago. In the present thesis it is postulated that the atmosphere is not mainly, but entirely, the accumulated product of the activity of organised life from time immeasurable.

*　　*　　*　　*　　*

The ancient rocks have been subject to many secondary forces, tending to change their appearance and even to alter somewhat their chemical character. Some of these changes have been referred to. Volcanic phenomena have played a part in some of these secondary transformations and therefore must be noticed. The belief that the seat of volcanic origin is in the " bowels of the earth " is not now generally held.

Laid down with the mineral matter is several times more weight of readily combustible material in a state of incipient and proceeding decomposition and spontaneous oxidation, aided by dampness and the presence and activity of bacteria. Many destructive fires have arisen in refuse heaps, slag heaps, stores of cotton, haystacks, etc., through this process of spontaneous oxidation, until the heat developed in the mass rises to ignition point and flames develop, which rapidly increases the rate of destructive oxidation. All the

---

* Those who have speculated on the subject, take the carbonic acid as being present when the world began, or assume that it has arisen from volcanic action and mineral springs. The breathing of simple animal life suffices, in this thesis.

elements for such reactions on a tremendous scale have been present even in the earliest deposits.

In the Mauvaises Terres (the Bad Lands) of Montana, U.S.A., the lignite deposits have been burning underground for centuries, and are still burning. Blue smoke issues from every fissure in the ground. There are abrupt sinks in the earth, sometimes miles in length, with steep banks often many feet in depth.

Even the outpourings of volcanoes, their fused lavas, their vast clouds of ashes that make day dark as night for many miles around and cover fertile plains with thick barren blankets and bury cities and their inhabitants; and their tremendous streams of mud, all bear witness on closer investigation of their ejecta, to their close association at the seats of their tumult with the remains of past life. The vitreous lavas, the light cellular pumice, the fine volcanic ash and the thick mud, or " Moya " as Humboldt named it, bear none of those peculiarities that we should expect to find in materials delivered out of the " womb of the earth ". Instead, they almost invariably bear the marks of their organic origin. Ehrenberg, reporting on some lava from the Isle of France, of a greyish white colour, found it to be in layers, with soft delicate plant fibres and roots permeating it throughout. (27)

During the voyage of the *Beagle* in 1836, a call was made at Ascension Island. Darwin describes its smooth conical hills of a bright red colour, with their summits generally truncated, rising separately out of a level surface of black rugged lava. One hill, formed of the older series of volcanic rocks, is remarkable from its slightly hollowed and circular summit having been filled up with many successive layers of ashes and fine scoriæ. These saucer-shaped layers crop out on the margin, forming rings of many different colours. " I brought away specimens of one of the tufaceous layers of a pinkish colour, and it is a most extraordinary fact, that Prof. Ehrenberg finds it almost wholly composed of matter which has been organised. He detects in it silicious shielded fresh-water Infusoria, and no less than twenty-five different kinds of the siliceous tissues of plants, chiefly of grasses. From the absence of carbonaceous matter Ehrenberg believes that these organic bodies have passed through the volcanic fire, and have been ejected in the state in which we now see them." (36)

English investigators are careful not to commit themselves. " A comparison of hand specimens of felsitic lavas of the hornstone type with altered slates and porcellanites of similar aspect shows that, without the microscope, we may meet with embarrassment." Even with the microscope, " when we consider the nature of the minute particles of which slates are composed, we need feel but little surprise at the occasional doubt which the microscopist experiences when dealing with rocks of this kind ". (51)

In the Basalt-Tuff of Cassel, Germany, a fine polishing or honestone is worked commercially, in which are leaf impressions and quite a score of different species of infusoria. (27) In the country surrounding the volcano " Hocksimmer " in the Eifel, fused Tuff deposits 180 feet thick lie, in which masses of Infusoria with their shields fritted by the heat are intimately mixed with the ejected scoriæ (see Plate II (b)). " It is so predominant in both older and more recent volcanic ejections, throwing out masses of life-forms, from many known and different formations or strata of the earth's crust, that a knowledge of the facts is no longer dependent on isolated opinion." (27) Ehrenberg made some experiments with diatom shields in a porcelain furnace, and found that they fritted and fused very like volcanic lava ; but if the shields were first freed from lime, and iron contamination, he could not entirely fuse them, even with an oxygen-gas flame.

Volcanic mud is even more surprising in its behaviour, and in its contents, which sometimes include countless fishes. The town of Pelileo, in Quito, South America, in 1797 was destroyed by an outflow of mud, which poured down on the plains from a height of 7,500 feet, carrying a Trap-Porphyry scoriæ of a green-grey colour, with glassy felspar, with it. " Many pieces are coloured black. These burn so well that the Indians of the interior still use the *Moya* to make their fires, and cook their food." (27) It burns like inferior peat, without flame. Prof. Ehrenberg recognised eight species of Infusoria, and eleven species of Phytolitharia. He pronounced the contents of the mud to be identifiable with the humus of high meadow land.

The Rare Earths and Minerals

In this thesis the aim has been to describe the source and origin of those elements that make up over 99 per cent of the earth's crust. It is suggested that it is incomplete if the remaining 1 per cent has no mention. What of gold and silver and precious stones?

Of some ninety elements forming part of the crust, two of them, oxygen and silicon, account for three-quarters of the whole. Six more, aluminium, iron, calcium, magnesium, sodium and potassium, account for nearly another quarter—23 per cent; carbon, hydrogen, phosphorus, sulphur and chlorine together amount to roughly 1 per cent. The other elements, nearly eighty all together, are so rare that they hardly constitute 1 per cent of the whole. Some sixty of them are so sparse that all of them only account for $\frac{1}{100}$ of 1 per cent, and may be regarded as not much more than chemical curiosities.

The analyses of organic residues have not been carried to that degree of accuracy needed to detect such mere traces of the rare elements as is indicated above, but sixty elements have been found. Besides the elements peculiar to coal as a fuel and the usual constituents of the ash of plant residues, there are also present titanium with smaller amounts of manganese, phosphorus, arsenic, barium, boron, cadmium and zinc, chlorine, gallium, germanium, iodine, lead, molybdenum, radium and thallium, silver, selenium, uranium and vanadium.

Spectrum analysis of flue dust show traces of lithium, rubidium and cæsium. (32)

Of the minerals or metallic ores, usually found in veins, pockets or layers, most of them are relatively heavy, and have suffered segregation from the lighter earths in the course of those secondary vicissitudes of weathering rock, and the carrying of detritus by rain and flood, while at or near the surface, and have been subjected to different but more varied and extreme processes of separation, when at great depths below the surface. " In the greater part of all subterranean action, temperatures are continually changing and therefore large masses of rock must be expanding and contracting, with infinite slowness, perhaps, and with infinite force. Changing temperatures must exert relatively changing forces of decomposition and recombination of the rock constituents, the whole mass of the rock in motion, either contracting itself and so gradually widening the cracks, or being compressed and thereby closing them, while water of every degree of heat and pressure congeals or drips, or throbs from pulse to pulse of foaming arteries, whose beating is felt through whole kingdoms, as your own pulses beat and surge." (46) The heated waters, heavily charged with mineral matter in solution (or in suspension) rise in the cracks and fissures and as the waters cool the minerals are deposited on the walls ; the duplication of the layers in mineral veins are evidence of this. The minerals may have been abstracted from the surrounding rocks by the percolating water, or they may have been brought by the subterranean stream from other strata within the crust, but many were, without doubt, originally widely diffused throughout the mass.

At the foot of the southern flank of the Harz Mountains lies a copper-bearing shale, under salt deposits, in a basin some 15 by 20 miles in extent. The shale is rich in fossils of the period—cephalopod life. The ore beds are made up of layers of bituminous copper shale. The ore contains 3–4 per cent of copper as sulphide, with 0·01–

0·02 per cent of silver. There are present also sulphides of lead and zinc, and traces of manganese, vanadium, nickel and chromium.

Copper ores in Northern Rhodesia and some of those in the Belgian Congo are bedded deposits, in form not unlike coal seams. The Northern Rhodesian deposit averages 4 per cent of copper.

When it is realised that copper is a constituent of the blood of the life forms that swarmed the seas of the period to which this deposit belongs, it is understandable that the copper is the accumulated debris of that life.

The blood of the crustacea—lobsters, crabs, etc., and of the mollusca, or shell-fish, contains a protein of a blue colour, known as *hemo-cyanin*. The copper content ranges from 0·33 per cent to 0·38 per cent. Wherever salt and gypsum deposits are found, copper ores may be located in the vicinity, as in the Harz, so it is in the Urals, in strata belonging to the same period, in a bituminous shale laid down under similar conditions. Instead of being a mere coincidence that they are found near each other, it is, again, " the necessary results of common causes ". The organic life that produced the calcareous shells from which, by chemical reactions and new combinations, the gypsum was derived, also produced the copper, while the soft body substance of the animals at their death became, by partial decomposition, the bitumen that bound the heterogeneous material together to form the black shale.

Copper is found in some species of plant life including spinach, potatoes and mushrooms. It is a constituent of the latex of the Indo-Chinese lacquer tree (*Rhus succedanea*), being present as a blue copper-protein oxidase to which the darkening and hardening of the latex is due. The oxidase portion contains 0·154 per cent of copper. The activity of the hardening process of lac is proportional to its oxidase content, that is, it is proportional to its copper content.

In this " Wonderland " of Nature, it should be no more surprising that a relatively few atoms are abnormal than that " freaks " are occasionally born among living creatures—heavy and misshapen. Atoms of higher mass may be pathological in origin, the result of sick conditions in organisms that produce them. A sick mollusc, for instance, may secrete strontium or barium in some proportion instead of pure lime, and exude it from its mantle or shell-forming gland, after the manner in which an irritated mollusc invests the cause of irritation (sand or other particle) in *nacre* or " mother of pearl ", by secreting a colloidal calcareous substance, film on film, which soon loses part of its moisture and hardens into a pearl. " In a recent spectroscopic analysis of molluscs, including the edible snail, strontium was found in fifty-one out of sixty-seven molluscs analysed." (5) Both these minerals, strontium and barium, are found associated with limestone ; barium is obtained from pockets in the limestone in Derbyshire, as barytes, and strontium is yielded by the limestone at Yate in Gloucestershire, as celestine. Both are sulphates of the mineral elements. Strontium is present in the ash of New Zealand coal.

Gold is found in rocks dating back to Silurian times, and was probably a constituent present in very low degree in algal forms of life. In a note on gold in plant ash (*Nature*, July 10, 1937) it is stated that *Equisetum palustre* is found to accumulate gold to the extent of 610 grains per ton of ash. It is also stated that the gold content is related to the silica content of the ash. Gold is the most widely diffused of all the rare metals. It is found in alluvial deposits and river sands in many parts of the globe ; the sands of California, the alluvium of the Yukon and Klondyke in Alaska, and the " diggings " of Victoria and South Australia have yielded much gold. Auriferous rocks are found in Mysore, in the Urals, in the United States, the Transvaal and other places. Some of the gold in alluvial deposits may have been derived from auriferous rocks by weathering. Metalliferous veins containing gold with other metallic ores, mainly iron pyrites, " streak " the rocks ; or a conglomerate of quartz pebbles cemented together by silica and iron oxide forms the rock mass, the gold being diffused chiefly in the cement. The " Rand " in South Africa is of this nature. In Transylvania, hydrothermal veins in limestone are worked, the sides of the fissures are thinly coated with gold in a felted mat of rod-like crystals, each infinitely small, but woven into tiny triangles, that glitter like leaves covered with frost. The gold usually contains some silver.

Gold combined with chlorine is soluble in water. The waters of the seas hold gold in solution in this form equal to one grain of gold to the ton of water. Gold readily enters the colloidal state ; " dissolved " in glass it gives the glass a deep red colour and is used in stained-glass window work.

Gold is found also combined with tellurium, the heaviest element of the sulphur group. Tellurium is only found in Nature associated with other elements of high atomic weights, as gold (at. wt. 197) and bismuth (at. wt. 208).

Selenium, another relative of sulphur, and found in very small proportions in native sulphur, is surely the " looking-glass " of this " Wonderland " of minerals, for it is sensitive to light. It is used in television—the electrical transmisssion of pictures. The electrical conductivity of selenium is increased by light waves falling on it, the conductivity being increased proportionally with the intensity of the light.

Silver is present in many lead ores, notably with galena, forming a double sulphide of silver and lead. It is also found combined with antimony (stibium) and with arsenic, as double sulphides respectively. Antimony, arsenic and bismuth belong to the phosphorus group of elements. Silver is a constituent of a few plant tissues, and many fungi. The edible mushroom *Agarious campestris* contains 0·02 per cent (of its dry weight) of silver, and of copper.

Arsenic is found in the soil in parts of Cornwall to the extent of 125 parts in a million. In the trial of Mrs.

Hearn (June 1931) for murder, a witness for the prosecu-
tion admitted that the arsenical content of the soil of
Lewannick churchyard was 240 times the amount found
in some of the organs of the victim whose body had been
exhumed there.   It was also admitted by expert analysts
that a little arsenic is often found with the phosphorus
in bone and tissue.   There is a growing body of evidence
that minute traces of arsenic are of widespread occurrence in tissues, and that the element is of physiological
importance.   It is always present in coal ash.

Lead and zinc resemble iron in that they are associated
with sedimentary formations.   Lead and zinc are found
in veins in limestone, often with calc-spar—a crystallised
(from water) form of carbonate of lime.   The concentration of lead and of zinc in certain seaweeds is over ten
thousand times that of these metals in sea-water.   The
constant decay of the vegetation has deposited these
minerals in the ooze, becoming part of the limestone
strata, and later, by those secondary vicissitudes to which
most rocks are subjected, the minerals have been
segregated.   Veins are worked in the Carboniferous
limestone in Derbyshire, and these minerals are mined
in the Mississippi Valley which at one time formed part
of the Cretaceous Sea.   Lead has been found in molluscs
and crustacea, as well as in sea-water.   Lead chloride
is soluble in hot water, and is precipitated by cooling.
An ore of lead (Matlockite) is combined with chlorine ;
hot subterranean waters would be able to cause the
segregation of Matlockite.

Lead appears to be the limit in stability among
Nature's elements.   The elements of higher atomic
weight than lead tend to break down spontaneously, lead
being formed in the process ; from thorium one form of
lead is produced, from radium another lead, from
actinium yet another lead, but the three leads, though
identical chemically, are yet different from one another,
and from common lead.   The " common " lead being,
of course, the organic product found in sea-weed debris.

The " White rabbits " among all this wonderland of

minerals are the radio-active elements—radium and uranium, cerium and thorium, and others. In some ores the active element appears to be associated with bismuth, and in others with barium. At this end of the atomic scale instability reigns like to that of living protoplasm—the greatest characteristic of which is changefulness. There is here among inert atoms of matter, something of a vital quality, a display of volition ; atoms possessing more of the attributes of will and energy than of matter. Uranium is being extracted from shale oil deposits in Sweden, the shale is the remains of small marine organisms. Russian scientists claim to have extracted radium and thorium from wheat and other growing plants. Prof. W. Vernadsky confirms this. A trace of uranium is found in yeast cells.

Manganese is found associated with iron widely spread in the plant kingdom and in many rocks. In land that was once bog-land, it is found in conditions similar to those of bog-iron ore deposition, and clay-iron stone. The water from moorland is often discoloured by manganese. In water conduits, microscopic growths containing manganese form fibrous gelatinous masses. *Sphagnum palustre* has 3·2 per cent of manganese oxide in its ash, and water-caltrop, *Trapa natans*, as much as 7 per cent. (7)

The blood pigment of the tropical mussel, *Pinna squamosa*, contains up to 0·81 per cent of manganese, and is brown in colour.

The metal vanadium is also a blood constituent, being present in the fluid of the marine creature called the sea-squirt. It is truly astonishing if these humble creatures can select from sea-water, what analysts fail altogether to find with all their modern methods and apparatus of precision. The metal occurs in a few rare mineral ores, associated with copper, and lead or iron.

Vanadium is commonly present in the ashes of coal, bitumen, asphalt and petroleum in varying amounts. The oxide is recovered from the soot that collects in the boiler flues of ships burning Mexican oil fuel.

Some Peruvian asphalts yield up to 30 per cent vanadic oxide from their ash, and a Peruvian ore occurs in black carboniferous rocks resembling coal, and yielding 10 to 13 per cent of vanadium.

Chromium is found associated with iron as chrome-iron ore, and with lead as the yellow lead chromate.

Cobalt and nickel are found as sulphides and arsenides. A rich cobalt-nickel deposit lies at Sudbury, Canada. A series of rocks about 2,000 metres thick rest on gneiss and granite, filling a basin 58 kilometres by 26. The upper part is acid granitic rock, the lower part being heavy basic diorite. At its base lie the nickel ore, with iron and copper, and traces of silver and platinum, a little gold, iridium and osmium, and smaller quantities of rhodium and palladium. " It is assumed that the nickel-bearing rocks at Sudbury were a molten mixture, which separated out according to the specific gravity of its components." (10) More probably they were water-separated, as has already been described. " Nickel is found in some plants, especially the tropical spices and alkaloids " (5), and in tea.

" Fluorspar ", a fluoride of calcium, is mined as " Blue John " in the limestone of Derbyshire, and a fluoride of lime and phosphate is found in some meta-morphosed rocks. There are traces of fluorine in bones and teeth. Buried bones accumulate fluorine in the metamorphosis and mineralising process. This provides a means of measuring the length of time the bones have been lying in the deposits.

Those crystallised earths known as precious stones are products of secondary transformations. Rubies and sapphires are indurated clay, crystallised, red in colour and blue and clear respectively. Emerald and beryl are double silicates of berylla, BeO and alumina ; the green stones are called emeralds, beryls are blue or yellowish. Amethyst consists of crystallised quartz of a blue-violet colour.

Perhaps the most curious thing in this world of wonderfully strange things is the diamond—a vegetable gum transformed.* If mucilaginous exudations of plant life become by transformation—by whatever forces it is brought about—the hardest thing in the world, it is surely the final satisfying proof of the soundness of the theory that the activities of organic life have created the hard rock masses of the earth.

In a world where all the rock masses are in the oxidised state it is in a high degree wonderful and strange that carbonaceous matter should escape either decomposition or oxidation anywhere. Yet it has been preserved as soft coal in Derbyshire, and as hard anthracite in South Wales; in Cumberland it is found simplified chemically, and changed to graphite; in India, South Africa and other regions it has become purified, and crystallised in the diamond state.

* See footnote, p. 82.

CHAPTER V

AIR AND WATER

> The laboratory of Nature is very different from that of experimental research. . . . With all our chemical knowledge we can never hope to rival in our laboratories the results that Nature has achieved. Nor can we hope to achieve through biological experiments in the laboratory what her silent processes have amounted to through millions of years.
>
> GENERAL SMUTS. *Holism and Evolution*, p. 211.

> Oxygen was discovered by Priestley in 1774. He described it as "the most extraordinary of all my unexpected discoveries".

AIR

THE atmosphere is the name applied to the gaseous mixture that envelops the earth and is commonly spoken of as the air we breathe. It consists of a mixture of gases, the two chief of which are oxygen and nitrogen in the proportion of 1 : 4. Both of them are colourless gases without taste or smell. Most of the chemical phenomena exhibited by the atmosphere are due to the oxygen, as when some metals tarnish or iron rusts, and when fuel is burned. Oxygen is the only gas that is capable of supporting respiration ; respiration may be said to be a relatively low-temperature combustion of animal matter.

It has already been suggested that the oxygen of the atmosphere is the accumulated end-product of physiological processes of plant life and radiant energy. From the evidence furnished by the geological record with its abundance of plant remains of ancient forests, of which the numerous coal seams bear witness, there can be no doubt that the vegetable kingdom has always had a preponderance over the animal kingdom. If the fundamental rocks were mainly of algal origin as has been suggested under the heading of granitic Rocks, the atmosphere must have increased from very small begin-

nings and in an attenuated state, to a density high
enough to form a respiratory medium suitable for the
kind of animal organisms that lived in Cambrian and
later times. Even after the development of the higher
orders of vertebrates, when the forests were inhabited by
the mylodon, and subsequently by the elephant and the
like, the same preponderance of the vegetable kingdom
persisted, as it persists to-day, over the consuming power
of the animal world. Throughout geological time,
therefore, oxygen has been constantly liberated and
diffused into the atmosphere, to the present time, and the
atmosphere continues to increase in volume and density.
To this changing environment all animals must adapt
themselves.

There are really two environments; an external
environment in which the animal lives and moves, and
an internal environment in which the cells of the
organism live. It is the efforts to maintain the internal
environment constant in changing external surroundings, that bring about the modification of structure.
Very slight variations from the normal in the composition of the internal medium induce appropriate
reactions. For instance, a very small increase in the
amount of carbon dioxide in the blood at once brings
about an increased activity of the breathing movements
by which this is corrected. The respiratory medium is
also a powerful factor affecting the internal environment, and this has slowly increased in density throughout
geological time. What would be the effects produced
by this changing external environment upon the animal
creation? Herbert Spencer (35) speculated on this
question one hundred years ago, and it would be useful
to restate his argument. Writing in the *Philosophical
Magazine*, February 1844, on the theory of Reciprocal
Dependence in the Animal and Vegetable Creations, he
stated: " . . . Superior orders of beings are strongly
distinguished from inferior ones by the warmth of their
blood. A low organisation is uniformly accompanied
by a low temperature, and in ascending the scale of

creation we find that, setting aside partial irregularities, one of the most notable circumstances is the increase of heat. It has been further shown, by modern discoveries, that such augmentation of temperature is a direct result of a greater consumption of oxygen ; and it would appear that a quick combustion of carbonaceous matter through the medium of the lungs is the one essential condition to the maintenance of that high degree of vitality and nervous energy without which exalted physical endowments cannot exist.

" Coupling this circumstance with the theory of a continual increase in the amount of atmospherical oxygen, we are naturally led to the conclusion that there must of necessity have been a gradual change in the character of the animate creation. If a rapid oxidation of the blood is accompanied by a higher heat and a more perfect *mental* and physical development, and if in consequence of an alteration in the composition of the air greater facilities for such oxidation are afforded, it may be reasonably inferred that there has been a corresponding advancement in the temperature and organisation of the world's inhabitants.

" Now this deduction of abstract reasoning is in exact accordance with geological observations. The records of creation demonstrate that there was an era in which the Earth was occupied exclusively by cold-blooded creatures requiring but little oxygen ; that it was subsequently inhabited by animals of superior organisation consuming more oxygen ; and that there has since been a continual increase of the hot-blooded tribe , and an apparent diminution of the cold-blooded ones.

" It would appear that there is some connection between the change in the vital medium and the increased intensity of life and superiority of construction which have accompanied it." (35)

It would be difficult to improve upon Herbert Spencer's argument, even after another hundred years of study of the subject. The change in the character and anatomical construction of the animal world was, of course, very gradual. Reptiles, though lung-breathers, suffered diminished metabolism as their surroundings were reduced in temperature, leading to hibernation and torpor. The increase of vitality that lung-breathing gave to them, led chiefly to increased glandular activity, stimulating the bone-forming apparatus. This is manifested in the giant reptiles of the Jurassic Period. It seems as though some trigger that restrains bone-formation within limits, had been released. Thick plates of osseous " armour " covered the bodies of these creatures, and this development appears to have been as much a hindrance to their activity as " shining armour " was to medieval knights-at-arms.

But this phase eventually passed and was followed by a tremendous advance in organic evolution, with a great acceleration in the rate of change ; cold-blooded animals, by structural changes in the glandular system, were succeeded by warm-blooded creatures ; and passed from the stage of propagation of their young by development of the egg, to placental development of the germ cell, and the suckling of the young.

The increase in nervous energy that brought about the transformation to placental development affected the whole glandular system, one gland stimulating the others, and these again, reacting, stimulating in turn. For instance, the super-renal gland is the principal organ of heat-control of warm-blooded life forms, including the human, in co-operation with the thyroid gland. Heat regulation is under the direct control of the sympathetic nervous system, but it is subject to the functional activity of the two glands named. In warm-blooded creatures the vascular arrangement of inter-renal tissue, and the para-sympathetic ganglion cells, are completely reversed from that of cold-blooded animals : in fishes, the groups of para-sympathetic cells are separate from the groups of inter-renal tissues ; in reptiles the groups have come in contact, the sympathetic cells having fused together, and surround the inter-renal tissue. In mammals, the cells are collected to form a mass in the centre of the inter-renal tissue. It is a complete reversal of the positions of the group of cells, and inter-renal tissue, the super-renal gland having been literally turned inside out; and the response to stimulus has been reversed also. (39) In mammals, exposure to cold is a powerful stimulus to glandular activity with increased metabolism and greater vitality. Mammals are able to adapt themselves quickly to changes of climate; their glandular system therefore plays an important role in evolution, acting along with and reacting to the respiratory factor, leading to still further development, and culminating in man.

It takes only a few seconds to read the above statement of the transformation from cold-blooded creatures to warm-blooded mammals that suckle their offspring, but the process itself required many millions of years for its accomplishment. Traces of untransformed relics still exist. Reptilian organisms propagated their young by laying eggs on the sand to hatch where deposited, and they still do so. Lizards lay an egg much like those laid by our domestic hen, which we value so much as a nourishing food. The domestic fowl, though a warm-blooded creature, still bears marks of its reptilian ancestry in the " lizard-skin " and scales on its legs.

It may then be accepted that the oxygen of our atmosphere is a product of the activity of organised life, but the other component of our atmosphere, nitrogen, is such an inert substance, combining only with great difficulty with only a few other elements, that it seems far removed from vital sources. Its inertness is more apparent than real; the molecule of nitrogen displays strong chemical affinity between the two atoms that constitute the molecule.

Nowhere is nitrogen found in combination with other than organic matter, only in animal organisms it is present in such combinations of vital import to the organism as muscle tissue, brain substance, and globin bodies of the blood, and in the waste excretory products of the metabolic processes—urea, hair and feathers, nails, hoofs, etc., and in plant tissues, in the gluten of wheat, the protein of peas, clover and the like.

The presence of combined nitrogen in the soil plays an important part in the fertility that is not well understood at present. From experiments made at the Rothamsted Agricultural Station it was estimated that as much as 80 lbs. of nitrogen is formed per acre, during a year's bare-fallow. (49) This nitrogen of a bare-fallow is formed or " fixed " in the soil by the vital activity of " nitrifying " bacteria that live in the soil. It is an end-product of the metabolic processes that take place in the life cycle of these microscopic organisms. It cannot be the free nitrogen of the atmosphere that is " fixed ", but it is derived from decomposing proteins on which the bacteria feed, and is the result of many and complex bio-chemical and physiological changes and exchanges.

There is a type of vegetation that these nitro-bacteria seem to prefer above all others, among the roots of which they thrive and multiply ; these are the leguminous plants—peas, beans, lupins, clover, etc. On the roots of these plants, when well advanced in growth, nodules or swellings can be seen, which are the homes of these bacteria. (44)

The nitro-organisms are active without the legumes, as is shown by the bare-fallow experiments, the legumes on the other hand are not nitrogen " fixers " in the absence of the nitro-organisms from the soil, but when both are present there is a great increase in activity, the symbiotic partnership produces a mutual stimulation to effort.

It has recently been found that many plants such as cereals and grasses, in addition to legumes, share symbiotic association with nitro-bacteria.

Nitrogen is the product of the life force and cosmic energy, as is oxygen, after the manner of the other admittedly organic elements. Through the agency of chlorophyll and sunlight, vegetable organisms build up proteins of a complex character, animals devour these vegetable products, and by the energy derived from their assimilation, living substances of a still higher order of complexity are formed—the animal proteins. On the death of both plants and animals these complex substances are broken down by the action of bacteria to carbon dioxide, nitrogen and water, leaving the mineral remains unchanged.

" The refuse in the drains of a large city is to that city a menace to health, and a potential danger on a large scale. It must be shunned as a virulent poison. Yet what is it but the worn-out remnant of what has been valuable ? This refuse consists of cast-off husks which were essential in their time to what is healthful ; they are not evil of themselves, they are only evil where they are. In due course they will be restored by the action of the healthful powers of a bacteriological nature to usefulness, fertility and a new generation ; they become part of the great scheme of the renewal of life. They are as essential to it as are old age and death to the eternal youthfulness of the world." (53)

Though the work of bacteria is often referred to as that of " breaking down " complex material to simple elements, the cells of bacteria possess remarkable synthetic capacities. Furnished only with water, glucose and simple sources of carbon and nitrogen, some types of bacteria can synthesise proteins, lipoids, growth accessories and enzymes, all organised into characteristic protoplasmic systems. The building-up of enzymes varies according to the environment. Such chemical synthesis and organisation cannot be imitated in the best-equipped laboratories.

The exact nature of enzymes is not yet known. Their reactions are veiled in mystery. The problem of synthesis and the accessory agents of the synthesis is an almost virgin field awaiting exploration. There are numerous names for enzymes and co-enzymes besides zymase, catalase, phosphatase, hydrolase, oxidase and others, but they serve to hide our own lack of knowledge more than they explain what they are. All enzymes, as far as we know, are produced inside living cells by secretion. In many reactions it may be the same activating enzyme operating under different conditions. The enzymatic processes of digestion always follow the same course, and the systems for activating cellular respiration have much in common *throughout* the animal kingdom.

Enzymes are colloidal particles of a protein nature carrying electric charges ; in the ultimate analysis enzymes may represent the behaviour of idio-electric charges " positioned " on the most unstable vibrating particles of a delicately-poised complex. Modern thought should therefore turn in the direction of an electronic interpretation. The activated complexes represent intermediate stages in hydrogen transfers and oxidations, and, in the state of electronic resonance, are of great significance in physiological processes and as precursors of the creative synthesis that is displayed in the growth and development of the organism.

WATER

Joseph Priestley discovered oxygen in 1774, while experimenting with the red oxide of mercury by exposing it to the sun's rays passed through a powerful lens ; he found later that when hydrogen and oxygen were mixed and ignited in a suitable glass vessel, moisture was produced. Cavendish, who first established the existence of hydrogen as an individual substance, showed that the water was actually the product of the chemical union of the two gases, hydrogen and oxygen.

Water is a tasteless and odourless liquid that in moderate amounts appears colourless also. The manufacture of water in the laboratory is not a practical economical proposition. In nature it enters into all the photo-chemical and physiological reactions ; sugar, for instance, is a definite number of carbon atoms combined with a definite number of water molecules arranged in a definite structural arrangement. Water constitutes over 90 per cent of the matter that is contained in most fruits and vegetables.

The hydrosphere, a term invented to cover the totality of water on the earth whether gathered together in the ocean, or invisible in the air, or condensed in the clouds, or falling as rain or snow, or creeping down in the glaciers, or coursing down in the rivers, or percolating underground, or rising in the sap of plants and trees, or circulating in the veins and arteries of animals—this hydrosphere is functionally one, for given sufficient time every drop might successively take the place of every other drop, passing from the ocean back into the ocean. Obviously life, and not least human life, is possible only within the bounds of the hydrosphere. (52)

" What is it that gives a unique interest to the surface of the earth ? Not its dead features, surely ; there are mountains also on the moon, ruins of a live past. No, it is the fluid envelopes of water and of air, that by their circulations, their physical and chemical reactions, and their relation to life, impart to the earth's surface an activity almost akin to life itself. Which is the fundamental—the living palpitating being, or the dead skeleton which it shapes and leaves behind as a monument ? " (52)

Water seems to possess something like vital qualities in that when the two gases are compounded and united, a new characteristic emerges ; water possesses qualities and powers that its constituents do not possess, it has something of that emergent quality that typifies the living organism.

There is evidence that in Nature's workshop hydrogen and oxygen are being united, thereby incrementing the liquid portion of the globe, so that the volume of the seas tends to increase from age to age. For this reason, among others, water possesses great geological interest apart from the fact that the seas provide the environment for innumerable creatures that possess in themselves geological interest. Water determines or changes the outlines of continents by the greater or lesser submersion of lowlands by the changes of the reciprocal relations of height between the higher land surfaces and the fluid portions of the earth's surface. The time within which man has been observing and recording the changes of the earth's surface forms but an insignificant fraction of the ages through which geological history has been in progress. Man has never witnessed the uplifting of a mountain-chain. Submerged forests have been found, and this at first sight seems to furnish evidence of subsidence of that portion of the land on which the trees have grown.

Again and again, in geological works, reference is made, in accounting for some formation such as slate deposits, to gradual subsidence of the land during its deposition. Giekie (14), writing of the deposition of the Carboniferous limestone, states : " The sea bottom was sinking, but so slowly that the growth of limestone, and the deposit of sediment probably on the whole, kept pace with it. The actual depth of the water may not have varied greatly even during a subsidence of several thousand feet. That this was the case may be inferred from the structure of the limestone itself. . . . Had there been no subsidence of the sea-floor during the accumulation of so thick a mass of organic debris, it is evident that the first beds of limestone must have been begun at a depth of 6,000 feet below the surface of the sea." He then gives several reasons why this could not have been the case. " We are led to conclude that the bottom slowly subsided until its original level, on which the limestone began to form, had sunk at least 6,000 feet."

Again—" In the successive strata of a coal field we are presented with the records of a prolonged period of subsidence, probably marked by longer or shorter periods of rest. These more stationary periods are indicated by the coal-seams, while the sandstones, shales and other strata record the downward movement."

Hugh Miller confirms this : " For repeated alternations in strata—repetitions carried on for hundreds of feet in vertical extent—we require that condition of gradual subsidence in the general crust that can alone account for the fact so often pressed upon the geologist in exploring the coal measures, that in deposits thousands of feet in thickness, each stratum had been laid down in shallow waters." (43) Charles Darwin, likewise, was " convinced that nearly all our ancient formations which for the greater part are rich in fossils, have been deposited during subsidences. Since publishing my views on this subject, I have watched the progress of Geology, and have been surprised to note how auther after author, in treating of this or that formation, has come to the conclusion that it was accumulated during subsidence." (37)

On the other hand, Alexander Agassiz, commenting on the formation of the Florida reefs, states that these " cannot be explained by subsidence " to which cause Darwin had ascribed the formation of barrier reefs in general. The evidence gathered by Murray, Semper and Agassiz shows that great submarine plateaux have gradually been built up in the Gulf of Mexico and the Caribbean by the decay of animal life upon banks which lie directly in the path of the great Atlantic equatorial currents. These currents carry along their course an immense amount of pelagic life, serving as food for the animals living on the bottom. Here we have all the elements of the gradual accumulation of submerged land, which, when it rises to a certain level, becomes the foundation upon which reefs are formed. (8)

Observations made on the Cocos-Keeling Atoll convinced Prof. F. Wood Jones (50) that organic debris falling from the ocean surface has a large share in the production of the several stages of atoll history. " Somewhere between the surface of the ocean and the bottom the action of waves prevents sediment coming to rest. Where sediment is at all times liable to settle on the living zooids, reef corals will not flourish. In the wave-stirred area above this line, reef corals can and do flourish. . . . (50)

" The surface waters continue to drop their burden of suspended matter over the reef, and where the surface of the reef becomes sufficiently flat, it settles on the zooids and their activity begins to wane. At length the central area dies—the zooids choked with sediment—and a raised ring of active living zooids surrounds a central depressed area. This leads to the development of a basin-shaped reef. . . .

" The basin-shaped reef continues to grow upwards until tide limit arrests the growth of its margins. At this stage the enclosed lagoon tends to become the resting-place of a vast amount of sediment, formed by the disintegration of coral fragments by the force of the waves. . . .

" This debris becomes cemented into solid breccia by the deposition of calcium carbonate around the particles that compose it. . . .

" In the Cocos-Keeling Atoll, the history shows a steady filling-in of the lagoon. No sinking has taken place in the whole history from the time when the wave-washed reef first reached the surface of the sea." (50)

The continual rain of tests, shells, skeletons and other organic mineral remains to the ocean floor, thereby raising it, will displace the volume of water so that its surface level is also raised, and low-lying land will tend to become submerged ; but not in the sense that the geologists use the term, notwithstanding that the relation of height of land surfaces to that of sea-level have been altered.

There are regions where the total thickness of the deposits of a particular period is as great as the mean depth of the oceans at the present day. In a few instances the deposits are even greater ; in Pennsylvania the thickness of deposits from the top of the Alleghany River coal series to the Trenton limestone is over 18,000 feet. " How great ", says Eduard Suess (10), " on prevailing assumptions, must the subsidence of a district once have been, where not merely the depth of the ocean covering it, but even the thickness of sediment deposited have attained so great a magnitude."

But is " subsidence of a district " a fair description

of phenomena that submerge nearly the whole globe ? Eduard Suess (10) makes this interrogation—" How is it to be explained that the very earliest of the geological formations, the Silurian, recurs in parts of the earth so widely removed from one another—from Ladoga to the Argentine Andes, and from Arctic America to Australia —and always attended by the same characteristic features ? " May it not be more probable that, notwithstanding the gradual growth in thickness of the earth's crust and increasing diameter of the globe in consequence of continuous deposits of mineral debris, other natural forces have been more than keeping pace with these, and caused the waters of the earth to increase at a yet more rapid pace ; and that these have produced this apparent subsidence that has been observed so generally by geologists ?

Man has, so far, not learnt from Nature all the mechanisms of synthesis that she uses. Organic chemistry, with all its achievements, cannot yet claim to rival the synthetic activities of the cell. Even respiration of plant cells is not fully understood. It has been too readily assumed to be the converse of photosynthesis, whereby the simple sugars are broken down to carbon dioxide and water, with the release of energy. Yet the process is not quite so simple, carbon is not consumed to the same extent as that primary element, or simple structure, hydrogen. Hydrogen is the primary fuel of the cell, just as it is the synthesised " brick " of which the complex elements are built up in the photosynthesis phenomena.

" The sugar molecule is a group of hydrogen atoms held in a carbon lattice. The hydrogen is ' burned ' by combining with oxygen activated by the ' enzyme of respiration ', forming water." (57) In doing so, the oxygen leaves an electron charge which " excites " the cell respiratory pigment—a further step in a chain of respiratory reactions. Throughout the life-cycle of plants water is continually being manufactured, or rather created, from a new combination of elements that have been condensed by the transformation of light waves as before described.

When life has passed out of plant and animal organisms active secretion of water ceases, but bacteria set to work to break down the complex material—cellulose, carbohydrates, proteins and tegument, by feeding thereon, and so releasing the nitrogen and the water, which bacteria have not the power to split up. Thus the incessant life and incessant death in both kingdoms of created things, with the co-operation of bacteria, contribute to the volume of the waters of the earth. Not only are the rocks " the daughters of Time ", but the atmosphere and the seas also.

Everywhere, the land is only able to keep its surface above the ever-increasing waters by constant strain and stress. Were it not for the constant growth and accumulation on the land surface combined with the continuous side-thrust and tremendous pressure of the mobile waters, slowly squeezing the land surfaces upwards and inland, the seas would long ago have drowned the land surfaces of the globe. The oldest crystalline rocks, while still showing evidences of having been laid down in seasonal courses, often display intensely crumpled and puckered wave-forms, arising from this side thrust.

Some idea of this lateral thrust, puckering the land nearest to the line of action, can be obtained by studying a topographical map of North and South America, noting the coast ranges of mountains stretching along the Pacific coast line from the extreme north of North America, practically without a break, to the extreme south of the Southern Continent. The Rocky Mountains, the North and South Cascades and the Sierra Nevada form a series of " hunches " a thousand miles across at its widest part. The present foldings are young, but probably older foldings lie deeper. The ranges are double, and in some regions three-fold ; in these the eastern ranges are the oldest, and when they had become too heavy, or had lost plasticity to thrust further upwards, younger folds developed and continued to mount still higher than the older range. From the consideration of this well-known example, the dependence of the processes of mountain formation on lateral pressure and deflection becomes more evident. (10) The Rocky Mountains have risen so slowly and quietly, that the Green River, which flowed across the site of the Uinta range, has not been deflected, but has actually been able to deepen its cañon as fast as the mountain range has been pushed upward. (14a) The comprehensive biological work of the *Challenger* expedition led Murray to conjecture that in Palæozoic times the oceans were not so deep as at present, and were strewn with numerous islands ; later, the continental land became more continuous and higher, while the oceans were more circumscribed and grew deeper. (24). Folded or puckered elevated land constitutes the greatest chains and ranges of mountains. The Alps, Pyrenees, Carpathians and the Himalayas are all folded. The Alps consist of crumpled heights ninety miles wide from north to south, which, before the land was squeezed and elevated, must have been more than seven hundred miles wide. This range and the Himalayas are the youngest mountains ; the latter are still being pushed up. The sheets of sedimentary rocks of which these ranges consist were being deposited in a sea much larger than the Mediterranean of the present time, and long after the Caledonian Mountain system had been developed.

All types of rock-folding can be resolved ultimately into wave patterns of crests and troughs, due to this lateral pressure. Where naturally hard rocks have been crumpled without fracture, the inference is that they were subjected to very steady and prolonged compression. In parts of Southern England vast wrinkling has been imposed on hard rocks (see Fig. 2, p. 113).

The land areas of the globe thus thrust up amount to little more than one quarter of the total surface of the sphere. The average elevation of the continental

land is estimated at 900 to 1,000 feet above sea-level, and
the mean depth of the oceans at 13,000 feet.   One-half
of the total area of the planet is an abyssal region from
two to five miles deep.

If, in Archæan times, with the earth 30 to 40 miles
less in diameter than now, with consequent lesser area,
the waters of the earth were confined in very wide but
shallow seas—a not improbable supposition according
to Murray (24)—the waters have since than time
increased so much more rapidly in bulk than the land,
for all the earth's increase in diameter in the long

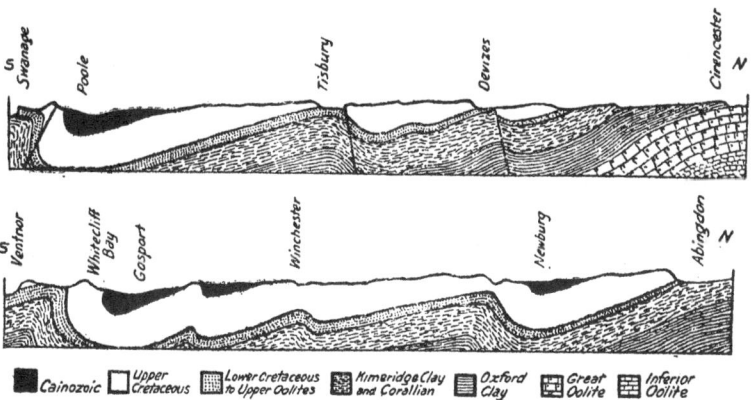

FIG. 2.—Simplified sections (thirty miles apart) across parts of Southern England

interval, that the cubic contents of the oceans is more than
tenfold that of the land projecting above.

There is geological evidence of several oceanic trans-
gressions, and these have been accepted as indications
of continental depression.   It appears more reasonable
now to assume, that the level of the sea has persistently
risen, and that the surfaces of the land areas have
become elevated, as here suggested, by intermittent
uplift.   This idea, however startling it may at first
appear, is quite in accordance with the analogy of
changes now going on in various regions of the globe ;
and is very necessary to a biological conception for all
natural phenomena.

Were it not for the heaped up masses of frozen water

H

at the two poles, the increase of the waters would be much more apparent. The Greenland ice-cap alone is rather larger than England, France, Germany, Austria, Spain and Portugal together, and is at the minimum upwards of 2,500 feet thick and may even reach 7,000 feet over a large part of it. The crevasses are often hundreds of feet in depth. In the Antarctic, the great Ice Plateau of South Victoria Land is 7,000 to 8,000 feet high, and that of King Edward VII Land attains to 11,000 feet. Sir Douglas Mawson, in a lecture to the Royal Institution, stated that Antarctica appears to be an immense plateau of ice, larger than Europe, and over 6,000 feet high, unbroken by mountains. He said the ice may be even 10,000 feet thick, probably resting on islands dotted under its area, like a blanket. This ice flows off into the sea as an immense glacier and breaks into icebergs, some of which are as large as, say, the island of Corsica. There is a barrier ice sheet at each of the poles nearly two thousand miles across. In the Arctic this great mass of ice slowly drifts south in the Central Arctic Basin, carried along by the ocean current southward along the coast of Greenland. Four Russian scientists recently set up their station on the ice-floe near the North Pole and were carried or drifted over nine hundred miles in seven months. Should climatic conditions return to the poles similar to those that reigned there in the Cretaceous period, there would be vast areas of lowland on every continent submerged by the release of the waters now imprisoned in these heaped-up masses of ice.

*   *   *   *   *

Earth, air, water—lithosphere, atmosphere, hydrosphere—each the outcome or product of biological processes energised by sunbeams ! There was more of the substance of reality than of peradventure in Erasmus Darwin's declaration—" the World itself might have been generated rather than created ; that is, it might have been produced from very small beginnings, increasing by the activity of its inherent principles, rather than by a sudden evolution of the Whole by the Almighty fiat ".

Dr. Darwin consistently put forward a well-rounded theory regarding the development of the living world. In *The Temple of Nature* he wrote :

> Age after Age expands the peopled plain,
> The tenants perish, but their cells remain.

and

> How life increasing peoples every clime,
> And young renascent Nature conquers Time.

The existence of life on the earth is regarded as an accident by many prominent scientists—" a unique phenomenon in the whole Universe ", as Sir James Jeans suggests. If life is so regarded as the rarest of accidents in a lifeless Universe, it seems rash to suggest that biology may become the normative science of the Cosmos. Yet the biological conception of the process of generation presents the possibility of a counter-movement to that of the degradation of energy, or " running down " of the Universe, that is so much in the minds of astronomers and physicists at the present time. A way of escape from the inexorable second law of thermo-dynamics, as is here suggested, should be welcomed and explored. Living things are constantly building up new substances, and thus hold the second law at bay. The late Sir Oliver Lodge claimed " as a physicist, that too much attention has been paid to this second law of thermo-dynamics, and that the final and inevitable increase of entropy to a maximum is a bugbear, an idol, to which philosophers need not bow the knee ". (55) " The thermo - dynamic theorist makes sweeping generalisations upon insufficient knowledge. He has, so far, left cosmic radiation unconsidered in determining the origin and destiny of the universe," says Prof. Millikan. (56) Sir James Jeans admitted later that " living organisms succeed in evading the law in varying degrees.   In fact it would seem reasonable to define life as being characterised by a capacity for evading the law." (23)

The second law of thermo-dynamics is " the nigger in the woodpile " of physical science.   The law, as stated in the handbooks, runs—" It is impossible by means of any automatic process to transfer heat from a cold body to a hot body without the assistance of some external force."   The law is otherwise called the law of increasing entropy.   The working hypothesis of thermo-dynamics seems to be that some billions of years ago the Creator wound up the material universe, and has left it to run down of itself ever since.

Now it is readily admitted that the quantitative aspect of the second law of thermo-dynamics is of practical importance in engineering science, especially in engine design.   The engineer's unit—the horse-power —is a measure of work done or energy spent.   Just how efficient the horse is, was not generally realised until motor-cars became the mode of transport.   Think what loads were carried on the old four-in-hand coach, what hills were climbed.   Then consider the motor-car with every device to reduce friction—ball-bearings and the like—and compare the horse-power that has to be installed to propel it.   Yet the engineers fixed their unit after tests carried out with the best cart-horses.   The horse does not obey the second law of thermo-dynamics ; on the other hand, with but little stimulus of the whip, he reverses the law.

So securely did the conservation of matter and of energy seem to be established that they have become accepted as universal laws governing the whole of creation, and have received unquestioning assent form-erly accorded to some theological propositions.   In his recent work *Human Knowledge : Its Scope and Limits*, Bertrand Russell makes the assertion that there is " considerable reason to think that everything in the behaviour of living matter is theoretically explicable in terms of physics and chemistry " ; but he uses a very cautious phrase: "which scientific investigation has made *probable* ".

The facts of biology lead us to the conclusion that the physical and chemical interpretation of the world is fundamentally imperfect. An acorn has in itself the potentiality not of one oak-tree alone, but of a forest of oak-trees in unlimited numbers and generations, as if it had the resources of the whole universe to draw upon.

There is no law of conservation imposed here.

Facts are but half of truth, interpretation is the other half. The facts herein recorded have been stated before, and there have been various interpretations of them. There seems to be some agreement that the earth's atmospheric envelope has been developed by the activities of plant life, as here described, and that the gradually increasing density of the air has played a part in biological evolution. A few observers have recognised evolution at work in the inanimate physical environment of living things, as a consequence of this. Some have even gone further in their interpretation and recognised a general law in Nature, wherein the waste products of the metabolic activities of one species are a source of energy to other species, and that none carry on a solitary one-way traffic with others, or even with their inanimate surroundings.

To the writer it seems logical that cosmic and biological evolution are one interlinked, inter-related, and united Whole, nor is this interpretation so new as perhaps it seems to the reader. Plato regarded the Logos as the agent of Creation, the sustainer *in being* of all created things. The prologue to the fourth gospel is a rhapsody on the cosmic principle. A liberal translation of the prologue would read: " In the beginning was the Logos, in Whom was Life. Through Life all things were made ; and without Life was not anything made that was made." But amid the dazzling triumphs of materialism, and the feverish exploitation of the earth's resources, such concepts have been almost totally obscured.

REFERENCES

(1) COMBER, PROF. N. M. "Pedology". Review of, *Science Progress*, July 1937, p. 115.
(2) LODGE, SIR OLIVER. *Ether and Reality*, 1925.
(3) SMUTS, GENERAL J. *Holism and Evolution*, 1925.
(4) BARTON-WRIGHT, E. C. *Recent Advances in Plant Physiology*, 1930.
(5) RAMAGE, HUGH. *Application of the Spectroscope to Biology*, 1933.
(6) HOPKINS, SIR FREDERICK. "Some Chemical Aspects of Life", Pres. Add. Brit. Assoc., 1933.
(7) WOLFF, DR. EMIL. *Aschen Analysen*, 1871.
(8) AGASSIZ, ALEXANDER. *Three Cruises of the Blake*, 1888.
(9) SCROPE, G. P. *The Geology and Extinct Volcanoes of Central France*, 1858.
(10) SUESS, EDUARD. *The Face of the Earth*, English trans., 1904.
(11) LODGE, SIR O. "Interaction of Life and Matter", article in *Hibbert Journal*, April 1931.
(12) SPOEHR, H. A. *Photosynthesis*, 1926.
(13) BROGLIE, LOUIS DE. *Matter and Light*, 1939.
(14) GEIKIE, SIR ARCHIBALD. *Class Book of Geology*, 3rd Ed., 1892.
(14a) GEIKIE, SIR ARCHIBALD *Text-Book of Geology*, 3rd Ed., 1893.
(15) HUMBOLDT, ALEX VON. *Cosmos*, Vol. I, 1845.
(16) LODGE. SIR O. *My Philosophy*, 1933.
(17) KRAUSE, ERNST. *Erasmus Darwin, Life of*, 1879.
(18) LYELL, SIR CHARLES. *Principles of Geology*, XIth Ed., 1875
(19) SWAINE, A. T. *The Earth—Its Genesis and Evolution*, 1913.
(20) GARDINER, PROF. S. *Coral Reefs and Atolls*, 1931.
(21) MILLIKAN, PROF. R. A. Discussion on "The Evolution of the Universe", Brit. Assoc. Mtg., 1931.
(22) LYELL, SIR CHARLES. *Elements of Geology*, 3rd Ed., 1878.
(23) JEANS, SIR JAMES. *The New Background of Science*, 1933.
(24) MURRAY, JOHN. Report of the Scientific Results of H.M.S. *Challenger*, Summary II, 1895.
(25) LODGE, SIR O. *Life and Matter*, 2nd Ed., 1905, p. 104.
(26) MILLER, HUGH. *The Old Red Sandstone*, 1858.
(27) EHRENBERG, PROF. C. *Mikro-Geologie*, 1845.
(28) "Lens", articles by, *New Statesman*, March 26, 1927, March 10, 1930.
(29) NEWTH, G. S. *Inorganic Chemistry*, 1905.
(30) LIBBY, PROF. W. T. Report in *Science To-day*, June 26, 1947.
(31) CLEMENTS, FRANK ED. *Plant Physiology and Ecology*, Nebraska, 1907.
(32) Fuel Research. Phys. and Chem. Survey of National Coal Reserves, No. 28, 1933.
(33) KAY, PROF. H. D. Article in *Nature*, Aug. 11, 1945.
(34) JEANS, SIR JAMES. *The Universe Around Us*, 1933.

REFERENCES

(1) COMBER, PROF. N. M. " Pedology ". Review of, *Science Progress*, July 1937, p. 115.
(2) LODGE, SIR OLIVER. *Ether and Reality*, 1925.
(3) SMUTS, GENERAL J. *Holism and Evolution*, 1925.
(4) BARTON-WRIGHT, E. C. *Recent Advances in Plant Physiology*, 1930.
(5) RAMAGE, HUGH. *Application of the Spectroscope to Biology*, 1933.
(6) HOPKINS, SIR FREDERICK. " Some Chemical Aspects of Life ", Pres. Add. Brit. Assoc., 1933.
(7) WOLFF, DR. EMIL. *Aschen Analysen*, 1871.
(8) AGASSIZ, ALEXANDER. *Three Cruises of the Blake*, 1888.
(9) SCROPE, G. P. *The Geology and Extinct Volcanoes of Central France*, 1858.
(10) SUESS, EDUARD. *The Face of the Earth*, English trans., 1904.
(11) LODGE, SIR O. " Interaction of Life and Matter ", article in *Hibbert Journal*, April 1931.
(12) SPOEHR, H. A. *Photosynthesis*, 1926.
(13) BROGLIE, LOUIS DE. *Matter and Light*, 1939.
(14) GEIKIE, SIR ARCHIBALD. *Class Book of Geology*, 3rd Ed., 1892.
(14a) GEIKIE, SIR ARCHIBALD *Text-Book of Geology*, 3rd Ed., 1893.
(15) HUMBOLDT, ALEX VON. *Cosmos*, Vol. I, 1845.
(16) LODGE. SIR O. *My Philosophy*, 1933.
(17) KRAUSE, ERNST. *Erasmus Darwin, Life of*, 1879.
(18) LYELL, SIR CHARLES. *Principles of Geology*, XIth Ed., 1875
(19) SWAINE, A. T. *The Earth—Its Genesis and Evolution*, 1913.
(20) GARDINER, PROF. S. *Coral Reefs and Atolls*, 1931.
(21) MILLIKAN, PROF. R. A. Discussion on " The Evolution of the Universe ", Brit. Assoc. Mtg., 1931.
(22) LYELL, SIR CHARLES. *Elements of Geology*, 3rd Ed., 1878.
(23) JEANS, SIR JAMES. *The New Background of Science*, 1933.
(24) MURRAY, JOHN. Report of the Scientific Results of H.M.S. *Challenger*, Summary II, 1895.
(25) LODGE, SIR O. *Life and Matter*, 2nd Ed., 1905, p. 104.
(26) MILLER, HUGH. *The Old Red Sandstone*, 1858.
(27) EHRENBERG, PROF. C. *Mikro-Geologie*, 1845.
(28) " Lens ", articles by, *New Statesman*, March 26, 1927, March 10, 1930.
(29) NEWTH, G. S. *Inorganic Chemistry*, 1905.
(30) LIBBY, PROF. W. T. Report in *Science To-day*, June 26, 1947.
(31) CLEMENTS, FRANK ED. *Plant Physiology and Ecology*, Nebraska, 1907.
(32) Fuel Research. Phys. and Chem. Survey of National Coal Reserves, No. 28, 1933.
(33) KAY, PROF. H. D. Article in *Nature*, Aug. 11, 1945.
(34) JEANS, SIR JAMES. *The Universe Around Us*, 1933.

cautious phrase : " which scientific investigation has made *probable* ".

The facts of biology lead us to the conclusion that the physical and chemical interpretation of the world is fundamentally imperfect. An acorn has in itself the potentiality not of one oak-tree alone, but of a forest of oak-trees in unlimited numbers and generations, as if it had the resources of the whole universe to draw upon.

There is no law of conservation imposed here.

Facts are but half of truth, interpretation is the other half. The facts herein recorded have been stated before, and there have been various interpretations of them. There seems to be some agreement that the earth's atmospheric envelope has been developed by the activities of plant life, as here described, and that the gradually increasing density of the air has played a part in biological evolution. A few observers have recognised evolution at work in the inanimate physical environment of living things, as a consequence of this. Some have even gone further in their interpretation and recognised a general law in Nature, wherein the waste products of the metabolic activities of one species are a source of energy to other species, and that none carry on a solitary one-way traffic with others, or even with their inanimate surroundings.

To the writer it seems logical that cosmic and biological evolution are one interlinked, inter-related, and united Whole, nor is this interpretation so new as perhaps it seems to the reader. Plato regarded the Logos as the agent of Creation, the sustainer *in being* of all created things. The prologue to the fourth gospel is a rhapsody on the cosmic principle. A liberal translation of the prologue would read : " In the beginning was the Logos, in Whom was Life. Through Life all things were made ; and without Life was not anything made that was made." But amid the dazzling triumphs of materialism, and the feverish exploitation of the earth's resources, such concepts have been almost totally obscured.

(35) SPENCER, HERBERT. " Reciprocal Dependence in the Animal and Vegetable Creations ", *Phil. Mag.*, Feb. 1844.

(36) DARWIN, CHARLES. *Journal of Voyage of the Beagle*, 1845.

(37) DARWIN, CHARLES. *Origin of Species*, VIth Ed., 1885.

(38) *London Clay Flora*, British Museum Nat. Hist., Reid & Chandler, 1934.

(39) CRAMER, W., Ph.D. *Fever, Heat Regulation, and Thyroid-Adrenal Apparatus*, 1928.

(40) *Handbook of the Geology of Great Britain*, edited by J. W. Evans, 1929.

(41) DALY, PROF. R. A. *Amer. Journal of Science*, 1907, p. 93.

(42) PERRIER, ED. *The Earth before History*, 1925.

(43) MILLER, HUGH. *Testimony of the Rocks*, 1857.

(44) NICOL, HUGH. *Microbes by the Million*, Penguin, 1939.

(45) WARMING, PROF. E. *Oecology of Plants*, 1909.

(46) RUSKIN, JOHN. *Ethics of the Dust*, 2nd Ed., 1877

(47) THOMSON, SIR J. A. *The Gospel of Evolution*, 1925.

(48) BAILEY, PROF. J. W. *Microscopic Observations in S. Carolina and Florida*, 1851.

(49) FREAM. *Elements of Agriculture*, 10th Ed., 1918.

(50) WOOD JONES, PROF. F. *Coral and Atolls*, 1910.

(51) Memoir, Geol. Survey. *The Felsitic Lavas of England and Wales*, 1885.

(52) MACKINDER, SIR H. J. " The Human Habitat ", Pres. Add. Geography Sect. Brit. Assoc. Mtg., 1931.

(53) GRAHAM, JOHN W. *The Faith of a Quaker*, 1920.

(54) LODGE, SIR O. *Life and Matter*, 2nd Ed., 1905, pp. 148–9.

(55) LODGE, SIR O. } Discussion on " The Evolution of the
(56) MILLIKAN, PROF. } Universe ", Brit. Assoc., 1931.

(57) SZENT-GYÖRGI, PROF. ALBERT VON. Essay on " Oxidation and Fermentation ", in *Perspectives in Bio-Chemistry*, 1937.